Springer Proceedings in Earth and Environmental Sciences

Series Editors

The series Springer Proceedings in Earth and Environmental Sciences publishes proceedings from scholarly meetings and workshops on all topics related to Environmental and Earth Sciences and related sciences. This series constitutes a comprehensive up-to-date source of reference on a field or subfield of relevance in Earth and Environmental Sciences. In addition to an overall evaluation of the interest, scientific quality, and timeliness of each proposal at the hands of the publisher, individual contributions are all refereed to the high quality standards of leading journals in the field. Thus, this series provides the research community with well-edited, authoritative reports on developments in the most exciting areas of environmental sciences, earth sciences and related fields.

João Rafael da Costa Sanches Galvão ·
Paulo Brito · Filipe dos Santos Neves ·
Henrique de Amorim Almeida ·
Sandra de Jesus Martins Mourato · Catarina Nobre
Editors

Proceedings of the 3rd International Conference on Water Energy Food and Sustainability (ICoWEFS 2023)

Springer

Editors
João Rafael da Costa Sanches Galvão
Instituto Politécnico de Leiria
Escola Superior de Tecnologia e Gestão
Leiria, Portugal

Paulo Brito
Instituto Politécnico de Portalegre
Escola Superior de Tecnologia e Gestão
Portalegre, Portugal

Filipe dos Santos Neves
Instituto Politécnico de Leiria
Escola Superior de Tecnologia e Gestão
Leiria, Portugal

Henrique de Amorim Almeida
Instituto Politécnico de Leiria
Escola Superior de Tecnologia e Gestão
Leiria, Portugal

Sandra de Jesus Martins Mourato
Instituto Politécnico de Leiria
Escola Superior de Tecnologia e Gestão
Leiria, Portugal

Catarina Nobre
Instituto Politécnico de Portalegre
Escola Superior de Tecnologia e Gestão
Portalegre, Portugal

ISSN 2524-342X ISSN 2524-3438 (electronic)
Springer Proceedings in Earth and Environmental Sciences
ISBN 978-3-031-48531-2 ISBN 978-3-031-48532-9 (eBook)
https://doi.org/10.1007/978-3-031-48532-9

This Springer imprint is published by the registered company Springer Nature Switzerland AG
The registered company address is: Gewerbestrasse 11, 6330 Cham, Switzerland

Paper in this product is recyclable.

Organization

Conference Co-chairs

João Rafael da Costa Sanches Galvão
Leiria Polytechnic, Portugal

Paulo Sérgio Duque de Brito
Portalegre Polytechnic, Portugal

Organizing Committee

Ana Carolina Assis
Portalegre Polytechnic

Catarina Nobre
Portalegre Polytechnic

Filipe Neves
Leiria Polytechnic

Flávio Craveiro
Leiria Polytechnic

Henrique Almeida
Leiria Polytechnic

Joel Vasco
Leiria Polytechnic

Luis Calado
Portalegre Polytechnic

Pedro Monteiro
Portalegre Polytechnic

Ricardo Gomes
Leiria Polytechnic

Roberta Panízio
Portalegre Polytechnic

Sandra Mourato
Leiria Polytechnic

Vânia Ribeiro
Leiria Polytechnic

Scientific Committee

Luísa Scmidt, ICS/Lisbon University, Portugal
Madalena Moreira, Évora University, Portugal
Manuel Jesús Hermoso Orzáez, Universidad de Jaén, Spain
Margarida Gonçalves, Nova University, Portugal
María Hernández, Extremadura University, Spain
Marika Tossavainen, HAMK University, Finland
Marina Antunes, ISCAL/Lisbon Polytechnic, Portugal
Marritta Kymalainen, HAMK University, Finland
Milica Kasanin-Grubin, University of Belgrade, Serbia
Mohammed Benhachmi, Faculté des Sciences et Techniques de Mahammedia, Marocco
Nicolau Almeida, Portalegre Polytechnic, Portugal
Noel Perera, Birmingham City University, UK
Nuno Simões, Coimbra University, Portugal
Ramón Sanguino, Extremadura University, Spain
Remco Kranendonk, Wageningen University and Research, The Netherlands
Ricardo Chacartegui, Sevilha University, Spain
Rodrigo Moruzzi, UNESP/São Paulo State University, Brazil
Rui Ganhão, Leiria Polytechnic, Portugal
Saifur Rahman, Virginia Tech Advanced Research Institute, USA
Sérgio Leitão, UTAD/Trás-os-Montes e Alto Douro University, Portugal
Sónia Pereira, Leiria Polytechnic, Portugal
Susana Neto, IST/Lisbon University, Portugal
Teresa Mouga, Polytechnic of Leiria, Portugal
Valentim Realinho, VALORIZA/Portalegre Polytechnic, Portugal
Zahra Kalantari, Stockholm University, Sweden

Keynote Speakers

Eduardo Zarza-Moya
Plataforma Solar de Almería—CIEMAT—Spain

Eduardo Zarza-Moya is a researcher with 37 years of experience in the field of concentrating solar thermal (CST) technologies. At present, he is working at the Plataforma Solar de Almería (PSA) as Technical Coordinator and Head of the R+D Unit online-Focus CST technologies. He has coordinated several national and International R+D projects related to CST technologies (the projects: DISS, Solar Thermal Desalination, CAPSOL, PREDINCER, DETECSOL, etc.). He is a member of the AEN206/SC117 and IEC/TC117 standardization committees for solar thermal power plants, the Scientific and Technical Committee of the European Association of Solar Thermal Electricity (ESTELA), and the Spanish representative at the Executive Committee of SolarPACES.

Concentrating Solar Thermal (CST) Technologies for Sustainability

Abstract

It is estimated that the global energy consumption will reach 18,608 Mtoe by 2035, and it has become evident that fossil fuels must be replaced by renewable energies in order to reduce the global warming and the climate change that are already underway and are clearly endangering the future of the mankind. Due to its unlimited nature, solar energy must play a significant role in this effort to replace fossil fuels. At present, there are different technologies to make use of the solar radiation and some of these technologies are already well known because they have achieved a wide commercial deployment worldwide. Photovoltaic (PV) systems are a good example of these successful solar technologies. However, Concentrating Solar Thermal (CST) technologies are also a very promising option to achieve a fully decarbonized energy sector. The commercial deployment of CST technologies is still small, with a total installed power of less than 8 GW if both power and heat applications are taken into account, and less than 7 GW if only CST power plants are considered. Nevertheless, CST technologies have many socio-economic and technical benefits that must be considered (i.e., they are suitable for both power and heat generation, there are many countries with a high level of direct solar radiation, excellent dispatchability due to their thermal energy storage systems, significant job creation in places where they are implemented, high local content and excellent complementarity with PV plants), and these benefits altogether make CST technologies an excellent option for clean and sustainable energy supply. The reasons why CST technologies must play a significant role in the decarbonization of the energy market will be further explained in this invited talk.

Fernando José Cebola Lidon
Faculty of Sciences and Technology of Universidade Nova de Lisboa (FCT/UNL)—Portugal

Fernando José Cebola Lidon is a full professor at the Faculty of Sciences and Technology of Universidade Nova de Lisboa (FCT/UNL) and a coordinator of the Research Center "Geobiosciences, Geoengineering and Geotechnologies"—GeoBioTec—Polo FCT/UNL and graduated in Biology and Geology from the University of Évora and in Biochemistry from the Faculty of Sciences of the University of Lisbon, with a Ph.D. in Biology—specialty Plant Biochemistry from FCT/UNL and recognition by ECE/USA. Develops is work in the field of Agroindustrial Production and Transformation Technologies, focusing its activity on the development of functional foods following sustainable perspective, considering principles of eco-efficiency with the adoption of last generation technologies and aiming at the promotion of economic added value. Thus, articulation with the agro-industrial industry is developing techniques for the biofortification of edible parts of foods of vegetable origin, equating with the introduction of technical itineraries to increase productivity indices, adoption of crop monitoring techniques with the introduction of edaphoclimatic sensors and definition of food and nutritional quality standards.

Biofortification of Pear Rocha with Calcium: Production Workflow and Quality Definition of the Final Product

Abstract

Low dietary intake of Ca in humans has been epidemiologically linked to various diseases (such as osteoporosis), which can have serious health consequences over time. Accordingly, the development of an agronomic workflow for Ca biofortification of Rocha pears (Pyrus communis L.) and the assessment of physicochemical deviations was carried out in an orchard located in the West region of Portugal. During fruits development, leaves were sprayed a total of seven times with $Ca(NO_3)_2$ (with concentrations between 0.1–0.6 kg ha^{-1}) and/or $CaCl_2$ (with concentrations between 0.4–8 kg ha^{-1}). During fruits development, chlorophyll a parameters, as well as leaf gas exchange parameters such as net photosynthesis, stomatal conductance, transpiration rate, instantaneous and water use efficiency, only showed minor deviations, which indicated that the threshold of toxicity was not surpassed. Calcium contents varied during fruits development and at harvest the average biofortification index varied between 47 and 63%. Besides, the equatorial region of the fruits showed for all treatments (substantially in Ca treated samples) higher Ca contents in the epidermal and in the central regions. It is concluded that, although prevailing a heterogeneous distribution of Ca in fruit tissues, high indexes of biofortification in Rocha pears can be prompt in the orchards, without substantial physicochemical changes. Accordingly, agronomic biofortification with Ca can be used as a strategy for benefiting consumer's health.

Wolf-Gerrit Früh
School of Engineering and Physical Sciences, Institute of Mechanical, Process and Energy Engineering, Heriot-Watt University, Edinburgh, Scotland

After completing a first degree in Physics at the University of Freiburg in Germany and a D.Phil. in Atmospheric Physics at the University of Oxford, Wolf-Gerrit Früh took up a position in 1997 teaching Mechanical and Energy Engineering at Heriot-Watt University in Edinburgh. His research covers a wide range from fundamental fluid dynamics experiments and simulations through wind turbine aerodynamics to energy systems modeling and the application of data science for prediction and optimisation for energy systems design and scheduling.

Recent work relevant to the energy-water-food nexus include a study to reduce the carbon footprint of whisky distillation through Renewable Energy and thermal storage, heat-recovery in commercial kitchen, aquaponics, vertical farming, and desalination through solar-powered reverse osmosis. Most of these projects included collaboration with industry.

The Role of Low-Carbon Energy Sources and Storage in the Water-Food-Energy Nexus

Abstract

Water, food, and energy are absolutely essential. Providing all three in a way where supply of one does not endanger any of the other is a key requirement for sustainable development. This contribution identifies key challenges of, and opportunities for, providing energy for water and food provision. A brief overview over the energy requirements for water treatment and food production will identify some of the main challenges, such as water purification to drinking water quality, or growing crops for the increasing population. Having identified some specific challenges, this contribution will illustrate opportunities for more sustainable water and food production through renewable energy systems with a particular focus on local solutions. Examples include water purification powered by solar energy, energy recovery from water treatment, and design of renewable energy systems required for controlled agriculture or aquaponics. The examples will highlight that many opportunities exist but that the energy system to support the water or food production must be carefully designed and optimized to improve the sustainability of water, food, and energy provision. In many cases, the optimization of an energy system for a particular task in a specific location will point toward a hybrid system, combining a range of energy generation technologies as well as some form of energy storage.

Antje Disterheft
NOVA School of Science and Technology, NOVA University Lisbon, Portugal

Antje Disterheft is a postdoctoral researcher and an integrated member at Center for Environmental and Sustainability Research (CENSE), from NOVA School of Science and Technology, NOVA University Lisbon. She is enthusiastic about exploring transformative processes that can support keeping human-nature systems in balance. Her research focuses on capacity building and new approaches to collaboration and co-creation for sustainability in Higher Education Institutions. In the past years, she started to engage deeper with the lens of care when thinking about sustainability and founded The CareLab for People and Planet at her faculty. The CareLab is a new space of collaboration, co-creation, and transformative learning where the interlinkages of inner and outer sustainability are explored.

She holds a Ph.D. in Social Sustainability (Universidade Aberta) and is an alumni of the Postdoc Academy for Transformational Leadership. German by nationality, but living in Portugal for more than 16 years, she enjoys being in nature, biking, and cooking.

Sustainability and Care—Why Our Inner Worlds Matter for Transformative Change

Abstract

Sustainability is inherently linked to questions of relationship: How do we relate to ourselves and to the world around us, and how can we enhance our transformative capacity to thrive within the planetary boundaries? In the current times of multiples crises, e.g., the climate crisis, the effects of the recent pandemic due to COVID-19, as well a global crisis of trust, the concepts of intertwined inner and outer care are highly relevant: Personal care for ourselves (physical, emotional, and mental care) will impact and reflect in care for our communities and environment and draws on the social-ecological system perspective. Besides progress and growing awareness of sustainability, the world continues on a rather unsustainable path. While the Sustainable Development Goals (SDGs) offer us a roadmap of sorts toward a more sustainable future by 2030, it is becoming more apparent that without deeper transformation the goals might not be reached. What is missing in a world of increasing knowledge and sophisticated technology, where knowledge and technology might not hold the only solutions to solve complex unsustainability?

A growing number of scientists regard the inclusion of inner transformation as a deep leverage point for our sustainability endeavors. In this speech, I will explore why sustainability and care start with us and how our inner transformation can contribute to achieving the SDGs and help implement transformative change.

Shaul Sorek
Ben-Gurion University of the Negev
Jacob Blaustein Institutes for Desert Research, Zuckerberg Institute for Water Research, Department of Environmental Hydrology and Microbiology, Sde Boker Campus, Israel

Graduated as B.Sc. in mechanical engineering at the Ben-Gurion University of the Negev and received his M.Sc. and Ph.D. degrees at mechanical engineering of the Technion—Israel Institute of Technology. His research interests are in the development of fundamental modeling for biomedical and environmental systems. These address theoretical formulation for macroscopic transport phenomena of multiphase and multi-components driven through heterogenous media by inertia to drag dominant forces, and related numerical methods implementing Eulerian and Lagrangian concepts. Concerning quantitative decision making for water related resources management, he aims on causality-driven directed information in addition to spatiotemporal measures.

Policy Directed Information Technology Governing Network Welfare by Balancing the Water-Energy-Food Nexus and Human Perceptions (WEFH) Measures

Abstract

The water-energy-food (WEF) nexus represents the most fundamental requirements for human existence. It is essential for social welfare and thus is a top strategic priority in the management of water related stakeholders, which enables security and sustainability of the WEF limited resources. In previous publications, following Boltzmann entropy, we prove that the WEF resources are linked by information, across the regional administrative nodal network (municipalities, districts, and states) system. We suggest the Volume (normalized WEF expenditures product) as the nexus holistic measure from high to low Volume nodes. The node's Volume with its household income combined density allows for a decision-making policy tool on a node's Welfare Mass (WM—product of Volume and density). This enables the aiming at balancing regulation of the system welfare. The intervention to balance household WEF consumption inequality, in conjunction with the subjective attitude uses big data mining and analytics approach vis a vis water availability, yields a computer-based decision-support system (DSS) for defunding/subsidizing actions to over/under respective consumption leads to activating regulatory policy instrument.

This WEFH input across a region welfare imbalanced nodal network is quantified and visualized to derive recommendations for policy-driven interventions to regional ascribed nodes' WM source/sink terms.

Our hypothesis of causality-driven directional information is exemplified by a sharp price increase in wheat and rice, for USA and Thailand, respectively, that manifests its impact on the temporal trend of Israel's administrative districts of the WEF expenditures.

Dr. Maria Dermiki
Atlantic Technological University, Ireland

Dr. Maria Dermiki has an interdisciplinary background in scientific research with a degree in Chemical Engineering from Aristotle University of Thessaloniki, M.Sc. in

Food Science and Nutrition from University of Ioannina, Ph.D. in Food Biosciences from the University of Reading, postdoctoral research experience in Sensory Science and Dairy Science and finally industrial experience in Food Product Development. She has a holistic approach in product development, starting from Food Chemistry, Food Processing, Sensory Evaluation and Consumer Science, Packaging and Legislation, while adhering to sustainable practices. She is interested specifically in understanding the food choices of different population groups to develop nutritious foods that address the three pillars of sustainability and contribute to sustainable diets. Finally, she applies the UN Sustainable Development Goals to all the modules she teaches at undergraduate and postgraduate level, with the aim to help students apply sustainable practices in their professional and everyday life.

"What Would You Eat to Save the Earth?," Challenges and Opportunities for Sustainable Food Product Development

Abstract

Nowadays, the inadequate natural resources, climate change, and the negative impact of food production on the environment call for more sustainable food production practices that will ensure there will be enough food to feed the growing world population. Numerous solutions are being explored to overcome the multiple challenges that current food production systems are facing. For example, researchers are exploring the application of protein sources alternative to animal, such as insects or plants that could have a lower environmental impact compared to livestock production. Adding to the growing challenge of food security, almost a third of food produced worldwide is lost or wasted along the supply chain. In order to achieve the UN Sustainable Development Goal 12.3 that aims to "halve global food waste at consumer and retailer level and reduce food losses at the primary and secondary production stages" multiple ways to reduce food waste are currently investigated. One way to achieve this and retrieve nutrients from food waste is through the concept of upcycled foods which is gaining research and commercial interest.

Despite the promising outcomes of the shift to more sustainable food products, there are numerous challenges still to be addressed. For example, there is a need for changes in the current legislation that hinders innovation toward the use of alternative ingredients such as insects or underutilized ingredients and food processing by-products. Another challenge is consumer acceptability of such alternative ingredients and products that contain them. This presentation will explore the potential of sustainable food product development while also discussing the challenges industry faces.

Conference Sponsors

The conference co-chairs and organizing committee wish to acknowledge the support and sponsorship given in the organization of the ICOWEFS 2023—International Conference on Water Energy Food and Sustainability, held at the Polytechnic Institute of Leiria, Portugal.

Official Sponsors

POLITÉCNICO DE PORTALEGRE

POLITÉCNICO DE LEIRIA

RUN REGIONAL UNIVERSITY NETWORK EUROPEAN UNIVERSITY

ICoWEFS 2023 is Organized in Cooperation with these Research Units

Event Sponsors and Exhibitors

Preface

The 3rd edition of **International Conference on Water Energy Food and Sustainability (ICoWEFS 2023)**, taking place in Leiria (May 10–12, 2023) Portugal, aims to be a major event to foster innovation and exchange knowledge in the water-energy-food nexus, embracing the Sustainable Development Goals (SDGs) of the United Nations, building a better future for all bringing together leading academics, researchers, and industrial experts.

Taking into account that the EU has announced several actions for the implementation of the SDGs, such as reducing emissions to achieve climate neutrality by establishing a goal of zero net emissions by 2050, but there will be an interim goal of reducing emissions by 55% by 2030. In addition, the EU required an increase to 45% of renewable sources in the energy mix by 2030, as well as that all new buildings in the EU produce zero emissions from 2030 linked to the plan that in the future, energy consumption should be smaller. The conference will also focus on interconnected areas such as waste and effluent recovery, renewable gases, renewable energies, carbon capture, efficient use of water, and sustainable agrifood technologies.

The conference expects to foster networking and collaboration among participants to advance the knowledge and identify major trends in the above mentioned fields.

This conference is the 3rd co-organization between two Portuguese Polytechnics from Leiria and Portalegre and is open to new partners in the future events.

This edition counted with 87 presentations and 6 keynote speakers from 23 nationalities, that enriched the debate and evolution of these four scientific areas, namely AgriFood, Energy, Sustainability and Water. And as a result of the submitted papers, this book proceedings contains 70 papers.

Finally, the technical visits to a Smart Management Water System and a Biogas Power Plant in the surroundings of the conference city are contributors to promoting the dissemination of knowledge and the interaction among the academy, research centers, and the industry.

Welcome to this 3rd edition of the ICoWEFS 2023 Conference in Leiria, Portugal.

ICoWEFS
2023

Leiria, Portugal
Portalegre, Portugal

João Rafael da Costa Sanches Galvão, Ph.D.
Paulo Brito, Ph.D.
Conference Chairs

Contents

Agrifood

The following part contains papers related to "Agrifood".

The editors had assistance from the following conference organizing committee member in the editing of this part:

- Vânia Sofia Santos Ribeiro

Comparison of Two Organically Produced Tomato Cultivars After Micronutrients Enrichment

Cláudia Campos Pessoa, Diana Daccak, Inês Carmo Luís, Ana Coelho Marques, Ana Rita F. Coelho, Manuela Simões, Fernando Reboredo, Maria Manuela Silva, Maria Fernanda Pessoa, Mauro Guerra, Roberta G. Leitão, Paula Scotti-Campos, Isabel P. Pais, José C. Ramalho, and Fernando C. Lidon

Abstract Iron (Fe) and zinc (Zn) are micronutrients commonly linked to nutritional deficits. Inadequate intakes of Fe can be linked to impaired cognition, weakness and fatigue or pathologies such as anemia. As for Zn, cell damages, stunting or increases in infection susceptibility can occur. In Portugal, an enrichment workflow with foliar sprays was performed, to increase Fe and Zn contents in two organically produced tomato cultivars. A total of 4 foliar sprays with a mixture of Maxiblend (concentrations of 1–7 kg/ha) and Zitrilon 15% (concentrations of 0.40–2.0 kg/ha), were applied during the production cycle. At harvest, mineral enrichment was assessed by µ-EDXRF, while total soluble solids (TSS) were measured with a refractometer. At harvest, Apple cultivar showed increases of 1.17–1.27 fold for Fe and 1.33–1.38 fold for Zn. For Chucha, Fe values increased by 1.10–2.05 fold, while Zn increased by 1.51–1.99 fold. Brix values at harvest ranged between 4.6 and 5.0° Brix and 4.2 and 4.9° Brix for Apple and Chucha cultivars respectively. Significant differences

C. C. Pessoa (✉) · D. Daccak · I. C. Luís · A. C. Marques · A. R. F. Coelho · M. Simões · F. Reboredo · M. M. Silva · M. F. Pessoa · F. C. Lidon
Earth Sciences Department, Faculdade de Ciências e Tecnologia, Universidade Nova de Lisboa, Caparica, Portugal
e-mail: c.pessoa@campus.fct.unl.pt

C. C. Pessoa · D. Daccak · I. C. Luís · A. C. Marques · A. R. F. Coelho · M. Simões · F. Reboredo · M. M. Silva · M. F. Pessoa · P. Scotti-Campos · I. P. Pais · J. C. Ramalho · F. C. Lidon
GeoBioTec Research Center, Faculdade de Ciências e Tecnologia, Universidade Nova de Lisboa, Caparica, Portugal

M. Guerra · R. G. Leitão
LIBPhys, Faculdade de Ciências e Tecnologia, Universidade Nova de Lisboa, Caparica, Portugal

P. Scotti-Campos · I. P. Pais
INIAV, Instituto Nacional de Investigação Agrária e Veterinária, Oeiras, Portugal

J. C. Ramalho
PlantStress and Biodiversity Laboratory, Centro de Estudos Florestais (CEF), Instituto Superior Agronomia (ISA), Universidade de Lisboa (ULisboa), Oeiras, Portugal

J. R. da. C. S. Galvão et al. (eds.), *Proceedings of the 3rd International Conference on Water Energy Food and Sustainability (ICoWEFS 2023)*, Springer Proceedings in Earth and Environmental Sciences, https://doi.org/10.1007/978-3-031-48532-9_1

occurred only in Chucha. In conclusion, the applied agronomic approach can increase both minerals in both cultivars. For both cultivars T3 surpassed the toxicity threshold, limiting a continuous application. Complementary, both cultivars presented different responses to the applied concentrations, with Apple cultivar needing higher concentrations than Chucha to present mineral increases. The applied workflow did not negatively impact Brix values. Overall cultivar variability in organically produced tomatoes must be taken into consideration to maximize results.

Keywords Foliar applications · Iron enrichment · *Lycopersicon esculentum* · Total soluble solids · Zinc enrichment · μ-EDXRF

1 Introduction

According to the Food and Agriculture Organization [1], the tomato crop assumes a prominent position as the second most important vegetable crop worldwide. Furthermore, according to FAOSTAT [2], this fruit production worldwide reached 186,821,216 tonnes in 2020. Of these, 1399,210 tonnes were produced in Portugal [2], the vast majority being exported (as sauces, paste, or canned tomatoes) to international destinations such as the United Kingdom or Japan (among others) [3]. Most of this production comes from the conventional way, compared to tomatoes grown in organic production mode. In fact, given the expected population increase in the upcoming years [4], the conventional mode can facilitate the fulfilment of the food needs of the majority of the population. In turn, organic farming emerges as a potential market (currently in growth) based on pillars of quality assurance, food safety and improvement of organoleptic characteristics. A concept with which a growing number of consumers identify and thus, are willing to pay the price difference, seeking to invest on sustainable production [5].

Mineral elements such as zinc (Zn) and iron (Fe) perform various functions that ensure the proper functioning of the human body, and their deficits can promote the development of pathologies. Iron is incorporated into proteins (hemoglobin and myoglobin), which enable the transport of oxygen to the different tissues of the human body [6]. In turn, Zn is part of different metabolic processes, due to the performance of regulatory and catalytic functions [7]. Iron and Zn deficits are common in human diets namely when considering that one half of the world's population is affected by mineral malnutrition [8]. In fact, according to the National Program for the Promotion of Healthy Eating—PNPAS [9], the nutritional deficit of Fe is the most common in the world. In more severe cases, deficits of this mineral element can trigger anemia [9], while Zn can affect memory, immune system, cell division and growth, or influence protein and DNA synthesis [10, 11]. According to the National Institute of Health [6], the Recommended Daily Amount (RDA) of Fe for adults (ages 19 and up), can range from 8 to 18 mg. Furthermore, it is noted that women need a higher daily amount (up to 27 mg during pregnancy), however a daily amount of 45 mg should not be exceeded. In the case of Zn, the RDA for adults (ages 19 and up) can vary between 8 and 11 mg, but a daily intake of 40 mg should not be exceeded [11].

Biofortification can be used as a strategy to counter these mineral elements deficits in diets, since it aims to increase the levels of chosen nutrients in the edible parts of target crops [12]. In this context, the agronomic route consists in the use of fertilizers directly applied in the soil or sprays to the aerial part of plants, followed by its absorption and further translocating to fruits/grains [12].

In this context, agronomic biofortification through foliar applications (with products in accordance with organic farming) of different tomato cultivars can be considered, as it aims to increase the levels of selected mineral elements (such as Fe and Zn) in the fruits. Agroindustries can thus provide to consumers products with functional properties, namely in a preventive perspective of pathologies such as the ones previously mentioned.

2 Materials and Methods

2.1 Enrichment Workflow with Fe and Zn

An experimental field was selected in Ourém (Portugal), after taking into consideration edaphoclimatic parameters and protocols for tomato growth according to organic farming practices.

Later, an agronomic enrichment workflow was implemented to increase the levels of Zn and Fe in two tomato cultivars (*Lycopersicon esculentum*, namely Apple-Edwin and Chucha-Hedvig), produced in organic mode. Thus, a total of four foliar sprays of a mixture of Zitrilon 15% and Maxiblend, spaced 10 to 16 days apart were performed during the productive cycle (from August to October). Both products can be used in organic farming and can be applied by foliar spray in tomato crops. Zitrilon 15% is a concentrated Zn fertilizer, with 15% Zn in chelated form (EDTA), and Maxiblend is majorly constituted of Fe (soluble in water, 5.25% and chelated, 4.75 + 0.50%) as well as other mineral elements (such as Mn (2%) > Zn (1%) > B (0.4%) > Cu (0.15%) > Mo (0.1%)). Beyond Control (0 kg/ha Zitrilon 15% + 0 kg/ha Maxiblend), a total of three treatments with different concentrations were applied: T1—0.4 kg/ha Zitrilon 15% + 1.0 kg/ha Maxiblend; T2—1.2 kg/ha Zitrilon 15% + 4.0 kg/ha Maxiblend; and T3—2.0 kg/ha Zitrilon 15% + 7.0 kg/ha Maxiblend. It should be noted that in treatment T3, there was only one foliar application due to a noticeable evolution of toxicity symptoms.

2.2 Mineral Content and Total Soluble Solids Assessment

After a brief cleaning of the tomatoes with deionized water, a longitudinal cut was performed to obtain slices that were subsequently dehydrated (at 60 °C) until constant weight. A micro-energy X-ray scattering fluorescence system (M4 Tornado™, Germany) was used to determine the levels of Zn and Fe fruits as described in a similar study [13]. The total soluble solids (TSS) were determined by using a refractometer (Atago, Japan), with values expressed in ºBrix, as mentioned in a similar study [13].

2.3 Statistical Analysis

A One-Way ANOVA ($P \leq 0.05$) was performed for both analyses, to assess statistical differences between all treatments for each variety and a Tukey test was applied for mean comparison (confidence level of 95%).

3 Results

3.1 Mineral Content in Fruits

Regarding Zn content of Apple cultivar (Fig. 1), treatments T2 and T3 were significantly superior to the Control, while T1 was significantly inferior to the Control. Regarding Zn content in Chucha cultivar (Fig. 1), all treatments were significantly superior to the Control, and T1 presented the highest value, being significantly higher than T2 e T3. Zinc increases in Apple cultivar varied between 1.33 and 1.38 fold, and for Chucha, the increases reached 1.51–1.99 fold.

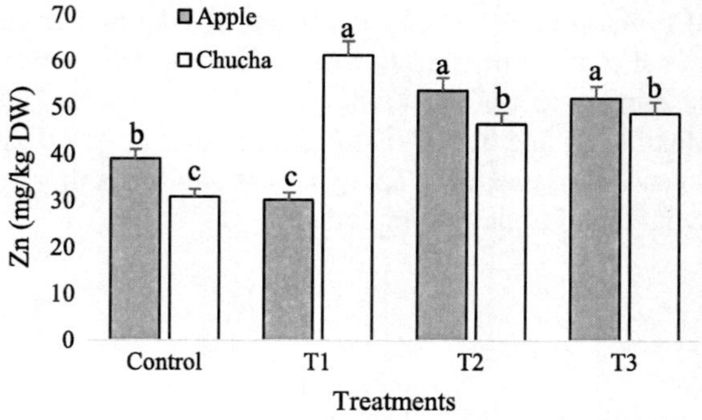

Fig. 1 Average values of Zn (mg/kg dry weight) content at harvest, of two tomato cultivars organically produced

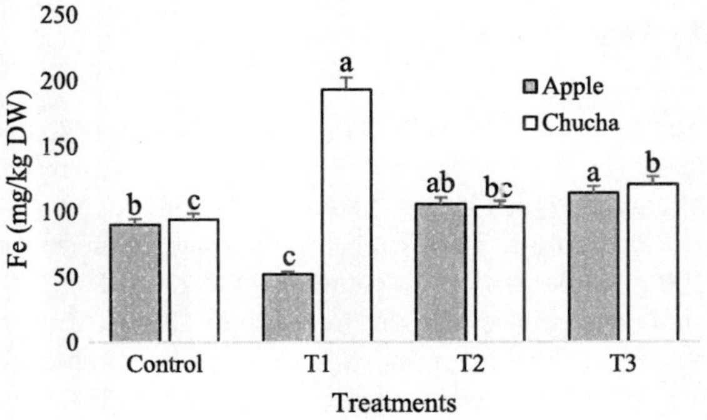

Fig. 2 Average values of Fe (mg/kg dry weight) content at harvest, of two tomato cultivars organically produced

Regarding Fe content of Apple cultivar (Fig. 2), treatments T2 and T3 presented higher values than the Control, significantly only in T3, while T1 was significantly inferior to the Control. Regarding Fe content in Chucha cultivar (Fig. 2), all treatments were superior to the Control, significantly only for T3 and T1, the latest presenting the highest value. Thus, at harvest, Fe content in sprayed fruits increased by 1.17–1.27 fold in Apple cultivar and for Chucha, values increased by 1.10–2.05 fold. For all treatments of both cultivars, Fe values were superior to Zn values. When comparing only the controls, Apple cultivar presents higher values of Zn, while higher values of Fe prevailed in Chucha cultivar.

3.2 *Total Soluble Solids*

For Apple cultivar, values ranged between 4.6 and 5.0° Brix, and for Chucha cultivar, values varied between 4.2 and 4.9° Brix. Significant differences were only observed in Chucha cultivar, in which T3 was significantly higher than the Control and T1 (Fig. 3).

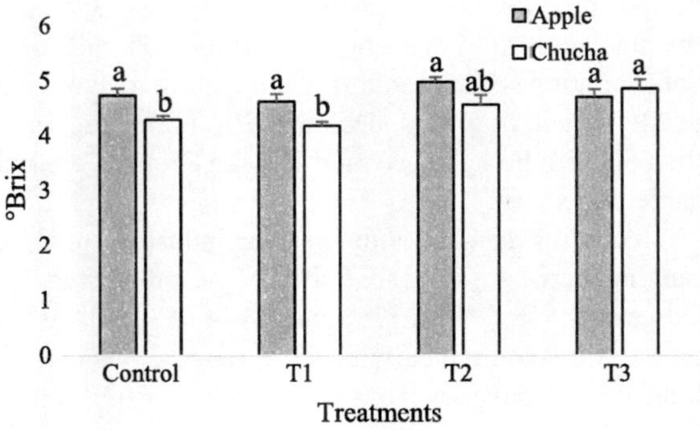

Fig. 3 Average values of total soluble solids content at harvest (°Brix), of two tomato cultivars organically produced

4 Discussion

Agronomic biofortification in tomato crops has been performed in different cultivars, in conventional or organic producing modes with different mineral elements such as Mg and Se [13, 14] and in some cases with Fe and Zn [15, 16].

According to PortFir [17], raw tomatoes present higher values of Fe (0.7 mg/100 g edible portion) in comparison to Zn (0.1 mg/100 g edible portion). Our data are in accordance with this trend, since all treatments of both cultivars revealed this tendency. The same trend was observed by another study [18] performed in Spain, where Fe values were superior to Zn in two distinct cultivars (Lladó and Antillas), produced both in organic and conventional mode.

Regarding mineral increases, T1 was not effective in incrementing Fe and Zn contents in Apple cultivar, but T2 and T3 led to increases of both mineral elements. However, although T3 presented significant increases of Fe and Zn with a single foliar spray, a continuous application is unfeasible since toxicity signs were evident. This suggests that the adequate concentration threshold was surpassed, and further applications could permanently damage the plant normal functioning. For Chucha cultivar, all treatments led to increases of both minerals, significantly in T1 and T3. Contrarily to Apple cultivar, mineral rises were achieved with lower concentrations (T1), but similarly, higher concentrations (T3) also surpassed the toxicity threshold for this cultivar, limiting further sprays. Thus, our study further enforces that agronomic workflow outcomes differ between cultivars. This is in accordance with another author [18], that points "cultivar" as the main aspect for variability in micronutrient content of tomato fruits. Furthermore, interactions between both Fe and Zn are reported, mainly due to chemical similarity and transport proteins [19]. In this context, an antagonistic interaction between both minerals (after simultaneous application) was not observed in this study. Accordingly, in cherry tomatoes (in soilless system), Fe applications let to synergic increases in other mineral contents such as Zn, while Fe increases in fruits were also reported [16]. Regarding the effects of sprays with higher concentrations, it is known that micronutrients such as Fe and Zn, when present in supra-optimal levels become toxic for plants, impairing their normal functioning (by instigating situations such as oxidative stress) [19]. Thus, the management of concentrations and number of foliar sprays need to be taken into consideration while monitoring the culture to avoid reaching toxicity levels that can result in yield or growth declines [20]. In this regard, concentrations slightly higher than 500 mg Fe kg^{-1} DW and 300 mg Zn kg^{-1} are indicated as critical in leaves of some crops [21].

According to [22], tomatoes for industrial processing can present TSS values ranging between 4.5 and 6.5%, while other studies reported values that ranged between 4.5–5.1 ºBrix [23] and 4.1–4.8% [24] for different tomato cultivars. Our data are in accordance with these ranges. Besides color, Brix serves as a quality indicator in tomatoes. However, this parameter is more strictly controlled in fruits from conventional methods rather than those organically produced, since the later are meant for direct consumption while the others are intended for industrial processing (where lower energy spending during evaporation processes can be achieved with higher Brix values) [24]. Nevertheless, Brix values can be associated to sweetness, a sensorial parameter often highly regarded by consumers, namely affecting subsequent purchases [25]. Additionally, TSS values are known to vary with maturation stages, cultivars, edaphoclimateric conditions, and agronomic practices [24]. In this context, our data suggest that both cultivars present similar Brix values, with Apple cultivar presenting slightly higher values than Chucha. Furthermore, the applied agronomic workflow did not present negative impacts in both cases, since the values either did not change or increased, being in accordance with industry and consumers preferences.

5 Conclusion

Both tomato cultivars organically produced (for direct consumption) showed positive results regarding the potential for biofortification by foliar route with the Fe and Zn. It should be noted, however, that mineral increases are cultivar dependent, with Chucha cultivar showing a greater accumulation of Fe and Zn. Besides, there was only a higher content of TSS in Chucha cultivar submitted to the enrichment workflow, which indicated the absence of negative impacts.

References

1. FAO - Food and Agriculture Organization: Land & Water – Crop Information – Tomato. https://www.fao.org/land-water/databases-and-software/crop-information/tomato/en/. Last accessed 2022/10/27.
2. FAO - Food and Agriculture Organization: FAOSTAT. https://www.fao.org/faostat/en/#data/QCL. Last accessed 2022/10/27.
3. GPP - Gabinete de Planeamento, Políticas e Administração Geral: Informação sobre produtos. https://www.gpp.pt/index.php/produtos/produtos. Last accessed 2022/10/28.
4. FAO - Food and Agriculture Organization of the United Nations.: The future of food and agriculture—Trends and challenges. FAO, Roma (2017). ISBN 978-92-5-109551-5.
5. Maples, M. Interis, M.G., Morgan, K.L., Harri, A.: Southeastern consumers' willingness to pay for environmental production attributes of fresh tomatoes. Journal of Agricultural and Applied Economics 50, 27–47 (2018). https://doi.org/10.1017/aae.2017.18.
6. NIH - National Institute of Health: Iron. https://ods.od.nih.gov/factsheets/Iron-Consumer/. Last accessed 2022/10/27.
7. Grüngreiff, K., Gottstein, T., Reinhold, D.: Zinc deficiency—An independent risk factor in the pathogenesis of haemorrhagic stroke? Nutrients 12, 3548 (2020). https://doi.org/10.3390/nu12113548.
8. Sharma, D., Jamra, G., Singh, U.M., Sood, S., Kumar, A.: Calcium biofortification: Three pronged molecular approaches for dissecting complex trait of calcium nutrition in Finger Millet (*Eleusine coracana*) for devising strategies of enrichment of food crops. Frontiers in Plant Science 7, 2028 (2017). https://doi.org/10.3389/fpls.2016.02028.
9. PNPAS - National Program for the Promotion of Healthy Eating: Iron. https://alimentacaosaudavel.dgs.pt/nutriente/ferro/. Last accessed 2022/10/27.

10. PNPAS - National Program for the Promotion of Healthy Eating: Zinc. https://alimentacaosaudavel.dgs.pt/nutriente/zinco/. Last accessed 2022/10/27.
11. NIH - National Institute of Health: Zinc. https://ods.od.nih.gov/factsheets/Zinc-Consumer/. Last accessed 2022/10/27.
12. Garg, M., Sharma, N., Sharma, S., Kapoor, P., Kumar, A., Chunduri, V., Arora, P.: Biofortified crops generated by breeding, agronomy, and transgenic approaches are improving lives of millions of people around the world. Frontiers in Nutrition 5, 12 (2018). https://doi.org/10.3389/fnut.2018.00012.
13. Pessoa, C.C.; Lidon, F.C.; Coelho, A.R.F.; Marques, A.C.; Daccak, D.; Luís, I.C.; Caleiro, J.C.; Kullberg, J.C.; Legoinha, P.; Brito, M.G.; Ramalho, J.C.; Silva, M.J.; Rodrigues, A.P.; Guerra, M.; Leitão, R.G.; Campos, P.S.; Pais, I.P.; Semedo, J.N.; Silva, M.M.; Galhano, C.; Leal, N.; Reboredo, F.H.; Pessoa, M.F.; Simões, M.: Magnesium accumulation in two contrasting varieties of *Lycopersicum esculentum* L. fruits: Interaction with calcium at tissue level and implications on quality. Plants 11, 1854 (2022). https://doi.org/10.3390/plants11141854.
14. Narváez-Ortiz, W.A., Becvort-Azcurra, A.A., Fuentes-Lara, L.O., Benavides-Mendoza, A., Valenzuela-García, J.R., González-Fuentes, J.A.: Mineral composition and antioxidant status of tomato with application of selenium. Agronomy 8, 185 (2018). https://doi.org/10.3390/agronomy8090185.
15. Coelho, A.R.F., Daccak, D., Luís, I.C., Marques, A.C., Pessoa, C.C., Silva, M.M., Simões, M., Reboredo, F.H., Pessoa, M.F., Legoinha, P., Ramalho, J.C., Campos P.S., Pais, I.P., Semedo, J.N., Lidon, F.C.: Can natural fortification increase Fe and Zn content in organically grown tomatoes? Biology and Life Sciences Forum 16, 26 (2022). https://doi.org/10.3390/IECHo2022-12504.
16. Buturi, C.V., Coelho, S.R.M., Cannata, C., Basile, F., Giuffrida, F., Leonardi, C., Mauro, R.P.: Iron biofortification of greenhouse cherry tomatoes grown in a soilless system. Horticulturae 8, 858 (2022). https://doi.org/10.3390/horticulturae8100858.
17. PortFir Homepage. http://portfir.insa.pt/. Last accessed 2022/11/10.
18. Ordóñez-Santos, L.E., Vázquez-Odériz, M.L., Romero-Rodríguez, M.A.: Micronutrient contents in organic and conventional tomatoes (*Solanum lycopersicum* L.). International Journal of Food Science & Technology 46(8), 1561–1568 (2011). https://doi.org/10.1111/j.1365-2621.2011.02648.x.
19. Rai, S., Singh, P.K., Mankotia, S., Swain, J., Satbhai, S.B.: Iron homeostasis in plants and its crosstalk with copper, zinc, and manganese. Plant Stress 1, 100008 (2021). https://doi.org/10.1016/j.stress.2021.100008.
20. Taiz, L., Zeiger, E.: Plant physiology. 3rd edn. Sinauer Associates, Sunderland, UK (2002).
21. Marschner, P.: Marschner's mineral nutrition of higher plants. 3rd edn. Academic Press, Cambridge, USA (2011). ISBN 9780123849052.
22. Garcia, E., Barrett, D.M.: Evaluation of processing tomatoes from two consecutive growing seasons: Quality attributes, peelability and yield. Journal of Food Processing and Preservation 30, 20–36 (2006). https://doi.org/10.1111/j.1745-4549.2005.00044.x.
23. Peixoto, J.V.M, Garcia, L.G.C., Nascimento, A.R., Moraes, E.R., Ferreira, T.A.P.C, Fernandes, M.R., Pereira, V.A.: Post-harvest evaluation of tomato genotypes with dual purpose. Food Science and Technology 38, 255-262 (2018). https://doi.org/10.1590/1678-457X.00217.
24. Rodrigues, P.M.C.S.F.: Comparação de cultivares de tomate de indústria em modo de produção biológico. Master thesis. University of Lisbon (2017). http://hdl.handle.net/10400.5/15122.
25. Magwaza, L.S., Opara, U.L.: Analytical methods for determination of sugar and sweetness of horticultural products - A review. Scientia Horticulturae 184, 179–192 (2015). https://doi.org/10.1016/j.scienta.2015.01.001.

Foliar Application of $ZnSO_4$ and ZnO on a *Vitis Vinifera* Cv. Moscatel in Portugal: Assessment of Quality Parameters

Diana Daccak, Inês Carmo Luís, Ana Coelho Marques, Ana Rita F. Coelho, Cláudia Campos Pessoa, José C. Ramalho, Mauro Guerra, Roberta G. Leitão, Paula Scotti-Campos, Isabel P. Pais, Maria Manuela Silva, Paulo Legoinha, Maria Fernanda Pessoa, Manuela Simões, Fernando H. Reboredo, and Fernando C. Lidon

Abstract Zinc deficiency affects the population worldwide, being developed several strategies to mitigate this health problem, such as agronomic biofortification. The absence of an adequate amount of Zn affects growth, immunity, and reproductive system. Accordingly, a technological itinerary for Zn agronomic biofortification was outlined using grapes *Vitis vinifera* cv. Moscatel (i.e., with irrigation), aiming the optimization of the Zn content, and the determination of potential implications in some quality parameters. Along the production cycle, three foliar applications of $ZnSO_4$ and ZnO fertilizers were performed at the concentrations of 0, 450 and 900 g ha^{-1}

D. Daccak (✉) · I. C. Luís · A. C. Marques · A. R. F. Coelho · C. C. Pessoa · M. M. Silva · P. Legoinha · M. F. Pessoa · M. Simões · F. H. Reboredo · F. C. Lidon
Earth Sciences Department, Faculdade de Ciências E Tecnologia, Universidade Nova de Lisboa, Campus da Caparica, 2829-516 Caparica, Portugal
e-mail: d.daccak@campus.fct.unl.pt

D. Daccak · I. C. Luís · A. C. Marques · A. R. F. Coelho · C. C. Pessoa · J. C. Ramalho · P. Scotti-Campos · I. P. Pais · M. M. Silva · P. Legoinha · M. F. Pessoa · M. Simões · F. H. Reboredo · F. C. Lidon
Faculdade de Ciências E Tecnologia, GeoBioTec Research Center, Universidade Nova de Lisboa, Campus da Caparica, 2829-516 Caparica, Portugal

J. C. Ramalho
Plant Stress & Biodiversity Laboratory, Centro de Estudos Florestais (CEF), Instituto Superior Agronomia (ISA), Universidade de Lisboa (ULisboa), Quinta Do Marquês, Oeiras and Tapada da Ajuda, Av. República, 2784-5051349-017 Lisboa, Portugal

M. Guerra · R. G. Leitão
Physics Department, Faculdade de Ciências E Tecnologia, LIBPhys, Universidade Nova de Lisboa, Campus da Caparica, 2829-516 Caparica, Portugal

P. Scotti-Campos · I. P. Pais
Instituto Nacional de Investigação Agrária e Veterinária, I.P. (INIAV), Avenida da República, Quinta do Marquês, 2780-157 Oeiras, Portugal

J. R. da. C. S. Galvão et al. (eds.), *Proceedings of the 3rd International Conference on Water Energy Food and Sustainability (ICoWEFS 2023)*, Springer Proceedings in Earth and Environmental Sciences, https://doi.org/10.1007/978-3-031-48532-9_2

(with the control being sprayed with water). At harvest, all biofortified grapes showed an increase in Zn content (greater at maximum concentrations of 900 g ha^{-1}), with ZnO-grapes achieving a 4.2-fold increase. Through the tissue analyses (μ-EDXRF) a Zn accumulation in the seeds was found after the application of Zn fertilizers. Moreover, sugar analyses presented mainly glucose and fructose within the range of 7.70–13.22% and 8.26–14.09% respectively, with no significant differences between the biofortified and control grapes. Also, in the colorimetric parameters accessed, no significant changes were observed in Moscatel grapes. It was concluded, that Zn biofortification was effective increasing the Zn amount with both fertilizers and additionally without negatively impacting the quality parameters.

Keywords Agronomic biofortification · *Vitis Vinifera* cv. Moscatel · Zn · Zn-fertilizers

1 Introduction

One third of the population suffers from zinc (Zn) deficiency, mainly in developing countries, of which 60–70% in Asia and Sub-Saharan Africa are susceptible to this health problem [1, 2]. Zinc is acquired through dietary intake, being found in foods such as red meat, some seafoods (e.g., oysters and crab), nuts and whole grains cereals [3, 4]. In developing countries, low dietary diversity of foods is one of the main reasons contributing to Zn deficiency, along with reduced intake, low absorption and/or a higher rate of Zn consumption by the body, which can affect zinc content [1, 3]. Dietary Zn reference for population is 9.4 and 7.5 mg per day for adult males and females, respectively (i.e., being higher depending on the consumption of more phytate through food such as unrefined cereals, legumes, and oleaginous seeds) [4, 5].

Zinc has several roles in biological processes, with prosthetic groups of Zn being found in about 2800–3000 proteins in the organism and required for the normal functioning of at least 300 enzymes (i.e., involved in all six classes of enzyme—hydrolases, lyases, ligases, isomerases, oxidoreductases, and transferases). Therefore, it influences the physical growth, development, immunity and reproductive function and neurobehavioral activity, linked to his action in the regulation of gene expression, signal transduction, neuronal transmission, and others [2, 6]. Zinc deficiency in the human body, leads to a retardation in physical growth, problems with immunity system, DNA damage, risk of cancer, infections, and problems in reproductive system [1, 7]. These conditions can cause other physiological problems, being responsible for several disorders namely: in the respiratory system, malaria, diarrhea, hair loss and skin lesions [7].

In this context, resulting of a global partnership the HarvestPlus program was created in 2003, where the main purpose is to mitigate micronutrient deficiencies (i.e., namely Zn) via crops biofortification [8]. Agronomic biofortification is the process of increasing a nutrient content in edible parts of plants (e.g., Zn), through soil and/or foliar application of fertilizers during the productive cycle [9]. In fact, the application of Zn fertilizers in crops is being study and his efficiency depends on some factors that influence the crop response such as: method of application, source of Zn, time of the application, genetic and environment conditions [2]. Also, it was reported that foliar application of Zn-fertilizers demonstrated to be efficient in cereals grains and in some fruit trees an increase in micronutrient content was observed along with a higher yield [10, 11].

Considering the above-mentioned problematic, this experimental study comprises the foliar application of $ZnSO_4$ and ZnO in grapes of *Vitis vinifera* cv. Moscatel with the aim of increasing the Zn content and additionally accessing the influence on some quality parameters.

2 Materials and Methods

2.1 *Experimental Itinerary*

Agronomic biofortification of a *Vitis vinifera* cv. Moscatel was performed at Setubal/Portugal (GPS coordinates 38° 35′ 47.113″ N; 8° 40′ 46.651″ W), through foliar application. After flowering, three pulverizations with $ZnSO_4$ and ZnO were applied during the productive cycle, at the concentrations of 0, 450 and 900 g ha^{-1} (i.e., the control vines were pulverized with water). Harvest was performed by 25th September of 2018.

2.2 *Orthophotomap and Surface Water Drainage*

A Drone (DJI Phantom 4 Pro+) equipped with high-definition and multi-sector RGB (i.e., with three Electromagnetic Spectra Bands—Red, Green and Blue) and Parrot Sequoia (i.e., with five Electromagnetic Spectra Bands—NIR, REG, Green, Red and RGB) cameras, was used to collect data on 1 August 2018. Parrot Sequoia multi-spectral camera calibration was performed considering the environmental brightness conditions. Images collected were then processed in a Workstation (AORUS, GIGA-BYTE Technology Co., Ltd—2019), to produce the final orthophotomap. Following [12], the general morphology and surface water drainage areas of the experimental fields were assessed using Agisoft PhotoScan Professional (Version 1.2.6, Software of 2016 and the ESRI of 2011 and ArcGIS Desktop—Release 10 from Redlands, CA: Environmental Systems Research Institute).

2.3 Zn Content in Grapes Analysis

After the 3rd foliar application, Zn content was quantified in randomized grapes (dried at 60 °C, until constant weight, grounded and processed into pellet) according to [13] with minor changes, using a XRF analyzer (model XL3t 950 He GOLDD+) under He atmosphere.

At harvest, the Zn content and location in the skin and seeds of grapes, was accessed with the μ-EDXRF system (M4 Tornado™, Bruker, Germany), according to [14]. The X-ray generator was operate at 50 kV and 100 μA without the use of filters, to enhance the ionization of low-Z elements. A set of filters between the X-ray tube and the sample, composed of three foils of Al/Ti/Cu (with a thickness of 100/50/25 μm, respectively), was applied for a better quantification. All the measurements with filters were perform with 600 μA current. Detection of fluorescence radiation was carried out by an energy-dispersive silicon drift detector, XFlash™, with 30 mm^2 sensitive area and energy resolution of 142 eV for Mn Kα. Measurements were carried out under 20 mbar vacuum conditions. These point spectra were acquired during 200 s.

2.4 Sugar Analysis

Grapes sugar extraction was carried out in 40 g of pulp (i.e., from several grape berries at harvest) per sample (n = 3), following [15] with minor changes. To perform the quantification and analyze of sugars data, software Breeze was used, based on the calibration curves of sucrose, glucose and fructose.

2.5 Colorimetric Analysis

According to [16], colorimetric parameters were accessed in the pulp of three randomized fresh grapes per sample (n = 3) with a scanning spectrophotometric colorimeter (Agrosta, European Union). The sensor provides a 40 nm full-width half-max detection, covering the visible region of the electromagnetic spectrum. This sensor has 6 phototransistors with sensibility in a specific region of the spectrum (380 nm—violet; 450 nm—blue; 500 nm—green; 570 nm—yellow; 600 nm—orange; 670 nm—red). Light was furnished by a white light-emitting diode (LED) covering all the visible region.

2.6 *Statistical Analysis*

Data were statistically analyzed using a One-Way ANOVA ($p \leq 0.05$) to evaluate differences between treatments. After, a Tukey's for mean comparison was performed (95% confidence level was adopt for all tests).

3 Results

3.1 *Meteorological Conditions*

Considering the field location, the meteorological conditions of 2018 were defined from the national meteorological network that is located at the Montijo Air Base (N 38°43′28.7″ O 9°0′40.114″). Data reported in the experimental period, was characterized by a maximum average temperature of 28 °C and a minimum average of 16.6 °C, with the maximum and minimum recorded was 44 °C and 11 °C respectively. According to the report by the Portuguese Institute of the Sea and the Atmosphere—IPMA [17] for the summer period, it refers to the coldest July since 2000 and August the 2nd warmest in the last 88 years.

3.2 *Surface Drainage Network*

The acquired data suggest a field morphology with small slope and surface drainage with water flow in the NE-SW direction (Fig. 1).

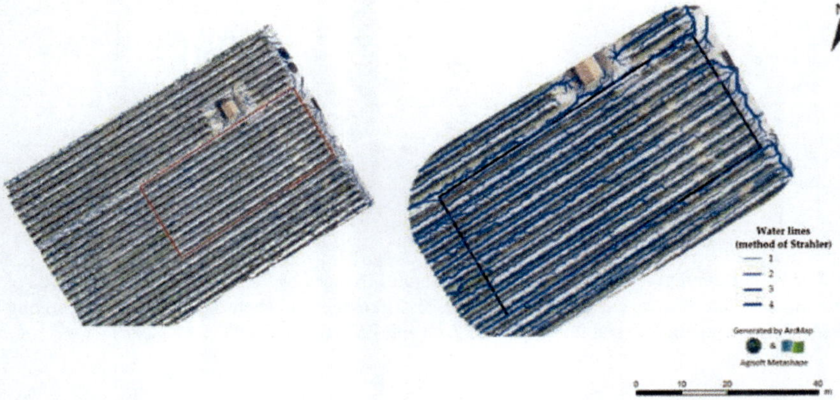

Fig. 1 Orthophotomap and map of the surface drainage network (order of water lines according to Strahler) of the vineyard Lau Novo-cv. Moscatel

3.3 Zn Content in Grapes

After the 3rd application of the Zn-fertilizers, an increase in Zn concentration was found in all grapes compared to control (Table 1). Indeed, Zn biofortification indexes varied between a 1.5 to 4.2-fold increase (with $ZnSO_4$- 450 g ha^{-1} not being significant) (Table 1). The highest concentrations of both treatments (900 g ha^{-1}) led to a higher accumulation of Zn in the Moscatel grapes (being the increase in ZnO-treated grapes significatively different regarding to the others) (Table 1).

At harvest, it was observed an accumulation of Zn in the seeds with Zn-fertilizer application, in relation to the skin (Fig. 2). Seeds analyze, showed a significant increase of Zn concentration with the application of $ZnSO_4$ and with the higher concentration of ZnO fertilizer 900 g ha^{-1} (although this last not being significative) (Fig. 2). As for the skin tissue, with Zn application it was found a decrease in Zn content, being observed significative higher values for control grapes regarding the others (Fig. 2).

Table 1 Average + S.E. (n = 3) of Zn amount in grapes of *Vitis vinifera* variety Moscatel in mg/kg after the 3rd foliar application with ZnO and $ZnSO_4$

	Zn (mg/kg)
Control	6.69 ± 0.01d
ZnO (450 g ha^{-1})	14.60 ± 1.77bc
ZnO (900 g ha^{-1})	28.13 ± 1.03a
$ZnSO_4$ (450 g ha^{-1})	9.74 ± 0.77 cd
$ZnSO_4$ (900 g ha^{-1})	15.07 ± 1.22b

The letters a, b, c, d indicates significative differences between the treatments (P < 0.05)

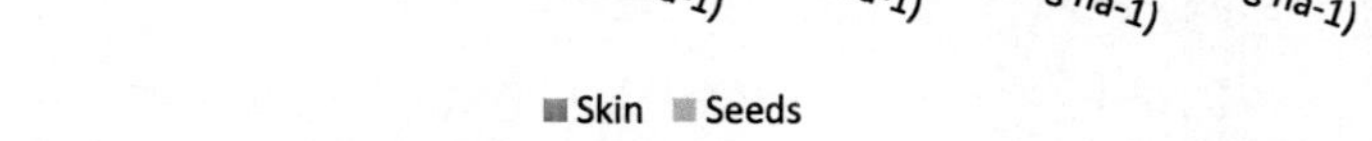

Fig. 2 Average content + S.E. (n = 3) of Zn amount in grapes of *Vitis vinifera* variety Moscatel in mg/kg. The letters a, b, c indicates significative differences between the seeds or skin among different treatments (0, 450 and 900 g ha^{-1} of ZnO and $ZnSO_4$) (P < 0.05)

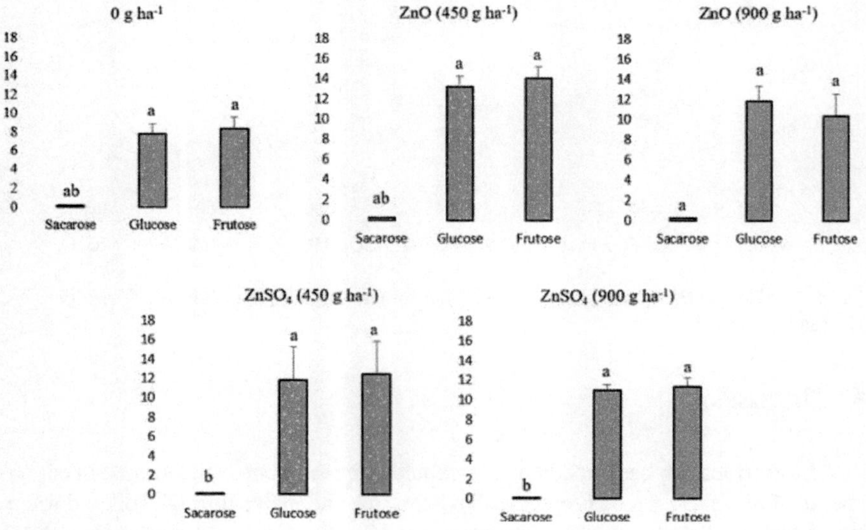

Fig. 3 Average content + S.E. (n = 3) of sugar (Sucrose, Glucose and Frutose) in grapes of *Vitis vinifera* L., variety Moscatel in % of fresh weight. The letters a, b indicates significative differences between the same sugar among different treatments (0, 450 and 900 g ha^{-1} of ZnO and $ZnSO_4$) ($P < 0.05$)

3.4 Sugar Composition in Grapes at Harvest

Moscatel pulp analysed showed three soluble sugars (sucrose, glucose and fructose) where a greater contribution of glucose and fructose was observed. The range for glucose and fructose was between 7.70–13.22% and 8.26–14.09% respectively (Fig. 3). In general, with Zn applications with both treatments, the sugar content didn't demonstrated significative differences (except in sucrose with treatment $ZnSO_4$ showing significantly lower values) (Fig. 3). Although, with Zn-treated grapes the sugar concentration was higher compared to grapes not biofortified.

3.5 Colorimetric Parameters of Grapes at Harvest

Grapes skin analysis in the visible spectral region (450–650 nm), showed maximum values of transmittance at 550 and 650 nm (Fig. 4). Transmittance values of Moscatel grapes didn't present differences with the application of $ZnSO_4$ and ZnO.

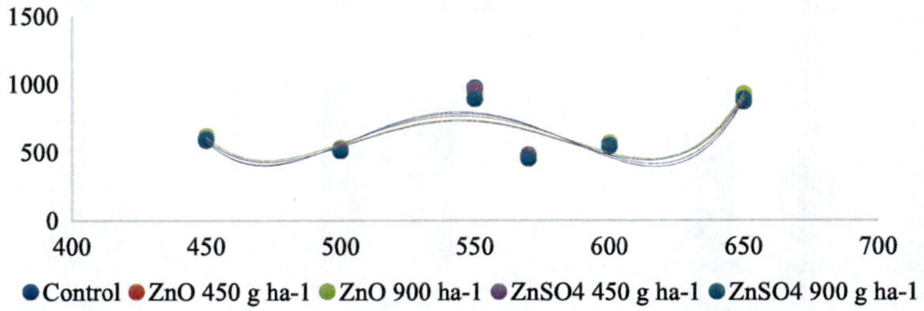

Fig. 4 Visible spectra showing the average of transmittance (n = 3) in the grapes *Vitis vinifera* cv. Moscatel at harvest

4 Discussion

Zinc biofortification has already demonstrated to be efficient as mentioned in experiments of the HarvestZinc project with wheat, rice and maize through soil and foliar applications [8]. As known [2] crops response to biofortification depends on genetic, time of application, environmental conditions, method of application and also the type of fertilizer. Accordingly, this experiment with grapes *Vitis vinifera* cv. Moscatel, demonstrated an increase of Zn amount in all grapes subjected to foliar application of ZnO and $ZnSO_4$ (Table 1), which is the most effective method applied compared to the soil application (i.e., since it prevents the fixation in the soil and the nutrient antagonistic effect in the Zn uptake, and also requires a low fertilizer application) [2, 8, 18]. Furthermore, the highest concentrations of the Zn-fertilizers (900 g ha^{-1}) applied led to a higher accumulation of Zn in the grapes (Table 1). Although $ZnSO_4$ is known to be more soluble, efficient, and additionally a source of sulfur for crops (required for an adequate growth and development), ZnO is cheaper [19–22] and reached a higher increase (i.e., 4.2-fold increase) in Moscatel grapes (Table 1). Thus, pointing to different responses of crops and genotypes under distinct edaphoclimatic conditions, to Zn-fertilizers. However, even under similar conditions, plants can display a different tissue concentration of Zn [23], being observed after the Zn biofortification in Moscatel grapes, a greater accumulation in the seeds (Fig. 2) (which according to [24] contributes to a sustainable crop growth and increase yield). In fact, it was already observed in previous studies [25], that the increase in Zn supply in pea plants (*Pisum sativum* L.) led to a higher concentration of bioavailable Zn in the seeds.

Regarding the quality parameters, the sugars of Moscatel grapes at harvest didn't show significant changes with the use of Zn-fertilizers (Fig. 3), which is important for the shape of berries, sensory properties, alcohol concentration and synthesis of organic acids, phenolics and aroma compounds [26]. Considering the sugars of Sonaka, Flame seedless and Red globe grapes (i.e., HPLC analysis) found in [27], our data showed higher levels (reaching a maximum of 13.22% and 14.09% for glucose and fructose respectively (Fig. 3)) compared to the range between 8.25–10.56% of fructose and 6.25–9.78% of glucose found in these varieties. Other sugars were not detected, as observed in [27] where maltose, xylose and lactose were below the detection limit, with fructose and glucose being the main sugars present in all the samples analyzed. Also, the field used in this study is irrigated (with surface drainage with water flow in the NE-SW direction although with a small slope which mitigates the differences in the water supply of the Moscatel grapes (Fig. 1)), which has an impact on the sugar accumulation, with the amount of glucose and fructose increasing in irrigated vines compared to non-irrigated vines [28].

As for the colorimetric parameters of the grapes biofortified with Zn, despite the climatic conditions of 2018 (i.e., year of the experiment), no changes were found (Fig. 4). In addition, this indicates the absence of sunburn damage, such as: loss of chlorophyll and consequently the green color [29].

5 Conclusions

Agronomic zinc biofortification led to a higher Zn concentration with both ZnO and $ZnSO_4$ (with the highest concentration (900 g ha^{-1}) showing the maximum increase). Furthermore, although the known solubility of $ZnSO_4$, the ZnO was more effective in Moscatel grapes, which indicates different responses of cultures to Zn fertilizers. Also, Zn biofortification led to a greater accumulation of Zn in the seeds. Moreover, color and sugars analyses were not affected after the Zn applications. Accordingly, this data suggests that the applied Zn-fertilizers has no negative impact on the physicochemical characteristics of grapes.

References

1. Dapkekar, A., Deshpande, P., Oak, M.D., Paknikar, K.M., Rajwade, J.M.: Zinc use efficiency is enhanced in wheat through nanofertilization. Sci Rep 8, 6832 (2018). https://doi.org/10.1038/s41598-018-25247-5
2. Praharaj, S., Skalicky, M., Maitra, S., Bhadra, P., Shankar, T., Brestic, M., Hejnak, V., Vachova, P., Hossain, A.: Zinc biofortification in food crops could alleviate the zinc malnutrition in human health. Molecules 26, 3509 (2021). https://doi.org/10.3390/molecules26123509
3. Wang, J., Um, P., Dickerman, B.A., Liu, J.: Zinc, magnesium, selenium and depression: A review of the evidence, potential mechanisms and implications. Nutrients 10, 584 (2018). https://doi.org/10.3390/nu10050584
4. Shkembi, B., Huppertz, T.: Influence of dairy products on bioavailability of zinc from other food products: A review of complementarity at a meal level. Nutrients 13, 4253 (2021). https://doi.org/10.3390/nu13124253
5. Gibson, R.S., Raboy, V., King, J.C.: Implications of phytate in plant-based foods for iron and zinc bioavailability, setting dietary requirements, and formulating programs and policies. Nutr Rev. 76, 793–804 (2018). https://doi.org/10.1093/nutrit/nuy028
6. Ackland, M.L., Michalczyk, A.A.: Zinc and infant nutrition. Arch Biochem Biophys 611, 51–57 (2016). https://doi.org/10.1016/j.abb.2016.06.011
7. Younas, N., Fatima, I., Ahmad, I.A., Ayyaz, M.K.: Alleviation of zinc deficiency in plants and humans throught anb effective technique; biofortification: A detailed review. Acta Ecologica Sinica (2022). https://doi.org/10.1016/j.chnaes.2022.07.008.

8. Harvestplus Homepage. https://www.harvestplus.org/. Last accessed 2022/11/15
9. Pal, V., Singh, G., Dhaliwal, S.S.: Agronomic biofortification of chickpea with zinc and iron through application of zinc and urea, Commun. Soil Sci. Plant Anal. 50, 1864–1877 (2019). https://doi.org/10.1080/00103624.2019.1648490
10. Nissar, R., Zahida, R., Kanth, R. H., Manzoor, G., Shafeeq, R., Ashaq, H., Bhat, R.A., Bhat, M.A., Tahir, S.: Agronomic biofortification of major cereals with zinc and iron-A review. Agricultural Reviews, 40, (2019). https://doi.org/10.18805/ag.R-1860
11. Rugeles-Reyes, S. M., Cecílio, A. B., Lopez Aguilar, M. A., Silva, P. H. S.: Foliar application of zinc in the agronomic biofortification of arugula. Food Sci. Technol. 39, 1011–1017 (2019). https://doi.org/10.1590/fst.12318
12. Pessoa, C.C., Lidon, F.C., Coelho, A.R.F.; Marques, A.C.; Daccak, D.; Luís, I.C.; Caleiro, J.C.; Kullberg, J.C.; Legoinha, P.; Brito, M.G., Ramalho, J.C., Silva, M.J., Rodrigues, A.P., Guerra, M., Leitão, R.G., Campos, P.S., Pais, I.P., Semedo, J., Silva, M.M., Galhano, C., Leal, N., Reboredo, F.H., Pessoa, M.F., Simões, M.: Magnesium accumulation in two contrasting varieties of *Lycopersicum esculentum* L. fruits: Interaction with calcium at tissue level and implications on quality. Plants 11, 1854 (2022). https://doi.org/10.3390/plants11141854
13. Luís, I.C., Lidon, F.C., Pessoa, C.C.; Marques, A.C.; Coelho, A.R.F.; Simões, M.; Patanita, M.; Dôres, J.; Ramalho, J.C.; Silva, M.M., Almeida, A.S., Pais, I.P., Pessoa, M.F., Reboredo, F.H., Legoinha, P., Guerra, M., Leitão, R.G., Campos, P.S.: Zinc enrichment in two contrasting genotypes of *Triticum aestivum* L. grains: Interactions between edaphic conditions and foliar fertilizers. Plants 10, 204 (2021). https://doi.org/10.3390/plants10020204
14. Coelho, A.R.F.; Ramalho, J.C.; Lidon, F.C.; Marques, A.C.; Daccak, D.; Pessoa, C.C.; Luís, I.C.; Guerra, M.; Leitão, R.G.; Semedo, J.M.N., Silva, M.M., Pais, I.P., Leal, M., Galhano, C., Rodrigues, A.P., Legoinha, P., Silva, M.J., Simões, M., Campos, P.S., Pessoa, M.F., Reboredo, F.H.: Foliar spraying of *Solanum tuberosum* L. with $CaCl_2$ and $Ca(NO_3)_2$: interactions with nutrients accumulation in tubers. Plants 11, 1725 (2022). https://doi.org/10.3390/plants11131725
15. Pessoa, C.C., Lidon, F.C., Daccak, D., Luís, I.C., Marques, A.C., Coelho, A.R.F., Legoinha, P., Ramalho, J.C., Leitão, A.E., Guerra, M., Leitão, R.G., Campos, P.S., Pais, I.P., Silva, M.M., Reboredo, F.H., Pessoa, M.F., Simões, M.: Calcium biofortification of Rocha pear fruits: Implications on mineral elements, sugars and fatty acids accumulation in tissues. Sci 4, 35 (2022). https://doi.org/10.3390/sci4040035
16. Daccak, D., Lidon, F.C., Luís, I.C., Marques, A.C., Coelho, A.R.F., Pessoa, C.C., Caleiro, J., Ramalho, J.C., Leitão, A.E., Silva, M.J., Rodrigues, A.P., Guerra, M., Leitão, R.G., Campos, P.S., Pais, I.P., Semedo, J.N., Alvarenga, N., Gonçalves, E.M., Silva, M.M., Legoinha, P., Galhano, C., Kullbergm J.C., Brito, M., Simões, M., Pessoa, M.F., Reboredo, F.H.: Zinc biofortification in *Vitis vinifera*: Implications for quality and wine production. Plants 11, 2442 (2022). https://doi.org/10.3390/plants11182442
17. IPMA Homepage. https://www.ipma.pt/pt/index.html. Last accessed 2022/11/22
18. Ram, H., Rashid, A., Zhang, W., Duarte, A. Á., Phattarakul, N., Simunji, S., Kalayci, M., Freitas, R., Rerkasem, B., Bal, R.S., Mahmood, K., Savasli, E., Lungu, O., Wang, Z.H., de Barros, V.L.N.P., Malik, S.S., Arisoy, R.Z., Guo, J.X., Sohu, V.S., Zou, C.Q., Cakmak, I.: Biofortification of wheat, rice and common bean by applying foliar zinc fertilizer along with pesticides in seven countries. Plant and soil 403, 389–401 (2016). https://doi.org/10.1007/s11104-016-2815-3
19. Umar, W., Hameed, M. K., Aziz, T., Maqsood, M. A., Bilal, H. M., Rasheed, N.: (2021). Synthesis, characterization and application of ZnO nanoparticles for improved growth and Zn biofortification in maize. Arch. Agron. Soil Sci 67, 1164–1176 (2021). https://doi.org/10.1080/03650340.2020.1782893
20. Szerement, J., Szatanik-Kloc, A., Mokrzycki, J., Mierzwa-Hersztek, M.: Agronomic biofortification with Se, Zn, and Fe: an effective strategy to enhance crop nutritional quality and stress defense—a review. J. Soil Sci. Plant Nutr 22, 1–31 (2022). https://doi.org/10.1007/s42729-021-00719-2

21. Narayan, O. P., Kumar, P., Yadav, B., Dua, M., Johri, A. K.: Sulfur nutrition and its role in plant growth and development. Plant Signal. Behav., 2030082 (2022). https://doi.org/10.1080/15592324.2022.2030082
22. Cakmak, I.: Enrichment of fertilizers with zinc: An excellent investment for humanity and crop production in India. J. Trace Elem. Med. Biol. 23, 281–289 (2009). https://doi.org/10.1016/j.jtemb.2009.05.002
23. White, P. J., Broadley, M. R.: Biofortification of crops with seven mineral elements often lacking in human diets–iron, zinc, copper, calcium, magnesium, selenium and iodine. New Phytol. 182, 49–84 (2009). https://doi.org/10.3389/fpls.2011.00080
24. Noulas, C., Tziouvalekas, M., Karyotis, T.: Zinc in soils, water and food crops. J. Trace Elem. Med. Biol. 49, 252–260 (2018). https://doi.org/10.1016/j.jtemb.2018.02.009
25. Sharma, P., Aggarwal, P., Kaur, A.: Biofortification: A new approach to eradicate hidden hunger. Food Rev. Int. 33, 1–21 (2017). https://doi.org/10.1080/87559129.2015.1137309
26. Trad, M., Boge, M., Ben Hamda, H., Renard, C. M., Harbi, M.: The Glucose-fructose ratio of wild Tunisian grapes. Cogent food agric. 3, 1374156 (2017). https://doi.org/10.1080/23311932.2017.1374156
27. Lodaya, J. D., Gotmare, S.: Determination of sugars in different grapes using high performance liquid chromatography-ELSD. JETIR 5 (2018). www.jetir.org (ISSN 2349–5162)
28. Parker, A. K., de Cortázar-Atauri, I. G., Gény, L., Spring, J. L., Destrac, A., Schultz, H., Molitor, D., Lacombe, T., Graça, A., Monamy, C., Stoll, M., Storchi, P., Trought, M.C.T., Hofmann, R.W., Van Leeuwen, C.: Temperature-based grapevine sugar ripeness modelling for a wide range of *Vitis vinifera* L. cultivars. Agric For Meteorol 285, 107902 (2020). https://doi.org/10.1016/j.agrformet.2020.107902
29. Gambetta, J. M., Holzapfel, B. P., Stoll, M., Friedel, M.: Sunburn in grapes: A review. Front. Plant Sci. 11, 2123 (2021). https://doi.org/10.3389/fpls.2020.604691

Influence of Knowledge, Motivations and Behaviors on Consumer Food Waste—Results from a Survey in Portugal

Joana Simões, Ana Carvalho, Tânia Pinto-Varela, and Margarida Gaspar de Matos

Abstract The aim of this study is to understand the influence of knowledge, motivations and behaviors on food wasted by households in Portugal and to provide suggestions for a communication strategy to raise awareness to this issue and promote food waste reductions. Data was collected through a survey made available online and distributed using the snowball sampling technique yielding a final sample of 376 answers. Three hypotheses were formulated to test the correlation of the amount of food wasted with knowledge about the impacts of food waste, and motivations and behaviors to reduce food waste. The results show that the amount of food wasted is not significantly correlated with knowledge but is moderately correlated with motivations and behaviors. Most respondents have already adopted behaviors to reduce food waste to some extent. However, weekly planning and the use of the freezer can be further incentivized. Motivations to reduce food waste include social, financial and environmental concerns, by this order, and the majority of respondents feels guilty when wasting food. Communication strategies may be more effective if they focus on promoting anti-food waste behaviors, involve all stakeholders of the food supply chain and provide positive messages instead of guilty feelings. The results apply to a segment of the population living in Portugal characterized by having a university degree, no children in the household and living mostly in the Lisbon Metropolitan Area. Further research is needed to segment consumers according to their socio-demographic characteristics, motivations, and behaviors regarding food waste to enable the design of tailored communication strategies.

Keywords Food waste · Survey · Consumer knowledge · Motivations and behaviors

J. Simões (✉) · A. Carvalho · T. Pinto-Varela
Centre for Management Studies (CEG-IST), Instituto Superior Técnico, University of Lisbon, Lisbon, Portugal
e-mail: simoesjoana@edu.ulisboa.pt

M. G. de Matos
Faculty of Medicine, ISAMB, University of Lisbon, Lisbon, Portugal

J. R. da. C. S. Galvão et al. (eds.), *Proceedings of the 3rd International Conference on Water Energy Food and Sustainability (ICoWEFS 2023)*, Springer Proceedings in Earth and Environmental Sciences, https://doi.org/10.1007/978-3-031-48532-9_3

1 Introduction

Around one third of the food that is produced globally is either lost or wasted every year [4], which translates into resources (water, energy, soils, labor) being wasted. Springmann et al. [22] estimate that the 50% reduction goal of food wastage set by the Sustainable Development Goals (SDG) would reduce the amount of resources used by the food system by a magnitude comparable to the reductions resulting from widespread dietary changes and technological changes in agricultural production.

In Portugal, the estimated food wastage is of about 1.89 million tons per year [3], with households representing the highest share (67%), followed by retail, with 12%. These shares are in line with other European countries [3], where EU-27 households' contribution is above 50%.

Given the importance of the consumption stage in the generation of food wastage, it is crucial to understand consumer's knowledge, motivations and behavior regarding it. Previous studies have been conducted in Poland [2, 19], USA [9], United Kingdom [18], Italy [20], Ireland [5], Denmark [24], Greece [16] and several European countries [6], suggesting that, although the causes of food waste may be common, cultural factors need to be considered. To the best of the authors' knowledge the only studies that have been conducted to understand consumer behavior towards food waste in Portugal were from Baptista et al. [1], where the amount of household food waste and reasons for wasting were determined through a survey with 804 valid responses and 41 semi-structured interviews, from Portugal et al. [17] that developed a questionnaire to understand major determinants of household food waste and collected 438 valid answers from families living in the North of Portugal that had a good knowledge of the issue and, from Pires et al. [15] that analyzed the food waste behavior during lockdown caused by COVID-19 using a questionnaire with 841 valid answers and reported that 36% threw away less food than before, with vegetables, milk and fruit having the highest reduction of food waste. Besides adding knowledge to the consumer food waste field in Portugal, this paper analyses the influence of knowledge, motivations and behaviors on the amount of food wasted, which have not been studied thoroughly before in Portugal.

The aim of this paper is thus to understand the influence of knowledge, motivations and behaviors on the amount of food wasted in households in Portugal and to provide suggestions for a communication strategy that raises awareness to this issue.

This paper is structure as follows: Sect. 2 presents the data and methodology, Sect. 3 provides the results and Sect. 4 discuss them and identifies limitations and future research. The paper ends with main conclusions.

2 Method

To achieve the aim of this paper, the following hypothesis were formulated and tested using Pearson correlations of factors representing knowledge, motivations and behaviors obtained in a factor analysis with Principal Component Analysis as the extraction method:

First Hypothesis: H_o: The quantity of food wasted is independent of the knowledge about the impacts of food waste;
H_1: The quantity of food wasted is not independent of the knowledge in the impacts of food waste;
Second Hypothesis: H_o: The quantity of food wasted is independent of the motivations to reduce food waste;
H_1: The quantity of food wasted in the past week is not independent of the motivations to reduce food waste;
Third Hypothesis: H_o: The quantity of food wasted is independent of the behaviors to reduce food waste;
H_1: The quantity of food wasted in the past week is not independent of the behaviors to reduce food waste.

A structured survey was designed based on other surveys available in the literature. Table 1 presents the questions translated from Portuguese to English, the type of answer requested and identifies the references from which the question was adapted. Food waste was clearly defined in the introductory text of the survey and examples of edible food waste were provided.

The survey comprised eight sections, but only six are presented (the other two are about information and drivers and barriers, which will not be analyzed in this paper). The first section addresses socio-demographic data. The second section inquiries about leftover use, shopping and cooking habits, excluding from the analysis the respondents that shop and cook less than half the times. Section 3 quantifies food wasted in the past week using the validated survey developed by Herpen et al. [8]. The following section addresses the knowledge about the impacts of food waste on sustainability. The 5th section inquiries about motivations to reduce food waste. The last section inquiries about the frequency of certain behaviors to reduce food waste.

Most questions request the level of agreement on a 5-point Likert type scale (1 = completely disagree to 5 = completely agree) or the frequency respondents think about or do something (1 = never; 2 = a few times; 3 = sometimes; 4 = many times; 5 = always). Respondents could also answer that they did not know, or they rather not say.

A pre-test of the survey was made using a small sample, to clarify its understandability and duration. The survey was distributed using the snowball sampling technique [7]. This is a nonprobability, convenience method of sampling used when it is difficult to access subjects with the targeted characteristics and it consists of recruiting participants among the acquaintances. These, in turn, recruited other participants until data saturation [11]. The main advantages of this method are its cost-effectiveness

Table 1 Constructs, questions, type of answer and reference from which the question was adapted from

Construct	Question	Type of answer	Reference
N/A	Socio-demographic data	Multiple	Adapted from Geffen et al. [6]
N/A	How often do you do the groceries in your household?	3-point scale frequency scale	Geffen et al. [6]
	How often do you do the cooking in your household?	3-point Likert type scale	Geffen et al. [6]
	How often do you use leftovers to prepare new meals?	5-point frequency scale	New
N/A	Food waste measurement in the past week	Multiple	Herpen et al. [8]
Knowledge	Food waste reduction is necessary for sustainable development	5-point Likert type scale	New
	More food is wasted at home than in supermarkets and restaurants	5-point Likert type scale	New
	Food waste is inevitable because only fresh fruit and vegetables provide the necessary nutrients for a healthy diet	5-point Likert type scale	Adapted from National Academies of Sciences, Engineering, and Medicine [13]
	I know what the impacts of food waste in sustainability are	5-point Likert type scale	New
	When I waste food, I'm… wasting water; contributing to climate change; causing food insecurity; contributing to an increase in food prices	5-point Likert type scale	New
Motivations	When I waste food I think about… Hungry people; Money spent; Food safety; Environmental impacts; Extra time needed to not waste any food	5-point frequency scale	Adapted from Geffen et al. [6]

(continued)

Table 1 (continued)

Construct	Question	Type of answer	Reference
	I feel guilty when I waste food	5-point Likert type scale	Stefan et al. [25]
Behaviors	Frequency with which the person adopts several behaviors: Shopping list; Storage management; Meal planning; Leftovers management; Storage good practices; Hierarchy in consumption; Self control; Clean plate	5-point frequency scale	Adapted from Geffen et al. [6]

Legend: N/A – not applicable

and the networking it may create for subsequent studies or follow-up activities, since the first sample is somehow linked to the researcher.

3 Results

3.1 Respondents' Characterization

The survey timespan was from February to April 2022, and it had a total of 440 responses, remaining 376 responses after validation. The socio-demographic characteristics are shown in Table 2. As in other similar questionnaires available in the literature, there are more female respondents (78%) than male. Respondents have between 18 and 76 years old, with an average age of 42 years old, the majority residing in Lisbon Metropolitan Area (65%). Additionally, more than half of the respondents do not have children in the household (66%) and the majority has a university degree (92%).

From the 376 valid responses, 51% wasted food, 78% of which were females (Table 3). 49% of respondents stated they frequently use leftovers, while 19% always use leftovers (Fig. 1). The number of people responding not to have wasted food in the past week increases with age among the respondents and does not seem to be influenced by the level of education (Fig. 2).

The products most wasted were fresh fruit (completely unused), bread (partly used), pasta, rice and other grains (leftovers) and soup (leftovers) (Fig. 3).

Most respondents already perform behaviors to reduce food waste. However, weekly planning and the use of the freezer can be incentivized (Fig. 4).

Table 2 Socio-demographic characteristics of the sample

Characteristic	Sample distribution
Gender	Female: 291 (78%) Male: 81 (22%) Other: 1 (0%) Rather not say: 1 (0%)
Age	Mean: 42, from 18 to 76 18–29: 71 (19%) 30–39: 104 (28%) 40–49: 90 (24%) 50–59: 73 (20%) 60–69: 29 (8%) More than 70: 7 (2%)
Area of residence	Metropolitan area of Lisbon: 243 (65%) Metropolitan area of Porto: 41 (11%) North: 20 (5%) Center: 42 (11%) South: 20 (5%) Madeira and Azores Islands: 8 (2%)
Total number of people in the household	Mean: 2.7, max: 6 1: 50 (13%) 2: 129 (34%) 3: 80 (21%) 4 or more: 115 (31%)
Number of children	Mean: 0.5, max: 4 0: 245 (66%) 1: 75 (20%) 2: 48 (13%) 3: 4 (1%) 4: 2 (1%)
Level of education	Secondary education: 21 (6%) University: 264 (70%) Technical-professional course: 4 (1%) PhD or higher: 86 (23%)

Table 3 Food wasted by gender

Food wasted	Female	Male	Other	Rather not say	Total
Yes	149	41	0	1	191
No	143	41	1	0	185
Total	292	82	1	1	376

3.2 *Data Analysis*

The data analysis was performed in SPSS version 28.0.1.1 for Mac with an alpha of 0.05. To find the latent structure that explains the observed correlations among

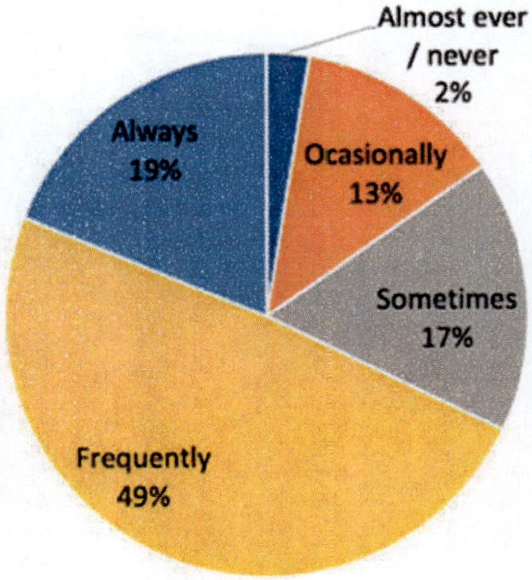

Fig. 1 Leftover use

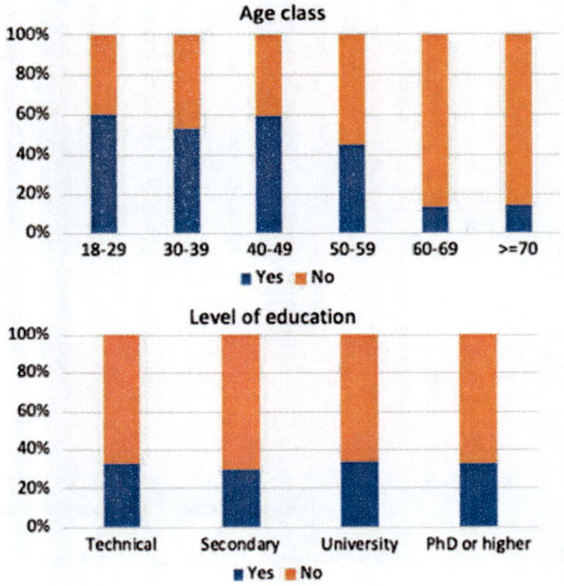

Fig. 2 Food wasted by age and level of education

the original items and since Likert type scale questions had a reduced number of classes that prevented them to be treated as quantitative variables, items with Cronbach's alpha of 0.883 (good reliability) were subjected to a factor analysis on the correlation matrix obtained using Spearman correlation matrixes for ordinal items as stated in Marôco [10]. The Kaiser–Meyer–Olkin measure is 0.869 (good reliability). The factor structure after Varimax rotation explains more than 52% of the observed information among the ordinal items. Five components were extracted: one component represents knowledge, two components represent behaviors related to planning

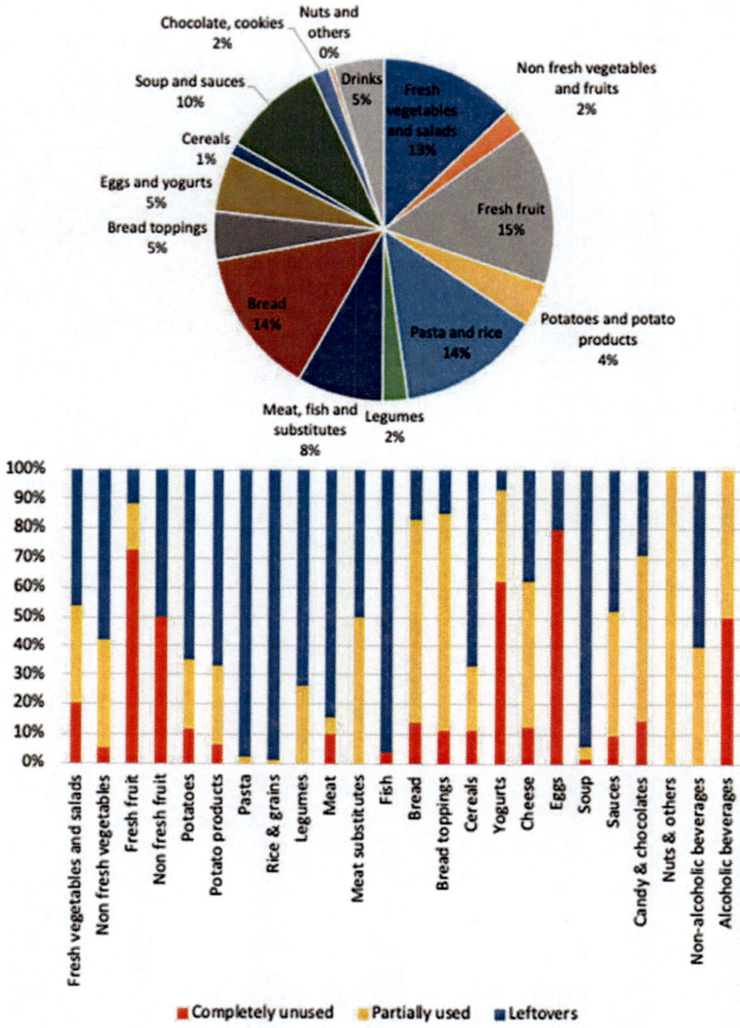

Fig. 3 Products most wasted and state they are most wasted

and shopping, cooking and leftover use, one component represents motivations, with social motivations having the highest weight (0.785), followed by financial (0.676) and environmental motivations (0.542), and one component represents information, which will not be analyzed in this paper.

Table 4 presents the results of the hypothesis tests. Pearson correlation (ρ) was selected to test the hypothesis since the independent and dependent variables are quantitative and may be linearly related.

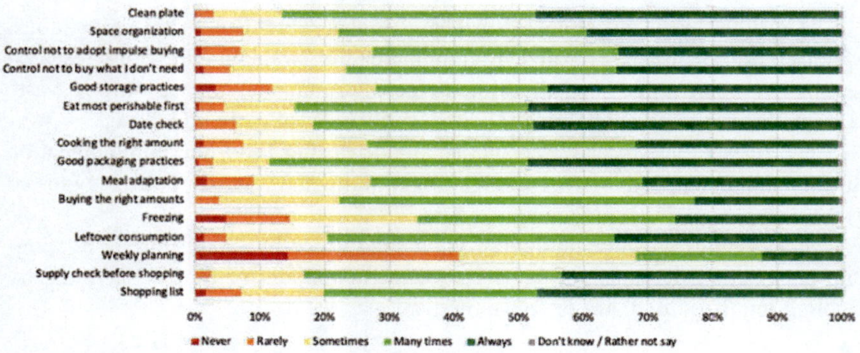

Fig. 4 Anti food waste behaviors already adopted by respondents

Table 4 Results of the hypothesis tests

Hypothesis	Pearson correlation	p value	Conclusion
First	$\rho = -0.055$	0.289	Correlation not significant H_0 cannot be rejected for $\alpha = 0.05$
Second	$\rho = -0.429$* $\rho = -0.407$*	< 0.001	*Correlation is significant Negative and moderate correlation H_0 can be rejected for $\alpha = 0.05$
Third	$\rho = -0.284$*	< 0.001	*Correlation is significant Negative and moderate correlation H_0 can be rejected for $\alpha = 0.05$

4 Discussion

The results show that the correlation of the amount of food wasted with knowledge about its impacts is not significant, but it is moderately significant with motivations and behaviors. For these two last constructs, the Pearson coefficients are both negative, meaning that motivations and behaviors to reduce food waste contribute to lower amounts of food wasted.

These results suggest communication strategies may be more effective if they focus on promoting action-related information (actions to reduce food waste) as suggested by Neubig et al. [14] and anti-food waste behaviors. Although most respondents already adopt anti-food waste behavior as depicted in Fig. 4, weekly meal planning and the use of the freezer can be further incentivized.

Due to their direct contact with consumers, retailers can play an important role in the implementation of a communication strategy through apps, social media and/or newsletters that stimulates the use of shopping lists and provides recipes with leftovers and tips on how to best freeze leftovers and fresh products to extend their shelf-life at home while maintaining their nutritional value and sensory properties.

The results reveal respondents attribute a higher weight to social motivations, which could be further explored in communication strategies. Previous research focused on saving money and appealing to the environmental impacts of food waste [23], but having think about hungry people could be even more effective.

Since respondents already feel guilty about wasting food, the communication strategy should deliver positive messages highlighting the value of food as suggested by Narvanen et al. [12].

These findings advance the literature on food waste by establishing a relationship between motivations and behaviors with the amount of food wasted and by highlighting the importance of social motivations.

4.1 Limitations and Future Research

These findings need to be interpreted carefully since food waste is the result of many drivers and barriers, of which routines and habits play a key role [21]. The amount of food wasted was measured indirectly using a validated survey [8], where respondents estimate the percentage of food their households wasted the week prior to the survey, and this is subject to bias and relies on memory. More accurate methods could have been used (direct weighing, volume assessing or waste composition analysis), but they are time consuming and are not feasible for large samples and for large geographical areas. Additionally, the sample is not representative of the Portuguese consumers because the sampling selection of the snowball technique is biased, and because most respondents have a university degree, are female and live in the Lisbon Metropolitan Area. Other similar studies are needed to validate these findings and to focus on the underrepresented groups.

Communication strategies should be tailored to different consumer segments, aligned with different motivations and disseminated in different channels. Further research should focus on segmenting consumers according to their socio-demographic characteristics and motivations and behaviors regarding food waste.

5 Conclusions

This paper studied the influence of knowledge about the impacts of food waste, motivations, and behaviors to reduce food waste on the amount of food wasted in Portuguese households. To this extent, three hypotheses were tested with a survey made available online and using Pearson correlation on factors representing knowledge, motivations and behaviors obtained with factor analysis with Principal Component Analysis as the extraction method. The total number of valid answers was 376.

The results showed no significant correlation between the amount of food wasted and knowledge about the impacts of food waste and moderate significant correlations between the amount of food waste and motivations and behaviors.

Given these results, communication strategies should focus on promoting anti-food waste behaviors and on motivating consumers with positive messages that address the social impacts of food waste.

References

1. Baptista, P., Campos, I., Pires, I., Vaz, S. (2012), *Do Campo ao Garfo, Desperdício alimentar em Portugal*, Lisboa, CESTRAS
2. Bilska, B., Tomaszewska, M., Kolozyn-Krajewska, D. (2019), Analysis of the behavior of Polish consumers in relation to food waste. Sustainability, 12, 304. https://doi.org/10.3390/su12010304
3. Eurostat homepage, https://ec.europa.eu/eurostat/statistics-explained/index.php?title=Food_waste_and_food_waste_prevention_-_estimates. Last accessed December 2022
4. FAO (2011), *Global food losses and food waste – Extent, causes and prevention*, Rome
5. Flanagan, A., Priyadarshini, A. (2021), A study of consumer behavior towards food-waste in Ireland: Attitudes, quantities and global warming potentials, *Journal of Environmental Management*, 284, 112046, https://doi.org/10.1016/j.jenvman.2021.112046
6. Geffen, L., Herpen, E., Trijp, H. (2017), *Quantified consumer insights on food waste, pan-European research for quantified consumer food waste understanding*, REFRESH
7. Goodman, L. (1961), Snowball sampling, *Annals of Mathematical Statistics*, 32, 148–170, https://doi.org/10.1214/aoms/1177705148
8. Herpen, E., Geffen, L., Nijenhuis, M., Holthuysen, N., Lans, I., Quested, T. (2019), A validated survey to measure household food waste, *MethodsX*, 6, 2767–2775, https://doi.org/10.1016/j.mex.2019.10.029
9. Kavanaugh, M., Quinlan, J. (2020), Consumer knowledge and behaviors regarding food date labels and food waste, *Food Control*, 115, 107285, https://doi.org/10.1016/j.foodcont.2020.107285
10. Marôco, J. (2021), *Análise Estatística com o SPSS Statistics*, 8.ª edição, Report Number, ISBN: 978-989-96763-7-4
11. Naderifar, M., Goli, H., Ghaljaie, F. (2017), Snowball sampling: a purposeful method of sampling in qualitative research, *Strides in Development of Medical Education*, in press, https://doi.org/10.5812/sdme.67670
12. Narvanen, E., Mesiranta, N., Sutinen, U., Mattila, M. (2018), Creativity, aesthetics and ethics of food waste in social media campaigns, *Journal of Cleaner Production*, 195, 102–110, https://doi.org/10.1016/j.jclepro.2018.05.202
13. National Academies of Sciences, Engineering, and Medicine (2020), *A National Strategy to Reduce Food Waste at the Consumer Level*, Washington, DC: The National Academies Press, https://doi.org/10.17226/25876
14. Neubig, C., Vranken, L., Roosen, J., Grasso, S., Hieke, S., Knorpfle, S., Macready, A., Masento, N. (2020), Action-related information trumps system information: influencing consumers' intention to reduce food waste, *Journal of Cleaner Production*, 261, 121126, https://doi.org/10.1016/j.jclepro.2020.121126
15. Pires, I, Fernández-Zamudio, M., Vidal-Mones, B., Martins, R. (2021), The impact of COVID-19 lockdown on Portuguese households' food waste behaviors, *Human Ecology Review*, 26 (1), 59–69, https://doi.org/10.22459/HER.26.01.2020.06
16. Ponis, S., Papanikolaou, P., Katimertzoglou, P., Ntalla, A., Xenos, K. (2017), Household food waste in Greece: a questionnaire survey, *Journal of Cleaner Production*, 149, 1268–1277, https://doi.org/10.1016/j.jclepro.2017.02.165

17. Portugal, T., Freitas, S., Cunha, L., Rocha, A. (2020), Evaluation of determinants of food waste in family households in the Greater Porto Area based on self-reported consumption practices. Sustainability, 12, 8781, https://doi.org/10.3390/su12218781
18. Quested, T., Marsh, E., Stunell, D., Parry, A. (2013), Spaghetti soup: the complex world of food waste behaviors, Resources, Conservation and Recycling, 79:43–51, https://doi.org/10.1016/j.resconrec.2013.04.011
19. Radzyminska, M., Jakubowska, D., Staniewska, K. (2016), Consumer attitude and behavior towards food waste, *Journal of Agribusiness and Rural Development*, 1, 39, 175–181, https://doi.org/10.17306/JARD.2016.20
20. Setti, M., Falasconi, L., Segrè, A., Cusano, I., Vittuary, M. (2016), Italian consumers' income and food waste behavior, *British Food Journal*, 118, 7, 1731–1746, https://doi.org/10.1108/BFJ-11-2015-0427
21. Simões, J., Carvalho, A., Matos, M. (2022), How to influence consumer food waste behavior with interventions? A systematic literature review, *Journal of Cleaner Production*, 373, 133866, https://doi.org/10.1016/j.jclepro.2022.133866
22. Springmann, M., Clark, M., Mason-D'Croz, D., Wiebe, K., Bodirsky, B., Lassaletta, L., Vries, W., Vermeulen, S., Herrero, M., Carlson, K., Jonell, M., Troell, M., DeClerck, F., Gordon, L., Zurayk, R., Scarborough, P., Rayner, M., Loken, B., Fanzo, J., Godfray, H., Tilman, D., Rockstrom, J., Willett, W. (2018), *Options for keeping* the food system within environmental limits, Nature, Vol 562, https://doi.org/10.1038/s41586-018-0594-0
23. Soma, T, Li, B., Maclaren, V. (2020), Food waste reduction: a test of three consumer awareness interventions, *Sustainability*, 12, 907, https://doi.org/10.3390/su12030907
24. Stancu, V., Haugaard, P., Lahteenmaki, L. (2016), Determinants of consumer food waste behavior: two routes to food waste, *Appetite*, 96, 7–17, https://doi.org/10.1016/j.appet.2015.08.025
25. Stefan, V., van Herpen, E., Tudoran, A., Lähteenmäki, L. (2013). Avoiding food waste by Romanian consumers: The importance of planning and shopping routines, *Food Quality and Preference*, 28, 375–381, https://doi.org/10.1016/j.foodqual.2012.11.001

Acceptance of Insect-Containing Foods: A Pilot Study Exploring the Use of Insects as a Meat Alternative

Leocardia Ranga, Jennifer Vance, and Maria Dermiki

Abstract Entomophagy is currently increasing in popularity due to its potential as a sustainable alternative to conventional meat. However, acceptability of entomophagy amongst European consumers is still low. The aim of this study was to determine the factors affecting the acceptability of insect burgers (relative to plant and beef burgers) amongst consumers in Ireland through an affective sensory test. A total of 29 naïve consumers participated in a blind liking test which was conducted in isolated sensory booths. The beef burger received the highest overall liking score (7.03 ± 1.84) followed by the plant (6.31 ± 1.69) and lastly the insect burger (6.24 ± 1.66). Qualitative findings, however, revealed that the most negative comments were made for the plant burger. Participants' meat attachment and perception of entomophagy had a significant effect ($p < 0.05$) on the acceptance of the insect burgers. The lower the meat attachment and the more positive participants' entomophagy perception was, the more inclined they were to accept the insect burger. The taste, texture, and appearance of the burgers also significantly affected participants' overall liking of them. Future interventions should focus on improving the sensory attributes of insect-containing foods and expanding the study to a representative sample of the population.

Keywords Entomophagy · Sensory attributes · Meat attachment · Perception of entomophagy

L. Ranga · J. Vance · M. Dermiki (✉)
Department of Health and Nutritional Sciences, Atlantic Technological University, Ash Lane, Sligo F91 YW50, Ireland
e-mail: Maria.Dermiki@atu.ie

L. Ranga
e-mail: Leocardia.Ranga@research.atu.ie

J. R. da. C. S. Galvão et al. (eds.), *Proceedings of the 3rd International Conference on Water Energy Food and Sustainability (ICoWEFS 2023)*, Springer Proceedings in Earth and Environmental Sciences, https://doi.org/10.1007/978-3-031-48532-9_4

1 Introduction

The unsustainability of the current global food production systems has been widely reported [1, 2]. Meat production in particular, accounts for most of food productions' greenhouse gas (GHG) emissions [2]. These GHG emissions from the food chain, if not mitigated, could hamper the Paris Agreement's aim towards curbing the increase of global average temperatures to abate climate change [3]. Aside from the GHG emissions, meat production also makes use of a considerable amount of land and water [4]. Consequently, with the world population set to increase to nine billion by 2050 [5], expanding meat production to meet the subsequent increase in demand for protein sources [6], will not be sustainable [7]. Thus, exploring sustainable protein alternatives is imperative to maintain food security [8]. This will also facilitate progress towards achieving the United Nations' sustainable development goals (SDGs) particularly SDG 12 which is sustainable production and consumption.

Insects have been generally found to be a potentially suitable alternative to meat due to their comparably high biological value protein content [9] and relatively lower environmental footprint [10–12]. Though the human consumption of insects (entomophagy) has been around for centuries in South America, Africa, and Eastern Asia [13], it recently garnered interest in the western world due to its potential as a sustainable protein source [14]. In 2021, the European Food Safety Authority (EFSA) conducted a risk assessment of yellow mealworms [15], migratory locusts [16] and house crickets [17] substantiating their safety for human consumption leading to the approval of these three insect species as novel food across Europe [18–20]. However, entomophagy acceptance amongst consumers in Europe is still relatively low [21].

Various consumer-related factors such as age, gender, education background, and perception towards entomophagy have been shown to have an impact on consumer willingness to accept insects as novel food [22–24]. Since insects are still a novel food in Europe, food neophobia and disgust are said to be some of the main barriers to entomophagy acceptance in this region [22, 23]. A suggested resolution has therefore been to introduce insects into food products which are already familiar to the consumer [25]. Likewise, familiarity [23, 26] and previous experience [22] with entomophagy has been linked with an increase in entomophagy acceptance amongst consumers. Nonetheless, the extent to which consumers are attached to meat could also influence their acceptance of meat alternatives. Meat attachment scores have thus been used as an indicator to determine consumer acceptance of meat alternatives, with higher scores associated with lower acceptance rates [27].

Ireland identifies as a farming country with about seven million cows reported in 2020 [28] and €3.5 billion worth of meat and livestock exports in 2021 [29]. Previous studies have investigated the meat attachment scale [27], however, it has not been investigated in Ireland, along with its effect on entomophagy acceptance, while the acceptance of insect snack foods in Ireland has been reported [30]. Nevertheless, as of 2011, 41% of Irish adult's protein intake came from meat and the overall meat consumption in Ireland has significantly increased ever since [31, 32]. Despite being relatively satisfactory in its overall SDG performance, Ireland, reportedly still faces major challenges in achieving SDG number 12 of responsible consumption and production and SDG number 13 of climate action by 2030 [33, 34]. It could be worth exploring insects as a meat alternative in Ireland and determine how the consumers' farming background and level of meat attachment translates into their willingness to accept the change.

Aside from the consumer-related factors, factors related to the characteristics of the insect-containing foods (product-related) such as its nutritional value [22] and sensory attributes [22], play a major role in consumer acceptance [35]. Problems concerning the challenging task of replicating the sensory properties of meat such as its appearance, flavor, texture, and palatability in meat alternatives have been reported [36]. This can lead to consumer rejection of such products based on undesirable or unfamiliar properties [35, 36]. This study aims to determine the consumer-related and product-related factors affecting the acceptability of insect-based burgers (comparing them with plant-based and beef-based burgers) amongst consumers in Ireland.

2 Materials and Methods

2.1 Data Collection and Participants' Profile

This pilot study received approval from the ethics committee of the Department of Health and Nutritional Sciences, Atlantic Technological University, Sligo (ATU Sligo), Ireland. Recruitment for participants took place in October 2022 via word of mouth and the email network of ATU Sligo. The exclusion criteria were, being a vegetarian or vegan, having any food allergies, being allergic to dust mites and being younger than 18 years of age. A total of 29 staff and students from ATU Sligo participated in the study experiment which was conducted in isolated sensory booths. At the beginning of the experiment, each participant was given a questionnaire to complete. This questionnaire consisted of seven questions relating to the participants' age, gender, field of study, farming background, meat attachment, previous experience, and perception of entomophagy. The meat attachment scale was used to measure participants' level of meat attachment [27], whilst their perception of entomophagy was determined using the 'specific beliefs and attitudes about insect consumption scale' [24].

Upon completing the questionnaire, participants conducted a single-blind sensory acceptability test of three burgers (insect-based, plant-based, and beef-based) which were presented to them monadically using three-digit codes. Samples were presented in a random balanced order except for the beef burger which was provided last to avoid any carry-over effects recorded during the piloting of the test. Participants were asked to rate overall liking, liking of external appearance, internal appearance, aroma, taste, and texture of each burger on 9-point hedonic scales, as well as to provide a reason for each rating. They were also requested to cleanse their pallets with the provided water and cream crackers before tasting each sample. Among the participants, there were slightly more males (51.7%) than females (48.3%) and there was a representation from all age groups (37.9% in the 21–25 age range, 17.2% in the 31–35 age range, 24.1% in the 41–50 age range and 20.7% over 50 years old). Participants' respective fields of education were grouped into science and non-science fields and those with a science background were more (75.9%) than those from non-science background (24.1%). About a third of the participants had previous experience with entomophagy (34.5%) and only 27.6% had a farming background.

2.2 Sample Attributes and Preparation

The beef and plant burgers were obtained from a local supermarket in Sligo, Ireland while the insect-based burgers were imported from Switzerland. Samples were chosen based on the protein content (closer to 16 g) and based on their resemblance with the appearance of the insect burgers after preparation as seen in Fig. 1. Sample handling and preparation was done in accordance with HACCP principles [37]. All samples were kept frozen at −18 °C until the day of the experiment when they were cooked from frozen in a convection oven at 220 °C for a period of 20 min. The core temperature of the cooked samples was measured using a temperature probe to ensure it was at least 75 °C and safe for consumption [37]. Once cooked, the samples were kept outside the temperature danger zone at 64 °C during the sensory evaluation session [37].

The day following the sensory evaluation sessions, the color, texture, and moisture content of the three samples was measured. A Lovibond LC-100 colorimeter was used to measure L/lightness, a*/green–red axes, and b*/blue–yellow axes at three different points of each sample. Texture was measured using a TA.XT texture analyzer set up with a Warner Bratzler blade and a slotted blade insert using the following parameters: 1.5 mm/s pre-test speed, 10 mm/s post-test speed, 30 mm distance, 40 g trigger type and 200 pps data acquisition rate. Moisture content was measured using an OHAUS MB23 moisture balance.

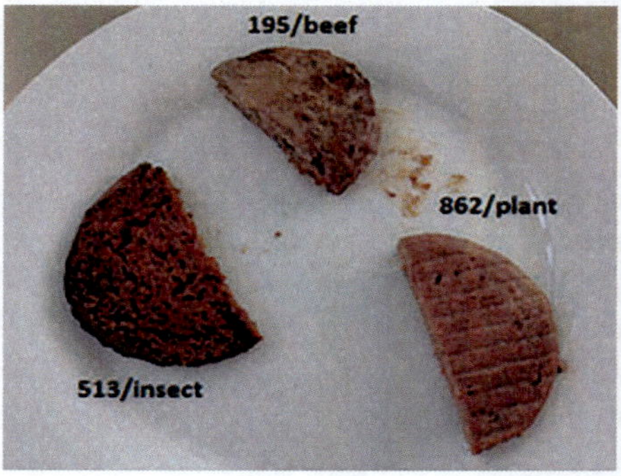

Fig. 1 The 3 burgers (insect, plant, and beef) used in this study and the 3-digit code assigned to each sample

2.3 Data Analysis

The reasons provided by the participants (qualitative data) justifying the rating for overall liking of the burgers were first coded and then grouped into themes. The quantitative data was analyzed using IBM SPSS V28 software. Non-parametric statistical tests (Kruskal Wallis and Mann Whitney U tests) were used to determine the effect of participants' age, field of study, gender, farming background and previous experience of entomophagy on the overall liking of the burgers. One-way analysis of variance (ANOVA) was employed to identify statistical differences between the overall liking, and liking of external appearance, internal appearance, aroma, taste, and texture of the three burgers. Spearman's correlation was used to determine the effect of participants' perception of entomophagy and meat attachments on their overall liking of the burgers, and Pearson's correlation was used to find any significant correlations between liking of sample sensory attributes and overall liking. Significance was established at $p < 0.05$.

3 Results and Discussion

3.1 Acceptance of Insect-Based Burgers

Participants' ($N = 29$) overall liking of the three burgers (insect, plant, and beef) and the liking of the respective sensory attributes can be seen in Fig. 2. Generally, the beef burger was the most liked of the three, attaining an overall liking score of 7.03 ± 1.84, followed by the plant (6.31 ± 1.69) and then the insect burger (6.24 ± 1.66). Although the insect burger received the lowest overall liking score, qualitative data revealed that the most negative comments were linked with the plant burger whilst the beef burger received the most positive comments. The latter validates further the highest liking scores of the beef burgers. Participants ($N = 79$) in a previous study also preferred beef burgers over insect and plant-based alternatives [35]. This could be due to the fact that in both studies participants were more familiar with the beef burgers compared to the insect or the plant burgers.

3.2 Factors Affecting Acceptance

Product-Related Factors. Pearson's correlation revealed positive correlations between overall liking scores and liking scores of each of the respective sensory attributes evaluated (external appearance, internal appearance, aroma, taste, texture). When comparing the three burgers (insect, plant, and beef), one -way ANOVA showed that there were significant differences ($p < 0.05$) in the liking scores of the external and internal appearances of the burgers. The beef burger had the highest

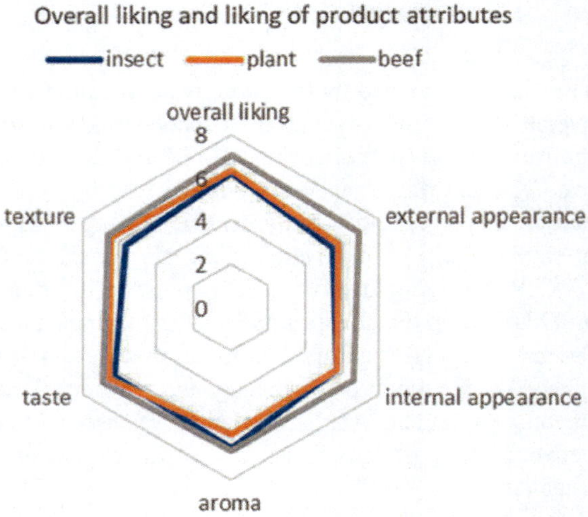

Fig. 2 Overall liking of the three burgers (insect, plant, and beef) and liking of the sensory attributes of each burger

liking score for external (6.93 ± 1.41) and internal appearance (6.62 ± 1.52) compared to the plant (5.79 ± 1.9 and 5.72 ± 1.53, respectively) and the insect burgers (5.48 ± 1.95 and 5.68 ± 1.41, respectively). Qualitative analysis from the reasons participants provided in support of their rating of each sensory attribute, identified taste and texture as the factors that mostly influenced their acceptance of the burgers. The beef burger was rated highest for taste (6.93 ± 1.79) and texture (6.62 ± 1.65) followed by the plant (6.55 ± 1.66 and 6.48 ± 1.66, respectively) and then the insect burger (6.27 ± 1.88 and 5.65 ± 2.12, respectively). There were some variations in the attributes of the burger samples (colour, texture, moisture content and nutritional content) used in this study as shown in Table 1.

As seen, the insect burger had a deep reddish colour that was different from the brown colour of the beef burger, with which participants are familiar. Moreover, the moisture content of the insect burger was two times lower compared to the beef burger, which could have affected its texture. Possibly if the insect burgers closely resembled the colour, texture and taste of the beef burgers, their acceptance would have increased. The insect burgers used in this study also had the highest salt and sugar content compared to the plant and beef burgers which could have had an impact on the overall taste of the burger. Past studies have also found the taste and appearance of insect containing foods to have an impact on consumers' acceptability of these products [22, 30, 35]. Future studies could focus on implementing processing techniques to improve the moisture content and reduce the salt and sugar content of the insect burgers. Apart from mimicking the sensory attributes of beef burgers when developing insect burgers, perhaps the more consumers get an opportunity to taste the insect burgers through free sample tastings, the more they can get acquired to the taste of the insect burgers and become familiar with those. Further research may be needed to test this notion.

Table 1 Color, texture, moisture, and nutritional composition of the three burgers (insect, plant, and beef) used in this study

Sample attributes	Sample type		
	Insect	Plant	Beef
Colour			
L	34.17 ± 1.36	**38.9 ± 2.81**	30.5 ± 3.24
a*	**14.53 ± 0.15**	10.93 ± 1.25	8.67 ± 1.91
b*	14.2 ± 1.11	**15.67 ± 1.29**	8.87 ± 2.44
Texture			
Firmness (g)	491.217 ± 8.95	754.392 ± 65.85	**1196.33 ± 235.82**
Toughness (g/sec)	2519.8 ± 587.51	4012.86 ± 111.34	**5459.09 ± 992.53**
Moisture			
Weight (g)	0.94	0.83	0.76
Moisture content (%)	16	**54.7**	33.3
Nutritional composition (Per 100 g)			
Protein (g)	**16.6**	16	16.2
Fat (g)	13.5	8.7	**19.5**
Saturated fat (g)	2.4	3.8	**9.1**
Sugar (g)	**3.8**	< 0.1 g	< 0.5
Salt (g)	**1.3**	1.1	0.65

Colour measurement: L = lightness, a* = red-green axes, b* = blue-yellow axes
Nutritional composition was obtained from nutrition facts table presented on the labels of the burger (Regulation EU No 1169/2011)
Bold font: highest measurement within an attribute

Consumer-Related Factors. The effect of participants' age and field of education on their overall liking of the burger samples, was determined using the Kruskal Wallis test whilst the effect of their gender, farming background and previous experience of entomophagy was determined through a Mann Whitney U test. Results showed that participants' age, gender, field of education, farming background and previous experience of entomophagy had no significant ($p > 0.05$) effect on their overall liking of the three burgers. Several past studies, nonetheless, did find age [23], gender [22, 23, 35] and previous experience with entomophagy [22, 26] to have a significant effect on consumers' acceptance of entomophagy. Perhaps significant differences in this regard would have been detected in this study if a larger sample size was used. Also, since this was an experimental pilot study, it could be that those who indicated interest in participating were those willing to try insect-containing foods in the first place. A mixture of experimental and non-experimental designs could help attract those not willing to taste insect-containing foods at the non-experimental stage, to elicit their views on entomophagy and obtain significant differences between the two groups (those willing to try insects and those not willing).

Spearman's correlation on the other hand revealed a significant effect ($p < 0.05$) of participants' perception towards entomophagy and their overall liking of the three burgers (insect, plant, and beef). Significantly positive correlations ($p < 0.05$) were detected between the level of agreement on the statement "eating insects is disgusting" and overall liking of the beef burger (r (27) = 0.421) along with "eating insects is good for the environment" and overall liking of the plant burger (r (27) = 0.382) and between "it is not natural for humans to eat insects" and overall liking of the beef burger (r (27) = 0.366). The more the participants found insects to be disgusting and 'not natural for humans to eat', the more they preferred the beef burgers. This study used a single-blind experimental test which could explain why participants thought the plant burger was the insect burger, causing them to like the plant burger more when they perceived insects to be good for the environment. These results align with that of past studies which found entomophagy perception and environmental impact of insects as food to have an effect on entomophagy acceptance amongst consumers [22–24].

Participants' level of meat attachment was also found through a Spearman's correlation to have a significant effect ($p < 0.05$) on their overall liking of the burgers. Significantly positive correlations ($p < 0.05$) were identified between participants' level of agreement with the statement "I love meals with meat" and overall liking of the beef burger (R = 0.373), and between "I'm a big fan of meat" and overall liking of the beef burger (R = 0.397). The more the participants were attached to meat, the more they preferred the beef burger over the alternatives. A previous study also reported the effect of meat attachment on meat consumption [27]. Educating consumers on the benefits of insects as food and the environmental impacts associated with meat consumption could tilt their perception of entomophagy on the positive side. This change in perception might reduce the level of meat attachment amongst consumers and increase the acceptance of insects as a meat alternative. Further studies could be done to investigate whether educating consumers in this regard will change their perception of entomophagy and level of meat attachment.

4 Conclusions

The aim of this pilot study was to determine the product-related and consumer-related factors affecting the acceptability of insect-based burgers amongst consumers in Ireland. Despite the limitation of the small number of participants, the findings of this pilot study corroborated with previous research in the field. The consumer-related factors that significantly affected participants' acceptance of the insect burgers were meat attachment and perception of entomophagy. The lower the meat attachment and the more positive the entomophagy perception a consumer had, the more likely they were to accept the insect burgers tested. When considering the product-related factors, appearance, taste, and texture significantly affected acceptance of the insect burgers. Future interventions should focus on improving these sensory attributes of insect-containing foods and determine how this improvement affects consumer attachment to conventional meat products. Since insect burgers contained higher levels of salt and sugar compared to the other burgers tested, future studies should focus on decreasing the levels of these ingredients making insect-containing products more nutritious. Finally, since the participants who agreed with the statement "eating insects is good for the environment" gave higher scores to the insect burgers, it is suggested to observe how educating consumers about the sustainability benefits of entomophagy could affect their perception and acceptance of it. The latter could be achieved by exploring the views of a representative sample of the Irish population or by expanding this work to more countries in Europe.

References

1. Ritchie, H., Roser, M.: Environmental Impacts of Food Production. Our World in Data, (2021).
2. Poore, J., Nemecek, T.: Reducing food's environmental impacts through producers and consumers. Science 360(6392), 987–992 (2018).
3. Clark, M., Domingo, N., Colgan, K., Thakrar, S., Tilman, D., Lynch, J., Azevedo, I., Hill, J.: Global food system emissions could preclude achieving the 1.5 °C and 2 °C climate change targets. Science 370(6517), 705–708 (2020).
4. Steinfield, H., Gerber, P., Wassenaar, T., Castel, V., Rosales, M., De Haan, C.: Livestock's long shadow: Environmental issues and options. FAO, Rome (2006).
5. United Nations.: World Population Prospects 2022. Department of Economic and Social Affairs Population Dynamics (2022).
6. Henchion, M., Hayes, M., Mullen, A.M., Fenelon, M., Tiwari, B.: Future protein supply and demand: strategies and factors influencing a sustainable equilibrium. Foods 6(7) 53 (2017).
7. Van Huis, A., Dicke, M., Van Loon, J.J.A.: Insects to feed the world. Journal of Insects as Food and Feed 1(1) 3–5 (2015).
8. Van Huis, A.: Edible Insects contributing to food security? Agriculture & Food Security 4(20) 1–9 (2015).
9. Rumpold, B.A., Schlüter, O.K.: Nutritional composition and safety aspects of edible insects. Molecular Nutrition & Food Research 57(5) 802–823 (2013).
10. Kusch, S., Fiebelkorn, F.: Environmental impact judgements of meat, vegetarian, and insect burgers: Unifying the negative footprint illusion and quantity insensitivity. Food Quality and Preference 78 103731 (2019).
11. Smetana, S., Mathys, A., Knock, A., Heinz, V.: Meat alternatives: life cycle assessment of most known meat substitutes. The International Journal of Life Cycle Assessment 20(9) 1254–1267 (2015).
12. Oonincx, D.G., De Boer, I.J.: Environmental impact of the production of mealworms as a protein source for humans-a life cycle assessment. PLoS ONE 7(12) e51145 (2012).
13. Gahukar, R.T.: Entomophagy and human food security. International Journal of Tropical Insect Science 31(3) 129–144 (2011).
14. Liceaga, A.M.: Approaches for utilizing insect protein for human consumption: Effect of enzymatic hydrolysis on protein quality and functionality. Annals of the Entomological Society of America 112(6) 529–532 (2019).
15. EFSA Panel on Nutrition, Novel Foods and Food Allergens (NDA), Turck, D., Bohn, T., Castenmiller, J., De Henauw, S., Hirsch-Ernst, K.I., Maciuk, A., Mangelsdorf, I., McArdle, H.J., Naska, A., Pelaez, C.: Safety of frozen and dried formulations from whole yellow mealworm (Tenebrio molitor larva) as a novel food pursuant to Regulation (EU) 2015/2283. EFSA Journal *19*(8) e06778 (2021).

16. EFSA Panel on Nutrition, Novel Foods and Food Allergens (NDA), Turck, D., Castenmiller, J., De Henauw, S., Hirsch-Ernst, K.I., Kearney, J., Maciuk, A., Mangelsdorf, I., McArdle, H.J., Naska, A., Pelaez, C.: Safety of frozen and dried formulations from migratory locust (Locusta migratoria) as a Novel food pursuant to Regulation (EU) 2015/2283. EFSA Journal *19*(7) e06667 (2021).
17. EFSA Panel on Nutrition, Novel Foods and Food Allergens (NDA), Turck, D., Bohn, T., Castenmiller, J., De Henauw, S., Hirsch-Ernst, K.I., Maciuk, A., Mangelsdorf, I., McArdle, H.J., Naska, A., Pelaez, C.: Safety of frozen and dried formulations from whole house crickets (Acheta domesticus) as a Novel food pursuant to Regulation (EU) 2015/2283. EFSA Journal *19*(8) e06779 (2021).
18. Commission implementing regulation (EU) 2022/169.: Authorising the placing on the market of frozen, dried and powder forms of yellow mealworm (*Tenebrio molitor* larva) as a novel food under Regulation (EU) 2015/2283 of the European Parliament and of the Council and amending Commission Implementing Regulation (EU) 2017/2470. Official Journal of the European Union (2022).
19. Commission implementing regulation (EU) 2021/1975.: Authorising the placing on the market of frozen, dried and powder forms of *Locusta migratoria* as a novel food under Regulation (EU) 2015/2283 of the European Parliament and of the Council and amending Commission Implementing Regulation (EU) 2017/2470. Official Journal of the European Union (2021).
20. Commission implementing regulation (EU) 2022/188.: Authorising the placing on the market of frozen, dried and powder forms of *Acheta domesticus* as a novel food under Regulation (EU) 2015/2283 of the European Parliament and of the Council and amending Commission Implementing Regulation (EU) 2017/2470. Official Journal of the European Union (2022).
21. Naranjo-Guevara, N., Fanter, M., Conconi, A.M., Floto-Stammen, S.: Consumer acceptance among Dutch and German students of insects in feed and food. Food Science & Nutrition *9*(1) 414–428 (2021).
22. Kane, B., Dermiki, M.: Factors and conditions influencing the willingness of Irish consumers to try insects: a pilot study. Irish Journal of Agricultural and Food Research 60 1–16 (2021).
23. Verbeke, W.: Profiling consumers who are ready to adopt insects as a meat substitute in a Western society. Food Quality and Preference 39 147–155 (2015).
24. Ruby, M.B., Rozin, P., Chan, C.: Determinants of willingness to eat insects in the USA and India. Journal of Insects as Food and Feed. 1(3) 215–225 (2015).
25. Barton, A., Richardson, C.D., McSweeney, M.B.: Consumer attitudes toward entomophagy before and after evaluating cricket (Acheta domesticus)-based protein powders. Journal of Food Science 85(3) 781–788 (2020).
26. Lensvelt, E.J., Steenbekkers, L.P.A.: Exploring consumer acceptance of entomophagy: a survey and experiment in Australia and the Netherlands. Ecology of Food and Nutrition *53*(5) 543-561 (2014).
27. Graça, J., Calheiros, M.M., Oliveira, A.: Attached to meat? (Un) Willingness and intentions to adopt a more plant-based diet. Appetite *95* 113–125 (2015).
28. Census of Agriculture 2020 Detailed Results Homepage, Census of Agriculture 2020 Detailed Results - CSO - Central Statistics Office, last accessed 2023/01/10.
29. Bord Bia.: Meat & Livestock Review 2020/21. Bord Bia (2021).
30. Herbert, M., Beacom, E.: Exploring Consumer Acceptance of Insect-based Snack Products in Ireland. Journal of Food Products Marketing 27(6) 267–290 (2021).
31. Irish Universities Nutrition Alliance.: National Adult Nutrition Survey. Ireland (2011).
32. Shahbandeh, M.: Consumption of meat in Ireland 2010–2021. Statista (2022).
33. Sachs, J., Lafortune, G., Kroll, C., Fuller, G., Woelm, F.: From Crisis to Sustainable Development: the SDGs as a Roadmap to 2013 and Beyond. Sustainable Development Report 2022. Cambridge University Press, Cambridge (2022).
34. Lafortune, G., Fuller, G., Bermont-Diaz, L., Kloke-Lesch, A., Koundouri, P., Riccaboni, A.: Achieving the SDGs: Europe's Compass in a Multipolar World. Europe Sustainable Development Report 2022. SDSN and SDSN Europe. France, Paris (2022).

35. Megido, R.C., Gierts, C., Blecker, C., Brostaux, Y., Haubruge, É., Alabi, T., Francis, F.: Consumer acceptance of insect-based alternative meat products in Western countries. Food Quality and Preference *52*, 237–243 (2016).
36. Lee, H.J., Yong, H.I., Kim, M., Choi, Y.S., Jo, C.: Status of meat alternatives and their potential role in the future meat market—A review. Asian-Australasian Journal of Animal Sciences 33(10) 1533 (2020).
37. Food Safety Authority of Ireland Homepage, Temperature Control I FAQs I The Food Safety Authority of Ireland (fsai.ie) and Domestic Practice I FAQ's I The Food Safety Authority of Ireland (fsai.ie), last accessed 2023/01/11.

Apple Puree Pasteurization by Pulsed Electric Fields: From Microbial Safety to Financial Viability Assessment

Enrique Pino-Hernández, Paula Valério, Sara Dias, Marta B. Evangelista, and Marco Alves

Abstract Food industries have been developing more sustainable preservation processes that maintain the nutritional value of foods and meet consumers' demands by providing safe and healthy products with a long shelf-life. Thus, it was necessary to study and take on challenges throughout the production process. The goal of this work was to evaluate the effect of pulsed electric fields (PEF) on the apple puree quality and to present a model that supports investment decisions taken by Portuguese agri-food companies on using PEF equipment to pasteurize the fruit product. PEF treatment ($E = 10$ kV/cm and $q = 644$ kJ/kg) was applied to apple puree at a laboratory scale. The processing conditions established resulted in 5-log CFU/g reduction of *Escherichia coli*. Puree PEF-treated and stored at 4 °C for 30 days was compared to puree without treatment regarding aerobic bacteria and yeast and mold. Moreover, financial viability assessment was developed using *Banco de Portugal* data (year 2021) for the classification of economic activities, namely 10320. The results of the technical and financial analysis indicate PEF pasteurization is suitable for upgrading pasteurization process. The investment required on PEF equipment for upgrading pasteurization process under study showed an internal rate of return of 21.54% and a payback of 4 years. In what concerns the final product, PEF treatment ensures the microbial stability of puree for at least 30 days. In conclusion, this study established the PEF conditions for puree non-thermal pasteurization and also showed a suitable model to support a decision investment on PEF technology.

Keywords Decision investment · Non-thermal pasteurization · Reineta Apple

These authors (Enrique Pino-Hernández and Paula Valério) contributed equally to this study and share first authorship.

E. Pino-Hernández · S. Dias · M. B. Evangelista · M. Alves (✉)
INOV.LINEA, Agri-Food Technology Transfer and Valorization Center, TagusValley Science and Technology Park, 2200-062 Abrantes, Portugal
e-mail: marco_alves@tagusvalley.pt

P. Valério
INOV.POINT, Business Hosting and Entrepreneurship, TagusValley Science and Technology Park, 2200-062 Abrantes, Portugal

J. R. da. C. S. Galvão et al. (eds.), *Proceedings of the 3rd International Conference on Water Energy Food and Sustainability (ICoWEFS 2023)*, Springer Proceedings in Earth and Environmental Sciences, https://doi.org/10.1007/978-3-031-48532-9_5

1 Introduction

The current global commitment to reduce climate change and food waste have been promoting the need for finding more sustainable technological processes [1]. Consumers are increasingly concerned about having a balanced diet, reducing consumption of unhealthy and ultra-processed products, and increasing daily consumption of fruits and vegetables. These habits could result in important benefits for consumers' health, with a remarkable impact on several food production systems and the environment [2]. Therefore, new studies on sustainable and financially viable processes for food valorization are necessary.

Apple (*Malus domestica*) is one of the most popular fruits in the world [3]. According to data from 2021, the global production of apples reached 93.1 million tons [4]. In European Union, apple production represents 20% of the global market. In Portugal specifically, apple was the most harvested fruit in 2021, with a total of 368,230 tons produced [4].

Post-harvested apples are separated and categorized according to size and quality. Fruits with visual defects or smaller sizes have lower economic value than optimal fresh fruits and are normally used for the production of different products such as juices, concentrates, or purees. Traditionally, thermal treatments have been used by the food industry to process fruit products. Raw apples are precooked at temperatures between 93 and 98 °C for 4 to 5 min. This is an important step for enzymatic inactivation and fruit tissue softening. Next, pre-treated fruits are pulped and pasteurized at 100 °C for 10–15 min to ensure the destruction of spoilage microorganisms [5]. However, this process has some negative effects, such as loss of some nutrients, formation of unhealthy chemical compounds and the modification of some organoleptic characteristics [5, 6].

The drawbacks of current thermal treatment, because of its long processing times and negative effects on the final product, have prompted the exploration of new technologies for fruit products disinfection or pasteurization such as Ohmic heating (OH), ultrasound (US), high-pressure processing (HPP), and pulsed electric fields (PEF) [2, 7, 8]. Specifically, HPP and PEF are alternative technologies that can produce products with higher quality in comparison to traditional thermal treatments. However, there are other aspects to consider when choosing those technologies, such as equipment costs, energy requirements, processing time, batch or continuous operation mode, and carbon footprint, among others. In what concerns PEF, this processing technology could offer some advantages when compared to HPP.

PEF promotes cell electroporation (non-thermal effect that creates pores in cell membranes), and shows great efficiency in microbial inactivation and as a processing technology. The application of short-duration PEF (μs–ms) leads to a total treatment time that is very short, which results in saving energy and operational costs, and in food results in minimally processed products, with sensory and nutritional characteristics similar to their fresh equivalents [9]. For this reason, PEF can contribute to sustainable food processing while promoting innovation and economic advantage for the food sector [10]. In spite of these benefits, PEF technology is highly dependent on the food electrical conductivity, because it influences the process's effectiveness. On the other hand, the lack of information and data reported in the literature prevents the replicating of processes and the development of comparative experiments, delaying the widespread use of this technology on an industrial scale [11]. Nonetheless, an industrial scale PEF equipment is currently available in TAGUSVALLEY and has been successfully used to conduct proof-of-concepts.

Considering the advantages of PEF technology, a study on the techno-financial viability of using PEF at industrial level was conducted to guide and evaluate the applicability of using PEF equipment and to support investment decisions taken by Portuguese agri-food companies. It is important to highlight that when this type of study is carried out, possible changes in the operating capacity, production levels, operating and maintenance costs, sales, income and outlays must be taken into account.

2 Material and Methods

2.1 *Raw Material*

Reineta Parda apples (*Pyrus Malus*) were obtained from an industrial unit (Cooperfrutas, S.A.) of Alcobaça, Portugal. Fresh fruits with some visual defects and small sizes to be sold in the fresh market were used in this study in order to fruit valorization. After reception, fruits were stored at 3 ± 1 °C until being used for their processing.

2.2 *Chemical Characterization*

To evaluate the nutritional composition of fresh apple fruits, analyses of protein (infrared spectrophotometry), lipids (Soxhlet extraction), carbohydrate (was obtained by the difference between the sum of protein, lipids, moisture, ash, and fiber contents), total sugars (Lane-Eynon method), fiber (sulphuric acid solution and loss in mass on incineration), and total energetic value were conducted based on the [12]. The proximate composition was expressed as g/100 g in dry weight.

2.3 *Product Preparation*

Fresh apples were washed, cut into 4 pieces using a sterile stainless knife and pretreated at 100 °C for 2 min. Next, fruits were blended using a kitchen robot (Vorwerk, Thermomix TM31, DE), and fresh lemon juice (2.5% w/w) as a natural acidulant was added [5]. Then, the puree was sieved due to the high amount of cored and pieces

Fig. 1 Sample preparation for PEF processing at laboratory scale. **1** Washing and cutting, **2** Pre-treatment, **3** Mash, **4** Refining, **5** PEF pasteurization, **6** Storage

of peel and to also avoid the formation of air bubbles in the puree product. After that, puree was immediately placed in the PEF treatment chamber between the electrodes and pasteurized. The PEF treatments were conducted at laboratorial scale in an PEF generator model EPULSUS_LBM3B-15 (EnergyPulse Systems, Lisbon, Portugal) using bipolar mode. Finally, the treated product was stored under refrigerator (see Fig. 1).

2.4 PEF Pasteurization Conditions

PEF pasteurization conditions were selected on preliminary tests considering *Escherichia coli* (ATCC 25922) as reference bacteria (data not shown). The pre-treated puree product (100 g) was inoculated with the bacteria at an approximate concentration of 10^7 CFU/g. *E. coli* determination was performed according to the method previously described in [9]. The PEF processing conditions used are reported in the Table 1.

Table 1 PEF pasteurization conditions

Processing parameters	Conditions
Electrical conductivity of puree at 25 °C (mS/cm)	≈4
PEF treatment time (μs)	1120
Electric filed strength, *E* (kV/cm)	10
Pulse duration rate (μs)	8
Specific energy, *q* (kJ/kg)	644
Current (A)	115

2.5 Storage Study

Before and immediately after PEF treatment samples were microbiologically analyzed. The effects of PEF pasteurization on microbial quality of puree product were compared with unprocessed control puree. Samples were analyzed at 0–10–20–30th days of storage at 3 ± 1 °C to determine the microbial stability.

Total aerobic plate counts and yeast and mold counts were determined on plate count agar (PCA, BBL/Difco Laboratories, Sparks, MD, USA) and Dichloran-rose engal chloramphenicol agar (DRBC, Merck, Germany), respectively, using 100 μL surface plating method [6]. PCA plates were incubated at 37 °C for 24 h, and RBC plates at 25 °C for 5 days. Results were expressed as log CFU/g.

2.6 Financial Viability Assessment

The financial feasibility study is focused on the apple puree pasteurization by PEF instead of conventional thermal pasteurization. For a closer approximation to the industry reality, it was considered as the basis of this study the cost and income structure of the average of all Portuguese companies with economic activities code (EAC) 10320 (manufacture of fruit and vegetable products), using for this data from the *Banco de Portugal*, for the year 2021. The replacement of conventional pasteurization equipment depends on economic considerations rather than the total physical uselessness of the equipment. Production capacity, new incomes derived from changes in the quality and price of the product, the use of new commercial possibilities were also considered to support the investment decision.

The equipment purchases cost and its specifications were taken from quotes provided by national and international suppliers, as well as information from manufacturers' websites.

PEF Equipment Specifications. The PEF equipment allows for laboratory and pilot scale experiments, including a bipolar solid-state Marx Modular generator, safety box, and treatment chamber to work in continuous mode. The PEF equipment specifications are showed in the Table 2.

Table 2 PEF equipment specifications

Parameters	Output
Maximum voltage	15,000 V
Pulse polarity	Positive and negative
Maximum frequency	200 Hz
Pulse duty cycle	220
Pulse width	2 μs to 200 μs
Efficiency	90%

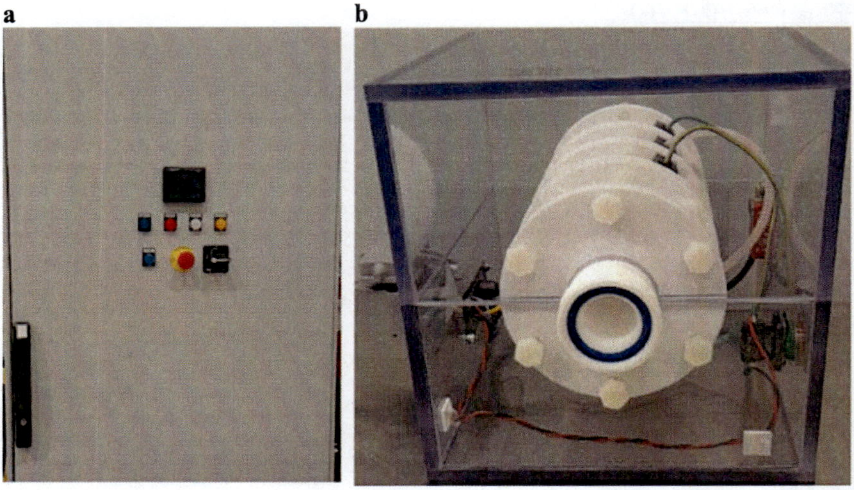

Fig. 2 PEF equipment (**a**) and continuous processing chamber DN10 (10 mm diameter) (**b**)

PEF Equipment Investment. Considering that the entire manufacturing process of the apple puree product using PEF is similar to the conventional production process, with the exception of the pasteurization operation where PEF equipment is introduced, an investment value of € 104,000 was considered. The equipment is shown in the Fig. 2. Regarding the remaining investment/equipment already used for production, it was considered the average tangible net assets of the companies with EAC 10320 previously chosen.

Financial Model. The financial feasibility study was carried out with the support of the project evaluation tool of Institute for Supporting Small and Medium-Sized Companies and Innovation (IAPMEI). To support investment decision from the perspective of apple puree product pasteurized by PEF instead of conventional treatment, a set of assumptions for the model were considered taken as basis the EAC 10320 data from *Banco de Portugal* (see Table 3).

The companies considered as samples are distributed throughout the Portuguese territory and have the developing of fruit products as their principal economic activity. The remaining useful life of the replaced asset (conventional equipment) will not be considered as income. In what concerns to the market segment and distribution channels, they are similar to the conventional product. However, consumers who seek foods that are minimally processed and without added sugars and preservatives could be reached.

Table 3 Summary of the financial assessment

Item	Scenario
Year of analysis	2021
1st year of project	2023
Project life (years)	5
Companies' samples	15 micro and 5 small companies
Conventional treated product price (Kg)	€ 5.75 + VAT
Annual product price increase (per year)	1%
Gross margin of cost of goods sold	42.77%
External supply and services (2023)	€ 241.800
Personnel expenses (2023)	€ 118.620
Quantities sold (2023)	185.687 kg
Growth rate of quantities sold (per year)	1%
Investment value (PEF Eq.)	€ 104.000
Annual amortization rate	12.5%
Tangible fixed assets (2023)	€ 643.300

3 Results and Discussion

3.1 Chemical Characterization

The proximate composition of the analyzed fresh apples is presented in Table 4 (dry weight). Fruit samples are high in fiber. According to nutritional claim, a food high in fiber means that the product contains at least 6 g per 100 g of this component [13]. The amount of fiber in some fruit products are dramatically reduced after processing. The beneficial effect of fruit products on human health is increased when consuming the products like puree due to higher fiber content when compared to liquid products [2].

Carbohydrates and total sugars are the most abundant components in Reineta Parda apple. Lipids and protein represent less than 4 g/100 g in dry weight of the

Table 4 Fresh apple nutritional characterization

Nutrition facts	Value (g/100 g in dry weight)
Lipids	0.50
Protein	3.73
Carbohydrate	80
Of which sugars	29
Fiber	7.8
Total energetic value (Kcal/100 g)	352

fruit. Usually, the predominant sugar in apples is fructose accounting for ~60% of the total sugar content [14]. Total sugars, defined as mono and disaccharides, play a crucial role in fresh apple's sensory characteristics because they turn them sweeter, which increases their commercial attractiveness [15].

In what concerns to possible changes induced by PEF pasteurization, it was not the aim of this study due to the short processing time. Previous studies reported that emerging thermal and non-thermal technologies have a minimal impact on nutritional and sensorial fruit attributes [6, 14–16].

3.2 Microbial Quality of Apple Puree Product

Total aerobic microorganisms cause spoilage of fruit during storage under refrigeration while molds and yeasts limit its shelf-life as previously reported [6, 7, 15]. At the beginning of the product storage study (day 0), untreated samples revealed total aerobic bacteria counts of 2.6 ± 0.1 log CFU/g and molds and yeasts values of 2.9 ± 0.2 log CFU/g despite of visual molds was not observed. Nonetheless, the microbial load increased during the storage period reaching its maximum value (total aerobic bacteria 3.7 ± 0.1 and molds and yeasts 6.5 ± 0.3 log CFU/g) after 30 days. Molds contamination was visible after 10 days of storage.

In what concerns PEF-treated samples, immediately after the process (day 0), high microbial reductions were obtained, since total aerobic bacteria and molds and yeasts counts were below the detection limit (<1 CFU/g). Selected PEF processing conditions previously reduced 5.8log CFU/g of inoculated *E. coli*. Moreover, PEF remarkably delayed the microbial growth population. Microbial counts remained below the detection limit for the whole storage period analyzed. The PEF treatment effects together with the storage conditions were an effective method to reach a higher microbial stability of the apple puree product for at least 30 days. Similar results were previously reported when apple juice and strawberry puree were processed by PEF and stored at 4 °C. The PEF treatments ensured microbial stability for at least at 3 and 28 days, respectively [7, 17].

From a technical context, apple puree product pasteurized by PEF technology has unique characteristics when compared to conventional one, namely: shorter processing time, minimally processed product, clean label, non-thermal pasteurization, energy efficiency in process, and ecofriendly technology.

3.3 Project Evaluation

The results obtained suggest that the payback time occurs in 4 years. The project proved to be quite attractive with an internal rate of return on investment of 21.54% and a net present value of € 355,083. As the selling price of the product has an important effect on the productivity and profitability of a project [18], the selling price of the product considered in this study was € 6.33 + VAT per kg. The net return on sales varies between 7 and 8%. According to several data-bases, this value is also similar to the selling price of analogous products processed by HPP. Thus, it is easily accepted that the final price of this apple puree product is slightly more expensive than the conventional one. Nevertheless, when PEF technology is introduced into a production process, the company´s characteristic must be considered because its economic and financial indicators would change depending on its resources, assets, productivity, commercial structure, among others aspects.

3.4 Operational Breakeven Point

The breakeven point for the 1st year of project is reached when 163.500 kg of product are sold or when revenue is 1.03 M€, which means that these are the quantities needed to cover all costs (fixed and variable). Any extra amount sold translates into positive net income for the company. From a techno-economic point of view, the integration of PEF equipment for pasteurization can be viable as a strategy to improve product quality and safety while increasing industrial sustainability and competitiveness.

4 Conclusions

This study showed that the use of PEF technology is a novel alternative to apple puree non-thermal pasteurization, avoiding the use of thermal pasteurization. The results obtained suggest that PEF pasteurization is a profitable investment for current or new companies that want to follow the technological changes in order to maintain their competitiveness in market with a more sustainable production process and product. In addition to being profitable, it also allows companies to keep up with market trends and meet the demands of consumers who are continuously looking for healthy and minimally processed foods. Moreover, The PEF equipment can be integrated into the food processing plant like any other piece of equipment and its installation is not expensive. With the implementation of the PEF equipment, a slightly redesign of the plant is recommended in order to allow a better use of the area and lower energy consumption.

Acknowledgements We would like to thank Cooperfrutas CRL for providing the raw material for the experiments performed for this work.

Funding This work was financially supported by the projects TAGUSVALLEY2030 RHaq (CENTRO-04–3559-FSE-000143), TAGUSVALLEY2030 IT (CENTRO-04–3559-FSE-000143), and WINBIO (POCI-01–0246-FEDER-181335) under European Social Fund from the European Union managed by CENTRO 2020 and PORTUGAL 2020.

References

1. Hassoun, A.; Bekhit, A.E.D.; Jambrak, A.R.; Regenstein, J.M.; Chemat, F.; Morton, J.D.; Gudjónsdóttir, M.; Carpena, M.; Prieto, M.A.; Varela, P.; et al. The Fourth Industrial Revolution in the Food Industry—Part II: Emerging Food Trends. *Crit. Rev. Food Sci. Nutr.* **2022**, 1–31, https://doi.org/10.1080/10408398.2022.2106472.
2. Rinaldi, M.; Langialonga, P.; Dhenge, R.; Aldini, A.; Chiavaro, E. Quality Traits of Apple Puree Treated with Conventional, Ohmic Heating and High-Pressure Processing. *Eur. Food Res. Technol.* **2021**, *247*, 1679–1688, https://doi.org/10.1007/s00217-021-03738-6.
3. Shah, Z.A.; Dar, M.A.; Dar, E.A.; Obianefo, C.A.; Bhat, A.H.; Ali, M.T.; El-Sharnouby, M.; Shukry, M.; Kesba, H.; Sayed, S. Sustainable Fruit Growing: An Analysis of Differences in Apple Productivity in the Indian State of Jammu and Kashmir. *Sustainability* **2022**, *14*, 14544, https://doi.org/10.3390/su142114544.
4. FAOSTAT Crops.
5. Landl, A.; Abadias, M.; Sárraga, C.; Viñas, I.; Picouet, P.A. Effect of High Pressure Processing on the Quality of Acidified Granny Smith Apple Purée Product. *Innov. Food Sci. Emerg. Technol.* **2010**, *11*, 557–564, https://doi.org/10.1016/j.ifset.2010.09.001.
6. Pino-Hernández, E.; Pereira, R.N.; Ballesteros, L.F.; Vicente, A.A.; Abrunhosa, L.; Teixeira, J.A. Effects of Moderate Electric Fields on the Post-Harvest Preservation of Chestnuts. *Food Bioprocess Technol.* **2021**, *14*, 920–934, https://doi.org/10.1007/s11947-021-02612-3.
7. Yildiz, S.; Pokhrel, P.R.; Unluturk, S.; Barbosa-Cánovas, G. V. Shelf Life Extension of Strawberry Juice by Equivalent Ultrasound, High Pressure, and Pulsed Electric Fields Processes. *Food Res. Int.* **2021**, *140*, https://doi.org/10.1016/j.foodres.2020.110040.
8. Timmermans, R.A.H.; Nederhoff, A.L.; Nierop Groot, M.N.; van Boekel, M.A.J.S.; Mastwijk, H.C. Effect of Electrical Field Strength Applied by PEF Processing and Storage Temperature on the Outgrowth of Yeasts and Moulds Naturally Present in a Fresh Fruit Smoothie. *Int. J. Food Microbiol.* **2016**, *230*, 21–30, https://doi.org/10.1016/j.ijfoodmicro.2016.04.014.
9. Yildiz, S.; Pokhrel, P.R.; Unluturk, S.; Barbosa-Cánovas, G. V. Identification of Equivalent Processing Conditions for Pasteurization of Strawberry Juice by High Pressure, Ultrasound, and Pulsed Electric Fields Processing. *Innov. Food Sci. Emerg. Technol.* **2019**, *57*, 102195, https://doi.org/10.1016/j.ifset.2019.102195.
10. Arshad, R.N.; Abdul-Malek, Z.; Roobab, U.; Munir, M.A.; Naderipour, A.; Qureshi, M.I.; El-Din Bekhit, A.; Liu, Z.W.; Aadil, R.M. Pulsed Electric Field: A Potential Alternative towards a Sustainable Food Processing. *Trends Food Sci. Technol.* **2021**, *111*, 43–54, https://doi.org/10.1016/j.tifs.2021.02.041.
11. Raso, J.; Frey, W.; Ferrari, G.; Pataro, G.; Knorr, D.; Teissie, J.; Miklavčič, D. Recommendations Guidelines on the Key Information to Be Reported in Studies of Application of PEF Technology in Food and Biotechnological Processes. *Innov. Food Sci. Emerg. Technol.* **2016**, *37*, 312–321, https://doi.org/10.1016/j.ifset.2016.08.003.
12. AOAC. (1998). Official methods of analysis (Vol. 16). Association of Official Agricultural chemists.
13. EU Nutrition and Health Claims Made on Foods. *Off. J. Eur. Union* **2006**.
14. Wibowo, S.; Essel, E.A.; De Man, S.; Bernaert, N.; Van Droogenbroeck, B.; Grauwet, T.; Van Loey, A.; Hendrickx, M. Comparing the Impact of High Pressure, Pulsed Electric Field and Thermal Pasteurization on Quality Attributes of Cloudy Apple Juice Using Targeted and Untargeted Analyses. *Innov. Food Sci. Emerg. Technol.* **2019**, *54*, 64–77, https://doi.org/10.1016/j.ifset.2019.03.004.
15. Pino-Hernández, E.; Pinto, C.A.; Abrunhosa, L.; Teixeira, J.A.; Saraiva, J.A. Hydrothermal and High-Pressure Processing of Chestnuts - Dependence on the Storage Conditions. *Postharvest Biol. Technol.* **2022**, *185*, 111773, https://doi.org/10.1016/j.postharvbio.2021.111773.
16. Abid, M.; Jabbar, S.; Hu, B.; Hashim, M.M.; Wu, T.; Lei, S.; Khan, M.A.; Zeng, X. Thermosonication as a Potential Quality Enhancement Technique of Apple Juice. *Ultrason. Sonochem.* **2014**, *21*, 984–990, https://doi.org/10.1016/j.ultsonch.2013.12.003.

17. Dziadek, K.; Kopeć, A.; Dróżdż, T.; Kiełbasa, P.; Ostafin, M.; Bulski, K.; Oziembłowski, M. Effect of Pulsed Electric Field Treatment on Shelf Life and Nutritional Value of Apple Juice. *J. Food Sci. Technol.* **2019**, *56*, 1184–1191, https://doi.org/10.1007/s13197-019-03581-4.
18. Coral, D.A.; Leonardo, V.; Óscar, F.C.; James, J.S. Process Design and Techno - Economic Assessment of Cellulolytic Enzymes Production from Coffee Husk through Process Simulation. *Biomass Convers. Biorefinery* **2022**, https://doi.org/10.1007/s13399-022-03130-8.

Bibliometric Analysis of Sustainable Indicators in Agribusiness

Sónia Monteiro, Patrícia Quesado, Verónica Ribeiro, Maria Eduarda Fernandes, Teresa Eugénio, João Costa, Hugo Rodrigues, and Marina Tomás

Abstract The aim of this paper is to analyze the scientific production in the last decade (2011–2022) related to sustainable indicators (environmental, economic, social and ESG—Environmental, Social and Governance) in agribusiness. As methodology, a bibliometric study was carried out on articles from Scopus database. Afterwards, the most cited articles, journals and countries were analyzed, as well as a thematic analysis of the keywords. Results indicate that research on sustainable indicators of agribusiness has increased over time. In last years, the number of scientific papers on these topics has been increasing, probably associated with the approval of the United Nations for Sustainable Development (known as 2030 Agenda, approved in 2015). Italy is the country with most publications, the highest number of citations, and with the largest collaboration network. Most papers are published in 2 journals of higher impact and LCA (Life Cycle Assessment) is cited as the most usual method to assess the impact of business. Regarding the indicators to evaluate the sustainability of agribusiness, there is a greater predominance of environmental indicators compared to economic and social indicators.

Keywords Sustainability · Indicators · Agribusiness · Bibliometric study

S. Monteiro (✉) · P. Quesado · V. Ribeiro · J. Costa
CICF-Research Centre on Accounting and Taxation, Polytechnic University of Cávado and Ave, Barcelos, Portugal
e-mail: smonteiro@ipca.pt

M. E. Fernandes · T. Eugénio
ESTG-CARME–Centro de Investigação Aplicada em Gestão e Economia, Polytechnique Institute of Leiria, Leiria, Portugal

H. Rodrigues · M. Tomás
Campotec, Portugal

J. R. da. C. S. Galvão et al. (eds.), *Proceedings of the 3rd International Conference on Water Energy Food and Sustainability (ICoWEFS 2023)*, Springer Proceedings in Earth and Environmental Sciences, https://doi.org/10.1007/978-3-031-48532-9_6

1 Introduction

A sustainable agribusiness model is one that produces sufficient food to meet the demand, now and in the future, thus contributing to achieve the Sustainable Development Goals (SDGs) in the United Nations Agenda for Sustainable Development (known as 2030 Agenda, approved in 2015). Although there is broad literature on sustainability in agribusiness, no study offers a comprehensive overview of the research about the sustainable indicators for measuring sustainability in the agribusiness. Barth et al. [1] developed a systematic literature review to understand sustainable business model innovation in the agri-food sector, regarding the three pillars of sustainability but do not address the issue of measurement and indicators. Silvestri et al. [11] suggest that Triple Bottom Line (TBL) approach of indicators (environmental, social, and economic) offers the best solution to ensure an easier transition to sustainability in the agri-food sector. This study considers the TBL approach suggested by Silvestri et al. [11] to define the keywords search strings in Scopus database. Using a bibliometric approach, the aim of this paper is to analyze the scientific production in the last decade (2011–2022) related to sustainable indicators (environmental, economic, social and ESG) in agribusiness.

Section 2 provides a brief literature review related to sustainability in the agribusiness industry. Section 3 describes the methodology underlying this study. The results are presented in Sect. 4 and Sect. 5 highlights the main conclusions and suggestions for future research.

2 Literature Review

Agriculture is one of the sectors with the greatest impact on climate change, as it is one of the largest contributors of GHG emissions, and also one of the most affected sectors by climate change [5]. Devitalized soils, increased pest and disease infestations, and decreased pollination are some examples of the climate changes' impacts on agriculture [12]. Further, much of the food produced is wasted either by producers or consumers (13.3% of food is wasted after harvest and before it reaches retail markets and 17% of food is wasted by the consumer). The recent impact of crises, such us COVID-19 pandemic and war in Ukraine, leads to a decrease in production and consequent increase in food prices, and exponentially in the years 2020 and 2021, bringing additional challenges to sustainability [3, 12].

Adopting a Circular Economy (CE) approach offers solutions towards a more sustainable agricultural system, aiming to use raw materials first as products and then as reused/recycled products, to reduce inputs to the minimum possible and to reduce energy use [6]. A shift to a CE model in the agriculture sector could reduce CO_2 emissions by 60%, water consumption by about 70% and fertiliser use by 80% by 2050 [2]. So, the agricultural sector is one of the key sectors in the transition towards a more sustainable development oriented towards EC principles [9].

To move towards a sustainable agricultural system, it is necessary to measure and quantify the impact of agricultural business on the economy, the environment and society, therefore, specific methods and indicators are used.

Bibliometric studies, in the field of sustainability and agriculture, address sustainability as a whole in agriculture [10] and more specifically the product value chain [7, 8], and indicators to assess sustainability in agribusiness [11, 13, 14].

Sustainable indicators in agribusiness can have the following purposes: to assess the sustainability of agribusiness through Life Cycle Assessment (LCA), to identify best sustainable agricultural practices and to assist in decision making [11]. The biggest limitations in the sector of agribusiness indicators, is the lack of access to information from companies. Understanding how companies use the indicators is essential because it allows to understand which are the advantages and disadvantages of the application of certain indicators [11].

Despite the increased interest in this area, the studies of evaluation of sustainable agricultural development are still immature and the collaboration between authors and institutions is not strong [14]. The current bibliometric study aims to contribute with additional data to this area.

3 Methodology

Bibliometric studies are based on statistical analysis method emerged in the 1950s [4]. This type of study provides information on various subjects. It allows, for example, to study the scientific impact, the collaboration between countries, authors, and universities, on a certain topic. There is several software that helps the researcher to perform a bibliometric analysis. In this article, we use the Biblioshiny software, one of the most used software in this area [10], to make a bibliometric analysis of articles on sustainable indicators (economic, social, and environmental) related to agribusiness. For such purpose, the methodology described in Fig. 1 was followed.

The research was carried out in the Scopus database, one of the best databases to produce a reliable bibliometric survey [11]. The research criteria were "Title, Keywords, Abstract." The search strings used are "Environmental indicators" OR "Social indicators" OR "Economic indicators" OR "Sustainability indicators" OR "ESG" OR "Environmental Social Governance". The string "AND" was also used to restrict the search of the indicators to the area of agribusiness, ESG and report. The search was limited to documents of the type "article" and "review" (in English language), in the period 2011–2022, and to the areas of "Business, Management and Accounting", "Econometrics and Finance".

By June 2022, a total of 157 documents (148 scientific articles and 9 reviews) were identified. The documents are from 63 different sources, with a total of 643 authors, as shown in Table 1.

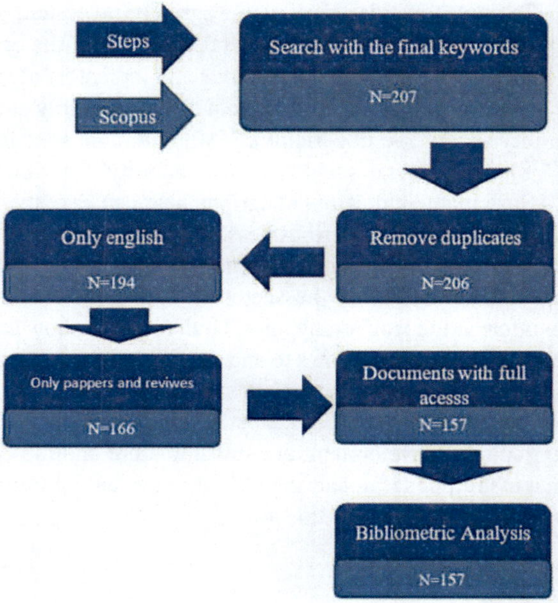

Fig. 1 Steps followed until final sample of articles

4 Results

4.1 Total Citations Per year

As we can observe in Table 2, the number of articles produced per year increased from 2015 onwards, particularly in 2018 (17 documents) and specially in 2021 (37 documents). However, regarding the average number of citations per article, the articles produced in 2012, 2014 and 2017 had on average over 30 citations per article. Recent years register fewer average citations per article, which is normal since they are more recent articles. In the column of _MeanTCperYear, it is possible to observe that the number of citations per year has been growing, which shows interest of researchers in these themes.

4.2 Most Cited Articles—Top 20

In Table 3 we can observe the 20 most cited articles related to sustainable indicators in agribusiness. The most cited article is by Yakovleva N. with a total of 158 citations, and an average of 14.36 citations per year. With a total citation above 100 are the articles by Egilmez G., Scholz K. and Banasik A., respectively with 116, 112 and 107 citations. We also highlight the articles by Petit-Boix A., Abdella Gm., Schmitt E.,

Table 1 Database description n = 157

Description	Results
Main information about data	
Timespan	2011:2022
Sources (Journals, Books, etc.)	63
Documents	157
Annual Growth Rate %	8.69
Document Average Age	3.94
Average citations per doc	18.89
References	9144
Document contents	
Keywords Plus (ID)	967
Author's Keywords (DE)	637
Authors	
Authors	643
Authors of single-authored docs	9
Authors collaboration	
Single-authored docs	10
Co-Authors per Doc	4.38
International co-authorships %	35.67
Document types	
Article	148
Review	9

Source R Bibliometrix outputs

Table 2 Publications and citations per year

Year	No.	Mean Tc Per Art	Mean Tc Per Year	Citable years
2011	4	23.50	2.14	11
2012	7	39.57	3.96	10
2013	4	22.25	2.47	9
2014	6	30.83	3.85	8
2015	10	26.40	3.77	7
2016	15	27.40	4.57	6
2017	15	34.80	6.96	5
2018	17	20.41	5.10	4
2019	16	22.50	7.50	3
2020	16	13.69	6.84	2
2021	37	5.27	5.27	1
2022	10	0.30		0

Source R Bibliometric outputs

Table 3 The 20 most cited articles

Paper	Total Citations	Tc Per Year
Yakovleva N, 2012, Int J Prod Res	158	14.36
Egilmez G, 2014, Resour Conserv Recycl	116	12.89
Scholz K, 2015, Resour Conserv Recycl	112	14.00
Banasik A, 2017, Int J Prod Econ	107	17.83
Castellini C, 2012, J Clean Prod	83	7.55
Petit-Boix A, 2017, J Clean Prod	78	13.00
Schmitt E, 2017, J Clean Prod	73	12.17
Kalmykova Y, 2016, J Clean Prod	68	9.71
Naidoo M, 2018, J Clean Prod	64	12.80
Montgomery H, 2011, World Dev	59	4.92
Reutter B, 2017, J Clean Prod	57	9.50
Rothwell A, 2016, J Clean Prod	55	7.86
Strezov V, 2013, J Clean Prod	48	4.80
Ahmad S, 2019, J Clean Prod	46	11.50
Nicoletti Junior A, 2018, J Clean Prod	46	9.20
Cristóbal J, 2016, J Clean Prod	42	6.00
Jane Dillon E, 2016, Int J Agric Sustainability	42	6.00
Goggins G, 2016, J Clean Prod	40	5.71
Raman S, 2014, J Clean Prod	40	4.44
Abdella Gm, 2020, J Clean Prod	39	13.00

Source R Bibliometric outputs

Naidoo M., Ahmad S., and Devkota Kp. Which have an average number of citations per year higher than 10.

4.3 Higher Impact Journals—Top 20

In this section the impact of the journals was analyzed (Table 4). Regarding the h_index, g_index and m_index the *Journal of Cleaner Production* stands out from the others, with a m_index of 2.364. Then, we have the journal *Resources, Conservation and Recycling* with an m_index of 0.667, and a TC of 333. In third place is World Development with a m_index of 0.417 and 141 TC. We also highlight the *International Journal of Production Economics*, with a total of 115 citations, despite having a h_index and g_index of 2. The *Journal of Cleaner Production* and *Resources, Conservation and Recycling* publish about a third of the documents of the entire collection. Furthermore, according to the Scimago Journal & Country Rank, these two journals are in the first quartiles, with a H index of 232 and 150, respectively.

Table 4 The 20 Journals with the highest impact

Journal	H_Index	G_Index	M_Index	Tc	Np	Py_Start
Journal of Cleaner Production	26	36	2.364	1436	47	2012
Resources, Conservation and Recycling	6	10	0.667	333	10	2014
World Development	5	6	0.417	141	6	2011
British Food Journal	3	6	0.273	52	6	2012
Ecological Economics	3	3	0.300	56	3	2013
Food Policy	3	3	0.333	45	3	2014
Internat. J. of Agricultural Sustainability	3	3	0.273	79	3	2012
Marine Policy	3	3	0.429	24	3	2016
Agricultural And Food Economics	2	2	0.286	19	2	2016
Agricultural Economics (Czech Republic)	2	3	0.222	18	3	2014
Agris On-Line Papers in Econ.and Inform	2	2	0.200	15	2	2013
Business Strategy and The Environment	2	2	0.250	49	2	2015
Custos e Agronegócios	2	3	0.250	15	4	2015
Environment, Devel. and Sustainability	2	2	0.182	32	2	2012
Internat. J.of Production Economics	2	2	0.333	115	2	2017
Studies In Agricultural Economics	2	2	0.286	30	2	2016
Accounting Perspectives	1	1	0.333	3	1	2020
Asian Social Science	1	2	0.100	15	2	2013
Business: Theory and Practice	1	1	0.500	3	1	2021
Canadian J. of Agricultural Economics	1	1	0.091	12	1	2012

Source R Bibliometric outputs

Table 5 The 20 countries with the most publications

Country	Freq	Country	Freq
Italy	44	Australia	13
Spain	29	Serbia	12
France	26	Sweden	12
Usa	25	Czech Republic	10
Germany	23	Chile	8
Uk	22	Kazakhstan	8
Ukraine	22	Denmark	6
Brazil	21	India	6
Netherlands	17	Pakistan	6
China	14	Qatar	6

Source R Bibliometric outputs

4.4 Country Analysis

In this section, first, we analyze the publications produced in each country on sustainable indicators for agribusiness (Table 5). Italy is the country with the most publications (44), followed by Spain, France, and USA with 29, 26, and 25, respectively. All the countries in the top 20 have at least six publications. Some countries have joint publications with other countries.

As we can see in Table 6, the most cited papers are from USA and Italy with a total of more than 300 citations, followed by Sweden and Australia with more than 200 citations, and, finally, Spain, Netherlands, Germany, Japan, Ireland and UK with more than 100 citations. However, regarding the average number of citations per article, the country that stands out on top is Japan with about 60 citations per article, followed by Malaysia, Sweden, and USA with more than 40 citations per article. Italy has an average of 22 citations per article.

Countries network was analysed, showing Italy as the country with the largest collaboration with other countries. The one with the smallest collaboration network is Qatar. Most collaborations occur between European countries such as Italy, Spain, France, Poland, Ireland, Hungary, Serbia, Netherlands, and Germany. The United Kingdom, despite being in another collaborative network, has some collaborations with countries such as Italy and Spain. This shows a great tendency and interest in this theme on the part of European countries, compared to countries from other continents.

4.5 Word Map, Trend Topics and Thematic Maps

By the word map analysis of the frequency of keywords plus, it is concluded that the food industry theme (33), environmental indicators (30), environmental impact

Table 6 The 20 countries with the most citations

Country	Total citations	Average article citations
Italy	330	22.00
USA	313	44.71
Sweden	234	46.80
Australia	201	28.71
Spain	194	24.25
Netherlands	186	31.00
Germany	151	16.78
Japan	123	61.50
Ireland	118	39.33
United Kingdom	118	19.67
China	84	16.80
Switzerland	76	38.00
Brazil	66	16.50
France	50	8.33
Malaysia	47	47.00
Qatar	46	23.00
Czech Republic	44	7.33
Philippines	34	34.00
Mexico	27	27.00
Canada	24	12.00

Source R Bibliometric outputs

(28), LCA (24), life cycle (22), sustainability (19), sustainability indicators (18), are present in the researched articles. Despite not appearing in the words map, sustainable development (46) is also widely addressed in the analyzed documents. This analysis demonstrates a greater approach to the environmental dimension of sustainability when faced with the economic and social dimensions.

The word map analysis of the frequency of the authors' keywords shows that the most frequent word is LCA (21), followed by sustainability indicators (11), environmental indicators (6), food security (6), food waste (6), nutrition (6) and supply chain (6). The most frequent word is sustainability (24) although it is not found on the map. This suggests a strong link between the use of LCA in assessing the sustainability of agribusiness.

As we can observe, LCA stands out as the most used methods to quantify a service or product in sustainable terms. Regarding the indicators, the most commonly used are related to the environmental dimension of sustainability, namely carbon footprint, gas emissions, greenhouse gases, land uses, food waste.

5 Conclusions and Future Research

This study addresses the scientific production, from the last decade, on the indicators to evaluate the sustainability performance in agribusiness. A large part of the studies is conducted in Europe, namely in Italy. These results are in line with the findings of Silvestri et al. [11], Wohlenberg et al. [13] and Yu & Mu [14]. The largest collaborations occur in Europe, namely between Italy, Spain, France and the UK. The European institutions are the ones that have the largest number of publications, highlighting Wageningen University (Netherlands). This institution is highlighted in studies on sustainable development [14] and on sustainability of the value chain [10]. Also, Asia presents studies in this area. China has a relevant role in studies on sustainable development, being one of the first countries to propose and implement a sustainable development strategy [14].

The interest of researchers in assessing the sustainability of agribusiness has grown in recent years, and although sustainable development, sustainability in agriculture and agribusiness are a "hot topic" [10, 11, 14] there is not much collaboration among authors. Our results indicate the existence of a week network of collaboration between authors, which corroborates the results of previous studies (e.g., Yu and Mu [14]). Also, we conclude that previous research mainly addresses the following areas: food industry, life cycle, sustainability, sustainability indicators and LCA, since these are the themes that stand out. Economic and environmental indicators and LCA are the driving themes of the research. Therefore, we conclude that the society pillar of sustainability does not have the same relevance/weight as the others.

Through the structural analysis, it is possible to conclude that the studies are related to sustainable developments in agribusiness, agriculture and value chain as well as product life cycle and environmental impact and environmental indicators. The three pillars of sustainability are mentioned as well as some methods to evaluate the performance of a product. LCA is one of the tools used to assess sustainability in agribusiness. It is also possible to realize some of the environmental indicators used to evaluate agribusiness, such as water footprint, carbon footprint, land use and greenhouse gas emissions. Regarding the indicators, a higher prevalence of economic and environmental indicators is observed in the sample compared to social indicators. This may reveal that there is still a focus on the environmental dimension of sustainability.

The research carried out with this bibliometric study highlighted that the transformation of agriculture and agribusiness into more sustainable sectors is essential, and there is a growing interest of the scientific community in these topics.

One of the limitations of the study was the use of a single database, in this case Scopus. Further research should include Web of Science database. Another limitation concerns the research areas: we only focus the areas of "Business, Management and Accounting" and "Economics, Econometrics and Finance".

It will be interesting to later conduct a content analysis of the most relevant articles of the sample and understand which indicators are used to measure the sustainable performance of agribusiness.

Acknowledgements The authors acknowledge the financial support of FCT—Fundação para a Ciência e a Tecnologia (Portugal) for projects funded through national funds UIDB/04928/2020 (MEF and TE) and (POCI-01-0247-FEDER-070155).

This work is financed by national funds through FCT—Foundation for Science and Technology, I.P., within the scope of multi-annual funding UIDB/04043/2020.

References

1. Barth, H, Ulvenblad P. & Ulvenblad P. (2017). Towards a conceptual framework of sustainable business model innovation in the agri-food sector: systematic literature review. *Sustainability*, 9
2. Ellen MacArthur Foundation. (2015). Growth within: a circular economy vision for a competitive Europe. in https://www.mckinsey.com/
3. FAO. (2021). World Food and Agriculture – Statistical Yearbook 2021. In World Food and Agriculture – Statistical Yearbook 2021
4. Godin, B. (2006). On the origins of bibliometrics. Scientometrics, 68(1), 109–133
5. IPCC. (2021). Climate Change 2021: The Physical Science Basis. Contribution of Working Group I to the Sixth Assessment Report of the Intergovernmental Panel on Climate Change. Cambridge University Press, Cambridge
6. Jurgilevich, A., Birge, T., Kentala-lehtonen, J., Korhonen-kurki, K., Pietikäinen, J., Saikku, L., & Schösler, H. (2016). Transition towards Circular Economy in the Food System. *Sustainability*, 8(1), 69
7. Luo, J., Ji, C., Qiu, C., & Jia, F. (2018). Agri-food supply chain management: Bibliometric and content analyses. Sustainability, 10(5)
8. Popa, D. (2022). Bibliometric Analysis of Agri-Food Supply Chain and Short Agro-Food Chain. Review of International Comparative Management, 23(2), 242–256
9. Poponi, S., Arcese, G., Pacchera, F., & Martucci, O. (2022). Evaluating the transition to the circular economy in the agri-food sector: Selection of indicators. *Resources, Conservation & Recycling*, 176, 1–13
10. Sarkar, A., Wang, H., Rahman, A., Memon, W. H., & Qian, L. (2022). A bibliometric analysis of sustainable agriculture: based on the Web of Science (WOS) platform. *Environmental Science and Pollution Research*, 29(26), 38928–38949
11. Silvestri, C., Silvestri, L., Piccarozzi, M., & Ruggieri, A. (2022). Toward a framework for selecting indicators of measuring sustainability and circular economy in the agri-food sector: a systematic literature review. *International Journal of Life Cycle Assessment*. https://doi.org/10.1007/s11367-022-02032-1
12. United Nations (UN, 2022). The sustainable development goals report 2022. https://unstats.un.org/sdgs/report/2022/
13. Wohlenberg, J., Schneider, R. C. S., & Hoeltz, M. (2020). Sustainability indicators in the context of family farming: A systematic and bibliometric approach. Environmental Engineering Research, 27(1)
14. Yu, S., & Mu, Y. (2022). Sustainable Agricultural Development Assessment: A Comprehensive Review and Bibliometric Analysis. Sustainability, 14, 118824

Introduction of Macroalgae in Human Food: A Matter of Sustainability

Tiago Mendes, Maria de Fátima Machado, and Vânia Ribeiro

Abstract World population growth and climate change are two factors that are increasingly impacting on the availability of nutritious and value-added food for human nutrition. One of the many causes of climate change is greenhouse gases, many of which come from the area of meat production for human consumption. On the other hand, the production of pastures to feed animals makes heavy use of water resources. As such, it is necessary to find alternatives to animal protein production, and macroalgae, among others, have been consumed worldwide for a long time, especially in the East. However, in western countries the growing interest in macroalgae and its introduction into the diet has been increasing, most likely because of the growing adherence to "plant-based diets". This article aims to review the potential of macroalgae in human food, its health benefits, and risks, as well as to demonstrate its potential as a sustainable animal protein substitute.

Keywords Macroalgae · Human food · Sustainability

1 Introduction

The world population, which was 7.8 billion in 2021, will continue to grow and is estimated to reach 9.7 billion by 2050 and 10.3 billion by 2100 [1]. Thus, it is suggested that reducing animal protein consumption is a solution to achieve a more

T. Mendes (✉) · M. de Fátima Machado · V. Ribeiro
Centre for Innovative Care and Health Technology, School of Health Sciences, CiTechCare, Polytechnic Institute of Leiria, 2411-901 Leiria, Portugal
e-mail: tgsmendes@gmail.com

M. de Fátima Machado
Ernesto Morgado, S. A. Rua Prof. Casimiro de Oliveira, 21, 3090-833 Barra, Alqueidão, Figueira da Foz, Portugal

V. Ribeiro
GeoBioTec, GeoBioSciences, GeoTechnologies and GeoEngineering, Earth Sciences Department, Universidade NOVA de Lisboa, 2829-516 Caparica, Portugal

J. R. da. C. S. Galvão et al. (eds.), *Proceedings of the 3rd International Conference on Water Energy Food and Sustainability (ICoWEFS 2023)*, Springer Proceedings in Earth and Environmental Sciences, https://doi.org/10.1007/978-3-031-48532-9_7

sustainable food supply chain [2–5]. To meet the goal of eradicating global hunger by 2030, governments must take immediate action to transform their agri-food networks [5]. Although the Sustainable Development Goals (SDGs) of the 2030 Agenda and its commitment to "leave no one behind (LNOB)" were outlined in 2015 with the agreement of all United Nations (UN) member states, they have been stalled in their progression, firstly due to COVID-19 (from which we had not yet fully recovered from the economic and social repercussions) and more recently the war in Ukraine. Insecurity and armed violence negatively influence the development of any country. Before the beginning of the war in Ukraine it was predicted that the main global economic variables would return to normal between 2022 and 2023, but with the war, the price of products and inflation have skyrocketed and Ukraine and Russia being the main producers and traders of cereals, energy and fertilisers, it will cause more inequality between the richest and poorest population, causing the latter to enter extreme poverty and hunger [6]. Considering climate change as one of many environmental concerns, greenhouse gases from livestock production account for 14.5% of global emissions. Many strategies to curb climate change have been adopted, such as: improving energy efficiency, various technological advances or reducing food waste, but it is recognized that they are insufficient to reduce greenhouse gas emissions and achieve their targets, so diets will have to be changed, particularly in reducing meat consumption. [4].

The scientific debate on the concept of sustainable diets has been increasingly on the agenda, and sustainable diets are understood as "nutritionally adequate and healthy diets, safe, with low environmental impact, economically fair, accessible and culturally acceptable" [4]. Thus, consciously, or unconsciously, currently, there is already an increase in adherence to "plant-based diets", which include vegetarianism, veganism or flexitarianism. These diets are based on the main consumption of plant derivatives but may include small amounts of animal foods (except for veganism). Thus, this accession has driven advances in the food industry with alternatives to food products of animal origin, based on plants, but with a flavour, texture or taste similar to those of animal origin. [6, 7]. There may be multiple reasons for this growing increase in the choice of plant-based diets, but among them, the following stand out: environmental factors (animal production is strongly criticised for its negative environmental impacts), greenhouse gas emissions, landscape degradation or excessive use of water resources. Or even nutritional factors, as according to the literature, plant-based diets tend to be associated with higher dietary quality and lower risk of chronic diseases [5, 6, 8, 9]. One of the concerns of plant-based diets is the source of protein, whether this is acquired from legumes, seeds, cereals or mycoproteins. Thus, it is considered an advantage to combine various ingredients to improve the composition of the product at nutritional, technological, functional and consumer preference levels [6, 8, 9]. Macroalgae thus appear as an alternative to be explored, whose sensory similarity with fish could be a point in favour among fish consumers, since it provides a sensation of familiarity to consumers, contributing to the reduction of a potential neophobia [5, 9, 10].

1.1 Macroalgae in Human Food

Based on several studies [2, 4] it is estimated that the products coming from the land, in the long term, may saturate the soils, with the need for new sources of vegetable protein. Macroalgae are an alternative that has been gaining prominence, since they are nutritious and sustainable, do not require arable land, are able to grow with few nutrients and are versatile in terms of use in various industries [4, 8]. Macroalgae are macroscopic (visible to the naked eye), multicellular algae that live in water and possess chlorophyll, thus carrying out photosynthesis and are considered to be the true "lungs" of the Earth. Their use stands out as an ingredient for human and animal food, as well as ingredients for the food, pharmaceutical and cosmetics industries, fertilisers for agriculture and others [3, 7, 11, 12]. Macroalgae are nutritionally composed of minerals (iron, calcium, iodine, selenium, potassium, magnesium) and vitamins (A, B1, B2, B6, B12, C and E), in addition to being sources of omega-3, soluble dietary fibre and some of them rich in protein. They are one of the few vegetable sources of vitamin B12 [10]. In addition to these nutritional characteristics, macroalgae normally grow quickly and do not require fertilisers or saturate the soil, making them an environmentally sustainable option for their more frequent inclusion in the diet [1, 8]. Thus, due to the increased demand for macroalgae, their sustainable production in aquaculture has increased, ensuring high quality food for the growing population [13]. Nowadays, we live in a deceptively abundant diet, surrounded by fast food, rich in calories and saturated fats and poor in essential nutrients. Macroalgae, however, are a natural food with high nutritional value: high mineral content (especially iron and calcium), the trace element iodine, vitamins, protein (with all the essential amino acids) and structural polysaccharides (fibre), and low fat content [3, 11, 14].

1.2 Benefits of Macroalgae on Human Health

In terms of benefits for human health, macroalgae have anti-inflammatory, antiviral, anti-tumour, anti-diabetic, antibacterial, anti-fungal, prebiotic or antioxidant properties [2, 3, 7–10, 13, 15]. Although macroalgae have long been included in the diet in the East, there are not many recent studies on their effect on humans. However, a recent intervention study in healthy Japanese children suggested that eating macroalgae, namely *Porphyra spp.*, appears to have long-term benefits on blood pressure [16]. Another recent study demonstrating the benefits of macroalgae consumption is that of Haskell-Ramsay et al. [17], in which they analysed the impact of brown macroalgae consumption on cognitive function when assessed postprandially after a high-carbohydrate meal, with the results appearing to be promising both in improving cognitive function and by decreasing postprandial insulin levels [17].

1.3 Potential for Cultivation and Marketing of Macroalgae for Human Consumption in the Central Portuguese Coast

All countries with a coastline, apart from the countries of the East, already have the habit of consuming macroalgae in their diet, out of necessity, because of periods of famine or war that they have gone through. However, Portugal, with a vast coastline, still does not have this habit of routine food inclusion (Tables 1 and 2) [15].

Thus, to highlight the edible macroalgae of central Portuguese coast: green macroalgae *Phylum Clorophyta,* brown macroalgae *Phylum Heterokontophyta* (or *Ochrophyta*) and all of them are grouped in the Class *Phaeophyceae* and red

Table 1 Historical perspective of the use of macroalgae worldwide [3]

Country/Region	Use	Year
Chile	Food and medicine	14,000 years ago
Korea and Japan	Food (nutritional properties and flavours)	Centuries ago
China	Food	500 BC
Islands	Food, due to soils not allowing conventional agriculture	Unknown

Table 2 Macroalgae use in Europe [3]

Country/Region	Macroalgae	Use
Wales	*Porphyra sp.*	Baked in various dishes such as laverbread
Portugal (Açores Island)	*Porphyra sp.*	Fried as an ingredient in omelets or pies ("Tortas de Erva Calhau")
Catania (Sicily)	*Ulva lactuca* *Chondracanthus teedei*	Eaten fresh, dried as a snack, mixed in salads, bread dough, curd, fish dishes
England Ireland Scotland	*Palmaria palmata*	Eaten raw, roasted, cooked in oat broth. Ingested for deworming dietary supplements Food additives (flavour enhancers, dyes, antioxidants, among others)
Norway France Ireland Great Britain	*Ascophyllum nodosum*	Extraction of phycocolloids to alginate protein for the food industry
France Great Britain Normandy Denmark	*Chondrus crispus* *Mastocarpus stellatus* *Furcellaria lumbricalis*	Carrageenan extraction for the food industry
Portugal Spain France	*Gelidium corneum* *Gracilariopsis longíssima*	Extraction of phycocolloids for alginate organics for the food industry

Table 3 Macronutrient composition (in percentage) of edible macroalgae and caloric values (kcal 100 g^{-1} in dry weight) [19]

Macroalgae	Protein (%)	Carbohydrates (%)	Lipids (%)	Crude fiber (%)	Energy (kcal 100 g^{-1})
Ulva rígida	15	62	1.4	17	255
Porphyra spp	24	52	2.1	17	256
Gracilaria gracilis	18	53	0.9	14	237
Osmundea pinnatifida	20	57	1.8	18	252

Table 4 Concentration of ash (expressed as %, w/w), macroelements (Ca, K, Mg, Na and P, expressed in mg g^{-1}) and trace elements (Cu, Fe, Mn, Se and Zn, expressed as μg g^{-1}) of edible macroalgae [19]

Macroalgae	Ash	Ca	K	Mg	Na	P	Cu	Fe	Mn	Se	Zn
	%	mg g^{-1}					μg g^{-1}				
Ulva rígida	21	6.9	20	30	74	2.5	3.6	383	18	15	10
Porphyra spp	22	2	29	6.3	42	6	5.9	110	53	2.9	61
Gracilaria gracilis	28	1.7	106	3.3	27	2.9	2.3	123	15	18	17
Osmundea pinnatifida	21	5.5	53	9.9	103	2.8	3.1	552	23	14	63

macroalgae Phylum Rhodophyta. Within the green macroalgae, Ulva rigid C. proliferates, which is a source of nitrate, sulfate and vitamins, having antimicrobial, antiviral, antihyperglycemic, antibacterial and antioxidant activities. In the red macroalgae, there is reproduction of *Osmundea pinnatifida* (quite aromatic), *Gracilaria gracilis* and *Porphyria*, rich in vitamins and minerals and omega 3 polyunsaturated fatty acids [7, 11, 14, 18]. The nutritional composition of macroalgae with potential interest for human consumption and which proliferate along the coastline of the central region of Portugal can be seen in Tables 3 and 4.

1.4 Macroalgae and Food Safety

In addition to all the benefits that macroalgae can offer as an alternative, within the vegetable base, to products of animal origin, the following food safety hazards must be taken into account: microbiological, chemical and allergenicity [1, 3, 10]. Within the microbiological field, microbial contamination can occur at any stage of macroalgae, from cultivation to storage. Although macroalgae can be reservoirs for *Vibrio parahaemolyticus* and *Vibrio vulnificus*, these bacterial species are sensitive to heating processes and therefore should not survive food processing. However, if macroalgae are eaten raw, the risk of being poisoned by those bacteria remains. On the other hand, even in aquaculture productions, if they do not have the appropriate hygiene and basic sanitation conditions or if they are in wildlife refuge areas, they can be contaminated. [1, 3, 10].

About chemical hazards, macroalgae can bioaccumulate heavy metals such as arsenic, lead, cadmium, and mercury. To reduce this risk, macroalgae should be harvested in uncontaminated areas at harvest time (younger leaves may not contain as much heavy metals compared to older leaves). Arsenic tends to be more accumulated by macroalgae belonging to the class *Phaeophyta*, followed by the *Phylum Rhodophyta* and *Cholophyta*. In the case of cadmium, it is found at higher levels in the red macroalgae, while mercury in the brown ones. But speaking about lead coming from macroalgae, human exposure to it can be considered minimal. Finally, in terms of iodine, macroalgae are considered a product rich in this mineral, since the concentration of iodine varies greatly depending on the species, and processing methods can influence its concentration and therefore human exposure [1, 3, 10].

Regarding allergenicity, allergic reactions to some red macroalgae have been described, as well as to certain proteins present in the macroalgae *Ulva spp.* and *porphyra* [1, 3].

2 Conclusions

Consumers' perception is influenced by the area where they live, as there are cultural, behavioural and habit differences. In Western Europe, more specifically in Mediterranean countries, there is a positive response to plant-based products, which may be related to the long-term importance of these foods in their diet, which could positively influence the introduction of macroalgae in conventional dishes.

More studies should be carried out in Western Europe to understand the knowledge and acceptance that the population has about macroalgae as a new food and, for this, it is necessary to increase knowledge of the benefits of its consumption, both for health and for environmental sustainability, as well as affordably priced and thus increase the likelihood of its consumption.

Furthermore, carrying out hedonic tests on innovative food products with the inclusion of macroalgae is essential to assess consumer acceptance and improve, if necessary, the sensory characteristics (aroma, texture, appearance, or general taste).

In addition to the above, attention must be paid to food safety for these plant-based alternatives, which will certainly be different from animal products and such a transition will require a rethink of food safety management. Other factors such as the price and cultural appeal of alternatives to animal products will need to be taken into consideration and it is expected that, due to increased consumer demand, the cost of plant based products will decrease and in addition to being more adapted to fast-food (hamburgers, nuggets, sausages), more adapted to traditional products from different regions, enhancing consumer acceptance.

Macroalgae constitute a valuable natural resource, still little studied and explored, mainly regarding its economic potential. It is necessary to have industries capable of investing and continuing to study macroalgae, so that there is greater scientific support. As it is a renewable natural resource, it is imperative that legislation regulate sustainable production and commercialization activities.

References

1. FAO. Thinking about the future of food safety – A foresight report. Retrieved from Rome:. https://doi.org/10.4060/cb8667en (2022)
2. Harrysson, H. *et al.* Production of protein extracts from Swedish red, green, and brown seaweeds, *Porphyra umbilicalis* Kützing, *Ulva lactuca Linnaeus, and Saccharina latissima* (Linnaeus) J. V. Lamouroux using three different methods. *Journal of Applied Phycology* 30, 3565–3580, https://doi.org/10.1007/s10811-018-1481-7 (2018)
3. Mendes, M. C. *et al.* Algae as Food in Europe: An Overview of Species Diversity and Their Application. *Foods (Basel, Switzerland)* 11, 1871, https://doi.org/10.3390/foods11131871 (2022)
4. Macdiarmid, J. I., Douglas, F. & Campbell, J. Eating like there's no tomorrow: Public awareness of the environmental impact of food and reluctance to eat less meat as part of a sustainable diet. *Appetite* 96, 487–493, https://doi.org/10.1016/j.appet.2015.10.011 (2016)
5. Palmieri, N., Nervo, C. & Torri, L. Consumers' attitudes towards sustainable alternative protein sources: Comparing seaweed, insects and jellyfish in Italy. *Food Quality and Preference* 104, 104735, https://doi.org/10.1016/j.foodqual.2022.104735 (2023)
6. Bin-Nashwan, S. A., Hassan, M. K. & Muneeza, A. Russia–Ukraine conflict: 2030 Agenda for SDGs hangs in the balance. *International Journal of Ethics and Systems* ahead-of-print, https://doi.org/10.1108/IJOES-06-2022-0136 (2022)
7. Pereira, L. Therapeutic and nutritional uses of marine algae: a pharmacy in the ocean. *Tradit Med Res* 7, 30, https://doi.org/10.53388/TMR20220310268 (2022)
8. Mellor, C. *et al.* Consumer Knowledge and Acceptance of "Algae" as a Protein Alternative: A UK-Based Qualitative Study. *Foods (Basel, Switzerland)* 11, 1703 https://doi.org/10.3390/foods11121703 (2022)
9. Anusha Siddiqui, S. *et al.* Consumer acceptability of plant-, seaweed-, and insect-based foods as alternatives to meat: a critical compilation of a decade of research. *Critical Reviews in Food Science and Nutrition*, 1–22, https://doi.org/10.1080/10408398.2022.2036096 (2022)
10. Wu, G. *et al.* Current Status and Future Trends in Removal, Control, and Mitigation of Algae Food Safety Risks for Human Consumption. *Molecules* 27, 6633, https://doi.org/10.3390/molecules27196633 (2022)
11. Gonçalves, A. M. M., García-Poza, S., Cotas, J., Marques, J. C. & Pereira, L. in *Proceedings of the 1st International Conference on Water Energy Food and Sustainability (ICoWEFS 2021).* (eds João Rafael da Costa Sanches Galvão et al.) 227–236 (Springer International Publishing), https://doi.org/10.1007/978-3-030-75315-3_27 (2021)
12. Pereira, L. Macroalgae. *Encyclopedia* 1, 177–188, https://doi.org/10.3390/encyclopedia1010017 (2021)
13. Al Khawli, F. *et al.* Aquaculture and its by-products as a source of nutrients and bioactive compounds. *Advances in Food and Nutrition Research* 92, 1–33, https://doi.org/10.1016/bs.afnr.2020.01.001 (2020)
14. Lomartire, S., Marques, J. C. & Gonçalves, A. M. M. An Overview to the Health Benefits of Seaweeds Consumption. *Marine Drugs* 19, https://doi.org/10.3390/md19060341 (2021)
15. Cascais, M. *et al.* Effects of Heat Treatment Processes: Health Benefits and Risks to the Consumer. *Applied Sciences* 11, 8740 https://doi.org/10.3390/app11188740 (2021)
16. Wada, K. *et al.* Effect of Dietary Nori (Dried Laver) on Blood Pressure in Young Japanese Children: An Intervention Study. *Journal of Epidemiology* 31, 37–42, https://doi.org/10.2188/jea.JE20190176 (2021)

17. Haskell-Ramsay, C. F. *et al.* Acute Post-Prandial Cognitive Effects of Brown Seaweed Extract in Humans. *Nutrients* 10, https://doi.org/10.3390/nu10010085 (2018)
18. El Zokm, G. M., Ismail, M. M. & El-Said, G. F. Halogen content relative to the chemical and biochemical composition of fifteen marine macro and micro algae: nutritional value, energy supply, antioxidant potency, and health risk assessment. *Environmental Science and Pollution Research International* 28, 14893–14908, https://doi.org/10.1007/s11356-020-11596-0 (2021)
19. Milinovic, J. *et al.* Nutritional Benefits of Edible Macroalgae from the Central Portuguese Coast: Inclusion of Low-Calorie 'Sea Vegetables' in Human Diet. *Int J Environ Sci Nat Res* 28, 556250, https://doi.org/10.19080/IJESNR.2021.28.556250 (2021)

Satellite Dataset Visual Analysis for Remote Soil Nutrient Estimation

Andrés Isaza-Giraldo, Manuel Pereira, Rafael Candeias, and Lucas Pereira

Abstract This paper proposes a methodology for visualizing satellite-based machine learning (ML) datasets to understand the visual components that will be used as inputs for developing ML models. The proposed methodology uses t-Distributed Stochastic Neighbor Embedding (T-SNE) methods to create visualizations of satellite images leveraging models that were pre-trained in ImageNet. T-SNE is a self-supervised learning tool used to transform high-dimensional spaces into two- or three-dimension embeddings, making it easier to visualize a broad dataset in a single image or space. The methodology is demonstrated using the LUCAS Topsoil Analysis dataset with satellite images from Sentinel-2. The dataset was constructed using the TerraSense Toolkit (TSTK). The T-SNE visualization tool aims to improve ML research by providing a clearer visual understanding of satellite-based datasets.

Keywords Satellite imagery · t-SNE (t-Distributed Stochastic Neighbor Embedding) · Soil sensing · Machine-learning · Agroindustry

---

A. Isaza-Giraldo
Faculdade de Belas-Artes, University of Lisboa, 1249-058 Lisbon, Portugal
e-mail: giraldo@edu.ulisboa.pt

M. Pereira · R. Candeias · L. Pereira (✉)
Instituto Superior Técnico, University of Lisboa, 1049-001 Lisbon, Portugal
e-mail: lucas.pereira@tecnico.ulisboa.pt

M. Pereira
e-mail: manuel.afonso.pereira@tecnico.ulisboa.pt

R. Candeias
e-mail: rafaelmcandeiras@tecnico.ulisboa.pt

M. Pereira
Instituto de Engenharia de Sistemas e Computadores—Investigação e Desenvolvimento, 1049-001 Lisbon, Portugal

A. Isaza-Giraldo · L. Pereira
Interactive Technologies Institute, LARSyS, 9050-100 Funchal, Portugal

J. R. da. C. S. Galvão et al. (eds.), *Proceedings of the 3rd International Conference on Water Energy Food and Sustainability (ICoWEFS 2023)*, Springer Proceedings in Earth and Environmental Sciences, https://doi.org/10.1007/978-3-031-48532-9_8

1 Introduction

Soil surveying is a process that reveals crucial information about the soil composition, such as nutrients, that could have a great impact on a more efficient and effective use of land. Traditionally this has been a time-consuming and expensive process that requires sampling on the field and intensive testing. Despite the progress made in the agricultural applications of ML using satellite images (e.g., [1, 2]), datasets are still a challenging matter as there are still few ground-truth available and because of specific limitations of satellite photography. According to Lillesand et al. [3] "When we look at aerial and space images, we see various objects of different sizes, shapes, and colors. (...) The images contain raw pixel data. These data, when processed by a human interpreter's brain, become usable information." Following this reasoning, it becomes clear that visually analyzing the dataset and trusting human capacity for interpretation is one of the paths toward a better dataset that could ultimately improve the performance of ML algorithms.

Against this background, this paper proposes a methodology to visualize satellite-based ML datasets to understand better the challenges associated with the visual components that utterly affect the performance of ML algorithms. More precisely, the proposed methodology uses t-Distributed Stochastic Neighbor Embedding (T-SNE) [4, 5] methods to create visualizations of the satellite images leveraging models that were pre-trained on ImageNet [6].

T-SNE methods have been used in many disciplines, for instance, data science, medicine, social sciences, media art. This self-supervised learning tool transforms high-dimensional spaces into two- or three-dimension embeddings, making it easier to visualize a broad dataset in a single image or space. The proposed methodology is demonstrated using the LUCAS Copernicus dataset with satellite images from Sentinel-2 [7–9]. We also analyze the dataset for two specific crops, maize, and common wheat, as these are among the most represented crops in the used dataset.

Although t-SNE has commonly been used to visualize the distance between satellite images according to their classification vectors, e.g., there are few examples of t-SNE used to plot satellite images themselves for the sake of visually embedding datasets altogether. One prominent example would be the process of creating the interactive platform Land Lines by Zach Lieberman and Google Data Arts Team for which they plotted several satellite images together in a grid before developing their interactive experience [10]. This later type of t-SNE is more commonly used in digital arts. Its application could have a significant impact on machine-learning dataset visualization, for example, for dataset cleaning purposes.

The remaining of this paper is organized as follows. Section 2 presents the datasets, tools, and methodology used to visualize the datasets. Section 3 presents and discusses the result of a simple case study that was developed to demonstrate the proposed method. The paper concludes in Sect. 4 with a summary of the results and future work directions.

2 Materials and Methods

This section first describes the datasets and tools used in this research. Then, the proposed methodology is described in detail.

2.1 *Dataset and Tools*

The dataset was constructed using information given in the LUCAS Survey, an effort of the European Soil Data Centre (ESDAC) for which there have been collected thousands of samples of soil in the European Union described some of its properties. The ground-truth taken were the ones intersecting three different surveys, as observed in Fig. 1: the LUCAS Topsoil Analysis, which contains information about soil composition, the LUCAS Copernicus which includes information on land use; and LUCAS 2018 that contains GPS coordinates. The LUCAS Copernicus survey is not a complete dataset per se, as it does not contain aerial images of the fields. The images were obtained from Sentinel-2 satellite, a multispectral satellite orbiting Earth since 2016.

The dataset construction workflow was introduced by author Manuel Pereira on the TerraSense Toolkit (TSTK) [11]. For each specific sample of the Lucas Survey, several images were taken in a range of five days prior to and five days after the date on which the ground truth was taken and as long as the maximum cloud coverage was less than 80%. This is because the satellite captures every point on earth only

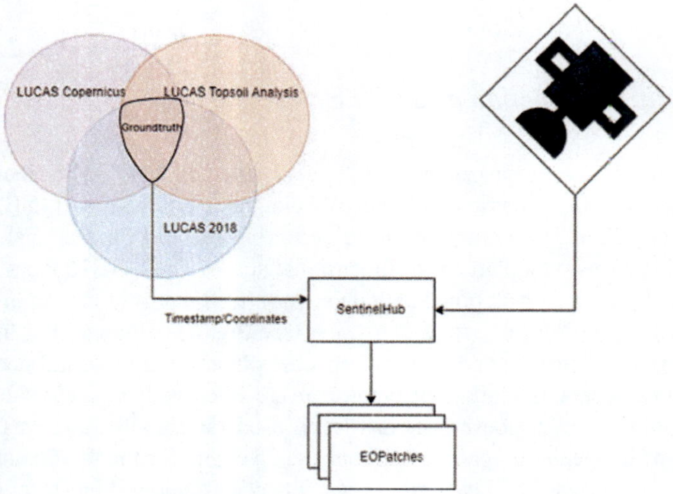

Fig. 1 Dataset construction workflow

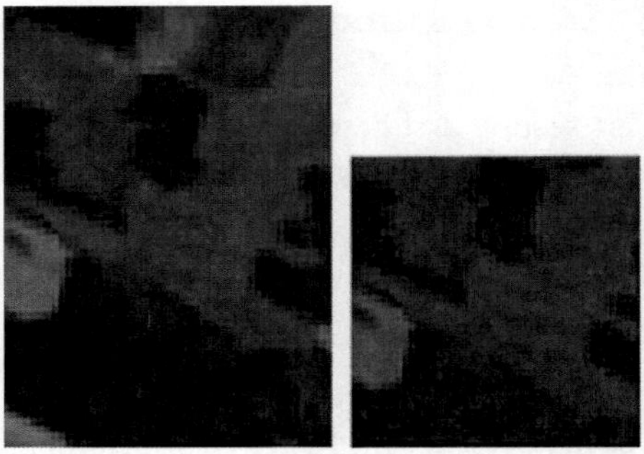

Fig. 2 RGB image with no background and image cropped to 100 × 100 px

twice or three times a week due to its orbit around Earth. In many cases, clouds might block the view of the direct soil rendering the images unusable.

The Sentinel-2 uses a multispectral camera with 12 bands ranging from the visual spectrum to short-wave infrared. The spatial resolution for the visual spectrum is 10 m, while it varies from 10 to 60 m on other bands. The LUCAS Survey also contains a polygon of up to 51 m defining an area around the ground truth where soil use does not change. The TSTK downloader downloads the polygon and an extra margin in all directions. The resolution of images may vary depending on the size of the polygon. Then images of the visible spectrum were created using red, green, and blue bands. The median resolution for the RGB images is 162 × 217 px. For the first tSNE presented in this paper, all of the images were exported with a white background and then cropped to 100 × 100 px resolution. For later tSNEs, images were exported with no background and cropped to 100 × 100 px, as seen in Fig. 2. Images with smaller resolution were upscaled to match this resolution.

2.2 Feature Extraction and t-SNE

An overview of the proposed methodology is given in Fig. 3. To analyze the high amount of data, t-Distributed Stochastic Neighbor Embedding (tSNE) images were created. tSNE is a commonly used clustering method for visualizing high-dimensional spaces on a plane or tridimensional space. The VGG16 keras classification model [12], a convolutional neural network (CNN) pre-trained on ImageNet, was utilized for high-level feature extraction (see Fig. 4). Although the ImageNet dataset consists of images and labels of specific objects, animals, and people, the current dataset contains a different type of image often with no clear subject and smaller resolution. Whatsoever, the last layer, used for classification on the 1000 categories of ImageNet is ignored, and only the weights for the 4096 connections on the last fully connected layer, whose weights correspond to high-level features. Then a Principal Component Analysis (PCA) extraction takes place, restricting the number of features to only the most relevant.

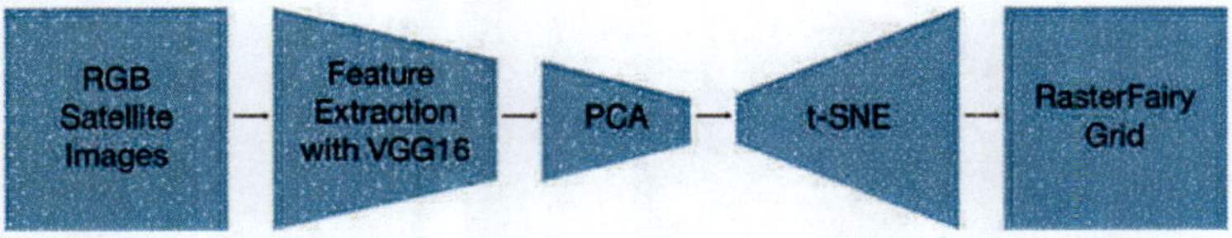

Fig. 3 Workflow diagram of the proposed methodology

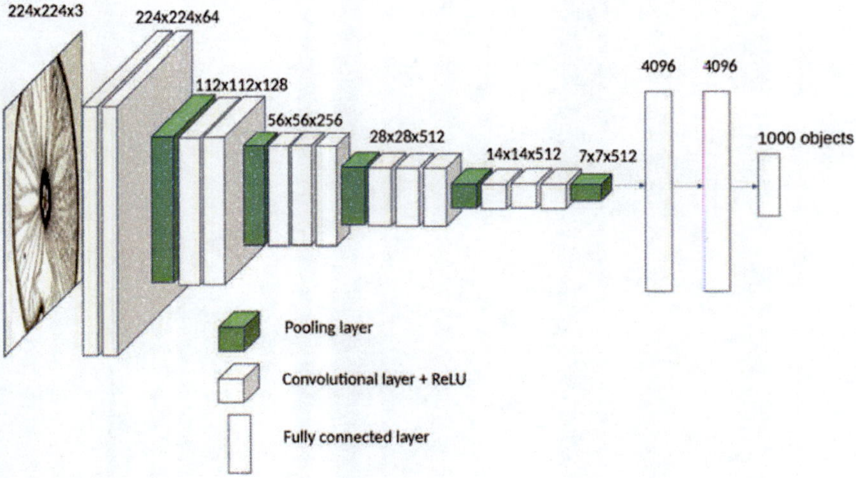

Fig. 4 VGG16 architecture

All the images are then plotted with t-SNE to a two-dimensional plane, where neighboring relationships indicate the similarity of features. This is done through an iterative random process that ultimately optimizes the probability of distribution through a stochastic process. In the last step, the images are replotted into a grid using RasterFairy, a tool created by Mario Klingemann [13]. This tool rasters the images into a regular structure while trying to preserve the neighboring relationship.

3 Results and Discussion

This section presents and discusses the obtained results. The first result is the generated tSNE for the entire dataset, it is shown on Fig. 5. This image is by its visual properties a result on itself. It is a tool to better visualize our own data, like a map that reveals the potential knowledge for the machine. The image is then described in visual terms on detail to point out the generated clusters and what it can say about our data.

Fig. 5 t-SNE representation for the entire dataset

In a third part, two specific crops, common wheat, and maize, are explored in more detail.

3.1 Entire Dataset t-SNE

The whole dataset consists of over 50,000 captures of soil for over 15,809 different points or soil ground truth. Meaning there might be more than one image for any point. The images were created using only the red, green, and blue channels of the patches. Then they were automatically cropped to 100 × 100px due to the capture size.

The t-SNE shown in Fig. 5 was made with a random sample of 10,000 images taken from the general set of images. Primary Component Analysis (PCA) was restricted to 300 features.

Some obvious conclusions can be drawn from looking at the t-SNE. The first is the persistence of clouds on the dataset; Even though clouds were already maxed out, they still took about 20% of the dataset. In the blue-shaded section of the t-SNE at the bottom, it could be observed that heavy moisture due to atmospheric conditions blurs the view of the soil. These images add much random noise to the dataset as no specific features can be extracted from them.

On the lower right side, there are a set of images that have white lines on their margins. These images had a smaller resolution than 100 px wide because the polygon provided by the LUCAS Copernicus was smaller than average. Thus, when applied the center crop, an extra white margin was left. In future experiments, images with less than 100 px wide were upscaled.

It is also observed some little clusters of black images or partially black images on the bottom of the tSNE, towards the center and center-right. The absence of visual information is due to the satellite camera not covering the whole of the requested area during its orbit.

There are also some reddish images clustered on the top left. Whatsoever, some other reddish images were grouped alongside clouds on the lower center-right. Further analysis indicates that some of these images are also brighter than some of the clouds. It is possible that VGG16 assigned similar features to dim clouds and bright reddish soil, which ended up grouping them after the stochastic clustering.

The rest of the dataset seems to be clear and distinguishable. A wide variety of greens and browns is observed. But also, there is a high presence of distinct forms due to cultivation area; the presence of roads, houses, and other structures; and topographical features.

3.2 Common Wheat and Maize Case Study

The TSTK was used to train individual models for each crop present on the LUCAS Survey, being that each crop has different visual properties related to soil nutrients. From the previous analysis, it was established that the model trained on Common Wheat was the most accurate one to predict soil nutrients, while the one trained on Maize was the least accurate [11]. The most significant difference for each crop is

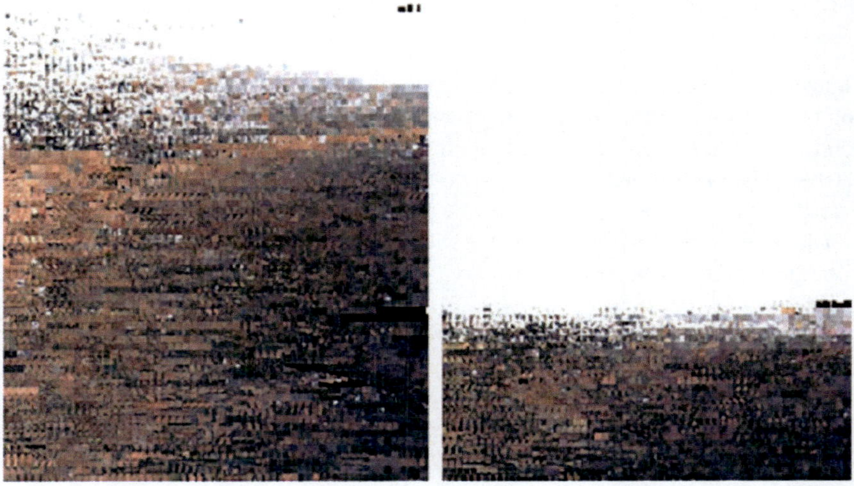

Fig. 6 t-SNE representation for individual crops: common wheat (left), maize (right)

the size of the dataset. As seen in Fig. 6, the number of samples for common wheat is 2.5 times larger than that for maize (4.133 and 1.796, respectively [7]).

Furthermore, the t-SNE differentiates two different types of images for each crop, possibly for when the plant is fully grown, and for when it is harvested or recently planted as observed on the detail shown in Fig. 7. This distinction is much clearer on wheat than maize. It could be hypothesized for a study that a machine trained on these images might be able to understand these differences very easily and make separate calculus for when there is leaf presence and when there are no leaves.

4 Conclusions and Future Works

To understand what is happening inside a satellite images dataset, it is essential to look at the visual information and thus to find ways to organize it to make it easier to comprehend by the human eye. Visual t-SNE help to cluster information altogether to better explore relationships within images but also to understand the possible challenges driven out of the dataset creation and further growth. It shows how task related to computer vision grow in the interdisciplinary efforts across the computational sciences, the multimedia practices and the agronomical field.

Satellite images, such as those provided by Sentinel-2, are a cost-efficient way to obtain visual information from the soil. Nonetheless, information regarding soil composition is still limited, and more robust datasets are needed to make a big picture of soil nutrients all across the European Union and the whole of the world.

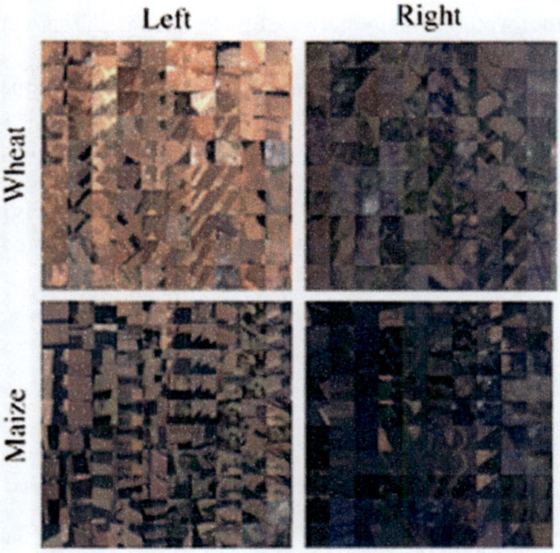

Fig. 7 Details of different clusters generated for each crop showing the different stages of plant growth

4.1 Future Work

The current dataset is still subject to different types of visual analysis. On one hand it is needed to study the visual spectrum for each of the crops that are not cited on this article to understand the how much visual information is actually present on the dataset.

On the other hand, it is needed to be understood which of the bands of the Sentinel-2 satellite are the most helpful for the study of each crop. As suggested by Wand and Wei [14, 15], certain bands might be more valuable than others in estimating the presence of chlorophyll in leaves which can be an indication of specific nutrients on the soil, specifically nitrogen for the study. The bands depend on the specific crop but some common bands might also be determinant. Lu et al. experiments propose that for the estimation of potassium from plant canopy, certain near-infrared and shortwave infrared are more adequate for such task [16]. It would be interesting to see t-SNE experiments with the different bands provided by the satellite to actually understand how it clusters differently from the visual spectrum cluster.

According to Mandrake study to detect sulfur on the soil [17], some sulfur might be easy to differentiate through aerial imagery. Others might have similar visual properties to non-sulfur components, which might lead to false positive results. For such reason, it is necessary to study which visual characteristic might lead further algorithms to provide false positive results.

Upson visualizing the results of the t-SNE for the present study, it has been suggested several times that this kind of image clustering could be applied through an interactive interface for both dataset analysis and dataset cleaning. This kind of tool would be extremely useful for cleaning larger datasets avoiding going through thousands of files.

However, what could have the most significant impact on the success of the TSTK would be an improved dataset of soil ground truth alongside crop type and GPS coordinates. Although the effort of the LUCAS Survey has been extremely useful, the dataset is not big enough for machine-learning algorithms result to be fully trusted. Further improvement and refinement of our dataset is going to be performed, but the help of agricultural institutions might play a big role in the further development on this cost-efficient technology.

References

1. Karthikeyan N, Shashikkumar M, Ramanamurthy J (2010) A study on vegetation vigour as affected by soil properties using remote sensing approach. https://doi.org/10.1109/RSTSCC.2010.5712811
2. Walshe D, McInerney D, De Kerchove RV, Goyens C, Balaji P, Byrne KA (2020) Detecting nutrient deficiency in spruce forests using multispectral satellite imagery. Int J Appl Earth Obs Geoinformation 86:101975. https://doi.org/10.1016/j.jag.2019.101975
3. Liu JG, Mason PJ (2016) Image Processing and GIS for Remote Sensing: Techniques and Applications, 2nd edition. Wiley-Blackwell
4. Cai TT, Ma R (2022) Theoretical Foundations of t-SNE for Visualizing High-Dimensional Clustered Data
5. Laurens van der Maaten and Geoffrey Hinton. Visualizing data using t-SNE. Journal of Machine Learning Research, 9(Nov):2579–2605, 2008
6. Deng J, Dong W, Socher R, Li L-J, Kai Li, Li Fei-Fei (2009) ImageNet: A large-scale hierarchical image database. In: 2009 IEEE Conference on Computer Vision and Pattern Recognition. IEEE, Miami, FL, pp 248–255
7. d'Andrimont R, Verhegghen A, Meroni M, Lemoine G, Strobl P, Eiselt B, Yordanov M, Martinez-Sanchez L, van der Velde M (2021) LUCAS Copernicus 2018: Earth-observation-relevant in situ data on land cover and use throughout the European Union. Earth Syst Sci Data 13:1119–1133. https://doi.org/10.5194/essd-13-1119-2021
8. Tóth G, Jones A, Montanarella L (2013) The LUCAS topsoil database and derived information on the regional variability of cropland topsoil properties in the European Union. Environ Monit Assess 185:7409–7425. https://doi.org/10.1007/s10661-013-3109-3
9. European Commission. Joint Research Centre. (2020) Assessment of changes in topsoil properties in LUCAS samples between 2009/2012 and 2015 surveys. Publications Office, LU
10. Liebermann, Zach (2016). Land Lines. https://zachlieberman.medium.com/land-lines-e1f88c745847
11. Pereira MAS (2022) TerraSenseTK: a toolkit for remote soil nutrient estimation. Master Thesis
12. Simonyan K, Zisserman A (2015) Very Deep Convolutional Networks for Large-Scale Image Recognition
13. Klingemann M (2022) About RasterFairy-Py3. https://github.com/Quasimondo/RasterFairy
14. Wang W, Yao X, Tian Y, Liu X, Ni J, Cao W, Zhu Y (2012) Common Spectral Bands and Optimum Vegetation Indices for Monitoring Leaf Nitrogen Accumulation in Rice and Wheat. J Integr Agric 11:2001–2012. https://doi.org/10.1016/S2095-3119(12)60457-2
15. Wang L, Wei Y (2016) Revised normalized difference nitrogen index (NDNI) for estimating canopy nitrogen concentration in wetlands. Optik 127:7676–7688. https://doi.org/10.1016/j.ijleo.2016.05.115

16. Lu J, Eitel JUH, Jennewein JS, Zhu J, Zheng H, Yao X, Cheng T, Zhu Y, Cao W, Tian Y (2021) Combining Remote Sensing and Meteorological Data for Improved Rice Plant Potassium Content Estimation. Remote Sens 13:3502. https://doi.org/10.3390/rs13173502
17. Mandrake L, Wagstaff KL, Gleeson D, Rebbapragada U, Tran D, Castaño R, Chien S, Pappalardo RT Hyperspectral Sulfur Detection Using An Svm With Extreme Minority Positive Examples Onboard EO-. 12

Agronomic Benefits of Biochar Application in Sandy Soils

Carolina Morim, Márcia Santos, Pollyanna Mendoza, Luís Tarelho, and Flávio Silva

Abstract Soil degradation is being exacerbated not only by population growth but also by the increasing frequency of forest fires that are recurrent in Portugal. There is a need to remediate degraded soils by using sustainable approaches, such as the valorization of biomass into biochar. To address this need, a soil incubation assay was performed with biochar produced at two pyrolysis temperatures (450 and 550 °C) in four application rates (0, 3, 6 and 10%) and three mean particle sizes ($S < 0.5$ mm, $M = [0.5;3.15]$ and $L > 3.15$ mm). The pH, electrical conductivity (EC), water holding capacity (WHC) and soil respiration were examined. The pH increased for all treatments to neutral values (5–7) and the water holding capacity increased by 69–225% when compared to the control soil. The soil respiration increased in all treatments, being directly proportional to the application rate of biochar thus showing positive microbial responses. Biochar application significantly improved key-physicochemical properties of the soil aiming at reduced irrigation demands and higher agronomic quality.

Keywords Biochar · Soil quality · pH · Water holding capacity · Electric conductivity

1 Introduction

Soil is an essential resource for human life as it provides a wide range of goods and services, such as carbon storage, water mass regulation, food production and shelter provision [1, 2]. An important threat to the quality and functions of this resource are forest wildfires that degrade the environmental compartments due to loss of vegetation, biodiversity, soil quality and fertility, erosion and many off-site effects such as ashes runoff into water bodies [3]. In Portugal, forest wildfires are systemic, and

C. Morim (✉) · M. Santos · P. Mendoza · L. Tarelho · F. Silva
Department of Environment and Planning, Centre for Environmental and Marine Studies (CESAM), University of Aveiro, Aveiro, Portugal
e-mail: carolinapedrosamorim@ua.pt

J. R. da. C. S. Galvão et al. (eds.), *Proceedings of the 3rd International Conference on Water Energy Food and Sustainability (ICoWEFS 2023)*, Springer Proceedings in Earth and Environmental Sciences, https://doi.org/10.1007/978-3-031-48532-9_9

their frequency tend to increase due to climate change projections [4]. Therefore, its necessary to implement methods that allow the conservation of soils while complying with the goals of Circular Economy. One way is the application of biochar as amendment for degraded soils [5–8]. Besides, an important wildfire prevention method is the appropriate management and removal of the residual forestry biomass (RFB) on site, otherwise its accumulation leads to fire propagation due to the high flammability of the species existing in Mediterranean forests. The use of RFB for energy production is not efficient due to its ash content and associated down-stream issues, therefore its valorization into biochar is a viable alternative [9].

Biochar is a carbon rich material (65% to 90%) that results from the thermochemical conversion of biomass under low-oxygen conditions at relatively low temperatures (pyrolysis, 300–700 °C). Its structural matrix and physicochemical properties may improve the quality of soils by increasing its organic matter (SOM), water holding capacity (WHC), cation exchange capacity (CEC) and nutrient content, as well as correcting pH and electrical conductivity (EC) to suitable agronomic ranges [8, 10–16]. The physicochemical properties and structure of biochar vary with the pyrolysis operating conditions such as heating rate, peak temperature and reaction time, as well as the biomass used as feedstock [11, 17].

The WHC is also dependent on the bulk density (BD), since the BD of biochar is substantially lower than that in most soils (<0.6 vs. 1.25 g cm^{-3}). Biochar application leads to a decrease in this BD and thus in soil compaction allowing for an increase in its porosity and fluids movement, therefore enhancing its capacity to retain water [14, 18–21].

Other biochar physicochemical parameters dependent on the pyrolysis process are the pH and electrical conductivity (EC) [2]. Typical values of pH for biochar are in the range 6.4–9.3 tending to increase with pyrolysis temperature, due to the ash and alkaline inorganic compounds. This feature allows to design biochar as a liming agent [2]. Higher superficial area and pore volume of biochar have proven to enable higher organic matter content since it allows for the allocation of microbial communities and microorganisms that otherwise would not be able to thrive in compacted soil [11, 14, 18, 19, 22].

This work was developed in the scope of BioValChar project (Reference: PCIF/GVB/0034/2019) which studies the valorization of forestry biomass residues into biochar and further application as a soil amendment. Therefore, the objective of our research was the valorization of residual forest biomass into biochar, its application in degraded soils and the evaluation of the physicochemical and biological alterations that occur in them.

2 Methodology

2.1 Soil Sampling and Characterization

Sandy soil was collected in Estarreja (north-central Portugal) from a forest area in the vicinity of industrial sites. The soil was sieved to a particle size <2 mm to remove coarse materials, and air dried (≈20 °C) for 5 days. The organic matter content was determined by the loss on ignition (LOI), in which the sample was ignited to 550 °C for 4 h in a muffle furnace [29].

2.2 Biochar Preparation and Characterization

Biochar was produced from acacia wood residues (*Acacia longifolia*) in a lab-scale auger-type continuously-fed pyrolysis reactor at the Department of Environment and Planning of the University of Aveiro, Portugal. The pyrolysis was conducted in two different temperatures of 450 and 550 °C, and the biochar was collected unsorted and then sieved in three particle size classes classified as: S < 0.5 mm, M = [0.5;3.15] and L > 3.15 mm. The proximate analysis was performed following the CEN/TS 14774-3:2004 (E), CEN/TS 14775:2004 (E) and CEN/TS 15148:2005 (E) to determine moisture, ash and volatile matter contents, respectively. The content of fixed carbon was calculated through the difference between the volatile matter and the ash content.

2.3 Pot Incubation Assay

The pot incubation assay was designed under a factorial basis with three independent variables (factors): three particle sizes, S < 0.5 mm, M = [0.5;3.15] and L > 3.15 mm; two pyrolysis temperatures 450 and 550 °C; and four application rates of 0, 3, 6 and 10% (w/w). To each pot 200 g of the mixture of soil and biochar were added and 3 repetitions were performed (n = 57). The incubation of the biochar in the soil performed for 50 days, in which the water holding capacity (WHC) of the treatments was maintained in a range of 20–85%, through regular weighing and watering with distilled water.

2.4 Analytical Methods

Water Holding Capacity (WHC)
The WHC was determined by adding 50 g of the treatment mixtures in filter paper (Whatman no. 42) into a cup with holes in the bottom. The mixtures were saturated for a period of 8 h, afterwards they were placed in a deep platter to allow the water to leave the bottom of the cups due to gravity and covered in the top, so that there was no evaporation, and left to stand for 24 h. The cups were then weighed and the WHC was calculated through the reason between the difference of the weight before and after saturation and their dry weight after 24 h in the furnace at 105 °C.

pH and Electrical Conductivity (EC)
The pH was analyzed following the ISO 10390 with an extraction rate of 1:5 (v/v) with $CaCl_2$, shaken for 1 h and analyzed. The electrical conductivity was analyzed following the ISO 11265 with an extraction rate of 1:5 (w/v) with distilled water, after which the solution was shaken for 30 min at 300 rpm and then filtrated with a 45 μm nitrocellulose filter.

Soil Respiration
The soil respiration was analyzed following the ISO 16072:2002(E), through the decrease of pressure recorded in respirometry equipment (Oxitop). 50 g of sample were incubated at 20 °C for a period of 19 days at a water holding capacity of 50% [23].

Statistical Analysis
The three factors under study were examined through variance analysis (ANOVA) at a confidence level of 95%. The significance of pyrolysis temperature, biochar particle size and application rate were tested on the dependent variables pH and electrical conductivity. The values of skewness and kurtosis between −2 and 2 were verified to support the assumption of normality when the Shapiro–Wilk test was significant, and homogeneity of variances was tested through Levene's test and verified for all cases. Statistical analysis was performed in the software IBM SPSS.

3 Results and Discussion

3.1 Soil and Biochar Characterization

The physicochemical characterization of the soil is shown in Table 1. The biochar used was pyrolyzed from acacia wood (*Acacia longifolia)* at two temperatures of 450 and 550 °C, being characterized through proximate analysis, bulk density, pH and electrical conductivity. The immediate analysis consisted in the ash, volatile and fixed carbon content as shown in Fig. 1.

Table 1 Organic matter (OM), pH, electrical conductivity (EC), bulk density (BD), water holding capacity (WHC) and bioavailable nutrients of the Estarreja sandy soil

Parameters	Soil
pH	3.8 ± 0.1
EC ($\mu S\ cm^{-1}$)	6.8 ± 0.4
OM (%)	$1.0 \pm (4.0 \times 10^{-2})$
WHC (%)	16 ± 3
BD ($g\ cm^{-3}$)	$1.5 \pm (4.5 \times 10^{-2})$
P ($mg\ kg^{-1}{}_{bs}$)	4.3 ± 0.1
Ca ($mg\ kg^{-1}{}_{bs}$)	108 ± 8
Mg ($mg\ kg^{-1}{}_{bs}$)	20 ± 1
K ($mg\ kg^{-1}{}_{bs}$)	9.6 ± 0.5
Na ($mg\ kg^{-1}{}_{bs}$)	11 ± 4

Fig. 1 Proximate analysis of biochar produced at two pyrolysis temperatures

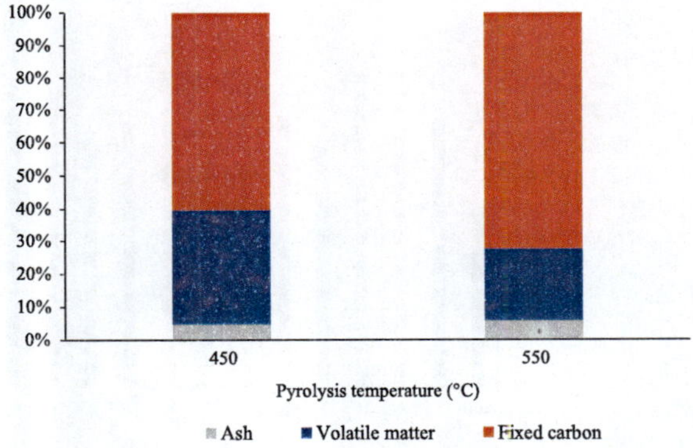

The biochar produced at 450 °C had a higher amount of volatile matter (35%) and lower ash and fixed carbon content (5 and 60%) when compared with the biochar produced at 550 °C (21%, 6% and 71% respectively).

For the preparation for the incubation assay three granulometric classes of biochar were made: S < 0.5 mm, M = [0.5;3.15] and L > 3.15 mm and their distribution was briefly examined. Higher pyrolysis temperatures and the shear stress of the process, usually promotes lower particle sizes [14]. Therefore, a higher amount of L particles was observed in the biochar produced at 450 °C (29%) when compared to that produced at 550 °C (13%). In both biochars the class M presented the higher number of particles within a range of 48–52%.

The bulk density of the biochar did not vary with the pyrolysis temperature. However, the difference was substantial in the particle size S, 0.33 $g\ cm^{-3}$, when compared to the other particle sizes, 0.10 to 0.13 $g\ cm^{-3}$. This was similar to the bulk density registered by Burrell et al., 2016, that observed BD values of 0.22, 0.47, 0.44 and 0.36 $g\ cm^{-3}$ from the pyrolysis of straw, woodchip and vineyard at 525 °C and vineyard at 400 °C, respectively [16].

The pH and EC of the biochar produced from both pyrolysis temperatures had the highest values in the 550 °C biochar, 7.0 and 516 μS cm^{-1}, respectively, when compared with the biochar produced at 450 °C, 6.2 and 379 μS cm^{-1}. The increase of pH and EC with higher pyrolysis temperature is expected due to the higher number of alkaline inorganic compounds present in the ash content. The values of pH obtained are similar to the typical reported values that range from 6.4 to 9.3 [2, 14].

3.2 pH and EC

The pH and EC were measured in the treatments of the incubation assay, after an extraction with $CaCl_2$ and H_2O, respectively, and the results are depicted in Fig. 2.

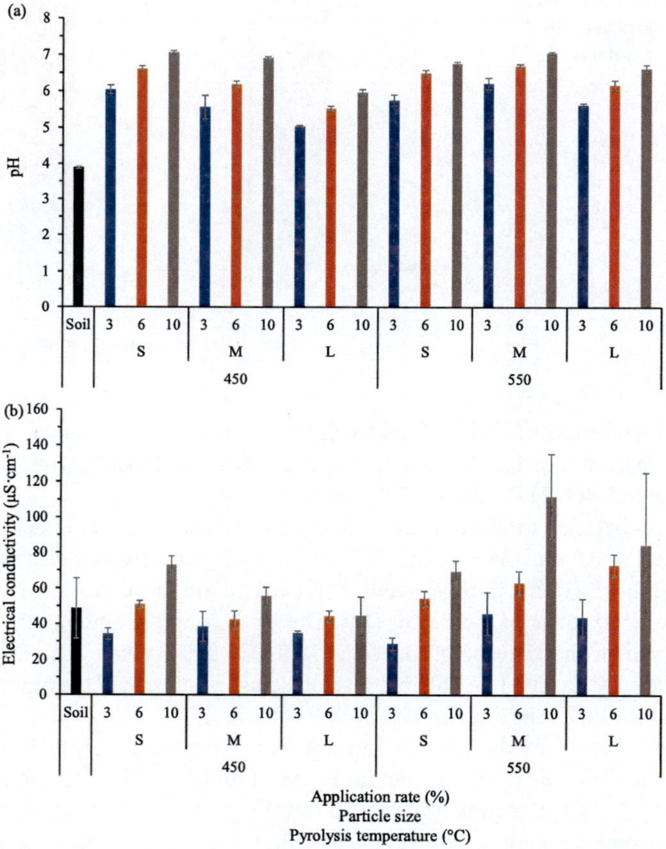

Fig. 2 **a** pH and **b** electrical conductivity of the treatment mixtures, with the control soil as a reference (black column)

The pH increased for all treatments with biochar regardless of pyrolysis temperature, being the increase directly proportional to the application rate and indirectly proportional to the particle size. For the pH, all factors (pyrolysis temperature, application rate and particle size) were statistically significant (*p-value* < 0.05), moreover the mean difference between the particle size L and the remaining particle sizes S and M was statistically significant while the difference between the particles S and M was not (*p-value* $= 1.00$).

The pH increased from 3.9 in the control soil to values in the range 5–7 for the 10% application rate of particle L and 3% application rate of particle S with the biochar produced at 450 °C. This increase may be due to the inorganic alkaline compounds that alter the pH of acidic soils or to a poor buffering capacity of the sandy soil [2]. These results are similar to those obtained by Watzinger et al., 2014, who observed an increase in the pH of an acidic soil after an application rate of 3% of biochar produced from wheat and willow bark [18]. In turn, our results were different to those obtained by La Rosa, de et al., 2015, and Zheng et al., 2018, who observed no substantial changes in the soil pH after biochar application rates of 10% to 20% and 0.5% to 10%, respectively, being that attributed to the alkaline nature and high buffering capacity of the soil in their studies [8, 24]. This fact highlights the importance of driving more soil-specific studies towards comprehensive information about valorization of biomass into biochar to provide effective agronomic gains.

The EC values were within the range of 28.6 to 112 $\mu S\ cm^{-1}$, corresponding to the application rate of 3% of particle size S and 10% application rate of particle size M of the biochar produced at 550 °C, respectively. The EC generally increased upon biochar application, with the exception of the 3 and 6% application rates of all particle sizes of the 450 °C biochar and the 3% application rate of the 550 °C biochar. The pyrolysis temperature and application rate were significant on the EC, and the increase observed may be due to the inorganic content in the ashes or to the higher ratio of superficial surface area to volume that allows for higher ion retention [25]. These results were similar to those obtained by Farhain et al. [20], that reported increases in pH and EC in the range 7.3–11 and to a maximum of 2.5 $dS\ m^{-1}$ in the soil after biochar application. Burrel et al. [16], also observed an increase in the EC in three soils after 3% application rate of biochar produced at 525 °C [16, 20].

3.3 WHC

The WHC values are shown in Fig. 3. All the treatments with biochar presented an increase in the WHC of the soil to values in the range 23–68%, which corresponds to an increase of 69% to 225% when compared with the 16% WHC of the control soil. This increase was directly proportional to the application rate, and the highest values were attained in the particle size M. The increase in this parameter can be due to the decrease in the BD, an increase in the soil storage pores and/or pore rearrangement between aggregates of the soil after biochar application which leads to an increase in its porosity [14, 26, 27].

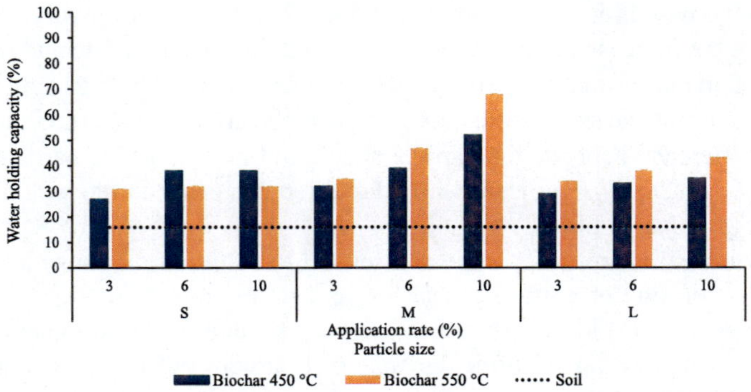

Fig. 3 Water holding capacity of the treatment mixtures, with the control soil as a reference (dotted line)

The M size particles induced a higher WHC than the size L, which could be due to the filling of the soil pores by small biochar particles, thus leading to a decrease in its permeability [28]. This effect was not observed in the S particle sizes, however that could be explained by a lesser change in the BD of the soil by the latter particles. Overall, the WHC was higher in the treatments containing biochar produced at pyrolysis temperature of 550 °C, being the exceptions the application rates of 6 and 10% of the particle size S. This could be explained by a higher superficial surface area of biochar with the increase of pyrolysis temperature which leads to a better pore structure and consequently to a higher WHC [29].

Similar results were obtained by Fu et al., 2021, who studied the hydraulic properties of desert soils after biochar application (produced at 500 °C) at rates of 0, 0.5, 1, 2, and 4% and particle sizes in the ranges of 0–0.25 mm, 0.25–1 mm, and 1–2 mm. They observed an increase in water content with the higher application rates, besides an increase of 65.1, 115, and 51.7%, for each particle size, with the application rate of 4% biochar [26]. Verheijen et al., 2019 and Laghari et al., 2015 observed an increase of 48 and 6.9% in the WHC of amended soils after a biochar application rate of 20% and 0.6%, respectively, being this increase explained by the retention of water against evaporation and gravity [14, 28].

3.4 *Soil Respiration*

The soil respiration was calculated through the pressure drop after an incubation period of 19 days, and the results are shown in Fig. 4. The soil respiration increased from 0.0032 mg O_2 g^{-1} d^{-1} to a range of values from 0.0095 to 0.0286 mg O_2 g^{-1} d^{-1}, which corresponds to an increase of 195–783% to the 3% application rate of the L particle for the biochar produced at 450 °C and 10% application rate of particle size S of biochar produced at 550 °C. The increase is directly proportional

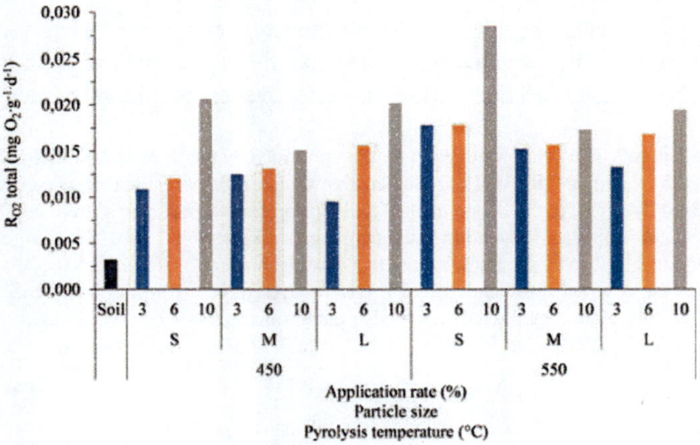

Fig. 4 Soil respiration of the treatment mixtures, with the control soil as a reference (black column)

to the application rate of biochar which corresponds to the increase in the water holding capacity that allows for a lower nutrient run-off and consequently to a higher available concentration of nutrients that leads to an increase in microbial activity [30]. Xu et al., 2016 observed that higher respiration rates for soils with biochar can be expected due to an improvement in the soil structure that leads to higher aeration [31].

4 Conclusion

The application of biochar significantly increased key-physicochemical properties of the soil that have agronomic relevance. The water holding capacity increased to values between 27 and 68% consisting of an improvement of 69–225% when compared to the control soil (WHC of 16%). This allows for a higher amount of available water to plants and consequently to a reduced demand for irrigation, which is critical under the current climate change scenarios. The increase in WHC also decreases the nutrient leaching and consequently improves the microbial activity, as observed by the increase in soil respiration within a range of 195–783% when compared with control. The pH was corrected to neutral values (5–7) in all the treatments due to the liming capacity of the biochar.

The medium particles, M = [0.5;3.15], induced the higher values for the WHC, while the smaller particles, S < 0.5 mm, promoted higher pH values. In turn, the larger particles, L > 3.15 mm, did not induced significant advantages in the parameters studied, which is of high importance since the M particles are the predominant range of particles for both pyrolysis temperatures (48–52.4%).

The improvements provided by the biochar application are strongly related to the type of soil that it is applied in, therefore study of the application of biochar in different types of soil is needed to assess the effectiveness of this amendment method.

Acknowledgements We acknowledge the funding provided by the Science and Technology Foundation (FCT) to the project "BioValChar- Sustainable valorization of residual biomass into biochar" (PCIF/GVB/0034/2019), in the scope of the Competition for Scientific Research and Technological Development Projects – Prevention and Action against Forest Fires- 2019, funding exclusively by national funds. F.C. Silva acknowledges the funding through national funds (OE) under the framework of the contract foreseen in art. 23° DL57/2016 (n° 4, 5, 6) of august 29th, rectified by DL57/2017 of July 19th. Acknowledges are also made for the financial aid provided by CESAM through the FCT/MCTES (UIDP/50017/2020+UIDB/50017/2020+ LA/P/0094/2020), and national funding.

References

1. Khorram, M.S., Zhang, Q., Lin, D., Zheng, Y., Fang, H., Yu, Y.: Biochar: A review of its impact on pesticide behavior in soil environments and its potential applications. J. Environ. Sci. (China). 44, 269–279 (2016). https://doi.org/10.1016/j.jes.2015.12.027
2. Neina, D.: The Role of Soil pH in Plant Nutrition and Soil Remediation. Hindawi. 2019, 9 (2019). https://doi.org/10.1155/2019/5794869
3. Lourenço, L., Fernandes, S., Castro, A.: Causas de incêndios florestais em Portugal continental. Análise estatística da investigação efetuada no último quindénio (1996 a 2010). In: Cadernos de Geografia. pp. 61–80 (2011)
4. Marques, S., Borges, J.G., Moreira, F., Carreiras, J.M.B., Oliveira, M.M., Cantarinha, A., Botequim, B., Pereira, J.M.C.: Characterization of wildfires in Portugal. 775–784 (2011). https://doi.org/10.1007/s10342-010-0470-4
5. Enaime, G., Baçaoui, A., Yaacoubi, A., Lübken, M.: Biochar for Wastewater Treatment—Conversion Technologies and Applications. Appl. Sci. 2020, Vol. 10, Page 3492. 10, 3492 (2020). https://doi.org/10.3390/APP10103492
6. Yuan, J., Wen, Y., Dionysiou, D.D., Sharma, V.K., Ma, X.: Biochar as a novel carbon-negative electron source and mediator: electron exchange capacity (EEC) and environmentally persistent free radicals (EPFRs): a review. Chem. Eng. J. 429, (2022). https://doi.org/10.1016/J.CEJ.2021.132313
7. Pituya, P., Sriburi, T., Wijitkosum, S.: Properties of biochar prepared from acacia wood and coconut shell for soil amendment. Eng. J. 21, 63–76 (2017). https://doi.org/10.4186/ej.2017.21.3.63
8. Zheng, H., Wang, X., Chen, L., Wang, Z., Xia, Y., Zhang, Y., Wang, H., Luo, X., Xing, B.: Enhanced growth of halophyte plants in biochar-amended coastal soil: roles of nutrient availability and rhizosphere microbial modulation. Plant Cell Environ. 41, 517–532 (2018). https://doi.org/10.1111/pce.12944
9. Departamento de Ambiente e Ordenamento, Centro de Estudos do Ambiente e do Mar: BioValChar -Valorização sustentável de biomassa residual para biochar, https://biovalchar.web.ua.pt/?targetPage=readArticle&categoryId=103&articleId=72
10. Baronti, S., Alberti, G., Vedove, G.D., di Gennaro, F., Fellet, G., Genesio, L., Miglietta, F., Peressotti, A., Vaccari, F.P.: The biochar option to improve plant yields: First results from some field and pot experiments in Italy. Ital. J. Agron. 5, 3–11 (2010). https://doi.org/10.4081/ija.2010.3

11. Verheijen, F., Jeffery, S., Bastos, A.C., Van Der Velde, M., Diafas, I.: Biochar Application to Soils: A Critical Scientific Review of Effects on Soil Properties, Processes and Functions. (2010)
12. Qambrani, N.A., Rahman, M.M., Won, S., Shim, S., Ra, C.: Biochar properties and eco-friendly applications for climate change mitigation, waste management, and wastewater treatment: A review. Renew. Sustain. Energy Rev. 79, 255–273 (2017). https://doi.org/10.1016/J.RSER.2017.05.057
13. Yang, X., Zhang, S., Ju, M., Liu, L.: Preparation and Modification of Biochar Materials and their Application in Soil Remediation. Appl. Sci. 2019, Vol. 9, Page 1365. 9, 1365 (2019). https://doi.org/10.3390/APP9071365
14. Verheijen, F.G.A., Zhuravel, A., Silva, F.C., Amaro, A., Ben-Hur, M., Keizer, J.J.: The influence of biochar particle size and concentration on bulk density and maximum water holding capacity of sandy vs sandy loam soil in a column experiment. Geoderma. 347, 194–202 (2019). https://doi.org/10.1016/j.geoderma.2019.03.044
15. Graber, E.R., Meller Harel, Y., Kolton, M., Cytryn, E., Silber, A., Rav David, D., Tsechansky, L., Borenshtein, M., Elad, Y., Tsechansky, L., Meller Harel, Y., Kolton, M., Rav David, D., Borenshtein, M., Elad, Y.: Biochar impact on development and productivity of pepper and tomato grown in fertigated soilless media. Plant Soil. 337, 481–496 (2010). https://doi.org/10.1007/s11104-010-0544-6
16. Burrell, L.D., Zehetner, F., Rampazzo, N., Wimmer, B., Soja, G.: Geoderma Long-term effects of biochar on soil physical properties. Geoderma. 282, 96–102 (2016). https://doi.org/10.1016/j.geoderma.2016.07.019
17. Gascó, G., Cely, P., Paz-Ferreiro, J., Plaza, C., Méndez, A.: Relation between biochar properties and effects on seed germination and plant development. Biol. Agric. Hortic. 32, 237–247 (2016). https://doi.org/10.1080/01448765.2016.1166348
18. Watzinger, A., Feichtmair, S., Kitzler, B., Zehetner, F., Kloss, S., Wimmer, B., Zechmeister-Boltenstern, S., Soja, G.: Soil microbial communities responded to biochar application in temperate soils and slowly metabolized 13C-labelled biochar as revealed by 13C PLFA analyses: Results from a short-term incubation and pot experiment. Eur. J. Soil Sci. 65, 40–51 (2014). https://doi.org/10.1111/EJSS.12100
19. Ali, S., Rizwan, M., Qayyum, M.F., Ok, Y.S., Ibrahim, M., Riaz, M., Saleem Arif, M., Farhan Hafeez, &, Al-Wabel, M.I., Shahzad, A.N.: Biochar soil amendment on alleviation of drought and salt stress in plants: a critical review. Environ. Sci. Pollut. Res. 24, 12700–12712 (2017). https://doi.org/10.1007/s11356-017-8904-x
20. Farhain, M.M., Cheema, M., Katanda, Y., Nadeem, M., Javed, B., Thomas, R., Saha, R., Galagedara, L.: Potential of developing podzolic soil-based potting media from wood ash, paper sludge and biochar. J. Environ. Manage. 301, (2022). https://doi.org/10.1016/J.JENVMAN.2021.113811
21. Razzaghi, F., Obour, P.B., Arthur, E.: Does biochar improve soil water retention? A systematic review and meta-analysis. Geoderma. 361, (2020). https://doi.org/10.1016/J.GEODERMA.2019.114055
22. Zhang, B., Zhang, L., Zhang, X.: Bioremediation of petroleum hydrocarbon-contaminated soil by petroleum-degrading bacteria immobilized on biochar. R. Soc. Chem. 35304–35311 (2019). https://doi.org/10.1039/c9ra06726d
23. Bloem, J., Hopkins, D.W., Benedetti, A.: Microbial biomass and numbers. Microbiol. methods Assess. soil Qual. 73–113 (2009). https://doi.org/10.1079/9780851990989.0073
24. de la Rosa, J.M., Paneque, M., Miller, A.Z., Knicker, H.: Relating physical and chemical properties of four different biochars and their application rate to biomass production of Lolium perenne on a Calcic Cambisol during a pot experiment of 79 days. Sci. Total Environ. 499, 175–184 (2014). https://doi.org/10.1016/j.scitotenv.2014.08.025
25. Ullah, N., Ditta, A., Khalid, A., Mehmood, S., Shahid Rizwan, M., Ashraf, M., Mubeen, F., Imtiaz, M., Mazhar Iqbal, M.: Integrated Effect of Algal Biochar and Plant Growth Promoting Rhizobacteria on Physiology and Growth of Maize Under Deficit Irrigations. J. Soil Sci. Plant Nutr. 20, 346–356 (2020). https://doi.org/10.1007/s42729-019-00112-0

26. Fu, G., Qiu, X., Xu, X., Zhang, W., Zang, F., Zhao, C.: The role of biochar particle size and application rate in promoting the hydraulic and physical properties of sandy desert soil. Catena. 207, 105607 (2021). https://doi.org/10.1016/j.catena.2021.105607
27. Agegnehu, G., Bass, A.M., Nelson, P.N., Bird, M.I.: Benefits of biochar, compost and biochar-compost for soil quality, maize yield and greenhouse gas emissions in a tropical agricultural soil. Sci. Total Environ. 543, 295–306 (2016). https://doi.org/10.1016/J.SCITOTENV.2015.11.054
28. Laghari, M., Mirjat, M.S., Hu, Z., Fazal, S., Xiao, B., Hu, M., Chen, Z., Guo, D.: Effects of biochar application rate on sandy desert soil properties and sorghum growth. CATENA. 135, 313–320 (2015). https://doi.org/10.1016/J.CATENA.2015.08.013
29. Yang, Y., Ye, S., Zhang, C., Zeng, G., Tan, X., Song, B., Zhang, P., Yang, H., Li, M., Chen, Q.: Application of biochar for the remediation of polluted sediments. J. Hazard. Mater. 404, (2021). https://doi.org/10.1016/J.JHAZMAT.2020.124052
30. Mitchell, P.J., Simpson, A.J., Soong, R., Simpson, M.J.: Shifts in microbial community and water-extractable organic matter composition with biochar amendment in a temperate forest soil. Soil Biol. Biochem. 81, 244–254 (2015). https://doi.org/10.1016/j.soilbio.2014.11.017
31. Xu, N., Tan, G., Wang, H., Gai, X.: Effect of biochar additions to soil on nitrogen leaching, microbial biomass and bacterial community structure. Eur. J. Soil Biol. 74, 1–8 (2016). https://doi.org/10.1016/j.ejsobi.2016.02.004

Effect of High-Pressure Processing Combined with Propolis Extract on the Quality and Microbiological Safety of Apple Puree

Diogo A. Gonçalves, Enrique Pino-Hernández, Jorge A. Saraiva, Carlos A. Pinto, Marta B. Evangelista, and Marco Alves

Abstract The aim of this study was to evaluate the influence of high-pressure processing (HPP) combined with natural extract on physical–chemical (Total soluble solids (TSS), moisture, pH, water activity (a_W), color parameters) and microbiological characteristics (Total mesophilic aerobic microorganisms count at 30 °C, molds and yeast count at 37 °C) of apple puree, during 90 days under refrigerated conditions. Three processes were applied: HPP (600 MPa, 3 min) with and without the addition of propolis extract (2.5% w/w); and conventional pasteurization (75 °C, 10 min), as a reference method in the industry. The results showed that immediately after the application of the three processes, the load of aerobic microorganisms decreased from 3.40 to < 1 Log CFU/g. At the end of the shelf-life study, in all treatments, the count of aerobic microorganisms and of molds and yeasts was less than 2 and 1 Log CFU/g, respectively. Likewise, in the evaluation of TSS, moisture and a_W no significant differences were found between processes. In the color analysis, HPP increased the stability over the shelf-life but promoted a higher initial change. The addition of extract enhanced color and pH stability, being the only treatment in which the product's color did not change significantly during shelf-life. In conclusion, HPP is a good alternative to conventional pasteurization in the production of apple puree. The addition of propolis enhances the quality of puree while providing differentiation to the product.

Keywords High hydrostatic pressure · Propolis extract · Development of food products · Food safety

---

D. A. Gonçalves · E. Pino-Hernández · M. B. Evangelista · M. Alves (✉)
INOV.LINEA - Agri-Food Technology Transfer and Valorization Center, TAGUSVALLEY—Science and Technology Park, 2200-062 Abrantes, Portugal
e-mail: marco_alves@tagusvalley.pt

J. A. Saraiva · C. A. Pinto
Chemistry Department, LAQV-REQUIMTE, University of Aveiro, Campus Universitário de Santiago, 3810-193 Aveiro, Portugal
e-mail: jorgesaraiva@ua.pt

J. R. da. C. S. Galvão et al. (eds.), *Proceedings of the 3rd International Conference on Water Energy Food and Sustainability (ICoWEFS 2023)*, Springer Proceedings in Earth and Environmental Sciences, https://doi.org/10.1007/978-3-031-48532-9_10

1 Introduction

The production of apple is one of the most important economic activities in the primary sector, being estimated that between 2010 and 2020 the volume of production grew from 750.5 million to 887 million metric tons. China and Europe are the biggest producers in the world, producing 46 and 12.28 million metric tons in 2021/2022 crop year, respectively [1]. After harvesting, the fruit is separated and categorized based on the caliber and defects to determine sales value and final application. One of the main application is the production of apple puree.

In the industry, the production of apple puree starts with preparatory operations, as suitable apples are washed and selected, while the ones with non-conformities are rejected. Bleaching is usually done after the preparatory steps to decrease raw material losses, energy requirements and mechanical stress in the crushing operation. Additionally, bleaching is essential to guarantee the polyphenol oxidase enzyme (PPO) inactivation, responsible for the colour changes by enzymatic browning [2]. After crushing the bleached apples are pasteurized to inactivate microorganisms and subsequently ensure the safety of the final product throughout the extended shelf-life. One of the biggest disadvantages of the traditional method applied to transform apple into puree is the usage of high temperatures in successive operations. Submitting food to high temperatures for a long period of time alters the nutritional value and greatly modifies the final product sensory and physical–chemical properties.

In order to minimize the disadvantages of high temperatures, companies are increasing investments in studying and using low-impact preservation methods, such as high-pressure processing (HPP). HPP is a food processing method based on the isostatic and *Le Chatelier's* principles. The isostatic principle implies that pressure is applied equally in all directions, therefore is distributed homogeneously and instantaneously, regardless of the composition, size or geometry of the food. This way there is a guarantee that no pressure gradient occurs in the process and that non-porous foods will keep their original shape after depressurization [3]. *Le Chatelier's* principle states that any phenomenon (molecular configuration, phase transition and chemical or biochemical reaction) that results in a volume reduction is favored by pressurization. For this reason, covalent bonds are not broken, and consequently, small and low molecular weight molecules are rarely affected in HPP. On the other hand, complex macromolecules that are united by *Van der Waals* and hydrogen bonds can undergo changes in their quaternary, tertiary and secondary structures [3]. Therefore, with HPP, it is possible to affect the viability of microorganisms without destroying important components such as vitamins, pigments and flavouring agents.

In the food industry, it is usual to combine HPP with different preservation methods to ensure the safety and stability of the final product (hurdle technology). Nowadays refrigerated storage combined with HPP is the golden standard in the food industry, but additives are increasingly being added to give more assurance in the quality maintenance of the product, especially in the stability of pH and color [4]. Additives may also be added to improve functionality or differentiate the product in the market. Propolis, a resin-like substance produced by bees from tarry and balsamic substances, might be used as an additive to improve product stability and functionality in the apple puree processed by HPP. It has been documented that propolis has nearly 420 chemical substances identified to date, the majority being flavones, flavanones, flavonols and dihydroflavonoids [5]. Those substances provide numerous beneficial effects for food quality and, consequently, for the consumer wellbeing. It has been documented that adding propolis to food matrices might provide antifungal, antiviral, antiparasitic, antioxidant effects, with additional in vitro studies showing that the consumption of propolis might have an inflammatory, anticancer and antidiabetic effects [6–8].

The aim of this study was to compare and evaluate the impact of three process methods in the physical–chemical (total soluble solids (TSS), moisture, water activity (a_W), colour parameters, pH, antioxidant capacity) and microbiological (total aerobic count of mesophilic microorganisms, moulds and yeasts) characteristics of Fuji apple puree. The applied processing methods were the conventional pasteurization (75 °C, 10 min) and HPP (600 MPa, 3 min) with and without the addition propolis extract (2.5% w/w). The processed apple puree was stored at 3 ± 1 °C for 90 days, with periodic monitoring of 15 days in the first month and 30 days in subsequent months.

2 Material and Methods

2.1 *Raw Material*

Fuji apples were obtained from Cooperfrutas, a Cooperative of fruit and vegetable producers from Alcobaça, Portugal. Apples were 2nd grade, not suitable for the fresh market. They appeared mature, but not overripe, and free from decay. Some defects were visible such as bruises, mild deformations, and brown surface discolorations. Propolis was purchased raw, without any transformation, from local beekeepers in Abrantes, Portugal. After reception, apples were stored in refrigerated conditions at 2 ± 1 °C, and propolis was frozen at −19 ± 1 °C.

2.2 *Extract Preparation*

Raw propolis was blended in a kitchen robot (Bimby, Vorwerk, Thermomix TM31, Germany) until a coarse powder was obtained. The propolis extract was made by mixing the powder with an aqueous solution of ethanol (70% v/v), followed by extraction by pulse electric fields (PEF). The PEF system is an EPULSUS_LBM3B-15 (EnergyPulse Systems, Lisbon, Portugal) capable of bipolar pulses, with different pulse width, frequency and up to 15 kV of power. The generator was connected to a batch cell with 2 parallel electrodes with an area of 100 cm^2 and 5 cm apart. The parameters during the extraction were the following: field strength of 2 kV/cm, pulse frequency of 60 Hz, pulse width of 25 μs and 19 kJ/Kg of specific energy.

After PEF, the extract was centrifugated (Nahita 2615/1, Auxilab, Spain) at 3000 rpm for 5 min and immediately vacuum filtrated using a 2 μm cellulose membrane filter. The extract was frozen at -19 ± 1 °C until use.

2.3 *Puree Preparation*

Apples were washed, peeled and bleached. Bleaching was done by placing the apples in a boiling pan with water for 2 min. Afterwards, the apples were blended in a kitchen robot (Bimby, Vorwerk, Thermomix TM31, Germany) and the extract (2.5% w/w) was added in half of the pure intended to be processed with HPP. Puree was immediately vacuum sealed in metallic opaque bags (bags PA/PE MET 2 sides 200X300X90M, 4Pack, Portugal) using a packaging machine (Henkelman, Marlin 42, Netherlands) in portions of 50 g. Three replicates for each process and monitoring time were produced and three replicates were also prepared for purée without any treatment (control sample). In total 54 samples were produced and stored under refrigeration (3 ± 1 °C) until HPP or pasteurization were applied.

2.4 *Preservation Methods*

2.4.1 HPP Process

The puree samples were pressurized at 600 MPa for 3 min, using water at 17 °C as the medium for transmitting pressure. The industrial high-pressure processing was done using a 55 L high-pressure pilot unit (Hiperbaric 55, Hiperbaric S.A, Burgos, Spain) with a maximum capacity of 55 L and a maximum pressure of 600 MPa. After treatment, the apple puree was stored at a temperature of 3 ± 1 °C.

2.4.2 Pasteurization Process

The puree packaged in the bags was placed in a glass flask (H11cm, Ø7.5 cm) and processed in an autoclave (Autester ST, J.P Selecta S.A, Barcelona, Spain) set to 75 °C for 10 min. The process was monitored using a datalogger (IFC400, MadgeTech, USA) inserted inside one flask with puree made for this propose. After the process the flasks were stored in a fridge at 3 ± 1 °C.

2.5 *Physicochemical Measurements*

Physicochemical characteristics of treated puree were monitored after each treatment (0 day), and after 15, 30, 60 and 90 days of refrigerated storage at 3 ± 1 °C. Samples of puree without any treatment were only measured 1 time after preparation. Each sample replica was measured 3 times to determine the values of °Brix, pH Moisture and a_W. In colour 9 measurements for each replica were performed.

2.5.1 Moisture, TSS, pH and a_W Measurement

Total soluble solids (TSS) were measured using a refractometer (HI 96,801, Hanna, Portugal), temperature during measurement was 20 ± 1 °C. The pH and a_W were measured using a pH meter and a water activity meter (C931, Consort, Belgium and a Aqualab 4TE, Decagon Devices, USA) respectively.

To quantify the moisture content, 5 ± 1 g of sample was weighed and placed in a crucible. The crucible was put in an oven (ED 23, Brinder, Deutschland) at 105 °C for 24 h. After drying, the samples were placed in a desiccator for 30 min and then reweighed.

2.5.2 Color Measurement

The hue alteration of apple puree during shelf-life was determined using a (Minolta, CR-310, JP). The apparatus measures the light received and converts it into spaces coordinates. These values are translated to a $L^*a^*b^*$ standard, where L^* represents lightness, a^* indicates the hue from green (−) to red (+), and $b*$ the hue from blue (−) to yellow (+). To determine the magnitude of color alteration, the overall color deviation was calculated by the following equation:

$$\Delta E^* = \left[(\Delta L^*)^2 + (\Delta a^*)^2 + (\Delta b^*)^2\right]^{0.5}$$

2.6 *Microbiological Determination*

Total mesophilic aerobic, molds and yeasts count were performed according to EN ISO 4833-1:2013 for total aerobic mesophilic and NP 3277-1:1987 for molds and yeast. The results were presented in terms of the logarithm of colony forming units per gram (Log CFU/g), and the measurements were carried out with the same frequency and length as the physical and chemical measurements.

2.7 Statistical Analysis

The data presented in this study was treated by the graph prism 9 program. Statistical significance was performed by applying a one-way ANOVA and post hoc test (Tukey's Test) with a significance level of $P < 0.05$. Data graphic representations also contain the standard deviation.

3 Results and Discussion

3.1 Moisture, TSS and a_W Analysis

The untreated puree had a TTS of 13.6 ± 0.1 °Brix, 88.0% moisture and 0.994 ± 0.001 a_W (Table 1). Pasteurization and HPP with and without propolis resulted in no significant changes in moisture and a_W. The values of these parameters were constant along the shelf-life period of 90 days. a_W is a major factor in preventing or limiting microbial growth, the vast majority of food microorganisms grow between 0.85 and 1, including pathogenic bacteria [9]. Between 0.85 and 0.61 a_W is also suitable for molds and yeast growth [9]. Thus, the observed a_W values of all samples are considered ideal for microbial development. Nonetheless pasteurization and HPP should be able to reduce microbial loads to secure levels for most bacteria, molds and yeasts presented in the raw apple puree, thus the microbiological growth is only a problem after opening the package/jar. The only exception are microbial spores, this biologic structure is not affected by HPP or pasteurization and can regrow into bacteria after the treatment is applied [10]. The high a_W, mild temperature and low °Brix are favorable conditions for spores regrow, therefore apple puree processed by mild pasteurization and HPP should be stored in refrigerated conditions.

TSS slightly changed after the application of each treatment. Puree treated with HPP and pasteurization had a °Brix of 12.5 and 12.8 respectively. The addition of extract, as it is an ingredient with very low solid content, resulted in lower values of °Brix and, similarly to a_W and moisture, it did not change during shelf-life. The recorded values of °Brix are lower than typical levels found in commercially available products, which can range from 20 to 25 °Brix [2]. Increasing the concentration of solid compounds is done in industry by adding sucrose or by concentrating the pure using heat, both methods can be applied in the same product [2]. The described methods have some disadvantages; the addition of sucrose contradicts new dietary trends and concentration reduces process yield and degrades the most heat-sensitive compounds. For these reasons, and to produce a puree with lower energy density, the described methods were not applied.

Table 1 °Brix, Moisture and a_W values of untreated puree and puree treated with pasteurization, HPP and HPP with extract, throughout 90 days of shelf-life

Treatment	Day	°Brix	Moisture (%)	a_W
Fresh apple	0	13.6 ± 0.1^{a}	$88.0 \pm 0.1^{a,A}$	$0.994 \pm 0.001^{a,A}$
Pasteurization	0	$12.8 \pm 0.4^{b,A}$	$87.9 \pm 0.3^{a,A}$	$0.996 \pm 0.001^{a,A}$
	15	$12.8 \pm 0.4^{a,A}$	$88.2 \pm 0.3^{a,A}$	$0.996 \pm 0.001^{a,A}$
	30	$13.0 \pm 0.1^{a,A}$	$87.9 \pm 0.2^{a,A}$	$0.997 \pm 0.001^{a,A}$
	60	$12.8 \pm 0.3^{a,A}$	$88.0 \pm 0.3^{a,A}$	$0.994 \pm 0.001^{a,A}$
	90	$13.0 \pm 0.3^{a,A}$	$87.9 \pm 0.2^{a,A}$	$0.993 \pm 0.002^{a,A}$
HPP	0	$12.5 \pm 0.4^{bc,AB}$	$87.9 \pm 0.2^{a,A}$	$0.996 \pm 0.001^{a,A}$
	15	$12.5 \pm 0.1^{ab,AB}$	$88.5 \pm 0.3^{a,A}$	$0.991 \pm 0.001^{a,B}$
	30	$12.6 \pm 0.2^{ab,AB}$	$87.8 \pm 0.4^{a,A}$	$0.991 \pm 0.001^{b,B}$
	60	$12.4 \pm 0.1^{a,B}$	$87.6 \pm 0.7^{a,A}$	$0.992 \pm 0.001^{a,AB}$
	90	$12.9 \pm 0.1^{ab,A}$	$87.8 \pm 0.3^{a,A}$	$0.992 \pm 0.001^{a,AB}$
HPP + Extract	0	$12.2 \pm 0.1^{c,AB}$	$88.4 \pm 0.2^{a,A}$	$0.996 \pm 0.001^{a,A}$
	15	$12.1 \pm 0.7^{b,AB}$	$89.0 \pm 0.1^{a,A}$	$0.993 \pm 0.001^{a,A}$
	30	$12.3 \pm 0.1^{b,AB}$	$88.3 \pm 0.2^{a,A}$	$0.993 \pm 0.001^{ab,A}$
	60	$12.7 \pm 0.3^{a,B}$	$88.0 \pm 0.3^{a,A}$	$0.992 \pm 0.001^{a,A}$
	90	$12.5 \pm 0.1^{b,A}$	$88.2 \pm 0.2^{a,A}$	$0.993 \pm 0.004^{a,A}$

Different lower-case letters represent significant differences between treatments in the same time point. Different capital letters represent significant differences in the same treatment in different time points ($\alpha = 5\%$)

3.2 pH Analysis

The pH is a very important characteristic in the quality of products as small differences in this parameter are easily detected by consumers and are usually associated with spoiled or expired products. Therefore, maintaining a consistent pH throughout the shelf-life is crucial. The results show that the Fuji apple puree without treatment has a pH value of 4.90. HPP treatment did not significantly modify puree pH. On the other hand, the addition of extract caused a decrease in the pH of the puree treated with HPP. The extract, being a matrix with a lower pH and representing a considerable portion of the apple puree (2.5% w/w), directly influenced the pH reading in the puree. The pasteurization process also significantly decreased the puree pH, one of the reasons may be the degradation of cell walls by heat, releasing acidic components into the matrix [11] (Fig. 1).

Throughout shelf-life, the samples processed by pasteurization presented the greatest stability in pH, between day 0 and day 90, the pH decreased by 1.86%. In pasteurization, the pH progressively decreased, in the last analysis (day 90) it was noticeable a continuation in the trend. This lack of stabilization might indicate further decreases in pH if longer shelf-life studies were performed (Fig. 2). The puree processed by HPP presented the greatest variation in pH throughout the shelf-life,

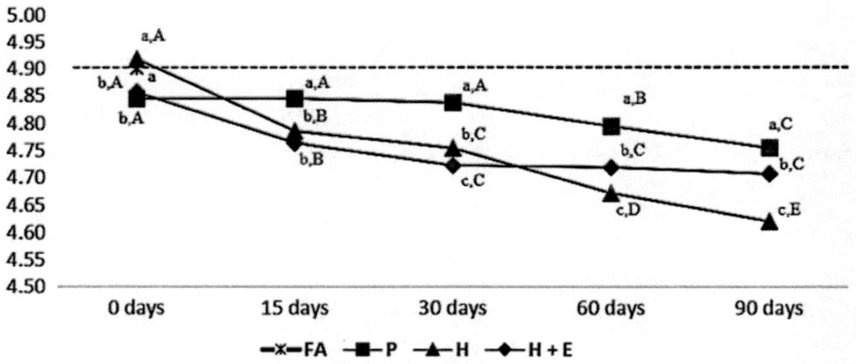

Fig. 1 pH values of fresh apple (FA) and puree treated with pasteurization (P), HPP(H) and HPP with extract (H + E), throughout 90 days of shelf-life. Different lower-case letters represent significant differences between treatments in the same time point. Different capital letters represent significant differences in the same treatment in different time points. ($\alpha = 5\%$)

with a decrease of 6.10%. In a similar manner as pasteurization, the last analysis shows an evident trend in the decrease of pH that might continue if a longer shelf-life evaluation took place. The addition of extract moderated the decrease in the puree pH processed by HPP, with a decrease of 3.09%. Despite the pH total variation being greater in relation to pasteurization, in the last 30 days of analysis the pH did not vary significantly. The lack of a downward trend suggests that in higher shelf-life studies, the puree with propolis treated with HPP has the greatest stability.

The pH variation during shelf-life can be explained by organic processes that slowly occur in the food matrix. Some organic process might release acidic by products to the matrix and reduce pH [12]. The propolis extract is a matrix that contains more than 50% of flavonoids and phenolic compounds, family of compounds with high scavenger capacity of hydroxyl radicals [13]. This capacity can reduce or delay the organic reactions of the different compounds of the matrix.

3.3 Color Analysis

The initial $L^*a^*b^*$ values of untreated puree were 71.50, -3.99 and 24.46, respectively. All applied treatments considerably modified the $L^*a^*b^*$ chromatic coordinates. HPP, compared to mild pasteurization, caused a more expressive increase in the a^* and b^*, which translates into a more reddish and yellowish puree. The difference between processes may have come from the possible inability of HPP to inactivate the biological activity of polyphenol oxidase (PPO) and peroxidase (POD), enzymes responsible for enzymatic browning [14].

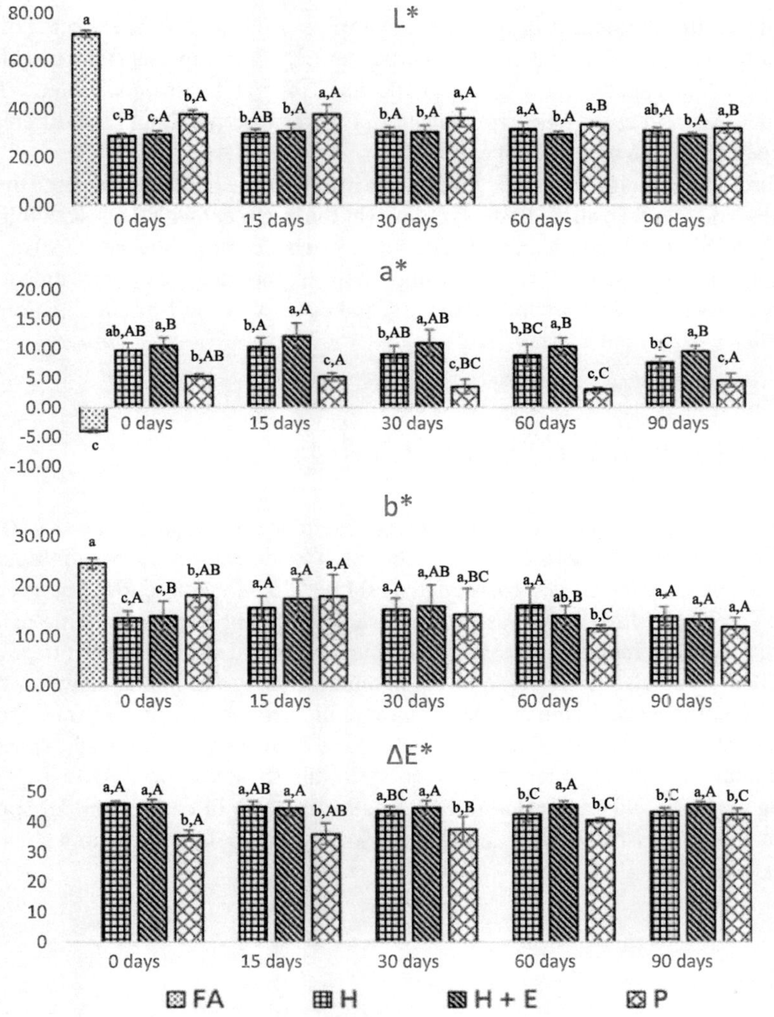

Fig. 2 Colour values of fresh apple (FA) and puree treated by HPP (H) and HPP with extract (H + E) and, pasteurization (P), throughout 90 days of shelf-life. Different lower case letters represent a significant difference between treatments in the same time point. Different capital letters represent significant differences in the same treatment in different time points ($\alpha = 5\%$)

Pasteurization treatment also increased a^* and b^* values compared with fresh apple, which can be justified by non-enzymatic browning triggered by high temperature. Non-enzymatic browning is caused by Maillard reactions, caramelization and/or destruction of pigments. Considering that the applied temperature was 75 °C, the destruction of pigments is the most plausible cause, since at these temperatures the Maillard and browning reactions are negligible [4].

The addition of extract in HPP treatment increased b^* value, as expected due to the different color of the extract. The parameter L^*, which represents the brightness of the product, experienced a significant decrease after all treatments. Pasteurization had the least effect on puree brightness and there were no significant differences between HPP with and without extract.

Throughout shelf-life significant changes in color were registered in puree treated by HPP and Pasteurization. Pasteurization was the treatment with the less stability in L^*, a^* and b^* values, considering that it progressively decreased during the 90 days of storage. The puree treated by HPP with propolis extract was the only treatment that did not vary significantly between the first and last day of shelf-life in all chromatic coordinates analyzed.

3.4 Microbiological Safety and Shelf-Life

Initially the apple contained 3.4 Log CFU/g of aerobic microorganisms and <1.0 Log CFU/g of molds and yeasts (Table 2). After each treatment, the count of mesophilic aerobic microorganisms was reduced to <1.0 Log CFU/g. In pasteurized puree, after 90 days of shelf-life, there was no increase in the number of mesophilic aerobic microorganisms, molds and yeasts. In the purée with and without extract treated by HPP, there was also no increase in the number of yeast and molds, however there was a small increase in the number of mesophilic aerobic microorganisms. From a food safety point of view, the magnitude of the increase is not relevant, since the maximum number recommended for safety in this food category is 3 Log CFU/g, 2 Log above the 90 days value in both treatments [15]. In conclusion, all applied treatments successfully maintained the food safety of apple puree with a shelf-life of 90 days.

Table 2 Microbiological analysis of fresh apple puree and puree treated with pasteurization, HPP, and HPP with extract

Mesophilic Aerobic Count at 30 °C by the pour plate technique (Log UFC/g)					
Treatment	0 days	15 days	30 days	60 days	90 days
Fresh Apple	3.40	–	–	–	–
Pasteurization	1.0	1.0	1.0	1.0	1.0
HPP	1.0	1.0	1.0	1.5	1.4
HPP + Extract	1.0	1.0	1.0	1.7	1.4

4 Conclusion

The results of this study demonstrated that HPP as a non-thermal pasteurization is a safe and effective alternative to conventional thermal pasteurization. During the 90 days of shelf-life, puree treated with HPP was able to maintain counts of mesophilic aerobic microorganisms, molds and yeast similar to pasteurization and well below the recommended maximum count.

The addition of propolis extract in the puree treated with HPP improved the product quality attributes by enhancing pH and colour stabilization throughout the shelf-life study. Additionally, propolis extract differentiates the product on the market by modifying the organoleptic characteristics and providing compounds that might improve consumer's health. However additional research in this area is needed to study if the new organoleptic characteristics are pleasant and to confirm the presence and bioavailability of nutraceutical compounds.

Acknowledgements and Funding We would like to thank Cooperfrutas for providing the raw material necessary for our study. This work was financially supported by the project TAGUS-VALLEY2030 RHaq (CENTRO-04–3559-FSE-000143) and TAGUSVALLEY2030 IT (CENTRO-04–3559-FSE-000143), under the European Social Fund of the European Union managed by CENTRO 2020 and PORTUGAL 2020, and by FCT/MCTES through UIDB/50006/2020 and UIDP/50006/2020.

References

1. Statista "Global leading apple producing countries in 2021/2022", https://www.statista.com/statistics/279555/global-top-apple-producingcountries/ Last accessed 2022/12/20
2. Barta, J., Cano, M., Gusek, T., Sidhu, J., Siha, N. "Handbook of Fruits and Fruit Processing". 1nd edn. Blackwell Publishing, USA (2006)
3. Balasubramaniam, VM., Barbosa-Cánovas, G., Lelieveld, H. "High Pressure Processing of Food". 1nd edn. Springer, Germany (2016)
4. Landl, A., Abadias, M., Sárraga, C., Viñas, I., & Picouet, P. A. "Effect of high pressure processing on the quality of acidified Granny Smith apple purée product." Innovative Food Science & Emerging Technologies, 11(4), 557–564 (2010).
5. Pobiega K, Kraśniewska K, Derewiaka D, Gniewosz M. "Comparison of the antimicrobial activity of propolis extracts obtained by means of various extraction methods." J Food Sci Technol, 56(12),5386–5395 (2019).
6. Natsir, M.; Kurniawati, D.; Kurniasih, Y. "Study of the utilization of propolis local propolis from Konawe, Southeast Sulawesi, as preservative natural preservative for beef." AIP Conference Proceedings 1823, 020049 (2017).
7. Bahtiti, N. "Study of Preservative Effect of "Propolis" on the Storage Quality of Mashed Potatoes." Food Science and Technology, 1(2), 17–20 (2013).
8. Pasupuleti VR, Sammugam L, Ramesh N, Gan SH. "Honey, Propolis, and Royal Jelly: A Comprehensive Review of Their Biological Actions and Health Benefits. Oxid Med Cell Longev" SI(2017), 5–26 (2017).
9. – FDA "Water Activity (aw) in Foods", https://www.fda.gov/inspections-compliance-enforcement-and-criminal-investigations/inspection-technical-guides/water-activity-aw-foods Last accessed 2022/12/20

10. Zhang, H., & Mittal, G. S. "Effects of High-Pressure Processing (HPP) on Bacterial Spores: An Overview." Food Reviews International, 24(3), 330–351 (2008).
11. – Dash, K. K., Fayaz, U., Dar, A. H., Shams, R., Manzoor, S., Sundarsingh, A., Deka, P., & Khan, S. A. "A comprehensive review on heat treatments and related impact on the quality and microbial safety of milk and milk-based products." Food Chemistry Advances, 1, 100041 (2022).
12. Techakanon, C., & Sirimuangmoon, C. "The Effect of Pasteurization and Shelf Life on the Physicochemical, Microbiological, Antioxidant, and Sensory Properties of Rose Apple Cider during Cold Storage." Beverages, 6(3), 43 (2020).
13. Treml, J., & Šmejkal, K. "Flavonoids as Potent Scavengers of Hydroxyl Radicals." Comprehensive Reviews in Food Science and Food Safety, 15(4), 720–738 (2016).
14. Arnold, M., Michałowska, A. "Enzymatic browning in apple products and its inhibition treatments: A comprehensive review" Comprehensive Reviews in Food Science and Food Safety, 21(6), 11 (2022)
15. Instituto Ricardo Jorge. "Interpretação de resultados de ensaios microbiológicos em alimentos prontos para consumo e em superfícies do ambiente de preparação e distribuição alimentar Valores-guia" (2019).

Infusion of Macroalgae: A Promise Methodology for Obtaining a Healthy Food Ingredient

Miguel Caetano, Mariana Gomes, Joaquina Pinheiro, and Rui Ganhão

Abstract Nowadays, the consumer demands for foods more natural, with reduced or without synthetic additives in their formulation and with health benefits, improving their life quality. In response, the Food Industry has been investigating a wide range of natural ingredients/compounds that bring not only a nutritional but also a food improvement. Some of these benefits compounds have been studied in a natural matrix, the macroalgae. The present study aimed to evaluate the impact of infusion in one brown and green macroalgae, *Fucus vesiculosus* and *Ulva* sp., respectively. In both macroalgae, the ratio of solid:liquid (w:v) applied were 1:16 and 1:50 and two time of infusion (5 and 15 min), was also studied. After the infusion total phenolic content (TPC), antioxidant capacity expressed by the scavenging of the free radical DPPH and the reducing power capacity of Fe (III), were also determined. By the obtained results in both macroalgae infusion extracts, the solid–liquid ratio of 1:16 (w:v) for 15 min, contributed to extract a higher content of TPC (36.52 mg GAE/100 g and 157.64 mg GAE/100 g in *Ulva* sp. *and Fucus*, respectively). Also, *Fucus*, showed a superior value of DPPH and FRAP, 62.8 2% and 41.77 mM FeSO4/g respectively. This indicated that using a simple treatment as infusion, an interesting and potential extraction of benefits compounds, as antioxidants and the phenolic it achieves. Thus, allied to the nutritional composition, the macroalgae is an alternative of using the "bioactive rich infusion" as a potential food and/or ingredient to be included in Mediterranean Food.

Keywords Seaweeds · Infusion · Antioxidants · Food industry · Bioactive

M. Caetano · M. Gomes · J. Pinheiro (✉) · R. Ganhão
MARE—Marine and Environmental Sciences Centre/ARNET—Aquatic Research Network, ESTM—School of Tourism and Maritime Technology, Polytechnic of Leiria, 2520-614 Peniche, Portugal
e-mail: joaquina.pinheiro@ipleiria.pt

J. R. da. C. S. Galvão et al. (eds.), *Proceedings of the 3rd International Conference on Water Energy Food and Sustainability (ICoWEFS 2023)*, Springer Proceedings in Earth and Environmental Sciences, https://doi.org/10.1007/978-3-031-48532-9_11

1 Introduction

In recent years we have seen an increase in average life expectancy around the world, have consequently also increased the risk of age-related diseases, such as cardiovascular and neurodegenerative diseases [1]. New studies have been looking for new compounds and foods with benefits for human health, in response to the problems associated with increasing average life expectancy.

For thousands of years the macroalgae have been used for various activities, for example in cooking and agriculture. In Asia, several macroalgae species are used for human consumption, namely species such as *Undaria* (Wakame), *Laminaria* (Kombu) and *Porphyra* (Nori). The Increase demand for healthy practice eating has led to elevate the scientific research on the nutritional value of macroalgae and their application in food products [2].

The incorporation and consumption of macroalgae in a varied and healthy diet brings health benefits, resulting from the consumption of foods rich in fiber, vitamins, and bioactive compounds, associated with the prevention of diseases such as cardiovascular diseases, degenerative diseases, and diabetes [3]. These properties are attributed to foods that have antioxidants such as carotenoids and phenolic compounds [4].

Oxidative damage at the cellular level is a major factor in premature aging, as well as in the development of neurodegenerative, cancer, and cardiovascular diseases. Free radical scavenging is important for decreasing oxidative damage through bioactive compounds with antioxidant properties.

The macroalgae have a diversity of colors, shapes, and sizes, and can be classified as follows: *Chorophyceae* (green algae), *Rhodophyceae* (red algae), and *Phaeophyceae* (brown algae) [5]. *Ulva sp.*, also known as sea lettuce, is a species of green macroalgae that can be found on rocky shores anywhere in the world. *Ulva* species have been on the market for many years all over the world. They are usually consumed as sea vegetables or used as biostimulants [3, 5]. *Fucus* spp. is an abundant and widely distributed genus of brown macroalgae and have been used in human nutrition for many years, mainly in Asia. This group of seaweeds has high fiber, vitamin, and mineral content and is low-fat, making them interesting on a nutritional level [6].

The main goal of the present study was to evaluate the impact of infusion treatment on quality and antioxidant capacity of two macroalgae, one brown and other green, *Fucus vesiculosus* and *Ulva sp.*, respectively. The infusion of macroalgae was performed using the two mixtures of solid–liquid ratio (1:16 and 1:50, w:v) and two period time (5 and 15 min). After treatment the obtained extracts were analyzed regarding the bioactive composition expressed by total phenolic content (TPC) and two antioxidant methodologies (DPPH and FRAP).

2 Materials and Methods

2.1 *Materials*

The macroalgae used in the present study, *Fucus vesiculosus* and *Ulva sp.* were acquired in Algaplus (Ílhavo, Portugal), and transport to the Research Laboratory of Cetemares at Peniche (Portugal).

The reagents used were Folin-Ciocalteu's phenol reagent (Merck, Germany); Gallic acid - analytical reagent (BIOCHEM Chemopharma, France); sodium carbonate anhydrous (BIOCHEM Chemopharma, France); 1,1-diphenyl-2-picrylhydrazyl Free Radical, DPPH (TCI, Laborspirit, Japan); 2,4,6-Tris(2-pyridyl)-s-triazine (for spectrophotometric det. Of Fe > 99%, HPLC (Sigma-Aldrich, Switzerland); Iron (III) chloride hexahydrate (Carlo Erba, France); sodium acetate 3-hydrate for analysis ACS, ISSO (Panreac Applichem, Germany); Iron (II) sulfate 7-hydrate (Reag. USP). for analysis, ACS (Panreac Applichem, Germany); acetic acid (VWR, France); hydrochloric acid (ANALAR NORMAPUR, VWR, France).

2.2 *Infusion Treatment*

Infusion is a methodology for extraction of add-value compounds from the macroalgae when exposed to water and heat treatment. In this study, the infusion was occurred during 5 and 15 min after the initial water boiling moment. Another studied variable was the impact of two proportion ratio of macroalgae and water, 1:16 and 1:50 (w:v, g:mL) [5].

2.3 *Methods*

2.3.1 Determination of Total Phenolic Content (TPC)

Total phenolic content (TPC) was determined by the adapted method of Folin–Ciocalteau [7]. Briefly, in a 96-well microplate, 20 μL of sample/standard was added followed by 100 μL of Folin-Ciocalteu (1/10; v/v). After 4 min of reaction, 80 μL of Na_2CO_3 (7.5%, w/v) was mixed. After 2 h of reaction in the dark at room temperature, the results were analyzed in a microplate reader (Bioteck, SynergyH1, Thermo-Scientific, Winooski, USA) at 750nm. Gallic acid was used as standard for the curve of calibration line using the concentration between 0.05 and 0.25 mg/mL ($y = 6.2845x + 0.0696$; $R^2 = 0.9948$). The obtained results were expressed as mg gallic acid equivalents per 100 g of macroalgae (mg GAE/100 g).

2.3.2 Determination of Antioxidant Capacity Expressed by the DPPH Radical Scavenging Activity

DPPH radical scavenging activity was determined according to the modified method of Brand-Williams (1995) [7]. Briefly, in a 96-well microplate, 50 μL of the sample was added to 150 μL of previously prepared DPPH solution with 80% of methanol. The reaction was left to stand for 30 min in the dark at room temperature, and then the absorbance was measured at wavelength of 517 nm in a microplate reader (Bioteck, SynergyH1, Thermo-Scientific, Winooski, USA). Finally, the antioxidant capacity of the samples was calculated using the following equation:

$$Antioxidant\ capacity\ (RSA, \%) = \frac{Abs_{DPPH\ initial} - Abs_{sample}}{Abs_{DPPHinitial}} \times 100 \quad (1)$$

2.3.3 Determination of Antioxidant Capacity Expressed by the Ferric Ion Reducing Antioxidant Power (FRAP)

The ferric ion reducing antioxidant power (FRAP) was performed according to the modified method described by [8]. This method is based on the reduction of Fe (III) to Fe (II) by electron transfer that occurs in the presence of antioxidants. The FRAP reagent was prepared according to the mixture of solutions 2,4,6-Tris(2-pyridyl)-s-triazine (TPTZ) making up with hydrochloric acid, the ferric solution using $FeCl_3$ and an acetate buffer solution. A mixture of 2.7 mL of FRAP, 90 μL of infusion extract and 270 μL of H_2O were performed. Then, the mixture was heated at temperature of 37 °C for 30 min. In the 96-well microplate, 200 μL of mixture/reaction was added to each well and the absorbance was measured at 593nm. Standard curve was performed using the $FeSO_4$ solution (75–1500 μM) (y = 0.0007x – 0.0107; R^2 = 0,9989), and results were expressed as mM $FeSO_4$/g.

2.4 *Statistical Analysis*

The analysis of variance (ANOVA) of data was carried out using the Statistica™v.8.0 Software from Statsoft [9]. The obtained results were considered statistically significant at a level of significance of 5% (p-value < 0.05), following the Tukey HSD (Honestly Significant Difference) test.

3 Results and Discussion

3.1 *Impact of Infusion on Total Phenolic Content and Antioxidant Capacity of Macroalgae*

The effect of infusion performed at different conditions (treatment time and proportion of macroalgae:water) on total phenolic content (TPC) of *Fucus vesiculosus* and *Ulva* sp. can be observed at Fig. 1.

Comparing the phenolic content of extracts without infusion (CTR) and after infusion (INF), an increase of this compounds was observed in brown and green macroalgae after the first 5 min of treatment. Regarding the phenolic content of *Fucus vesiculosus* extracts, the infusion ratio of 1:50 (w:v) and the treatment time of 15 min conducted to high value compared to ratio of 1:16 (w:v) and infusion time of 5 min. On the contrary, the infusion of *Ulva* sp. at proportion of 1:16 (w:v) after 5 and 15 min did not contribute to an efficient extraction of phenolic content, since no significant difference was denoted in all infusion extracts. Also, when the ratio increase (1:50, w:v), it was not possible to observe a positive results (TPC <

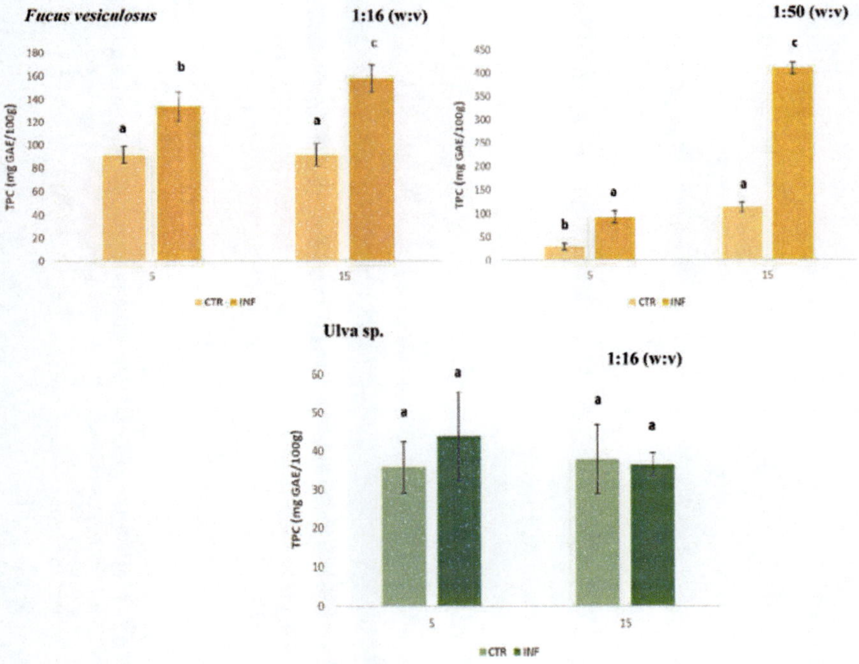

Fig. 1 Total phenolic content (TPC, mg GAE/100 g) of *Ulva* sp. and *Fucus vesiculosus* extracts without (CTR) and with infusion (INF) (*Fucus vesiculosus at 1: 16 (w:v) and 1:50 (w:v)* and *Ulva* sp. at 1:16 *(w:v)*). The vertical lines represent the standard deviation and different letters indicates significant differences at *P*-value < 0.05, Tukey test

0 mg GAE) of phenolic content in the obtained *Ulva* sp. extracts. This could be due several factors as the low content of these compounds in green macroalgae, the low extraction efficacy by using the water as solvent and heat as extraction method [10, 11]. So, in the future another assay will be realized to validate the effect of infusion in *Ulva* sp. Comparing the phenolic content resulted in both macroalgae after infusion, it is evident that this is influenced by the macroalgae type which demonstrates the importance of their choice to the success of infusion as extraction process.

The antioxidant capacity expressed by DPPH radical scavenging activity of the infusion extracts of *Ulva sp.* and *Fucus vesiculosus* after treatment at different period time and proportion of macroalgae:water ratio (1:16 and 1:50 (w:v)) is shown at Fig. 2.

The antioxidant capacity of *Ulva* sp. extracts without and with infusion was similar in ratio of 1:50 (w:v) ($P > 0.05$, Tukey test) ranged from 54.93% ± 5.60 to 57.16% ± 6.45, in CTR-15min and CTR-5min, respectively. Despite the low value observed in macroalgae extracts of *Fucus vesiculosus*, the impact of infusion as extraction method was better after treatment with ratio of 1:50 (w:v) and 15 min, with a significant increase ($P < 0.05$, Tukey test) of 65%. By observation of antioxidant capacity of *Ulva* sp. extract, no significant difference ($P > 0.05$, Tukey test) was denoted between the treatment conditions, as ratio of macroalgae:water and infusion time. The higher

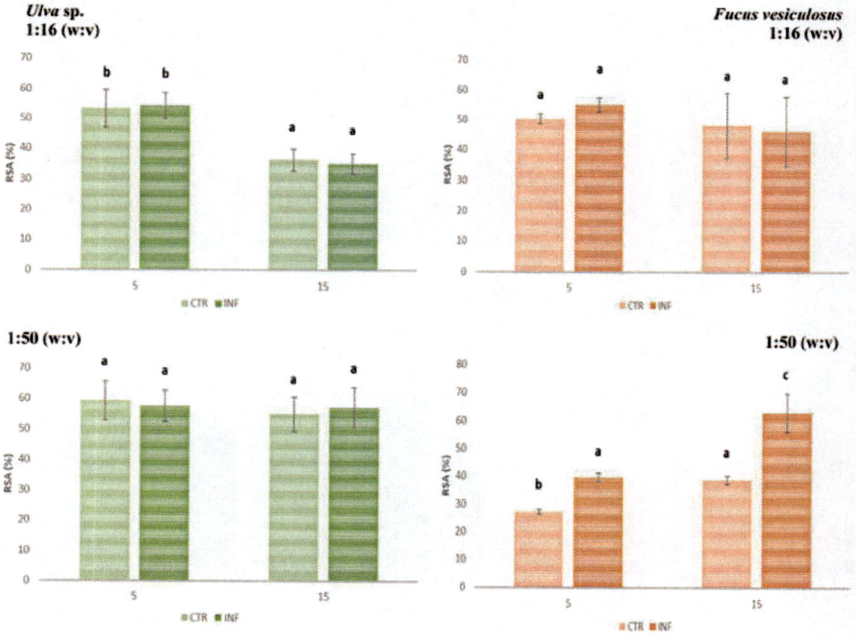

Fig. 2 Antioxidant capacity expressed by DPPH radical scavenging activity (RSA, %) of *Ulva* sp. and *Fucus vesiculosus* extracts without (CTR) and with infusion (INF). The vertical lines represent the standard deviation and different letters indicates significant differences at *P*-value < 0.05, Tukey test

value of DPPH radical reducing capacity on *Ulva* extracts was observed after 5 min of infusion, where value was high than 50 %, which was not observed for the infusion time of 15 min with values lower than 50 %. This indicates that over the infusion time the DPPH radical scavenging capacity decreases. The antioxidant capacity of *Fucus vesiculosus* extracts was demonstrated higher than 50% of capacity for a treatment time of 15 min, the highest value recorded in this study. Comparing the ratio of 1:50 (w:v) with the ratio of 1:16 (w:v), lower values for a time of 5 min and ratio of 1:50 (w:v), and a higher capacity for a time of 15 min, was found. Looking at both macroalgae extracts and considering the high values, we found a better capacity of reducing the DPPH radical for a ratio of 1:50 (w:v), in the case of *Fucus vesiculosus* specifically for treatment time of 15 min. Our study is in the line with previous reports that highlighted the antioxidant richness of macroalgae [12]

The assessment of antioxidant capacity expressed by ferric ion reducing antioxidant power (FRAP) of the studied infusion extracts of macroalgae can be observed at Fig. 3.

A similar behavior of antioxidant capacity by DPPH and FRAP methodology was observed in all macroalgae extracts. In green macroalgae extracts, no significant differences ($P > 0.05$, Tukey test) between the studied ratio was registered. In the

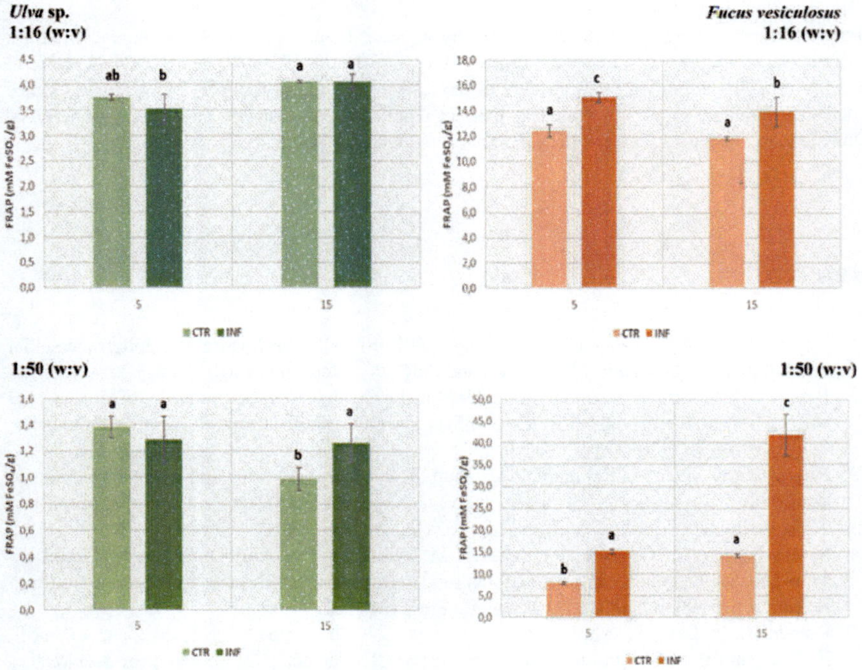

Fig. 3 Antioxidant capacity expressed by ferric ion reducing antioxidant power (FRAP, mM $FeSO_4$/g) of *Ulva* sp. and *Fucus vesiculosus* extracts without (CTR) and with infusion (INF). The vertical lines represent the standard deviation and different letters indicates significant differences at *P*-value < 0.05, Tukey test

brown macroalgae extracts an increase of antioxidant capacity from 14.10 to 41.77 mM $FeSO_4$/g was detected after the infusion during 15 min at ratio of 1:50 (w:v). This fact indicates that *Fucus vesiculosus* demonstrates an interesting content of antioxidant capacity when the applied infusion treatment was the ratio of macroalgae:water of 1:50 (w:v). Through the FRAP methodology it is possible to denote elevated values in both macroalgae extracts than verified in the literature. This demonstrates the richness potential of antioxidant capacity of both infusion extracts [13].

4 Conclusions

Nowadays, consumers have a growing preference for natural products with health benefits leading to alternative and innovative food source research for achieving these goals. In this study, the brown macroalgae *Fucus vesiculosus* showed the potential, after the infusion treatment, for obtaining an interesting and potential ingredient to be included in the food products for a healthy diet and consequent well benefits. Nonetheless, future studies will be needed to investigate the effect of the infusion macroalgae as natural and healthy ingredients incorporated in food product.

Acknowledgements This work was funded by national funds through FCT - Fundação para a Ciência e a Tecnologia, I.P., under the project MARE (UIDB/04292/2020 and UIDP/04292/2020) and the project LA/P/0069/2020 granted to the Associate Laboratory ARNET. Mariana Gomes thanks the financial support of Fundação para a Ciência e Tecnologia (FCT) with collaboration of Direção Geral do Ensino Superior (DGES), by "Verão com Ciência 2022".

References

1. Dominguez, L. J., di Bella, G., Veronese, N., & Barbagallo, M. Impact of mediterranean diet on chronic non-communicable diseases and longevity. Nutrients, 13(6), (2021).
2. Bayomy, H. M. Effects of culinary treatments on the physicochemical properties of Ulva lactuca collected from Tabuk coast of Red sea in Saudi Arabia. Saudi Journal of Biological Sciences, 29(4), 2355–2362 (2022).
3. Pangestuti, R., Haq, M., Rahmadi, P., & Chun, B. S. Nutritional value and biofunctionalities of two edible green seaweeds (Ulva lactuca and caulerpa racemosa) from indonesia by subcritical water hydrolysis. Marine Drugs, 19(10) (2021).
4. Santos-Buelga, C., González-Paramás, A. M., Oludemi, T., Ayuda-Durán, B., & González-Manzano, S. Plant phenolics as functional food ingredients. In Advances in Food and Nutrition Research, Vol. 90, pp. 183–257. Academic Press Inc (2019).
5. Roleda, M. Y., Lage, S., Aluwini, D. F., Rebours, C., Brurberg, M. B., Nitschke, U., & Gentili, F. G. Chemical profiling of the Arctic sea lettuce Ulva lactuca (Chlorophyta) mass-cultivated on land under controlled conditions for food applications. Food Chemistry, 341 (2021).
6. Catarino, M. D., Silva, A. M. S., & Cardoso, S. M. Phycochemical constituents and biological activities of Fucus spp. In Marine Drugs, Vol. 16, Issue 8, MDPI AG (2018).

7. Pinheiro, J.; Rodrigues, S.; Mendes, S.; Maranhão, P.; Ganhão, R. Impact of Aqueous Extract of Arbutus unedo Fruits on Limpets (Patella spp.) Pâté during Storage: Proximate Composition, Physicochemical Quality, Oxidative Stability, and Microbial Development. Foods, 9, 807 (2020). https://doi.org/10.3390/foods9060807.
8. Ganhão, R.; Pinheiro, J.; Tino, C.; Faria, H.; Gil, M.M. Characterization of Nutritional, Physicochemical, and Phytochemical Composition and Antioxidant Capacity of Three Strawberry "Fragaria × ananassa Duch." Cultivars ("Primoris", "Endurance", and "Portola") from Western Region of Portugal. Foods, 8, 682. (2019) https://doi.org/10.3390/foods8120682.
9. StatSoft Inc. STATISTICA (data analysis software system), version 8; StatSoft Inc.: Tulsa, OK, USA, 2007.
10. Generalić Mekinić, I.; Skroza, D.; Šimat, V.; Hamed, I.; Čagalj, M.; Popović Perković, Z. Phenolic Content of Brown Algae (Pheophyceae) Species: Extraction, Identification, and Quantification. Biomolecules, 9, 244 (2019). https://doi.org/https://doi.org/10.3390/biom9060244.
11. Pappou, S.; Dardavila, M.M.; Savvidou, M.G.; Louli, V.; Magoulas, K.; Voutsas, E. Extraction of Bioactive Compounds from Ulva lactuca. Appl. Sci., 12, 2117 (2022). https://doi.org/https://doi.org/10.3390/app12042117.
12. Corsetto, P.A.; Montorfano, G.; Zava, S.; Colombo, I.; Ingadottir, B.; Jonsdottir, R.; Sveinsdottir, K.; Rizzo, A.M. Characterization of Antioxidant Potential of Seaweed Extracts for Enrichment of Convenience Food. Antioxidants, 9, 249 (2020). https://doi.org/https://doi.org/10.3390/antiox9030249.
13. Silva, A. M., Pinto, D., Fernandes, I., Gonçalves Albuquerque, T., Costa, H. S., Freitas, V., Rodrigues, F., & Oliveira, M. B. P. P. Infusions and decoctions of dehydrated fruits of Actinidia arguta and Actinidia deliciosa: Bioactivity, radical scavenging activity and effects on cells viability. Food Chemistry, *289*, 625–634 (2019).

Extraction of *Moringa oleifera* Leaves to Obtain Antioxidants and Vitamins

Olívia Gomes, Anabela Leitão, Hermínio C. de Sousa, Mara Braga, and Licínio Gando-Ferreira

Abstract Moringa leaves have become popular as a dietary supplement and herbal medicine. Therefore, it would be of nutritional benefit to determine the appropriate extraction method and solvent that would promote the extraction with high biological activity and high yield of active compounds to produce nutraceuticals. In this study, an evaluation of two methods of extracting bioactive compounds from *Moringa oleifera* leaves was proposed. In this evaluation, some parameters were compared, such as the yield, the content of extracted phenolic compounds, the antioxidant activity and the quantity of vitamins extracted and preserved in the extracts. Three types of solvents: water, ethanol (EtOH) and ethyl acetate (Act) were used at two different temperatures and different extraction times. The average extraction yield in the hot bath shows an increase of 13.2% when compared to vacuum pressure bath extraction. In terms of phenolic compound content and antioxidant activity, the average increase was 4.9% and 4.3%, respectively. Vitamins were detected in the samples and, due to their polarity, they were mostly found in the solvent with the presence of ethanol (50% and 100% of EtOH). The solvent with the presence of ethyl acetate (50% and 100% of Act) showed low yield and low solubility with the compounds of interest in the extractions, such as phenolic compounds and vitamins. The positive effect of low pressure on the extraction of fat-soluble vitamins was observed.

Keywords *Moringa oleifera* leaves · Extraction · Phenolic compounds · Antioxidant activity · Vitamins

O. Gomes (✉) · H. C. de Sousa · M. Braga · L. Gando-Ferreira
Department of Chemical Engineering, Faculty of Sciences and Technology, University of Coimbra, CIEPQPF, Rua Sílvio Lima, 3030-790 Coimbra, Pólo II, Portugal
e-mail: olivia@eq.uc.pt; lferreira@eq.uc.pt

A. Leitão
Faculty of Engineering, LESRA—Laboratory of Separation, Reaction and Environmental Engineering, Agostinho Neto University, Av. Ho Chi Min No 201, Luanda, Angola

J. R. da. C. S. Galvão et al. (eds.), *Proceedings of the 3rd International Conference on Water Energy Food and Sustainability (ICoWEFS 2023)*, Springer Proceedings in Earth and Environmental Sciences, https://doi.org/10.1007/978-3-031-48532-9_12

1 Introduction

In recent years, moderate and severe hunger has affected about two billion people worldwide [1], which drastically impacts the high rates of malnutrition observed in the world. About 21%, 7% and 6% of children worldwide are stunted, wasted and overweight, respectively. Furthermore, the 2020 global nutrition report shows that one in nine and one in three people are hungry or undernourished and overweight or obese, respectively [2].

The medicinal characteristics of distinct plant matrices are credited mainly to the presence of secondary metabolites, which are used in pharmaceutical or nutraceutical articles as food complements [3, 4]. Recently, *Moringa oleifera* leaves have become popular as a nutritional and herbal supplement. Its leaves are edible and rich in protein, with a well-balanced essential amino acid profile. At the same time, it contains high levels of vitamins A and C, as well as powerful natural antioxidants [5]. Moringa has long been used as a nutritional supplement to reduce malnutrition [3].

To obtain the compounds present in Moringa leaves, extraction methods must be optimized that make it possible to control the properties of the final extract. Solid–liquid extraction methods are generally used for the extraction of bioactive compounds and, specifically, for the extraction of phenolic compounds from different plant matrices [6, 7]. There are reports of conventional methods used to get these compounds, such as Soxhlet, hot bath with agitation or maceration. This step is very important because these compounds can be chemically altered or degraded during the extraction process. Theoretically, the ideal method should be simple, fast, economical and applicable to a large industry [8].

In this study, an evaluation of two methods of extracting bioactive compounds from *Moringa oleifera* leaves was proposed. In this evaluation, some parameters were compared, such as the yield, the content of extracted phenolic compounds, the antioxidant activity and the amount of vitamins extracted and preserved in the extracts. Three types of solvents (water, ethanol and ethyl acetate) were used at two different temperatures and different extraction times.

2 Materials and Methods

2.1 Plant Material

Small branches were harvested from three *Moringa oleifera* trees in Luanda, Angola (8°57′24.9″S, 13°13′02.9″E), with fresh mature leaves of three different colors from dark green to green yellowish. Upon arrival at the laboratory, the branches were washed with running tap water, the leaves removed and separated into the three colors (Fig. 1). The drying of the leaves spread out on trays began at room temperature (about 30 °C) in a well-ventilated place and ended in an oven at 50 °C until constant weight. After this drying process, which lasted about two weeks, the leaves were

Fig. 1 Image of leaves harvested for characterization analysis

crispy and ready for grinding. The grinding was done in a coffee grinder, using small amounts of leaves and the same grinding speed to obtain a fine powder with the same granulometry. Powdered samples were stored in sealed bags kept away from light and humidity.

2.2 *Extraction of Bioactive Compounds from* Moringa Oleifera *Leaf Powder*

Hot bath and vacuum pressure bath extraction methods were chosen to perform the extractions to compare different effects on the extracts. These are simple, low cost and fast processes.

For hot bath extraction, a thermostatic stirred water bath of 18L VWR (1.05 kW, 60 Hz) (United States) was used with agitation at 100 rpm and temperature controlled at 60 °C using a 200 ml glass vials for the extractions. The extractions were performed in two different times: 3 and 6 h. On the second extraction method, a rotary evaporator (Büchi Rotavapor R-210) (United States) was used, with stirring at 50 rpm and temperature controlled at 50 °C. A pressure of 0.5 mbar was applied. The test was performed in 1 h and 2 h. The solid:solvent ratio used for both extraction methods was 1:200 (g/mL).

Once completed the extractions, the extract was centrifuged at 5.000 rpm for 15 min. The supernatant was added to a volumetric flask and then, the organic solvents were removed in a rotary evaporator and the aqueous extract was subjected to lyophilization for 48 h. Different solvents mixtures were used to evaluate the effect of liquid phase polarity on the extraction of different bioactive compounds. These solvent and their mixtures are:

- 100% distilled water (100% H_2O).
- 50% ethanol/distilled water (50% $EtOH/H_2O$).

- 100% ethanol (100% EtOH).
- 50% ethyl acetate/distilled water (50% Act/H_2O).
- 100% ethyl acetate (100% Act).

The extracts were solubilized in solvents used on the extraction and analyzed in terms of phenolic compounds, antioxidant activity and vitamins on HPLC.

2.3 *Characterization of* Moringa Oleifera *Leaf Extracts*

2.3.1 Total Phenolic Content

For the analysis of the total phenolic content, the method by Sengev [9] modified for use in microplates was used. The mixture for the reaction is composed of 100 μL of sample (distilled water for the blank), 100 μL of Folin-Ciocalteu 1N solution and 800 μL of Sodium Carbonate (Na_2CO_3) at 75 g/L. After an incubation period of 30 min heated to 40 °C, the absorbance was read at 750 nm in a spectrophotometer. The analysis was performed at least in duplicate. The result was calculated from a standard curve of gallic acid and expressed in equivalents of gallic acid, mg EGA/L.

2.3.2 Antioxidant Capacity (DPPH)

Initially, for evaluating the DPPH of the samples, 2.5 ml of sample was diluted in ethanol (or at most a solution with 50% water). Subsequently, except for the blank tests, 1 ml of an alcoholic solution of 2,2-diphenyl-1-picrylhydrazyl (DPPH) (0.3 mM) prepared daily and protected from light was added. In blank tests, 1 ml of solvent is added to the 2.5 ml of sample dilution to prepare the solutions. After 30 min in the dark, the absorbance was measured at 517 nm. The percentage of antioxidant activity (AA, %) is presented in relative proportion. For each extract, the extract solution concentration curve was plotted as a function of absorbance. With the interpolation equation, IC_{50} values were calculated, defined as the concentration of the extract that shows 50% of the antioxidant activity through the DPPH method.

2.3.3 Vitamins

Chromatographic analysis of water-soluble vitamins (vitamin C and complex B- B_1, B_2 and B_9) was performed following the method described by Seal et al. [10] with minor modifications. The extracts were analyzed at a concentration of approximately 5 mg/ml in a 10% ethanol in water solution. The C18 column (5 μm) Phenomenex (4.6 × 250 mm) was used. The mobile phase contains acetonitrile (Solvent A) and aqueous trifluoroacetic acid (TFA, 0.01% v/v) (Solvent B). The elution gradient was 1% A and 99% B with a flow rate of 0.5 ml/min over 5 min, from 1 to 25%. HPLC chromatograms of all vitamins were detected using a photodiode array UV detector at four different wavelengths (210, 245, 275 and 290 nm) according to the absorption maxima of the analyzed compounds.

For the chromatographic analysis of fat-soluble vitamins (vitamin A and E), extracts were performed following the method described by Garai [11]. Using reverse phase C18 column (5 μm), 95:5 methanol:water as mobile phase and flow rate 1 ml/min. The total analysis time per sample was 25 min.

The injection volume was 20 μl for all samples. The quantification of the vitamins content in the samples was done by the calibration curve.

3 Results and Discussions

The extraction yields, total phenolic content and antioxidant activity in the different extracts of *Moringa oleifera* leaves are also shown in Fig. 2. The extraction yields varied between 4.96 and 41.22% for hot bath and 3.89–35.64% in vacuum pressure bath. The hot bath extraction method (Fig. 2a) gave the highest average vacuum pressure extraction yield (Fig. 2b). It was due to the higher temperature (60 °C) that was used during the extraction process. Cell membranes were easily ruptured and untargeted substances such as carbohydrates were also released into the solution, which was able to increase the yield of the aqueous extract.

Regarding the yield of 100% ethyl acetate and 50% solution with water extractions were significantly lower than the other solvents, which was also confirmed by the HPLC analysis. Usually, ethanol would be blended with leaves to infuse and dissolve phenolic compounds based on the "like dissolves like" concept [12], with the help of vacuum pressure, low pressure has great effect on cell membranes and forced phenolic compounds are easily released [12, 13]. In addition, the partial removal of the air inside the flask indicated that the negative pressure in atmosphere had a positive effect on the preservation of more sensitive compounds. Due to the wide range of polarity, ethanol extracts compounds of high or low polarity that represent a dark green sticky substance in the final product. On the other hand, the aqueous one required a shorter extraction time because it allowed the samples to be boiled for several hours to break the cell membranes and then the phenolic compounds were released into the water.

With the results obtained in Fig. 2, the average extraction yield in the hot bath shows an increase of 13.18% when compared to vacuum pressure bath extraction. In terms of phenolic compound content and antioxidant activity, the average increase was 4.90% and 4.28%, respectively.

A yield greater than 35% was obtained by the extraction process carried out in a rotary evaporator at 50 °C. The growth can also be observed in the values obtained for the phenolic compounds (on average, 13.31%). On the other hand, the mean IC_{50} value of antioxidant activity decreased by 1.58%. This demonstrates that phenolic

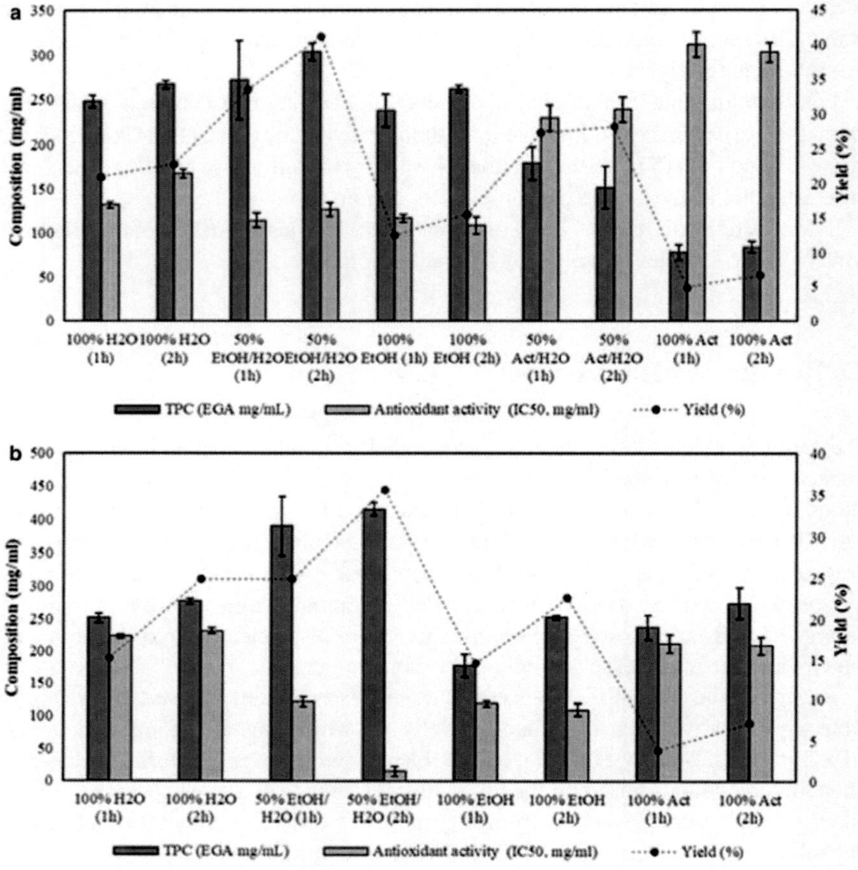

Fig. 2 Extraction yield (% dry mass), total phenolic content (mg/g GAE dry weight), antioxidant activity (IC_{50}, mg/ml) from hot bath (**a**) and from vacuum pressure bath (**b**) extraction with *Moringa oleifera* leaf

compounds and antioxidants that can be degraded in oxidative environments (presence of oxygen) were preserved with the decrease in pressure inside the extraction container.

Table 1 shows the concentrations of vitamins A and E in the hot bath and vacuum pressure bath extractions. In the vacuum pressure bath, with a pressure of 0.5 mbar, concentrations of vitamin A were obtained in a mixture of water and ethanol, pure ethanol and pure ethyl acetate. Vitamin E was not found in pure water or in mixtures with water, mainly because it is not soluble in aqueous solutions. It is also possible to observe the effect of the low oxidative atmosphere under the extraction in a vacuum pressure bath. It was observed that vitamin A appears in greater quantity when extraction is performed using this method.

Table 1 Identification and quantification of vitamins A and E in Moringa leaf powder

Extraction method	Time (h)	Vitamins (mg/ml)	Solvent				
			H_2O 100%	EtOH/H_2O 50%	EtOH 100%	Act/ H_2O 50%	Act 100%
Hot bath	3	A	0	0	0.32 ± 0.02	0	0
		E	0	0	7.32 ± 0.32	0	5.56 ± 0.13
	6	A	0	0	0	0	0
		E	0	0	2.89 ± 0.09	0	1.31 ± 0.23
Vacuum pressure bath	1	A	0	0	1.77 ± 0.06	0	2.27 ± 0.83
		E	0	0	0.89 ± 0.11	0	0
	2	A	0	291.57 ± 1.23	23.83 ± 1.18	0	33.81 ± 2.46
		E	0	0	3.01 ± 0.21	0	1.43 ± 0.31

The quantification of vitamin C and some of the B complex in the extracts led to the concentration values shown in Table 2. It is possible to notice that vitamin C is a very temperature sensitive molecule and is degraded during its extraction. Since ethanol is highly polar, it is likely that it has a strong connection to the molecules of the B complex vitamins analyzed here [11]. For this reason, it was possible to identify the molecules in solvents with the presence of water, especially in the mixture with ethanol.

Table 2 also shows the quantification of vitamin C and some of the B complex in the extracts obtain in a rotary evaporator, with a pressure of 0.5 mbar. As with the extractions carried out in the vacuum pressure bath extraction, it was not possible to detect the presence of vitamin C, except in the one containing distillate water as a solvent. In these tests, it was possible to perceive that the vitamins again tend to solubilize in solvents with the presence of ethanol.

4 Conclusions

The extraction yields, total phenolic content and antioxidant activity in the different extracts of *Moringa oleifera* leaves were obtained. The average extraction yield in the hot bath shows an increase of 13.18% in total dry mass, 4.90% in terms of phenolic compound content and 4.28% in terms of antioxidant activity, when compared to vacuum pressure bath extraction. Vitamins were detected in the samples and, due to their polarity, they were mostly found in the solvent with the presence of ethanol (50% and 100% EtOH). The solvent with the presence of ethyl acetate (50% and 100% Act) showed low yield and low solubility with the compounds of interest in the extractions, such as phenolic compounds and vitamins. The positive effect of low pressure on the extraction of fat-soluble vitamins was observed.

Table 2 Identification and quantification of vitamins C and some of complex B in Moringa leaf powder

Extraction method	Time (h)	Vitamins (mg/ml)	Solvent				
			H_2O 100%	EtOH/H_2O 50%	EtOH 100%	Act/H_2O 50%	Act 100%
Hot bath	3	C	0	0	0	0	0
		B_1	3.20 ± 0.12	7.20 ± 0,83	1.25 ± 0.08	0	0
		B_2	3.82 ± 0.13	4.05 ± 0.08	0.47 ± 0.02	0	0
		B_9	13.22 ± 0.21	71.30 ± 1.47	0	0	0
	6	C	0	0	0	0	0
		B_1	7.11 ± 0.41	8.59 ± 0.82	6.46 ± 0.61	3.37 ± 1,21	0
		B_2	11.14 ± 2.04	1.43 ± 0.31	2.89 ± 0.38	0.89 ± 0,11	0
		B_9	23.43 ± 2.10	4.62 ± 0.82	21.33 ± 1.33	8.59 ± 0,82	0
Vacuum pressure bath	1	C	0	0	0	0	0
		B_1	4.51 ± 0.38	1.43 ± 0.31	8.59 ± 0.82	0	0
		B_2	83.63 ± 3.67	13.36 ± 1.01	6.27 ± 0.83	0.83 ± 0,02	0
		B_9	143.59 ± 10.31	7.20 ± 0.38	2.80 ± 0.27	0	0
	2	C	7.90 ± 1.24	0	0	0	0
		B_1	38.60 ± 1.45	24.29 ± 4.12	3.20 ± 0.12	5.61 ± 0.87	1.61 ± 0.34
		B_2	101.42 ± 3.64	65.72 ± 5.24	6.46 ± 0.61	1.63 ± 0.68	2.39 ± 0.13
		B_9	63.22 ± 2.10	0	8.10 ± 0.87	0	3.03 ± 0.24

Acknowledgements This work was carried out under the Project MORfood (Ref^{a}. 541163254/2019) – *Microencapsulation of Moringa oleifera extracts and their application in functional foods to mitigate children malnutrition in developing countries*, funded by Foundation for Science and Technology (FCT) and Aga Khan Development Network (AKDN), and the Strategic Projects FCT-MEC of The Chemical Process Engineering and Forest Products Research Centre (CIEPQPF): UIDB/04539/2020 and UIDP /04539/2020.

References

1. FAO; IFAD; UNICEF; WHO. The State of Food Security and Nutrition in the World 2020. Transforming Food
2. Systems for Affordable Healthy Diets; FAO: Rome, Italy, 2020.
3. DevelopmentInitiatives. 2020 Global Nutrition Report: Action on Equity to End Malnutrition; Development Initiatives: Bristol, UK, 2020.
4. Zhao, J.; Davis, L.C.; Verpoorte, R. Elicitor signal transduction leading to production of plant secondary metabolites. Biotechnol. Adv. 2005, 23, 283–333.
5. Olson, M.E.; Fahey, J.W.; Jed,W. Moringa oleifera: A multipurpose tree for the dry tropics. Rev. Mex. Biodiv. 2011, 82, 1071–1082.
6. González de Peredo, A.V.; Vázquez-Espinosa, M.; Espada-Bellido, E.; Jiménez-Cantizano, A.; Ferreiro-González, M.; Amores-Arrocha, A.; Barbero, G.F. Development of New Analytical Microwave-Assisted Extraction Methods for Bioactive Compounds from Myrtle (Myrtus communis L.). Molecules 2018, 23, 2992.
7. Vázquez-Espinosa, M.; González-de-Peredo, A.V.; Espada-Bellido, E.; Ferreiro-González, M.; Toledo-Domínguez, J.J.; Carrera, C.; Barbero, G.F. Ultrasound-Assisted Extraction of Two Types of Antioxidant Compounds (TPC and TA) from Black Chokeberry (Aronia melanocarpa L.): Optimization of the Individual and Simultaneous Extraction Methods. Agronomy 2019, 9, 456.
8. Fombang, E.N.; Saa, W.R. Production of a Functional Tea from Moringa oleifera LAM Leaf Powder: Optimization of Phenolic Extraction Using Response Surface Methodology. J. Nutr. Food Sci. 2016, 6, 556.
9. Sengev, A.I.; Abu, J.O.; Gernah, D.I. Effect of Moringa oleifera Leaf Powder Supplementation on Some Quality Characteristics of Wheat Bread. Food Nutr. Sci. 2013, 4, 270.
10. Seal T, Chaudhuri K, Pillai B. A rapid high performance liquid chromatography (HPLC) method for the simultaneous estimation of water-soluble vitamin in ten wild edible plants consumed by the tribal people of North-Eastern Region in India. Pharmacognosy Magazine 2018. 14:72-77.
11. Garai L. Improving HPLC Analysis of Vitamin A and E: Use of Statistical Experimental Design. Procedia Computer Science 108C (2017) 1500–1511.
12. Chinma, C.E.; Abu, J.O.; Akoma, S.N. Effect of Germinated Tigernut and Moringa Flour Blends on the Quality of Wheat-based Bread. J. Food Process. Preserv. 2014, 38, 721–727.
13. Páramo-Calderón, D.E.; Aparicio-Saguilán, A.; Aguirre-Cruz, A.; Carrillo-Ahumada, J.; Hernández-Uribe, J.P.; Acevedo-Tello, S.; Torruco-Uco, J.G. Tortilla Added with Moringa Oleifera Flour: Physicochemical, Texture Properties and Antioxidant Capacity. LWT 2019, 100, 409–415.

Effects of Autohydrolysis on the Phenolic Extraction and Antioxidant Capacity of *Camellia* sp. Seed Husk Extracts

R. M. Ferreira, B. Piñeiro-Lago, J. A. Saraiva, S. M. Cardoso, M. D. Torres, and H. Domínguez

Abstract The *Camellia* genus stands out as one of the most prevalent plant groups across Asia. Its exquisite flowers, valued as adornments, and its role in tea production have led to a surge in cultivating plants from this genus, notably focusing on *Camellia sinensis*, in recent years. However, this escalated cultivation has inadvertently led to a rise in by-product generation, potentially exerting detrimental effects on the environment. To mitigate this, strategies for waste reduction are imperative. Thus, it becomes essential to embrace novel principles of bio-refinery and circular bioeconomy, prioritizing the utilization of natural resources like eco-friendly solvents and residual biomass. This study homes in on the husk of *Camellia sp.* seeds, selected for its abundant phytochemical content and potent antioxidant properties. The study characterized the fundamental composition of the husk used as the primary material and through hydrothermal extraction (autohydrolysis), a sustainable approach applying subcritical water conditions, the husk samples were processed for the solubilization of different compounds. Towards understanding the technique's impact, varying solid-to-liquid ratios—1:8 (w:v) and 1:30 (w:v)—were employed, while maintaining a constant temperature of 160 °C. The resulting extracts underwent different analyses via distinct spectrophotometric methodologies. Total phenolic compound content (TPC) is assessed using the Folin-Ciocalteu method, while TEAC, DPPH, and FRAP assays gauge extract antioxidant capacities. The findings emphasized that elevating the solid-to-liquid ratio from 1:8 to 1:30 (w:v) correlates with heightened phenolic compounds and enhanced antioxidant activity.

R. M. Ferreira · J. A. Saraiva · S. M. Cardoso
LAQV-REQUIMTE, Department of Chemistry, Campus Universitário de Santiago, University of Aveiro, 3810-193 Aveiro, Portugal

B. Piñeiro-Lago
Department of Innovation, B. PIÑEIRO-LAGO, Pazo Quinteiro da Cruz. A Cruz 88-Lois, 36635 Ribadumia, Pontevedra, Spain

M. D. Torres (✉) · H. Domínguez
Department of Chemical Engineering, Edificio Politécnico, CINBIO, Universidade de Vigo (Campus Ourense), As Lagoas, 32004 Ourense, Spain
e-mail: matorres@uvigo.gal

J. R. da. C. S. Galvão et al. (eds.), *Proceedings of the 3rd International Conference on Water Energy Food and Sustainability (ICoWEFS 2023)*, Springer Proceedings in Earth and Environmental Sciences, https://doi.org/10.1007/978-3-031-48532-9_13

Keywords *Camellia* sp. · Autohydrolysis · Polyphenols · Antioxidant

1 Introduction

The *Camellia* genus of the family *Theaceae* is the most traditional and famous plants in Asia, being found in eastern and southern Asia, from the Himalayas to Japan and Indonesia [1]. Different species and cultivars are prevalent in many countries across the world, including Spain and Portugal. Some species belonging to the *Camellia* genus have a high economic and commercial value due to their subproducts, which include, tea, seed oils, floral extracts, flowering shrubs, and ornamental plants [2]. The high abundance and economic value have boosted interest in the biological properties of this plant genus and several studies were undertaken to evaluate its antioxidant, antimicrobial, and antitumor activity. Over the years a huge amount of work was published demonstrating that these species are an enormous source of phenolic and antioxidant compounds with proven biological activity against bacteria, fungi, and virus as well as antitumor potential [3–5]. Most of the studies were focused on the *Camellia* genus of three specific species; namely, *C. sinensis* (the tea plant), *C. oleifera* (used to obtain edible oil), and *C. japonica* (ornamental flowers), however, there are among the other species from this genus many others with biological properties.

The growing plantation of these species has spurred interest in studying their subproducts that otherwise would be discarded. The *Camellia sp.* fruit shell is an agricultural waste from farming and an industrial by-product of seed oil processing, usually destined for composting, burning, and for the production of fertilizers [6, 7]. In some cases, such as for *Camelia oleifera*, the shells account for 54% of the underutilized plant components, therefore it is attracting increasing attention and could be proposed as a source of different active phytochemicals [8].

To manage with the great quantities of agricultural by-products and industrial waste produced by the plantation of plants of the genus *Camellia*, many countries around the world, particularly those where environmental concerns, have been, in recent years, developing and using different techniques to obtain greener products. The use of the remaining residues exerts a favorable influence on the diminution of solid remnants fostering the creation of economically viable materials imbued with substantial intrinsic worth. Furthermore, this practice contributes to the safeguarding of the environment [9]. Thus, the use of agro-industrial by-products, such as *Camellia* sp seed husk, needs to be extensively studied to know all its constituents, and to originate new products or applications that have gained more and more importance over the years.

In this context, the application of pressurized hot water using subcritical conditions operating with conventional heating (CH) and using water as the only solvent has gained importance in the last years due to the efficient extraction and production of extracts rich in many bioactive compounds [10]. The main goal of this study is the valorization of *Camellia sp.* seed husk by-products using an eco-friendly technology to produce an extract rich in phenolic and antioxidant compounds to explore different alternative valorization schemes into commercially interesting products.

2 Material and Methods

2.1 Raw Material

Camellia sp. fruit husks were collected in Pazo Quinteiro da Cruz Ribadumia (Pontevedra, Spain) in the Summer of 2022. The husks were dried, ground to a particle size under 0.5 mm, and stored in darkness at room temperature, in hermetic plastic bags until further analysis.

2.2 Extraction Method

Extracts from *Camellia* sp. Seed husk (CSH) were obtained through the process of autohydrolysis, utilizing conventional heating (CH) methodology. This procedure was executed within a 100 mL extractor (Parr Instrument Company, Moline, IL, USA). The milled CSH material was brought into contact with distilled water, in a liquid-to-solid mass ratio (LSR) (w:w) of 8:1 and 30:1, under non-isothermal conditions, up to a heating temperature of 160 °C. Then, the suspension was immediately cooled with water through a stainless-steel coil. After extraction, filtration with a vacuum pump to separate solid and liquid phases was carried out.

In order to compare the effect of the operational conditions in the equipment, minimizing the effect of the different heating–cooling profiles, a severity concept incorporating the effects of both time and temperature was used. Usually the severity, S_0, is expressed as the decimal logarithm of R_0, defined as:

$$\begin{aligned}\log R_0 &= \log(R_{0,heating} + R_{0,cooling}) \\ &= \log\left\{\left[\int_0^{t_{\max}} \exp\left(\frac{T(t) - T_{ref}}{\omega}\right)dt\right] + \left[\int_{t_{\max}}^{t_{final}} \exp\left(\frac{T\prime(t) - T_{ref}}{\omega}\right)dt\right]\right\}\end{aligned} \quad (1)$$

where R_0 is the severity factor, t_{max} (min) is the time to reach the target temperature T_{max} (°C), t_{final} (min) is the time for the heating–cooling, and T(t) and T'(t) the temperature profiles in the heating and cooling stages, respectively. ω (14.75 °C), T_{ref} (100 °C).

2.3 Analytical Techniques

The moisture content was assessed gravimetrically through oven-drying at 105 °C until a consistent weight was achieved after 24 to 48 h. For the determination of ash content, gravimetric analysis was employed following calcination in a muffle furnace at 575 °C for 6 h. Mineral content analysis was carried out via acid digestion using a Marsxpress apparatus (CEM). This involved combining 0.3 g of ash with 10 mL of nitric acid and 1 mL of hydrogen peroxide. The treatment process involved exposure to 1600 W for 15 min until reaching 200 °C, followed by a 10 min maintenance at this temperature. Protein content was computed using the standard conversion factor of 6.25 [34]. Carbohydrates within CSH were evaluated post-acid hydrolysis utilizing 72% sulfuric acid at 30 °C for 1 h and 4% sulfuric acid at 121 °C for 60 min. The analysis occurred using an HPLC system (Model 1200, Agilent Technologies) employing Aminex HPX-87H and HPX-87P columns with specific temperature and flow rate conditions. Fatty acid methyl esters (FAME) were prepared following the UNE-EN ISO 12966–3:216 protocol and subjected to GC–MS QP 2010 analysis (Shimadzu) with a prescribed temperature profile. Identification of FAMEs utilized authentic standards and the NIST MS Search 2.0 library. The average outcomes from two independent assessments were presented as a percentage of the total FAME. Lignin contents and extractives within Camellia sp. shells were determined using analytical methods developed by the National Renewable Energy Laboratory (NREL).

2.4 Total Phenolic Content

Total phenolic content was determined by the Folin–Ciocalteu method with gallic acid as the standard. Briefly, 0.5 mL of extract was mixed with 3.75 mL of water, 0.25 mL of Folin–Ciocalteu's reagent (diluted 1:1 with water), and 0.5 mL of sodium carbonate (10%, w/v). Samples were incubated for 1 h in the absence of light at room temperature before absorbance readings at 765 nm [12]. All assays were determined in triplicate.

2.5 Antioxidant Capacity

The radical scavenging against 2,2-diphenyl- 1-picrilhydrazil (DPPH) was studied by the inhibition percentage calculated as the relative decrease in absorbance of the extract (50 μL) at 515 nm after 16 min in 2 mL of methanolic solution of DPPH (3.6×10^{-5} M). The 2,2′-azinobis-(3-ethyl-benzothiazoline-6-sulfonate) (ABTS) radical scavenging capacity was expressed as Trolox equivalents. Absorbance readings at 734 nm were recorded after 6 min at 30 °C for a mixture of diluted radical solution (1.0 mL) and extract (10 μL). The diluted radical solution was composed of a mixture of $ABTS^+$ reagent (0.3840 g of ABTS and 0.0662 g of potassium persulfate) and phosphate buffer (PBS). The mixture was ready to use when an absorbance of 0.700 was reached. The reducing power assay (FRAP) was based on the reduction of Fe^{3+} to Fe^{2+}. Briefly, 1 mL of sample was placed in a test tube, 2.5 mL phosphate buffer (0.2 M and pH 6.6), and 2.5 mL of potassium ferricyanide (1%) were also added and mixed with a vortex. After incubation at 50 °C for 30 min, 2.5 mL of trichloroacetic acid (10%) was added, and then a centrifugation step for 10 min and 4500 rpm was needed. The supernatant was mixed with distilled water (1:1, v/v) and 0.5 mL of ferric chloride (0.1%, w/v). The standard curve was made with ascorbic acid, and the absorbance was read at 700 nm in an (Evolution 201, Thermo Scientific, Shanghai, China) spectrometer [12, 13]. All assays were determined in triplicate.

3 Results and Discussion

3.1 *Elemental and Proximal Composition of* Camellia sp. *Seed Husk*

Table 1 shows the elemental composition of *Camellia* seed husk, with almost 42% total carbon content, as the second major constituent. The carbon content obtained for the mix of plant seed husk is in the range of that found for different *Camellia* species such as *C. oleifera*, *C. mollissima*, and *C. cathayensis*, which possess 48.6, 38.5 and 45.4% of carbon respectively. Similarly, both nitrogen (N) and hydrogen (N) are in the range reported for these species [14].

The proximal composition of *Camellia* sp. husks has been analyzed and the results are summarized in Table 2. The ash content was found to be slightly higher than the previously reported values for *C. oleifera* seed husks, ranging from 2.2% to 3.1% [15, 16]. However, it is worth noting that different species of Camellia have been found to have varying ash content. For instance, Zhang et al. (2018) reported ash contents of 2.6% for *C. oleifera*, 3.2% for *C. mollissima*, and 6.9% for *C. cathayensis* [14]. Therefore, considering that the mix of seed shells used in this study is from a variety of Camellia species, the ash content values obtained are within the expected range.

It is important to note that ash content is a crucial parameter in evaluating the nutritional and industrial potential of *Camellia sp.* husks. This parameter provides

Table 1 Elemental composition of dry *Camellia* sp. seed husks

Element	Content (%)
N	0.7 ± 0.1
C	41.8 ± 0.2
H	5.3 ± 0.6
O	52.1 ± 0.3

Table 2 Composition of tea plant seed shell, expressed in dry basis, except for moisture

Component	Content (%, d.b.)
Moisture (%, w.b.)	11.6 ± 0.3
Ash (%)	4.4 ± 0.1
Protein (%)	4.3 ± 0.1
Lipids (%)	3.1 ± 0.1
Lignin and extractives (%)	22.7 ± 0.4
Carbohydrates (%)	65.5 ± 1.5

insights into the mineral makeup of the husks, serving as an indicator of both the soil conditions in which the plants thrived and the plants' efficiency in absorbing nutrients. Furthermore, the ash content bears relevance for industries utilizing plant biomass as raw material, particularly in sectors like biofuel production, where it can influence the material's combustion characteristics. Thus, the outcomes outlined in this study enhance the comprehension of the nutritional and industrial possibilities embedded in *Camellia sp.* husks which can have important implications for sustainable resource utilization and waste management.

The material under investigation is characterized by a low lipid content of 3%, with stearic acid and palmitic acid identified as the major components, accounting for 53 and 47% of the identified fatty acids, respectively. Stearic acid is well-known for its emollient, emulsifying, and lubricating properties, which make it a commonly used ingredient in cosmetic formulations aimed at maintaining the skin barrier. Similarly, palmitic acid is also useful in such applications due to its detergent, emulsifying, and emollient properties [17].

Moreover, the protein content of the seed shell husk is low, with a value of 4.3%. Similar results were reported by Sun et al. [18] for *Camellia oleifera Abel.* hull obtained from the National Engineering Research Center for Oil-tea *Camellia*, Hunan Academy of Forestry, China, where values ranging from 2 to 4% were obtained.

It is worth noting that the lipid and protein content of plant materials can have significant implications for their potential applications in various industries, including food, pharmaceuticals, and cosmetics. Therefore, understanding the chemical composition of such materials is crucial for identifying their potential applications and developing sustainable resource utilization strategies. The findings presented here contribute to the existing knowledge on the chemical composition of *Camellia sp.* seed shell husks and can serve as a basis for further research on their potential applications in different industries.

3.2 Polyphenolic Content and Antioxidant Activity

The effects of the different solid–liquid ratios on the TPC and antioxidant capacity of *Camellia sp.* seed husk are presented in Table 3.

Table 3 Effects of solid: liquid ratio on the extraction of total phenolic compounds (TPC) and on the antioxidant capacity given by the TEAC and DPPH antiridicalar methods as well as the ferric reducing power (FRAP) of *Camellia sp.* seed husk extracts obtained after autohydrolysis at 160 °C

Ratio solid: liquid	TPC ($g_{Gallic\ acid\ eq.}$/ $100g_{sample}$)	TEAC ($g_{Trolox\ eq.}$/ $100g_{sample}$)	DPPH EC_{50}(μg/ mL)	FRAP ($g_{BHA\ eq.}$/ $100g_{sample}$)
1:8	16.4 ± 0.2	28. 0 ± 4.5	60.4 ± 0.7	9.9 ± 0.1
1:30	16.7 ± 0.1	34.0 ± 0.1	52.3 ± 0.9	11.9 ± 0.2

The increase of the solid: liquid ratio from 1:8 to 1:30 leads to the increase in both the phenolic compounds extracted and as expected in the antioxidant capacity once part of this activity is a result of the phenolic compounds present [19, 20]. As expected, the addition of more water allowed the extraction of a higher quantity of extractives once it allows the compounds to be in contact with the solvent increasing their extraction [21].

The maximum extraction yield of phenolic obtained for 1:30 (s:L ratio) was 16.7 ± 0.1 which is around 8 times higher than the results obtained for commercially available Ethiopian tea from *Camellia sinensis* infusions [22] and 10 times higher than what was reported by Nugraheni et al. [23] for Black Tea extracts of *Camellia sinensis* L. showing the potential of this extraction method for the production of a polyphenolic rich extract.

As for the antioxidant capacity, it is noteworthy that the observed outcomes concerning the ability to neutralize $ABTS^{+\bullet}$ and $DPPH^{\bullet}$ radicals, along with iron reduction, notably surpassed the values documented in existing literature for *Camellia sinensis* L. teas [22, 23]. This enhancement in overall antioxidant capacity, subsequent to the rigorous heat treatment, might arise from the creation of novel neoantioxidants derived from Maillard, caramelization and thermoxidation influencing the overall antioxidant potential. Similar dynamics have been reported by Merichel Plaza et al. in their investigations of diverse macroalgae and plant materials post-Subcritical Water Extraction (SWE), thus underscoring the effect of extreme heat treatment on bolstering antioxidant capacity [24].

These findings provide valuable insights into the potential use of *Camellia sp.* seed shell husks as a source of natural antioxidants and highlight the importance of optimizing the extraction conditions to maximize the yield of phenolic compounds and antioxidant activity. Such studies have significant implications for developing sustainable resource utilization strategies for plant-based waste materials and improving their value-added applications in various industries.

4 Conclusions

A mix of *Camellia sp.* seed husks was used as a raw material for the autohydrolysis process to evaluate the extraction yield and to maximize the total phenolic compounds extracts and the antiradical properties of the extracts. The optimal conditions were previously established at 160 °C and a solid: liquid ratio of 1:30 with a total yield of 8.9 mg of extractables per mL of water used. This condition reproduced in total phenol content of 16.7 g GAE/100 g extract and antiradical capacity of 34.0 g Trolox eq./100 g extract and an EC_{50} of 52.3 μg/mL measured against TEAC and DPHH radicals. Moreover, the extracts presented a capacity to reduce iron of 11.9 g BTH eq./100 g extract. To conclude, autohydrolysis technology allowed to obtain high amounts of bioactive compounds with high biological activities (such as high antioxidant capacity) from *Camellia sp.* seed husks, selecting water as a solvent and solid: liquid ratio of 1:30 as the most suitable operation conditions to recover bioactive compounds, reducing the generated environmental impact, when compared with other conventional methods.

References

1. Páscoa, R.N.M.J.; Teixeira, A.M.; Sousa, C. Antioxidant capacity of Camellia japonica cultivars assessed by near- and mid-infrared spectroscopy. *Planta* **2019**, *249*, 1053–1062.
2. Yang, C.; Liu, X.; Chen, Z.; Lin, Y.; Wang, S. Comparison of Oil Content and Fatty Acid Profile of Ten New Camellia oleifera Cultivars. *J. Lipids* **2016**, *2016*, 1–6.
3. Ye, Y.; Xing, H.; Chen, X. Anti-inflammatory and analgesic activities of the hydrolyzed sasanquasaponins from the defatted seeds of Camellia oleifera. *Arch. Pharm. Res.* **2013**, *36*, 941–951.
4. Chitsazan, A. Anti-cancer properties of green tea probed viaquantum mechanics calculations. *Orient. J. Chem.* **2015**, *31*, 393–408.
5. Teixeira, A.M.; Sousa, C. A review on the biological activity of camellia species. *Molecules* **2021**, *26*, 1–25.
6. Yan, C.; Liu, Y.; Cao, L.; Xia, M.; Zhang, Q.; Li, C.; Ruan, R. Oligosaccharide preparation from microwave-ethanol pretreated Camellia oleifera seed shell by enzymolysis of Agrocybe aegerita. *Ind. Crops Prod.* **2021**, *161*, 113155.
7. Tang, N.; Tan, X.; Cai, Y.; He, M.Y.; Xiang, Z.Y.; Ye, H.; Ma, J.L. Characterizations and application potentials of the hemicelluloses in waste oil-tea camellia fruit shells from Southern China. *Ind. Crops Prod.* **2022**, *178*, 114551.
8. Sharma, R.; Verma, S.; Kumar, D. Polyphenolics and therapeutic insights in different tissues extract and fractions of Camellia sinensis (L.) Kuntze (Kangra Tea). *Food Biosci.* **2021**, *42*, 101164.
9. Morales-Contreras, B.E.; Flórez-Fernández, N.; Dolores Torres, M.; Domínguez, H.; Rodríguez-Jasso, R.M.; Ruiz, H.A. Hydrothermal systems to obtain high value-added compounds from macroalgae for bioeconomy and biorefineries. *Bioresour. Technol.* **2022**, *343*.
10. Esteban-Lustres, R.; Torres, M.D.; Piñeiro, B.; Enjamio, C.; Domínguez, H. Intensification and biorefinery approaches for the valorization of kitchen wastes – A review. *Bioresour. Technol.* **2022**, *360*.
11. Overend, R.P.; Chornet, E. Fractionation of lignocellulosics by steam-aqueous pretreatments. *Philos. Trans. R. Soc. London. Ser. A, Math. Phys. Sci.* **1987**, *321*, 523–536.

12. Órbenes, G.; Rodríguez-Seoane, P.; Torres, M.D.; Chamy, R.; Zúñiga, M.E.; Domínguez, H. Valorization of artichoke industrial by-products using green extraction technologies: Formulation of hydrogels in combination with paulownia extracts. *Molecules* **2021**, *26*.
13. Flórez-Fernández, N.; Ferreira-Anta, T.; Torres, M.D.; Domínguez, H. Valorization of arnica montana wastes after extraction of the ethanol tincture: Application in polymer-based matrices. *Polymers (Basel).* 2021, *13*.
14. Zhang, J.; Wang, Z.; Du, M.; Yao, X.; Hu, L. Physicochemical properties of Camellia nut shell and its thermal degradation characteristics. *BioResources* **2015**, *10*, 647–659.
15. Zhang, L.; Zhang, W.; Zhang, F.; Jiang, J. Xylo-oligosaccharides and lignin production from Camellia oleifera shell by malic acid hydrolysis at mild conditions. *Bioresour. Technol.* **2021**, *341*, 125897.
16. Xu, W.; Zhang, W.; Han, M.; Zhang, F.; Lei, F.; Cheng, X.; Ning, R.; Wang, K.; Ji, L.; Jiang, J. Production of xylooligosaccharides from Camellia oleifera Abel fruit shell using a shell-based solid acid catalyst. *Bioresour. Technol.* **2022**, *365*.
17. Baumann, L. *Cosmeceuticals and Cosmetic Ingredients*; 2015; Vol. 1; ISBN 9788578110796.
18. Sun, F.F.; Tang, S.; Liu, R.; Tang, Y.; Wang, R.; Zhang, Z.; Gao, Z.; Li, H.; Li, C.; Xiao, Z. Biorefining fractionation of the Camellia oleifera Abel. hull into diverse bioproducts with a two-stage organosolv extraction. *Ind. Crops Prod.* **2016**, *94*, 790–799.
19. Sawicki, T.; Wiczkowski, W. The effects of boiling and fermentation on betalain profiles and antioxidant capacities of red beetroot products. *Food Chem.* **2018**, *259*, 292–303.
20. Zeb, A. Concept, mechanism, and applications of phenolic antioxidants in foods. *J. Food Biochem.* **2020**, *44*, 1–22.
21. Che Sulaiman, I.S.; Basri, M.; Fard Masoumi, H.R.; Chee, W.J.; Ashari, S.E.; Ismail, M. Effects of temperature, time, and solvent ratio on the extraction of phenolic compounds and the anti-radical activity of Clinacanthus nutans Lindau leaves by response surface methodology. *Chem. Cent. J.* **2017**, *11*, 1–11.
22. Bizuayehu, D.; Atlabachew, M.; Ali, M.T. Determination of some selected secondary metabolites and their invitro antioxidant activity in commercially available Ethiopian tea (Camellia sinensis). *Springerplus* **2016**, *5*, 1–9.
23. Nugraheni, Z. V.; Gunawan, W.D.; Laili, H.N.; Putro, H.S.; Wahyudi, A. Analysis of antioxidant activity of black tea extract (Camellia sinensis L.) and the effect on organoleptic properties of cassava getuk. *AIP Conf. Proc.* **2021**, *2349*.
24. Plaza, M.; Amigo-Benavent, M.; del Castillo, M.D.; Ibáñez, E.; Herrero, M. Facts about the formation of new antioxidants in natural samples after subcritical water extraction. *Food Res. Int.* **2010**, *43*, 2341–2348.

Oxidative Stability of Sunflower Oil Enriched with Fig Tree Leaves Bioactive Compounds at Frying Temperature

Susete Pinteus, Marta Neves, Carla Tecelão, Susana Silva, Alexandra Cruz, Susana Bernardino, Vasco Mendes, Vânia Ribeiro, Daniela Vaz, and Maria Jorge Campos

Abstract The pressure imposed by the increase of world's population and climate change makes it imperative to find sustainable strategies to optimize food resources and decrease waste production based on circular economy principles. This work evaluated the potential of fig tree leaves to reduce the oxidation of sunflower oil when exposed to frying temperatures. The bioactive compounds extracted directly from fig tree leaves were incorporated into the oil at a ratio of 12.5% (m/m), with the assistance of ultrasounds. The supplemented oil was characterized in terms of its total phenolic content (TPC), which was determined using the Folin-Ciocalteu method,

---

S. Pinteus
MARE—Marine and Environmental Sciences Centre & ARNET—Aquatic Research Network Associated Laboratory, Polytechnic of Leiria, 2520-614 Peniche, Portugal

M. Neves · C. Tecelão · S. Silva · A. Cruz · S. Bernardino · M. J. Campos (✉)
MARE-Marine and Environmental Sciences Centre & ARNET—Aquatic Research Network Associated Laboratory, ESTM, Polytechnic of Leiria, 2520-641 Peniche, Portugal
e-mail: mcampos@ipleiria.pt

V. Mendes
Quinta da Mó de Cima, 1600-131 Lisboa, Portugal

V. Ribeiro
Centre for Innovative Care and Health Technology, School of Health Sciences, CiTechCare, Polytechnic of Leiria, 2411-901 Leiria, Portugal

Faculdade de Ciências E Tecnologia, Earth Sciences Department, GeoBioTec, GeoTechnologies and GeoEngineering, Universidade NOVA de Lisboa, 2829-516 Caparica, GeoBioSciences, Portugal

D. Vaz
School of Health Sciences, Polytechnic of Leiria, 2411-901 Leiria, Portugal

Coimbra Chemistry Centre, Institute of Molecular Sciences, University of Coimbra, 3004-535 Coimbra, Portugal

LSRE-LCM—Polytechnic of Leiria, 2411-901 Leiria, Portugal

ALiCE—Associate Laboratory in Chemical Engineering, University of Porto, 4200-465 Porto, Portugal

J. R. da. C. S. Galvão et al. (eds.), *Proceedings of the 3rd International Conference on Water Energy Food and Sustainability (ICoWEFS 2023)*, Springer Proceedings in Earth and Environmental Sciences, https://doi.org/10.1007/978-3-031-48532-9_14

and its antioxidant capacity, which was evaluated using both the 2,2-Diphenyl-1-picrylhydrazyl (DPPH) assay and the ferric reduction antioxidant power (FRAP) assay. Also, the color of the oil and the oxidation products (K232 and K270) were analyzed.

Supplementing the oil with fig tree leaves resulted in a higher antioxidant capacity compared to the control, as demonstrated by the DPPH assay (5128 $\pm$ 466 vs. 1106 $\pm$ 407 μM Trolox eq/kg oil), as well as lower levels of primary and secondary oxidation products. The results suggest that fig tree leaves can be a valuable source of antioxidant compounds with the capacity to improve the nutritional value and oxidative stability of sunflower oil under frying temperatures.

Keywords Antioxidant activity · Circular economy · Vegetable oils oxidation · Figs' orchards · Food quality · Orchards sub-products · Sustainable food · Waste valorization

1 Introduction

World's population increase and the growing awareness for sustainability issues are imposing new challenges on the food industry. Research for new food sources and strategies to increase food products shelf-life, allied with new technological approaches to deliver safer food products, are essential for this new paradigm. Edible vegetable oils play a key role in industrial food production but also in domestic use. In 2021–2022 it was estimated that around 200 million metric tons of vegetable oils were consumed worldwide, being the sunflower oil the third most consumed (17.88 million metric tons), after palm (71.12 million metric tons), soybean (59.3 million metric tons) and rapeseed (29.44 million metric tons) oils [1]. Although the use of vegetable oils is inevitable, some sustainability issues have been raised due to the high environmental impacts of intensive culture of plants for oil extraction, such as canola, sunflower, soybean, and palm. These cultures are amongst the most destructive crops in the world, given their association to global deforestation, accounting for 20–30% of global crop lands [2, 3]. Thus, it is essential to find sustainable strategies for reusing vegetable oils, without jeopardizing food quality and safety. Several plant extracts were already evaluated for their potential to decrease the oxidative degradation of vegetable oils during frying, showing promising results [4]. However, there is still a need to create strategies based on circular economy approaches to develop low-cost, sustainable, and environmentally friendly options for food markets and domestic use. In line with this view, the main objective of this work was to understand the potential use of *Ficus carica* L. (fig) tree leaves, resulting from the figs production industry, to preserve sunflower oil quality during heating at frying temperature, adding value to a natural product that is currently considered a waste. Figs production has a high economic relevance in several European countries, with Mediterranean region accounting for over 50% of total global production (https://www.atlasbig.com/en-gb/countries-by-figs-production). Normally, fig leaves are a neglected waste, however, it is already known that this biological material contains valuable bioactive compounds including antioxidant substances [5]. This work aimed to screen the potential of fig tree leaves to preserve sunflower oil quality during heating at frying temperature.

2 Material and Methods

2.1 Chemicals and Reagents

Gallic acid (97.5% purity), Folin-Ciocalteu reagent, DPPH, and FRAP reagents were purchased from SigmaAldrich.com. *n*-hexane (≥97%), methanol (≥99.8%), isooctane (≥99.8%), $FeSO_4$ were obtained from VWR International, LLC, Carnaxide, Portugal.

2.2 Sample Preparation

Fig tree leaves (Dauphine variety) were supplied from Quinta da Mó in July 2021, transported in cooled boxes. These were then immediately oven dried (40 °C), grinded to a fine powder and stored in vacuum at room temperature until further use.

2.3 Sunflower Oils Enrichment with Fig Tree Leaves

Grinded fig tree leaves were added to sunflower oil in a proportion of 12.5% (m/m) in 500 mL glass bottles. Samples were then submitted to ultrasounds (VWR Ultrasssonic Cleaner USC-TH, 45 kHz), for 20 min at room temperature. The mixture was then filtered with gauze and centrifuged (5 min, 4000 rpm, 15 °C) to remove solid plant residues from the supplemented oil. Filtered sunflower oil was kept at −20 °C, protected from light until further analysis [6]. Sunflower oil, without plant biomass incorporation was submitted to ultrasounds and filtered in the same conditions described above and was used as control.

2.4 Heating Assays

Control and supplemented oil (5 × 20 g) were placed in 50 mL goblets and heated at frying temperature (180 °C) in an oven for periods of 6 h per day, over 4 days, totaling 24 h of heating. Between heating cycles, the oil samples were maintained overnight at room temperature and protected from light [6]. The samples submitted to different heating times (0, 6, 12, 18 and 24 h), were stored at −20 °C until further analysis.

2.5 Evaluation of Total Phenolic Content and Antioxidant Activity of Oil Samples

2.5.1 Preparation of Extracts

Control and supplemented oil (10 g) were dissolved with 10 mL of hexane, in Falcon tubes, and the mixture was extracted with 20 mL of methanol:water (80:20) under vortex stirring during 5 min. After centrifugation at 7000 rpm during 5 min, the methanolic phase was collected in a round bottom flask. This procedure was repeated two more times. Then, the recovered phases were combined and dried in a rotary evaporator at 150 rpm, 70 °C. The dried residue was then solubilized in 5 mL methanol and kept at −20 °C until further analysis concerning total phenolic content and antioxidant activity [7].

2.5.2 Total Phenolic Content Analysis

Total phenolic content was determined by the Folin-Ciocalteu method [8] and the results expressed as mg of gallic acid equivalents per kg of oil.

2.5.3 Evaluation of Antioxidant Capacity of Supplement Sunflower Oil Extracts

The antioxidant capacity was evaluated by the 2,2-Diphenyl-1-picrylhydrazyl (DPPH) method [9] and by the ferric reducing antioxidant power (FRAP) [10].

2.6 Determination of Color Changes

Changes in the color of oil along heating assays were analyzed according to the Portuguese standard NP 937 [11].

2.7 *Quantification of Conjugated Dienes and Trienes*

The oxidation products of oil samples were determined spectrophotometrically as described in Sousa et al., 2021 [12]. Briefly, oil samples (5–10 mg) were diluted in 25 mL of isooctane. Then, the absorbance of oil solutions was measured at 232 nm (related to the presence of primary oxidation products, conjugated dienes) and at 270 nm (related to the presence of secondary oxidation products, conjugated trienes) using quartz cells. The results of K232 and K270 were obtained according to Eq. 1.

$$K\lambda = \frac{E\lambda}{c \times b} \tag{1}$$

where Kλ is the specific extinction coefficient at the respective wavelength; Eλ is the absorbance at wavelength λ; c is the concentration of the oil solution (g/100 mL) and b is the path-length (cm).

3 Results

3.1 *Evaluation of Antioxidant Potential*

The total phenolic content (TPC) and the antioxidant capacity, of sunflower oil samples are shown in Fig. 1 and Table 1, respectively.

The supplemented oil had a higher content of phenolic compounds than the control oil (0 h), as shown in Fig. 1. However, no significant differences were observed

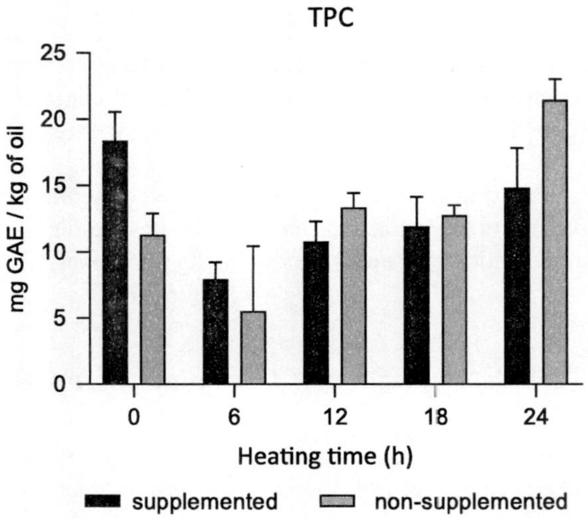

Fig. 1 Total phenolic content (TPC) of sunflower oil samples (control and oil supplemented with fig tree leaves) determined during heating times at 180ºC. The bars represent the mean ± standard error of the mean (SEM). Control—original sunflower oil not submitted to heating at frying temperature

Table 1 Antioxidant potential of sunflower oil submitted to heating at 180 °C along time. The numbers correspond to the mean ± standard error of the mean (SEM) of two independent experiments with four replicates. (Nd – not detected)

Heating (180 °C) Time (h)	DPPH (μM Trolox eq/Kg oil)		FRAP (μM $FeSO_4$ eq/Kg oil)	
	Supplemented	Non-supplemented	Supplemented	Non-supplemented
0	5128 ± 466	1106 ± 407	156 ± 76	Nd
6	590 ± 146	567 ± 99	145 ± 90	Nd
12	670 ± 113	127 ± 170	147 ± 120	Nd
18	Nd	440 ± 82	Nd	Nd
24	Nd	292 ± 50	Nd	Nd

between the supplemented oil and the control in samples subjected to frying temperature, suggesting that these compounds were degraded by high temperature. The unexpected increase in TPC values observed after 24h of heating may be due to interferences caused by oxidation products in the analytical method used. In fact, Folin reagent showed to be reactive towards several compounds, particularly unsaturated fatty acids, aldehydes, and ketones [13].

The antioxidant capacity of the supplemented oil was clearly higher than that of the control oil at time zero, suggesting that some antioxidant compounds had been transferred from the fig tree leaves to the oil during ultrasound extraction. However, as observed for TPC values, the heating process led to a pronounced decrease in the antioxidant capacity of the oil samples (Table 1). These results may be sustained by the thermolability of the antioxidant compounds.

3.2 *Evaluation of Quality Parameters of Sunflower Oil*

Sunflower oil oxidation status along heating at frying temperature was estimated spectrophotometrically by the formation of conjugated dienes (K232) and trienes (K270), which are primary and secondary oxidation products, respectively. The results are shown in Fig. 2.

The results suggest that the supplemented sunflower oil suffered less primary and secondary oxidation, than the control. These results may suggest the protective effect of the antioxidants provided by the fig tree leaves.

3.3 *Determination of Chromatic Characteristics*

The color is an indicative of oil quality. In the present study, the color of supplemented oil ranged between 48.7 and 87.5%. The non-treated oil showed 87.7% transparency.

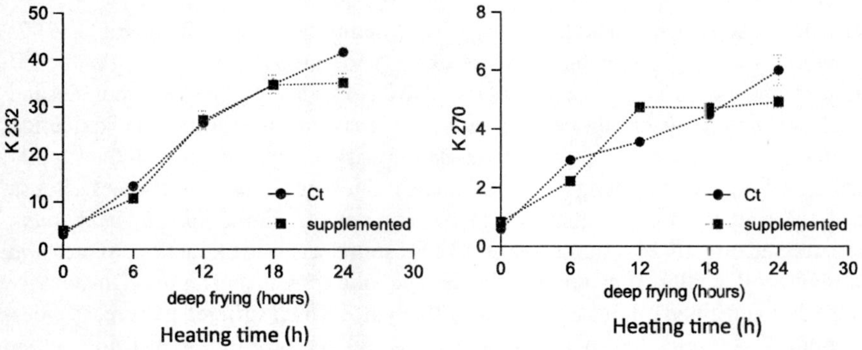

Fig. 2 Evaluation of sunflower oil quality along heating at frying temperature (180 °C). Each point represents the mean ± standard error of the mean (SEM) of two independent experiments. Ct-sunflower oil without supplementation

The samples enriched with fig tree leaves without heating showed the lowest transparency (48.7%) since the oils were enriched with fig tree leaves pigments, presenting a greenish color. This color disappeared when the oils were subjected to thermic treatment.

4 Discussion and Conclusions

Although being commonly associated with a negative connotation for having a high caloric content, vegetable edible oils are in the basis of a healthy diet, when consumed in moderation, and if of high quality. However, vegetable oils are highly susceptible to degradation through oxidative processes, due to their high content in polyunsaturated fatty acids [14]. Oils undergo several reactions (polymerization, hydrolysis, oxidation, etc.) promoted by heating that generate undesirable substances such as aldehydes, acids, ketones, partial acylglycerols, hydroperoxides, among others, leading to the development of undesired odors and flavors, such as rancidity. These substances and their subproducts are also associated with toxic effects in the human body. Therefore, the reduction of frying cycles is crucial for maintaining oil quality and food safety [14].

One of the strategies adopted by food industry is the partial or total replacement of used oil by new one, thus decreasing the concentration of noxious compounds [15]. However, this routine brings additional costs to the food industry, and is not sustainable or environmentally friendly. So, it is important to develop sustainable strategies to maintain the oil quality during deep-frying. This work aimed to evaluate the potential of fig tree leaves to maintain vegetable oil quality over time during heating at frying temperature. Fig three leaves are neglected waste products derived from figs farming, that contain valuable bioactive compounds, including antioxidant compounds. Previous works on this biological material identified three flavonoids with high antioxidant capacity, namely 3-O-(rhamnopyranosyl-glucopyranosyl)-7-O-(glucopyranosyl)-quercetin, isoschaftoside, and rutin [5]. n this work, the oil supplemented with fig tree leaves was found to have higher total phenolic content and antioxidant capacity than the control. Moreover, analyzing the formation of oxidation products (K232 and K270), it was possible to verify a slight decrease in the supplemented oil suggesting that fig tree leaves could reduce the oil oxidation process. Sousa et al. [6] evaluated the potential of bioactive compounds from *Crithmum maritimum* L, extracted directly into sunflower oil by ultrasound-assisted extraction, to preserve sunflower oil quality. They reported higher phenolic content and antioxidant activity in the supplemented oil, in comparison with control which suffered a decrease when submitted to accelerated oxidation conditions (60 °C for 12 days) [6]. In another work, Sousa et al. [6] evaluated the potential of extracts obtained from the brown seaweed *Pelvetia canaliculata,* to increase sunflower oil oxidative stability under accelerated oxidation conditions at 60 °C/12 days. They observed that the antioxidant activity of supplemented oil was sixfold higher than the control, concluding that algae extracts contributed to an increase of sunflower oil stability [12]. Other study carried out by Neves et al. [16] reported that supplementation of sunflower oil with ethanolic extracts of *Pterospartum tridentatum* clearly improved oil oxidative stability under frying conditions and at room temperature [16]. Kreivaitis et al. [17] evaluated the capacity of sage and thyme extracts to preserve rapeseed oil quality during deep-frying and storage, verifying an improvement in the oxidation stability of this vegetable oil [17].

The increase of oxidative stability of vegetable oil, under frying conditions and at room temperature (shelf life), can significantly reduce environmental constraints from waste oils. To create more sustainable products, further research focusing on circular economy is essential. This study found that vegetable oil supplemented with fig tree leaves had a higher antioxidant capacity and lower levels of primary and secondary oxidative products after being exposed to frying temperature for 24 h. These results suggest that fig tree leaves, which are currently overlooked, could be a valuable source of antioxidant compounds that can help to extend the shelf life of vegetable oils.

Further studies are required to fully assess the potential of fig tree leaves in enhancing the antioxidant capacity of vegetable oils. This may involve exploring various green extraction methods, such as microwave-assisted extraction, to obtain concentrated extracts for supplementing oils.

Acknowledgements This work was funded by national funds through FCT—Fundação para a Ciência e a Tecnologia, I.P., under the project MARE (UIDB/04292/2020 and UIDP/04292/2020), the project LA/P/0069/2020 granted to the Associate Laboratory ARNET and Internationalization Operational Program for financial support to the project 100%Figo (POCI- 01-0247- FEDER-064977) and the grant awarded to Susete Pintéus.

References

1. https://www.statista.com/statistics/263937/vegetable-oils-global-consumption
2. Pendrill, F., Persson, U. M., Godar, J., & Kastner, T. (2019). Deforestation displaced: trade in forest-risk commodities and the prospects for a global forest transition. *Environmental Research Letters*, *14*(5), 055003.
3. Meijaard, E., Brooks, T. M., Carlson, K. M., Slade, E. M., Garcia-Ulloa, J., Gaveau, D. L. A., Lee, J. S. H., Santika, T., Juffe-Bignoli, D., Struebig, M. J., Wich, S. A., Ancrenaz, M., Koh, L. P., Zamira, N., Abrams, J. F., Prins, H. H. T., Sendashonga, C. N., Murdiyarso, D., Furumo, P. R., Macfarlane, N., … Sheil, D. (2020). The environmental impacts of palm oil in context. *Nature plants*, *6*(12), 1418–1426.
4. Blasi, F., & Cossignani, L. (2020). An overview of natural extracts with antioxidant activity for the improvement of the oxidative stability and shelf life of edible oils. *Processes*, *8*(8), 956.
5. Li, C., Yu, M., Li, S., Yang, X., Qiao, B., Shi, S., Zhao, C., & Fu, Y. (2021). Valorization of Fig (*Ficus carica* L.) Waste Leaves: HPLC-QTOF-MS/MS-DPPH System for Online Screening and Identification of Antioxidant Compounds. *Plants (Basel, Switzerland)*, *10*(11), 2532.
6. Sousa, G., Alves, M. I., Neves, M., Tecelão, C., & Ferreira-Dias, S. (2022). Enrichment of Sunflower Oil with Ultrasound-Assisted Extracted Bioactive Compounds from *Crithmum maritimum* L. *Foods (Basel, Switzerland)*, *11*(3), 439.
7. Janu, C., Kumar, D. S., Reshma, M. V., Jayamurthy, P., Sundaresan, A., & Nisha, P. (2014). Comparative study on the total phenolic content and radical scavenging activity of common edible vegetable oils. *Journal of food biochemistry*, *38*(1), 38-49.
8. Ainsworth, E. A., & Gillespie, K. M. (2007). Estimation of total phenolic content and other oxidation substrates in plant tissues using Folin-Ciocalteu reagent. *Nature protocols*, *2*(4), 875–877.
9. Brand-Williams, W., Cuvelier, M. E., & Berset, C. L. W. T. (1995). Use of a free radical method to evaluate antioxidant activity. *LWT-Food science and Technology*, *28*(1), 25-30.
10. Benzie, I. F., & Strain, J. J. (1996). The ferric reducing ability of plasma (FRAP) as a measure of "antioxidant power": the FRAP assay. *Analytical biochemistry*, *239*(1), 70-76.
11. *Norma Portuguesa. NP 937 (1987) Gorduras e Óleos Comestíveis—Determinação da Cor Dos óleos e Suas Características Cromáticas. Processo Corrente*; Instituto Português da Qualidade: Lisboa, Portugal, 1987; p. 11.
12. Sousa, G., Trifunovska, M., Antunes, M., Miranda, I., Moldao-Martins, M., Alves, v.D., Vidrih, R., Lopes, P.A., Aparício, L., Neves, M., Tecelão, C., Ferreira-Dias, S (2021). Optimization of ultrasound-assisted extraction of bioactive compounds from *Pelvetia canaliculata* to sunflower oil. *Foods*, *10*(8), 1732.
13. Everette, J. D., Bryant, Q. M., Green, A. M., Abbey, Y. A., Wangila, G. W., & Walker, R. B. (2010). Thorough study of reactivity of various compound classes toward the Folin-Ciocalteu reagent. *Journal of agricultural and food chemistry*, *58*(14), 8139–8144. https://doi.org/https://doi.org/10.1021/jf1005935
14. Zhang, Q., Saleh, A. S., Chen, J., & Shen, Q. (2012). Chemical alterations taken place during deep-fat frying based on certain reaction products: a review. *Chemistry and physics of lipids*, *165*(6), 662–681.
15. Firestone, D. (2007). Regulation of frying fat and oil. In *Deep frying* (pp. 373–385). AOCS press.
16. Neves, M., Miranda, A., Lemos, M. F. L., Silva, S., & Tecelão, C. (2020). Enhancing oxidative stability of sunflower oil by supplementation with prickled broom (*Pterospartum tridentatum*) ethanolic extract. *Journal of food science*, *85*(9), 2812–2821.
17. Kreivaitis, R., Gumbytė, M., Kazancev, K., Padgurskas, J., & Makarevičienė, V. (2013). A comparison of pure and natural antioxidant modified rapeseed oil storage properties. *Industrial Crops and Products*, *43*, 511-516

Energy

The following part contains papers related to "Energy".

The editors had assistance from the following conference organizing committee member in the editing of this part:

- Roberta Mota Panizio

Barriers Found in the Integrated Implementation of Energy and Environmental Management Systems Through ISO 50001 and ISO 14001

Fabio Daniel Chaves Almanza and Gabriel Leon de los Santos

Abstract The relationship between energy and environment has been known for a long. Nevertheless, the synergy among energy and environmental management systems is still few explored in the literature. In this sense, the objective of the present work is to shed light on the barriers found in the literature for the integrated implementation of the ISO 50001 and ISO 14001 international standards. A systematic search of scientific papers was developed in several databases, such as Scopus, ScienceDirect and Web of Sciences among others. After discarding book chapters and articles non-related to ISO 50001 and ISO 14001 integrated management systems, 14 papers were selected, in which a total of 30 barriers were mentioned. Classifying similar barriers, 13 categories were obtained. With 20%, the most referred barrier was the scarcity of resources, including financing, data, and time; followed by deficient leadership strategies and/or commitment, with 13%. The results of this research show that management teams need to become aware that the synergy between energy and environmental management systems can bring competitive advantages in the medium and long term, by optimizing resources, increasing efficiency, and achieving sustainability. The findings presented can also contribute to policymakers and certification bodies for promoting sustainable development.

Keywords Barriers · Energy and environmental management systems · ISO 50001 · ISO 14001 · Synergy · Sustainability · Optimization

---

F. D. Chaves Almanza (✉) · G. Leon de los Santos
Faculty of Engineering, Universidad Nacional Autónoma de México, UNAM, Mexico City, Mexico
e-mail: fabiodaniel@comunidad.unam.mx

G. Leon de los Santos
e-mail: tesgleon@comunidad.unam.mx

J. R. da. C. S. Galvão et al. (eds.), *Proceedings of the 3rd International Conference on Water Energy Food and Sustainability (ICoWEFS 2023)*, Springer Proceedings in Earth and Environmental Sciences, https://doi.org/10.1007/978-3-031-48532-9_15

1 Introduction

Energy and environmental management systems have relatively little time to be coexisting together in the practice of industrial, commercial and other organizations. Thus, there are few studies to analyze the optimization procedures of both standards, basically because the practical experiences and results of implementation, operation and administration of ISO 50001 began to give opportunities since its early arrival around 10 years ago. Therefore, the analysis of the barriers to optimize the joint implementation of the standards via the synergies between them, is very relevant these days. The present article delves into this topic.

Sustainable development is based on achieving a balance between three main dimensions: environment, society and economy [1–4], so that the needs of the present are met without jeopardizing the capacity of future generations to meet their needs [3, 5–7]. In an effort to accomplish this objective, the United Nations adopted in 2015 the 2030 Agenda with 17 Sustainable Development Goals, which moreover intent to cover all areas of human activity [8, 9]. In relation to energy, Goal 7 considers: "Ensure access to affordable, reliable, sustainable and modern energy for all". And regarding environment, Goal 13 proposes: "Take urgent action to combat climate change and its impacts" [10]. Therefore, a close relationship between sustainability, energy and environment is observed. In this sense, energy and environmental management systems assume a relevant role to encourage organizations worldwide seeking sustainable performance [11]. On one hand, ISO 14001 explicitly states that a systematic approach to environmental management can provide information to senior management to generate long-term success and create options to contribute to sustainable development [6]. On the other hand, energy management systems are also a proven methodology of best practices that ensure sustainability [12–14].

In the last years, organizations all over the world that adopted ISO 50001 have experienced energy performance improvements [15]. Despite being a recent standard coming to light in 2011, in 2018 more than 22,000 facilities worldwide had implemented ISO 50001 [16]. The ISO 14001 was first published in 1996 and it has received significant attention as well, since it is the second most implemented ISO standards in the world after ISO 9001 [17, 18]. The relevance of ISO 50001 and ISO 14001 is growing more and more, however still few studies consider them in an integrated management system [19]. The complementary role between energy and environmental management has been known for a long time [20]. In the early years of the past decade, some authors already reported that energy management and efficiency were drawing attention in relation to environmental management [21–24]. And so today the integration among ISO 50001 and ISO 14001 is just starting to be explored in the literature [20]. In fact, considering the compatibilities between both standards, a potential sector is identified for the integration of ISO 50001 and ISO 14001 as an important synergy [15]. The lack of research in this issue, clearly involves little empirical work reported [20]. Consequently, the present study aims to shed light on the barriers faced by organizations during the integrated implementation of the ISO 50001 and ISO 14001 standards, according to the literature.

2 Background

2.1 *Energy Management Systems*

Based on the systemic approach [25, 26], an energy management system is defined by ISO (International Organization for Standardization) as "the set of interrelated or interacting elements to establish an energy policy and energy objectives, and processes and procedures to achieve those objectives" [27]. As noted, the standard ISO 50001 for the implementation of energy management systems emerged in 2011, responding to the need to make efficient use of energy, which has recently aroused great interest [28]. In fact, rising energy costs due to global demand created the need for a global standard for energy use management [29]. Before ISO 50001 there were approximately 14 regional and/or local energy management standards [20], but it is currently one of the best known and most adopted globally. The second and current version of this standard was published in 2018, and it continues to support organizations to continually improve their energy performance [16]. In this way, the adoption of ISO 50001 is accelerating as companies incorporate energy management requirements for supplier and sustainable corporate strategies [30, 31].

2.2 *Environmental Management Systems*

On the same systemic basis that characterizes all ISO management systems standards, ISO 14001 provides organizations with a framework for protecting the environment and responding to changing environmental conditions [6]. Globally, ISO 14001 is the most widely adopted standard for environmental management systems [32–35]. Extensive research and empirical evidence show that ISO 14001 have great impact on organizations performance, not only achieving waste minimization as a key environmental indicator [36] but also driving benefits such as reduced waste management cost, lower distribution costs, better corporate image, and savings in energy consumption and materials [37]. Furthermore, implementation of ISO 14001 involves the development of cleaner, safer and healthier products and workplaces [38–40].

2.3 *Synergy of Energy and Environmental Management Systems*

Sustainability and market dynamics are increasingly demanding more robust and integrated management systems for organizations. In this sense, ISO standards have evolved through different versions published over several decades. Currently, there is a clear correspondence between the sections of the standards associated with different management systems [19, 41–43]. Moreover, another common principle in all ISO standards for management systems is the continuous improvement transversal approach. Known as the PDCA cycle, it is an acronym for plan, do, verify and act. This methodology is a continuous iterative process to be incorporated into everyday organizational practices [11, 20, 44]. Now, regarding energy and environment synergy, it has been recognized that ISO 50001 shares several compatible requirements related to ISO 14001 [15]. Some authors have highlighted the advantages of integrating ISO 50001 and ISO 14001, emphasizing the common elements shared through comparisons between the standards [15, 20, 45–47].

An integrated energy and environmental management system leads to the optimization of resources, especially time, financing and documentation [11, 48]. But the main synergy is that energy efficiency is related to better environmental performance, as energy consumption generally means environmental impact [20, 49]. Energy management is considered to have a significant role in mitigating climate change and environmental pollution [47]. For instance, energy efficiency is linked to the savings obtained by reducing CO_2 emissions [50, 52]. However, it has been argued that both ISO 50001 and ISO 14001 are weak in promoting the use of renewable energy [47].

3 Methods

To achieve the research objective, it was defined a rigorous methodology to identify the barriers presented in the integrated implementation of ISO 50001 and ISO 14001, which led to a systematic literature review procedure that minimized the risk of bias. Based on the approach of Rampasso et al. [19], the methodological procedure followed is shown in Fig. 1.

Fig. 1 Methodological procedure

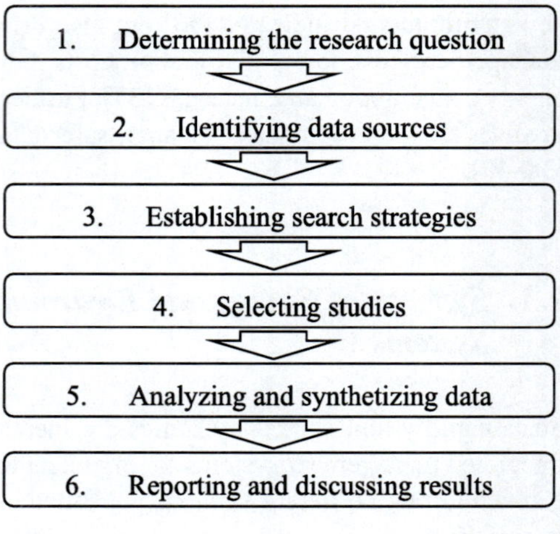

According to this procedure, the first stage of the study was to determine the research question. Given the objective of the research, the related question to be answered was: What are the barriers faced by organizations during the integrated implementation of the ISO 50001 and ISO 14001 standards? Bearing in mind the research question, a systematic search of scientific papers was developed in several databases, such as Scopus, ScienceDirect, VirtualE, Scirus, Web of Sciences, and others related to providers Elsevier and Clarivate Analytics. The election of these databases considered that they are representative from all scientific works published around the world, especially regarding the topic of interest. Thus, the second stage of the study was satisfied.

To ensure that suitable articles were found for the study, the following strategies were established:

- **Type of articles**: Articles from scientific journals were mainly considered, but also proceeding papers as well as review articles. Book chapters were discarded.
- **Key phrases for database search**: For the search of articles in databases, the phrases "ISO 50001 and ISO 14001" and "Energy and Environmental Management Systems" were used, for the specific issue of interest.
- **Abstract review**: It was important to confirm that the articles were related to the combination of both standards, by reviewing the abstracts.
- **Keywords for articles review**: In candidate articles, the words "Barriers", "Limitations", "Gaps", "Lacks", "Difficulties" and "Problems" were searched in any part of the contents. Also, all fragments related to the combination of ISO 50001 and ISO 14001 were reviewed along the contents.

The search in databases led a result of 103 candidate articles that fulfilled the methodological criteria. However, after reviewing their contents only 14 articles were related to some extent to the barriers in the integrated implementation of ISO 50001 and ISO 14001. This first result shows that the synergy among energy and environmental management systems is still few explored in the literature, as it is confirmed by Laskurain et al. [20]. The development of stages 5 and 6 of the methodology is presented in the next section.

4 Results and Discussion

The graphic in Fig. 2 shows the year of publication for the 14 studies analyzed. As mentioned, the ISO 50001 standard was first published in 2011; the second and current version came to light in 2018. This is closely related to the distribution of publication years, displaying the highest number of studies in 2019 with 3 articles.

After reviewing each of the 14 selected articles, a list of 30 barriers was obtained. Then, similar barriers were classified as common content letting a result of 13 categories. According to the number of times the barriers were mentioned, they were sorted in descending order. Figure 3 presents this result, and the articles where each barrier is mentioned as well.

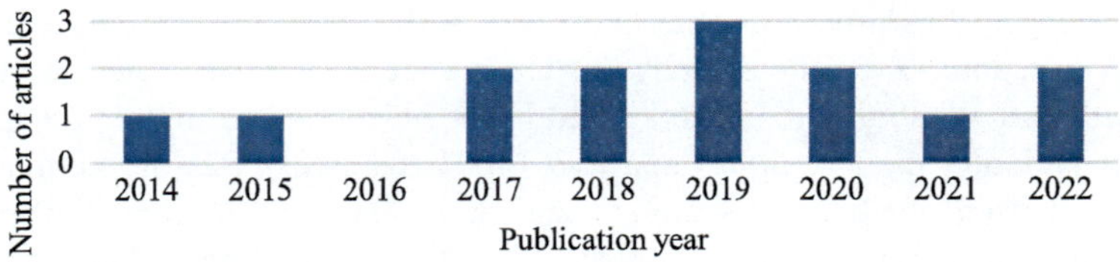

Fig. 2 Publication year of the articles reviewed

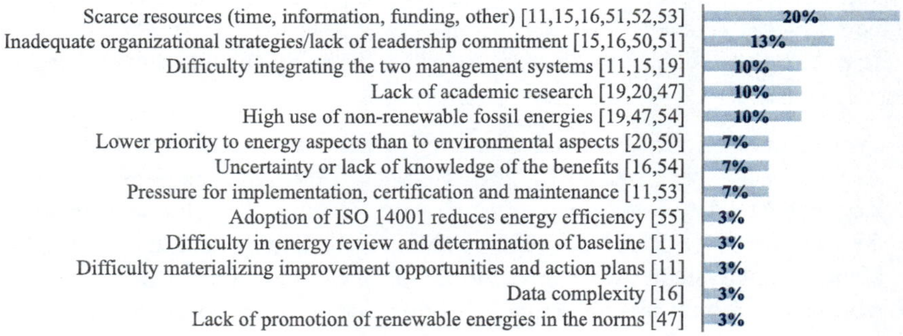

Fig. 3 Barriers in the integrated implementation of ISO 50001 and ISO 14001

It is important to note that Iker Laskurain from Spain is the researcher most involved in this topic, being the author of 3 of the 14 articles selected for the analysis. The most cited barrier in the integrated implementation of ISO 50001 and ISO 14001 is the "scarcity of resources", with 20% of the total barriers mentioned. Gunarathne and Lee [51] focus on the lack of systematic energy information organizations deal, particularly in the early stages of environmental strategy implementation. Chrysikopoulos and Chountalas [52] states that an important barrier is the lack of appropriate financing mechanisms, for energy and environmental efficiency projects. Fuchs et al. [16] refer the elevated certification cost of the management systems. Similarly, the high cost of the development and implementation of the Energy Management System in organization with an ISO 14001 certification is highlighted by Uriarte et al. [15]. These authors, also emphasize that time spent on this objective is one of the most important obstacles. The lack of resources is then usual in organizations facing constraints to achieve the implementation of projects related to both energy and environmental issues [11, 53].

The "inappropriate organizational strategy or the lack of leadership commitment" is as well an important barrier to consider, scoring 13%. The absence of leadership is one of the main challenges [16], while the non-well oriented strategies are generating lack of vision and cultural change [15, 50]. Regarding these most cited barriers, reacting policies to achieve energy efficiency for environmental improvement, are identified as one of the main causes [51]. Hence, the role of policymakers in the establishment of directives for organizations to prior energy and environmental efficiency, is now a key issue. In addition, certification bodies in charge of issuing new standards can play a major role in fostering this synergy.

The 10% of the barriers found refer to the "difficulty integrating the two management systems". For companies just starting in this area, it may not be feasible to target more than one standard at a time [11]. In this sense, Uriarte et al. [15] note that there is currently no methodology that provides a path to accomplish the integration of ISO 50001 and ISO 14001. Despite the relevance of ISO 50001, few studies consider it in an integrated management system with ISO 14001 [19]. Therefore, the absence of an integrating method is associated to the "lack of academic research", a barrier mentioned in 3 studies. Thus, the synergy of energy and environmental management systems is undervalued. For Laskurain et al. [20, 47] this research avenue could certainly provide fertile ground, not only for academic purposes but also in organizational fields.

Likewise, 10% of the barriers analyzed regards the "high use of non-renewable fossil energies". The use of non-renewable sources of energy has serious consequences for the environment and society, which is especially worrying considering that non-renewable sources are still the main source of energy generation worldwide [19]. Even, Daoud et al. [54] observe that oil and gas companies make large investments to devote considerable resources in pursuit of energy savings in the sector, instead promoting renewable energy sources. In no sector, the use of renewable energy represents up to 20% of total energy use [47]. Here, there is also a great opportunity to establish policies that promote the integration of energy and environmental management that necessarily led to greater use of renewable energy.

The "pressure for implementation, certification and maintenance of the integrated management system", the "uncertainty or lack of knowledge of the benefits" and the "lower priority to energy aspects than to environmental aspects" are barriers with 7% of citations. For the first one, Brem et al. [11] note that the pressure could be significant for certain organizations or projects that want to achieve these goals with minimal resources. This is also referred by Patel et al. [53], especially in small and medium-sized enterprises, which aims to identify and adopt energy efficient technologies, production processes to improve environmental as well as financial performance. On the second hand, the uncertainty of the benefits is an important challenge in this integrated implementation, according to Fuchs et al. [16]. Indeed, Daoud et al. [54] state that every managerial team needs to understand that nowadays the low carbon economy is becoming a need. Third, however, for ISO 14001 certified organizations, the standard makes little reference to the use of energy, so energy aspects are often overlooked and organizations concentrate on the more obvious environmental elements [50], which has an indirect effect on environmental improvement [20].

The last barriers analyzed scored 3% each one, corresponding to a single mention in the articles reviewed. According to Jeong and Lee [55] the adoption of ISO 14001 results in approximately 6% to 12% lower energy efficiency compared with non-adoption, in a trade-off relationship. For Brem et al. [11] conducting a baseline analysis is one of the most labor-intensive aspects of implementing and maintaining a management system based on ISO standards. And such difficulty is also observed in materializing improvement opportunities and action plans [11]. Fuchs et al. note that for a successful implementation, the complexity of the data is essential [16]. Finally, Laskurain et al. [47] highlight the lack of promotion of renewable energies in both ISO 50001 and ISO 14001 standards, which is closely related to the high use of non-renewable energy, discussed above. Thus, the synergy of both standards could be optimized not only by disrupting policies, but also by integrated standards, for pursuing sustainability.

5 Conclusions

The integration of energy and environmental management systems is an issue with still great research potential, even when energy and environment are closely linked. ISO 50001 and ISO 14001, the most relevant standards worldwide respectively, are carefully approached and facilitate the integration, however, organizations face different barriers that hinder the implementation of this integrated management system. Since the scarcity of resources and the lack of managerial vision, understanding and commitment are the most frequent limitations, organizations need to deeply examine their realities to correctly plan the standards integrated implementation [19]. It is essential that managers are aware of the benefits of this initiative, not only within the organization including significant financial savings, but also abroad in favor of other stakeholders, especially the environment and society.

In this sense, the organizations that accept this challenge are broadly supporting sustainable development. Therefore, policymakers may take part on the promotion of this kind of initiatives. Similarly, standardization bodies could also encourage this sustainable course of action by integrating the standards. Furthermore, these regulatory actors could take advantage of this conjuncture to drive an extended use of renewable energy.

References

1. Riestra, L.: Las dimensiones del desarrollo sostenible como paradigma para la construcción de las políticas públicas en Venezuela [Sustainable development dimensions: a paradigm for the construction of public policies in Venezuela]. Tekhné. Vol. 21, Núm 1, 024–033 (2018).
2. Gómez, C.: Objetivos de Desarrollo Sostenible (ODS): una revisión crítica [Sustainable Development Goals (SDGs): a critical review]. Papeles de relaciones ecosociales y cambio global, 140, 107-118 (2018).
3. Artaraz, M.: Teoría de las tres dimensiones de desarrollo sostenible [Theory of the three dimensions of sustainable development]. Revista Ecosistemas, 11(2), (2002).
4. Armindo, J., Fonseca, A., Abreu, I., Toldy, T. Perceived importance of sustainability dimensions in the Portuguese metal industry. Int. J. Sustain. Dev. World Ecol., 26, 154–165 (2019).
5. Gómez, I.: Desarrollo sostenible [Sustainable development]. Editorial Elearning, SL. Primera edición (2020).

6. ISO. ISO 14001:2015, Environmental Management Systems—Requirements with Guidance for Use; ISO/IEC: Geneva, Switzerland, 2015.
7. Brundtland, G.H. Report of the World Commission on Environment and Development: Our Common Future; United Nations: Oslo, Norway, 1987.
8. Zenchanka, S., Malchenka, S.: "Three "Gs" for campus sustainability development." Handbook of Theory and Practice of Sustainable Development in Higher Education, pp. 291–302. Springer, Cham (2017).
9. Zhao, X., Castka, P., Searcy, C.: ISO Standards: A Platform for Achieving Sustainable Development Goal 2. Sustainability, 12(22), 9332 (2020).
10. UN Sustainable Development Goals (SDGs). Available online: https://www.un.org/sustainabledevelopment/sustainable-development-goals/, last accessed 2022/10/3.
11. Brem, A., Cusack, D. Ó., Adrita, M. M., O'Sullivan, D. T., Bruton, K. How do companies certified to ISO 50001 and ISO 14001 perform in LEED and BREEAM assessments?. Energy Efficiency, 13(4), 751-766 (2020).
12. Medojevic, M., Medic, N., Marjanovic, U., Lalic, B., Majstorovic, V.: Exploring the impact of industry 4.0 concepts on energy and environmental management systems: Evidence from Serbian manufacturing companies. In IFIP International Conference on Advances in Production Management Systems (pp. 355–362). Springer, Cham (2019).
13. Petrovic, J.; Medojevic, M.: Importance and role of the state, ISO 50001 standards and regulations in the introduction of the energy management system. IEEP 2015, Zlatibor (2015).
14. UNIDO: Achieving impact and market credibility - Policy and conformity assessment frameworks for EnMS/ISO 50001 – Expert Group Meeting report, Vienna, 8–10 (2014).
15. Uriarte-Romero, R., Gil-Samaniego, M., Valenzuela-Mondaca, E., Ceballos-Corral, J.: Methodology for the successful integration of an Energy Management System to an Operational Environmental System. Sustainability, 9(8), 1304 (2017).
16. Fuchs, H., Aghajanzadeh, A., Therkelsen, P.: Identification of drivers, benefits, and challenges of ISO 50001 through case study content analysis. En. policy, 142, 111443 (2020).
17. Domingues, J.P.T.; Sampaio, P.; Arezes, P.M.: Analysis of integrated management systems from various perspectives. Total Qual. Manag. Bus. Excell., 26, 1311–1334 (2015).
18. Hernandez-Vivanco, A.; Domingues, P.; Sampaio, P.; Bernardo, M.; Cruz-Cázares, C.: Do multiple certifications leverage firm performance? A dynamic approach. Int. J. Prod. Econ., 218, 386–399 (2019).
19. Rampasso, I. S., Melo Filho, G. P., Anholon, R., de Araujo, R. A., Alves Lima, G. B., Perez Zotes, L., & Leal Filho, W.: Challenges presented in the implementation of sustainable energy management via ISO 50001: 2011. Sustainability, 11(22), 6321 (2019).
20. Laskurain, I.; Heras-Saizarbitoria, I.; Casadesús, M. Do energy management systems add value to firms with environmental management systems? Environmental Engineering & Management Journal, 18, 17–30 (2019).
21. Vikhorev K., Greenough R., Brown N.: An advanced energy management framework to promote energy awareness, Journal of Cleaner Production, 43, 103-112 (2013).
22. Böttcher C., Müller M.: Insights on the impact of energy management systems on carbon and corporate performance. An empirical analysis with data from German automotive suppliers, Journal of Cleaner Production, 137, 1449-1457 (2014).
23. Zhu Q., Lujia F., Mayyas A., Omar M.A., Al-Hammadi Y., Al Saleh S.: Production energy optimization using low dynamic programming, a decision support tool for sustainable manufacturing, Journal of Cleaner Production, 105, 178-183 (2014).
24. Aughney N., O'Donnell G. E.: The energy saving opportunity in targeting non-value add manufacturing activities–a structured approach, Journal of Cleaner Production, 86, 191-200 (2015).
25. Bertalanffy, L. V.: General systems theory. The science of synthesis: exploring the social implications of general systems theory, 103 (2010).
26. Florkowski, W. J., Banks, N., Shewfelt, R. L., & Prussia, S. E. (Eds.).: Postharvest handling: a systems approach. Academic press (2021).

27. ISO. ISO 50001:2018. Energy Management Systems—Requirements with Guidance for Use; ISO/IEC: Geneva, Switzerland, 2018.
28. Jovanović, B., & Filipović, J.: ISO 50001 standard-based energy management maturity model–proposal and validation in industry. Journal of Cleaner Production, 112, 2744-2755 (2016).
29. Gopalakrishnan, B., Ramamoorthy, K., Crowe, E., Chaudhari, S., & Latif, H.: A structured approach for facilitating the implementation of ISO 50001 standard in the manufacturing sector. Sustainable Energy Technologies and Assessments, 7, 154-165 (2014).
30. Curkovic, S., Sroufe, R.: Using ISO 14001 to promote a sustainable supply chain strategy. Business Strategy and the Environment, 20 (2), 71–93 (2011).
31. Perkins, R., Neumayer, E.: Geographic variations in the early diffusion of corporate voluntary standards: comparing ISO 14001 and the Global Compact. Environment and planning A, 42(2), 347-365 (2010).
32. Bugdol, M., Goranczewski, B., Kądzielawski, G.: Systemic support and environmental awareness in a normalised environmental management system consistent with ISO 14001. Management of Environmental Quality: An International Journal, 32(5), 949-969 (2021).
33. Székely, F., Knirsch, M.: Responsible leadership and corporate social responsibility: Metrics for sustainable performance. European Management Journal, 23(6), 628-647 (2005).
34. Bracke, R., Albrecht, J.: Competing environmental management standards: how ISO 14001 outnumbered EMAS in Germany, the UK, France, and Sweden. Environment and planning C: government and policy, 25(4), 611-627 (2007).
35. Montiel, I., Husted, B. W., Christmann, P.: Using private management standard certification to reduce information asymmetries in corrupt environments. Strategic Management Journal, 33(9), 1103-1113 (2012).
36. Singh, M., Brueckner, M., & Kumar, P.: Environmental management system ISO 14001: effective waste minimisation in small and medium enterprises in India. Journal of Cleaner Production (102), 285-301 (2015).
37. Grandic, D.: The impact of environmental management systems (EMSs) and ISO 14001 standards on corporate performance: a systematic literature review. Eurasian Journal of Business and Management, 5(1), 21-34 (2017).
38. Schilizzi, S., Lesourd, J. B.: Accounting systems and environmental decision making: what costs, what benefits? No. 412–2016–25786 (2001).
39. Melnyk, S. A., Sroufe, R. P., & Calantone, R.: Assessing the impact of environmental management systems on corporate and environmental performance. Journal of operations management, 21(3), 329-351 (2003).
40. Tarí, J. J., Molina-Azorín, J. F., Heras, I.: Benefits of the ISO 9001 and ISO 14001 standards: A literature review. Journal of Industrial Engineering and Management (JIEM), 5(2), 297-322 (2012).
41. Nunhes, T.V.; Ferreira Motta, L.C.; de Oliveira, O.J.: Evolution of integrated management systems research on the Journal of Cleaner Production: Identification of contributions and gaps in the literature. J. Clean. Prod., 139, 1234–1244 (2016).
42. Wilson, J.P.; Campbell, L. Developing a knowledge management policy for ISO 9001: 2015. J. Knowl. Manag., 20, 829–844 (2016).
43. Wilson, J.P.; Campbell, L. ISO 9001:2015: The evolution and convergence of quality management and knowledge management for competitive advantage. Total Qual. Manag. Bus. Excell. (2018).
44. Comoglio, C., Botta, S.: The use of indicators and the role of environmental management systems for environmental performances improvement: a survey on ISO 14001 certified companies in the automotive sector. Journal of Cleaner Production, 20, 92–102 (2012).
45. Chiu, T.Y., Lo, S.L., Tsai, Y.Y.: Establishing an Integration-Energy-Practice Model for Improving Energy Performance Indicators in ISO 50001 Energy Management Systems. Energies, 5, 5324–5339 (2012).
46. Zobel, T., Malmgren, C.: Evaluating the Management System Approach for Industrial Energy Efficiency Improvements. Energies, 9, 774 (2016).

47. Laskurain, L., Saizarbitoria, I.H., Casadesús, M.: Fostering renewable energy sources by standards for environmental and energy management. Renew. Sustain. Energy Rev., 50, 1148–1156 (2015).
48. Zeng S.X., Xie X.M., Tam C.M., Shen L.Y.: An empirical examination of benefits from implementing integrated management systems (IMS), Total Quality Management, 22, 173-186 (2011).
49. Lefebvre H.: On common ground ISO/IEC take on terminology. ISO Focus+, 2:5 (2011).
50. Laskurain, I., Ibarloza, A., Larrea, A., Allur, E.: Contribution to energy management of the main standards for environmental management systems: The case of ISO 14001 and EMAS. Energies, 10(11), 1758 (2017).
51. Gunarathne, N., Lee, K. H. The link between corporate energy management and environmental strategy implementation: Efficiency, sufficiency and consistency strategy perspectives. Journal of Cleaner Production, 293, 126082 (2021).
52. Chrysikopoulos, S., Chountalas, P.: Integrating energy and environmental management systems to enable facilities to qualify for carbon funds. Energy & Environment, 29(6), 938-956 (2018).
53. Patel, J. D., Shah, R., Trivedi, R. H.: Effects of energy management practices on environmental performance of Indian small-and medium-sized enterprises. Journal of Cleaner Production, 333, 130170 (2022).
54. Daoud, M., Zamzam, M., Helmy, A., Rageh, A.: ISO 50001-Energy Management Systems Accreditation, ADCO Safeguards the Environment with Efficient Energy Utilization. In Abu Dhabi International Petroleum Exhibition and Conference. OnePetro (2014).
55. Jeong, S., Lee, J.: Environment and Energy? The Impact of Environmental Management Systems on Energy Efficiency. Manufacturing & Service Operations Management (2022).

Impact of Urban Morphology on Energy Consumption—An Approach Towards Sustainability

Babak Farham, Flávio Craveiro, António Brandão Moniz, and Andreas Seebacher

Abstract Political and economic crises and issues such as global warming, decreased fossil fuel reserves, environmental pollutions and acid rains are just some of the problems caused by improper energy consumption. Energy efficiency optimization with the right policy and definition of accurate scientific methods proven in energy consumption includes planning, enforcement, organizing permanent control, and continuously reviewing technology upgrades. Additionally, developing legal solutions to implement these programs prevents the destruction and waste of energy resources and reduces the side effects of improper energy consumption. The shape of the city and its components can affect the city's quality and its spaces, as well as the air quality change, creating a particular climate in their area, causing phenomena such as heat islands, dust, rain, albedo, atmospheric inversion, and some others. By making changes in the shape of the city, both on a large scale and a micro-scale, atmospheric currents and, consequently, air quality in cities can be improved. Urban morphological indicators, including building height, building type, occupancy level, segmentation, block size, block shape, density, and urban green, can affect the weather and climate conditions outside and inside the building and, thus, the energy demand in the areas. By creating effective reforms in the model of configuration of the urban regions, including the pattern of land fragmentation, the physical characteristics of buildings, the network of passages and open spaces, it is possible to increase the energy efficiency of buildings in the long term in the regional(synoptic) climatic conditions of the city and the neighbourhood.

B. Farham (✉) · A. B. Moniz
Nova School of Science and Technology, Nova Lisbon University, Lisbon, Portugal
e-mail: b.farham@campus.fct.unl.pt

F. Craveiro
IAUD, Universidade de Lisboa, Lisbon, Portugal

School of Technology and Management, Politechnic of Leiria, Leiria, Portugal

A. Seebacher
Institute for Technology Assessment and System Analysis (ITAS), Karlsruhe Institute of Technology, Karlsruhe, Germany

J. R. da. C. S. Galvão et al. (eds.), *Proceedings of the 3rd International Conference on Water Energy Food and Sustainability (ICoWEFS 2023)*, Springer Proceedings in Earth and Environmental Sciences, https://doi.org/10.1007/978-3-031-48532-9_16

Keywords Energy · Urban morphology · Sustainability · Urban planning

1 Introduction

Nowadays, energy has a special place and importance for all countries, and finding the process of energy consumption in different countries for many economic and political planning is critical. Political and economic crises and issues such as global warming, limited fossil fuel reserves, environmental concerns, acid rain, overcrowding, economic growth and consumption are all broad topics that are widely considered by thinkers and governments. They are trying to find appropriate solutions to reduce the problems caused by improper energy consumption in the world, especially environmental crises, which have led to extensive efforts [22].

The existence of undesirable culture (unfair, unreasonable, etc.) and inappropriate patterns in energy consumption, lack of proper management and policy (including pricing policy), and use of weak technologies to convert and distribute and use energy are effective factors of improper energy consumption [16].

Modern methods of architectural design and urban planning can be used to save building energy consumption, which urban designers have so far paid less attention to and are more focused on issues such as aesthetics, land use and transportation [1–3].

In terms of economics, energy loss means the depletion of resources that could be used for the welfare of individuals in society and the detrimental effects of energy loss on the planet. In the design of buildings and cities, all aspects of human social, economic, psychological and individual aspects should be considered. Considering the energy loss rate in buildings to be at least 30%, every one percent saving and preventing its loss has a lot of financial and environmental benefits for every country and the world [11].

Energy efficiency optimization, which is a fundamental issue and a constant need for any country, with the right policy and definition of accurate scientific methods proven in energy consumption, ensures the energy sustainability of that country. This issue prevents the destruction and waste of energy resources and reduces the side effects of improper energy consumption. These policies include planning, organizing permanent control, continuously reviewing technology upgrades, and developing legal solutions to implement these programs.

2 Urban Morphology

The word “morphology” comes from the Greek root “morphé” meaning form, and “logos”, meaning study. The term morphology is attributed to Johann Wolfgang von Goethe, a German poet, novelist, playwright, and philosopher who invented it in the early nineteenth century (1749–1832) in a biological context [4]. Urban morphology studies urban forms as human habitation [15].

Urban morphology has been developed since the late nineteenth century by researchers with different backgrounds in language, culture, and disciplines as an interdisciplinary knowledge in collaboration with experts in various fields of science. As an example: Urban morphology, in simple terms, is the study of the form of cities. Urban morphologists agree on what this discipline is, but there is disagreement over how to study the form of cities [13].

Urban morphology is a systematic study of the form, structure and functions of urban texture and the origin and methods of the evolution of these textures, as well as the study of plans, buildings, uses, streets and future perspectives in urban texture and analysis of results and side effects of these items over time [18].

An energy-efficient form of urban morphology considers the effect and amount of natural ventilation in summer and the protection of buildings against cold winds in winter. Additionally, the amount of solar radiant energy received (light and heat) and the absorption of this solar thermal energy by the walls of a building complex that affects its cooling, heating and lighting energy consumption at different times of the year is also taken into account.

2.1 *Morphological Indicators Affecting Energy Consumption*

Each built-in urban element creates a particular artificial climate around and above it and can affect the weather around it and constantly interact with it. The shape, height and size of the buildings, the direction of the streets and buildings, the level of the spaces, and the geometric shape of the city affect the climate of the urban area. The shape of the city and its components can affect the quality of the city and its spaces, as well as change the air quality of the city [23].

The shape of the city causes the concentration or dispersal of air pollutants, which is one of the problems of the world's major cities, depending on the air flows and formation of thermal islands within the city. It can be said that by making changes in the shape of the city, both on a large scale and on a micro scale, atmospheric currents and, consequently, air quality in cities can be improved [5].

Accordingly, urban morphological design and planning are considered based on the structure, context, and form in relation to culture and human behaviour [9]. The relationship between sustainable energy development, energy system, planning, and urban morphology can be explained by a framework. It can be said that the spatial configuration of elements in an urban area significantly impacts ecosystem dynamics. Generally, it can be argued from a morphological point of view that urban development directly affects an urban settlement configuration, the leading indicators of which are the types of buildings, patterns of streets, and blocks forming it [10].

Urban morphological indicators, including building height, building type, occupancy level, segmentation, block size, block shape and density, can affect the weather conditions outside and inside the building and, thus, the amount of energy demand in the areas. They affect the city, and every building in terms of energy consumption must be considered on an urban scale, considering its location and conditions [8]. By creating effective reforms in the model of configuration of urban areas, including the pattern of land fragmentation, the physical characteristics of buildings, the network of passages and open spaces, it is possible to increase the energy efficiency of buildings in the long term in the regional (synoptic) climatic conditions of the city and the neighbourhood (Fig. 1).

The urban morphological limits of land areas for certain functions can be adjusted based on the significance of the influencing factors, to achieve building energy conservation. For example, the building height limits of hotels, educational buildings, and offices can be appropriately raised [21].

Making changes in the morphological configuration of urban textures and achieving an optimal pattern is a very effective way (passive system) to increase energy efficiency in urban textures [1–3]. Making changes in the morphological configuration of urban textures and achieving an optimal model to combine the layout of density and physical features of buildings, land use, street networks, and open spaces can reduce a significant part of the energy consumption in buildings.

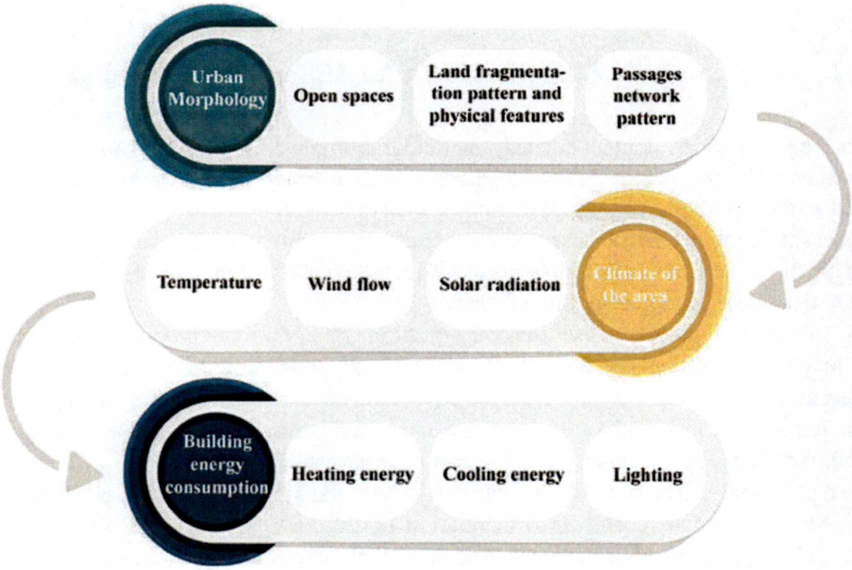

Fig. 1 Morphological indicators affecting energy consumption

2.2 Energy Consumption in Buildings

The main factors in the amount of energy consumption in buildings are climate, morphology, building design, productivity system and behaviour of residents, which are effectively related to each other [6]. Studies conducted in 2009 focusing on urban morphology in energy efficiency in different parts of Paris showed that energy consumption in modern textures is 1.8 times that of traditional, contemporary Parisian textures [19]. Urban textures consist of two major parts of the mass, including buildings and space, public and private open spaces such as passages, squares, and courtyards [14].

The combination of volatile spaces and the physical properties of masses and spaces can effectively reduce energy demand. These factors can potentially reduce half of the energy used in buildings. Although the morphology parameter has less effect than other parameters, it significantly reduces energy consumption in the long term [1–3]. Other indicators such as orientation, building elongation, number of floors, area of windows, type of canopy and form, density, orientation, and height of buildings as architectural and urban planning indicators affect the energy consumption of buildings. These factors can be optimized to reduce about 50% of energy consumption [1–3].

The energy performance of buildings (i.e., the energy demand of buildings) at the urban scale can be assessed in the scope of urban morphology. The influence of urban form parameters such as green area, building height and building density on energy consumption showed that the higher buildings surrounding a highly dense green area reduce the external temperature and the cooling loads of the building by around 5%.

The following chart (Fig. 2) shows how the urban parameters impact the urban microclimate and residential energy use:

Various research illustrated urban morphological indicators such as density, rugosity, sinuosity, contiguity and solar admittance [1, 2]. The relationship between built forms, density and solar potential (i.e., sky view factor at ground level and daylight availability on building façade) can provide beneficial information for planning solar cities [7]. Measuring the heating and cooling demands of the building revealed the significant role of urban settings in solar access [12]. The layout and position of open spaces with buildings and street networks, due to the impact on access to sunlight and wind (natural energy and ventilation), affect energy consumption in buildings.

Moreover, calculating the renewable energy potential shows the construction standard's importance. The radiation received on the roofs of the neighbourhoods with the best orientation and dense and sparse models can not cover more than 0.6 times the energy needed. On the other hand, if the best construction standard (beyond the energy codes) is used, and the neighbourhoods are oriented toward an angle with minimum radiation in both dense and sparse scenarios, 0.9 to 1.5 times of energy will be possibly covered by renewable sources [20].

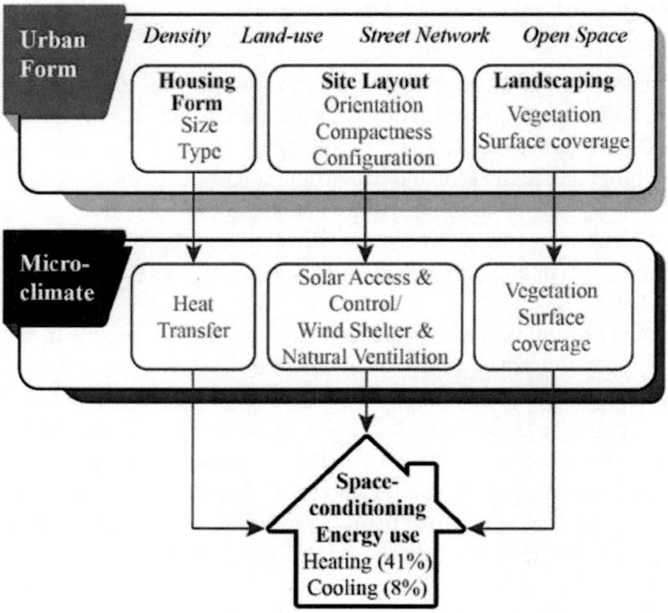

Fig. 2 Flow chart of how urban form variables impact urban microclimate and residential space-conditioning energy use [17]

Appropriate tree planting in the prevailing wind directions during winter and along the banks of the river can block the cold winds blowing into the city and reduce heating energy demand [21]. Changing the construction standards in different scenarios decreases heating loads by up to 70 and 85% in different case studies. A comparison between scenarios with different S/V (density, as it is the most effective morphological parameter) and construction standards show that regardless of the morphological feature, all neighbourhoods can reach a heating demand of less than 40 kWh/m^2 by changing the construction type [20].

3 Conclusions

Current urban planning in the cities is not sufficient for global energy consumption. Therefore, it is essential to establish a rough strategy towards energy consumption on the urban scale based on the current energy consumption methods and the ways to combine with urban form parameters to improve and integrate those systems into the urban planning and energy efficiency policies.

This research indicates that the shape and placement of buildings in urban blocks can affect air quality and reach a comfortable temperature in different urban areas in hot and cold seasons. By making effective reforms in the configuration of cities, it is possible to take a significant and effective step in reaching a pleasant temperature in every season of the year and energy consumption efficiency.

References

1. Adolphe, L. (2001a): "A simplified model of urban morphology: application to an analysis of the environmental performance of cities". Environment and Planning B: Planning and Design, (28), pp. 183–200.
2. Adolphe, L. (2001b): "Modelling the link between built environment and urban climate: towards simplified indicators of the city environment". Paper presented in: Seventh International IBPSA Conference, Rio de Janeiro, Brazil, 13–15 August 2001, pp. 13–15.679–684.
3. Adolphe, L. (2009): "Morphologic integrators of a sustainable city". Paper presented in: PLEA 2009: 26th conference on passive and low energy architecture, Quebec, Canada, 22–24 June 2009.
4. Aronoff, M. and Fudeman, K., (2005). "What is Morphology?" [pdf.] Oxford: Blackwell.
5. Bahraini, S.H. (2012). Urban Design Process, Eighth Edition. Tehran, Iran: University of Tehran Press.
6. Baker, N. and Steemers, K. (2000). Energy and Environment in Architecture: A Technical Design Guide. London, England: Taylor & Francis.
7. Cheng, V., Steemers, K., Montavon, M. and Compagnon, R. (2006). 'Urban form, density and solar potential', PLEA 2006: The 23rd Conference on Passive and Low Energy Architecture, Geneva, Switzerland, 6–8 September 2006.
8. Chrysoulakis, N., Grimmond, S., Feigenwinter, C. et al. (2018) 'Urban energy exchanges monitoring from space'. Scientific Reports 8, 11498. Available at: https://doi.org/10.1038/s41598-018-29873-x.
9. Delzendeh, E., Wu, S., Lee, A. and Zhou, Y. (2017). 'The impact of occupants' behaviours on building energy analysis: A research review', Renewable and Sustainable Energy Reviews, December 2017 (80), pp. 1061–1071.
10. Esmaeilpour Zanjani, N., Ziyari, Y. A., Zarabadi, Z. S., Sabbaghi, H. R., (2021). 'Comparative Study of Energy Role in Urban Morphology with an Emphasis on the Formation of Spatial Structure (Case Studies: Tehran and Berlin)'. J. Urban Manage Energy Sustainability, 3(1), Summer and Autumn 2021, pp. 43 – 56, Available at: http://www.ijumes.com/article_249028_8e509735cb0b9069958cb8890cce3dca.pdf.
11. Fong,W.K., Matsumoto, H., Lun, Y.F. and Kimura, R. (2007). "System Dynamic Model for the Prediction of Urban Energy Consumption Trends". Proceeding of the 6th International Conference on Indoor Air Quality, Ventilation & Energy Conservation in Buildings (IAQVEC 2007), 28–31 October 2007, Sendai, Japan, pp. 762–769.
12. Futcher, J.A., Kershaw, T.J. and Mills, G. (2013) 'Urban form and function as building performance parameters', Building and Environment, April 2013 (62), pp. 112-123.
13. Gauthier, P. and Gilliland, J. (2006). "Mapping Urban Morphology: a classificathion scheme for interpreting contributions to the study of urban from", Urban Morphology, 10(1), pp. 41-50.
14. Hong, T., Chen, Y., Belafi, Z. and D'Oca, S. (2018). 'Occupant behavior models: A critical review of implementation and representation approaches in building performance simulation programs', Building Simulation. 11, 1–14 [Online] Available at: https://doi.org/10.1007/s12273-017-0396-6.
15. Jones, A.N. and Larkham, P.J. (1991). "A Glossary of Urban Form", Historical Geography Research Series no.26, Norwich: Geo books.
16. Kariuki, D. (2018) energytoday Available at: https://www.energytoday.net/economics-policy/barriers-renewable-energy-technologies-development/.
17. Ko, Y.K. (2012). 'The energy impact of urban form: an approach to morphologically evaluating the energy performance of neighborhoods' (Thesis), UC Berkeley, Spring 2012, p. 9. Available at: https://escholarship.org/uc/item/6zd36454.

18. Madanipour, A. (1999). Urban space design, an attitude to social and spatial process, Tehran, Iran: Urban Planning and Processing Company.
19. Salat, S. (2010). 'Energy loads, CO2 emissions and building stocks: morphologies, typologies, energy systems and behaviour', Building Research & Information, 37:5–6, 598–609, Research on Building Stocks [Online]Available at: https://doi.org/10.1080/09613210903162126.
20. Sanei, A. (2022). The Impact of Urban Morphology and Construction Standards on the Energy Consumption of Neighborhoods, Montreal, Quebec, Canada. Presented in Partial Fulfilment of the Requirements for the Degree of Master of Applied Science in Building Engineering at Concordia University, July 2022.
21. Song, S., Leng, H., Xu, H., Guo, R., Zhao, Y. (2020). 'Impact of Urban Morphology and Climate on Heating Energy Consumption of Buildings in Severe Cold Regions', International Journal of Environmental Research and Public Health, 11 November 2020, 17(22), 8354; [Online]Available at: https://doi.org/10.3390/ijerph17228354
22. Shukla, V. and Kumar, N. (Eds.). (2020). Environmental Concerns and Sustainable Development, Singapore: Springer.
23. Ünlü, T. (2011). 'Towards the Conceptualization of Piecemeal Urban Transformation: The Case of Mersin, Turkey', Built Environment, (37) Number 4, December 2011, pp. 445–461(17) Alexandrine Press [Online] Available at: https://www.ingentaconnect.com/ contentone/alex/benv/2011/00000037/00000004/art00005

Syngas Application in Dual-Fuel Engines: A Brief Overview

Vítor Matos, Catarina Nobre, and Paulo Brito

Abstract Synthetic gas (syngas) can be generated from various carbon-rich sources, such as biomass and waste materials, using a process known as gasification. Syngas is a relatively low-energy fuel gas suitable for use in compression ignition (diesel) internal combustion engines (ICE). Utilizing syngas in engines holds promise for lowering greenhouse gas and hydrocarbon emissions. For instance, it can be combined with diesel in dual fuel engines (DFE). If implemented in heavy machinery, this approach can have a significant impact on reducing the demand for fossil fuels across various economic sectors and lessening their usage in engines. This study aims to provide a concise overview of the potential applications of syngas in DFE, addressing challenges related to syngas impurities and the emissions associated with this utilization.

Keywords Syngas · Dual fuel engines · Syngas contaminants · Diesel/syngas · Emissions

1 Introduction

Today, the primary objective of humanity is to attain independence from the reliance on fossil fuels. The escalating expenses related to energy and raw materials are compelling various sectors including industry and commerce to establish sustainable methods of functioning. A crucial focus lies in the advancement and transformation of sustainable types of biomass into feedstock for biorefineries, biofuels, biobased

V. Matos (✉) · C. Nobre · P. Brito
VALORIZA—Research Centre for Endogenous Resource Valorization, Portalegre Polytechnic University, Campus Politécnico 10, 7300-555 Portalegre, Portugal
e-mail: 12773@ipportalegre.pt

C. Nobre
e-mail: catarina.nobre@ipportalegre.pt

P. Brito
e-mail: pbrito@ipportalegre.pt

J. R. da. C. S. Galvão et al. (eds.), *Proceedings of the 3rd International Conference on Water Energy Food and Sustainability (ICoWEFS 2023)*, Springer Proceedings in Earth and Environmental Sciences, https://doi.org/10.1007/978-3-031-48532-9_17

products, and bioenergy. This emphasis stems from the imperative to ensure energy and environmental security, promote rural economic growth, and broaden the array of products and markets within the forestry and agricultural sectors.

For a considerable duration, diesel has functioned as the established fuel within the transportation and agricultural domains [1]. According to Sharma et al. [2], the predominant portion of emissions produced during agricultural activities originates from the direct combustion of diesel [2]. Consequently, addressing the imperative to diminish or enhance the efficiency of fuel utilization emerges as a potential resolution for decreasing emissions in these sectors. A strategy to accomplish this objective involves adopting alternatives with a reduced carbon footprint, such as employing a blend of diesel and syngas in engines designed for dual fuel usage—Dual-Fuel Engines (DFEs).

Syngas constitutes a gaseous amalgamation primarily of hydrogen (H_2), carbon monoxide (CO), accompanied by modest quantities of methane (CH_4), carbon dioxide (CO_2) and other minor hydrocarbons. This renewable gas is acquired through the gasification of solid biomass, waste materials, or coal [2, 3]. At present, syngas is widely acknowledged as an optimal fuel source for generating electricity, particularly in smaller-scale applications. Furthermore, on a larger scale, it proves valuable in processes like Fischer–Tropsch synthesis, employed for producing liquid fuels, and syngas methanation, utilized in biomethane generation. Within these applications, substantial volumes of hydrocarbons and impurities are undesirable, and the preference lies in a gas composition featuring solely CO and H_2. This underscores the necessity for comprehensive gas purification procedures [4].

Within DFEs, the heat generated through combustion is a result of a combination of diesel fuel and an alternative fuel, like syngas. The objective of dual-fuel mode (DFM) is to curtail diesel fuel usage, driven by both environmental and consumption considerations. For this purpose, it becomes essential to ascertain the optimal quantity of diesel injection across the complete spectrum of engine operational scenarios. This optimization is crucial to lower specific fuel consumption and attain peak performance across various engine loads [5]. DFEs can be categorized into mobile or stationary, contingent on their intended applications. These engines hold a significant role in generating power and heat due to their adaptable nature that allows variable power output. Additionally, they offer advantages such as economical capital investment, reliability, commendable performance under partial loads, efficient operation, and safety [6].

Recognizing the considerable potential of syngas applications, especially within engines applicable to various high-emission sectors such as agricultural machinery, the objective of this work is to offer a brief summary of syngas utilization in DFEs. This overview encompasses aspects like fuel characteristics, contaminants present in syngas, operational procedures, fuel usage patterns, and emissions associated with employing syngas in this context.

Table 1 Physical–chemical properties of syngas and diesel [10]

Parameter	Syngas[1]	Diesel ($C_{12}H_{26}$)
LHV (MJ/kg)	10.1–119.8	42.0
Stoichiometric Air/Fuel	2.5–34.3	14.9
Density (kg/m^3)	0.09–1.9	850
Ignition temperature (°C)	873–923	477–533

2 Syngas as Dual Fuel in Compression Ignition Engines

The blend of syngas alongside diesel in DFEs is regarded as a viable alternative from both energy and environmental perspectives. Within diesel engines, syngas serves as a partial substitute for traditional fossil fuels, enabling cost reduction, emissions mitigation, and improved engine performance within specific load ranges [7]. While syngas possesses a relatively lower calorific value (LHV) in comparison to conventional fossil fuels, any resultant decline in performance is somewhat mitigated due to the syngas-air mixture's low air-to-fuel ratio under stoichiometric conditions. Consequently, the energy density of the mixture closely approaches stoichiometric levels, rendering it nearly equivalent for both syngas and regular fossil fuels [8].

In the context of ignition, syngas lacks the capability for auto-ignition due to its ignition temperature, necessitating the presence of diesel to initiate combustion. This contrasts with diesel engines where such temperatures are naturally achieved [9]. In a study conducted by Sahoo et al. (2011), various syngas compositions were experimented with alongside diesel in DFEs. The physical–chemical properties of their syngas and diesel samples are presented in Table 1, reflecting the shared attributes typical to these fuels [10].

The chemical composition of syngas exhibits several distinct properties compared to conventional fossil fuels, resulting in favorable effects on engine performance and pollutant generation, including but not limited to [10–14]:

- A relatively elevated laminar flame velocity stands out as a significant influencer of engine performance. This parameter is critical for both engine design and the formation of pollutants, and its determination is notably impacted by factors such as mixture composition, pressure, and temperature. Its calculation often relies on the H_2/CO ratio.
- Syngas boasts a broad flammability range, a characteristic of considerable importance for engine applications. This attribute permits the utilization of the gas even in scenarios deviating from stoichiometric conditions, establishing an operational fuel range spanning from 20 to 60% in air.
- The high knock resistance of syngas empowers the application of elevated compression ratios, which, combined with the gas's extensive flammability

[1] For various syngas compositions: 100% H_2 + 0% CO; 75% H_2 + 25% CO, 50% H_2 + 50% CO, 0% H_2 + 100% CO.

range, proves advantageous for challenging combustion conditions. Nonetheless, achieving a balance between NO_x and CO emissions necessitates the implementation of fresh combustion strategies.
- H_2 plays a pivotal role in enhancing mixture reactivity, while CO exerts an inhibitory influence on combustion chemistry by extending the ignition time.

Overall, varying compositions of syngas lead to diverse outcomes in terms of engine performance.

3 Syngas Contaminants

The use of syngas invariably necessitates a cleaning process, a complexity that parallels the specific requirements outlined in Sect. 1. The removal of contaminants from syngas derived from biomass and waste sources encompasses various constituents, including particulate matter, tars (condensable hydrocarbons), nitrogen compounds, sulfuric compounds, alkali metals (primarily potassium and sodium), and hydrogen chloride (Cl) [15]. Levels of these contaminants exhibit considerable variation, significantly influenced by source material impurities and the methods employed in syngas production [16]. Drawing upon the extensive literature data referenced by Dziubak and Dziubak [17], it becomes apparent that abrasive wear arises from dust particles of specific dimensions, typically within the range of 5–20 μm. The severity of wear is contingent upon the particle's hardness and the sulfur content within the fuel. Empirical findings presented by these researchers reveal that a mineral dust layer on the airflow meter's measurement component leads to a 17% reduction in voltage. Similarly, the combination of dust and oil layers results in a more significant reduction of 46.7% in output voltage. This translates to lower sensor output voltage values, implying decreased air mass measurement and subsequently dispensing a smaller quantity of fuel. These factors collectively contribute to a reduction in engine power [17]. Post the Euro VI Regulatory measure introduced after 2013, a maximum sulfur content of 10 ppm in diesel fuel was mandated [18].

3.1 Particulate Matter

Within a gasifier, particulate matter spans a range of sizes, encompassing dimensions from under 1 μm to surpassing 100 μm. The composition of these particles exhibits substantial variability contingent upon the feedstock utilized and the specific gasification process employed [19]. Predominantly, inorganic compounds and solid residual carbon originating from biomass gasification constitute the primary components of these particulate materials. However, there's also a possibility for bed materials or catalysts to be separated out. The inorganic fraction encompasses alkaline-earth metals (primarily calcium), alkali metals (potassium and sodium), silica (SiO_2), and other metallic elements such as iron and magnesium [20]. For numerous syngas applications, the demand necessitates the removal of over 99% of particulates. Even in cases of direct combustion processes, which tend to tolerate some level of particulate matter, there exists a general requirement to lower particulate concentrations to levels below 50 mg/m^3 [21].

3.2 Tars

Tars are classifiable as condensable organic compounds and encompass a spectrum extending from primary oxygenated substances to more substantial deoxygenated hydrocarbons and polycyclic aromatic hydrocarbons (PAHs) [22]. Irrespective of their quantity or composition, tar presents a pervasive challenge in gasification processes due to their propensity to clog filters, pipelines, and engines. Moreover, tars can incapacitate catalysts in purification systems or downstream procedures [23].

3.3 Sulfur

Sulfur-based impurities, exemplified by H_2S, can vary in concentration within the range of 0.1 mL/L to exceeding 30 mL/L. The specific concentration depends on the feedstock used. These sulfur compounds have the potential to cause corrosion on metal surfaces [24]. In scenarios where the syngas is subjected to combustion, the sulfur contaminants undergo oxidation, forming sulfur dioxide (SO_2), which is subject to regulatory control as a pollutant. Remarkably, even minute quantities of sulfur can ultimately impair the effectiveness of catalysts intended for syngas purification, thus obstructing subsequent downstream applications [25].

3.4 Nitrogen Compounds

Numerous biomass and waste feedstocks contain nitrogen content that can yield notable quantities of ammonia (NH_3). Nevertheless, around 2/3 of this ammonia undergoes decomposition into molecular nitrogen (N_2) within the parameters of a typical gasification temperatures. Consequently, the presence of ammonia in syngas typically remains limited to several hundred to a few thousand parts per million. Despite these relatively low concentrations, there exist certain applications, such as chemical synthesis, where even such levels of ammonia can prove detrimental [26].

4 Internal Combustion Engines in Dual-Fuel Diesel/Syngas Mode

The ignition characteristics of internal combustion engines (ICE) are typically gauged through the cetane number (CN) or octane number (ON). In diesel engines, ignition quality relies not only on CN but also on factors encompassing energy content, density, cold flow properties, and sulfur content [27].

Syngas serves as a fuel in ICE under DFM, where diesel operates as the pilot fuel. The syngas is injected within the air through the engine's air intake. While the properties of diesel remain unaltered, the fuel blend consisting of diesel and syngas displays unique characteristics. The presence of the pilot fuel (diesel) is imperative to initiate the combustion of syngas. It's worth noting that the auto-ignition temperature of syngas (500 °C) is higher than the temperature reached by the fuel mixture in the diesel engine during the compression stroke [28].

Additionally, it's essential to consider that dual fuel injection feasibility in diesel engines with a compression ratio exceeding 17:1 might be constrained [29]. The requisite quantity of diesel fuel for pilot ignition varies and is largely contingent upon the quality and energy content of the syngas. Nevertheless, syngas can replace 60–90% of the diesel fuel needed to operate a diesel engine at a specific power output [29, 30]. The electronic control unit (ECU) of the diesel engine in DFM adjusts the injected diesel fuel amount as required to sustain engine power, adapting to shifts in the energy content of the syngas.

Emissions resulting from the dual-fuel mixture of syngas and diesel in ICE generally exhibit lower levels than when solely running on diesel. The reduction in fossil CO_2 emissions is proportionate to the extent by which biomass-based syngas replaces conventional diesel, given the carbon–neutral status attributed to biomass [30].

4.1 Combustion in Dual Fuel Analysis

In DFEs, combustion occurs through six key stages, in contrast to the five stages observed in traditional diesel engines, as depicted in Fig. 1.

The progression from phases A to F in DFEs can be elucidated as follows: [32]:

- Following diesel injection, Phase A-B ensues, marked by the postponed ignition of the pilot fuel. This stage is characterized by intricate chemical reactions that precede the actual combustion's heat release. Typically, the ignition delay duration is extended in dual-fuel combustion. Gaseous fuel, due to its elevated inert component, experiences even further prolonged ignition, owing to its slow-burning characteristics. At the conclusion of this phase, self-ignition of the injected pilot fuel (diesel) occurs.
- Preceding combustion, the initiation of multiple ignition cores takes place, followed by a premixed combustion stage, culminating in a specific pressure

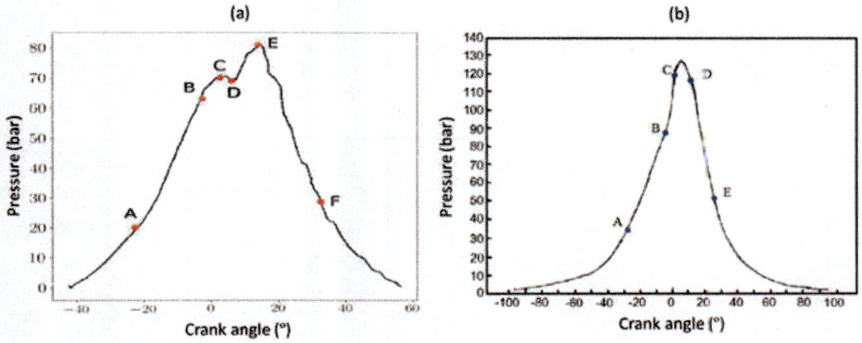

Fig. 1 Combustion phases of the engine: **a** Dual fuel; **b** Diesel [31]

surge (B-C phase). This pressure escalation is linked to the combustion of the air-diesel mixture encompassing the ignition point.

- The subsequent C-D phase is distinguished by a pressure reduction attributed to the delay in injecting the primary fuel (diesel).
- Consequently, the D-E phase is marked by the swift combustion of diesel, leading to a secondary pressure increase.
- As combustion progresses, both fuels contribute to the process amid turbulent flame propagation (E–F phase).

Hence, the combustion process observed in DFEs exhibits the same fundamental characteristics commonly associated with ICEs. This involves an initial premixed combustion of diesel fuel followed by its controlled combustion, with the heat release rate amplified through concurrent premixed combustion of syngas [33]. However, it's important to note that the comparison of performance and emissions outcomes between diesel engines utilizing pure diesel fuel and those employing dual fuel (w/ syngas) can vary depending on factors such as engine specifications, operational conditions, and the blend ratio utilized. Nevertheless, in a general sense, numerous studies have highlighted that incorporating syngas in DFEs can yield enhancements in both performance and emissions, as compared to the use of diesel fuel alone.

4.2 Diesel Consumption

Numerous investigations have been conducted employing syngas as a partial substitute for diesel, yielding savings that can reach up to 80% of diesel consumption [33, 34]. Parikh et al. (1986) reported that direct and indirect injection methods in diesel engines can achieve diesel savings of 80% and 90%, respectively [35]. Attaining such notable reductions in diesel consumption is attributed to various factors within the syngas, including temperature and turbulence in the engine intake and pre-cleaning

Table 2 Maximum diesel savings in DFEs [5]

Rotation (rpm)	Cilinder	Cubic capacity (dm^3)	Power (kW)	Compression ratio	Max. diesel savings (%)	Refs.
650	1	1.428	4.4	22:1	90.0	[35]
1500	1	0.411	5.9	18:1	64.2	[37]
1200	1	0.319	4.0	17.6:1	74.2	[36]
1800	6	6.494	68.4	17.5:1	75.0	[38]
1500	6	6.614	77.0	15:1	86.0	[11]
1500	1	0.553	3.5	18:1	45.7	[39]
1500	1	0.553	5.0	18:1	50.0	[40]

processes. It's important to highlight that these minimal diesel consumption thresholds were realized under highly controlled conditions and optimal engine speed ranges.

It's important to note that the extent of diesel fuel substitution cannot exceed a certain threshold due to the inherent tendency of engine thermal efficiency to decline with increasing gaseous fuel content. As such, engine efficiency experiences improvements as the diesel fuel proportion increases. Consequently, challenges arise in simultaneously managing engine efficiency and diesel consumption due to the inversely proportional relationship between them [36]. In practice, the realized savings in diesel consumption can range from 30 to 90%, accompanied by variations in power reduction levels. The extent of power reduction is contingent upon factors such as syngas composition and specific engine operating parameters like injection timing and compression ratios, which can range from 14:1 to 22:1 (Table 2).

5 Emissions from Dual Fuel Combustion

Operating in DFM presents potential benefits with regard to exhaust emissions, contingent upon the specific engine operating conditions. Among the prominent exhaust pollutants, the focus lies on NO_x and CO emissions. CO levels are influenced by the composition of the injected syngas, while NO_x emissions are predominantly influenced by the engine's geometric configuration and operational parameters [5]. Notably, the presence of H_2 in the syngas composition contributes to this dynamic. A higher fraction of H_2 is associated with increased NO_x production, alongside a concurrent decrease in emissions of unburned CO and hydrocarbons. Furthermore, considering the inclusion of H_2 within the syngas composition, a higher fraction tends to enhance NO_x production while simultaneously reducing emissions of unburned CO and hydrocarbons [40, 41].

The incorporation of syngas can potentially lead to a reduction in the ignition delay time and an increase in the laminar flame speed. These combustion characteristics consequently exert a favorable influence on the engine's lean burn condition, contributing to a reduction in emissions [42].

5.1 *Exhaust Gas Return Valve (EGR)*

An approach employed by automobile manufacturers involves utilizing an exhaust gas recirculation valve (EGR), which redirects a portion of exhaust gases back into the engine. This technique facilitates the reduction of NO_x, unburned hydrocarbons, and particulate matter by mitigating the peak temperature attained during combustion [43]. The advantages of EGR extend beyond pollutant reduction; it also imparts a stabilizing impact on flame propagation by enlarging the flame's ignition point and diminishing resistance to hydrodynamic perturbations associated with thermal expansion [44]. Essentially, the exhaust gases function as energy absorbers, curbing flame velocity and adiabatic temperature while retarding the development of unstable molecular structures. EGR's utilization has been explored and investigated as a means to effectively and stably operate at combustion temperatures surpassing the dew point of tars. Consequently, this approach eliminates the necessity for a tar filtration system [44, 45]. While this methodology holds promise for lowering costs associated with filtration systems, comprehensive analysis is required to ascertain its cost-effectiveness in the context of sustained engine use under these conditions, considering potential challenges that might arise.

5.2 *Supercharging*

Numerous researchers have explored the application of supercharging (via turbochargers) to enhance the engine's volumetric efficiency, a challenge rooted in the relatively lower mass density of syngas in comparison to air [38, 46, 47]. The incorporation of turbocharging, coupled with advancements in injection timing, results in notable enhancements in combustion efficiency compared to the atmospheric premixed dual-fuel scenario. This advanced approach leads to a reduction in CO emissions, attributed to the heightened oxygen availability that facilitates nearly complete combustion during the process.

6 Conclusions

ICEs offer a reliable and valuable technology for power generation, transportation, and heat production. However, the environmental challenges stemming from their operation necessitate the implementation of novel fueling strategies. While DFE are more commonly employed in stationary applications due to fewer logistical constraints, they still present an effective means of converting biomass waste into energy.

The utilization of syngas as a fuel presents an environmentally conscious and sustainable approach for disposing of biomass waste or waste derivatives from various industrial sectors. In this context, syngas-powered engines can be viewed as green and sustainable energy units, providing an alternative to reliance on oil companies.

Furthermore, the integration of syngas into engines can substantially diminish greenhouse gas and hydrocarbon emissions. By skillfully adjusting engine parameters, it's possible to mitigate CO and NO_x emissions when selecting a gas composition tailored to this objective. Research on DFE demonstrates the potential to achieve noteworthy reductions in diesel consumption. Elevated efficiencies coupled with low emissions can also be attained across a wide operating range through the adoption of suitable strategies like exhaust gas recirculation and turbocharging. Recent efforts have been aimed at optimizing these engines using advanced numerical techniques. However, the complexity associated with the presence of two fuel supply lines poses challenges linked to storage and transportation. Additionally, the literature often neglects to account for scenarios involving failure, no-load situations, or abnormal conditions related to one of the fuel lines. These issues are quite plausible given the variable of syngas composition.

Acknowledgements The present work was funded by Fundação para a Ciência e a Tecnologia, I.P. under the project UIDB/05064/2020, and by project ALT20-03-0145-FEDER-039485, co-financed by the ERDF, through the Regional Operational Program of Alentejo.

References

1. R. Bates and K. Dölle, "Syngas Use in Internal Combustion Engines - A Review," *Adv. Res.*, vol. 10, no. 1, pp. 1–8, 2017, doi: https://doi.org/10.9734/air/2017/32896.
2. P. Sharma and B. B. Sahoo, "An ANFIS-RSM based modeling and multi-objective optimization of syngas powered dual-fuel engine," *Int. J. Hydrogen Energy*, vol. 47, no. 44, pp. 19298–19318, 2022, doi: https://doi.org/10.1016/j.ijhydene.2022.04.093.
3. S. M. Santos, A. C. Assis, L. Gomes, C. Nobre, and P. Brito, "Waste Gasification Technologies: A Brief Overview," *Waste*, vol. 1, no. 1, pp. 140–165, Dec. 2022, doi: https://doi.org/10.3390/waste1010011.
4. R. Thomson, P. Kwong, E. Ahmad, and K. D. P. Nigam, "Clean syngas from small commercial biomass gasifiers; a review of gasifier development, recent advances and performance evaluation," *Int. J. Hydrogen Energy*, vol. 45, no. 41, pp. 21087–21111, 2020, doi: https://doi.org/10.1016/j.ijhydene.2020.05.160.

5. M. Fiore, V. Magi, and A. Viggiano, "Internal combustion engines powered by syngas: A review," *Applied Energy*, vol. 276. 2020, doi: https://doi.org/10.1016/j.apenergy.2020.115415.
6. *Synthesis Gas Combustion: Fundamentals and Applications - Google Books*.
7. F. Y. Hagos, A. R. A. Aziz, and S. A. Sulaiman, "Trends of syngas as a fuel in internal combustion engines," *Adv. Mech. Eng.*, vol. 2014, 2014, doi: https://doi.org/10.1155/2014/401587.
8. A. Pradhan, P. Baredar, Kumar, and Anil, "Syngas as An Alternative Fuel Used in Internal Combustion Engines: A Review," *J. Pure Appl. Sci. Technol.*, vol. 5, no. 2, pp. 51–66, 2015, [Online]. Available: http://nlss.org.in/abstract/syngas-as-an-alternative-fuel-used-in-internal-combustion-engines-a-review-anushka-pradhan-prashant-baredar-anil-kumar-2.html.
9. R. Rabello de Castro, P. Brequigny, and C. Mounaïm-Rousselle, "A multiparameter investigation of syngas/diesel dual-fuel engine performance and emissions with various syngas compositions," *Fuel*, vol. 318, 2022, doi: https://doi.org/10.1016/j.fuel.2022.123736.
10. B. B. Sahoo, U. K. Saha, and N. Sahoo, "Theoretical performance limits of a syngas–diesel fueled compression ignition engine from second law analysis," *Energy*, vol. 36, no. 2, pp. 760–769, Feb. 2011, doi: https://doi.org/10.1016/J.ENERGY.2010.12.045.
11. R. Uma, T. C. Kandpal, and V. V. N. Kishore, "Emission characteristics of an electricity generation system in diesel alone and dual fuel modes," *Biomass and Bioenergy*, vol. 27, no. 2, pp. 195–203, Aug. 2004.
12. Y. F. Li, H. Z. Huang, H. Zhang, N. C. Xiao, and Y. Liu, "Fuzzy sets method of reliability prediction and its application to a turbocharger of diesel engines," *Adv. Mech. Eng.*, vol. 2013, 2013, doi: https://doi.org/10.1155/2013/216192.
13. E. Monteiro, M. Bellenoue, J. Sotton, N. A. Moreira, and S. Malheiro, "Laminar burning velocities and Markstein numbers of syngas-air mixtures," *Fuel*, vol. 89, no. 8, pp. 1985–1991, 2010.
14. M. Fossum, R. V Beyer, and I. Bioenergy, "Co-combustion: Biomass fuel gas and natural gas," *SINTEF Energy Res. Trondheim*, 1998.
15. A. Kéromnès *et al.*, "An experimental and detailed chemical kinetic modeling study of hydrogen and syngas mixture oxidation at elevated pressures," *Combust. Flame*, vol. 160, no. 6, pp. 995–1011, 2013, doi: https://doi.org/10.1016/j.combustflame.2013.01.001.
16. P. J. Woolcock and R. C. Brown, "A review of cleaning technologies for biomass-derived syngas," *Biomass and Bioenergy*, vol. 52, pp. 54–84, 2013, doi: https://doi.org/10.1016/j.biombioe.2013.02.036.
17. T. Dziubak and S. D. Dziubak, *A Study on the Effect of Inlet Air Pollution on the Engine Component Wear and Operation*, vol. 15, no. 3. 2022.
18. Yihao Xie, Francisco Posada, and Ray MinjareS, "Diesel sulfur content impactson Euro VI soot-free vehicles:Considerations for emergingmarkets.," *Front. Environ. Sci. Eng.*, vol. 10, no. 5, 2020.
19. A. C. Hoffmann and L. E. Stein, *Alex C. Hoffmann · Louis E. Stein Gas Cyclones and Swirl Tubes.* .
20. E. Gustafsson, M. Strand, and M. Sanati, "Physical and chemical characterization of aerosol particles formed during the thermochemical conversion of wood pellets using a bubbling fluidized bed gasifier," *Energy and Fuels*, vol. 21, no. 6, pp. 3660–3667, 2007, doi: https://doi.org/10.1021/ef7002552.
21. Seville JPK, *Gas Cleaning in Demanding Applications*. Dordrecht: Springer Netherlands, 1997.
22. D. J. Stevens, "Hot Gas Conditioning: Recent Progress with Larger-Scale Biomass Gasification Systems," *Natinal Renew. energy Lab.*, no. NREL/SR-510–29952, p. 103, 2001, doi: https://doi.org/10.2172/786288.
23. W. Torres, S. S. Pansare, and J. G. Goodwin, "Hot gas removal of tars, ammonia, and hydrogen sulfide from biomass gasification gas," *Catal. Rev. - Sci. Eng.*, vol. 49, no. 4, pp. 407–456, 2007, doi: https://doi.org/10.1080/01614940701375134.
24. C. A. P. R. J. Lovell, S. W. Dylewski, "Project Summary: Control of Sulfur Emissions from Oil Shale Retorts." 1982.

25. B. Dou, M. Zhang, J. Gao, W. Shen, and X. Sha, "High-temperature removal of NH_3, organic sulfur, HCl, and tar component from coal-derived gas," *Ind. Eng. Chem. Res.*, vol. 41, no. 17, pp. 4195–4200, 2002, doi: https://doi.org/10.1021/ie010913h.
26. K. M. Hansson, J. Samuelsson, C. Tullin, and L. E. Åmand, "Formation of HNCO, HCN, and NH_3 from the pyrolysis of bark and nitrogen-containing model compounds," *Combust. Flame*, vol. 137, no. 3, pp. 265–277, 2004, doi: https://doi.org/10.1016/j.combustflame.2004.01.005.
27. N. Bezaire, K. Wadumesthrige, K. Y. Simon Ng, and S. O. Salley, "Limitations of the use of cetane index for alternative compression ignition engine fuels," *Fuel*, vol. 89, no. 12, pp. 3807–3813, 2010, doi: https://doi.org/10.1016/j.fuel.2010.07.013.
28. R. G. Papagiannakis and D. T. Hountalas, "Combustion and exhaust emission characteristics of a dual fuel compression ignition engine operated with pilot diesel fuel and natural gas," *Energy Convers. Manag.*, vol. 45, no. 18–19, pp. 2971–2987, 2004, doi: https://doi.org/10.1016/j.enconman.2004.01.013.
29. S. Szwaja, V. B. Kovacs, A. Bereczky, and A. Penninger, "Sewage sludge producer gas enriched with methane as a fuel to a spark ignited engine," *Fuel processing technology*, vol. v. 110. Elsevier B.V., 2013.
30. J. D. Martínez, K. Mahkamov, R. V. Andrade, and E. E. Silva Lora, "Syngas production in downdraft biomass gasifiers and its application using internal combustion engines," *Renew. Energy*, vol. 38, no. 1, pp. 1–9, 2012, doi: https://doi.org/10.1016/j.renene.2011.07.035.
31. W. N. W. Mansor, "Dual Fuel Engine Combustion and Emissions – an Experimental Investigation coupled with computer simulation," *PhD Thesis*, vol. 1, 2014.
32. O. M. I. Nwafor, "Knock characteristics of dual-fuel combustion in diesel engines using natural gas as primary fuel," *Sadhana - Acad. Proc. Eng. Sci.*, vol. 27, no. PART 3, pp. 375–382, 2002, doi: https://doi.org/10.1007/BF02703658.
33. J. B. Heywood, *Internal combustion engine fundamentals*. McGraw-Hill Education, 2018.
34. A. A. DENNETIERE, F. LEORAT, and G. F. BONNICI, "Power generation from biomass residues using the gasifier/dual-fuel engine technique," ACS Publications, 1980.
35. P. P. Parikh, A. G. Bhave, D. V. Kapse, and Shashikantha, "Study of thermal and emission performance of small gasifier-dual-fuel engine systems," *Biomass*, vol. 19, no. 1–2, pp. 75–97, 1989.
36. B. Mahgoub, S. Sulaiman, and Z. A. Abdul Karim, "Performance study of imitated syngas in a dual-fuel compression ignition diesel engine," *Int. J. Automot. Mech. Eng.*, vol. 11, pp. 2282–2293, Jun. 2015, doi: https://doi.org/10.15282/ijame.11.2015.11.0192.
37. P. Sombatwong, P. Thaiyasuit, and K. Pianthong, "Effect of pilot fuel quantity on the performance and emission of a dual producer gas - Diesel engine," *Energy Procedia*, vol. 34, pp. 218–227, 2013.
38. S. Dasappa and H. V. Sridhar, "Performance of a diesel engine in a dual fuel mode using producer gas for electricity power generation," *Int. J. Sustain. Energy*, vol. 32, no. 3, pp. 153–168, 2013, doi: https://doi.org/10.1080/14786451.2011.605945.
39. H. Singh and S. K. Mohapatra, "Production of producer gas from sugarcane bagasse and carpentry waste and its sustainable use in a dual fuel CI engine: A performance, emission, and noise investigation," *J. Energy Inst.*, vol. 91, no. 1, pp. 43–54, Feb. 2018.
40. A. Malik and S. K. Mohapatra, "Biomass-based gasifiers for internal combustion (IC) engines-A review," *Sadhana - Acad. Proc. Eng. Sci.*, vol. 38, no. 3, pp. 461–476, Jun. 2013, doi: https://doi.org/10.1007/S12046-013-0145-1.
41. U. Azimov, E. Tomita, N. Kawahara, and Y. Harada, "Effect of syngas composition on combustion and exhaust emission characteristics in a pilot-ignited dual-fuel engine operated in PREMIER combustion mode," *Int. J. Hydrogen Energy*, vol. 36, no. 18, pp. 11985–11996, Sep. 2011.
42. S. Feng, "Numerical Study of the Performance and Emission of a Diesel-Syngas Dual Fuel Engine," *Math. Probl. Eng.*, vol. 2017, 2017, doi: https://doi.org/10.1155/2017/6825079.
43. J. A. Smith and G. J. J. Bartley, "Stoichiometric Operation of a Gas Engine Utilizing Synthesis Gas and EGR for NOx Control ," *J. Eng. Gas Turbines Power*, vol. 122, no. 4, pp. 617–623, 2000, doi: https://doi.org/10.1115/1.1289386.

44. K. Vien, *Experiments on the Effect of Exhaust Gas Recirculation Diluent on the Laminar Burning Speed and Stability of Syngas/Air flames*. Northeastern University, 2016.
45. S. Bhaduri, H. Jeanmart, and F. Contino, "Tar Tolerant HCCI Engine Fuelled with Biomass Syngas: Combustion Control Through EGR," *Energy Procedia*, vol. 105, pp. 1764–1770, 2017.
46. S. Bhaduri, H. Jeanmart, and F. Contino, "EGR control on operation of a tar tolerant HCCI engine with simulated syngas from biomass," *Appl. Energy*, vol. 227, pp. 159–167, Oct. 2018.
47. Z. A. Z. and M. A. M. S. Hassan, F. Mohd Nor, "Performance and Emission Characteristics of Supercharged Biomass Producer Gas-diesel Dual Fuel Engine," *J. Appl. Sci.*, 2011, doi: 11: 1606–1611.

Routes of Interest to Produce Sustainable Biofuels from the Promotion of H_2 and CO_2—An Alternative for the Transport Sector

Vitor Silva, Catarina Nobre, and Paulo Brito

Abstract Combining hydrogen and carbon dioxide to produce sustainable biofuels seems to be an important alternative for the decarbonization of the transport sector, which is highly dependent on fossil fuels and, consequently, has significant CO_2 emissions. Currently, much of the world's hydrogen production comes from steam reforming (SMR) of natural gas. However, interest in the production of hydrogen from a renewable source has grown and leveraged the improvement of technologies, such as water electrolysis and biomass gasification. Carbon capture, sequestration, use and storage (CCUS) technologies, in addition to limiting the concentration of CO_2 in the atmosphere, also contribute to the use of CO_2 as a raw material in the production of fuels. CO_2 can be captured from the exhaust gases of conventional power plants or non-energy sectors, as well as captured directly from the air (DAC). In this sense, this article describes some possibilities to produce sustainable biofuels (biokerosene, biohydrogen and biomethane) from a technological platform of thermochemical-electrochemical integration and the Power-to-Liquids (PtL) route. The article also shows an overview of the use of renewable energy in the transport sector, especially in the European Union (EU).

Keywords Palavra-chave · Sustainable biofuels · Transport · H_2 · CO_2

1 Introduction

The transport sector has the highest dependence on fossil fuels compared to other sectors and accounts for 37% of CO_2 emissions from end-use sectors. Although it was one of the sectors most affected by the Covid-19 pandemic, in 2021, global CO_2 emissions from the transport sector returned to their historical growth trend (Fig. 1). There is an 8% increase (approximately 7.7 Gt CO_2) as pandemic restrictions are

V. Silva (✉) · C. Nobre · P. Brito
VALORIZA - Research Center for Endogenous Resource Valorization, Campus Politécnico, 10, 7300-555 Portalegre, Portugal
e-mail: vme.silva@campus.fct.unl.pt

J. R. da. C. S. Galvão et al. (eds.), *Proceedings of the 3rd International Conference on Water Energy Food and Sustainability (ICoWEFS 2023)*, Springer Proceedings in Earth and Environmental Sciences, https://doi.org/10.1007/978-3-031-48532-9_18

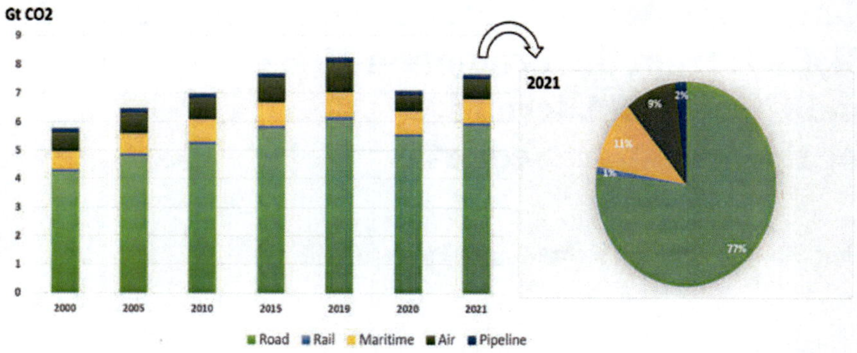

Fig. 1 Evolution of CO_2 emissions (Gt) in the transport sector (2000–2021). Adapted from IEA [8]

lifted and passenger and freight movements return to normal after the unprecedented decline that occurred in 2020 [8].

In 2021, the road mode of transport was responsible for 77% of CO_2 emissions, followed by 11% of the maritime mode of transport, 9% of the air mode of transport, 2% of the pipeline mode of transport and 1% of the mode of transport railway. There is a strong impact of the use of road transport on CO_2 emissions, but also on air quality in large urban centers, which profoundly affects the health of the population and the local environment.

To improve sustainability in the passenger and goods transport sector, it is necessary to apply a series of policies that associate the use of new forms of transport with energy efficiency measures and sustainable biofuels, to reduce carbon emissions, on all modes of transport.

Thus, in items 2 and 3, some possibilities of integrated systems for the sustainable production of biofuels will be presented, which combine technologies such as: biomass gasification, water electrolysis, methanation and anaerobic biodigestion of residues. Also noteworthy is the capture and use of CO_2, both from stationary sources and direct capture from the air (DAC).

1.1 Overview of the Use of Renewable Energy in the Transport Sector in the EU

According to the European Environment Agency (EEA), the share of energy from renewable sources used for transport in the EU has increased from less than 2% in 2005 to 10.2% in 2020. This means that, collectively, EU countries have reached the target of 10% for the share of energy from renewable sources in the transport sector. Preliminary data from the EEA indicate that, in 2021, the share of energy from renewable sources used for transport in the EU has stabilized around the same

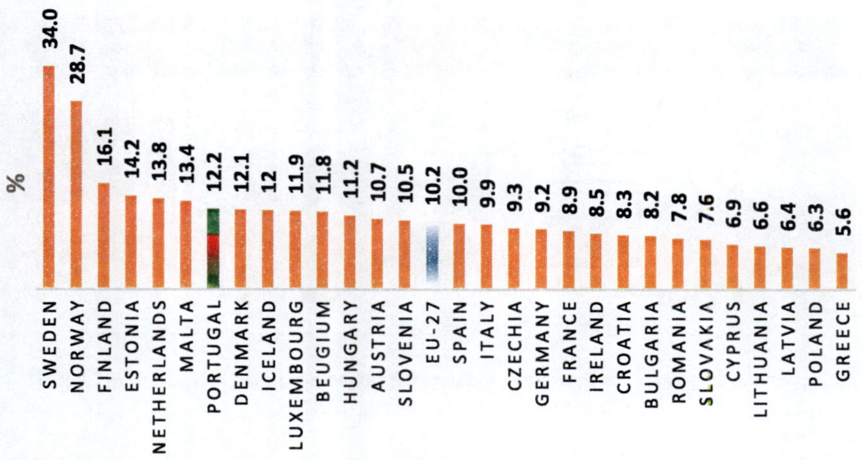

Fig. 2 Share of energy from renewable sources used in transport in Europe, by country, in 2021 (preliminary data). Adapted from EEA [3]

level (10.2%) (Fig. 2). Progress across EU Member States varies significantly, from 5.6% in Greece to 34% in Sweden. Portugal had a participation of 12.2% in 2021 [3].

2 The Power-To-Liquids (PtL) Technology Route to Produce Liquid Fuels

Within the scope of limiting the concentration of CO_2 in the atmosphere, the use of CO_2 as a raw material for the synthesis of different products, such as methanol, dimethyl ether, and liquid fuels, has been gaining a lot of interest. In particular, the Power-to-Liquids (PtL) concept, which combines the hydrogenation of CO_2 with the production of hydrogen from the electrolysis of water, powered by renewable sources, seems to be a very interesting solution to produce liquid fuels and chemicals [12].

PtL is a liquid hydrocarbon production route that uses electricity, water and CO_2 as resources (Fig. 3). The two main syntheses to produce renewable fuel by PtL are: Fischer–Tropsch Synthesis (FT) and Methanol Synthesis and Conversion (MSC). The production of liquid hydrocarbons via the PtL technological route comprises three main stages:

1. the production of hydrogen from renewable electricity using water electrolysis.
2. the supply of renewable CO_2 and its conversion.
3. synthesis to liquid hydrocarbons with further upgrading/conversion into refined fuels.

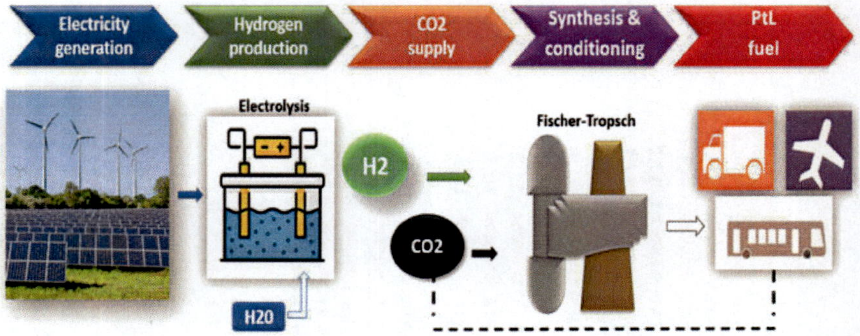

Fig. 3 Generic diagram of the Power-to-Liquids (PtL) technology route. Adapted from LBST [11]

The Fischer–Tropsch (FT) synthesis requires carbon monoxide. In synthesis routes such as BtL (Biomass-to-Liquid) and CtL (Coal-to-Liquid), this is provided from the gasification of biomass and coal, respectively. In the case of FT-PtL, CO_2 from concentrated sources or extracted from the air is used. According to Panzone et al. [12], the hydrogenation of CO_2 through the FT mechanism includes two reaction stages (Eqs. 2.1 and 2.2):

$$nCO_2 + nH_2 \leftrightarrow nCO + nH_2O \quad \delta H_r^{298K} = +41\,\text{kJ/mol} \tag{2.1}$$

$$nCO + 2nH_2 \leftrightarrow (-CH_2-)_n + nH_2O \quad \Delta H_r^{298K} = -166\,\text{kJ/mol} \tag{2.2}$$

In the first stage (Reaction 2.1) the conversion of CO_2 into CO occurs through the inverse process of the water–gas displacement reaction (RWGS). It is a reversible and endothermic reaction that quickly reaches equilibrium. It is favored at high temperatures and is not influenced by pressure as the number of moles does not change during the reaction. In the second step (Reaction 2.2), CO reacts in the Fischer–Tropsch reaction to produce organic products. This reaction is exothermic and involves a decrease in the number of product molecules, so it is favored at high pressures. It is slower than RWGS and generally has a kinetic control [12].

Renewable energy-based hydrogen production is referred to as green hydrogen. Water electrolysis is used to split water into its components of oxygen and hydrogen using renewable electricity (wind and solar power). The main types of electrolyzers used are proton exchange membrane electrolysis cell (PEMEC), solid oxide electrolysis cell (SOEC) alkaline water electrolysis cell (AWEC). At present, AWEC and PEMEC are proven as commercial technologies. However, SOEC has attracted many to bring this technology to market to achieve better energy efficiency. Based on operating temperature, the electrolysis technologies can be categorized into low-temperature electrolysis (LTE) and high-temperature electrolysis (HTE). Low-temperature electrolysis AWEC and PEMEC are operated below 100 °C, whereas SOEC operates at significantly higher temperatures (650–1000 °C). Despite their lower operating temperatures, high efficiency and technological maturity, low-temperature electrolysis cells cause high electrical energy consumption. Thermodynamically, the electrical energy demand to electrolyze water decreases as the operating temperature increases [15].

In the case of high-temperature electrolysis, the waste heat (220–250 °C) from the exothermic Fischer–Tropsch synthesis is used for steam generation, thus decreasing the electricity demand. High-temperature electrolysis can, in addition, allow the co-electrolysis of water vapor and CO_2, producing hydrogen and carbon monoxide in a single step (CoSOEC). In this case, a reverse step of the CO exchange process it's not necessary. However, coelectrolysis is subject to further research and technology validation [11].

Like any other renewable hydrocarbon synthesis process, production via PtL results in a mixture of gasoline, aviation kerosene, diesel, and other fuels. The product mix can be modified to at least a 50% share of jet fuel components. The conversion of the crude product derived from the FT route to aviation fuel and other hydrocarbons comprises several process steps, notably hydrocracking, isomerization and distillation. Currently, these processes are used on a large scale in oil refineries and in CtL and GtL plants. As environmental benefits of this route, in addition to the emission of greenhouse gases being practically neutral, the synthetic fuel produced offers better combustion with less pollutants. It should be noted that the demand for water used on the PtL route is almost insignificant and the land use requirements are lower compared to biofuels produced by other routes [11].

According to a study by the German Environment Agency [5], the main challenge for the short-term implementation of the PtL route is the production cost compared to conventional fuel. Cost reductions can be achieved through lower renewable electricity costs (wind and solar), increased efficiency through improved production of PtL processes (high temperature electrolysis, CO_2 extraction, etc.) and economies of scale. The energy efficiency of PtL route is significantly influenced by the type of CO_2 source used and by the applied electrolyzer technology.

2.1 *Real Cases of Biokerosene Production*

A real case[1] of CO_2 use, from the direct capture of the air, is what is happening in the Netherlands (Fig. 4). A consortium formed by Rotterdam The Hague Airport (part of the Royal Schiphol Group), SAF global leader SkyNRG and Swiss direct air capture pioneer Climeworks have agreed to build a demonstration plant producing sustainable aviation fuel (SAF) fully circular, with 100% CO_2 derived from air. Zenid Jet Fuel From Air aims to be the world's first industrial-size demonstration plant for

[1] Another case is that of the German aircraft manufacturer Deutsche Aircraft and the South African chemical and energy company—Sasol, which signed a memorandum of understanding (Jun/22) on the advanced technology PtL-SAF, based on hydrogen green and CO_2 capture, for the production of sustainable fuel for aviation, through the Fisher-Tropsh technological route [16].

Fig. 4 Modular sustainable aviation fuel (SAF) production plant [2, 6, 10]

sustainable aviation fuels (SAF) made from air. Direct air capture technology supplies CO_2 to a highly efficient co-electrolysis unit, which transforms CO_2 and added water into synthesis gas (H_2 and CO). Synthesis gas is transformed into liquid hydrocarbons by a modular Fischer–Tropsch reactor and then refined into sustainable jet fuel. The process runs entirely on renewable energy and uses only CO_2 from the air [17].

Another relevant example is the Brazil-German technical cooperation partnership, in the Project "Alternative Fuels without Climate Impacts—ProQR". The project comprises the construction of modular plants, located in remote areas, but close to airports (such as in the Brazilian Amazon region). In this way, they can be easily built, with their stations installed in containers, in addition to already supplying fuel directly at their point of consumption, at airports where transporting fuel was previously a challenge [14].

The transport of people and goods in the Brazilian Amazon region mainly involves road, river and air modes, however, there are periods of the year when river floods make road transport unfeasible and, at other times, the low volume in some rivers, prevent river transport, making air transport the best option. However, the continental dimensions of the country make aviation fuel distribution logistics difficult, making it quite complex, which increases the cost of fuel for airlines and, consequently, for the final consumer. Therefore, local production can reduce logistical costs, given the distance between the production hub and the consumer hub located in remote regions. Furthermore, the consumption of sustainable aviation fuel (SAF) will imply a reduction in fossil fuel consumption and will contribute to the reduction of CO_2 emissions in the aviation sector.

The design concept is presented in Fig. 5. The carbon dioxide (CO_2) is captured from the air or from stationary sources (industries). Using renewable energy (solar or wind), the electricity generated feeds a co-electrolysis system (CoSOEC), which

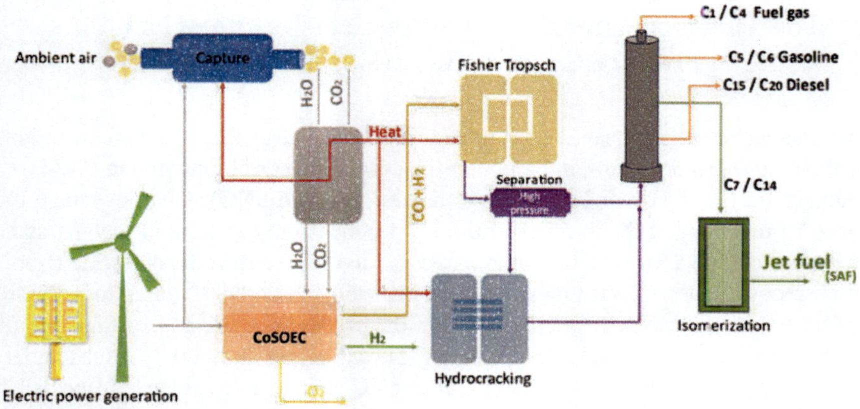

Fig. 5 Pro-QR project concept for sustainable aviation fuel production in Brazil. Adapted from PTX HUB [14]

separates the components of water (H_2 and O_2) and combines hydrogen with captured CO_2 to form the synthesis gas (H_2 and CO). The syngas goes to the Fisher-Tropsch process, where it is transformed into synthetic crude oil, then hydrocracked and separated into its main fractions: fuel gas, gasoline and diesel). The fraction corresponding to aviation fuel (C7/C14) is sent to the isomerization process, so that the jet fuel meets quality requirements and can thus supply aircraft.

It should be noted that in 2021, the airline sector was responsible for more than 2% of global energy-related CO_2 emissions, having grown faster in recent decades than road, rail, or maritime transport. As countries emerged from Covid-19 lockdowns, aviation emissions in 2021 reached around 720 Mt, recovering almost a third of the drop from 2019 levels seen in 2020 [9].

A growing number of regulatory frameworks, including RENOVABIO from Brazil (trade in certificates), California's Low Carbon Fuel Standard from the State of California/USA (based on carbon intensity) and ReFuelEU[2] Aviation proposed by the European Union (a combination mandate), include incentives for the SAF along with the US-proposed SAF Grand Challenge [9].

[2] The proposal introduces an obligation for aviation fuel suppliers to blend an increasing share of SAFs (advanced biofuels and e-kerosene) into fuel supplied to airports in Europe. This obligation covering both types of fuels would start in 2025 at 2% and increase to 5% in 2030. The regulation is currently under analysis/revision, for subsequent approval by the Council and the European Parliament.

3 The Thermochemical-Electrochemical Technological Route to Produce Gaseous Biofuels

Sustainable biofuels appear as an important source of energy for the decarbonization of the road transport sector, in particular the development of biohydrogen (bio-H_2), biomethane (bio-CH_4) or bio-synthetic natural gas (bio-SNG). One advantage of these biofuels is that the burning of bio-H_2 produces only water and oxygen and, in the case of bio-CH_4, in addition to its production from anaerobic digestion (from waste recovery), the CO_2 itself extracted from the biogas, could be reused to increase the production of biomethane, thus making the process circular and sustainable. It should be noted that the use of renewable gases can represent not only a substantial contribution to the climate, but also to greater EU energy independence (especially from Russian natural gas). In this context, the development of studies and projects of an integrated system is observed, with emphasis on the thermochemical (gasification and methanation), electrochemical (electrolysis) and biological (anaerobic digestion) platform, to produce renewable gaseous biofuels.

Giglio et al. [7] carefully analyzed the integration between biomass gasification, high temperature solid oxide electrolysis cell (SOEC) and catalytic methanation (Fig. 6). Electrolysis was employed to enrich the hydrogen content of the produced synthesis gas and to supply the gasification agent (made by a mixture of steam and oxygen). Oxygen (instead of air) allows the production of synthesis gas with higher calorific value and without nitrogen dilution. In this way, the oxygen (and not just the hydrogen) resulting from the splitting of water is also used, as it is considered a less valuable by-product than H_2. Thermal integration between hot and cold process streams is performed to increase overall plant efficiency. According to the authors of the article, the main objective of this research consists in estimating and optimizing the efficiency of the electrolysis-assisted bio-SNG production process, setting some important operational parameters and, through a reliable system configuration, guaranteeing the recovery adequate energy. Furthermore, the developed methodology can provide guidelines to investigate the production of other synthetic fuels (eg methanol, dimethyl ether) via biomass gasification and electrolysis fueled by renewable sources.

A project that has been gaining space is the use of biogas, replacing natural gas, to produce bio-H_2, through the steam reforming process. Biogas consists mainly of methane (CH_4) and carbon dioxide (CO_2) and is formed by the decomposition of organic matter (solid urban waste, industrial waste, agricultural waste, animal waste, among others), through the process of anaerobic digestion. Biogas can be used to generate energy or heat. However, when used as fuel for the transport sector, it must go through a purification stage to eliminate contaminants (CO_2 and H_2S) to acquire the specifications of vehicular natural gas (bio-CH_4).

Likewise, for steam reforming, it is indicated that the biogas contaminants are eliminated (mainly sulfur compounds), to avoid contamination of the catalysts. A basic design diagram comprises the anaerobic digestion (AD), purification, steam reforming (SMR), water–gas shift (WGS) and hydrogen purification (PSA) unit, as shown in Fig. 7.

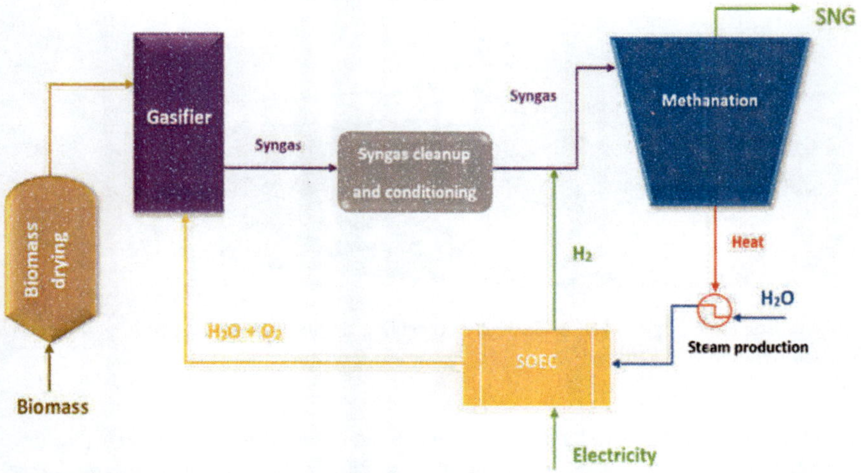

Fig. 6 Biomass process concept for electrolysis-assisted bio-SNG production. Adapted from Giglio et al. [7]

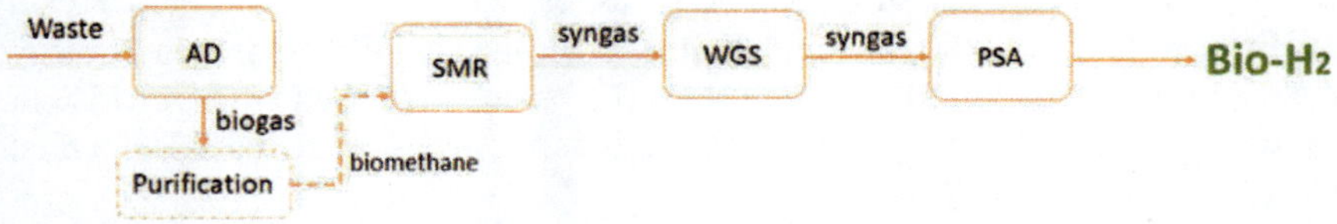

Fig. 7 Simplified diagram of the bio-H_2 production project from waste

In steam reforming (SMR), methane reacts with water steam, under conditions of high temperature and pressure, in the presence of a catalyst, to form syngas (H_2 and CO). Then, in the water–gas shift reactor (WGS), the CO reacts with the water vapor, at a lower temperature than that of the reformer, in the presence of a catalyst, forming more H_2 and also CO_2. The gas effluent will go to the H_2 purification step (PSA), for later use.

3.1 Real Case of Bio-H_2 Production from Gasification

With a view to combating climate change, the Municipality of Cascais (Lisbon, Portugal), in partnership with the companies Floating Particle and IPIAC, is developing a project to produce hydrogen from gasification, which will transform 50 tons of household waste into five tons of hydrogen a year. The hydrogen produced will supply two buses, which already circulate in the municipality of Cascais (Fig. 8a), and, in the future, garbage collection trucks. Named STELLA (Fig. 8b), the proof-of-concept gasification unit (a small reactor) for hydrogen production has a waste

Fig. 8 **a** Bus fueled by bio-H_2. **b** Container containing the gasification reactor to produce H_2. Adapted from Floating Particle [4]

conversion capacity of 250 kg/day and a hydrogen production capacity of 24 kg/day. The purity of hydrogen produced is 99.995% at 25 bar. The electrical load power of the station is 35 kVA [1, 4].

Therefore, this project fits into the Portuguese context for the energy transition, whose National Energy and Climate Plan (PNEC 2030), emphasizes the importance of the production and use of renewable gases, especially bio-H_2 and bio-CH_4, in the different sectors of the economy. Furthermore, the National Hydrogen Strategy (EN-H_2) sets targets for 2030 in the use of H_2, namely: the use of 10% to 15% of H_2 in natural gas networks, 2% to 5% H_2 in industry and 1.5% to 2.5% H_2 in final energy consumption [13].

4 Conclusion

In this article, it was found that fossil fuels are the dominant energy in the transport sector and the major contributors to the increase in greenhouse gases, which categorically shows that without a great acceleration in innovation and use of clean energy, it will be possible to achieve net zero emissions by 2050. Global interest in Hydrogen is growing more and more as a solution to an energy transition. Furthermore, the capture, sequestration, use and storage of carbon (CCUS) is an important alternative not only to limit the concentration of CO_2 in the atmosphere, but also a great opportunity for the use of CO_2, as a raw material and input, in the production of chemicals and fuels.

The Power-to-Liquids (PtL) technological route combines the hydrogenation of CO_2 with the production of hydrogen from the electrolysis of water (from renewable sources), to produce sustainable aviation fuel (SAF), in particular for use in air transport. In addition to this route, interest has been growing in integration projects between thermochemical (gasification and methanation) and electrochemical (electrolysis) processes, to produce gaseous biofuels, such as bio-H_2, bio-CH_4 or bio-SNG, to application in the road transport of people and goods.

It was also observed the importance of public policies and incentive programs and the use of renewable energy (sustainable fuels), technological development, energy efficiency and carbon credits to mitigate greenhouse gas emissions.

Acknowledgements This research was funded by Fundação para a Ciência e a Tecnologia, I.P. (Portuguese Foundation for Science and Technology) under the project UIDB/05064/2020 (VALORIZA – Research Centre for Endogenous Resource Valorization).

References

1. CASCAIS, 2021. From trash to hydrogen: Cascais will supply buses with domestic waste. Available in: https://www.cascais.pt/noticia/do-lixo-ao-hidrogenio-cascais-vai-abastecer-autocarros-com-residuos-domesticos (accessed 1.20.23).
2. CLEW, 2019. Clean Energy Wire. Start-up Sunfire's e-fuels can decarbonise industries most addicted to fossil fuels. Amelang S., 14 June 2019. Available in: https://www.cleanenergywire.org/news/start-sunfires-e-fuels-can-decarbonise-industries-most-addicted-fossil-fuels (accessed 1.2.23).
3. EEA, 2022. EUROPEAN ENVIRONMENT AGENCY. Use of renewable energy for transport in Europe. Available in: https://www.eea.europa.eu/ims/use-of-renewable-energy-for (accessed 1.9.23)
4. FLOATING PARTICLE, 2023. Stella Hydrogen Technology. Available in: https://www.floatingparticle.com/stella/ (accessed 1.20.23).
5. GERMAN ENVIRONMENT AGENCY, 2016. Power-to-Liquids: Potentials and Perspectives for the Future Supply of Renewable Aviation Fuel. September de 2016. Available in: https://www.umweltbundesamt.de/en/publikationen/power-to-liquids-potentials-perspectives (accessed 1.4.23).
6. GCC, 2019. Green Car Congress. Rotterdam The Hague Airport Initiates study for the production of renewable jet fuel from air.31 May 2019. Available in: https://www.greencarcongress.com/2019/05/20190531-rotterdam.html (accessed 12.8.22).
7. GIGLIO *et. al.*, 2021. Integration between biomass gasification and high-temperature electrolysis for synthetic methane production. Biomass and Bioenergy 148 (2021) 106117. https://doi.org/10.1016/j.biombioe.2021.106017. .
8. IEA, 2022a. INTERNATIONAL ENERGY AGENCY. Transport. Improving the Sustainability of passenger and freight transport. Available in: https://www.iea.org/topics/transport (accessed 12.10.22).
9. IEA, 2022b. INTERNATIONAL ENERGY AGENCY. Aviation. Tracking report — September 2022. Available in: https://www.iea.org/reports/aviation (accessed 12.10.22).
10. INERATEC, 2019. Thechnology. Compact Containerized Chemical Plants. Available in: https://ineratec.de/en/technology-2/ (accessed 1.2.23).
11. LBST, 2016. Ludwig bölkow systemtechink. Renewables in Transport 2050. Empowering a Sustainable Mobility Future With Zero Emissions Fuels From Renewable Electricity – Europe and Germany (accessed 12.5.22)
12. PANZONE *et al.*, 2020. Power-to-Liquid catalytic CO2 valorization into fuels and chemicals: focus on the Fischer-Tropsch route. Journal of CO2 Utilization. Volume 38, May 2020, Pages 314–347. https://doi.org/10.1016/j.jcou.2020.02.009
13. PORTUGAL ENERGIA, 2023. National Energy and Climate Plan 2021–2030; National Hydrogen Strategy (EN-H2). Available in: https://www.portugalenergia.pt/setor-energetico/bloco-3/ (accessed 8.18.23).
14. PTX HUB, 2023. ProQR. Climate Neutral Alternative Fuels. Brazil on the way to clean aviation. Available in: https://ptx-hub.org/proqr/ (accessed 1.11.23).

15. ROY & ETHAKOTA, 2022. H2TECH. Solid oxide electrolysis cell (SOEC): Potential technology for low-cost green H_2. Available in: https://h2-tech.com/articles/2022/q4-2022/special-focus-future-of-hydrogen-energy/solid-oxide-electrolysis-cell-soec-potential-technology-for-low-cost-green-h-sub-2-sub/ (accessed 1.17.23).
16. SASOL, 2022. Sasol eco FT and Deutsche Aircraft join forces to accelerate Power-To-Liquid (PTL) aiming for carbon-neutral flight. Available in: https://www.sasol.com/media-centre/media-releases/sasol-ecoft-and-deutsche-aircraft-join-forces-accelerate-power-liquid-ptl-aiming-carbon-neutral (accessed 1.9.23).
17. ZENID, 2021. Sustainable aviation fuel made from air. Fully circular. Available in: https://zenidfuel.com/ (accessed 1.2.23).

Valorisation of Leather Wastes from the Footwear Industry via Pyrolysis: Product Yields and Characterisation

F. Silva, D. Direito, R. Pilão, and A. M. Ribeiro

Abstract The footwear industry, although contributing positively to the Portuguese economy, generates a lot of residues including chromium tanned leather wastes. This study presents results on the slow pyrolysis of these residues at particle sizes between 5 and 6.3 mm and temperatures inside the reactor ranging from 444 to 875 °C. For this range of temperatures, char yield decreased from 36.2 to 28.1% and gas yield increased between 33.2 and 47.6%. The liquid phase yield was highest at 444 °C (30.6%) and then decreased to 24.3% at 875 °C. As the operating temperature increased, there was an increase in the ash content (15.8 to 21.6%) and a decrease in volatiles content (20.9–3.7%) of biochars. The higher heating values (HHV) of the chars were between 21.9 and 23.1 MJ/kg. Preliminary results show that an increase in particle size leads to a rise in char yield. The infrared analysis of the liquid phase indicated the presence of water, alcohols, phenols, carboxylic acids, alkanes, alkenes, and aromatics. The liquid phases had an average water content of 43%, densities between 1110 and 1120 kg/m^3 and the HHV between 12.7 and 15.1 MJ/kg. The gases produced were analysed by gas chromatography. For 538 °C, the production of H_2, CO, CH_4 and CO_2 was, respectively, 1.00, 0.37, 0.18 and 1.37 mol/kg of leather. At 875 °C, H_2 was the most produced gas (2.74 mol/kg leather), followed by CO, CO_2 and CH_4 with quantities of 1.40, 1.26 and 0.42 mol/kg leather.

Keywords Chromium tanned leather wastes · Pyrolysis · Biochar · Bio-oil

F. Silva · R. Pilão · A. M. Ribeiro (✉)
CIETI, ISEP, Polytechnic of Porto, Rua Dr. António Bernardino de Almeida, 4249-015 Porto, Portugal
e-mail: asr@isep.ipp.pt

D. Direito
BIOREF—Laboratório Colaborativo Para as Biorrefinarias, Rua Amieira, Apartado 1089, 4466-901 S. Mamede de Infesta, Portugal

J. R. da. C. S. Galvão et al. (eds.), *Proceedings of the 3rd International Conference on Water Energy Food and Sustainability (ICoWEFS 2023)*, Springer Proceedings in Earth and Environmental Sciences, https://doi.org/10.1007/978-3-031-48532-9_19

1 Introduction

In Portugal, the shoe manufacturing industry contributes significantly to the national economy. In 2017, 83.3 billion pairs of shoes were exported, generating around 1.96 billion euros. Furthermore, 87% of these exports correspond to leather footwear [1]. Nonetheless, the leather shoe industry is very pollutant, since it generates around 70.5 tonnes of residues per day, 53–67 tonnes of which correspond to chromium tanned leather wastes [2]. These are normally disposed in landfills without any treatment, which causes the release of chromium(III) into water streams, with consequent water contamination and pollution of the surrounding environment [3]. So, it is of the utmost importance to find efficient and low carbon footprint waste management techniques to deal with these toxic residues. One of those techniques and the focus of this study is pyrolysis. In this process, waste is thermally decomposed in an inert atmosphere at temperatures ranging from 400 to 1000 °C. The obtained products are biochar (the solid phase), bio-oil (the liquid phase) and the non-condensable gases mainly composed of carbon monoxide (CO), hydrogen (H_2), carbon dioxide (CO_2), methane (CH_4) and other gases in minor percentages [3]. All of these are value-added products that can be used as energy sources by other industries, thus contributing to a circular economy.

Several authors investigated the pyrolysis of leather wastes. Yilmaz et al. [4] studied the pyrolysis of various tanned wastes at temperatures of 450 and 600 °C. Marcilla et al. [5] investigated the fast and slow pyrolysis of chromium tanned leather wastes. Simioni et al. [6] used a fluidized bed reactor for the pyrolysis of chromed leather shavings at temperatures of 500 and 600 °C. Kluska et al. [7] and Fang et al. [8] studied the influence of temperature on the pyrolysis of chrome tanned leather residues. Van Rensburg et al. [9] investigated the pyrolysis of post-consumer leather shoe wastes. González-Lucas [10] studied the microwave-assisted pyrolysis of vegetable and chromium tanned leather residues.

In this work, the influence of temperature on the pyrolysis of leather trimmings from the footwear industry was investigated. Also, some preliminary results of the influence of particle size on char yield are presented. Biochars were characterised in terms of proximate analysis and higher heating value, HHV. The liquid phases were analysed by infrared spectroscopy (FTIR) with attenuated total reflectance (ATR) module. Density, water content and HHV of the liquid phases were also determined. The gas produced was analysed by gas chromatography (GC).

2 Experimental

2.1 Leather Waste Characterisation

The leather wastes used in this work come from a Portuguese footwear manufacture. The leather trimmings were characterised in terms of proximate analysis, namely, moisture (EN14774), ash (EN 15,148) and volatiles contents (EN 14,775). Chromium content was determined by standard ASTM D6714—01. The ultimate analysis was carried out according to standard CENT/TS 15,104. The HHV of the feed material, biochars and bio-oils produced during pyrolysis was measured using a Parr calorimeter 1451.

2.2 Experimental Apparatus and Procedure

Pyrolysis of the leather waste was performed in a refractory steel reactor, 78 cm in height and 9.7 cm in internal diameter. The pyrolyser was equipped with a vertical split furnace linked to a control system and with two thermocouples, one, the set-point control placed on the oven, and another located inside the reactor.

For each experiment a wire mesh basket was filled with a 100 g of pre-dried leather, which was then placed in the heating zone of the reactor. The carrier gas was nitrogen, fed at the top of the reactor at a flow rate of 0.105 kg/h.

The volatiles exiting the reactor flowed through a vertical double pipe heat exchanger, with water as cooling fluid, and a set of two trapping bottles immersed in an iced bath where the liquid phase was collected. After, the non-condensable gas flowed through a silica gel fixed bed to remove traces of water. During each experiment various samples of the gas phase were collected in Tedlar bags for further analysis by GC.

When each experiment ended, the heating system was switched off and the carrier gas was left to run for another two hours. After the system was cooled, the mass of biochar in the basket and the bio-oil collected in the trapping bottles were weighted. The amount of gas produced was determined by performing mass balances.

Pyrolysis experiments were carried at set point temperatures (T_{set}) in the range of 565–1030 °C (corresponding to final inside temperatures, T_i, in the range of 444 to 875 °C), with leather residues having a particle size between 5.3 and 6 mm. Also, some preliminary experiments were done to verify the effect of particle size on char yield. These were performed using leather squares with 10 mm side length, for $T_i =$ of 444 and 630 °C.

2.3 Characterisation of Solid, Liquid and Gas Phases

The biochar was characterised in terms of its proximate analysis (moisture, volatiles, ash, and fixed carbon contents) and HHV using the methods described in Sect. 2.1.

Samples of the liquid phase were analysed by FTIR with ATR module. The infrared spectra were collected in the 4000–400 cm^{-1} region, with a resolution of 4 cm^{-1}. The HHV of the liquid phases was measured using the Parr calorimeter. The bio-oil density was determined with a 10 mL glass pycnometer at 20 °C (ASTM D 1298). The water content was determined using a Karl Fisher apparatus. The gas fraction was analysed in a Dani 1000 DPC gas chromatograph equipped with an OPT333 injector, a 60/80 Carbonex 1000 column and a thermal conductivity detector TCD OPT266. Argon was used as the carrier gas and the gases identified were CO_2, CO, H_2, CH_4, O_2 and N_2.

3 Results and Discussion

3.1 Properties of Leather Waste

The moisture content, proximate and ultimate analysis on a dry basis (db), chromium oxide content and HHV on a dry basis of the leather wastes used in this work, together with data provided by Fang et al. [8] for chrome tanned leather shavings, and by González-Lucas et al. [10] for chrome leather trimmings, are listed in Table 1.

Table 1 Proximate and ultimate analysis, chromium oxide content and HHV of the residues

	This work	Fang et al. [8]	González-Lucas et al. [10]
Moisture (wt%)	12.3	9.2	36.9
Proximate analysis (db—wt%)			
Ash	6.4	10.32	7.16
Volatiles	73.1	74.60	–
Fixed carbon	20.5	15.08	–
Ultimate analysis (db—wt%)			
C	48.9	42.13	44.71
H	6.4	5.52	5.83
N	11.4	13.41	17.6
O	33.3	27.18	23.7
S	–	1.44	1.09
Chromium oxide (wt%)	2.5	2.18	–
HHV (MJ/kg)	20.3 (db)	17.2	12–14

The proximate analysis shows that this leather waste has a high value of volatiles (73.1%), close to the 74.6% presented by [8]. The volatile content is related to the production of bio-oil and non-condensable gas during the pyrolysis process. The ash content was 6.4%, lower than the data obtained by [8] and [10]. The percentage of chromium oxide (2.5%) is close to the value determined by [8] (2.18%). The fixed carbon content of leather waste was 20.5%, and the amount of fixed carbon content is positively correlated with biochar yield during pyrolysis process [11].

The mass percentage of carbon (C), hydrogen (H), nitrogen (N) and oxygen (O) in the present leather residues was of 48.9, 6.4, 11.4 and 33.3%, respectively. For C, H and O, these values are slightly higher than the data presented by [8] and [10], but the N content was lower. The nitrogen is derived from the protein, which is one of the main components of leather [12].

The higher heating value was 20.3 MJ/kg, higher than the values of 17.2 and 12–14 MJ/kg determined, respectively, by [8] and [10]. This difference could be due to the type of leather waste or to the moisture content of the raw materials. The HHV of the leather trimmings of this work was close to values presented by some lignocellulosic biomass, such as olive stones, 21.1 MJ/kg [13], and almond shells, 19.72 MJ/kg [14].

3.2 *Effect of Parameters on Pyrolysis Product Yields*

Effect of pyrolysis temperature. As mentioned before, pyrolysis experiments were carried at T_{set} in the range of 565 to 1030 °C, corresponding to T_i of 444 to 875 °C, respectively, using a particle size between 5.3 and 6 mm.

Figure 1 shows the trend of the yields of the char, liquid phase, and non-condensable gas as a function of T_i. In general, as T_i rose from 444 to 875 °C, char yield decreased from 36.2 to 28.1% and the gas yields increased from 33.2 to 47.6%. The highest liquid yield, 30.6%, was obtained at a temperature of 444 °C and then decreased to 24.3% at 875 °C.

Similar behaviour was reported by other researchers. Marcilla et al. [5] using fast pyrolysis and for temperatures of 450, 500 and 550 °C, obtained char yields of 38.5, 30.4 and 31%, respectively. Gas yields increased from 20.5 to 26.6% as the pyrolysis

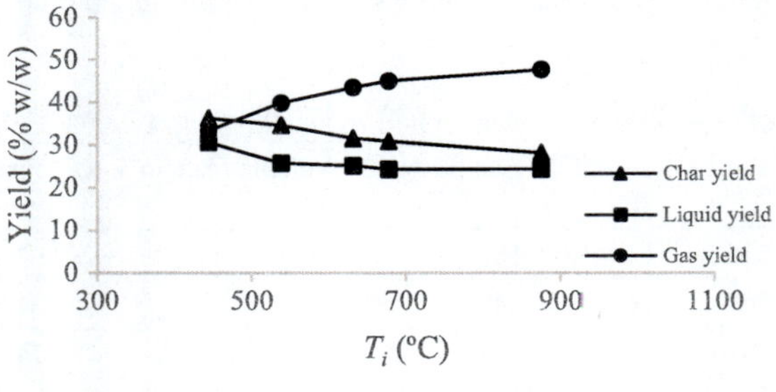

Fig. 1 Yields of solid, liquid and gas phases as a function of T_i

temperature was rose. Liquid yield presented a maximum of 44.5% at 500 °C. Kluska et al. [7] performed pyrolysis experiments in the range of 300 to 500 °C. The solid yield was about 33% in the range of 300 to 450 °C and 29% at 500 °C. For the liquid phase, the yield increased from 21 to 40%, presenting a maximum of 42% at 400 °C. The decrease in liquid yield at high temperatures could be related to the gas phase secondary cracking reactions that contribute to the release of low molecular weight compounds.

Effect of particle size on char. Some preliminary results were carried out to determine the influence of particle size on char yield using particle sizes of d_p = 5–6.3 mm and squares with a side length l = 10 mm. When changing from one particle type to the other at T_i = 444 °C, char yield increased from 36.2 to 39.4% and at T_i = 630 °C, the increase was from 31.5 to 32.9%. Similar behaviour was found by Somerville and Deev [15] when working with the pyrolysis of radiata pine wood, with nitrogen as the carrier gas. As particle size was increased from 0.5 to 1 mm to 67.5 mm, char yield rose from 16.6 to 25%. Ogunkanmi et al. [16], working with the pyrolysis of palm kernel shells at 550 °C, found a change in char yield from 19.33 to 70.67% as the particle size rose from 1.18 to 5 mm. Larger particles offer more resistance to the escape of condensable gases, leading to secondary cracking which results in an increase in solid yield [17].

3.3 Analysis of Solid, Liquid, and Gaseous Products

Proximate analysis and high heating value of solid products. The proximate analysis (db) and HHV of chars obtained during the pyrolysis experiments was carried out. Table 2 shows the results obtained for T_i = 444, 631 and 875 °C. As temperature was rose the ash content increased from 15.8 to 21.6%, volatiles decreased from 20.9 to 3.7% and the fixed carbon increased from 63.3 to 74.7%. Yilmaz et al. [4] obtained chars from chromium tanned leather with ash concentrations of 19.5 and 22.6%, at temperatures of 450 and 600 °C. Biochars produced by Gorshkov et al. [18] during the pyrolysis of several nuts at temperatures of 400 to 600 °C, showed an increase in the ash content and a decrease in the volatiles as the temperature was rose.

The HHV was of 21.9 MJ/kg for chars obtained at T_i = 444 and 630 °C and 23.2 MJ/kg for the char at T_i = 875 °C. The HHV of chars was higher than for the raw material (20.3 MJ/kg). Kluska et al. [7] produced chars from the pyrolysis of leather

Table 2 Proximate analysis (db) and HHV of chars

T_{set} (°C)	T_i (°C)	Ash (% w/w)	Volatile matter (% w/w)	Fixed carbon (% w/w)	HHV (MJ/kg)
565	444	15.8	20.9	63.3	21.9
765	630	18.8	9.4	71.8	21.9
1030	875	21.6	3.7	74.7	23.1

wastes at 300 to 500 °C, which presented HHV between of 18.55 to 21.50 MJ/kg. Van Rensburg et al. [9] found an HHV = 25.55 MJ/kg for chars obtained from used leather footwear at temperatures in the range of 409 to 690 °C. According to [9], the chars from leather wastes can be used in the metallurgical industry, as a fuel, in the production of activated carbon and as an adsorbent.

FTIR analysis, density, water content and HHV of liquid products. The liquid phases were analysed by FTIR-ATR, a technique which allows the identification of functional groups. As all the samples analysed presented similar curves, only the data obtained for the bio-oil from the pyrolysis experiment at $T_i = 444$ °C is shown in Fig. 2. Table 3 presents the peaks identified, the characteristic wavelength of each functional group and compounds associated: water, alcohols, phenols, carboxylic acids, alkanes, alkenes, and aromatics [19, 20].

The presence of these substances in bio-oils from the pyrolysis of chromium leather wastes was confirmed by Marcilla et al. [5], Simioni et al. [6] and Van Rensburg et al. [9], using gas chromatography-mass spectrometry (GC–MS) analysis.

Fig. 2 FTIR spectrum of the liquid phase obtained at $T_i =$ 444 °C

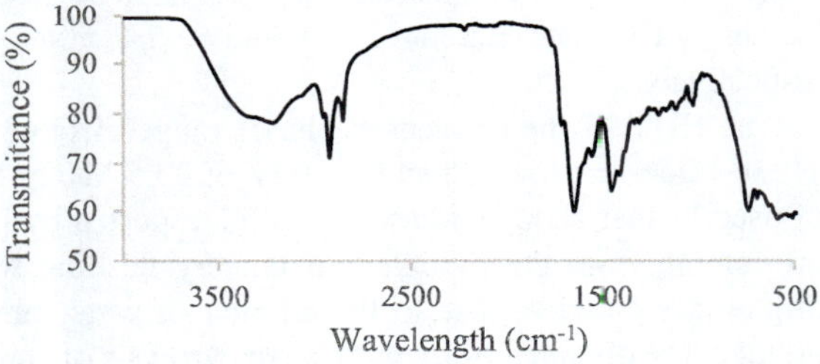

Table 3 Results obtained from the FTIR-ATR analysis at $T_i = 444$ °C

Peak wavelength (cm^{-1})	Wavelength (cm^{-1})	Functional group	Compounds
3388	3200–3700	O–H stretching vibrations	Water, alcohols, phenol, or carboxylic acids
2933	2800–3000	C-H stretching vibrations	Alkanes
2855			
1710	1550–1750	C = C stretching vibrations	Alkenes, carboxylic groups
1612			
1515	1600–1430	C = C stretching vibrations	Aromatics
1463			
1364	1350–1500	C-H deformation vibrations	Alkanes
1215	1275–1000	C-H deformation vibrations	Aromatics
1110			
1021			

Table 4 Density and heating values of the liquid phases

T_{set} (°C)	T_i (°C)	Density (kg/m^3)	HHV (MJ/kg)
565	444	1110	12.7
765	630	1120	13.2
1030	875	1120	15.1

Density and HHV of the liquid phase produced at T_i = 444, 630 and 875 °C are shown in Table 4. The density of the liquid phase varied between 1110 and 1120 kg/m^3 and was slightly higher than the value of 997 kg/m^3, reported by Amdouni et al. [21] for bio-oil from the pyrolysis of tannery fleshing. For lignocellulosic biomass, Shah et al. [22] compiled densities in the range of 951–1190 kg/m^3. The liquid phases exhibited an average water content of 43% (w/w). The presence of water in the liquid phase has a positive impact in lowering its viscosity (increases the fluidity), but a negative influence on the heating value. As the water content increases, there is a decrease in the HHV, and the possibility of phase separation rises [23]. According to Yilmaz et al. [4], the condensed phase produced by the pyrolysis of chromium-tanned leather had a water content of 65.9 and 67.5%, at temperatures of 450 and 600 °C, respectively.

The HHV of the condensed phases ranged from 12.7 to 15.1 MJ/kg (Table 4). These values are lower than the HHV of 33.6 MJ/kg for bio-oil from the pyrolysis of used leather shoe residues at 550 °C, reported by [9], and 39.36 MJ/kg for bio-oil coming from the pyrolysis of tannery fleshing wastes at 600 °C [21]. These differences could be due to the amount of water present in the bio-oils. For the liquid phase obtained from microwave pyrolysis of chromium tanned leather wastes, González-Lucas et al. [10] reported an HHV of 2 MJ/kg and for the respective heavy fraction, the HHV increased to 21.5 MJ/kg. The bio-oils obtained from pyrolysis could be used for energy production, as they can be fired in boilers and furnaces, as a source for obtaining chemicals or as a transport fuel after proper treatment [17].

Gas chromatography analysis. The non condensable gases produced during the pyrolysis of the leather residues were analysed by GC. Figure 3 shows the variation of the volumetric percentage of the different gases produced (CO_2, CO, CH_4 and H_2) with time, for T_i = 875 °C. For this temperature, the most released gas was H_2 with a maximum volumetric fraction of 15.2% at t = 2940 s and at a temperature of 600 °C. CO presented a maximum concentration of 8.5% at t = 2760 s and 502 °C, followed by CO_2 with a maximum volumetric fraction of 5.8% (t = 2250 s and 256 °C). The lowest maximum concentration, 2.9%, was presented by CH_4 at t = 2760 s and 502 °C. For the pyrolysis experiment run at T_i = 538 °C, the gas with the highest concentration was CO_2, with a maximum volumetric fraction of 6.8% at t = 2040 s and at 250 °C, followed by H_2 with maximum concentration of 5.5%, at t = 2760 s and 447 °C. The maximum volumetric fraction of CO was 2.2%, at t = 2040 s and at 250 °C. Finally, CH_4 presented the lowest maximum concentration of 1.0%, at t = 2760 and at 447 °C.

Table 5 lists the amount of H_2, CO, CH_4 and CO_2, in mol/kg of leather, released during the pyrolysis experiments at T_i = 536 and 875 °C and the HHV of the produced

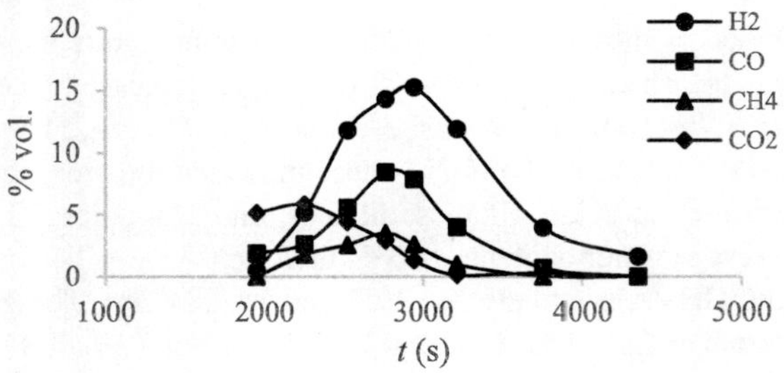

Fig. 3 Variation of the concentrations of CO_2, CO, CH_4 and H_2 with time, for $T_i = 875$ °C

Table 5 Total amount of H_2, CO, CH_4 and CO_2 produced during the experiments at $T_i = 538$ and 875 °C and HHV of the gas mixture

T_i (°C)	Gas production (mol/kg of leather)				HHV (kJ/Nm^3)
	H_2	CO	CH_4	CO_2	
538	1.00	0.37	0.18	1.37	279.3
875	2.74	1.40	0.42	1.26	731.5

gas (in kJ/Nm^3), calculated according to Eq. (1) [17]:

$$HHV = 10 \times (12.74 \times \%H_2 + 12.63 \times \%CO + 39.82 \times \%CH_4) \quad (1)$$

The total production of each individual gas was determined by considering the area under the curve of the function, molar flow rate of the gas versus time, and a constant mass flow of nitrogen equal to 0.105 kg/h. The HHV of the produced gas increased from 279.3 to 731.5 kJ/Nm^3 as the temperature rose from $T_i = 538$ to 875 °C. The presence of H_2, CO_2, CO and CH_4 in the produced gas was also reported by [7] and [8].

4 Conclusions

This study presents results on the pyrolysis of waste leather trimmings from the footwear industry, on a fixed bed reactor. The effect of operating temperature on product yields was studied for T_i ranging from 444 to 875 °C and using leather with particle size d_p = 5–6.3 mm. As the temperature rose, char yield decreased from 36.2 to 28.1%, gas yield increased from 33.2 to 47.6% and the liquid phase yield was highest at 444 °C (30.6%) and then decreased to 24.3% at 875 °C. Preliminary results showing the influence of particle size on char yield were carried out for T_i equal to 444 and 630 °C and for particle sizes of d_p = 5—6.3 mm and squares with $l = 10$ mm. For $T_i = 444$ °C, char yield changed from 36.2 to 39.4%, and for $T_i = 630$ °C, the increase in char yield was from 31.5 to 32.9%, as the particle size increased from d_p = 5.3–6.0 mm to squares with $l = 10$ mm. The proximate analysis of the produced biochars showed that, as the operating temperature increased, there was an increment on the ash content (15.8 to 21.6%) and a decrease in the percentage of volatiles (20.9 to 3.7%). The HHV of the chars was between 21.9 and 23.1 MJ/kg. The FTIR-ATR analysis of the liquid phase indicated the presence of water, alcohols, phenols, carboxylic acids, alkanes, alkenes, and aromatics. The liquid phases presented an average water content of 43%, their density was between 1110 and 1120 kg/m^3 and the HHV varied between 12.7 and 15.1 MJ/kg. The gas produced was analysed in terms of CO_2, CO, CH_4 and H_2 by CG. For $T_i = 538$ °C, the production of H_2, CO, CH_4 and CO_2 was, respectively, 1.00, 0.37, 0.18 and 1.37 mol/kg of leather. At $T_i = 875$ °C, H_2 was the most produced gas (2.74 mol/ kg of leather), followed by CO, CO_2 and CH_4 with quantities of 1.40, 1.26 and 0.42 mol/kg of leather, respectively.

Acknowledgements This work was supported by Fundação para a Ciência e Tecnologia—SFRH/BD/137913/2018.

References

1. APPICAPS Homepage, https://www.apiccaps.pt/getfilev2/download.aspx?f=/2018-facts--numbers.pdf, last accessed 2023/01/22.
2. CTCP Homepage, https://www.ctcp.pt/detalhe-publicacao/valorizacao-de-residuos-de-couro-curtido-com-cromiocalcado/373.html, last accessed 2023/01/22.
3. Jadhav, N.C., Jadhav, A.C.: Waste and 3R's in Footwear and Leather Sectors. In: Muthu, S. (eds): Leather and Footwear Sustainability. Textile Science and Clothing Technology. 1st ed. Springer, Singapore (2020).
4. Yılmaz, O., Kantarli, I.C., Yuksel, M., Saglam, M., Yanik, J. "Conversion of leather wastes to useful products," Resources, Conservation and Recycling 49(4), 436–448 (2007).
5. Marcilla, A., León, M., García, Á, N., Bañón, E., Martínez, P.: Upgrading of Tannery Wastes under Fast and Slow Pyrolysis Conditions. Industrial & Engineering Chemistry Research 51(8), 3246–3255 (2012).
6. Simioni, T., Matos, E., Bacca, E., Perondi, D., Godinho, M., Dettmer, A.: Pyrolysis of chrome leather waste shavings in fluidized bed. Journal of the American Leather Chemists Association 109, 342–352 (2014).
7. Kluska, J., Ochnio, M., Kardaś, D., Heda, L.: The influence of temperature on the physico-chemical properties of products of pyrolysis of leather-tannery waste. Waste Management 88, 248–256 (2019).
8. Fang, C., Jiang, X., Lv, G., Yan, J., Lin, X., Song, H., Cao, J.: Pyrolysis characteristics and Cr speciation of chrome-tanned leather shavings: influence of pyrolysis temperature. Energy Sources, Part A: Recovery, Utilization, and Environmental Effects 41(7), 881–891 (2019).
9. Van Rensburg, M., Nkomo, S., Mkhize, N.: Characterization and pyrolysis of post-consumer leather shoe waste for the recovery of valuable chemicals. Detritus 14, 92–107 (2021).
10. González-Lucas, M., Peinado, M., Vaquero, J., Nozal, L., Aguirre, J., González-Egido, S.: Microwave-assisted pyrolysis of leather waste. Energies 15, 1273 (2022).
11. Zaman, C., Pal, K., Yehye, W., Sagadevan, S., Shah, S., Adebisi, G., Marliana, E., Rafique, R., Johan, R.: Pyrolysis: A Sustainable Way to Generate Energy from Waste. In Samer, M. (eds): Pyrolisys, 1st edn. InTech (2017).
12. Cabeza, L., Taylor, M., DiMaio, G., Brown, E., Marmer, W., Carrió, R., Celma, P., Cot, J.: Processing of leather waste: pilot scale studies on chrome shavings. Isolation of potentially valuable protein products and chromium. Waste Management 18(3), 211–218 (1998).

13. Sanchez, F., San Miguel, G.: Improved fuel properties of whole table olive stones via pyrolytic processing. Biomass and Bioenergy 92, 1–11 (2016).
14. Din, M., Amanat, A., Hussain, Z., Khalid, R., Rauf, A.: Pyrolysis of almond shell biomass: effect of temperature and catalyst on product yield. International Journal of Environmental Analytical Chemistry, 1–12 (2021).
15. Somerville, M., Deev, A.: The effect of heating rate, particle size and gas flow on the yield of charcoal during the pyrolysis of radiata pine wood. Renewable Energy 151, 419–425 (2020).
16. Ogunkanmi, J., Kulla, D., Omisanya, N., Sumaila, M., Obada, D., Dodoo-Arhin, D.: Extraction of bio-oil during pyrolysis of locally sourced palm kernel shells: Effect of process parameters. Case Studies in Thermal Engineering 12, 711–716 (2018).
17. Basu, P.: Biomass Gasification, Pyrolysis and Torrefaction. Practical Design and Theory. 3rd edn. Elsevier Inc., London (2018).
18. Gorshkov, A., Berezikov, N., Kaltaev, A., Yankovsky, S., Slyusarsky, K., Tabakaev, R., Larionov, K.: Analysis of the physicochemical characteristics of biochar obtained by slow pyrolysis of nut shells in a nitrogen atmosphere. Energies 14(23), 8085 (2021).
19. Stuart, B: Infrared Spectroscopy: Fundamentals and Applications. John Wiley and Sons, Ltd. (2004).
20. González, F., Ramiro, A., González-Garcia, C., Gánan, J., Encinar, J., Sabio, E., Rubiales, J.: Pyrolysis of almond shells. Energy applications of fractions. Industrial & Engineering Chemistry Research 44, 3003–3012 (2005).
21. Amdouni, S., Hassen, A., Asmi, A., Chagtmi, R., Haddad, K., Jamaaoui, F., Khedhira, H., Chérif, C.: Tannery fleshing wastes conversion into high value-added biofuels and biochars using pyrolysis process. Fuel 294(7), 120423, (2021).
22. Shah, M., Khan, N., Kumar, V., Qurashi, A.: Pyrolysis of walnut residues in a fixed bed reactor: effects of process parameters, chemical and functional properties of bio-oil. Journal of Environmental Chemical Engineering 9(4), 105564, (2021).
23. Bardalai, M., Mahanta, D.: A review of physical properties of biomass pyrolysis oil. International Journal of Renewable Energy Research 5(1), 277-286 (2015).

Cashew Nutshell Waste as an Energy Feedstock: Analysis of the Potential for Combustion and Gasification

Raul Pereira Micena, Luís Filipe do Carmo Calado, Celso Eduardo Tuna, José Luz Silveira, and Paulo Sérgio Duque de Brito

Abstract Considered an agricultural product of great socioeconomic interest in northeastern Brazil, the cashew nut has as one of its main by-products the kernel. This inner part of the nut is industrialized through processes that consume, most of the time, environmentally aggressive energy resources, such as firewood and LPG fossil gas. Being the nutshell a bio-waste of these processes, we evaluated the potential application of this material to replace firewood and LPG, in its solid state in natura and in the form of syngas resulting from its thermal gasification process. The cashew nut processing has a thermal fuel consumption between 2 and 7 GJ per ton of processed cashew kernel, depending on whether the fuel is LPG or firewood. The resulting emissions reach values between 142 and 727 kg of CO_2eq per ton of processed kernel, respectively. Given the quantities of nutshell produced in the processing process, it was calculated that the nutshells have a potential to provide between 28.5 and 45.7 GJ per ton of kernel in the form of solid fuel. If gasified, they can provide up to 33.69 GJ per ton of kernel, values much higher than the energy needs of the processes involved. These results allow us to conclude that the nutshell has good

R. P. Micena · C. E. Tuna
Laboratory of Optimization of Energy Systems (LOSE), UNESP—São Paulo State University, Institute of Bioenergy Research (IPBEN-UNESP), São Paulo, Brazil
e-mail: raul.micena@unesp.br

C. E. Tuna
e-mail: celso.tuna@unesp.br

L. F. do Carmo Calado · P. S. D. de Brito (✉)
VALORIZA-Research Center for Endogenous Resource Valorisation, Polytechnic Institute of Portalegre, 7300-555 Portalegre, Portugal
e-mail: pbrito@ipportalegre.pt

J. L. Silveira
Laboratory of Optimization of Energy Systems (LOSE), UNESP—São Paulo State University, Institute of Bioenergy Research (IPBEN-UNESP), Núcleo de Revalorização de Resíduos (REVALORES), Universidade Federal Do ABC (UFABC), São Paulo, Brasil
e-mail: jose.luz@unesp.br
URL: https://www.feg.unesp.br/ipben

J. R. da. C. S. Galvão et al. (eds.), *Proceedings of the 3rd International Conference on Water Energy Food and Sustainability (ICoWEFS 2023)*, Springer Proceedings in Earth and Environmental Sciences, https://doi.org/10.1007/978-3-031-48532-9_20

potential for energy utilization in agro-industrial processes, both as a substitute for firewood and LPG gas.

Keywords Cashew nutshell · Gasification · GHG emissions

1 Introduction

In Brazil, one of the agro-industrial products of high economic and social relevance, especially in the northeastern region, is the cashew. The cashew tree, or cashew (*Anacardium occidentale*), is native to the northeastern region of Brazil and presents considerable adaptability to poorly fertile soils, hot climates, and water stress. Due to these particularities, cashew has stood out as a product of great economic importance for the northeastern region, especially where the semi-arid climate predominates. In addition, its production occurs during the dry season, in the off-season of annual crops, making it an agricultural product of socioeconomic value, in the field and in the industries.

Currently, the vast majority of Brazilian commercial plantations (about 95%) are located on small and medium-sized farms (<100 ha), cultivated by single producers or in associations in small communities or rural settlements. The rest of the production is exploited by large properties (>100 ha), the same as in the 1960's and 1980's with common cashew trees, with high heterogeneity among plants and gaps in the stands, which result in low productivity [1].

In 2021, cashew production in Brazil was 110.6 thousand of tons, grown on 427.6 thousand ha, grown mainly in the states of Ceará, Piauí and Rio Grande do Norte, corresponding to 91.0% of Brazilian production. In the same year, the average yield of the Brazilian production was 288 kg/ha [2].

In 2021 the world's 20 largest cashew producers produced 3.67 million tons of cashew nuts, with this ranking led by Ivory Coast (838,000 t), India (738,000 t) and Vietnam (399,000 t), according to data provided FAO [3].

1.1 Cashew, Its Parts and Products

The edible product of the cashew tree is composed of the nut, the true fruit, and the hypertrophied peduncle, the pseudo fruit. The mass of the nut corresponds to about 10% of the cashew mass. The nut is composed of the Cashew Nutshell (CNS, 65–70% of the nut mass), kernel skin (about 3%) and kernel (28–30%). From the CNS is extracted the Cashew Shell Liquid (CNSL), with which can be obtained paints, varnishes, plastics, phenolic resins and waterproofing agents. The CNSL represents about 20% of the mass of the whole nut. [4]. From the shell can be obtained tannins in general, while the Cashew Kernel is the edible part, which can be prepared for

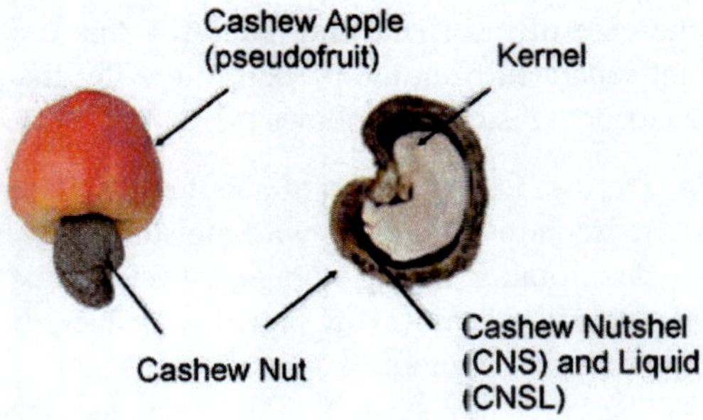

Fig. 1 Adapted from [5]

Table 1 Using data from [4]

Component	Mass (%)
Cashew nut	100
Deoiled cashew nutshell (CNS)	52
Cashew nutshell liquid (CNSL)	13
Pelicule	3.0
Kernel	32

consumption raw, with or without salt and or in the form of flour [1]. The main parts are shown in Fig. 1.

The cashew nut is composed of some main parts, which are briefly shown in Fig. 1. The gravimetric distribution of these parts is shown in Table 1.

From the residues derived from the processing of cashew nuts, the CNS can be highlighted and corresponds to more than half of the weight of the nut, as shown in Table 1. In the form of a bagasse after the extraction of the CNSL, it can be used for energy purposes in the form of a solid fuel, or in the form of a gaseous fuel, by means of biodigestion or gasification processes.

Besides other industrial applications, the CNSL itself has potential for energy utilization, having a lower calorific value equal to 36.26 MJ/kg, close to that of diesel oil [6]. The CNSL can be used as an additive or even as a raw material for biodiesel production [7].

Both the pseudo-fruit and the kernel contained in the nut are edible and economically important, giving origin to several derived products. In the case of the kernel, an industrial process of several steps is necessary for its preparation in the desired conditions. Such processes demand energy consumption, often through the consumption of fossil fuels or deforested firewood.

1.2 *Cashew Nut Processing*

Cashew nut processing is the process of transforming the raw cashew nuts into a product (kernel) ready for food consumption. On a small industrial scale, as is the case of most Brazilian facilities, this process starts with the reception of the nut separated from the pseudofruit. After that, the processing is performed with a sequence of steps, as follows [8].

- Drying: Laying layers of about 30 cm in terraces or cemented patios, with daily rotation to obtain nuts with moisture between 7 and 8% to avoid problems with deterioration during storage, especially due to the proliferation of fungi.
- Cleaning: removal of pieces of leaves, twigs, soil, sand and residues of the peduncle (pseudofruit).
- Classification: separation in perforated plates or sieves for separation according to the size of the nut.
- Storage: Storage in bags or piles of bags for the organization of production and transport during operations.
- Weighing: To control the production stock, as well as check the quality, and the unwanted rewetting can be observed.
- Autoclaving: In preparation for cutting, the nuts are cooked at 110°/10 min in LPG gas stove or in a simple cauldron for 30 min. More rustic equipment can be run on wood or chestnutshell.
- Cooling/Drying: cooled in a ventilated place for 20 min and then dried naturally to the weather.
- CNSL Extraction: Manual operation done with a knife and stylus and other instruments to open the bark and remove the kernel.
- Stuffing: To obtain kernels with humidity between 2.5 and 3.0% they are stored in greenhouses with hot air circulation between 60 and 70 degrees for 6–8 h.
- Cooling: In trays for about two hours at room temperature.
- Peeling: After the drying and cooling process, in general the peel is easily removed from the kernel and separated by hand. Hard beans must be returned to the parboiling operation.
- Selection/Sorting: At this stage the nut kernels are sorted according to size, integrity and color.
- Frying and salting: Bathed in hot oil for 3–6 min and then removed and freed of excess with absorbent material or, for best efficiency, a centrifuge. And then the nut kernels can be salted according to the desired product.
- Raw or fried cashew nut kernels packaging.

The present work is focused on the survey of greenhouse gas emissions associated with this industrial process, as well as the survey of a potential use of its own waste as an environmentally more appropriate alternative. In this sense the CNS is analyzed as a possible fuel to replace firewood, as well as a gaseous fuel to replace LPG, if used as a raw material in a gasification process.

2 Methodology

2.1 *Survey of the Energy Consumption in Cashew Nut Processing*

Once the industrial process of the cashew nut core is described, a survey of the energy consumption involved, and their respective emissions is carried out. The potential of waste from this process is evaluated, precisely CNS, for replacing current non-renewable fuels, since many times, in Brazil, firewood is obtained from deforestation. The direct combustion of CNS is evaluated as well as its gasification to obtain syngas.

Among all those mentioned processes, those that consume energy in some way, it was calculated that for each 550 kg of cashew nuts processed, 26 kg of LPG gas and 16.15 kWh of electricity are needed. Thus, in specific terms, the LPG consumption is 47.27 kg/ton, while the electrical consumption is 29.36 kWh/ton [9, 10].

Larger processing systems often use a boiler for cooking the nuts. In these cases, for production above 2 tons/day of nuts, 2 m^3/day of firewood is required [9].

Table 2 details the energy consumption of cashew nut processing, typical for a small to medium sized processing plant. Energy consumption is given per ton of processed nut kernel, with data from [11, 12].

From Table 2, with the product between the LHV and the consumption of each fuel, one can see an energy consumption that varies between 2.2 and 7.4 GJ/ton, if LPG gas or firewood is used, respectively, and 29.4 kWh/ton.

The thermal energy consumed in the CNS beneficiation is given by Eq. (1):

$$Ec = Ce \times LHV \tag{1}$$

where E_c is the energy consumption (GJ) per ton of processed kernel, C_e is the specific energy consumption (kg/t) and *LHV* is the Lower Calorific Value (MJ/kg) of the fuel.

The greenhouse gases emissions ($kgCO_2eq/t$) produced from processing per ton of cashew nut kernel processed are calculated with Eq. (2):

$$G_{CO2} = Ec \times EF \tag{2}$$

where E_c is the energy consumed, calculated with Eq. (1) and *EF* is the emission factor ($kgCO_{2eq}/t$) of each fuel used.

Table 2 Energy consumption of cashew nut processing

Energy input	Consumption	LHV (MJ/kg)	Emission factors (EF)	
LPG	47.3 kg/ton	46.3	2.92	$kgCO_2/MJ$
Firewood	500.0 kg/ton	14.7	1.4	$kgCO_2/MJ$
Electricity	29.4 kWh/ton	–	0.126	$kgCO_2/kWh$

Table 3 Energy available from cashew nut processing residues

Waste	Production (t/t_{pk})	Density (kg/m^3)	LHV (MJ/kg)	Available energy (MJ/kg_{pk})
CNS without CNSL	1.625	207	17.5	28.5
CNS with CNSL	2.03	350	22.5	45.7
CNSL	0.406	982	36.26	14.7

2.2 *The Use of CNS as a Gasification Fuel*

Cashew Nutshell can be used as fuel in the gasification process and SHRIRAME et al. (2022) studied the gasification of CNS with and without the presence of CNSL, whose properties are shown in Table 3:

The production of CNS and CNSL are shown in Table 3 as a function of the mass of processed nut kernel. To calculate the potential production of syngas from the gasification of CNS, a production of 5.91 Nm^3/kg is considered in relation to the total gasified biomass. It is also considered for the resulting syngas a LHV of 3.51 MJ/m^3. These values were obtained experimentally with the gasification of CNS by [13].

Thus, the energy potential (E_g) per kg of processed nut kernel (MJ/kg) of the syngas produced from the deoiled CNS gasification calculated with the Eq. 3:

$$E_g = V_s \times LHV_s \tag{3}$$

where V_s is the assumed syngas production potential in relation to the mass of CNS (m^3/kg) and LHV_s is the assumed lower heating value of the produced syngas (MJ/m^3).

3 Results and Discussion

With the emission factors of these energy sources in Table 2, a total of between 141.7 and 727.4 kg of CO_2 equivalent per ton of processed cashew kernel is reached.

Regarding the greenhouse gas emissions resulting from the cashew nut processing, with Eq. (1) we obtained values of thermal energy (LPG and firewood) consumed per ton of kernel produced and with Eq. (2) we calculated the total emissions. The results are shown in Table 4.

Consumption of electricity and LPG, and electricity and firewood, leads to total emissions equivalent to 141.74 and 727.42 kg of CO_{2eq} per ton of processed nut kernel, respectively. These values are obtained with the sum of the total emissions resulting from the use of LPG plus electricity and firewood plus electricity, respectively, with values according to Table 4.

Table 4 Specific energy consumption and total emissions per ton of kernel

Energy input	Energy consumption (GJ/t_{pk}) or (kWh/t)	Total emissions (kg of CO_2eq/t_{pk})
LPG	2.19	138.03
Firewood	7.35	723.71
Electricity	29.36	3.71

Table 3 shows that the CNS without CNSL has an energy potential of 28.5 MJ/kg_{pk}, or 28.5 GJ/t_{pk}, significantly higher than the energy consumption of firewood or LPG, shown in Table 4. This allows to state that the CNS presents itself as a viable substitute for such feedstocks.

According to Table 2, the processing of 1 kg of kernel results in 1.625 kg of chestnut shell. With the considered potential of 5.91 Nm3/kg of chestnut shell, there is therefore a potential of 9.60 Nm^3 of syngas for each kg of kernel. From Eq. 3 we have an energy potential of 33.69 MJ available for each kg of kernel processed.

The replacement of firewood by CNS can be a more easily implemented solution in small installations. While gasification would require capital investments, this deserves further analysis regarding the best scaling of this process.

4 Conclusions

A survey of energy consumption of the cashew nut processing process reveals values between 2.19 and 7.35 GJ per ton of processed cashew kernel, depending on whether the fuel is LPG or firewood. The emissions reach values between 141.74 and 727.42 $kgCO_{2eq}$ per ton of processed kernel, respectively. These values alone lead at first sight to the recommendation of the preferential use of LPG over firewood, also due to the greater energy efficiency of the use of the former, as well as the smaller carbon footprint.

As a fossil fuel, LPG gas also results in greenhouse gas emissions, and as a gaseous fuel it can also be replaced by a biogenic alternative, such as syngas produced by gasification of the nutshells themselves.

Finally, it is concluded that cashew nutshells have sufficient potential to be used as a substitute for both forest firewood and LPG gas, if gasified.

Future studies may point out pre-treatments and other necessary actions in order to make the energy content of the CNS available in the best way, both in direct combustion and gasification.

Acknowledgements This study was financed in part by the *Coordenação de Aperfeiçoamento de Pessoal de Nível Superior—Brasil (CAPES)—Finance Code 001.*

This work was funded by FCT—Fundação para a Ciência e a Tecnologia within the R&D Units Project Scope UIDP/04077/2020, VALORIZA—Research Center for Endogenous Resource Valorization.

References

1. L. A. L. Serrano and P. F. A. P. Pessoa, *Sistema de Produção do Caju Aspectos econômicos da cultura do cajueiro*, 2nd ed., no. 2. Fortaleza: Infoteca Embrapa (2016).
2. Companhia Nacional de Abastecimento, "Análise Mensal—Castanha de Caju—Julho de 2022," Brasília—DF, 2022. [Online]. Available: http://www.conab.gov.br/OlalaCMS/uploads/arquivos/12_09_20_16_03_35_castanhadecajuagosto2012.pdf.
3. Food and Agriculture Organization (FAO), "FAOSTAT," *Food and Agriculture Statistics*, 2022. https://www.fao.org/faostat/en/#rankings/countries_by_commodity, last accessed 01/12/2023.
4. D. C. Ike, M. U. Ibezim-Ezeani, and O. Akaranta, "Cashew nutshell liquid and its derivatives in oil field applications: an update," *Green Chem. Lett. Rev.* , vol. 14, no. 4, pp. 618–631 (2021).
5. Cardolite, "Cashew Nutshell Liquid (CNSL) Technology I Cardolite," 2023. https://www.cardolite.com/technology/, last accessed 12/01/2023.
6. H. Y. Shrirame, A. G. Mohod, Y. P. Khandetod, K. G. Dhande, and J. S. Dhekale, "Cashew Nutshell a Green Energy Option for Gasification," *IARJSET Int. Adv. Res. J. Sci. Eng. Technol. ISO*, vol. 9, no. 10 (2022).
7. M. L. Adekanbi and T. T. Olugasa, "Utilizing cashew nutshell liquid for the sustainable production of biodiesel: A comprehensive review," *Clean. Chem. Eng.*, vol. 4, p. 100085 (2022).
8. Francisco Fábio de Assis Paiva, Deborah dos Santos Garrutti, and Raimundo Marcelino da Silva Neto, *Aproveitamento industrial do caju*, 1st ed., vol. 1. Fortaleza: Embrapa Agroindústia Tropical—SEBRAE/CE (2000).
9. F. Fábio De Assis *et al.*, *Processamento de Castanha de Caju.* Brasília (2006).
10. Francisco Fábio de Assis Paiva, Raimundo Marcelino da Silva Neto, and Pedro Felizardo Adeodato de Paula Pessoa, *Minifábrica de processamento de castanha de caju.* Fortaleza (2000).
11. CETESB, "Inventário de emissões das fontes estacionárias do Estado de São Paulo ," São Paulo, 2009. Last accessed: 01/12/2023. [Online]. Available: https://sistemasinter.cetesb.sp.gov.br/inventariofontes/manual_de_preenchimento.pdf.
12. Prefeitura de Porto Alegre, "Inventário de emissões de gases de efeito estufa de Porto Alegre," Porto Alegre—RS, 2021. Last accessed: 01/12/2023. [Online]. Available: https://americadosul.iclei.org/wp-content/uploads/sites/78/2021/08/relatorio-inventario-poa-web.pdf.
13. N. Tippayawong, C. Chaichana, A. Promwangkwa, and P. Rerkkriangkrai, "Gasification of cashew nutshells for thermal application in local food processing factory," *Energy Sustain. Dev.*, vol. 15, no. 1, pp. 69–72 (2011).

A Case Study of Analyzing Energy Use and Equivalent Carbon Footprint in an Educational Building in Cairo, Egypt

Mohamed A. Afifi, Fatma Fathy, and Hussein Hamza

Abstract With natural catastrophes such as storms, floods, and wildfires occurring throughout the globe as a direct result of climate change, the whole built environment industry must undergo a drastic transformation. In addition, human activities have caused many communities to struggle with challenges such as energy security, resource depletion, air quality, and public health. This paper aims to investigate the energy performance of a lecture Hall on a university campus in Cairo, Egypt. The energy consumption is monitored using a Fluke power quality analyzer over a period of time in the winter season. Results are analyzed to find potential opportunities to minimize the building's energy consumption and carbon footprint. This paper highlights the potential to mobilize university buildings towards reaching Net zero carbon campuses by 2050 while producing a ripple effect that can result in adopting climate actions within the local communities and governments.

Keywords Energy use · Carbon footprint · Educational building · Energy monitoring

1 Introduction

Energy consumption reduction in educational buildings has become an increasingly important topic in recent years due to the need to reduce emissions and environmental impacts and create a more sustainable environment [1]. Various strategies

M. A. Afifi (✉)
Department of Electrical Power and Machines, Faculty of Engineering, Ain Shams University, Cairo, Egypt
e-mail: mohamed.afify01@eng.asu.edu.eg

F. Fathy · H. Hamza
Architecture Department, Faculty of Engineering, Ain Shams University, Cairo, Egypt
e-mail: fatma.fathy@eng.asu.edu.eg

H. Hamza
e-mail: hussein.a.faried@eng.asu.edu.eg

J. R. da. C. S. Galvão et al. (eds.), *Proceedings of the 3rd International Conference on Water Energy Food and Sustainability (ICoWEFS 2023)*, Springer Proceedings in Earth and Environmental Sciences, https://doi.org/10.1007/978-3-031-48532-9_21

are used to reduce energy consumption in educational buildings, including building design, energy efficiency measures, and renewable energy technologies and their effectiveness [2]. An improved design is one of the most common strategies for reducing energy consumption in educational buildings [3]. This includes incorporating passive design features such as natural ventilation [4] and daylighting [5]. Other strategies for reducing energy consumption include using renewable energy sources such as solar or wind power and implementing building automation systems to control lighting, heating, and cooling systems [6]. In addition to improved building design and renewable energy sources, several other technologies can be used to reduce energy consumption in educational buildings. For example, occupancy sensors can automatically turn off lights when a room is unoccupied [7]. Additionally, building materials with high thermal mass can help regulate indoor temperatures and reduce the need for air conditioning [8]. Additionally, using natural ventilation systems can reduce cooling costs by up to 30%, while using passive solar design techniques can reduce heating costs by up to 50% [9]. Moreover, reducing water consumption is another important strategy for reducing a building's carbon footprint [10]. Additionally, retrofitting existing buildings with insulation materials can improve their thermal performance and reduce energy consumption. Overall, previous studies have highlighted several strategies and technologies that have been used successfully to reduce energy consumption in educational buildings. By utilizing these strategies and technologies, it is possible for educational institutions to significantly reduce their overall energy costs while also improving their environmental sustainability efforts. In this paper, an investigation of energy use is carried out in an educational building at Ain Shams University campus in Cairo, Egypt, to identify energy performance in relation to space occupancy and user behaviour. The analyzed data provides the potential saving opportunities and the associated carbon reduction.

2 Carbon Footprint in Egypt

In 2019, "Global Warming" reached a record high, with energy-related CO_2 emissions reaching 33 Gt globally. Egypt contributed 250 million tonnes to this figure in 2018, placing it 27th in the world rankings. 40% of Egypt's CO_2 emissions come from its electricity production industry, which relies on oil and gas (80%), , with renewables (solar, wind and hydroelectric) making up the remaining 10% [11]. Egypt plans to construct new power plants with a combined capacity of 30GW by 2030. This will add to the country's already sizeable producing capacity of 60GW, bringing total energy generation to almost 90 GW [11]. The power-generating mix was proposed in Vision 2030 for Egypt. This idea aims to employ a diverse (open to all possible energy sources) energy mix by 2030. In this scenario, the projected energy mix would include 29% coal, 27% oil and gas, 16% solar, 14% wind, 9% nuclear, and 5% hydro [12]. The amount of CO_2 released to generate 1kWh of electrical energy depends on the carbon content in the fuel; according to [11], and [13], the current CO_2 emission per kWh generation of electricity in Egypt is $464gCO_2e$. This value is used in this study to determine the amount of reduction in CO_2 footprint of the building if the proposed saving opportunity is implemented. E is the energy consumption in (Kilowatt Hour)

$$E = \frac{\sqrt{3} * V_{ab} * I_a * cos(\phi) * (\text{ TimeStep })}{1000 * 60 * 60} kwh \tag{1}$$

where V_{ab} is the recorded line to line voltage, I_a is the line current, $\cos(\phi)$ is the power factor and Time step of 1 s which is the resolution of the data logger.

$$\text{CO}_2 e = \frac{E * 464}{1000} kg\text{CO}_2 \tag{2}$$

The energy consumption can be calculated using (1) and the CO_2 equivalent can be calculated using (2).

3 Methodology

3.1 Building Description

Physical characteristics. The case study was selected to be a lecture hall building at Ain shams university campus, in Cairo. The building shown in Fig. 1 contains two identical lecture halls having a foot print area about 430 m^2 with a capacity of 300 students per hall during the semester, and the two halls are used with a reduced capacity of 150 students per hall during the exams. The two hall energy consumption depends mainly on the lighting and the air conditioning systems. The building is manually operated, with one operator available for the two lecture halls; no occupancy or temperature sensors are installed. The operator manually turns on and off the lights, and the air conditioner is operated without changing the temperature set point; it is operated for air conditioning during the summer and used for ventilation during winter. The air conditioning system is operated during the day only, or when the halls are utilized, the system is not working 24 h. Moreover, the air conditioning system works with a fixed set temperature and is neither controlled by a building management system (BMS) nor changed by the operator, and is set to 18C°.

Electrical Characteristics. The single line diagram of the building under study shown in Fig. 2 and the description of items are shown in Table 1. The building under study is fed from 3 phase 400 V transformer, feeding a main distribution panel located at Hall A, where the operator uses the panel to manage the facility's electric loads. The main circuit breaker (C.B) (Q0) is the 400 A moulded case C.B, which is normally closed; the main C.B feeds a distribution bus bar that feeds several C.Bs, Q1 and Q2 are 125 A-C.Bs is used to control the air conditioning system of each hall; the air conditioning system consists of one condenser unit and four air handling units, each lecture hall is equipped with two handling units. Q3 and Q4 are feeding

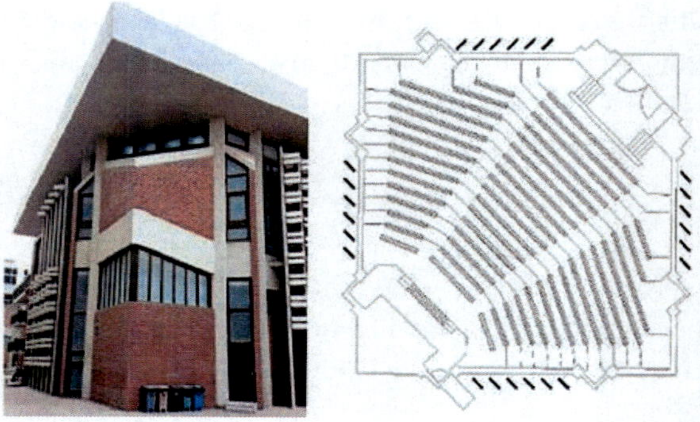

Fig. 1 The building under study

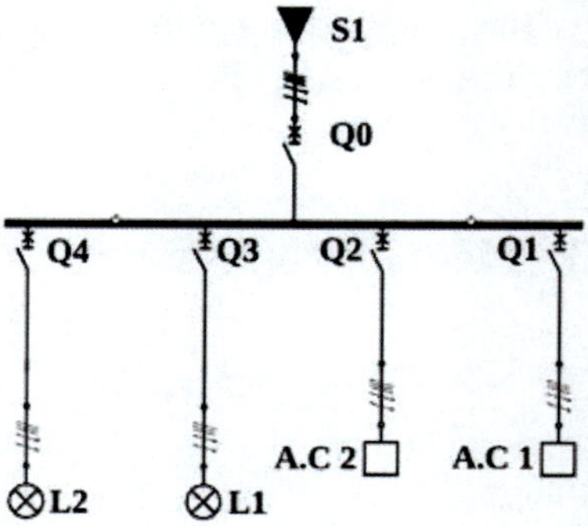

Fig. 2 The SLD of the electric installation

the lighting loads of each hall; the lighting loads are led panels 60 cm*60 cm each of 54 watts, and each hall contains 50 panels. The current drawn for each lecturing hall for the lighting is 10 amperes. The current drawn for each lecturing hall for the air conditioning system is 30 amperes, indicating that 25% of the loading of each lecture hall accounts for lighting and the other 75% for the air conditioning system; other auxiliary loads are of neglected consumption.

3.2 *Energy Monitoring*

The energy monitoring is done on the building using Fluke 435 power quality analyser shown in Fig. 3; the measurements are done on the main distribution panel where the three current probes are used on each phase of C.B Q0, the main C.B of the two halls, the voltage measurement is taken phase-phase. The data logging period was two weeks, with a resolution of one second. The first week was a typical study week where the two halls were not equally occupied by the same number of students or the same occupancy period. The second week was the final exams week, where the two halls were approximately occupied by the same number of students for two

Table 1 SLD description of the electric installation

Item	Description
S1	3-phase 400 V—4 wire source
Q0	3-phase 400 A main C.B
Q1	3-phase 125 A for A.C of hall A
Q2	3-phase 125 A for A.C of hall B
Q3	3-phase 100 A lighting of hall A
Q4	3-phase 100 A lighting of hall B
L1	All the lighting loads of hall A
L2	All the lighting loads of hall B
A.C 1	Air condition load of hall A
A.C 2	Air condition load of hall B

exam periods each day. The second week's data can be used to analyze the operator's behaviour as the same temperature conditions as shown in Fig. 4, and the same number of students were met. Hence we can assume a normalized condition where all the variables are fixed except the behaviours of the operator.

The outside temperature values were obtained from [14], which acquires the data from Cairo International Airport station. The temperature curve for the second week of energy monitoring shows an approximately constant temperature over the week and during the two exam periods with ±3 °C. The period of the second week, from the 9th of January to the 13th of January and the outside temperature ranged from 14 to 20 °C in this week, is considered an average week for the winter season in Cairo.

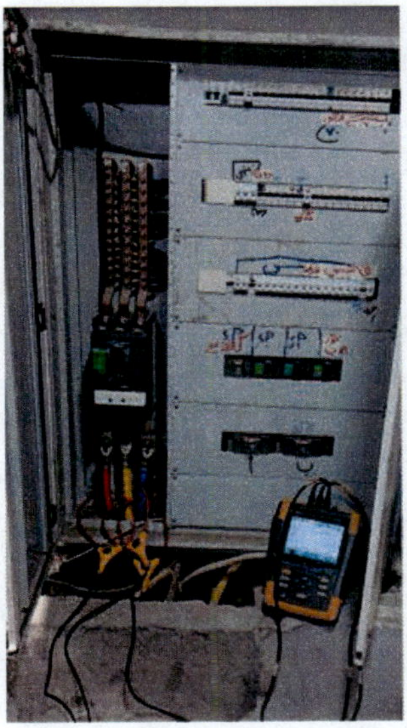

Fig. 3 The fluke measurement on the main C.B Q0

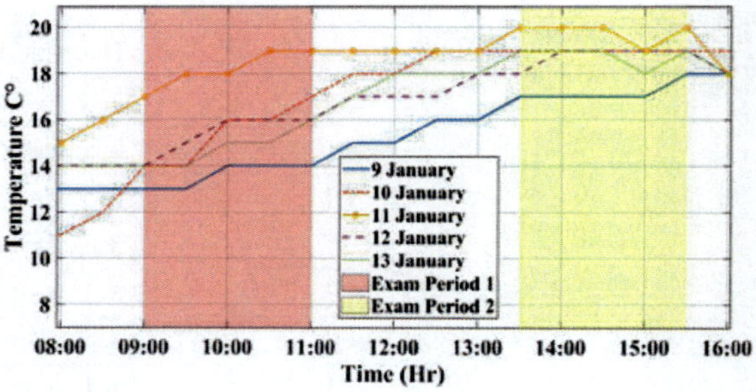

Fig. 4 The temperature recorded during the measurement period

4 Results

Figure 5 depicts the recorded waveform of the line-to-line voltage over the period of twenty-four hours for the second week of energy monitoring in addition to one day from the first week. The voltage waveform indicates an increase of the voltage over the night and a reduction during the day; the voltage is within the nominal range of ±5%, which has no significant effect on the system regarding energy consumption. However, the measured data helped the faculty determine the over-voltage problem, which occurred during the night and was the cause of equipment failure in another building. The line-to-line voltage is be used in calculations to determine the power and energy consumption value. The building is occupied within the two exam periods, which last 2 h, from 9:00 to 11:00 and 14:00 to 16:00. The temperature rises about 3° C^o from an average of 16° in the first period to 18.5 in the second period across the week. This rise caused a difference in energy consumption and reflected in cooling loads. The current consumption every day is depicted in Fig. 6 over the two exam periods to compare the behavioural changes that influence the current and the power consumption and lead to the detection of the saving opportunities.

The current consumption depends on the lighting in the two lecture halls and the air condition operating status. The optimum energy consumption is shown in Fig. 7 on the 9th and 11th of January, where only the lighting was used in both halls, and the air condition was not used in the two days in accordance with the recorded temperature.

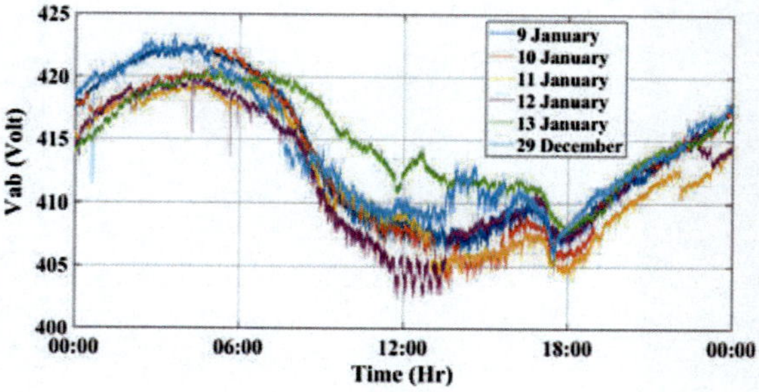

Fig. 5 Line to line voltage over 24 h period

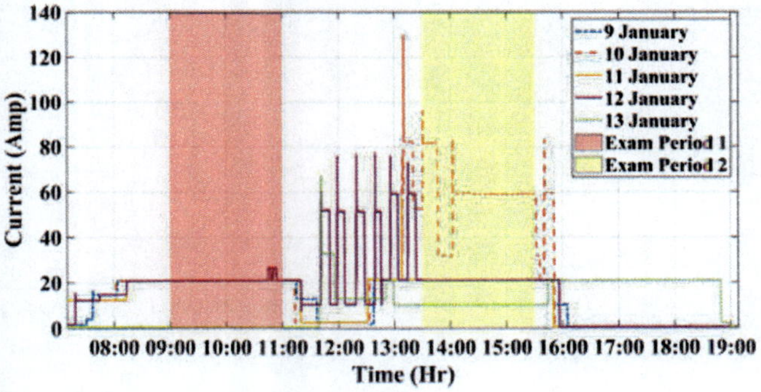

Fig. 6 Current of phase A over 24 h period

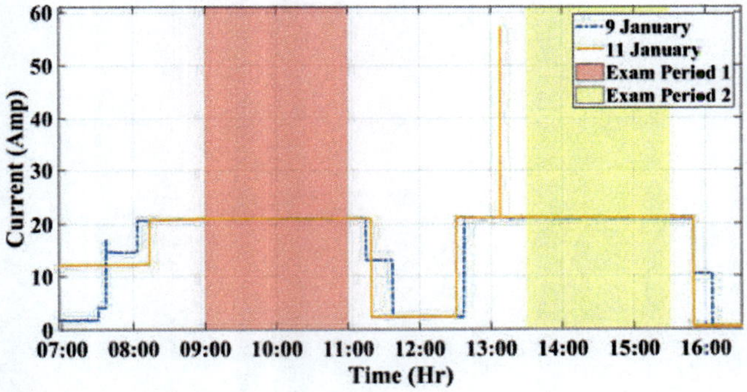

Fig. 7 Current of phase A 9th and 11th of January

Figure 8 depicts the current consumption on the 10th of January when the air conditioners were used without an essential need when the temperature was lower than the other days in the study. The air conditioner's condenser unit turned off and on several times because the temperature set point was attained. This caused the current to spike several times. This proves that the air conditioning was used for ventilation, not to decrease the interior temperature. The air condition operational in the period shown in Area 2 in Fig. 8 results in the consumption of an extra 70.894kwh and a 32.895 kg of CO_2.

Figure 9 depicts the current consumption on the 12th of January when the air conditioners were turned on and experienced a multiple on and off where no students were present. The air condition operational in the period shown in the energy consumption area in Fig. 9 results in the consumption of an extra 15.605 kwh and a 7.2406 kg of CO_2.

Figure 10 shows the current consumption on the 13th of January, where a late exam period was organized; the lighting was used during the night period and was left turned on after the exam. The late exam period resulted in an extra 45.634kwh

Fig. 8 Current of phase A 10th of January

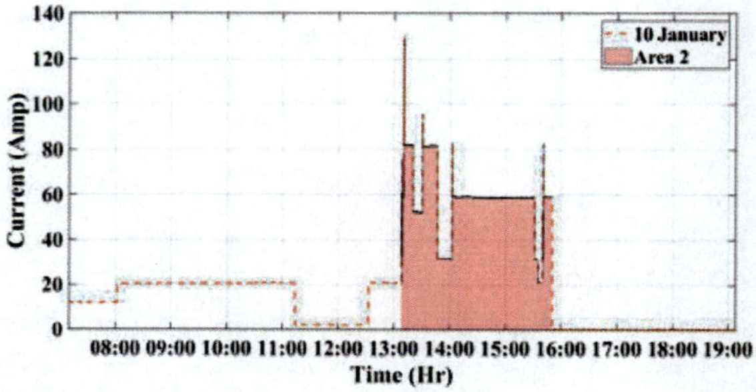

Fig. 9 Current of phase A 12th of January

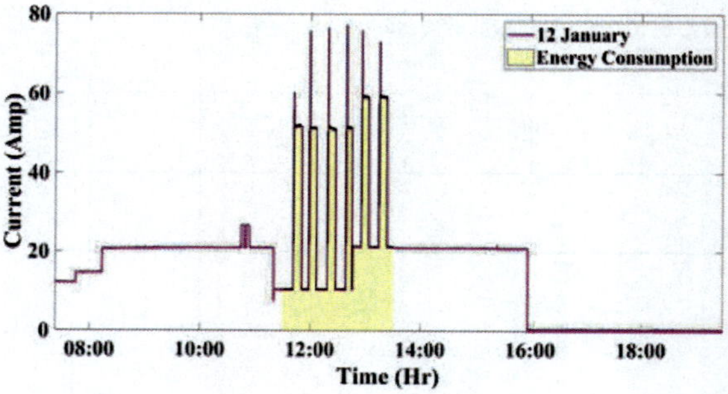

Fig. 10 Current of phase A 13th of January

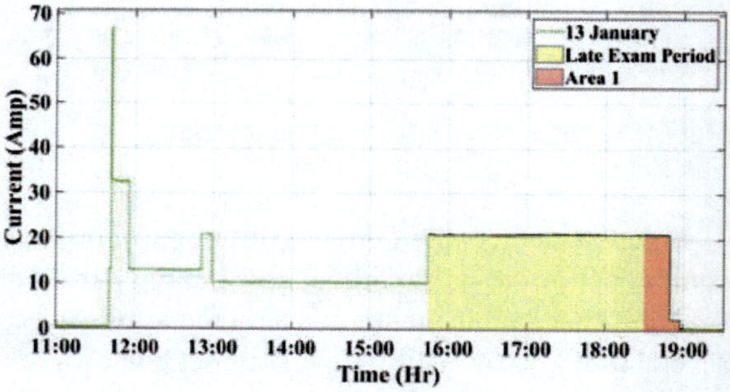

and a 21.174 kgCO_2. The excess energy consumption and the CO_2 can be avoided if the exam is rescheduled for a morning period when natural daylighting can be used.

5 Discussion and Recommendations

Develop a long-term strategic plan for future net zero campuses; this study can be considered the first step. Install a permanent energy monitoring system with more extended monitoring periods and different seasons to see if there are ways to save energy that could be put into place. Encourage energy-efficient behaviours and engage the public through awareness activities. Optimize the building's heating, ventilation, and air conditioning (HVAC) system by adjusting the thermostat, replacing the air filters regularly, and doing regular maintenance to ensure the system runs well. As a passive solution, the operator's actions can be optimized. Implementing Building Automation Systems (BAS) can help monitor, analyze and control various systems in the building to reduce energy consumption. Install solar panels on the roof of the building to reduce energy consumption. Passive solutions can include but are not limited to designing the building to take advantage of natural ventilation. Use operable windows, vents, or skylights to allow for natural airflow. Incorporate daylight harvesting techniques like light shelves to take advantage of natural light. This can reduce the need for artificial lighting and improve visual comfort. Active solutions can include but are not limited to automatic lighting control systems to save energy and make it easier on the eyes. The use of Variable Speed Drives for HVAC systems.

6 Conclusion

This paper investigated the energy performance of a lecture Hall on a university campus in Cairo, Egypt. The energy consumption is monitored using a Fluke power quality analyzer over a period of time in the winter season. Results are analyzed, and potential opportunities to minimize the building's energy will be achieved if simple behaviour solutions were implemented, like using natural ventilation instead of the air conditioner for ventilation; also, adjusting the set point of the Heating, ventilation, and air conditioning (HVAC) system can reduce the use of energy consumption. Rescheduling of using the halls in the evening to early periods and using natural daylight can also significantly impact energy consumption reduction and, consequently, the carbon footprint of the building. Future work will consider monitoring the inside temperatures, lighting levels, and energy use to correlate the required comfort levels for thermal and visual aspects with power and lighting loads. Also, the proposed recommendation to implement solutions that save energy will be studied in depth in the future from both a technical and an economic point of view. Moreover, monitoring will include a more extended period of analysis to compare the periods where the hall is used in the study with the exams; also, the summer season can be compared with the winter, where the average temperature levels are higher.

Acknowledgements The authors would like to acknowledge the British Council Going Global Partnerships for funding the project: Towards Net Zero Carbon Campuses for Climate Change Mitigation and Resilience, which enabled the publication of this study.

References

1. P. Marrone, P. Gori, F. Asdrubali, L. Evangelisti, L. Calcagnini, and G. Grazieschi, "Energy benchmarking in educational buildings through cluster analysis of energy retrofitting," Energies, vol. 11, p. 649, mar 2018.
2. M. Khoshbakht, Z. Gou, and K. Dupre, "Energy use characteristics and benchmarking for higher education buildings," Energy and Buildings, vol. 164, pp. 61-76, apr 2018.
3. H. Ibrahim, M. S. Elsayed, W. S. Moustafa, and H. M. Abdou, "Functional analysis as a method on sustainable building design: A case study in educational buildings implementing the triple bottom line," Alexandria Engineering Journal, vol. 62, pp. 63-73, jan2023 .
4. L.-R. Jia, J. Han, X. Chen, Q.-Y. Li, C.-C. Lee, and Y.-H. Fung, "Interaction between thermal comfort, indoor air quality and ventilation energy consumption of educational buildings: A comprehensive review," Buildings, vol. 11, p. 591, nov 2021.
5. V. Costanzo, G. Evola, and L. Marletta, "A review of daylighting strategies in schools: State of the art and expected future trends," Buildings, vol. 7, p. 41, may 2017
6. F. Asdrubali, L. Calcagnini, L. Evangelisti, C. Guattari, and P. Marrone, "Effectiveness of materials, technologies, and renewable energy in educational buildings through cluster analysis of energy retrofitting," in Sustainable Building for a Cleaner Environment, pp. 25-37, Springer International Publishing, jul 2018.
7. L. Martirano, "A sample case of an advanced lighting system in a educational building," in 2014 14 th International Conference on Environment and Electrical Engineering, IEEE, may 2014.
8. Z. S. Zomorodian, M. Tahsildoost, and M. Hafezi, "Thermal comfort in educational buildings: A review article," Renewable and Sustainable Energy Reviews, vol. 59, pp. 895–906, jun 2016.
9. J. Song, X. Huang, D. Shi, W. E. Lin, S. Fan, and P. F. Linden, "Natural ventilation in london: Towards energy-efficient and healthy buildings," Building and Environment, vol. 195, p. 107722, may 2021.
10. R. M. Almeida, N. M. Ramos, M. L. Simões, and V. P. de Freitas, "Energy and water consumption variability in school buildings: review and application of clustering techniques," Journal of Performance of Constructed Facilities, vol. 29, no. 6, p. 04014165, 2015.
11. L. Abdallah and T. El-Shennawy, "Evaluation of CO_2 emission from egypt's future power plants," Euro-Mediterranean Journal for Environmental Integration, vol. 5, sep 2020
12. "Sustainable development strategy: Egypt vision 2030," May 2016.
13. H. Ritchie, M. Roser, and P. Rosado, "Energy," Our World in Data, 2022. https://ourworldindata.org/energy.
14. "Cairo weather, cairo, egypt.," Jan 2023.

Energy Efficiency and Industrial Management in the Context of the Circular Economy

João Galvão, Filipe Neves, Marcelo Cardoso, and Manuel Galvão

Abstract The energy transition in the industrial sector, as well as the way energy is produced and consumed in areas heavily dependent on electricity, dictates that it is urgent to change this scenario by resorting to methodologies based on the circular economy, to mitigate waste and reducing greenhouse gas (GHG) emissions, with the aim of local energy production, in the context of applying the "United Nations Sustainable Development Goals". The research presented in this work intends to analyze the weaknesses and strengths that the plastic processing industry suffers, with the aim of becoming more competitive and efficient in the use of energy, as well as less waste generating, through the application of a methodology of management, such as the circular economy. It was through participation in an international project DEMOLA and applying its methodology based on management tools such as the Mind Map, PESTEL, and SWOT analysis supported by an interdisciplinary team of researchers, and they had a plastic company, as a case study. The results extracted should be disseminated throughout the industrial sector. Finally, conclusions are presented from the study carried out, achieving an improvement in the efficiency and sustainability of the company chosen as a case study and a contribution to sustainable development.

Keywords Energy efficiency · Sustainability · Circular economy · Recycling

J. Galvão (✉) · F. Neves · M. Cardoso
Polytechnic of Leiria, School of Technology and Management, 2411-901 Leiria, Portugal
e-mail: jrgalvao@ipleiria.pt

J. Galvão
INESCCoimbra, DEEC, Rua Sílvio Lima, Polo II, 3030-290 Coimbra, Portugal

F. Neves
CIIC—Computer Science and Communication Research Center, 2411-901 Leiria, Portugal

M. Galvão
Department of Mechanical Engineering, University of Minho, 4800-058 Guimarães, Portugal

J. R. da. C. S. Galvão et al. (eds.), *Proceedings of the 3rd International Conference on Water Energy Food and Sustainability (ICoWEFS 2023)*, Springer Proceedings in Earth and Environmental Sciences, https://doi.org/10.1007/978-3-031-48532-9_22

1 Introduction

The 2030 Agenda for Sustainable Development is composed of 17 Sustainable Development Goals (SDGs) and was approved in September 2015 by 193 members. It provides guidelines for the present and future societies based on a model of sustainable development. This document resulting from the joint work of governments and citizens around the world to create a new global model to end poverty, aims to promote prosperity and well-being of all, protect the environment and fight climate change [1]. In this scientific article, we intend to highlight the importance of five of these goals: the numbers 7, 8, 9, 11 and 12 (Fig. 1). This was achieved in the context of implementing the principles of circular economy [2, 3] in an industrial environment, having, as reference, one company in the plastics sector, which integrated a Demola International Project Studies that is one innovative way in co-creation work [4].

1.1 Objectives and Framework

One of the main objectives of this investigation is to initially improve energy efficiency in two companies that manufacture plastic components and are large energy consumers and waste generators, in a circular economy context and based on the methodology of the co-creation process. In this process, the Mind Map techniques, PESTEL analysis and SWOT analysis will structure the identification of opportunities and solutions to achieve improved performance in energy consumption and consequently in the management of manufacturing processes and reduction of waste [5–7].

Bearing in mind that a possible definition of circular economy could be according to Ellen MacArthur Foundation—an organization established in September 2010-, with the aim of accelerating the transition to a regenerative circular economy, the idea is "to shift from a linear model of resource consumption that follows a 'take- make-dispose' pattern, to an industrial economy that is 'restorative by intention'; i.e. that replaces the 'end-of-life' concept with restoration, shifts towards the use of renewable energy, eliminates *the use of toxic chemicals, which impair reuse, and aims for the elimination of waste through the superior design of materials, products, systems, and, within this, business models*". A possible comprehensive definition is related to several aspects and a circular economy could be more focused on materials and

Fig. 1 Five of SDGs of the 2030 agenda for sustainable development

resources, but on the other hand, it could be a transformation of the entire economy towards a specific goal of regeneration. On the other hand, another overview of this definition can be: *"A circular economy strives for the longest possible use of products and raw materials. In practical terms, this means avoiding waste through reuse or further use (recycling / reuse). Waste avoidance and reuse always come before recycling"* [8].

Relate as soon as possible the concept of circular economy with management policies committed to the sustainability of the planet and society, applying the objectives referred to in Fig. 1 and supported by the business processes of the Circular Plastics Alliance model, which is today a great movement of the European Union.

The Circular Plastics Alliance endorses the ambitious target that, by 2025, at least 10 million tonnes of recycled plastics must find their way into products and packaging in Europe each year, helping to implement the circular and lifecycle economy and 6 stages: Recycling; Sourcing; Manufacturing; Distribution; Use; Resource efficiency [9].

The incorporation of networked sensors in industrial processes increases productivity and avoids repetition in the circularity of parts and the sustainability of companies, as there is access to real-time information on the operation of processes. From here comes the ability to act directly and informedly, for corrections and adjustments in manufacturing or in the provision of services, resulting in increased efficiency, reduced waste and less material consumed [10].

According to F. Hofmann and Jaeger-Erben (2020, p. 2), a circular economy "connects post-use and process waste with production through processes such as recycling and repurposing of by-products and attempts to preserve the inherent value of products and product components by maximizing the number of consecutive use phases and use time in each of these phases via repair, maintenance, upgrade, resale, refurbishment and remanufacturing [11]. But the circular economy uses waste as an input in other parts of the value chain and widely focuses on closed-loop materials [12]. Subsequently, the circular economy research stream has generally started in emerging countries, finally all those circular economy dimensions can be tied directly to a sustainability context and achieve competitive low-carbon economy or a green environment in 2050 [13].

1.2 Motivation and Recruitment

The Demola International Project Studies is an international innovation co-creation program for collaborative work between companies and multidisciplinary teams of polytechnic students and other organizations based on Mind-Map, PESTEL and SWOT analysis diagnostic and management methodologies. It is aimed at promoting the entrepreneurship spirit through the empowerment by innovative and co-creation methodologies to find solutions for problems identified by organizations/companies.

The process of recruiting facilitators and students to form teams for the several challenges and works to be developed with factories/companies involved the dissemination in digital platforms or flyers. The team members will learn crucial skills that are relevant in working life, including interdisciplinary communication and project management skills. Participants will also get knowledge with strategic foresight tools and design thinking methodologies. In addition, they can access educational materials, and team management was conducted with the support of digital platforms.

The training of team facilitators was carried out weekly online by trainers from Finland, with the weekly support of the coordinators of the Demola Project at Polytechnic Leiria [14].

The innovation co-creation process consists in the constitution of teams made up of 5/6 students from multidisciplinary areas, one or two people from the company and the facilitator (Teacher), who meet weekly and work on the real cases/challenges of the companies. These "cases" are real problems/challenges that companies/ associations pose and, for which answers are sought, with a view to presenting suggestions and future strategic guidance.

Every day, companies face legal, economic, social, technological, environmental, and international challenges, in addition to uncertainty and change. It is in this context that the co-creation of innovation is important as a response to challenges/cases.

2 Case Study Characterization and Co-creation Team

The company chosen as a case study is located about 10 km west of the city of Leiria, with year-round sun exposure according to the GSA [15, 16]. The extrusion and injection activity, mainly directed to be applied on cooling/freezing equipment for food industry and stores, as well as construction and agriculture/garden are the main scope of the company's production. This company has a productive area of 3400 m^2, with all the means to manufacture technical profiles from small to large dimensions, for the widest areas and applications. Its diversity of equipment allows it to work with different types of material, resulting in the production of a distinct variety of products.

The range of manufactured products is in a small series of different types of products, where there is variation in the material to be injected, the design or the manufacturing process. The company's main raw material is thermoplastics. The company has the service of designing specifically the product to the customer needs. Faria & Mouroço attends to keep improving its design development of the product, to accomplish different client specific requirements. The vision of the company is to improve the manufacturing process, reusing all the raw materials more effectively, instead of wasting which results in a decrement in energy requirements.

Four students formed the team, from two different Polytechnics, two teachers/ facilitators and members of the company staff, with knowledge and know-how from several branches of engineering, such as Polymers, Energy and Environment, Product

Design, Electronics and Computing, Production Engineering. The innovation co-creation team met weekly and worked on the real problems/challenges/cases of the companies tracing a future strategic direction. The "cases" were real problems/challenges that companies/associations were posed with and for which answers were sought, with a view to presenting suggestions and future strategic guidance.

After the initial knowledge of the company and holding meetings with the administration of visits to the entire production area and collection information.

The methodologies of the following 3rd paragraph were applied in the analysis of these data and the results were presented for a future development strategy.

3 Features of Methodology Tools and Research Core

3.1 Methodology

The methodology applied to the case study was a simple energy audit process supported by the following three tools: Mind Map, PESTEL and SWOT analysis to detect relevant factors that condition the efficient management of the business and the energy area of this factory, with high level of consumption. In this work the Mind Map and the SWOT analysis are the tools that will be discussed in more detail.

Mind Maps are good to organize ideas, thoughts or concepts and see how they are interrelated. In the brainstorming sessions this technique is well accepted in problem solving workshops. The Miro Mind Map tool helps distributed teams to discuss ideas, map out information architecture and conceive charts for management matters (Fig. 2). It is expected that by searching for new methods and solutions to reduce the quantity of waste previously described, the company will be directed to a development based on circular economy and energy efficiency and the energy sources associated with the production process are broken down using partial Mind Map analysis (Fig. 3).

The PESTEL analysis is a powerful and widely used tool for understanding strategic risk. It identifies the changes and the effects of the external macro environment on a company´s competitive position, with greater detail and provides a view of the market, as it analyzes external changes to the company, namely political, economic, social, technological, ecological and legal factors, in the business environment. The use of the PESTEL matrix is essential to have a broad idea of the business area and the company's external environment and to support the detection of opportunities and threats relevant to the business activity carried out.

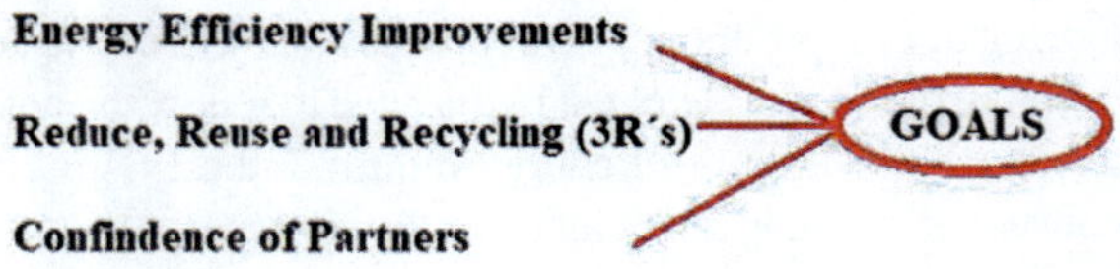

Fig. 2 Partial view of the mind map tool

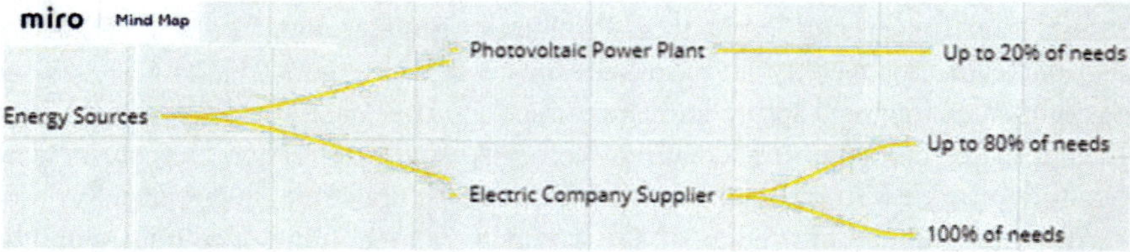

Fig. 3 Flow of energy sources present in the case study

The SWOT analysis makes it possible to assess the company's internal and external factors, while the PESTEL analysis only considers external aspects, but in a more comprehensive way. Therefore, the best approach is to use both analysis tools: carry out a PESTEL analysis and use the information collected in the SWOT analysis Opportunities and Threats.

3.2 *Research Core*

The methodology used by these three tools (Mind Map, PESTEL and SWOT analysis) applied in the case study aims to detect relevant factors that condition the efficient management of the business, in particular the fact that this company has a high consumption of electricity.

In the PESTEL analysis, the data of external factors and the information retained can lead to choosing the best strategy to achieve the company's growing goals. With this tool it can be study variables that influence the regular operations of Faria & Morouço company. To do so, a comprehensive literature review was performed by applying a Political, Economic, Social, Technological, Environmental, and Legal analysis. The information collected yielded an overview of the company and the factory, which serves as a basis for defining the models and tools necessary to support sustainable decision-making for better performance and efficiency of the company.

As previously stated, the SWOT analysis is a tool that allows the determination of some relevant details that influence the company's management. Some decisions are made at the strategic level for its growth, such as the internal factors: Strengths and Weaknesses and the external factors: Opportunities and Threats, Table 1.

4 Results and Proposals

During the simple energy audit process that was carried out, an opportunity for energy rationalization was detected in the need to renew the company's electrical installation to reduce the amount of energy required, the existing manufacturing processes and consequently their financial costs.

Table 1 Table of SWOT analysis

Strengths • Long-standing relationship with customers and partners; • Long-term jobs and relation employee/company; • More than four decades of experience and know-how; • Member of several R&D projects with companies, universities, and research centers; • Openness of management to new ideas; • Capacity to develop its own product/design with tooling	*Weaknesses* • The biggest volume of purges the highest energy and labor consumption to be regrind; • Age of the machines with consequent less efficient power consumption; • High power consumption with artificial lighting due to too narrow windows; • Huge amount of plastic waste on machine/production's purge; • Low power consumption from renewable energy source; • Old layout of the production area with many interruptions in raw materials distribution
Opportunities • Being part of EU policy of circular economy • The prohibition of single use plastics lead to the development of new design plastic parts and materials	*Threats* • Costs of equipment to replace and renew the existing machinery • Large number of local competitors onthe same market and area; • The impossibility to reflect the increase of the cost of energy on the costs of the sales

The existing production processes of injection and extrusion produce a large number of purges. The purge parts are too large and molded into a giant ball, making the process of cutting them. This forces the need to use equipment that demands a large amount of energy, undermining the company's electrical efficiency.

The semi-automatic feeding of injection molding machines results in loss of time and efficiency due to the distance between the machines and the raw material to be used in the injection process. The arrangement of equipment on the shop floor must be optimized and the feeding of extruders and injection machinery must be automated by network installation of several types of light sensors, flowmeters and other devices, all supported by a sensor network with data stored in the cloud.

The lack of windows in the storage area implies the need for installation of artificial lighting throughout the warehouse, working day and night, which results in high energy costs.

Although there are some sectors of the company where LED lighting fixtures already exist, there is a lack of use of natural light and an improvement in sustainability in the use of this natural resource.

There is already a renewable electrical energy power plant using photovoltaic technology, which covers about 20% of the company's current consumption, but should be expanded. It occupies the available space in the surrounding area, outside the main nave of the building, in the car park, taking into account the example of good energy sustainability practices in other companies and institutions [17–19] (Fig. 4).

The amount of material and raw material that was bought and wasted was estimated. Within the different materials bought, it was calculated that 5% of each

Fig. 4 Proposal for a new area of electrical energy power plant in car park and charging station for electric mobility

material is wasted and that from 431,300 kg of material acquired in total, 21,565 kg is wasted on different stages of the production process.

Comparing the total of material acquired with the wasted material, it can be concluded that the company is wasting a significant amount of the purchased material that ends up affecting its economy as well.

To reduce energy consumption, some equipment such as grinders and shredders that require high power-therefore large consumptions and having associated the price of the kwh, must have their operation relocated to the night period, when the kWh is cheaper.

To avoid price volatility in the purchase of raw materials, the company's stakeholders must assure the supply of their demands. The stock of raw materials is carried out in two ways: stock for production or by opportunity price and material they used.

In the context of the circular economy applied to this company as a case study, it should be considered that the short production series should be rethought, redesigned and recycled at any point in the chain of the process to reduce pressure on the environment, as in new materials, energy, water and natural resources.

In general, the company will benefit from the following: Creation of resilience and competitive advantages; New business and higher profit potential; New financing opportunities; Stimulus to innovation and eco-design; Reduction of costs with raw materials; Direct gains from recovery/recycling of materials; Certification of products with good environmental practices; Reduction of product complexity; Easier management life cycles of products.

Therefore, it is urgent to apply the management philosophy based on the principles of the circular economy to mitigate this waste and with this objective achieved, the five of SDGs of the 2030 Agenda for Sustainable Development, presented in Fig. 1, will be effectively fulfilled in this company's management mode.

5 Conclusions and Perspectives

During the period in which this work was carried out, the whole process was very intense, which included the training aspect for the application of the co-creation work methodology, gathering information with several weekly meetings, visits to the case under study, centers and lectures by specialists. There is a collaborative relationship between team members and the partner, resulting in mutual trust. Tasks were shared by the team, in an atmosphere of cordiality and sharing in the search for solutions to existing problems in the areas of recycling and energy, bearing in mind the circular economy methodology.

Members and researchers, in this co-creation process, used the tools and analyzes presented in point 3, having also sought answers to the threats and opportunities in this field of study/company with research centers and experts, involving the company's stakeholders and gathering insights that helped build ideas of possible responses to the difficulties encountered.

From the company's point of view, participating in the co-creation process was an opportunity for the company to think, reflect on the challenge ahead; gather important information within the company to understand more about plastic waste and expand the local electricity production mode by photovoltaic technology in the surrounding areas within the company perimeter.

In this project, working with a multidisciplinary team means that different points of view are shared, which lead to a greater ability to solve problems. Having a team that includes young people still in training mobilizes these future leaders of companies to find solutions within the scope of the circular economy, which is advantageous. Furthermore, the acquisition of skills by young people in the areas of energy efficiency, waste reduction and sustainability lead companies to become closer to the practice of the five objectives of the mentioned sustainable development principles.

The younger people go through this experience of co-creation in the aforementioned areas, more companies and society will be positively affected by these practices for sustainable development. Sharing experiences and working together helped to better understand and contribute to fighting waste, implementing circular economy.

As future perspectives, the company should monitor energy consumption, which is essential to define and adjust efficient energy management strategies. In particular, studying the evolution of consumption pattern behavior can allow defining energy policies according to the season of year [20].

Deepen the company's management actions in the context of the circular economy methodology and carry out the integration of resources and digital networks within the factory, in order to have cleaner final products, less CO_2 emitters, less energy consumers, always focusing on reuse within the circular economy.

Acknowledgements Study supported by UIDB/00308/2020 project.

References

1. The 2030 Agenda for Sustainable Development of United Nations see: https://sustainabledevelopment.un.org, last acess 17/1/2023
2. The Ellen MacArthur Foundation; see: https://ellenmacarthurfoundation.org/topics/circular-eT. *Bconomy-introduction/overview,* last acess 23/1/2023
3. Natur-Tec® Company, see: https://circulesolutions.com/circular-packaging/, last acess 10/10/2022.
4. DEMOLA International Project Studies, Demola:—Global community of co-creation experts, see https://www.demola.net .17/1/2023
5. Mind Mapping Tool for Teams—Miro, see: https://miro.com/
6. İ. Yüksel (2021), Developing a Multi-Criteria Decision Making Model for PESTEL Analysis, International Journal of Business and Management; Vol. 7, No. 24, DOI: https://doi.org/10.5539/ijbm.v7n24p52
7. T. Bonnici, D. Galea, (2015), SWOT Analysis, in book Wiley Encyclopedia of Management, editor J. Wiley & Sons, DOI:
8. Wasser 3.0 Organization, see: https://wasserdreinull.de/en/knowledge/water/circular-economy/?gclid=EAIaIQobChMI4M21_--9_AIV3IxoCR1yOgqCEAAYBCAAEgLif_D_BwE, last acess 4/1/2023
9. Circular Plastic Allinace Organization https://www.circular-plastics-alliance.com/en/welcome-to-the-circular-plastics-alliance/
10. U. Awan, R. Sroufe, M. Shahbaz, Industry 4.0 and the circular economy: A literature review and recommendations for future research, Research Article. Wiley Edition, 2021, DOI: https://doi.org/10.1002/bse.2731
11. Hofmann, F., & Jaeger-Erben, M. (2020). Organizational transition management of circular business model innovations. Business Strategy and the Environment, 29, 2770–2788. https://doi.org/10.1002/bse.2542
12. Hyun, Y., Hosoya, R., & Kamioka, T. (2018). The implications of big data analytics orientation upon deployment. In Proceedings of the 6th International Conference on Information Technology: IoT and Smart City (pp. 42–48). New York, NY, United States: Association for Computing Machinery. https://doi.org/10.1145/3301551.3301566
13. Barreiro-Gen, M., & Lozano, R. (2020). How circular is the circular economy? Analysing the implementation of circular economy in organisations. Business Strategy and the Environment, 29, 3484–3494. https:// doi.org/https://doi.org/10.1002/bse.2590
14. S. C. Rodrigues, Trainee Dossier, Curso: Aprendizagem com base em processos de co-criação, Número: "IPLeiria.2021.2.1", Ação: Curso 2—Ação 1, Número da Operação: POCH-04–5267-FSE-000814, June 2021.
15. GSA—Global Solar Atlas 2.0 Maps, Web-Based Application of Solargis Data with funding provided by the Energy Sector Management Assistance Program (ESMAP), see:: https://globalsolaratlas.info
16. D. Jorge, M. Cardoso, M. Cruz, P. Ferreira, A. Simões, J. Galvão, (2021), Final Report Look Inside of Factory Circular Economy and Energy Resources with Faria & Mouroço S.A., Project Demola/Link me Up.
17. J. Galvão, A. Nabais, M. Galvão, J. Candeias , T. Pereira, J. Ramos,,"Efficiency in an Intensive Energy Industrial Consumer", Renewable Energy and Power Quality Journal (RE&PQJ), http://www.icrepq.com/icrepq20/443-20-galvao.pdf; doi: https://doi.org/10.24084/repqj18.443
18. Galvão J., Ascenso R., Pereira S., Moreira L., Neves F. (2021) Smart Energy in Sustainable Services Building, Proceedings of the 1st International Conference on Water Energy Food and Sustainability (ICoWEFS 2021), Springer, Cham. doi:https://doi.org/10.1007/978-3-030-75315-3_47
19. SketchUp for Web | 3D Modeling, see: https://www.sketchup.com/pt-BR/ last access: 20/11/2022

20. J. Viera, J. Aguilar, M. R-Moreno, C. Quintero (2023) Analysis of the Behavior Pattern of Energy Consumption through Online Clustering Techniques, Journal Enegies (ISSN 1996–1073), Section: F5: Artificial Intelligence and Smart Energy, Special Issue: Machine Learning and Data Mining Applications in Power and Multi-Energy Systems.

Do Digital Finance and Energy Consumption Promote Sustainable Development in Emerging Economies?

Arindam Paul and Dukhabandhu Sahoo

Abstract In the era of new economy, digital finance, a technology-driven version of finance, is crucial to achieving sustainable development goals. Digital finance leads to an increase in the use of energy, which may be a threat to the environment if the sources of energy are dominated by non-renewable ones. Hence, it is crucial to study the nexus of digital finance, energy consumption, and sustainable development. The objective of the study is to investigate the impacts of digital finance, renewable energy, and non-renewable energy consumption on sustainable development in 13 emerging economies over the period of 2011–2021 using Driscoll-Kraay standard error method. The findings of the study reveal that digital finance enhances sustainable development, whereas non-renewable energy consumption hinders it. In addition, foreign direct investment inflows, natural resource rent, and unemployment have detrimental impact on sustainable development, while trade openness promotes it. Based on the findings, the study proposes that emerging economies should promote digital finance, reduce non-renewable energy consumption, and design environment-friendly foreign direct investment policies to foster sustainable development.

Keywords Digital finance · Energy consumption · Sustainable development · Emerging economies · Driscoll-Kraay standard error

1 Introduction

Sustainable development is first defined in the Brundtland report published in 1987 by the United Nations World Commission on Environment and Development as "development that meets the needs of the present, without compromising the ability of future generations to meet their own needs." Transitioning to a low-carbon economy is crucial to achieving sustainable development, which needs improving financial

A. Paul (✉) · D. Sahoo
School of Humanities, Social Sciences, and Management, Indian Institute of Technology Bhubaneswar, Bhubaneswar, Odisha, India
e-mail: ap59@iitbbs.ac.in

J. R. da. C. S. Galvão et al. (eds.), *Proceedings of the 3rd International Conference on Water Energy Food and Sustainability (ICoWEFS 2023)*, Springer Proceedings in Earth and Environmental Sciences, https://doi.org/10.1007/978-3-031-48532-9_23

services and resources [1]. During the 26th Conference of the Parties (COP26) of the United Nations Framework Convention on Climate Change (UNFCCC), finance is considered vital to achieving a sustainable environment, and developing countries can curb greenhouse gas emissions by developing their financial system. In recent years, information and communication technologies have been developed commendably, which introduces the concept of digital finance. Digital finance is a term used to describe the impact of new technologies on the financial service industries. Digital finance reduces the distance between customers and banks and thus includes more people in the financial system [2]. The development of digital finance makes financial services cheaper and easier to access for individuals as well as enterprises [3]. Digital finance is an energy-intensive industry. Therefore, with the progress of digital finance, the use of cleaner energy is also crucial to achieving a sustainable economy. However, achieving sustainable development is harder for emerging economies than developed countries due to their inherent structural characteristics.

Emerging economies are in their transition period. Rapid growth and industrialization can cause massive CO_2 emission growth in the near future. It is predicted that CO_2 emissions from emerging and developing countries will increase by five gigatonnes (GT) over the next two decades [4], which will hinder sustainable development. Therefore, the world's energy and climate future heavily depends on the decision made in emerging economies regarding sustainable development, and their failure leads to the failure of the global economy to reach the sustainable development goal. In this context, it is crucial to examine the dynamic among digital finance, energy consumption, and sustainable development in emerging countries.

Previous studies investigated the impact of digital finance either on environmental sustainability [5] or economic growth [6] and ignored the combination of economic, social, and environmental sustainability to represent sustainable development. These studies were also extensively based on China. Further, there are a limited number of studies investigating the impact of energy consumption on sustainable development [7, 8]. In addition, to the best of the authors' knowledge, there is hardly any literature analyzing the nexus among digital finance, energy consumption, and sustainable development in emerging countries. In order to bridge these gaps, the study aims to analyze the impacts of digital finance and disaggregate energy consumption on sustainable development in emerging economies over the period 2011–2021. To fulfill the objective, the fixed effect model with Driscoll-Kraay (D-k) standard error [9] is employed, which is robust to the presence of heteroskedasticity, autocorrelation, and cross-sectional dependence. The study reports that digital finance promotes sustainable development, whereas non-renewable energy consumption negatively affects it. The novelty of the study lies in introducing a new digital finance index covering access and usage, which can be used for more countries and longer time periods. Further, this is the first study investigating the linkage among digital finance, energy consumption, and sustainable development.

The rest of the paper is organized as follows. Section 2 provides a detailed discussion of the literature, and Sect. 3 provides the data sources and econometrics methods. Section 4 discusses the findings, while sect. 5 ends with conclusions and policy implications.

2 Literature Review

2.1 *Digital Finance-Sustainable Development Nexus*

There is a plethora of studies on financial inclusion. However, digital finance, a technology-driven version of finance, is an emerging concept. Digital finance can influence sustainable development through two channels. One channel is by affecting the environment. Zhao et al. [5] studied the impact of digital finance inclusion on CO_2 emissions in 30 Chinese provinces from 2011 to 2018. They found an inverted U-shaped relationship between digital finance and CO_2 emissions. A study by Zheng and Li [3] in 30 Chinese provinces during the period 2013–2020 reported that digital financial inclusion improved environmental quality by reducing CO_2 emissions. Shi et al. [10] examined the impact of digital financial inclusion and environmental regulation on SO_2 emissions per unit output in 30 Chinese provinces for the period 2011–2019. The findings revealed that digital financial inclusion mitigated SO_2 emissions per unit output, and with more stringent environmental regulation, the mitigating effect of digital financial inclusion was more profound. Furthermore, Wang et al. [11] analyzed the digital finance-environment nexus in 284 prefecture-level Chinese cities from 2011 to 2017. Using the spatial econometric model, they reported that digital financial inclusion increased CO_2 emissions of local cities but decreased those of neighboring cities. Yang et al. [2] investigated the impact of digital finance on $PM_{2.5}$ concentration in 283 prefecture-level Chinese cities over the period 2011–2018. They confirmed that digital financial inclusion raised $PM_{2.5}$ concentration but reduced it after reaching a certain threshold. However, Chang [12], in 31 Chinese provinces from 2011 to 2020, found that digital finance had a significant negative impact on CO_2 emissions after reaching a certain level, and its inhibitory impact increased with its further development. He also revealed that digital finance decreased agricultural carbon emissions via its mediating effect on farmer entrepreneurship and agricultural technology innovation. For 283 prefecture-level Chinese cities for the period 2011–2019, Xue et al. [13] documented that digital finance reduced CO_2 emissions, but its mitigating effect decreased with the increase of the digital threshold and green threshold.

Another channel is by affecting the economy. Jiang et al. [14] explored the digital finance-economic growth nexus in 30 Chinese provinces during the period 2011–2018. The findings indicated that digital finance enhanced economic growth. A similar result is obtained by Ahmad et al. [15] in 30 Chinese provinces for the period 2011–2018. Furthermore, a study by Khera et al. [16] in 52 emerging markets and developing economies investigated the impact of digital financial inclusion on economic growth by using the cross-sectional instrumental variable method. The study reported that digital financial inclusion increased economic growth. Shen et al. [17] examined the digital finance-economic growth nexus in 105 countries and found that digital finance positively impacted economic growth. They also confirmed that digital finance had a positive spillover impact on economic growth. However, Ding et al. [6], in 30 Chinese provinces from 2011 to 2019, reported that digital financial inclusion had a direct promotion effect and a negative spillover effect on economic growth.

2.2 *Energy-Sustainable Development Nexus*

Consumption of clean energy is crucial for sustainable development. Guney [7] examined the impact of renewable and non-renewable energy on sustainable development in 40 developed and 76 developing countries from 1990 to 2014. They found that both renewable and non-renewable energy consumption had a positive impact on sustainable development, but the impact of renewable energy was greater than that of non-renewable energy. However, another study by Guney [18] in 20 high-income countries over the period 1990–2015 reported that renewable energy consumption promoted sustainable development, whereas non-renewable energy consumption hindered it. Furthermore, Oke et al. [19] investigated the energy-sustainable development nexus in 51 African countries from 1990 to 2015 and found that renewable increased sustainable development. Islam et al. [8] analyzed the impact of disaggregate energy consumption on sustainable development in the Russian Federation over the period of 1990–2020. The findings revealed that renewable energy consumption escalated sustainable development, whereas non-renewable energy consumption declined it.

3 Data and Methods

3.1 *Data and Variables*

Explained variable: The explained variable of the study is sustainable development. Following Guney [7, 18] and Spaiser et al. [20], the study takes adjusted net saving rate (ANSR) as the proxy for sustainable development. The measure of adjusted net saving (ANS) is given in Eq. (1).

$$ANS = (GS - DEP) + E - CD - R \tag{1}$$

where GS refers to gross saving, DEP is depreciation rate, E represents public spending on education, CD indicates CO_2 and particulate emissions damage, and R includes net energy, mineral, and forest depletion. ANSR is measured as a percentage of gross domestic product and is a much better indicator of sustainable development compared to growth variables [7].

Core explanatory variables: Digital finance, renewable and non-renewable energy are the key explanatory variables of the study. Direct data for digital finance are not available. Hence, the study chooses mobile subscriptions, and internet uses as

Table 1 Definition and data sources of variables

Variables	Definition	Unit of measurements	Data sources
ANSR	Adjusted net saving rate	Percentage of GDP	WDI
DFI	Digital finance index	Index	Authors' construction
NREC	Non-renewable energy consumption	Quadrillion Btu	EIA
REC	Renewable energy consumption	Quadrillion Btu	EIA
INQ	Inequality (Gini coefficients)	Index	SWIID of Solt [25]
FDII	Foreign direct investment inflow	Percentage of GDP	WDI
UMP	Unemployment	Percentage of labor force	WDI
PCGDPG	Per capita gross domestic product growth	Percentage	WDI
NRR	Natural resource rent	Percentage of GDP	WDI
TO	Trade openness (sum of export and import)	Percentage of GDP	WDI

Notes WDI World Development Indicator, *EIA* Energy Information Association, *SWIID* Standardized World Income Inequality Database

the proxy of access to digital finance and the number of mobile and internet banking transactions (during the reference year) and the value of mobile and internet banking transactions as the proxy for the usage of digital finance. Then the study utilizes principal component analysis to formulate a digital finance index. The data of mobile subscriptions and internet users are collected from the World Development Indicator, and the data of the number and value of mobile and internet banking transactions are extracted from the International Monetary Fund.

Control variables: The study takes inequality (Gini index), foreign direct investment inflow, trade openness, unemployment, per capita gross domestic product growth, and natural resource rent as control variables that may affect sustainable development, and ignoring them may lead to a biased result. Control variables are chosen on the basis of literature [21, 22].

All the variables are transformed into natural logarithms to reduce the sharpness and outliers of the data as well as to express the estimated coefficients in the elasticities [23, 24]. The study employs annual data for 13 emerging economies[1] over the period of 2011 to 2021.[2] Table 1 provides variable descriptions and data sources, and Table 2 presents descriptive statistics of the variables.

[1] The countries are Argentina, Brazil, Chile, Colombia, Hungary, India, Indonesia, Korean Republic, Malaysia, Mexico, Russia, Thailand, and Turkey. Countries and time-period are chosen according to the availability of data.

[2] The study extrapolates missing data to make the data set balanced.

Table 2 Descriptive statistics

Variables	Obs	Mean	Std. Dev	Min	Max
LnANSR	143	2.451	0.525	0.308	3.201
LnDFI	143	2.691	0.479	0.750	3.438
LnNREC	143	1.581	1.064	−0.327	3.392
LnREC	143	−0.774	1.149	−3.182	1.536
LnINQ	143	3.702	0.173	3.311	3.924
LnFDII	143	3.801	0.329	0.652	5.300
LnUMP	143	1.632	0.731	−1.386	2.711
LnPCGDPG	143	2.590	0.376	0.211	3.140
LnNRR	143	0.655	1.538	−4.057	2.865
LnTO	143	4.109	0.560	3.113	5.126

3.2 Econometrics Methods

The study focuses on the following model to investigate the impact of digital finance and disaggregate energy consumption on sustainable development:

$$\begin{aligned} ANSR_{it} = f(&DFI_{it}, NREC_{it}, REC_{it}, INQ_{it}, FDII_{it}, {}_0 \\ &UMP_{it}, PCGDPG_{it}, NRR_{it}, TO_{it} \end{aligned} \tag{2}$$

where the variables are previously defined.

Equation (2) can be expressed in a more specific Cobb–Douglas form in the following way:

$$\begin{aligned} ANSR_{it} = {} & DFI_{it}^{\beta_{1i}} NREC_{it}^{\beta_{2i}} REC_{it}^{\beta_{3i}} INQ_{it}^{\beta_{4i}} FDII_{it}^{\beta_{5i}} \\ & UMP_{it}^{\beta_{6i}} PCGDPG_{it}^{\beta_{7i}} NRR_{it}^{\beta_{8i}} TO_{it}^{\beta_{9i}} e^{u_{it}} \end{aligned} \tag{3}$$

where u_{it} is assumed an independent and identically distributed (i.i.d.) disturbance term. Subscript i and t denote country and year. Now by taking the natural logarithm in the both sides and adding a constant on the right-hand side, we can get the following:

$$\begin{aligned} LnANSR_{it} = {} & \beta_{0i} + \beta_{1i} LnDFI_{it} + \beta_{2i} LnNREC_{it} + \beta_{3i} LnREC_{it} \\ & + \beta_{4i} LnINQ_{it} + \beta_{5i} LnFDII_{it} + \beta_{6i} LnUMP_{it} \\ & + \beta_{7i} LnPCGDPG_{it} + \beta_{8i} LnNRR_{it} + \beta_{9i} LnTO_{it} + u_{it} \end{aligned} \tag{4}$$

where β_{0i} is the constant term. β_{1i}, β_{2i}, β_{3i}, β_{4i}, β_{5i}, β_{6i}, β_{7i}, β_{8i}, and β_{9i} are the long-run coefficients which needs to be estimated.

First, we need to check cross-sectional dependence (CD), which is crucial in the study of panel data. Increasing globalization and trade liberalization have increased interdependence among countries, and ignorance of this may lead to misleading conclusions [26]. In this study, the Breusch and Pagan LM test [27], Pesaran scaled LM test, Pesaran CD test [28], and Bias-adjusted LM test [29] are employed to check the cross-sectional dependency among data. To check the time series properties of the variables, the cross-sectionally augmented Im-Pesaran-Shin (CIPS) unit root test [30] is employed, which considers the CD problem.

For diagnostic checking, the study employs the multicollinearity test, the Wooldridge test to test for serial correlation, the modified Wald test to test for the group-wise heteroskedasticity, the Hausman test [31] between fixed and random effect model, the Jarque–Bera (J-B) normality test, and previously mentioned four cross-sectional dependence tests. The results of these tests show the presence of serial correlation, group-wise heteroskedasticity, and cross-sectional dependence in data. Moreover, the Hausman test provides evidence for the fixed effect model, and the J-B test shows that the residuals are not normally distributed. Henceforth, we employ the fixed effect model with Driscoll-Kraay (D-k) standard error [9], which is robust to the presence of these problems [32].

4 Results and Discussion

4.1 Results of Diagnostic Tests

The results of the CD tests are given in Table 3, which confirm the existence of cross-sectional dependence among the panel. The results of the CIPS unit root test reported in Table 4 show that the variables are stationary at the first difference, i.e., I(1). The results of the diagnostic tests are presented in Table 5. The results show that the data have serial correlation, heteroskedasticity, and cross-sectional dependence. The result of the Hausman test shows the presence of a fixed effect in the model, and the result of the J-B test indicates the non-normality of the dataset. However, the Variance Inflation Factor (VIF) (in Table 6) of the independent variables shows no severe multicollinearity among the variables. So, D-K standard error method is appropriate for this study.

4.2 Estimation Result

The result of the D-K standard error estimation is reported in Table 7. The table shows that the estimated coefficient of digital finance is positively significant at 1% level of significance. A 1% increase in digital finance raises sustainable development by 0.408%. Digital finance increases economic efficiency and improves environmental quality by raising the availability of financial services and reducing transaction costs and asymmetric information. In this way, digital finance promotes sustainable development. This finding can be related to the findings of Jiang et al. [14] and Zheng and

Table 3 Results of cross-section dependence tests

Variables	Breusch-Pagan LM		Pesaran scaled LM		Bias-corrected scaled LM		Pesaran CD	
	Statistic	p-value	Statistic	p-value	Statistic	p-value	Statistic	p-value
LnANSR	197.679***	0.000	9.582***	0.000	8.932***	0.000	−0.497***	0.000
LnDFI	511.665***	0.000	34.721***	0.000	34.071***	0.000	15.181***	0.000
LnNREC	323.043***	0.000	19.619***	0.000	18.969***	0.000	10.088***	0.000
LnREC	355.047***	0.000	22.181***	0.000	21.531***	0.000	13.116***	0.000
LnINQ	442.066***	0.000	29.149***	0.000	28.499***	0.000	−0.018	0.985
LnFDII	199.483***	0.000	9.726***	0.000	9.076***	0.000	3.499***	0.000
LnUMP	328.540***	0.000	20.059***	0.000	19.409***	0.000	6.603***	0.000
LnPCGDPG	489.194***	0.000	32.922***	0.000	32.272***	0.000	21.343***	0.000
LnNRR	503.168***	0.000	34.041***	0.000	33.391***	0.000	18.9***	0.000
LnTO	318.918***	0.000	19.289***	0.000	18.639***	0.000	6.399***	0.000

Note ***denotes 1% level of significance

Table 4 Results of CIPS unit root test

Variables	With constant		With a constant and trend	
	Level	First difference	Level	First difference
LnANSR	−0.8	−1.740	−0.909	−2.925*
LnDFI	−1.372	−2.677**	−1.778	−2.954*
LnNREC	−1.248	−1.518	−1.251	−2.933*
LnREC	−1.599	−2.893***	−1.853	−3.546***
LnINQ	−2.605***	−3.309***	−2.714*	−3.272**
LnFDII	−0.877	−2.767***	−1.938	−2.888*
LnUMP	−1.202	−2.698**	−2.156	−2.934*
LnPCGDPG	−2.454**	−3.781***	−2.558	−4.002***
LnNRR	−0.371	−2.443**	−1.405	−3.332***
LnTO	−0.912	−2.543**	−2.270	−2.945*

Note *, **, and *** denote 10, 5, and 10% significance levels, respectively

Table 5 Results of diagnostic tests

Tests	Test statistics	p-value
Poolability test	F (12, 121) = 10.29***	0.000
Modified Wald test for groupwise heteroskedasticity	χ^2 (13) = 734.47***	0.000
Wooldridge test for autocorrelation in panel data	F (1, 12) = 83.034***	0.000
Hausman test	χ^2 (9) = 32.3***	0.000
Jarque–Bera normality test	45.972***	0.000
Breusch-Pagan LM test	153.21***	0.000
Pesaran scaled LM test	6.021***	0.000
Bias-corrected scaled LM test	5.371***	0.000
Pesaran CD test	−0.531	0.595

Note *** denotes 1% level of significance

Table 6 Variance inflation factor

Variables	VIF	1/VIF
LnREC	4.269	0.234
LnTO	3.272	0.306
LnNREC	3.205	0.312
LnINQ	2.762	0.362
LnUMP	2.455	0.407
LnNRR	1.521	0.658
LnDFII2	1.510	0.662
LnPCGDPG	1.089	0.918
LnFDII	1.011	0.989
Mean VIF	2.344	

Li [3]. The estimated coefficient of non-renewable energy is negatively significant. A perusal of the result reveals that a 1% increase in non-renewable energy consumption decreases sustainable development by 1.976%. Consumption of non-renewable energy increases the level of carbon emissions, which have a negative effect on the environment as well as sustainable development. This finding is similar to the finding of Guney [18]. Renewable energy has no significant impact on sustainable development, which indicates that sources of renewable energy in emerging economies are not sufficiently developed to promote sustainable development.

Furthermore, among the control variables, inequality positively impacts sustainable development. The result indicates that a 1% increase in inequality leads to 3.511% increase in sustainable development. Rising inequality leads to the concentration of wealth and income in the hand of a few, those have a higher marginal propensity to save. This will lead to higher savings in the economy and the availability of more loanable funds [33]. This may induce investment in environment-related technologies, which improve the quality of the environment as well as promote sustainable development. Foreign direct investment inflow has a negative impact on sustainable development. Sustainable development decreases by 0.99% due to a 1% increase in foreign direct investment inflow. However, trade openness enhances sustainable development. A 1% increase in trade openness raises sustainable development by 0.349%. Besides, natural resources rent and unemployment hinder sustainable development. It shows that a 1% increase in natural resource rent and unemployment lead

Table 7 Results of regression with Driscoll-Kraay standard errors

Variables	Coef	Std. Err	P-value
LnDFI	0.408***	0.097	0.002
LnNREC	−1.976***	0.406	0.001
LnREC	−0.180	0.103	0.111
LnINQ	3.511**	1.285	0.021
LnFDII	−0.099**	0.038	0.028
LnUMP	−0.430***	0.097	0.001
LnPCGDPG	−0.010	0.064	0.878
LnNRR	−0.195***	0.021	0.000
LnTO	0.349***	0.071	0.001
Constant	−8.858*	4.863	0.099
Number of Observations	143		
Number of groups	13		
F (11,30)	141.58		
Prob > F	0.000		
Within R^2	0.2281		

Note *, **, and *** denote 10, 5, and 10% levels of significance, respectively to a decrease in sustainable development by 0.195% and 0.43%, respectively. Excessive depletion of natural resources degrades environmental quality, which leads to a setback in sustainable development.

5 Conclusion and Policy Implications

The study aims to examine the impacts of digital finance, renewable, and non-renewable energy consumption on sustainable development in 13 emerging economies for the period 2011–2021. The results of the cross-sectional dependence tests assert the presence of cross-sectional dependency, and thus the cross-sectionally augmented Im-Pesaran-Shin unit root test is performed for all the variables before using the appropriate econometric tool to draw relevant implications. The result of the unit root test shows that all the variables are stationary at the first difference, i.e., I(1). The result of the Driscoll-Kraay standard error corrected method reveals that digital finance has a positive effect on sustainable development, whereas non-renewable energy consumption affects it negatively. However, the study finds no significant impact of renewable energy consumption on sustainable development. In addition, foreign direct investment inflows, natural resource rent, and unemployment hinder sustainable development, while trade openness and inequality promote it.

Based on the findings, the study proposes that emerging economies should promote digital finance. Including digital finance reduces transaction costs and asymmetric information, improves economic efficiency, and thus fosters sustainable development. On the other hand, these countries should reduce non-renewable energy consumption. Excessive use of non-renewable energy raises the threat of global warming and climate change and becomes a hindrance to sustainable development. Further, these economies should invest more in the development of renewable energy plants or sources to reap the benefits. Additionally, it is advised that these economies should be careful while inviting foreign direct investment to their countries as foreign investors may be attracted to these emerging economies to exploit the weak environmental regulations. Therefore, emerging economies need to design foreign investment policies that promote sustainable development.

References

1. Campiglio, E.: Beyond carbon pricing: The role of banking and monetary policy in financing the transition to a low-carbon economy. Ecological economics. 121, 220–230 (2016). https://doi.org/10.1016/j.ecolecon.2015.03.020
2. Yang, L., Wang, L., Ren, X.: Assessing the impact of digital financial inclusion on PM2.5 concentration: evidence from China. Environmental Science and Pollution Research. 29(15), 22547–22554 (2022). https://doi.org/10.1007/s11356-021-17030-3

3. Zheng, H., Li, X.: The impact of digital financial inclusion on carbon dioxide emissions: Empirical evidence from Chinese provinces data. Energy Reports. 8, 9431-9440 (2022). https://doi.org/10.1016/j.egyr.2022.07.050
4. International Energy Agency: Financing Clean Energy Transitions in Emerging and Developing Economies, https://www.iea.org/reports/financing-clean-energy-transitions-in-emerging-and-developing-economies, last accessed 2022/11/20
5. Zhao, H., Yang, Y., Li, N., Liu, D., Li, H.: How does digital finance affect carbon emissions? Evidence from an emerging market. Sustainability. 13(21), 12303 (2021). https://doi.org/10.3390/su132112303
6. Ding, R., Shi, F., Hao, S.: Digital Inclusive Finance, Environmental Regulation, and Regional Economic Growth: An Empirical Study Based on Spatial Spillover Effect and Panel Threshold Effect. Sustainability. 14(7), 4340 (2022). https://doi.org/10.3390/su14074340
7. Guney, T.: Renewable energy, non-renewable energy and sustainable development. International Journal of Sustainable Development and World Ecology. 26(5), 389-397 (2019). https://doi.org/10.1080/13504509.2019.1595214
8. Islam, M. M., Sohag, K., Shahbaz, M.: Assessment of Nexus between energy consumption and sustainable development in Russian Federation: A disaggregate analysis. World Development Sustainability. 1, 100027 (2022). https://doi.org/10.1016/j.wds.2022.100027
9. Driscoll, J. C., Kraay, A. C.: Consistent covariance matrix estimation with spatially dependent panel data. Review of economics and statistics. 80(4), 549-560 (1998). https://doi.org/10.1162/003465398557825
10. Shi, F., Ding, R., Li, H., Hao, S.: Environmental Regulation, Digital Financial Inclusion, and Environmental Pollution: An Empirical Study Based on the Spatial Spillover Effect and Panel Threshold Effect. Sustainability. 14(11), 6869 (2022). https://doi.org/10.3390/su14116869
11. Wang, X., Wang, X., Ren, X., Wen, F.: Can digital financial inclusion affect CO2 emissions of China at the prefecture level? Evidence from a spatial econometric approach. Energy Economics. 109, 105966 (2022). https://doi.org/10.1016/j.eneco.2022.105966
12. Chang, J.: The role of digital finance in reducing agricultural carbon emissions: evidence from China's provincial panel data. Environmental Science and Pollution Research. 29(58), 87730-87745 (2022). https://doi.org/10.1007/s11356-022-21780-z
13. Xue, Q., Feng, S., Chen, K., Li, M.: Impact of digital finance on regional carbon emissions: An empirical study of sustainable development in China. Sustainability. 14(14), 8340 (2022). https://doi.org/10.3390/su14148340
14. Jiang, X., Wang, X., Ren, J., Xie, Z.: The Nexus between digital finance and economic development: evidence from China. Sustainability. 13(13), 7289 (2021). https://doi.org/10.3390/su13137289
15. Ahmad, M., Majeed, A., Khan, M. A., Sohaib, M., Shehzad, K.: Digital financial inclusion and economic growth: Provincial data analysis of China. China Economic Journal 14(3), 291-310 (2021). https://doi.org/10.1080/17538963.2021.1882064
16. Khera, P., Ng, M. S. Y., Ogawa, M. S., and Sahay, M. R.: Is digital financial inclusion unlocking growth? (IMF Working Paper No. 2021/167). Washington, D.C., United States: International Monetary Fund (2021). https://doi.org/10.5089/9781513584669.001
17. Shen, Y., Hu, W., Hueng, C. J.: Digital financial inclusion and economic growth: a cross-country study. Procedia Computer Science. 187, 218-223 (2021). https://doi.org/10.1016/j.procs.2021.04.054
18. Guney, T.: Renewable energy consumption and sustainable development in high-income countries. International Journal of Sustainable Development and World Ecology. 28(4) 376-385 (2021). https://doi.org/10.1080/13504509.2020.1839807
19. Oke, D. M., Ibrahim, R. L., Bokana, K. G.: Can renewable energy deliver African quests for sustainable development? The Journal of Developing Areas. 55(1) (2021). https://doi.org/10.1353/jda.2021.0022
20. Spaiser, V., Scott, K., Owen, A., Holland, R.: Consumption-based accounting of CO2 emissions in the sustainable development Goals Agenda. International Journal of Sustainable Development and World Ecology. 26(4), 282-289 (2019). https://doi.org/10.1080/13504509.2018.1559252

21. Kaimuri, B., Kosimbei, G.: Determinants of sustainable development in Kenya. Journal of economics and sustainable development. 8(24), 17–36 (2017). Available at https://core.ac.uk/reader/234648100
22. Koirala, B. S., Pradhan, G.: Determinants of sustainable development: Evidence from 12 Asian countries. Sustainable Development. 28(1), 39-45 (2020). https://doi.org/10.1002/sd.1963
23. Kumari, N., Kumar, P., Sahu, N. C.: Do energy consumption and environmental quality enhance subjective wellbeing in G20 countries? Environmental Science and Pollution Research. 28(42), 60246-60267 (2021). https://doi.org/10.1007/s11356-021-14965-5
24. Behera, J., Sahoo, D.: Asymmetric relationships between information and communication technology (ICT), globalization, and human development in India: evidence from non-linear ARDL analysis. Journal of Economic Structures. 11(1), 1-25 (2022). https://doi.org/10.1186/s40008-022-00269-5
25. Solt, F.: Measuring income inequality across countries and over time: The standardized world income inequality database. Social Science Quarterly. 101(3), 1183-1199 (2020). https://doi.org/10.1111/ssqu.12795
26. Salahuddin, M., Gow, J., Vink, N.: Effects of environmental quality on agricultural productivity in sub Saharan African countries: A second generation panel based empirical assessment. Science of The Total Environment. 741, 140520 (2020). https://doi.org/10.1016/j.scitotenv.2020.140520
27. Breusch, T., Pagan, A.: The lagrange multiplier test and its applications for the error components model with incomplete panels. Review of Economic Studies. 47(1), 239-253 (1980). https://doi.org/10.2307/2297111
28. Pesaran, M. H.: General diagnostic tests for cross-sectional dependence in panels. Empirical economics. 60(1), 13-50 (2021). https://doi.org/10.1007/s00181-020-01875-7
29. Pesaran, M. H., Ullah, A., Yamagata, T.: A bias-adjusted LM test of error cross-section independence. The econometrics journal. 11(1), 105-127 (2008). https://doi.org/10.1111/j.1368-423X.2007.00227.x
30. Pesaran, M. H.: A simple panel unit root test in the presence of cross-section dependence. Journal of applied econometrics. 22(2), 265-312 (2007). https://doi.org/10.1002/jae.951
31. Hausman, J. A.: Specification tests in econometrics. Econometrica: Journal of the econometric society. 46(6), 1251–1271 (1978). https://doi.org/10.2307/1913827
32. Hoechle, D.: Robust standard errors for panel regressions with cross-sectional dependence. The stata journal. 7(3), 281-312 (1978). https://doi.org/10.1177/1536867X0700700301
33. Kaldor, N.: Alternative Theories of Distribution. Review of Economic Studies. 23 (2), 83-100 (1956). https://doi.org/10.2307/2296292

Post-treatment Modification Using Deep Eutectic Solvent in PES and PES/SAPO-34 Membranes for Gas Separation

Jonathan S. Cardoso, Zhi Lin, Paulo Brito, and Licínio Ferreira

Abstract Mixed matrix membranes (MMM) are heavily studied and the main concern by researchers is the difficulty to obtain a high selectivity membrane with low defects, mainly interfacial voids due to the poor interaction between polymer matrix and filler. The functionalization of zeolite surface in PolyEtherSulphone (PES)/Silico-AluminoPhosphate-34 (SAPO-34) led to membranes with higher compatibility. Recently, the use of ionic liquids in the particles surface to improve the gas permeability and separation efficiency has presented advanced results. Although, the cost and environmental issues on Ionic Liquids (ILs) still represent a limit to its use. The Deep Eutectic Solvents (DESs) present an alternative route to replace ILs, with low-cost and environmentally friend. This paper aims to study the potential use of choline chloride/urea, a DES mixture, in ratio 1:2, to replace ILs as a post-healing methodology, showing a 20% increase in CO_2 permeability in PES membranes and 15% increase in CO_2/N_2 selectivity in PES/SAPO-34 membranes due to an increase in compatibility between polymer and zeolite promoted by the post-treatment using DES.

Keywords SAPO-34 · Deep eutectic solvents · Mixed matrix membranes · Gas separation

J. S. Cardoso (✉) · L. Ferreira
CIEPQPF, University of Coimbra, 3030-790 Coimbra, Portugal
e-mail: jonathancardoso@ipb.pt

Z. Lin
CICECO, University of Aveiro, 3810-193 Aveiro, Portugal

P. Brito
CIMO, Polytechnic Institute of Bragança, 5300-253 Bragança, Portugal

J. R. da. C. S. Galvão et al. (eds.), *Proceedings of the 3rd International Conference on Water Energy Food and Sustainability (ICoWEFS 2023)*, Springer Proceedings in Earth and Environmental Sciences, https://doi.org/10.1007/978-3-031-48532-9_24

1 Introduction

Silico-aluminophosphates (SAPOs) are zeolites in the form TO_4 (T = Si, Al or P) and widely used in the catalysis and separation processes because their pore size is in dimensions near to most common molecules. Varied three-dimensional arrangements of the basic building units result in frameworks containing the channels of molecular dimensions that characterize the molecular sieves. Zeolites are classified according to size: small (less than 4 Å), medium, (4–6 Å), large, (6–8 Å), or super large (bigger than 8 Å) pore. Molecular sieves with pores larger than 20 Å, with amorphous walls, are classified as mesoporous [1]. SAPOs can be prepared basically through different methods which influence the crystal size and characteristics of the final material [2].

SAPO-34 is a solid with a chabazite structure (CHA) and it appeared as one of the main filler materials in mixed matrix membranes (MMM) used for post-combustion gas separation in comparison with other zeolites mainly due to its pore size (0.38 nm) is near the kinetic diameter of gases like H_2 (0.29 nm), CO_2 (0.33 nm), N_2 (0.36 nm), CO (0.37 nm), CH_4 (0.38 nm) [3, 4]. This means that the pore size of SAPO-34 is big enough to allow H_2, He, CO_2 to permeate and smaller enough to block the permeation of other gases.

Although, several studies in the production of MMM using SAPO-34 as a dispersed solid presented excellent separation result, the main problem in these membranes is the compatibility between particles, an inorganic component, and the polymer matrix, an organic component [5–9]. Surface functionalization of zeolites improves the zeolite-polymer matrix interaction. The functional groups function as a bound that promote a better link and diminishes interfacial problems [10–12]. Since deep eutectic solvents emerged as a low-cost alternative for others surface modifications, the use of those solvents in MMMs shows a new route that can be studied [13, 14]

2 Materials and Methods

2.1 Synthesis and Characterization of SAPO-34

To synthesize the SAPO-34 samples, a gel having the molar composition 1 Al_2O_3; 1 P_2O_5; 0.6 SiO_2; 1.5 Morpholine; 0.5 Tetraethylammonium Hydroxide (TEAOH); 70 H_2O was prepared. The gel was aged for 24 h under mixing and then dried at 90 °C for 24 h. The sample was prepared by adding water to dried gel in the mass ratio of 1:1 and heated at 200 °C with different reaction times. The solid obtained was centrifuged, washed, dried at 105 °C in an oven and calcined at 560 °C for 8 h.

The samples were characterized via Powder X-Ray Diffraction (PXRD) for the analysis of the crystalline structure of the synthesized materials, Dynamic Light Scattering (DLS) for particle size distribution, pore size analysis to determine the pore distribution and surface area to verify the textual properties of the samples after the structural modifications.

2.2 Synthesis and Characterization of Mixed Matrix Membranes

20% w/w Polyethersulfone (PES) membrane was prepared in N-Methyl-2-Pyrrolidone (NMP), 15% of total PES was added in NMP and mixed for 1 h at 80 °C. The remaining PES was added and mixed for 3 h at 80 °C, then the solution was stirred overnight at room temperature. For a 20% PES/SAPO-34 membranes, first 20% of SAPO-34, in relation to PES mass, was added in NMP, the solution was stirred for 1 h and kept in ultrasonic bath for 4 h to avoid particle agglomeration, then 15% of total PES was added in NMP/SAPO-34 solution and mixed for 1 h at 80 °C. The remaining PES was added and mixed for 3 h at 80 °C, then the solution was stirred overnight at room temperature. The resulting solutions were degassed for 4 h in ultrasonic bath. The membranes were casted with a knife of 0.3 mm, then dried at 80 °C for 2 h and at 200 °C for 20 h under vacuum. The samples were characterized by scanning electron microscopy (SEM) and Fourier-transform Infrared (FTIR) spectroscopy.

2.3 Gas Permeation Tests

Gas permeation experiments were conducted using pure gases to determine their permeability and diffusivity coefficients and the results were compared to each other in a permeability unit called Barrer ($10^{-10} cm^3_{STP} \cdot cm / cm^2 \cdot s \cdot cmHg$). A schematic representation of the permeation system is shown in Fig. 1 The membranes were cut into 47 mm disks and mounted on a steel permeation cell and tested at feed pressure of 10 bar at room temperature. The permeabilities were obtained according to Eq. 1.

$$P_i = \frac{V_s l}{T_{AMB} \Delta p} \frac{T_{STP}}{p_{STP} A} \frac{dp}{dt} \tag{1}$$

The permeability of a certain gas (P_i) is related to the thickness of the film (l), the fixed volume of the permeate (Vs) which is measured as standard. The ambient temperature (T_{AMB}), the difference of pressure between permeate and retentate (Δp), the permeation area (A) and temperature and pressure in normal conditions (T_{STP} and P_{STP}). The ideal selectivity results (α_{ab}) are represented by the ratio between the permeability (P) of two pure gases, a and b, respectively, measured separately under the same conditions, in the same membrane to evaluate the separation performance, as presented in Eq. 2.

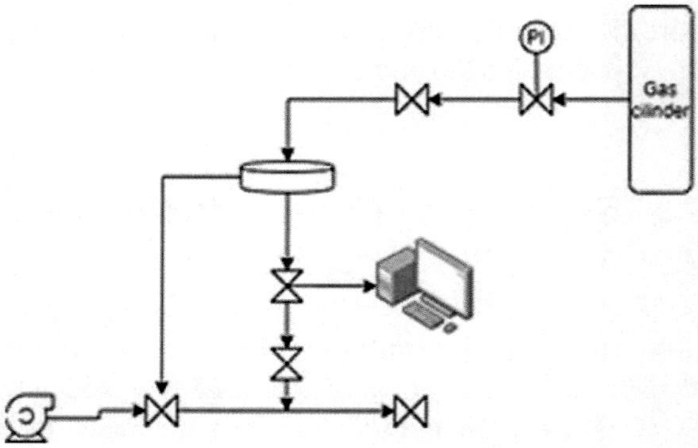

Fig. 1 Schematics of the permeation system

$$\alpha_{ab} = \frac{P_a}{P_b} \tag{2}$$

2.4 Post-impregnation

The membranes tested were placed in a kitasato, and vacuum was applied for a period of 1 h to remove the air from the pores. The additive DES (ChCl/Urea in mass ratio of 1:2) was then introduced with a syringe, through a septum and the membrane stays under vacuum for 1 h to allow the additive to fill the membrane pores. The membrane was then removed and dried using a tissue paper.

3 Results and Discussion

3.1 SAPO-34 Synthesis and Characterization

SAPO-34 was synthetized from the same batch, using different reaction times (12, 16, 24 and 48 h). The results showed that in all reaction times the SAPO-34 was obtained, although the samples with reaction time of 24 h or more presented a higher purity as presented in XRD spectra when compared to chabazite (CHA) standard x-ray dispersion analysis (see Fig. 2). The adsorption–desorption nitrogen isotherms are depicted in Fig. 3 and their textural properties were collected in Table 1.

The XRD patterns of different reaction times (12, 16, 24 and 48 h) are presented in Fig. 2 were compared to a standard XRD for the same structure where each peak represents a plain of diffraction which characterize the solid. The adsorption–desorption nitrogen isotherms in Fig. 3 exhibited a type I isotherm for all samples prepared, which is characteristic of well-defined microporous structure. The textural

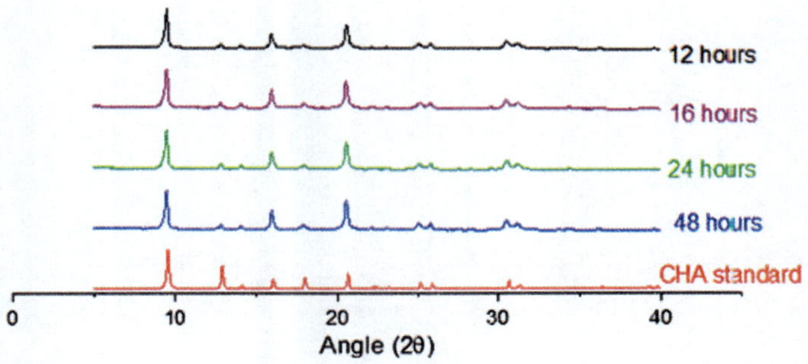

Fig. 2 PXRD patterns of samples with different reaction times and CHA standard

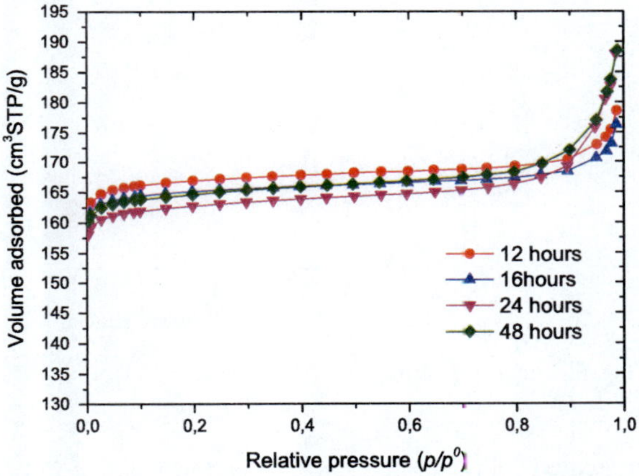

Fig. 3 Adsorption–desorption nitrogen isotherms of samples with reaction times of 12, 16, 24 and 48 h

Table 1 Textural properties of the materials

Sample	S_{BET} ($m^2\ g^{-1}$)	$S_{Langmuir}$ ($m^2\ g^{-1}$)	S_{ext} ($m^2\ g^{-1}$)	S_{mic} ($m^2\ g^{-1}$)	V_{mic} ($mm^3\ g^{-1}$)
12 h	477	725	6	471	0.257
16 h	470	718	5	465	0.254
24 h	464	707	9	455	0.249
48 h	470	715	10	460	0.251

properties show that all materials presented a high BET area (S_{BET}), which is related to the total area of the zeolite (internal and external) while presenting a small external area (S_{ext}) shows that the internal area is much larger that the external area due the presence of the pores, this is expressed by the microporous area (S_{mic}), presenting a large amount of microporous with a high surface area which is compared to those in literature. The sample with 24 h of reaction time was selected since it presented a higher purity even though with a bigger particle size. With that, another sample was

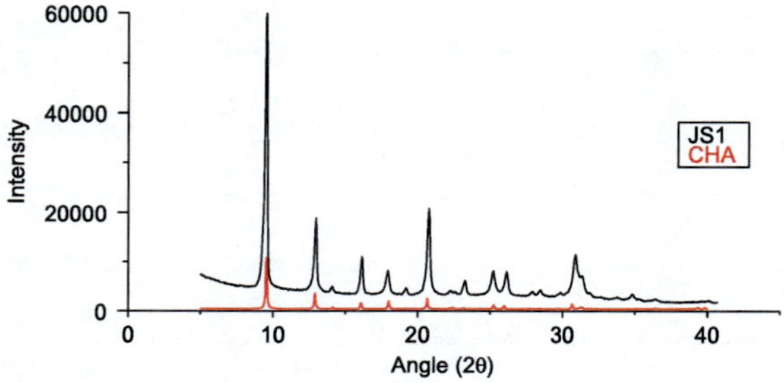

Fig. 4 XRD for JS1 sample

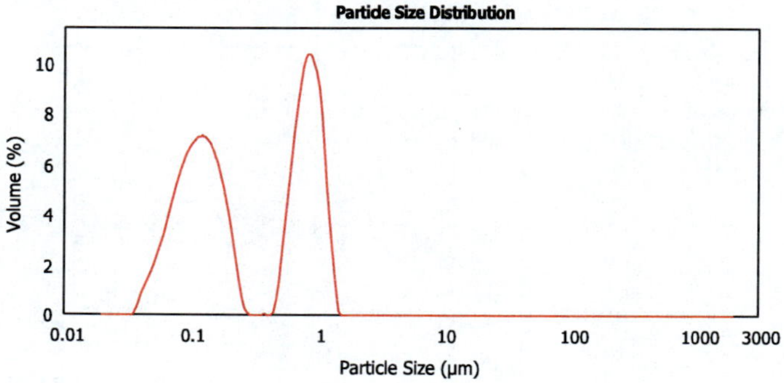

Fig. 5 LDS for JS1 sample

prepared following the same procedure, named JS1, although changing the silicon source from LUDOX AS-40 to LUDOX HS-40. The XRD and LDS analyses are presented in Figs. 4 and 5.

Sample JS1 presented the same purity as the samples from LUDOX AS-40 as obtained by comparing patterns in Figs. 2 and 4 with the standard CHA, however, the average particle size obtained was around 0.2 μm, as presented in Fig. 5, which is suitable to be used as the filler for the PES mixed matrix membranes.

3.2 *Membrane Characterization*

The PES and PES/SAPO-34 membranes were characterized by SEM to obtain images to evaluate the compatibility after the inclusion of SAPO-34 and DES in the membranes. Figures 6 and 7 presents the SEM analysis of a PES membrane before and after the DES post-treatment, respectively.

Figure 6 presents the neat PES membrane with a high density and no porosity, which is expected since the permeation of components is through the sorption–diffusion mechanism. After the post-treatment using DES, the neat PES membranes

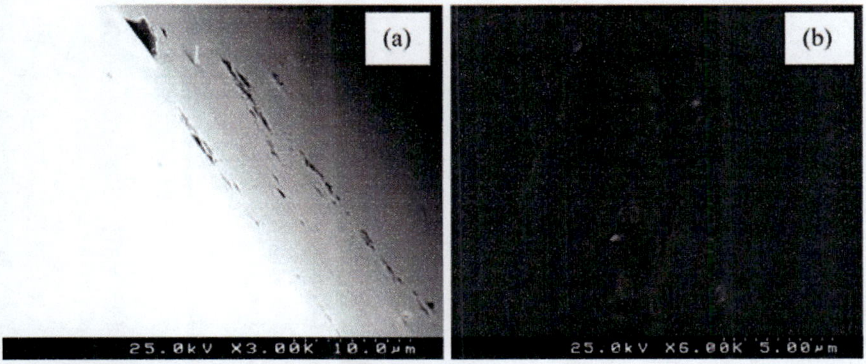

Fig. 6 Cross-section (**a**) and surface (**b**) of PES membrane

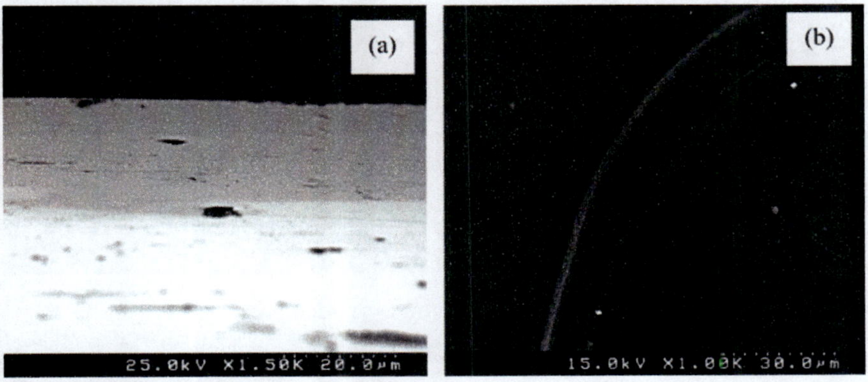

Fig. 7 Cross-section (**a**) and surface (**b**) of DES post-treated PES membrane

presented a similar morphology in SEM analysis (Fig. 7), showing that the post-treatment with DES does not affect the membrane structure and morphology.

Figure 8 shows the FTIR spectrums of the neat PES membranes with and without DES, the three peaks observed at 1576, 1482 and 1405 cm^{-1} indicate the presence of benzene rings. The two peaks at 1320 and 1296 cm^{-1} indicate the presence of ether function group. The two peaks at 1144 and 1100 cm^{-1} indicate the presence of sulfone group. Those spectrums are in accordance with FTIR spectrums for PES in literature [15, 16] and implies that the use of DES does not change the membrane matrix.

The PES/SAPO-34 membrane shows a high dispersion of SAPO-34, as presented in Fig. 9, although it is clear the presence of small gaps between some SAPO-34 crystals with absence of PES, which are characterized as interfacial voids. Those voids can be better noticed when comparing with the SEM of the cross-section obtained after the post-treatment using DES in Fig. 10. Post-treatment with DES clearly decreased the gaps between filler and matrix filling the voids in the membrane.

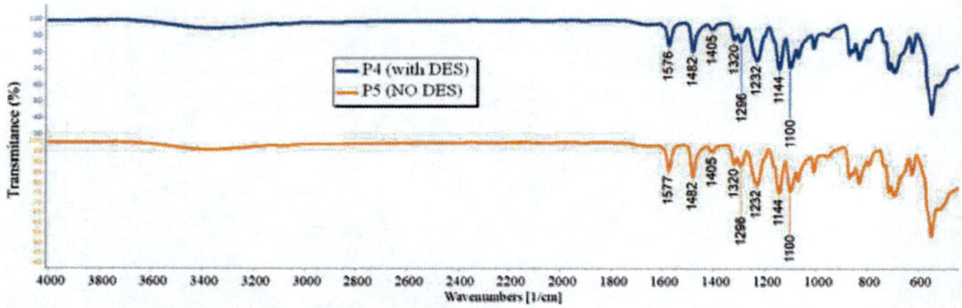

Fig. 8 FTIR spectrums of neat PES membranes

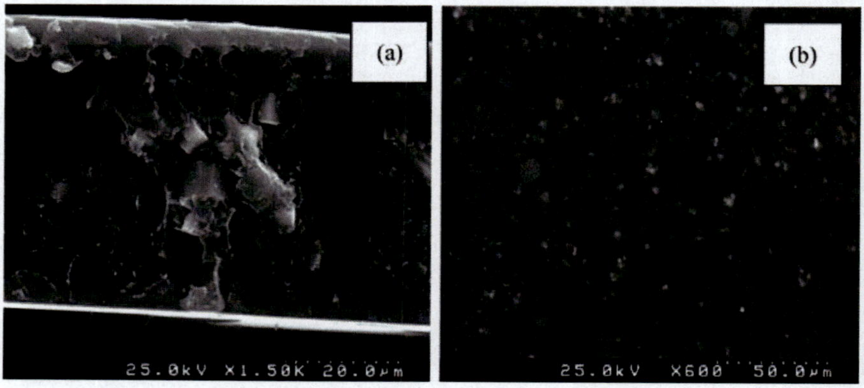

Fig. 9 Cross-section (**a**) and surface (**b**) of PES/SAPO-34 membrane

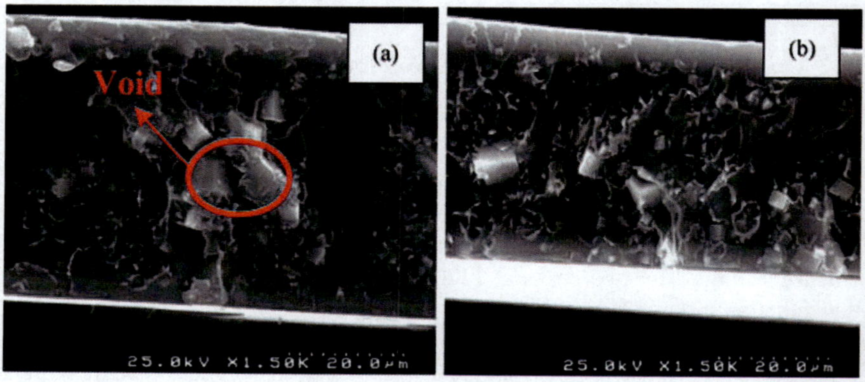

Fig. 10 Comparison in cross-section of pre-treated (**a**) and post-treated (**b**) PES/SAPO-34 membrane

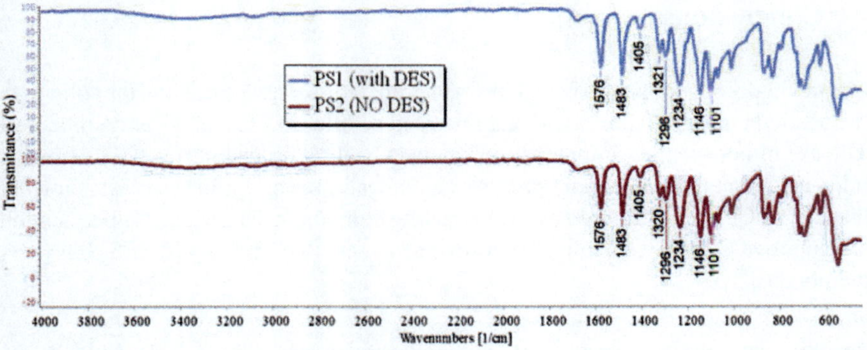

Fig. 11 FTIR spectrums for PES/SAPO-34 membranes

Figure 11 shows the FTIR spectrums of the PES/SAPO-34 membranes before and after the DES post-treatment, the three peaks observed at 1576, 1483 and 1405 cm^{-1} indicate the presence of benzene rings. The two peaks at 1320 and 1296 cm^{-1} indicates the presence of ether function. The two peaks at 1146 and 1101 cm^{-1} indicates the presence of sulfone group. Both spectrums are in accordance with FTIR spectrums for PES in literature [15, 16] and implies that the addition of SAPO-34 or the use of DES does not change the membrane matrix.

3.3 *Gas Permeation Tests*

The permeability obtained in neat PES membrane was 1.028 Barrer for CO_2 and 0.044 Barrer for N_2 with a selectivity of 23.26. While the permeability for neat post-treated PES membrane was 1.270 Barrer for CO_2 and 0.054 Barrer for N_2 with a selectivity of 24.65. The results show a small increase in selectivity mainly due to an increase in 20% of CO_2 permeability when compared to a non-treated membrane, which results an increase in permeability when compared to the initial membrane.

The permeability obtained in PES/SAPO-34 membrane was 0.97 Barrer for CO_2 and 0.07 for N_2 with a selectivity of 13.80. This result shows the presence of interfacial voids due to the low compatibility between PES and SAPO-34. Those voids can be noticed in SEM analysis presented in Fig. 10. Those imperfections allow gases to pass through without selectivity, which explains the increase in permeability and reduction in selectivity.

The permeability for CO_2, N_2 and selectivity obtained after the post-treatment using DES was 0.42 Barrer, 0.03 Barrer and 15.89, respectively. The results show a decrease in permeability due to the reduction of interfacial voids, DES replaced the air present in these gaps and improved the polymer-zeolite interaction.

4 Conclusions

The methodology for nano-sized SAPO-34 preparation was defined and the solid with the desired characteristics obtained to be used as filler in PES polymer membranes. The use of post-sealing impregnation can be a healing procedure for poor compatibility after synthesis. This work presents an innovation as preliminary results showing that 1:2 ChCl/Urea DES used as a post-sealing compound can reduce N_2 permeation and improve CO_2/N_2 separation performance in neat PES and MMM PES/SAPO-34 membranes.

Funding This work was supported by the Strategic Project of CIEPQPF (UIDB/00102/2020), CICECO-Aveiro Institute of Materials (UIDB/50 011/2020, UIDP/50 011/2020 & LA/P/0 0 06/ 2020), financed by Fundação para a Ciência e Tecnologia (FCT) through national funds. J. S. Cardoso was funded by the FCT through the PhD grant (SFRH/BD/148170/2019).

References

1. Sena, F.C., de Souza, B.F., de Almeida, N.C., Cardoso, J.S., Fernandes, L.D. Influence of framework composition over SAPO-34 and MeAPSO-34 acidity. Appl Catal A Gen 406:59–62 (2011). doi: https://doi.org/10.1016/j.apcata.2011.08.010
2. Askari, S., Halladj, R., Sohrabi, M. An overview of the effects of crystallization time, template and silicon sources on hydrothermal synthesis of sapo-34 molecular sieve with small crystals. Reviews on Advanced Materials Science 32:83–93 (2012)
3. Luna, F.J., Schuchardt, U. Modificação de zeólitas para uso em catálise. Quim Nova 24:885–892 (2001). doi: https://doi.org/10.1590/S0100-40422001000600027
4. Prakash, A.M., Unnikrirhnan, S. Synthesis of SAPO-34: High Silicon lncorporation in the Presence of Morpholine as Template. J Chem Soc, Faraday Trans, 90:2291–2296 (1994). doi: https://doi.org/10.1039/FT9949002291
5. Ahmad, N.N.R., Mukhtar, H., Mohshim, D.F., Nasir, R., Man, Z. Effect of different organic amino cations on SAPO-34 for PES/SAPO-34 mixed matrix membranes toward CO2/CH4 separation. J Appl Polym Sci 133:1–6 (2016). doi: https://doi.org/10.1002/app.43387
6. Junaidi, M.U.M., Khoo, C.P., Leo, C.P., Ahmad, A.L. The effects of solvents on the modification of SAPO-34 zeolite using 3-aminopropyl trimethoxy silane for the preparation of asymmetric polysulfone mixed matrix membrane in the application of CO2 separation. Microporous and Mesoporous Materials 192:52–59 (2014). doi: https://doi.org/10.1016/j.micromeso.2013.10.006
7. Junaidi, M.U.M., Leo, C.P., Ahmad, A.L., Ahmad, N.A. Fluorocarbon functionalized SAPO-34 zeolite incorporated in asymmetric mixed matrix membranes for carbon dioxide separation in wet gases. Microporous and Mesoporous Materials 206:23–33 (2015). doi: https://doi.org/10.1016/j.micromeso.2014.12.013
8. Li, S., Falconer, J.L., Noble, R.D. Improved SAPO-34 membranes for CO2/CH4 separations. Advanced Materials 18:2601–2603 (2006). doi: https://doi.org/10.1002/adma.200601147
9. Nasir, R., Ahmad, N.N.R., Mukhtar, H., Mohshim, D.F. Effect of ionic liquid inclusion and amino-functionalized SAPO-34 on the performance of mixed matrix membranes for CO2/ CH4separation. J Environ Chem Eng 6:2363–2368 (2018). doi: https://doi.org/10.1016/j.jece.2018.03.032
10. Huang, Y., Yu, M., Carreon, M.A., Song, Z., Zhou, R., Li, S., Feng, X., Zhou, S.J., Zong, Z., Meyer, H.S. SAPO-34 Membranes for N2/CH4 separation: Preparation, characterization,

separation performance and economic evaluation. J Memb Sci 487:141–151 (2015). doi: https://doi.org/10.1016/j.memsci.2015.03.078

11. Hudiono, Y.C., Carlisle, T.K., Bara, J.E., Zhang, Y., Gin, D.L., Noble, R.D. A three-component mixed-matrix membrane with enhanced CO2 separation properties based on zeolites and ionic liquid materials. J Memb Sci 350:117–123 (2010). doi: https://doi.org/10.1016/j.memsci.2009.12.018
12. Hudiono, Y.C., Carlisle, T.K., LaFrate, A.L., Gin, D.L., Noble, R.D. Novel mixed matrix membranes based on polymerizable room-temperature ionic liquids and SAPO-34 particles to improve CO2 separation. J Memb Sci 370:141–148 (2011). doi: https://doi.org/10.1016/j.memsci.2011.01.012
13. Craveiro, R., Neves, L.A., Duarte, A.R.C., Paiva, A. Supported liquid membranes based on deep eutectic solvents for gas separation processes. Sep Purif Technol 254 (2021). doi: https://doi.org/10.1016/j.seppur.2020.117593
14. Jiang, B., Zhang, N., Wang, B., Yang, N., Huang, Z., Yang, H., Shu, Z. Deep eutectic solvent as novel additive for PES membrane with improved performance. Sep Purif Technol 194:239–248 (2018). doi: https://doi.org/10.1016/j.seppur.2017.11.036
15. Alenazi, N., Hussein, M., Alamry, K., Asiri, A. Nanocomposite-Based Aminated Polyethersulfone and Carboxylate Activated Carbon for Environmental Application. A Real Sample Analysis. C (Basel) 4:30 (2018). doi: https://doi.org/10.3390/c4020030
16. Noel Jacob, K., Senthil Kumar, S., Thanigaivelan, A., Tarun, M., Mohan, D. Sulfonated polyethersulfone-based membranes for metal ion removal via a hybrid process. J Mater Sci 49:114–122 (2014). doi: https://doi.org/10.1007/s10853-013-7682-1

Pyrolysis and Combustion of Lemon Verbena and Oregano Biomass: Characterization and Thermodynamic Analyses

R. García-Mateos, S. Cuéllar-Borrego, J. A. Sandía-Manchado, and A. I. Parralejo-Alcobendas

Abstract The non-irrigated areas of the southwest of Spain need the introduction of profitable and diverse crop alternatives, as traditional crops are at the limit of their profitability. For this reason, the implementation of the cultivation of Medicinal and Aromatic Plants has been studied, as they improve the productive diversity of the agricultural sector in Extremadura, the capacity to obtain new products and indirectly increase the stability of the rural population. The exclusive use of one part of the plant leads to the generation of a high volume of plant waste, so the energetic use of this waste can be a key to competitive improvement. Specifically, an energy study of the pyrolysis and combustion process of waste from the production of lemon verbena (Aloysia citrodora) and oregano (Origanum virens) has been carried out by means of Thermogravimetric Analysis (TGA), Differential Thermogravimetry curve (DTG) and pyrolysis and combustion indexes, with the aim of comparing the results with other biomass waste.

Keywords Circular economy · Residue · Biofuel

1 Introduction

The southwest of Spain is characterised by a low population density and a high degree of ageing, with a regressive demographic trend in recent years. This problem is intensified in the most disadvantaged rural areas, such as non-irrigated lands. These areas need to implement profitable and diverse crop alternatives, as traditional crops are at the limit of profitability. The aim is to incorporate crops to produce sustainable products with high added value, based on the use and enhancement of high-quality local production, in order to be able to compete via differentiated quality.

R. García-Mateos (✉) · S. Cuéllar-Borrego · J. A. Sandía-Manchado · A. I. Parralejo-Alcobendas
Scientific and Technological Research Centre of Extremadura (CICYTEX), C/Pamplona 64, 06800 Mérida, Spain
e-mail: raquel.garcia@juntaex.es

J. R. da. C. S. Galvão et al. (eds.), *Proceedings of the 3rd International Conference on Water Energy Food and Sustainability (ICoWEFS 2023)*, Springer Proceedings in Earth and Environmental Sciences, https://doi.org/10.1007/978-3-031-48532-9_25

Furthermore, within the framework of the circular economy it is important to establish production systems that metabolise with the environment, that are self-sustainable through the continuous reuse of raw materials. This includes the reuse of waste generated in the area, to stimulate the transition towards a Circular Economy, which boosts global competitiveness, accelerates sustainable economic development, and creates new jobs.

In order to study the viability of different crops and their effect on the region, the Scientific and Technological Research Centre of Extremadura (CICYTEX), within the framework of the Extremadura ERDF Operational Programme 2014–2020, resolved to finance the project "Agro-livestock farming strategies for the economic and environmental sustainability of non-irrigated land in Extremadura GREENHOPE". Among the activities proposed in the project, it is concluded that an optimal crop to achieve these objectives is the cultivation of Medicinal and Aromatic Plants, as they improve the diversity of production in Extremadura's agricultural sector, the capacity to obtain new products and indirectly increase the stability of the rural population. The exclusive use of one part of the plant leads to the generation of a high volume of plant residue, generally fine wood from stems and young branches of annual growth. The study of the energetic use of the residues obtained after the harvesting of aromatic plants, as well as the residues generated after the extraction processes of compounds of interest, is key to the competitive improvement of this type of crops. Specifically, a study of residues from the production of lemon verbena (Aloysia citrodora) and oregano (Origanum virens) has been carried out. These are crops and residues that have hardly been studied in previous projects, which makes their study more interesting from an energy point of view.

Among the different thermal processes of biomass, pyrolysis and combustion are simple and direct processes for energy extraction. For the study of thermal processes, thermogravimetric analysis (TGA) is a basic tool, as it allows to evaluate the suitability of these crop residues as biofuels, as well as to determine the thermal process with the best performance. In addition, in order to carry out the thermogravimetric analysis correctly, the Differential Thermogravimetry (DTG) curve has been calculated, corresponding to the first derivative of the TG curve, in which the different stages of the pyrolysis and combustion processes and the speeds at which these stages occur can be observed. The pyrolysis and combustion indexes are also calculated, as a comparative parameter of the characteristics of the combustion process with other types of biomasses.

2 Materials and Methods

2.1 Sample Collection and Preparation

For this study, residues of lemon verbena (LV) and oregano (OR) grown in south-western Spain were selected. The samples were crushed in a hammer mill as a first process for proximate analysis, and then in a ball mill to obtain the optimum particle size for the thermogravimetric analysis.

2.2 Proximate Analysis and Higher Heating Value

Moisture was calculated according to the standard [1], by using a balance and a stove. Volatile compound percentage was determined by using an oven, according to the standard [2], whereas ash content was obtained by using an oven according to the standard [3]. Based on volatile matter and ash percentages, fixed carbon percentage was determined by subtraction of 100%. For higher heating value (HHV) analysis of the samples, a calorimetric pump was used according to the standard [4].

2.3 TG Analysis

The thermogravimetric analysis (TG) was carried out by using a thermobalance (NETZSCH STA 449F5). Different samples from each residue were selected, with variable weights from 21 to 37 mg in an aluminium crucible (Al2O3), undergoing temperatures between 30 and 900 °C with a temperature ramp of 20 °C/min in argon atmosphere, and a gas flow of 50 mL/min, and in air atmosphere (80/20) with a gas flow of 50 mL/min. The protective gas was argon (20 mL/min). The Difference Thermogravimetry curve (DTG) has been calculated, corresponding to the first derivative of the DT curve. All the samples were done in triplicate to assure a suitable reproducibility of results.

In order to determine the different temperature events the tangent method [5, 6] has been selected.

In the case of pyrolysis, the main characteristic temperatures of the process were determined, such as the starting temperature of fast pyrolysis Tm, the end temperature of pyrolysis, Tf, and the peak temperature Tp. The pyrolysis index, Pi, the end of pyrolysis index, Pf and the pyrolysis characteristic index P were calculated as well [7]:

$$P_i = \frac{DTG_{max}}{t_p t_m}$$

$$P_f = \frac{DTG_{max}}{\Delta t_{1/2} t_p t_f}$$

$$P = \frac{DTG_{max} DTG_{mean}}{T_m^2 T_f}$$

In the case of combustion, the ignition temperature, combustion temperature and maximum temperature, Ti, Tf, and Tp, respectively, were obtained. In addition, the ignition index Di, the combustion index Df, the devolatilization index D and the characteristic combustion index S were calculated [8–10].

$$D_i = \frac{DTG_{max}}{t_p t_i}$$

$$D_f = \frac{DTG_{max}}{\Delta t_{1/2} t_p t_f}$$

$$D = \frac{DTG_{max}}{T_p \Delta T}$$

$$S = \frac{DTG_{max} DTG_{mean}}{T_i^2 T_f}$$

3 Results and Discussion

3.1 Proximate Analysis and Higher Heating Value

The results concerning the proximate analyses, as well as the moisture, are shown in Table 1.

Considering the moisture data, in both cases the moisture content is not high (<10%), so no pre-drying of the biomass is necessary for its use as biofuel [11].

Regarding the proximate analysis, in the case of volatile matter, lemon verbena residues LV presented a higher value (76.64%), while oregano OR had a lower content

Table 1 Proximate analyses

Sample	Moisture % wb	Ash % db	Volatile Matter % db	Fixed Carbon % db	HHV MJ/kg db
LV	6.52	3.22	76.64	20.14	18.71
OR	7.33	4.67	73.77	21.66	18.15

db dry basis, *wb* web basis

(73.77%). A higher content of volatile matter is beneficial for the pyrolysis process of biomass.

In the case of ash content, lemon verbena (3.22%) presented lower values than oregano (4.67%), although in both cases it is a low percentage compared to other biomasses, whether of herbaceous origin or from wood residues [6, 12]. The importance of the ash content is particularly significant when biomass is used for combustion indomestic heating boilers. In this case the ash content is limited by the corresponding quality standard [13], which sets a maximum value of 6%.

Finally, as regards the higher calorific value, the values obtained are higher than those obtained in other studies of herbaceous waste, such as rice straw [14], but similar to those obtained for rice husk [15]. Considering the applicable standard [13], the value obtained is suitable for producing pellets for domestic combustion > 14.5%.

3.2 TG Analysis: Pyrolysis

Thermal analysis of pyrolysis is used to determine the ability of the samples to maintain self-energy balance. The thermal process curves of both residues, as well as the DTG curve, can be seen in Fig. 1. The main combustion parameters of both samples are included in Table 2.

The pyrolysis process occurs in 4 stages. The first stage is due to the elimination of moisture and the start of hemicellulose decomposition, until the start of fast pyrolysis Tm, which takes place at 255.40 °C for the LM sample and 251.80 °C for the OR sample, with both samples presenting very similar values. The second stage is where the main pyrolysis process takes place, from the onset of fast pyrolysis Tm

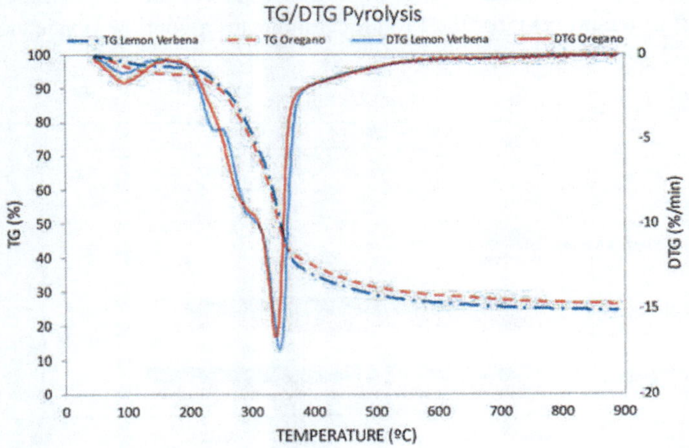

Fig. 1 TG and DTG pyrolysis profiles

Table 2 Main pyrolysis parameters

Parameter	Sample	
	LV	OR
Tm (°C)	255.40	251.80
Tp (°C)	344.00	338.40
Tf (°C)	372.00	376.00
DTG peak max (%/min)	17.41	16.66
DTG mean (%/min)	2.00	1.95
tm (min)	11.70	11.60
tp (min)	15.50	15.20
tf (min)	16.90	17.10
Δt1/2 (min)	16	15
Pi (%/min3 × 10^{-3})	96.00	94.49
Pf (%/min4 × 10^{-3})	85.31	89.11
P (%/C3 × 10^{-7})	14.35	13.60
Residual mass (%)	24.53	26.35

to the final pyrolysis temperature Tf. In Fig. 2, a first peak related to the decomposition of hemicellulose and a second main peak (344 ºC for LV and 338.40 for OR) corresponding to the decomposition of cellulose can be observed, while lignin was decomposed during the second and third stages without any characteristic peak [16]. In the third stage (carbonisation), a slow decomposition of the residues took place, generating ash and fixed carbon as final products, until the end of the process at 372.00 °C for the case of lemon verbena and 376.00 °C for the case of oregano. Finally, a last stage took place up to 900 °C, the temperature at which the test was completed.

As for the pyrolysis indexes (Fig. 2), a high value for the Pi index indicates an easier start of the process, for the Pf index it indicates a good pyrolysis yield in the

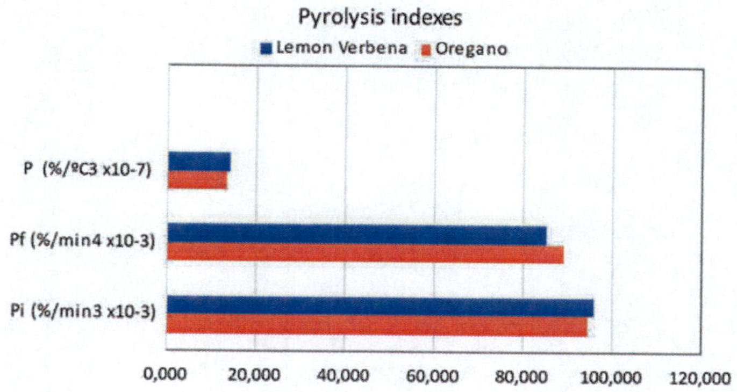

Fig. 2 Pyrolysis indexes

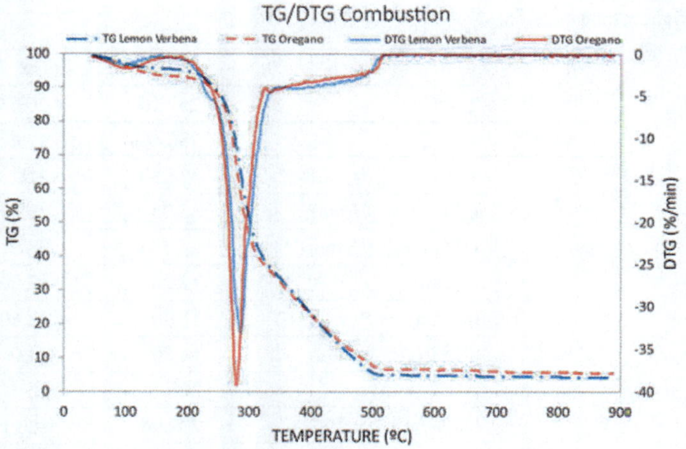

Fig. 3 TG and DTG combustion profiles

biomass and for the P index it indicates less energy needed for the pyrolysis process. The values obtained in both samples are very similar, with better values in the indexes for the case of lemon verbena.

3.3 *TG Analysis: Combustion*

The results obtained from the thermogravimetric analysis were used to compare the reactivity and combustibility of the lemon verbena and oregano residues. The thermal process curves of both residues, as well as the DTG curve, can be seen in Fig. 3. The main combustion parameters of both samples are included in Table 3.

According to the results obtained from the thermogravimetric tests, inferred from Fig. 3, four different areas were observed during the combustion of the samples. The first stage is related to moisture loss, from room temperature to the temperature at which a 5% weight loss occurs [17], resulting in 209 °C for lemon verbena and 129 °C for oregano. At higher temperatures, a faster stage started with the combustion of volatile compounds, with a peak temperature of 286.40 ºC for LV and 279.70 ºC for OR, due to the thermal decomposition of cellulose and hemicellulose, until the final temperature of this stage, where a change in the slope of the TG curve occurred. The third stage corresponds to lignin decomposition and charcoal combustion, ending at the combustion temperature, which was 328.30 and 318.30 ºC, for lemon verbena and oregano respectively. Finally, a last stage took place up to 900 °C, the temperature at which the test was completed. According to literature, the decomposition temperature of hemicellulose, cellulose and lignin under an oxidized atmosphere is within 220–315, 315–400 and 175–890 °C respectively [18]. This suggests that the temperature range identified for the tested samples is correct, and it is concluded that

Table 3 Main combustion parameters

Parameter	Sample	
	LV	OR
Ti (°C)	267.10	263.50
Tp (°C)	286.40	279.70
Tf (°C)	328.30	318.30
DTG max (%/min)	33.04	39.33
DTG mean (%/min)	2.73	2.73
ti (min)	12.20	12.00
tp (min)	12.90	12.60
tf (min)	14.50	14.00
Δt1/2 (min)	20	16
Di (%/min3 × 10^{-3})	209.94	260.12
S (%/C3 × 10^{-7})	38.57	48.64
D (%/C2 × 10^{-3})	5.98	8.68
Df (%/min4 × 10^{-3})	239.83	385.13
Residual mass (%)	4.24	5.40

the weight loss produced is due to the total thermal breakdown of hemicellulose and the transformation of cellulose and lignin into residual mass.

Table 3 shows the values of the combustion indexes. As for the Di-index, the higher the value of this index, the easier combustion occurs. The S-index helps to determine the ignition behaviour, and the higher the value of the S-index, the higher the combustion activity. As for the D index, it is related to the release of volatile compounds, and its value increases with the increase of the maximum rate of weight loss and the decrease of the temperature at which these volatile compounds evolve. Regarding the Df index, its value decreases with the increase of the Δt2 parameter, which implies higher combustion rates.

In all cases, the values of the indexes obtained for the oregano sample were higher than the values for the lemon verbena sample (Fig. 4), so it can be concluded that the waste from the oregano plant has better combustion qualities. On the other hand, it is observed that the oregano sample has a higher residual mass than the lemon verbena sample, which is detrimental to the combustion process.

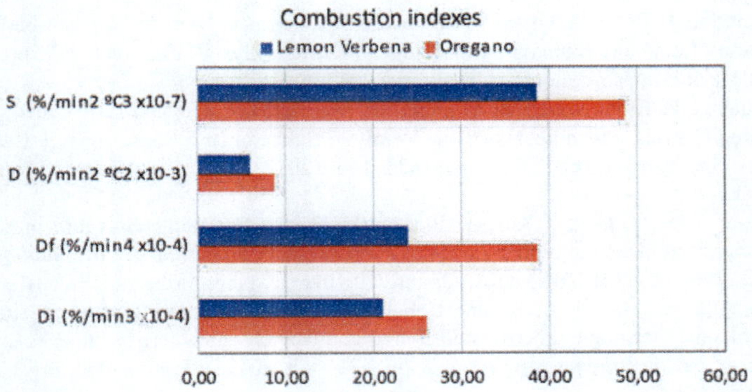

Fig. 4 Combustion indexes

4 Conclusion

Considering the above, it can be stated that both residues, lemon verbena and oregano, are biomass products suitable for both pyrolysis and combustion, and therefore present optimal behaviour for energy use.

In neither case is prior drying of the waste necessary, and both have optimum results in terms of ash content, volatile matter and higher heating value.

Two conclusions are drawn from the thermogravimetric study. The first is that for the use of the sample by pyrolysis process, lemon verbena has better qualities, a value in accordance with the higher content of volatile matter in this sample. The second conclusion is that oregano has better characteristics for combustion than lemon verbena, although the residual matter is slightly higher, which may be detrimental to the combustion process.

Acknowledgements This work was supported by the Junta de Extremadura, within the framework of the Extremadura ERDF Operational Programme 2014-2020. The authors would like to thank Ms. Teresa Miranda García-Cuevas for her help and collaboration.

References

1. "ISO 18134–2:2017: Solid biofuels. Determination of moisture content. Oven dry method. Part 2: Total moisture. Simplified method.," 2017.
2. "ISO 18123:2016. Solid biofuels. Determination of the content of volatile matter," 2015.
3. "ISO 18122:2015. Solid biofuels. Determination of ash content," 2015.
4. "ISO 18125:2018. Solid biofuels - Determination of calorific value," 2018.
5. M. T. Miranda, R. García-Mateos, J. I. Arranz, F. J. Sepúlveda, P. Romero, and A. Botet-Jiménez, "Selective use of corn crop residues: Energy viability," *Applied Sciences (Switzerland)*, vol. 11, no. 7, Apr. 2021, doi: https://doi.org/10.3390/app11073284.

6. M. Mureddu, F. Dessì, A. Orsini, F. Ferrara, and A. Pettinau, "Air- and oxygen-blown characterization of coal and biomass by thermogravimetric analysis," *Fuel*, vol. 212, pp. 626–637, Jan. 2018, doi: https://doi.org/10.1016/j.fuel.2017.10.005
7. S. Paniagua, L. Escudero, R. N. Coimbra, C. Escapa, M. Otero, and L. F. Calvo, "Effect of Applying Organic Amendments on the Pyrolytic Behavior of a Poplar Energy Crop," *Waste Biomass Valorization*, vol. 9, no. 8, pp. 1435–1449, 2018, doi: https://doi.org/10.1007/s12649-017-9885-1
8. C. Moon, Y. Sung, S. Ahn, T. Kim, G. Choi, and D. Kim, "Effect of blending ratio on combustion performance in blends of biomass and coals of different ranks," *Exp Therm Fluid Sci*, vol. 47, pp. 232–240, May 2013, doi: https://doi.org/10.1016/j.expthermflusci.2013.01.019
9. S. Paniagua, A. I. García-Pérez, and L. F. Calvo, "Biofuel consisting of wheat straw–poplar wood blends: thermogravimetric studies and combustion characteristic indexes estimation," *Biomass Convers Biorefin*, vol. 9, no. 2, pp. 433–443, 2019, doi: https://doi.org/10.1007/s13399-018-0351-5.
10. Y. Zhang, Y. Guo, F. Cheng, K. Yan, and Y. Cao, "Investigation of combustion characteristics and kinetics of coal gangue with different feedstock properties by thermogravimetric analysis," *Thermochim Acta*, vol. 614, pp. 137–148, 2015, doi: https://doi.org/10.1016/j.tca.2015.06.018.
11. R. K. Singh, T. Patil, and A. N. Sawarkar, "Pyrolysis of garlic husk biomass: Physico-chemical characterization, thermodynamic and kinetic analyses," *Bioresour Technol Rep*, vol. 12, no. September, p. 100558, 2020, doi: https://doi.org/10.1016/j.biteb.2020.100558.
12. C. Wang, F. Wang, Q. Yang, and R. Liang, "Thermogravimetric studies of the behavior of wheat straw with added coal during combustion," *Biomass Bioenergy*, vol. 33, no. 1, pp. 50–56, Jan. 2009, doi: https://doi.org/10.1016/j.biombioe.2008.04.013.
13. "ISO 17225–6:2021. Solid biofuels. Fuel specifications and classes. Part 6: Graded non-woody pellets," 2022.
14. D. Zinla, P. Gbaha, P. M. E. Koffi, and B. K. Koua, "Characterization of rice, coffee and cocoa crops residues as fuel of thermal power plant in Côte d'Ivoire," *Fuel*, vol. 283, p. 119250, 2021, doi: https://doi.org/10.1016/j.fuel.2020.119250
15. J. I. Arranz, M. T. Miranda, I. Montero, and F. J. Sepúlveda, "Thermal study and emission characteristics of rice husk using tg-ms," *Materials*, vol. 14, no. 20, 2021, doi: https://doi.org/10.3390/ma14206203.
16. C. Di Blasi, "Modeling chemical and physical processes of wood and biomass pyrolysis," *Prog Energy Combust Sci*, vol. 34, no. 1, pp. 47–90, 2008, doi: https://doi.org/10.1016/j.pecs.2006.12.001.
17. Z. Yang *et al.*, "Combustion behaviours of tobacco stem in a thermogravimetric analyser and a pilot-scale fluidized bed reactor," *Bioresour Technol*, vol. 110, pp. 595–602, Apr. 2012, doi: https://doi.org/10.1016/j.biortech.2011.12.119
18. H. Bi, C. Wang, Q. Lin, X. Jiang, C. Jiang, and L. Bao, "Combustion behavior, kinetics, gas emission characteristics and artificial neural network modeling of coal gangue and biomass via TG-FTIR," *Energy*, vol. 213, Dec. 2020, doi: https://doi.org/10.1016/j.energy.2020.118790.

Agreements, Energy Efficiency and Self-production in Building for Sustainability

João Galvão, Filipe Neves, Manuel Galvão, and Ricardo Gomes

Abstract A large amount of existing legislation, agreements and treaties, the main one being the 2030 Agenda for Sustainable Development, must be a permanent reference to reduce the consequences of the Urban Heat Island Effect that manifests itself in cities and allow populations to live more efficiently, healthier buildings and less waste of energy (heat, cold and electricity). In this context, a building will be presented as a case study and after carrying out a survey of inefficiency anomalies, both energetic and constructive, it will be possible to implement several actions that increase its energy efficiency. In addition, it will be a reference in reducing energy consumption and will be healthier and more energetically autonomous through renewable sources.

Keywords Green/landscape roof · Energy production · Energy efficiency · Sustainability

1 Introduction

The number of agreements, directives and plans coming from international and national institutions such as the goal 11—Make cities and human settlements inclusive, safe, resilient and sustainable of the 2030 Agenda for Sustainable Development of United Nations, provides guidelines for the present date and into the future society

J. Galvão (✉) · F. Neves · R. Gomes
Polytechnic of Leiria, School of Technology and Management, 2411-901 Leiria, Portugal
e-mail: jrgalvao@ipleiria.pt

J. Galvão · R. Gomes
INESCCoimbra, DEEC, Rua Sílvio Lima, Polo II, 3030-290 Coimbra, Portugal

F. Neves
CIIC—Computer Science and Communication Research Center, 2411-901 Leiria, Portugal

M. Galvão
Department of Mechanical Engineering, University of Minho, 4800-058 Guimarães, Portugal

J. R. da. C. S. Galvão et al. (eds.), *Proceedings of the 3rd International Conference on Water Energy Food and Sustainability (ICoWEFS 2023)*, Springer Proceedings in Earth and Environmental Sciences, https://doi.org/10.1007/978-3-031-48532-9_26

based on a model of Sustainable Development [1]. As an example, in December 2019, the European Commission announced its growth strategy based on combating climate change and promoting a new green economy—evidently, this strategy has a strong impact on the construction sector, in a particular context of UHIE—Urban Heat Island Effect. The European Green Deal thus brings the environment to the center of Europe's development policy, assuming Europe's ambition to be carbon neutral in 2050 [2].

The main objective of this research is applying some principles of sustainability in one existing government building and improved energy rating and reduced operating costs.

1.1 Treaties for Sustainability and New Buildings Concepts

Following the 2030 Agenda and the Green Deal, the government in Portugal approves the RNC 2050 (Roteiro para Neutralidade Carbónica/Roadmap for Carbon Neutrality) resolution nº 104/2019 [3]. By 2022, electricity production will already be around 60% from renewable sources and after the Climate Summit/COP 2022, the government of Portugal wants to reach 80% of electricity production from renewable sources by 2026. This goal is more ambitious than what is expressed in some plans proposed by the government.

This government provision brings forward this objective previously determined in the RNC 2050 by 4 years and consequently accelerates the implementation of goal 11 of the 2030 Agenda. Nevertheless, other actions—also supported by the PNEC 2030 (Plano Nacional para Energia e Clima/National Energy and Climate Plan), which is the main energy and climate policy instrument for the 2021–2030 decade, completed this objective towards a carbon neutral future. It also falls within the scope of the obligations established by the Regulation of Governance of the Energy and Climate Action Union, which requires all member states to prepare and submit to the European Commission their integrated energy and climate plans.

The PNEC 2030 sets ambitious but achievable national targets for the 2030 horizon to reduce greenhouse gas emissions, incorporating more renewable energy, increase energy efficiency and interconnections, the energy security, the research, the innovation and competitiveness, according to the guidelines proposed in the RNC 2050.

Urban areas require a significant amount of energy, water, and other natural resources, which have a strong environmental impact. One solution to ensuring sustainable urban development and environmental protection is to make buildings more sustainable, using new construction solutions, technologies, and materials. One can then think of achieving smart and progressively self-sustaining buildings, with autonomous energy production to achieve a maximum energy efficiency class, as well as a "nZEB" (nearly Zero Energy Building) label or be a "Passive House". All this globally, integrating the purpose of "Healthy Building" and digitalization target, with high comfort levels for its residents, also with impact on significant energy savings and being a contribution to decarbonization in heating and cooling situations [4–6].

One of the most interesting new features of buildings regarding its environmental and sustainability upgrading, concerns green/landscape roofs design and other actions to promote the local energy production or raise the energy efficiency of the cities, as well as decreasing the UHIE [7, 8].

According to Mohapatra et al. [9], the intensive use of green roofs permits an annual runoff reduction of 65–85% of the annual precipitation, while the corresponding value for extensive green roofs is in the range of 2–81%. Additionally, Li and Yeung [10] showed that the retention of semi-intensive vegetated green roof and intensive vegetated green roof depends on the coverage area of the roof.

The literature provides some disadvantages/advantages and others multifunctional benefits of green roof and in short:—Disadvantages: (i) structural overload limitations; (ii) possible damages arising from roof leaks with the potential to result in structural damages; (iii) limited choice of plants; and (iv) installation costs and periodic maintenance costs;—Advantages (i) reduction of in the urban heat island effect; (ii) reduction or delaying rainwater runoff; (iii) reducing energy requirements concerning building heating and cooling; (iv) improvement of carbon retention [11, 12].

The classification assigned is generally extensive or intensive, according to the substrate thickness [13–15]. Extensive green roofs are typically installed on existing buildings on the other side, intensive green roofs are more often installed on new buildings, because the additional weight must be initially designed in the building's structure.

Another concept linked to the mentioned technologies is "Healthy Construction" which is the result of the integration of materials, design and maintenance with a view to ensuring the health and well-being of the occupants. In this sense, the value of a healthy building encompasses many areas of building management, including physical, emotional, intellectual, spiritual, occupational, environmental, and social well-being, to improve outcomes for employers, developers, owners and investors.

Some features of a healthy building are associated with basic health and well-being factors that affect job performance, such as: Temperature, Moisture, Air movement, Noise, Lighting, and other Air pollutants. These factors are responsible for creating a healthy environment in all spaces of the building which will make it an ideal place for work and occupation in general. This is a process that requires a thorough assessment of the existing infrastructure with HVAC sensors, setpoint devices, and IoT sensors. Through a variety of platforms, it can help solve technical problems, collect data and analyze useful information to manage the health and ongoing well-being of occupants.

The main benefits of having a healthy building are: Maximization of space efficiency; Improvement of occupants' well-being; Improvement of the occupants' experience; Reduction of operating costs.

It reduces operating costs by obtaining information through coordinated and connected sensors to determine where and when to address problems.

Two of the main advantages of having a healthy build are as follows: Buildings empowered with this data have a decrease in complaints and increase in employee satisfaction; IoT health information can be used to automatically control the building and finances.

The platform to manage all this data should have the ability to combine all these aspects into a comprehensive and coordinated end-to-end solution, from electrical to mechanical areas; from connected IoT sensors that constantly capture building data to machine learning models. As an example, there is the external control of shading curtains, through their movement of descending and ascending according to the external luminosity consequently controlling the interior temperature of the building. The net sensors would drive predictive maintenance by the means of wearables that monitor the health of occupants, while putting them in control of their environment, so that autonomy in energy production and the sustainability of buildings and cities is achieved [16, 17].

2 Case Study Features, Research Core and Results

2.1 Energy Efficiency and Building Thermal Comfort

The International Energy Agency says that buildings are responsible for over 40% of world primary energy demand and 24% of greenhouse gas emissions [18]. As cities are growing, more urban areas are emerging, with higher temperatures than those of rural areas. Therefore, the urban heat island appears in cities [8, 19] due to the concentration of buildings, with high energy consumption and building thermal comfort requirements, particularly in residential and service buildings.

A good solution to decrease the heat flux released to the atmosphere and to mitigate the effect of high temperatures of the roof's surface is the technology of green or living roofs and consequently, the temperature gradient inside buildings will be lower [20, 21].

Some research from Quezada-Garcia et al. [22] said that the main objective of green roofs' contribution to energy efficiency and sustainability in cities is to reduce the urban heat island effect and consequently the reduction of energy, inherent costs, the increase of energy efficiency, maintaining levels of comfort inside the houses. On the other hand, the heat gain of the buildings will be reduced, with the consequence of a decrease in the consumption of air conditioning inside the building.

2.2 Methodology and Building Chosen

The methodology used was through a simple energy audit generic/deambulatory, following what is determined in the Portuguese Energy Audit—Decree-Law no. 68-A/2015. Initially it collected the constructive characteristics of the building and consumption data: electricity, gas, and water. Then, a breakdown of electrical consumption was carried out, through the installation of energy analyzers in the electrical panels, in order to identify opportunities to rationalize consumption.

The next step was the processing and analysis of the collected data and the consequent detailed characterization of the building's behavior.

In the last stage, a report was prepared in which the building was described, consumption was accounted for using graphs and tables, anomalous situations and opportunities for rationalization of consumption, technically and economically viable.

The building chosen to support this study is located in Campus 2, ESSLei Building—School of Health Sciences, in Leiria Polytechnic, Portugal [23–26].

The main objective of this research is to analyze the possibility of the School of Health Sciences building, Fig. 1, being classified as an energetically sustainable building, that is, with a balance neutral energy, becoming a standard building for the entire nZEB community from a perspective of sustainable development and energy efficiency, taking into account the following areas: (i) increase the comfort of life; (ii) reduce energy consumption; (iii) reduce fossil energies consumption; and (iv) reduce the tap water consumption and increase the water resources.

Bearing in mind that many energy losses (heat/cool) reach percentage values as shown in Fig. 2, an energy audit is required in order to evaluate what energy (thermal and electrical) is consumed in the building.

The building is about 20 years old and has a ground floor and 3 upper floors. The 3rd floor being the management area and teachers' offices. It has an auditorium on

Fig. 1 School of health sciences building

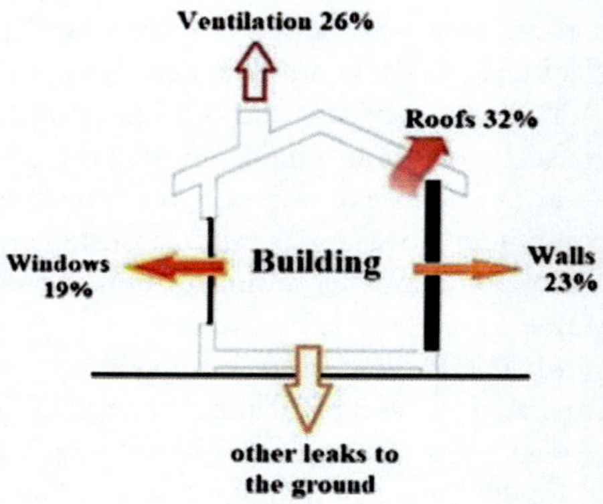

Fig. 2 Thermal energy losses of a building

the 3rd and 2nd floor, administrative services on the 1st floor and a bar on the ground floor. The remaining areas are teaching rooms and laboratories with equipment to support the classes.

There are several areas with stairs and one lift, with double glazed windows facing north and south. On the 1st floor several terraces can be found and there are no trees in the surrounding area. It is painted white and has black coating on the walls facing north, south and west outside the auditorium building and in the building top several tubes of air-conditioning system fill this roof.

2.3 *Research Core, Data Analysis and Efficiency Actions*

In the context of the study proposed to achieve the objectives presented in Sect. 2.2, the building energy audit began by collecting energies consumptions and water, by analyzing in-voices from previous years (2018–2021). Next, different load diagrams are designed, to understand the behavior of these variables during a month, week, and daily, Fig. 3. There was also a need to disaggregate some electrical consumption, having been placed for almost 2 weeks in electrical energy analyzer panels to collect consumption in greater detail and subject to future corrections.

Also, a thermographic survey was carried out on various building elements to identify thermal leaks, as well as any electrical equipment with high temperature that leads to loss of insulation and necessary replacement.

As an example, Fig. 4 represents an outside window, where the temperature varies between the frame 60.1 °F = 15.6 °C and the interior wall temperature 65.4 °F = 18.5 °C is representative of the energy loss able to be corrected.

At the end of the energy audit process, several opportunities for energy reduction were detected and energy efficiency technologies were proposed to this building such as these three main areas:

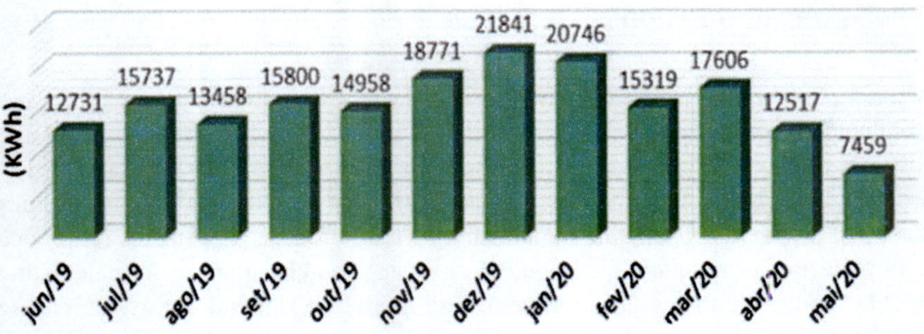

Fig. 3 Building electrical consumption diagram by year

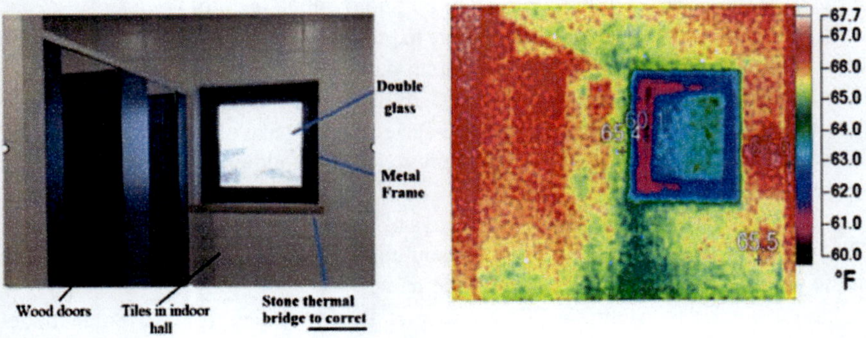

Fig. 4 Building window (left) and identification of thermal leaks through the frame (right)

1. **Energy efficiency**: Resort to local production of energy from a renewable source via photovoltaic technology; Decrease energy demands, make the insulate the building's exterior and reduce thermal leakage; Increase energy efficiency (HVAC systems);
2. **Thermal comfort**: Increase system efficiency; Reduce loads; Low Temperature Heating and High Temperature Cooling Systems; Use of building mass to reduce peaks (Thermo-Active-Building-Systems); Install landscape/green roofs;
3. **Indoor air quality**: Heat recovery; Increase system efficiency; Natural ventilation and hybrid ventilation; Air distribution and cleaning (contaminant removal) effectiveness; Reduce loads (pollution sources).

In this context, part of the energy production will be highlighted by resorting to the simulation of the local electricity production system, for satisfaction of the building's electricity supply and installation of chargers in the car park.

3 Results and Proposals

A local photovoltaic production system was designed, and several scenarios were simulated with the System Advisor Model (SAM). A SAM is a techno-economic software model that facilitates decision-making for people in the renewable energy projects, The photovoltaic model calculates a grid-connected photovoltaic system's electrical output using separate module and inverter models. It estimates losses due to the effect of temperature on module performance and has options for calculating shading and other losses in the system. The model also includes a system sizing assistant to help in project development for to determine the number of modules and inverters in the system [27].

The next step will be to present some inputs and outputs when designing and simulating the energy production system by photovoltaic technology:

Inputs: • As an initial step, it is necessary to provide the location of the photovoltaic system to be implemented, in this case the car park; • This software has in its database the atmospheric conditions of the same, within a radius of 10 km; • Define which type of photovoltaic module to use in the system; • The size of the installation area. These factors will determine how many photovoltaic modules will be needed in the system;

• Definition of the space where two arrays of photovoltaic panels will be installed. The first array in the parking row on the south side of the park and the second array in the parking row on the north side of the park.

Outputs: • This system has 208 monocrystalline modules from JA Solar model JAM72S20-460; • 100 modules in the first array and the remaining 108 in the second array, whose total module power will be of 95.7 kW; • It has an inverter model SUN2000-100KTL-USH0, from the company Huawei Technologies, with a power of 100 kW, in order to exceed the total power of the photovoltaic modules; • To connect all the components to each other and to the building, a cable path with a length of about 1000 m was determined, of which 500 m is used only for the connection of the modules to the inverter.

Regarding the economic analysis, this solution presents an initial investment of around 65 thousand euros, however, after the first year, this presents a net saving of around 17 thousand euros, where, consequently, it has a simple payback period of 3 years and 9 months. The 3D modeling of the case study was carried out in the Sketchup program, with a proposal for the location of the green roof and the local electricity production plant (with the cars taking advantage of its shading), Figs. 5 and 6 [28].

The main advantages of green roofs are: • Improved air quality; • Favoring biodiversity and promoting carbon retention; • Real estate valuation and improvement of the urban landscape; • Improved water quality with reuse potential; • Attenuation of the heat island effect; • Possibility of producing food on top of buildings; • Support for rainwater management by promoting precipitation retention; • Increased thermal and acoustic insulation of buildings; • Increased waterproofing lifetime.

Fig. 5 School of health sciences building, aerial view and park car

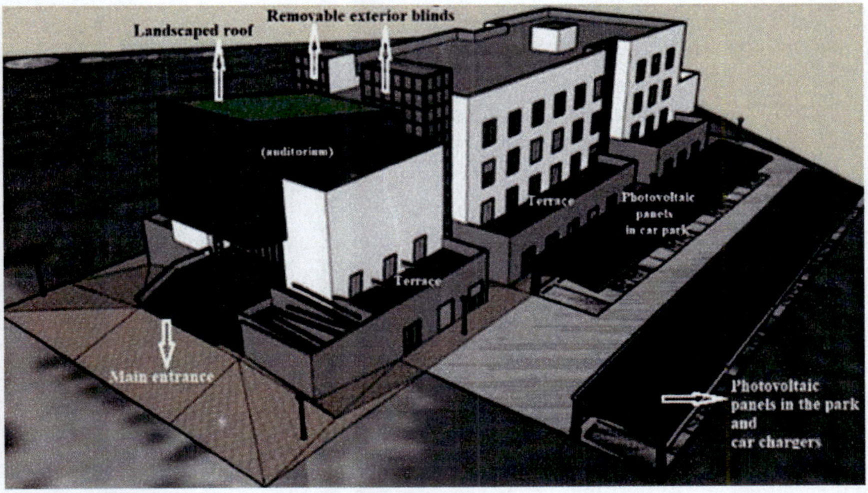

Fig. 6 3D modeling of school of health sciences building and the proposed photovoltaic system and building landscaped/green roof

In the final report presented to the management team of this building, several actions were proposed to improve its energy performance, such as: sensorized removable exterior blinds with an upward and downward movement according to the external luminosity and consequent control of the interior temperature of the building from the outside temperature, so that inside the building there is an average comfort temperature of around 21 degrees.

Additionally, it is recommended to install tap flow regulators and recovery roof rainwater for sanitary and irrigation of landscaped spaces. To install new generations of led lamps. Install thermal break frames on the windows, due to multiple anomalies in the existing frames. To eliminate thermal leaks in pillars, beams, and other areas, by improving thermal insulation. Finally, new software should be developed with algorithms supported by decision systems in the control and management of energy variables in the building, banishing all energy leaks in the building [29].

4 Conclusions and Perspectives

The need to carry out actions that make an effective application of objective 11—Make cities and human settlements inclusive, safe, resilient and sustainable, of the 2030 Agenda for Sustainable Development and the multiple agreements and treaties that the annual Climate Summits/COPs held in recent years are a global plan for the transformation to occur, towards a more sustainable society, with almost zero emissions and which manages to blur the UHIE. This transformation should rely on the involvement of governments and local authorities with a view to mobilizing populations towards the common goal.

The implementation of the best constructive practices in urban areas, is a process that requires investment and responsibility from every stakeholder, for the next goals to be achieved: (i) reduction of energy consumption, (ii) increase in local production of electrical and thermal energy, by renewable source; (iii) increase in the use of new materials and passive insulation technologies like vegetation facade or green/landscape roof; (iv) reduction of tap water demand and, (v) increase in the water recycling in urban area. To improve the energy efficiency of the analyzed building, it is necessary to: (i) replace the existing luminaires for LED lamps, (ii) replace the window frames and (iii) decrease thermal bridges.

A local electricity production plant in the car park is proposed, with the advantage of its shading for cars. A tank should be installed to retain roof rainwater for sanitary purposes and garden irrigation in the building's surroundings. With respect to shading of glazed areas, it is recommended to place the external control of shading curtains in their movement of descending and ascending according to the external luminosity and control of the interior temperature of the building. Lower energy needs for air conditioning as there is greater thermal stabilization inside the building both in the heating season/winter and in the cooling season/summer.

Finally, taking into account the aforementioned treaties, public and private institutions, as well as citizens, must make an effort to implement agreements and directives in order to effectively achieve the aforementioned objectives and have a relative sustainability and reduction of carbon levels.

Acknowledgements Study supported by UIDB/00308/2020 project.

References

1. The 2030 Agenda for Sustainable Development of United Nations" see: https://sustainabledevelopment.un.org, last access 17/12/2022.
2. Ursula Von Der Leyen, European Green Deal, in https://ec.europa.eu/info/strategy/priorities-2019-2024/european-green-deal_pt, last acess 20/11/2022.
3. APA/Portuguese Environment Agency - https://www.apambiente.pt/clima/plano-nacional-de-energia-e-clima-pnec, last acess 2/12/2022.
4. P. Torcellini, S. Pless, and M. Deru, Zero Energy Buildings: A Critical Look at the Definition, NREL - National Renewable Energy Laboratory D. Crawley U.S. Department of Energy, Conference Paper NREL/CP-550–39833, USA (2006).
5. PHI - Passive House Institute, Efficiency: The first Renewable Energy, in https://passivehouse-international.org/upload/20210215_Press_Release_Campaign_Efficiency_First.pdf, last acess 25/11/2022.
6. Gomes, R., Galvão, J., Gala, P., Prola, L., Ribeiro, V. (2021). An Overview of Green Roofs in Urban Areas: Impact on Buildings and Food-Energy-Water Nexus. In: , *et al.* Proceedings of the 1st International Conference on Water Energy Food and Sustainability (ICoWEFS 2021). ICoWEFS 2021. Springer, Cham. https://doi.org/10.1007/978-3-030-75315-3_67.
7. The Landscape Roofs, Published by Landscape Architecture and Urbanism at University od Greenwich London, see https://thelandscape.org/landscape-roofs/, last acess 2/10/2022.
8. Gago, E. J., et al.: The city and urban heat islands: A review of strategies to mitigate adverse effects. Renewable and Sustainable Energy Reviews, 25, 749–758 (2013).
9. Mohapatra, S., et al.: A critical appraisal of green vegetated roofs: Energy and environment in focus. Materials Today: Proceedings (2020).
10. Li, W. C., Yeung, K. K. A.: A comprehensive study of green roof performance from environmental perspective. International Journal of Sustainable Built Environment, 3(1), 127–134 (2014).
11. Cristiano, E., Deidda, R., Viola, F.: The role of green roofs in urban Water-Energy-Food-Ecosystem nexus: A review. Science of The Total Environment, 756, 143876 (2021).
12. Shafique, M., Xue, X., Luo, X.: An overview of carbon sequestration of green roofs in urban areas. Urban Forestry & Urban Greening, 47, 126515 (2020).
13. Cascone, S., et al.: The evapotranspiration process in green roofs: A review. Building and Environment, 147, 337–355 (2019).
14. Eksi, M., et al.: Effect of substrate depth, vegetation type, and season on green roof thermal properties. Energy and Buildings, 145, 174–187 (2017).
15. Cascone, S., et al.: A comprehensive study on green roof performance for retrofitting existing buildings. Building and Environment, 136, 227–239 (2018).
16. Katya Alejandra Manzano, Healthy Building, Manager Schneider Electric, in: https://www.voltimum.pt/artigos/artigos-tecnicos/, last acess 10/10/2022
17. Galvão, J., Ascenso, R., Pereira, S., Moreira, L., Neves, F. (2021). Smart Energy in Sustaiable Services Building. In: , et al. Proceedings of the 1st International Conference on Water Energy Food and Sustainability (ICoWEFS 2021). ICoWEFS 2021. Springer, Cham. DOI: https://doi.org/10.1007/978-3-030-75315-3_47.
18. IEA - International Energy Agency, Solar Heating & Cooling Programme, Homepage, https://iea-shc.org/, last acessed 1/11/2022.
19. 20. Santamouris, M.: Cooling the cities – A review of reflective and green roof mitigation technologies to fight heat island and improve comfort in urban environments. Solar Energy, 103, 682–703 (2014).
20. Gomes R., Galvão J., Gala P., Prola L., Ribeiro V. (2021) An Overview of Green Roofs in Urban Areas: Impact on Buildings and Food-Energy-Water Nexus. In: da Costa Sanches Galvão J.R. et al. (eds) Proceedings of the 1st International Conference on Water Energy Food and Sustainability (ICoWEFS 2021). ICoWEFS 2021. Springer, Cham, https://doi.org/10.1007/978-3-030-75315-3_67.

21. 22. Yang, J., et al.: Green and cool roofs' urban heat island mitigation potential in tropical climate. Solar Energy, 173, 597–609 (2018).
22. 23. Quezada-García, S., et al.: Heterogeneous model for heat transfer in Green Roof Systems. Energy and Buildings, 139, 205-213 (2017).
23. Barbeiro A., Antunes J. (2022), Energy Analysis to ESSLei and Renewable Energy Solutions, Final Report of the Project Curricular Unit, ESTG, Polytechnic Leiria.
24. 25. Galvão J., Moreira, L., Gaspar, G., Vindeirinho, S., Leitão, S., (2017), Energy System Retrofit in a Public Services Building, Management of Environmental Quality: An International Journal, Vol. 28 Issue: 3, pp. 302-314, DOI: https://doi.org/10.1108/MEQ-02-2014-0028 B23.
25. M. Almeida, R. Ascenso, J. Galvão, L. Moreira, S. Leitão, (2017), "Hybrid Energy System for NZEB", IN-TECH 2017 - International Conference on Innovative Technologies; Slovenia, ISSN 0184–9069; http://www.in-tech.info/.
26. 27. Galvão, J.R., Nabais, A.S., Galvão, M., Candeias, J., Pereira, T., & Ramos, J. (2020). Efficiency in an Intensive Energy Industrial Consumer. Renewable energy & power quality journal, 18, 599-602, DOI: https://doi.org/10.24084/repqj18.443.
27. Software SAM - System Advisor Model, NREL, USA. see https://sam.nrel.gov, last access 17/10/2021.
28. SketchUp for Web | 3D Modeling, see: ttps://www.sketchup.com/pt-BR/products/sketchup-for-web, last access: 20/2/2022.
29. Advance Housing Renovation with Solar and conservation, International Energy Agency, Solar Heating and Cooling Programme, Task 37, 2010, see https://www.iea-shc.org 22nov2022.

Preliminary Studies of Acorn Syngas Application in Dual-Fuel Diesel Engines

Nadezhda Silva Krop, Paulo Brito, and Luís Calado

Abstract Acorns are a nut that can be found across Portugal, particularly in the Alentejo region where the raising and feeding of Iberian Pigs is most present. Despite this contribution, they're often ignored by the general population and therefore go to waste. While there are different attempts at reviving its use among people, this paper aims to explore and demonstrate the results of using acorn syngas in a dual-fuel diesel engine, as an alternative option. To determine the effectiveness of acorn syngas in dual-fuel engine applications, the engine was first run using diesel exclusively in 1000, 1500, and 2000 rpm cycles for 20 min each, and then the same parameters were applied when running the diesel engine in dual-fuel mode using the produced acorn syngas. The results overall illustrate a decrease in the amount of diesel fuel consumed by the engine, as well as a decrease in the intake air flow and the exhaust gas flow. While the obtained results demonstrate a lot of potential for the use of this often-ignored biomass, more studies should be conducted to further determine its viability in dual-fuel engine applications.

Keywords Acorns · Gasification · Dual-fuel engines

1 Introduction

Besides its environmental benefits, gasification is also quoted as an economic technology, often favorable for implementation in rural or remote areas [1]. In addition to aiding in the development of these regions, this technology can also be seen as an interesting opportunity for the valorization of the region's local resources, therefore promoting autonomy. The Alentejo region of Portugal is primarily agrarian, and

N. S. Krop (✉) · P. Brito
VALORIZA—Research Center for Endogenous Resource Valorization, Polytechnic Institute of Portalegre, Campus Politécnico 10, 7300-555 Portalegre, Portugal
e-mail: 21273@ipportalegre.pt

L. Calado
Polytechnic Institute of Portalegre, Campus Politécnico 10, 7300-555 Portalegre, Portugal

J. R. da. C. S. Galvão et al. (eds.), *Proceedings of the 3rd International Conference on Water Energy Food and Sustainability (ICoWEFS 2023)*, Springer Proceedings in Earth and Environmental Sciences, https://doi.org/10.1007/978-3-031-48532-9_27

while it's not considered remote, a large portion of its territory consists mostly of *Montado*, a man-made landscape typical to the Iberian Peninsula [2].

With the Alentejo's large agrarian production, the presence of agricultural and agro-industrial wastes is to be expected. While most of these wastes are not taken advantage of for the most part, they do signify an interesting economic opportunity for the agro-industrial industry and all those others adjacent to it, regarding the production of energy. A somewhat often forgotten, albeit important, example of a biomass common in the Alentejo region is the acorn. Present in the human diet until recent times, this nut is now forgotten by the general public and is mostly used in the feeding of Iberian pigs, famous for their serrano ham. Despite their importance in the swine industry, more than half of all acorns produced every fall go to waste, either being consumed by the local wild fauna, or rotting away in the wilderness [3]. Nowadays, there are attempts at reeducating and reintroducing this nut into the Western world's diet, albeit these attempts are very slow and acorn-derived products are still not very accessible to the general public due to its high prices and its unavailability in mainstream supermarkets. While this attempt at revival is a great measure of reducing wastes and tackling issues such as food shortages—acorns have been humanity's reliable partner during times of famine—other methods of valorization can be explored, such as energy production. While gasification has several usages, this paper will focus on the application of acorn syngas in a dual-fuel engine at a laboratory-level to determine the efficiency of this biomass.

2 Dual-Fuel Engines

Dual-fuel engines can be defined as engines that have the ability to operate using 2 different types of fuel [4]. Usually, this consists of a gaseous fuel and a liquid fuel that often acts as a "pilot fuel", to help ignite the gaseous fuel and get the engine running. Some additional benefits these engines have is their ability to work solely with fossil fuel whenever gaseous fuel is unavailable, and the fact that these engines can be made by modifying pre-existing engines. There are other types of dual-fuel engines that are made from scratch such as HPDI (High Pressure Direct Injection), but nevertheless this doesn't take away from the possibility of using preexisting parts.

In literature, it's most common to see research pertaining to the use of diesel/natural gas mixtures in dual-fuel engines, however this doesn't mean there's no research available pertaining to the use of diesel/syngas mixtures. Most of these studies pertain to determining the effects of syngas mixtures on parameters such as flow rate, pollutant removals, and the performance of the engine per se [5–9]. Real life examples using syngas are difficult to come by, and this is made even more difficult when it's narrowed down to the applications of acorn syngas, as studies relating to acorns focus mostly on its nutritional value and its culinary properties [10–13]. Acorns are seldom considered for energy production, hence the lack of material on its uses for dual-fuel engines.

3 Methodology

The present study aims to determine the difference in parameters for a dual-fuel diesel engine, by performing tests using diesel fuel, and then using the engine in dual-fuel mode. First, two tests were conducted using only diesel as the fuel for the engine, meaning that the engine was set to work like a normal diesel engine. Both tests were performed at rotational speeds of 1000, 1500, and 2000 rpm. For temperature and flow measurements the first test used only an anemometer and the second test used both an anemometer and flowmeters. During each test, the engine was left running for 20 min for each rotation speed, meaning that each test had a total duration of 60 min. Parameters such as the intake air flow, exhaust gas flow, temperature, and consumption of fuel per hour were measured. The consumption of fuel per hour was determined by using a flask containing 1 L of diesel fuel, and then finding the difference between the final amount of fuel present in the flask and the initial amount of diesel. The flask was refilled to the 1 L limit every 20 min when the rotational speed was changed.

Once these tests were performed, two more rounds of tests were conducted in dual-fuel mode, measuring the same parameters as those measured in diesel-exclusive mode, with the added parameters of intake gas flow and fuel consumption difference. Two syngas samples from two separate acorn gasification tests were used. Measurements were taken using the help of an anemometer and flowmeters in both sets of tests. Like the first set of tests performed under diesel-exclusive mode, the engine was left running for 20 min for each rotation speed and had a total duration of 60 min. Different from the tests in diesel-exclusive mode, only one set of tests was done at rotational speeds of 1000 and 1500, the last rotation being of 2200 rpm instead of 2000 rpm. The other set of tests was done at 800, 1300, and 1500 rpm. The variations in rotations are the result of due to the increase in engine power and consequent increase in syngas flow and compression levels, which results in an increase in engine rotation speed. Increase in syngas flow demonstrates the relationship between compression levels and rotation speed in such conditions [14]. Since there was no storage unit available for syngas, the dual-fuel tests were performed while simultaneously gasifying the acorns. This means that as the syngas was being produced, it was directly injected into the engine.

4 Results

The following table illustrates the results obtained for the tests performed both in diesel-exclusive and dual-fuel mode, using two different samples of acorn syngas with a LHV (Lower Heating Value) of 5.2 and 4.6 MJ/kg respectively.

The results shown display the patterns of consumption for the engine in both diesel and dual-fuel mode. While these patterns are consistent, the intake airflow has demonstrated a significant decrease of 57.1% when a syngas/diesel mixture is used, with higher values of LHV being associated with lower values of intake air flow. The contrary can be said about intake gas flow, as lower values of LHV are associated with higher values according to Table 1. Regarding exhaust gas flow, Table 1 shows both an increase and decrease in the exhaust gas flow, with the higher LHV value being associated with the lower exhaust gas flow values of 64.8% on average. The temperature is displayed as a consistent value of 70 °C given the anemometer's limits, which meant that the maximum temperature the device was capable of reading was of 70 °C. However, it is more than likely that the actual temperatures from the exhaust gas were much higher than that since the plastic from the anemometer was showing signs of melting shortly after being bought and used.

The approaches taken in literature to the study of syngas/diesel mixtures in diesel engine functioning is generally done using other approaches. Olanrewaju examined the effects of syngas substitution in regard to the examined diesel engine's heat release rate and combustion rate [15], Tuan and Luong studied parameters pertaining more to exhaust emissions, specific energy consumption, diesel replacement rate, alongside the performance of the engine per se [9], Cacua et al. focused more on the substitution levels and the emissions of the diesel engine [16], and Fiore et al. studied the volumetric efficiency, heat losses, premixed flame propagation, compression ratio, knocking analysis, and engine performance and emissions of an HCSI engine [17]. These, among other papers, study and discuss many more parameters that vary from publication to publication and between authors. Naturally this also depends on the nature of the work being done and the objectives that are intended on being met.

Table 1 Diesel-fueled and Syngas/Diesel engine test results

Fuel type	Time (m)	Rotations (rpm)	Intake gas		Exhaust gas		Consumption (L/h)
			Air flow (Nm^3/h)	Gas flow (Nm^3/h)	Flow (Nm^3/h)	Temperature (°C)	
Diesel (No flowmeters)	20	1000	79.8	–	171.0	70.0	1.3
		1500	125.4	–	228.0		2.0
		2000	148.2	–	319.2		2.1
Diesel (Flowmeters)		1000	75.0	–	131.4		1.5
		1500	120.0	–	227.0		1.6
		2000	160.0	–	358.4		2.0
Syngas/diesel (LHV 4.6 MJ/kg)		800	30.0	20.0	160.0	70.0	0.8
		1300	85.0	35.0	358.2		1.6
		1500	90.0	45.0	255.5		1.5
Syngas/diesel (LHV 5.2 MJ/kg)	20	1000	17.5	14.0	71.8		1.3
		1500	21.5	16.0	86.6		2.0
		2200	60.0	27.5	114.0		4.3

Overall, despite their abundance, acorns are an often ignored and “obscure” biomass which is slowly being brought up to light again in different forms. Their use as gasification feedstock and subsequent application of the syngas in dual-fuel engines, can represent a valorization pathway with potential regarding energy recovery from wastes.

5 Conclusions

This paper explores the application of previously produced acorn syngas in a dual-fuel diesel engine to explore its functionality. During the engine tests, its intake air flow, gas flow, exhaust gas flow, temperature, and fuel consumptions were measured. Overall, the parameters observed showed that:

- The intake air flow decreased significantly with the presence of the higher LHV syngas with 5.2 MJ/kg.
- The intake gas flow increased with the presence of the lower LHV syngas with 4.6 MJ/kg.
- With the insertion of syngas, exhaust gas flow values have shown both an increase and decrease in comparison to the values displayed when only diesel was used.
- The fuel consumption rates in diesel and dual-fuel mode are overall consistent.

More tests and studies are recommended to further understand the properties of the engine, alongside with exploring other parameters and properties of the engine such as those pertaining to its emissions, brake thermal energy, fuel injection time, diesel replacement rate, specific energy consumption, among others, depending on the scope of research performed.

References

1. D. Masera and A. Faaij, “Renewable Energy for Inclusive and Sustainable Industrial Development. The Case of Biomass Gasification.,” p. 41, 2014.
2. Associação Bandeira Azul da Europa, “Montado,” 2015. https://desafiouhu.abae.pt/ecossistemas-de-portugal/montado/
3. M. Sottomayor, “Potencial Económico da Bolota em Portugal: Análise Exploratória.” pp. 1–14, 2015.
4. A. Yuksel, “What is a Dual Fuel Engine, and its Benefits for Oil and Gas Applications?,” 2021. https://www.cummins.com/news/2021/09/29/what-dual-fuel-engine-and-its-benefits-oil-and-gas-applications
5. H. Guo, W. S. Neill, and B. Liko, “The Combustion and Emissions Performance of a Syngas-Diesel Dual Fuel Compression Ignition Engine,” in *ASME 2016 Internal Combustion Engine Division Fall Technical Conference*, Oct. 2016. doi: https://doi.org/10.1115/ICEF2016-9367.
6. C. Garnier, A. Bilcan, O. Le Corre, and C. Rahmouni, “Characterisation of a Syngas-Diesel Fuelled CI Engine,” Apr. 2005. doi: https://doi.org/10.4271/2005-01-1731.

7. N. Kousheshi, M. Yari, A. Paykani, A. Saberi Mehr, and G. F. de la Fuente, "Effect of Syngas Composition on the Combustion and Emissions Characteristics of a Syngas/Diesel RCCI Engine," *Energies*, vol. 13, no. 1, p. 212, Jan. 2020, doi: https://doi.org/10.3390/en13010212.
8. B. K. M. Mahgoub, S. Hassan, S. A. Sulaiman, R. Mamat, A. A. Adam, and F. Y. Hagos, "Combustion and Performance of Syngas Dual Fueling in a CI Engine with Blended Biodiesel as Pilot Fuel," 2017.
9. L. A. Tuan and P. H. Luong, "Simulation of a syngas - diesel dual fuel engine for small-scale power generator," *J. Sci. Technol.*, no. April 2014, pp. 36–41, 2014, [Online]. Available: https://www.researchgate.net/profile/Tuan-Le-Anh-5/publication/340050764_Simulation_of_a_syngas_-_diesel_dual_fuel_engine_for_small-scale_power_generator/links/5e7488a5299bf1c76a20133c/Simulation-of-a-syngas-diesel-dual-fuel-engine-for-small-scale-power-g
10. R. A. Marc *et al.*, "Red oak (Quercus rubra L.) fruits as potential alternative for cocoa powder: Optimization of roasting conditions, antioxidant, and biological properties," *Forests*, vol. 12, no. 8, Aug. 2021, doi: https://doi.org/10.3390/f12081088.
11. A. Pasqualone *et al.*, "Effect of acorn flour on the physico-chemical and sensory properties of biscuits," *Heliyon*, vol. 5, no. 8, Aug. 2019, doi: https://doi.org/10.1016/j.heliyon.2019.e02242.
12. M. Taib and L. Bouyazza, "Composition, Physicochemical Properties, and Uses of Acorn Starch," *J. Chem.*, vol. 2021, 2021, doi: https://doi.org/10.1155/2021/9988570.
13. A. F. Vinha, J. C. M. Barreira, A. S. G. Costa, and M. B. P. P. Oliveira, "A New Age for Quercus spp. Fruits: Review on Nutritional and Phytochemical Composition and Related Biological Activities of Acorns," *Compr. Rev. Food Sci. Food Saf.*, vol. 15, no. 6, pp. 947–981, 2016, doi: https://doi.org/10.1111/1541-4337.12220.
14. M. Ajvad and H. Y. Shih, "Modeling syngas combustion performance of a can combustor with rotating casing for an innovative micro gas turbine," *Int. J. Hydrogen Energy*, vol. 45, no. 55, pp. 31188–31201, 2020, doi: https://doi.org/10.1016/j.ijhydene.2020.08.113.
15. F. O. Olanrewaju, H. Li, Z. Aslam, J. Hammerton, and J. C. Lovett, "Analysis of the effect of syngas substitution of diesel on the Heat Release Rate and combustion behaviour of Diesel-Syngas dual fuel engine," *Fuel*, vol. 312, p. 122842, Mar. 2022. https://doi.org/10.1016/j.fuel.2021.122842
16. K. Cacua, A. Amell, and F. Cadavid, "Effects of oxygen enriched air on the operation and performance of a diesel-biogas dual fuel engine," *Biomass and Bioenergy*, vol. 45, pp. 159–167, Oct. 2012, doi: https://doi.org/10.1016/j.biombioe.2012.06.003.
17. M. Fiore, V. Magi, and A. Viggiano, "Internal combustion engines powered by syngas: A review," *Appl. Energy*, vol. 276, no. July, p. 115415, 2020, doi: https://doi.org/10.1016/j.apenergy.2020.115415.

Pre-treatment and Valorization of Swine Farming Wastewater by Filtration with *Arundo Donax* Biomass

Bruna Petry, Catarina Nobre, Gonçalo Lourinho, Paulo Brito, and Margarida Gonçalves

Abstract The growing increase in intensive swine production has huge environmental impacts. In addition to the high organic load and nutrients, swine effluents have large amounts of suspended solids, which hinder and limit several treatment technologies. Solid–liquid separation using filtration or centrifugation is often used as pre-treatment to reduce suspended solids, also reducing other critical parameters. In this work, pig farming effluents were filtrated using a biological filter composed of *Arundo donax* biomass and the filtrated effluents were characterized to evaluate removal of contaminants. Total suspended solids (TSS) initially at 336.5 mg/L were eliminated, and biomass filtration was also able to remove 48% of chemical oxygen demand (COD), 49% of total nitrogen (TN) and 87% of total phosphorus (TP). The *Arundo donax* biomass after the effluent filtration showed an increase of ash, carbon and hydrogen contents. Calcium (Ca), magnesium (Mg) and phosphorus (P) were the mineral components strongly reduced in the filtered effluent and concentrated in the filter biomass. Through this pre-treatment, the effluent reached a composition more compatible to valorization by anaerobic digestion or production of microalgae, without requiring extensive dilution. The biomass filter can be often replaced and recycled through composting, to obtain biofertilizers for agronomical applications. This pre-treatment approach promotes circular economy in swine production units.

Keywords Pig farming effluents · Filtration · Pre-treatment · *Arundo donax*

B. Petry (✉) · C. Nobre · P. Brito · M. Gonçalves
VALORIZA—Centro de Investigação para a Valorização de Recursos Endógenos, Instituto Politécnico de Portalegre, Portalegre, Portugal
e-mail: b.petry@campus.fct.unl.pt

B. Petry · M. Gonçalves
MEtRICS—Mechanical Engineering and Resource Sustainability Center, NOVA School of Science and Technology, Caparica, Portugal

G. Lourinho
BIOREF—Laboratório Colaborativo para as Biorrefinarias, S. Mamede Infesta, Portugal

J. R. da. C. S. Galvão et al. (eds.), *Proceedings of the 3rd International Conference on Water Energy Food and Sustainability (ICoWEFS 2023)*, Springer Proceedings in Earth and Environmental Sciences, https://doi.org/10.1007/978-3-031-48532-9_28

1 Introduction

The growing increase of intensive swine production, as a response to the growing demand for food of animal origin, has significant environmental impacts, mainly due to the high volumes of effluents generated. Livestock residues stand out as major emitters of greenhouse gases (GHG), as well as potential polluters of groundwater and surface water. Due to the high concentration of solids, organic load, nitrogen, phosphorus, heavy metals present in the swine effluents, these are difficult to treat but they are also a potential source of nutrients for recovery or raw materials for bioenergy production. The pig farming effluents consisting of feces, urine and washing water are mostly sent to stabilization ponds and later applied to the soil as fertilizers. This type of treatment, in addition to being inefficient, leads to the proliferation of insects, gas emissions, odor, and the constant requirement to remove the mud that is deposited at the bottom of the pond.

Due to the complexity of this type of effluent, it is necessary to combine several treatment processes to improve treatment efficiency, as well as to promote the recycling of its nutrients. Pig farming effluents typically have large amounts of solids, which must be reduced before other tertiary treatments can be applied. The solid–liquid separation methods used to reduce TSS and nutrients concentrations in livestock wastewaters include gravitational, mechanical and chemical technologies [4]. Studies indicate large reductions in nutrients, COD, and TSS through the isolated and combined use of these methods [7], however with additional costs. Filtration is a mechanical process that can be performed with different types of devices and filtering materials, which will affect the efficiency of the process. The use of biomass filters has shown environmental and economic benefits, since the filter material can be recovered as a soil conditioner and to produce fuel, in addition to being biodegradable and available [1].

In this sense, this work seeks to present an alternative to the pre-treatment of swine wastewater by evaluating the performance of a green filter composed of *Arundo donax* biomass. *Arundo d.* is a perennial invasive species highly distributed in Portuguese territory, which, in addition to hampering the development of native species, has a high growth rate, therefore requires frequent cutting operations to prevent uncontrolled widespread. These cutting and uprooting operations generate a large amount of biomass waste, that should be recovered and directed to circular economy applications. Also considering the provisions of Directive 91/271/EEC, which states that sludge from wastewater treatment must be recovered and that its disposal in surface water must be progressively eliminated, the retention of suspended solids in a biomass filter contributes to sludge removal and effluent clarification.

The objective of this work is to promote circular economy in the pig production industry, by proposing economic and environmentally sustainable alternatives for the pre-treatment of pig farming wastewaters. Biomass filters containing milled *Arundo* biomass will be used to filter raw pig farming effluent to obtain a clarified effluent that may be used in anaerobic digestion or microalgae production. The filtered effluent will be characterized for various critical parameters to evaluate the pre-treatment efficiency. The used biomass filter will also be characterized for elemental and mineral composition to evaluate retention of nutrients that might be relevant for possible application in composting.

2 Material and Methods

For this work, vertical filters adapted in 1.5 L PET bottles were prepared with 100 g of previously dried, crushed and sieved *Arundo d.* biomass. The bottom of the PET bottle was removed, the lid was pierced with about 0.5 cm, and the mouth of the bottle was wrapped in fabric to retain the biomass inside the bottle. The bottle was closed with the cap, inverted, filled with biomass and compacted at the bottom of the bottle (mouth), and placed on a support to receive the effluent. The diluted effluent (1:1) was previously sieved to remove coarse solids, and 1 L of it was poured over the biomass for filtration. This sequence of procedures can be observed in the following diagram (Fig. 1).

2.1 *Effluent Characterization*

For this work, about 7 L of swine effluent were used, provided by the Aviliz Group, an important company in the agricultural sector that produces more than 15,000 pigs per year in different farms distributed in the regions of Rio Maior, Leiria and Marinha Grande (Portugal). The effluent is treated in stabilization ponds, and the sample was taken before the point of disposal in the pond, showing a dense, dark appearance and a strong odor. The raw effluent sample, as well as the filtered effluent, were analyzed for pH (Crison MicropH 2001 m), electrical conductivity (Mettler Toledo MC226 Conductivity Meter), COD, TN and TP using the Hanna Instruments test kits. Total solids (TS) and volatile (VS) and fixed solids (FS) were determined according to standard methods (2540B and 2540E, respectively). TSS was determined by vacuum filtration on a 0.45 μm cellulose acetate filter of known volume. The turbidity was determined according to the model proposed by [9] through the absorbance at 750 nm by the UV–VIS spectrophotometer (Biochorm Libra S4).

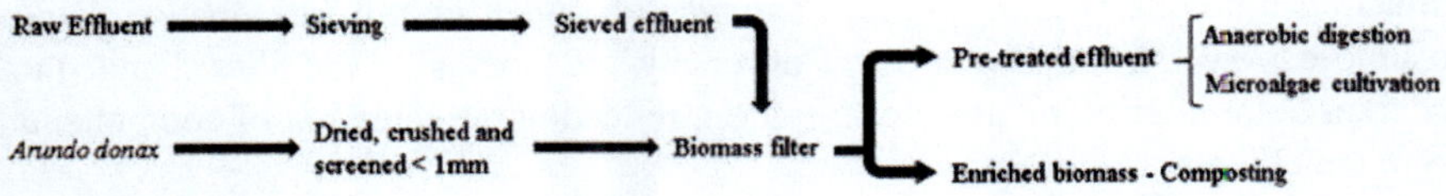

Fig. 1 Flowchart of the materials used in the study and their processes

2.2 Biomass Characterization

The *Arundo donax* canes were collected on November 10, 2022, in the Sintra region of Portugal. These were about 2.5 m high, were cut into smaller pieces, approximately 30 cm, and placed in an oven (Memmert, U-30) at 105 ± 2 °C for 48 h for drying. After drying, all constituent parts of the Arundo were crushed in a crossed beater mill (Retsch SK 100), and subsequently sieved below 1 mm for use. After filtering the effluent, the biomass used was poured into trays together with the sludge retained in the filter, homogenized and dried in an oven (Memmert, U-30) at 105 ± 2 °C for 48 h. The elemental composition of the original biomass and used biomass filter was determined using an elemental analyzer (Thermo Finnigan-CE Instruments Flash EA 1112 CHNS series). The ashes of these biomass samples were digested with 10 mL of diluted nitric acid (HNO_3), at 90 °C, and the solution was filtered, diluted to 100 mL final volume and analyzed by ICP-AED—Inductively Coupled Plasma – Atomic Emission Spectroscopy.

3 Results and Discussion

The filtration test took about 48 h to be complete at atmospheric pressure. The raw effluent was diluted (1:1), but still presented a large concentration of solids that tended to precipitate over the biomass filter, forming a sludge layer. In a simple sedimentation test, when pouring 300 mL of this raw effluent sample into a graduated cylinder, after 24 h of rest, it can be observed that 105 mL of sludge is deposited at the bottom of cylinder, that is, 35% of the effluent volume. Effluent flow through this layer was very slow but, on the other hand, also provided an additional filtering medium to retain suspended particles.

3.1 Effluent Characterization

The main characteristics of the raw effluent and of the filtered effluent are compared in Table 1.

Among the results obtained, what stands out are the TSS, for which it was possible to obtain a 100% removal. The difference in the appearance of the raw and filtrated effluent can be observed in Fig. 2, changing from a denser and dark opaque liquid to a more brownish and clearer one. Conversely, the biomass of the filter changed to a darker color after being used for filtration, reflecting the retention of components from the effluent.

Effluent filtration removed TS and VS contents by 48% and 60% respectively, while FS were only reduced by 26%, what confirms the organic character of most

Table 1 Result of the main parameters analyzed before and after effluent filtration

Parameter	Raw effluent	Filtered effluent
pH	8.43	7.78
TS (g/L)	47.45 ± 0.05	24.65 ± 0.05
FS (g/L)	17.80 ± 0.2	13.15 ± 0.15
VS (g/L)	29.09 ± 0.2	11.50 ± 0.15
TSS (mg/L)	336,54	0
CQO (mg/L)	348,100	181,500
N total (mg/L)	3950	2000
P total (mg/L)	116,687	151

Fig. 2 Figure on the left represents the effluent before and after filtration, and on the right the dry biomass used as a filter (on the left the crushed raw *Arundo d.* and on the right after filtration with the adhered sludge)

solids retained in the biomass filter. This reduction in organic and inorganic contaminants is also confirmed by the 48% COD removal and 81% reduction of turbidity for the filtered effluent. The parameters of TN and TP showed a reduction of 49% and 87% respectively.

The use of raw biomass as a filter medium for pig farming effluent pre-treatment was not found in the literature, but Thao and co-workers studied the filtration of these effluents using biochar produced from mango tree leaves. The swine effluent was pre-treated by anaerobic digestion and subsequently filtered by biochar showing reductions of 46.9% of COD, 78.1% of TSS, 2.13% of TN and 2.7% of TP [7]. The results obtained in the present work, using filtration with *Arundo d.* biomass, showed higher reductions of the effluent organic and inorganic loads in a single step, using an available and low-cost biomaterial.

The reduction of COD and TN can help to achieve an optimum C/N ratio for anaerobic digestion specially if the filtrated effluent is combined with a biomass source to adjust TS content [3].

In addition of the potential for biogas production, the pre-treated effluent can also be used to produce microalgae, what will further reduce its organic and inorganic load. Due to its high concentration of nitrogen and decrease of turbidity the filtered effluent can be used as culture medium for microalgae that might be subsequently

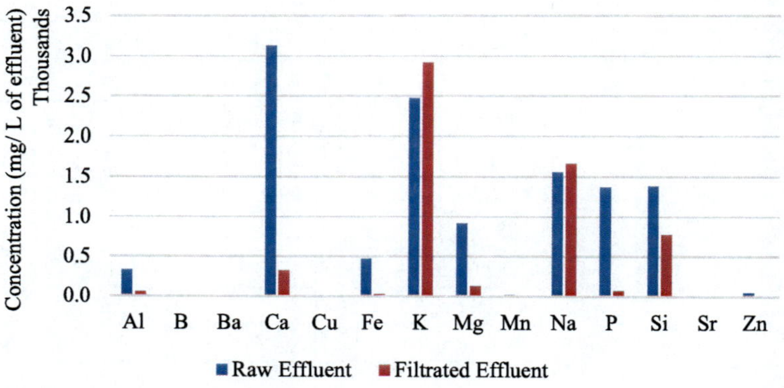

Fig. 3 Mineral composition of the raw and filtered pig farming effluents

used as agricultural fertilizers, creating a circular economy valorization pathway [8] (Fig. 3).

The mineral components present in the raw pig farming effluent were also strongly affected by the filtration process. The filtered effluent showed reduced concentrations of calcium (Ca), magnesium (Mg), phosphorus (P), iron (Fe) and aluminum (Al) relatively to the raw effluent, indicating the possible speciation of these minerals in the form of non-soluble species. Phosphorus often occurs in the form of phosphate groups connected to phospholipids, proteins or carbohydrates. All other minerals strongly reduced by filtration can exist as divalent or trivalent cations that can easily form complexes with organic or inorganic species to yield large aggregates with poor solubility in the aqueous phase [6].

3.2 Arundo Donax *Biomass Before and After Filtration*

The filtration with *Arundo* biomass, allows a fractionation of the effluent and biomass components, which is shown in the diagram below (Fig. 4).

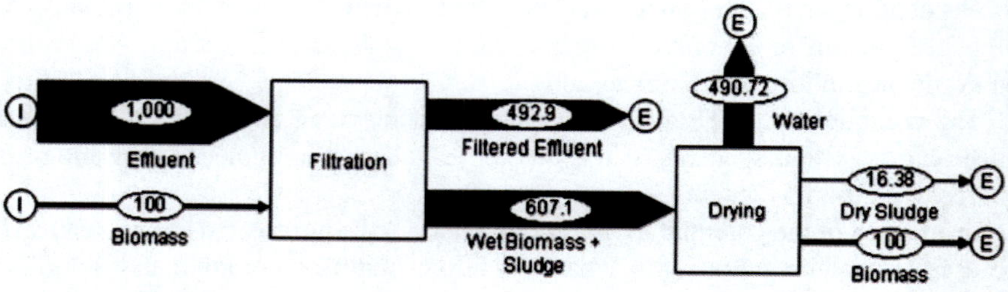

Fig. 4 Filtration test yield flowchart

As can be seen, the filtration efficiency is around 50%, and the remaining effluent mass is retained as sludge and water in the biomass filter. Considering that the biomass was milled and dried before use in the filter, it had a high capacity to retain water and suspended solids, what justifies this large reduction in the volume of filtered effluent. As the filtering biomass becomes saturated and less efficient in the effluent clarification, it can be replaced by fresh biomass, given the low cost and availability of this biomaterial. The used biomass saturated with water and retained solids can be directed to stabilization through composting. Composting is already known and used to treat the sludge from ponds that treat this type of wastewater, however, despite being and economical method, it requires the constant removal of the sludge that is deposited at the bottom of the ponds. Some studies shown favorable results in the use of *Arundo d.* as a bulking agent in composting. Pelegrín et al. [5] evaluated several co-composting scenarios of sewage sludge and agro-food sludge with different proportions of *Arundo d.* As a result, they found that the highest proportion of *Arundo d.* promoted a faster degradation of organic matter, in addition to being a strategy to mitigate the emission of GHG during the composting process. Several factors are important for the control of composting, among them the humidity of the compost. The use of *Arundo d.* biomass after filtration can also contribute to this factor since its high humidity reduces the need for irrigation of the compost piles.

The elemental composition and ash contents of the raw *Arundo d.* biomass and of the same biomass after use as filter medium are presented in Table 2.

The used biomass of the filtering medium (*Arundo d.*) containing the retained sludge, showed an increase in ash and carbon contents, relatively to the raw *Arundo* biomass. These modifications can be explained by retention of organic solids from the effluent in the biomass filter, namely lipid components, that have high carbon content and poor solubility in the water phase. On the other hand, ash increase is also expected by retention of inorganic components of the raw effluent, namely sand particles and precipitated compounds.

The mineral composition of the used biomass also showed significant differences relatively to the raw biomass, as showed in Fig. 5.

As expected, the minerals removed in a greater extent from the effluent were found to increase in the used biomass filter, designated as *Arundo* biomass + effluent

Table 2 Results of elemental analysis of *Arundo d.* biomass before and after filtration

	Arundo d. Biomass	Biomass + Sludge
Ash (%)	6.0	17.5
C (%)	40.35	42.63
H (%)	5.37	5.54
N (%)	1.74	0.28
O (%)	52.14	51.40
S (%)	0.40	0.15
O/C	0.97	0.90
H/C	1.60	1.65

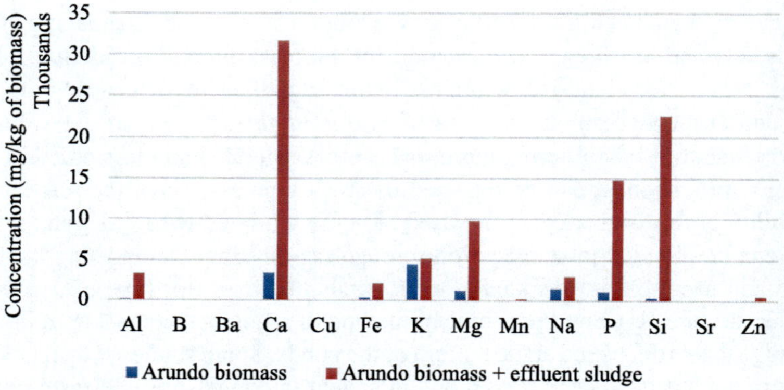

Fig. 5 Mineral composition of raw and after filtration biomass

sludge. Higher concentrations of Ca, Mg and P improve the quality of this material as a feedstock for composting. Silica (Si) also increased in the used biomass filter, reflecting the retention of soil particles removed from the effluent.

4 Conclusions

The wastewater from swine farming still has a great impact on the environment, being a great challenge to treat it. Treatment efficiency and costs involved are important factors for selection of the treatment method to be used. The results obtained in this work showed that it is possible to strongly reduce the load of organic and inorganic contaminants in pig farming effluents by filtration through milled *Arundo donax* biomass. This proposal promotes the recovery of the waste nutrients through a circular economy approach and optimizes the process of removing the sludge present in these wastewaters. Recycling the filter material for composting ensures the production of other valuable products while avoiding the creation of an additional wastestream.

Acknowledgements This work was supported by project PigWasteBiorefinery (ALT20-03-0246-FEDER-000054) and by national funds through the Fundação para a Ciência e Tecnologia, I.P. (Portuguese Foundation for Science and Technology) from FCT/MCTES UIDB/05064/2020 (VALORIZA—Research Centre for Endogenous Resource Valorization) and UIDB/04077/2020-2023 and UIDP/04077/2020-2023 (MEtRICs—Mechanical Engineering and Resource Sustainability Center).

References

1. Alvarez, X., Cancela, Á., Rodríguez, A., Valero, E., & Sanchez, Á. (2021). Green filters of Eucalyptus globulus for microalgae harvesting from freshwater reservoir and reuse of biomass harvested for pellet production. *Process Safety and Environmental Protection, 147*, 497–504. https://doi.org/10.1016/j.psep.2020.11.037.
2. Council Directive of 21 May 1991 concerning urban waste water treatment (91/271/EEC). (1991). *Official Journal of the European Communities*, (L), 40–52.
3. O'Flaherty, V., Collins, G., & Mahony, T. (2010). Anaerobic Digestion of Agricultural Residues. In R. Mitchell & J.-D. Gu (Eds.), *Environmental Microbiology* (Second Edi). Galway. https://doi.org/10.1002/9780470495117.ch11.
4. Osabutey, A., Haleem, N., Uguz, S., Min, K., Albert, K., Anderson, G., & Yang, X. (2023). Growth of Scenedesmus dimorphus in swine wastewater with versus without solid – liquid separation pretreatment. *Bioresource Technology, 369*(October 2022), 128434. https://doi.org/10.1016/j.biortech.2022.128434.
5. Pelegrín, M., Sáez-Tovar, J. A., Andreu-Rodríguez, J., Pérez-Murcia, M. D., Martínez-Sabater, E., Marhuenda-Egea, F., … Moral, R. (2018). Composting of the invasive species Arundo donax with sewage and agri-food sludge : Agronomic , economic and environmental aspects. *Waste Management, 78*, 730–740. https://doi.org/10.1016/j.wasman.2018.06.029.
6. Schott, C., Cunha, J. R., Weijden, R. D. Van Der, & Buisman, C. (2022). Phosphorus recovery from pig manure : Dissolution of struvite and formation of calcium phosphate granules during anaerobic digestion with calcium addition. *Chemical Engineering Journal, 437*(P2), 135406. https://doi.org/10.1016/j.cej.2022.135406.
7. Thao, N. T. P., Nga, N. T. T., Kim, H. T. T., Kien, T. T., Hieu, T. T., Thang, N. V., … Le, T. H. (2023). Combination of biochar filtration and ozonation processes in livestock wastewater treatment and application for soil cultivation. *Case Studies in Chemical and Environmental Engineering*, 100286 https://doi.org/10.1016/j.cscee.2022.100286.
8. Viegas, C., Gouveia, L., & Gonçalves, M. (2021). Bioremediation of cattle manure using microalgae after pre-treatment with biomass ash. *Bioresource Technology Reports, 14*(January), 100681. https://doi.org/10.1016/j.biteb.2021.100681.
9. Goodner, K.L.. Estimating Turbidity (NTU) From Absorption Data, *Synergy*, https://www.synergytaste.com/wp-content/uploads/sites/2/2022/05/Estimating-Turbidity-NTU-From-Absorption-Data.pdf (2009).

Residential District Heating Network with Peer-To-Peer Market Structure: The Case of Nordhavn District

António Sérgio Faria, Tiago Soares, and Linde Frölke

Abstract Over the last decades, district heating has been under development, especially the technologies like heat pumps, solar thermal and cogeneration. However, there is still a long way to go regarding regulation, legislation and market liberalization, which varies across countries and regions. The objective of this work is to investigate the potential benefits of decentralized district heating systems in residential areas. By studying a case study of EnergyLab Nordhavn, a residential area in Copenhagen, Denmark, the paper compares the market outcomes of decentralized systems such as community markets to the centralized pool market currently in practice, under the EMB3Rs platform. The study focuses on key market outputs such as dispatched production, revenues, and daily consumption patterns. Additionally, the paper examines the impact of advanced features such as flexible heat consumption and network awareness in the market. The results of this research suggest that decentralized district heating systems have the potential to improve market outcomes and increase energy efficiency in residential areas.

Keywords District heating · Residential heating · Decentralized market structures · Community · EMB3Rs

1 Introduction

District Heating (DH) systems play a crucial role in meeting energy demand for heating and cooling, and it is important to consider viability in their planning and design. The research in [1] aims to examine the impact of integrated planning on the environmental and economic sustainability of DH systems. A decision tool for local authorities is developed based on a modified Vester system analysis and

A. S. Faria (✉) · T. Soares
INESC TEC, Porto, Portugal
e-mail: antonio.s.faria@inesctec.pt

L. Frölke
DTU Electro, Kongens Lyngby, Denmark

J. R. da. C. S. Galvão et al. (eds.), *Proceedings of the 3rd International Conference on Water Energy Food and Sustainability (ICoWEFS 2023)*, Springer Proceedings in Earth and Environmental Sciences, https://doi.org/10.1007/978-3-031-48532-9_29

research in spatial, environmental, and energy planning. The main factors affecting DH feasibility include a mix of functions, density, building type, and heat source. Spatial planning policies must consider building density to ensure the long-term viability of DH systems. In [2] information is provided on how to design and plan DH systems in countries with low DH market share for heating and cooling supply. The article covers: building a heat atlas through essential techniques and methodologies for district thermal energy demand prediction and the integration of sustainable resources; Identifying technical barriers to the implementation of sustainable heat sources; Discussing technologies for the optimal planning of DH systems including current approaches, novel configurations, and upgrading measures for existing DH systems. Also, in [3] a novel 5th-generation DH model that implements the energy-prosumer concept of bilateral heat trading between the DH network and the building is proposed. The model utilizes the low temperature of the DH return pipe's water and its technical feasibility was evaluated through simulation analysis. Also, the replacement of old technology is assessed in [4] through the economic impact of converting six conventional DH systems in Poland to more sustainable and efficient systems. The new systems would include flat solar collectors, seasonal thermal energy storage, and gas heating plants as flexible heat sources.

Combined Heat and Power (CHP) plants can also play an important role in DH due to its flexibility. For instance, [5] presents simulations of a real CHP unit and DH network to show that this improves overall economic efficiency and increases the CHP unit's operating range in the electricity market. The proposed methodology provides logical results for calculating CHP efficiency indicators. While the work in [6] assesses the integration of pipeline energy storage in DH systems by adjusting CHP unit electricity production based on fluctuating electricity prices. However, the operational use of pipeline energy storage is challenging due to uncertainty, computational complexity, and limited sensors. To overcome these challenges, this study models CHP economic dispatch as a Markov decision process and uses reinforcement learning to estimate system dynamics.

The integration of solar energy into DH systems has become a crucial aspect in reducing greenhouse gas emissions and creating sustainable energy solutions for communities, [7] aims to determine the most cost-effective distribution system for a solar DH system by comparing the marginal life cycle cost of two systems. The simulations show that a novel distribution system with lower network temperatures and central domestic hot water preparation is the most efficient. The study found that the investment costs were more significant than operational costs in reducing life cycle costs. In [8] the use of solar energy transformation technologies in DH is also addressed. It examines the first year of operation of a large-scale solar collector field in the Baltic states. The study analyzes solar collector efficiency and factors such as temperature and flow. The benefits of optimal solar collectors' placement, storage space, and demand for thermal energy for solar energy production are also highlighted, [9] provides a case study to include solar energy in cooling systems in the Middle East. The work concludes that the availability of installation area and electricity tariff play a significant role in the economic competitiveness of different solar-cooling configurations. The optimal solution is the photovoltaic district cooling system configuration, offering the lowest operating cost and the highest environmental performance.

Thermal energy storage can also be used to balance energy demand and supply, ensuring stable and efficient operation of the system. This is supported by the study in [10] showing that using thermal energy storage, specifically chilled water, in a community of 2000 buildings, can have a positive impact on reducing the need for electricity generation and storage. This can lead to a self-sufficient community powered by renewable energy. The simulations indicate that the use of thermal storage reduces the required photovoltaics rating and the electric energy storage system capacity during the summer, [11] also provides an overview of thermal communities' research, but a lack of literature, directives and case studies is found. The authors state that further feasibility assessments of published studies are needed. Alternatively, [12] presents a model for assessing the economic viability of integrating a supermarket and power substation waste heat into existing DH systems.

In this work, the aim is to compare and analyze market outcomes of centralized and community schemes. In particular, it is relevant to address the effects of the choice of market mechanism on parameters such as the price of heat, generation and consumption schedules and settlement. The second aim of this case study is to validate and test the EMB3Rs platform. This is done by running a full simulation including the Core functionalities (CF) module, Geographical Interface System (GIS), Techno-Economic (TEO), Market Module (MM), and Business Module (BM).

2 Market Structures and EMB3Rs Platform

In this section, the market structures are explored, as well as the main features and capabilities of the EMB3Rs platform [13].

2.1 *Market Structures*

The main purpose of this subsection is to provide a set of market structures capable of simulating existing and future markets for DH systems. More precisely, three different market designs are available: Pool, Peer-to-Peer, and community-based structures.

1. Pool: In this design, the merit order mechanism is used (similar to the electricity market) to determine the order in which producers and consumers will be dispatched. This mechanism prioritizes the cheapest producers, as well as the highest-bidding consumers, to ensure that the most cost-effective energy sources are used. The market clearing price is determined by the price bid of the marginal producer or consumer, which is the agent that is only partially scheduled. All participants in the market will pay and receive the same price per unit of energy,

known as uniform pricing. The mathematical formulation for this market design can be found in [14];

2. P2P: This mechanism proposes that any pair of peers can trade heat on a bilateral basis, without third-party (coordinator) supervision. That is, each peer can exchange with another peer on an individual basis, defining the amount of energy to be bought or sold at a given price. Note that prosumers can trade with both producers and consumers, but such interaction is limited by the type of offer they submit on the market. The main advantage is that the privacy of peers is protected, as peers share limited information with those they want, taking full control of their facilities. A detailed description of this design can also be found in [14];
3. Community-based: The community-based market design intends to represent a more hierarchical structure of bilateral peer trades. In this semi-decentralized model, there is a community manager responsible for the community's energy management. In terms of privacy, it requires less exchange of information than the Pool market model but more than the P2P market model. This system is often seen as a compromising solution between the centralized and decentralized approaches. The community manager supervises all the trading activities within the community, as well as works as an intermediary in the heating trade with the main heat supplier [14].

2.2 EMB3Rs Platform

The EMB3Rs platform is a tool for assessing the feasibility of exchanging excess heat and cooling in several sites, including industrial operations and DH systems. Through simulating various supply–demand scenarios, network configurations, and commercial and market frameworks, the platform can assist users and stakeholders in comprehending the economic viability of investing in recovering waste heat and cooling as an energy source.

The five modules of the platform work together to provide a comprehensive understanding of the possibilities for recovering excess heat and cooling in various situations. The CF module enables the identification of possible sources and sinks of excess heat and cooling, as well as the associated costs of recovery and utilization. The GIS module helps to find and evaluate potential solutions for connecting sources and sinks, taking into account factors such as losses, costs, network length, and available pipeline capacity. The TEO module helps to identify the most cost-effective solutions for utilizing excess heat, considering factors such as regulations, available heat, load profiles, and technical and economic characteristics of technologies. The MM enables users to simulate current and future trends in the heat and cooling markets and assess the economic potential and savings of their investments. Using the MM, users can choose the best market framework for their specific economic, environmental, and social interests, helping them to make informed decisions about their investments in recovering excess heat and cooling. The BM is designed to consider

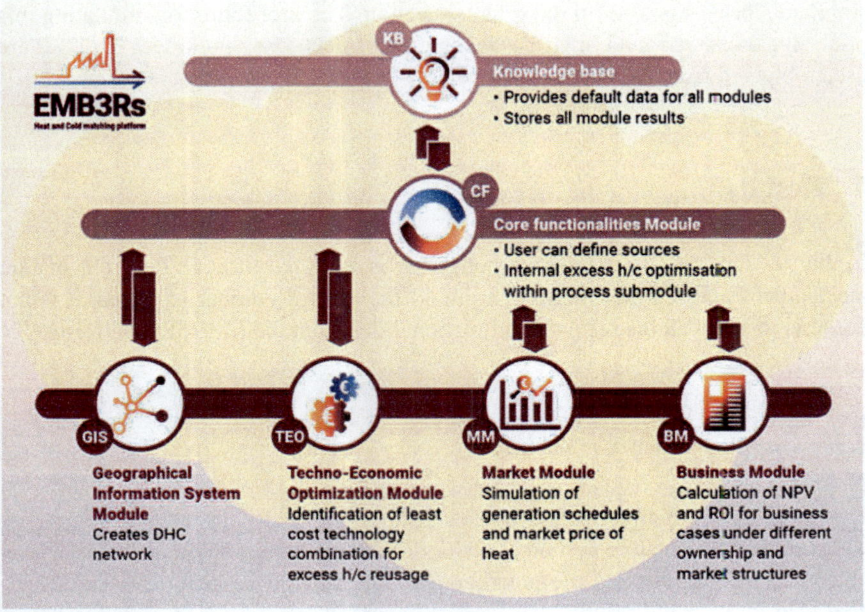

Fig. 1 EMB3Rs Platform modules functionalities

different ownership structures and market frameworks and evaluate key metrics such as net present value, levelized cost of heat, and internal rate of return (Fig. 1).

3 Case Study Description

This section provides a complete description of the case study, including goals for simulation under the EMB3Rs platform, as well as the collected data used in the case study.

The case study area is located in the Nordhavn neighbourhood in the city of Copenhagen. The area is linked to the EnergyLab Nordhavn (ELN) project [15]. It is composed of 30 residential consumers and a local supermarket. The residential buildings are supplied by the DH network operated by HOFOR. Note that the DH network in this area is already in place. Furthermore, the area is connected to the greater Copenhagen DH Network. The local supermarket from the chain Meny can provide excess heat that could inject back into the Nordhavn DH grid. The company Danfoss owns the heat recovery unit (HRU) of the Meny supermarket.

From this project, the total hourly heat consumption of 30 consumers is available. This consumption includes domestic hot water (DHW) and space heating (SH). Also, for the supermarket, the heat consumption is available, as well as an hourly production profile. In addition to the collected data, several open data sources have been used to simulate a price signal from the grid connection. The procedure for simulating this price signal can be found in [16]. Among others, electricity spot prices for DK2 are used, obtained from Nord Pool open data.

4 Results

In this section, the different agent behaviour is analyzed through different market mechanisms and features. Note that due to the limited number of pages, it is not possible to show all the results and functionalities from the EMB3Rs platform.

4.1 Consumption Flexibility

The simplest simulation of the Pool market is used, with and without the energy budget bidding format, to see the effect of flexible consumption on different market outcomes: the load profiles, the generation profiles, and the market prices. The scheduled load for each hour of the day averaged over the full simulated period is shown in Fig. 2. The solid red line shows the average pattern in the total scheduled load when there is no energy budget, so the load is more inflexible. The dotted line shows the effect of this energy budget, which allows the load to be shifted from one hour to another hour during the same day. It can be seen that a large share of the load during morning and evening hours is shifted to the night, and some of it is shifted to mid-day. This is the result of electricity prices normally being higher in the morning and evening so that the supermarket heat is more expensive there. To minimize cost, the load is scheduled to avoid those expensive periods of the day.

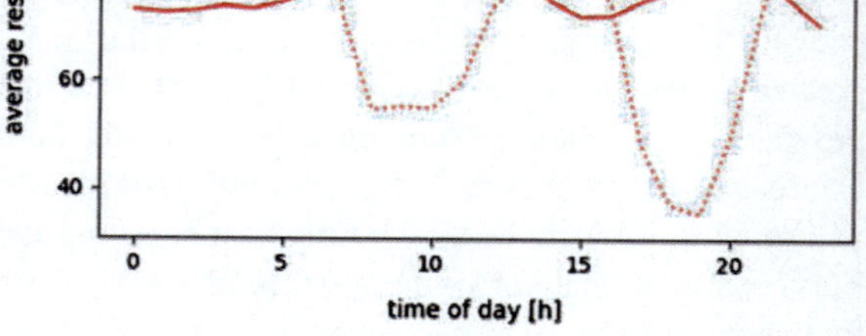

Fig. 2 Energy budget effect on the pool market consumption pattern

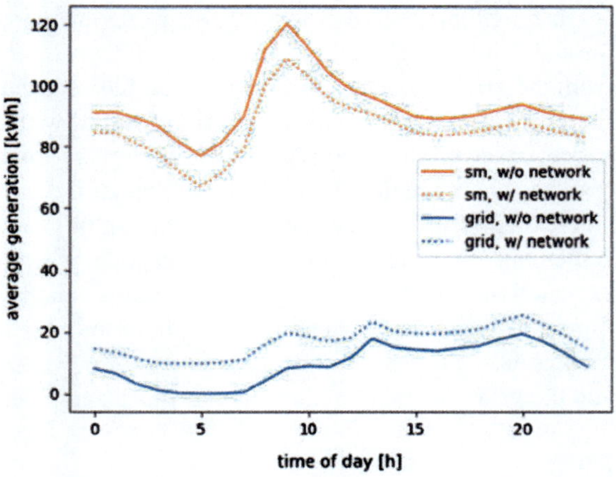

Fig. 3 The effect of network constraints on the daily generation patterns of the supermarket (sm) and grid

4.2 Network-Aware Mechanism

Generally, the different market designs are network independent. That is, the physical constraints to heat transport resulting from the district heating or cooling network layout are not considered. However, in the short-term pool market, there is an option to take network directions into account. As a result, the generation and consumption are scheduled such that the flow directions in the network are respected.

Figure 3 shows the effect of adding those network constraints on the scheduled generation of the supermarket (yellow) and grid (blue). The average daily generation pattern in a pool market without network (solid line) is compared with the network-direction option (dotted line). It can be seen that the supermarket is, on average, consistently scheduled less, and this is offset by increasing the grid import. The reason is that due to the network constraints, there are now several residential consumers that cannot be served by the (generally cheaper) supermarket excess heat. These consumers have no choice but to import heat from the grid instead.

4.3 Community Markets (Peak Shaving)

For the community market, the user must choose an objective for the community. The short-term MM provides two options for this: peak shaving and autonomy. For this market type, the user needs to define which agents are in the community, and which agents are not. In this case study, we consider that all agents except the grid connection are in the community. Energy Budget feature is activated in all simulated markets in Sects. 4.3 and 4.4.

First, the community market is compared with the peak-shaving penalty to the pool market. The daily scheduled generation patterns for the grid and the supermarket are shown in Fig. 4. Here, the average generation for each hour of the day is plotted, for different market types. It can be seen that the low peak-shaving penalty does not have a visible effect on these schedules. However, for the higher penalty, the import from the grid is affected, as shown by the blue dashed line. The import profile from the grid becomes more constant during the day, with a lower peak. To compensate, supermarket production is increased for several hours of the day, mostly during low-price hours in the night and the middle of the day.

Next, it is investigated how the community peak-shaving objective affects the settlement. From the form of the corresponding optimization problem, it is to be expected that prices increase as a result of this penalty, compared to the pool market. Figure 5 shows that in the high-load winter months from October to February, the high penalty on peak import leads to increased revenues for both the supermarket and the grid. The increased revenue is most significant for the supermarket.

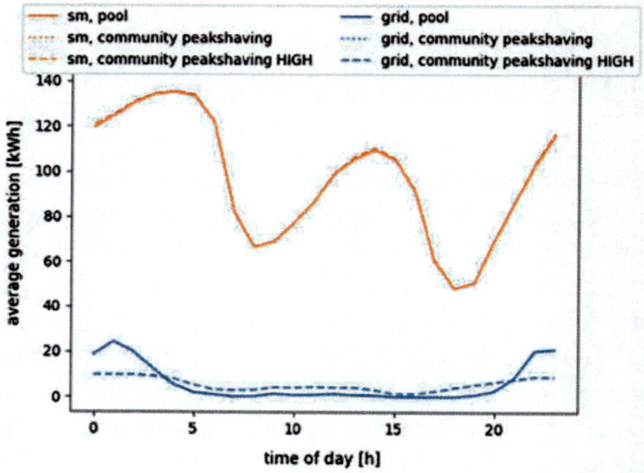

Fig. 4 The effect of peak-shaving objective on the average daily generation patterns

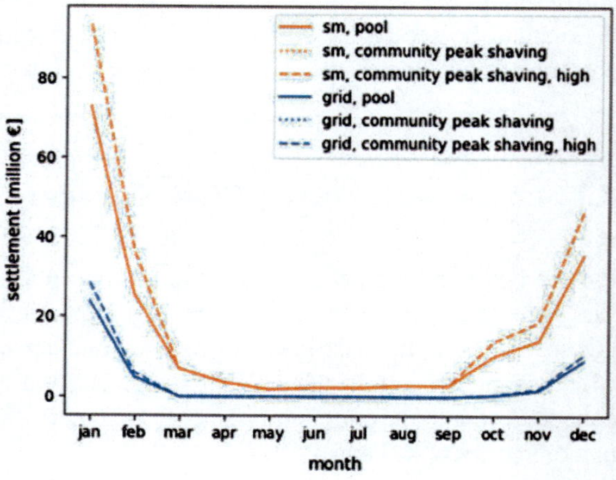

Fig. 5 The effect of a community with the peak-shaving objective on total settlement per month

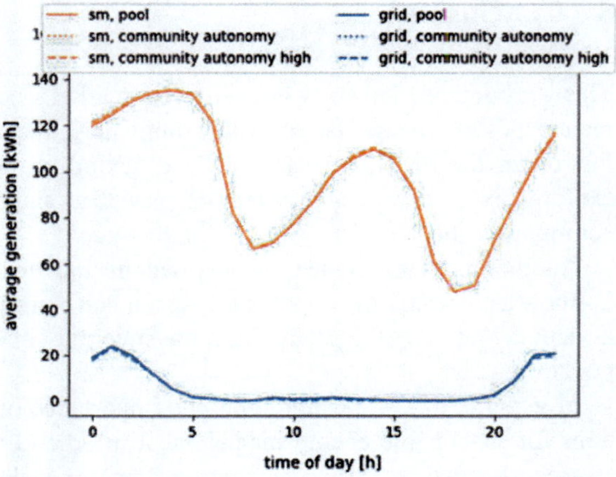

Fig. 6 The effect of a community market with autonomy objective on average daily generation profiles

4.4 *Community Market (Autonomy)*

In Fig. 6 the effect on the average daily generation pattern of the grid and supermarket is visualized. Again, the effect is very small for the low penalty. For the higher import penalty, the effect is still rather small. The only change is that the supermarket production increases in some hours, whereas the grid import on average is decreased in some hours of the day. In all, this means that the community has become more autonomous.

Regarding the settlement, the main effect is that the supermarket is earning more from October to February, because the supermarket is scheduled more, as Fig. 6 shows.

4.5 *Investment in a Supermarket Heat Pump*

Here, the goal is to investigate if the community of residential consumers and the supermarket together would benefit from investing in the supermarket heat pump. To analyze this, the pool market with Energy Budget is compared in two cases: (1) with the supermarket excess heat included and (2) without the supermarket excess heat. The average price per unit of heat without the supermarket excess heat is 784 DKK/MWh = 105 EUR/MWh, and 313 DKK/MWh = 41.7 EUR/MWh with the supermarket excess heat. Comparing the total settlement for the group of 30 consumers with and without the excess heat from the supermarket, the total savings in heat cost for one year is 675.000 €. This means that the consumers could invest this amount in the installation of the supermarket heat pump and earn their investment back within a year.

5 Conclusion

This work intended to show how different market mechanisms can affect the performance of certain agents, regarding both heat dispatched and economical benefits, under the EMB3Rs platform. The flexibility on the consumer side was also explored by the energy budget feature, as well as autonomy and peak-shaving on the community side.

Another finding was the monetary benefits that decentralized structures can bring and how a possible investment in this sector can come up with short payback periods, as demonstrated by the profits from the supermarket allowance to participate in the market.

In general, this work highlights the importance of considering innovative solutions for sustainable energy management in urban environments. The goal was to aware policymakers, urban planners and energy stakeholders about the benefits of integrating supermarkets or other industries in the district heating context. This inclusion can create a symbiotic relationship that benefits both parties while also reducing carbon emissions.

Acknowledgements This work is supported by the European Union's Horizon 2020 through the EU Framework Program for Research and Innovation, within the EMB3Rs project under agreement No. 847121. It is also supported by the Scientific Employment Stimulus Programme from the Fundação para a Ciência e a Tecnologia (FCT) under the agreement 2021.01353.CEECIND.

References

1. F. Zach, S. Erker, and G. Stoeglehner, "Factors influencing the environmental and economic feasibility of district heating systems - A perspective from integrated spatial and energy planning," *Energy. Sustain. Soc.*, vol. 9, no. 1, 2019, doi: https://doi.org/10.1186/s13705-019-0202-7.
2. M. Jiang, C. Rindt, and D. M. J. Smeulders, "Optimal Planning of Future District Heating Systems—A Review," *Energies*, vol. 15, no. 19. p. 7160, 2022, doi: https://doi.org/10.3390/en15197160.
3. Y. Youn and Y. Im, "Technical Feasibility Assessment for a Novel Fifth-Generation District Heating Model of Interconnected Operation with a Large-Scale Building," 2022.
4. B. S. Thomson, "Greening of the District Heating Systems—Case Study of Local Systems," pp. 1–43, 2022, doi: https://doi.org/10.1007/bfb0074381.
5. P. Ruševljuk, K. Lepiksaar, A. Siirde, and A. Volkova, "Economic dispatch of chp units through district heating network's demand-side management," *Energies*, vol. 14, no. 15, 2021, doi: https://doi.org/10.3390/en14154553.
6. K. Stepanovic, J. Wu, R. Everhardt, and M. de Weerdt, "Unlocking the Flexibility of District Heating Pipeline Energy Storage with Reinforcement Learning," *Energies*, vol. 15, no. 9, pp. 1–25, 2022, doi: https://doi.org/10.3390/en15093290.
7. M. Andersen, C. Bales, and J. O. Dalenbäck, "Economic Analysis of Heat Distribution Concepts for a Small Solar District Heating System," *Energies*, vol. 15, no. 13, 2022, doi: https://doi.org/10.3390/en15134737.

8. I. Polikarpova, R. Kakis, I. Pakere, and D. Blumberga, “Optimizing large-scale solar field efficiency: Latvia case study,” *Energies*, vol. 14, no. 14, 2021, doi: https://doi.org/10.3390/en14144171.
9. R. Ismaen, T. Y. Elmekkawy, S. Pokharel, A. Elomri, and M. Al-Salem, “Solar Technology and District Cooling System in a Hot Climate Regions: Optimal Configuration and Technology Selection,” *Energies*, vol. 15, no. 7, 2022, doi: https://doi.org/10.3390/en15072657.
10. R. E. Transition, “Thermal Storage for District Cooling—Implications for Renewable Energy Transition,” *Renew. Energy Transit. Asia*, 2021, doi: https://doi.org/10.1007/978-981-15-8905-8.
11. A. G. Papatsounis, P. N. Botsaris, and S. Katsavounis, “Thermal/Cooling Energy on Local Energy Communities: A Critical Review,” *Energies*, vol. 15, no. 3, 2022, doi: https://doi.org/10.3390/en15031117.
12. H. Dorotic, C. Kristijan, M. Josip, P. Tomislav, and D. Neven, “Technical and Economic Assessment of Supermarket and Power,” 2022.
13. “Heat and Cold matching platform.” https://www.emb3rs.eu/ (accessed Mar. 31, 2023).
14. A. S. Faria, T. Soares, J. M. Cunha, and Z. Mourão, “Liberalized market designs for district heating networks under the EMB3Rs platform,” *Sustain. Energy, Grids Networks*, vol. 29, p. 100588, Mar. 2022. https://doi.org/10.1016/J.SEGAN.2021.100588.
15. “EnergyLab Nordhavn.” http://www.energylabnordhavn.com/. (accessed Jan. 02, 2023).
16. Linde Frölke, Ida-Marie Palm, Jalal Kazempour, “Market integration of excess heat,” *Electr. Power Syst. Res.*, vol. 212, no. 108459, 2022. https://doi.org/10.1016/j.epsr.2022.108459.

Projection of the Climate Change Impacts on Electricity Production Derived from Renewable Sources

Pietro Campagna de Souza, Joao Sousa, and Sandra Mourato

Abstract The potential impact of climate change on the energy sector has prompted the engineering community to take a closer look. In view of this, it is inevitable to assume that the world faces the great challenge of investing in new alternative energy sources and trying to predict the next cycles in their production processes. The present work approaches this issue in order to project the production of this type of energy, using a standard emissions scenario and different time horizons. Based on the energy matrix and the energy supply in Portugal and on the climate projections, the work aims to verify the potential of energy production from renewable sources in future scenarios. The methods used were based on the analysis of hydro, wind and solar production cycles between the years 2008 and 2020. The comparison of these cycles with the climate data of the same period allows to obtain statistical projection models for the periods 2041–2070 and 2071–2100 for the emission scenario RCP8.5. At this stage, only the relationships between projections of future climate and water production are presented. The results indicate a reduction in hydropower production by 2100, which could be up to 12%.

Keywords Renewable energy · Climate change · Energy mix · Emission scenarios

---

P. C. de Souza (✉) · J. Sousa · S. Mourato
School of Technology and Management, Polytechnic Institute of Leiria, 2411-901 Leiria, Portugal
e-mail: 2200160@my.ipleira.pt

J. Sousa
INESC—Coimbra, DEEC, Polo II, University of Coimbra, 3030-790 Coimbra, Portugal

S. Mourato
MED—Mediterranean Institute for Agriculture, Environment and Development and CHANGE—Global Change and Sustainability Institute, University of Évora, Mitra Pole, Ap. 94, 7006-554 Évora, Portugal

J. R. da. C. S. Galvão et al. (eds.), *Proceedings of the 3rd International Conference on Water Energy Food and Sustainability (ICoWEFS 2023)*, Springer Proceedings in Earth and Environmental Sciences, https://doi.org/10.1007/978-3-031-48532-9_30

1 Introduction

1.1 Contextualization and Challenges Associated with the Insertion of Renewable Sources in the Energy Matrix

It is well known that the world's population is increasingly moving towards greater integration of renewable energy. This awareness plays a fundamental role in the development of new solutions and studies to mitigate the negative impact on the environment in which we live, as well as on such important political and economic sectors of society.

In recent years, there has been a growing concern about the climate and how uncertainties about the future of energy production are creating a need to re-evaluate the methods currently used. It is well known that many international commitments have been made to reduce the effects of global warming, which has greatly encouraged the study of renewable energy sources, but there are still future opportunities that have not been fully studied and explored, which will certainly allow better conditions and improvement of these capture operations.

In this context, it is important to understand the evolution of the climate in the past in order to anticipate decisions that would mitigate undesirable situations. In times of drought, for example, hydropower production is severely affected, increasing the vulnerability of basic services, particularly in certain regions of the world. This may be accompanied by other serious risks to human existence, such as disease outbreaks, increased poverty and poor infrastructure. Efforts to integrate renewable sources into the energy matrix must be structured in an integrated management of the available tools to achieve the optimal capacity of the matrix. A study carried out to investigate the impact of climate change on mountain hydrology in Peru showed that population growth and rising temperatures will increase per capita daily water consumption, but the use of existing technology will be able to minimize the gap between water supply and demand in the coming years [1]. This shows that with the tools available, it is possible to simulate the processes of the hydrological system, predict scenarios and thus outline strategies to minimize negative impacts in the future.

In the Sixth Assessment Report of the Intergovernmental Panel on Climate Change (IPCC), scientists from more than 60 countries were concerned to show that the world needs to reduce greenhouse gas (GHG) emissions on a large scale in the coming years in order to limit global warming to 1.5 °C (as set out in the Paris Agreement) [2]. The challenge is that while progress is being made, global efforts are still far from what is needed. For example, extreme temperatures (unbalanced heat or cold waves), constant changes in wind patterns, heavy rainfall or its absence, all these elements with sudden changes during the years end up generating indeterminate and non-standardized forecast variables [3].

It is difficult for people to accept that they can also do their bit, for example by using solar panels at home to produce solar energy [4]. In addition to investment in large solar farms, decentralized production should be promoted to ensure production closer to the point of consumption, thus avoiding investment in large infrastructure and losses in the distribution and transport of electricity over long distances.

This issue is relevant to society in general, which will benefit from increased production of resources that can be reused, and to decision-makers and companies that will work to ensure that new measures are constantly being considered and implemented to reduce uncertainty and increase the reuse of natural resources.

The aim of this article is to evaluate the impact of climate change on the renewable energy sector, on the supply side, relating the cycles of water, solar and wind production to climate projections, so that those responsible in this context can know the behavior of the future energy matrix and work on new actions to extract the maximum potential of renewable resources for the coming years. The application of the methodology is presented only for hydroelectric energy in mainland Portugal.

2 Characterization of the Renewable Energy Mix in Mainland Portugal

Renewable energy sources are those that are derived from resources that are not exhausted in nature and can therefore be used indefinitely if processed in the most appropriate way. Water, wind and solar energy are well-known examples of this type of source.

In Portugal, the government started in 2007 to promote the use of renewable energy in an undeniable way [5]. Figure 1 shows the Portuguese electricity generation mix based on renewable sources between 2000 and 2020.

As can be seen in Fig. 1, over the years there has been a variable supply of hydroelectricity, but based on a virtually fixed installed capacity, as there has been no new investment in this area. However, since 2004 there has been an increase in

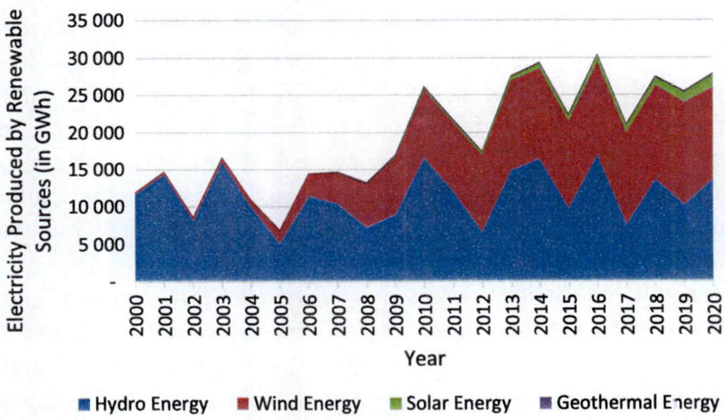

Fig. 1 Renewable energies contribution to the mainland portuguese electricity mix (GWh)

the supply of wind energy due to new investments in this sector. And from 2013, the supply of solar energy will increase as a result of increased investment in this area. Although energy availability depends on different sources, there is a similarity in peaks and troughs over the years, which may be related to energy imports and exports between Portugal and other European countries.

Portugal also has its own National Climate and Energy Plan (PNEC 2030), which aims to increase the share of renewable energy sources in total electricity consumption to more than 47% by 2030. By 2020, Portugal will rank 9th in the EU in terms of renewable electricity consumption [6].

Over the years, changes in the hydrological cycle have been observed, which may have a disproportionate impact on the population. Cases of flooding or drought, when they cannot be predicted, lead to a variety of circumstances that affect the needs of the population [4].

A study of Portugal's main river basins predicted a reduction in hydropower generation of between 17 and 44% for Portugal by 2050. The authors of the study expected changes in the mix and an increase in photovoltaic solar energy (PV), but this would not be enough to compensate for the reduction in hydroelectric power generation caused by climate change. It was assumed that hydroelectric generation could be improved if there were more significant changes in the operation of the main reservoirs in Portugal [7].

As mentioned above, being a country that is highly affected by the effects of climate change, Portugal is investing heavily in low-carbon energy and is one of the European countries that uses the most renewable energy. The development of hydroelectric power plants, for example, has been further exploited and the country remains steadfast in the fight against climate change. In 2023, according to REN (National Energy Networks), Portugal will have reached a new maximum in hydroelectric production of electricity from dams [8].

The impact of climate change on hydroelectric production is high, and therefore hydroelectric projections need to be accurately determined, considering the degree of uncertainty, since they can vary widely and do not depend only on precipitation. It is important to optimize the operation of the reservoirs so that the quality of the water produced can be used for different purposes, taking into account the different locations and seasons. This is urgent because energy demand in Europe, where water availability is likely to be lower, could increase by more than 25% by 2050 [9].

Climate change affects electricity production and demand, and thus the economic system. The intensification of natural phenomena such as floods, droughts, hurricanes, storms, heat waves, rising sea levels, melting polar ice and agricultural losses can ultimately cause fluctuations in energy prices, exchange rates, food production and other costs to society in general [10].

It is undeniable that political efforts to circumvent these situations and increase investment in renewable energy sources are greater in more developed countries, and that different ways of balancing supply and demand are providential in these countries. Economic and efficient technological solutions are still being studied to promote a more significant development for use on a systemic scale.

3 Methodology

In the specific objectives it is proposed to expose the results of some studies on the relationship between the energy matrix and the supply of renewable energy, not including the analysis of energy demand by the industrial, residential, services and agricultural sectors, to identify and analyze the crucial points related to the results of the evolution of these cycles for forecasts.

By analyzing the supply data provided by the DGEG (Directorate General of Energy and Geology) in the Energy Balance of Mainland Portugal between the years 2000 and 2020, it is possible to know the volume of imports, exports and primary energy consumption. Data was also obtained from the DGEG on the installed capacity of power plants. It was necessary to establish a relationship between these two data, through a normalization due to a very abrupt evolution of the integration of renewable generation, so as not to disturb the growth of the installed power in the electricity supply, according to Eq. 1:

$$Utilization\ Factor = \frac{Production\ of\ Energy}{Maximum\ Production\ Allowed\ With\ the\ Installed\ Power} \tag{1}$$

The climatic data considered were minimum temperature (TN), mean temperature (TG), maximum temperature (TX), precipitation (RR), wind speed (FG), radiation (QQ) and relative humidity (HU). The data were obtained from the E-OBS database, 0.25 degrees, between 1995 and 2022 [11].

MATLAB software was used to read, process and analyze these data. A pre-processing of the data was carried out and the database for the study was created. The annual, monthly and seasonal averages of all these climatic variables were calculated for comparison with the energy matrix supply. For winter, an average of all variables and energy sources for all study years was used for December, January and February. For spring an average of March, April and May was used, for summer June, July and August and for autumn September, October and November.

In order to verify that the climate data of the E-OBS were close to the observed values, the observed history of the available climate components, average, maximum, minimum and precipitation, between the years 2000 and 2020 was extracted from the Portuguese Institute of the Sea and Atmosphere (IPMA).

The climate data for the periods 2041–2070 and 2071–2100 for the RCP8.5 scenario were obtained from the Climate Portal [12], using the annual, monthly and seasonal time averages, the geographical area of mainland Portugal, the respective climate variable, the 30-year average statistics and the "ENSEMBLE" of the available models. A correction was made for the deviation from the data in the RCP8.5 scenario.

The correlations between historical data and E-OBS data were analyzed. The correlations between the E-OBS data and the Energy Matrix Supply data in the period 2000 to 2020 were analyzed and a statistical model was created to be used in the periods 2041–2070 and 2071–2100.

The data were stored in a database and divided into periods such as annual, monthly, winter, spring, summer and autumn. For each variable (TN, TG, TX, RR, FG, QQ and HU) the observed history, the climate model in the historical period (E-OBS), the climate portal data for the RCP8.5 scenario are available. The characterization of the electricity production on the non-standard scale and for the normalized scale of the water, solar and wind components is also available.

4 Results

In order to illustrate the methodology presented, we will present the results for hydro energy only. Figure 2 shows the dispersion between the historical observations (mm) and the climate model (E-OBS) of accumulated precipitation (mm) for the period 2008–2020, with its linear trend line $Y = 0.4329x + 40.919$ and its respective $R^2 = 0.176$, which in a way validates the data to be used for the future period.

Figure 3 shows the dispersion between the energy production supply (GWh) and the climate model (E-OBS) of accumulated precipitation (mm) for the period 2008–2020 with monthly data. Several regressions were calculated, with exponential trend lines, power and polynomial of up to six classes, but no improvement was observed and therefore we opted for a simpler study with a linear trend line with the equation $y = 3.1449x + 776.11$ and its corresponding $R^2 = 0.0948$. The data are not very similar to the trend line because there are other variables, such as the management of water reserves through reversible cycle plants in periods of drought, which affect electricity production, and which are not considered in this study.

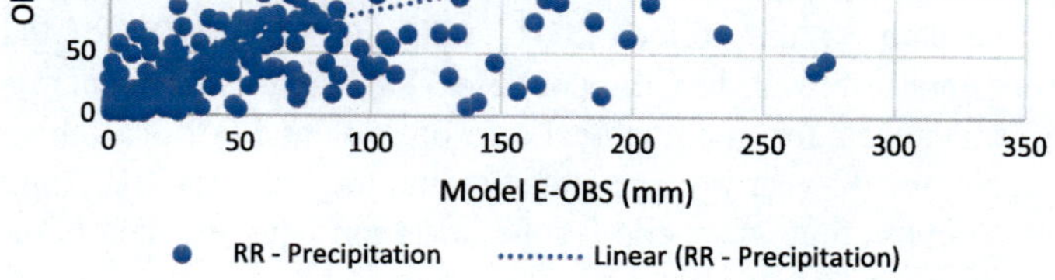

Fig. 2 Regression between the Observations (mm) and the E-OBS of the accumulated precipitation (mm) between the years 2008 to 2020 with monthly resolution

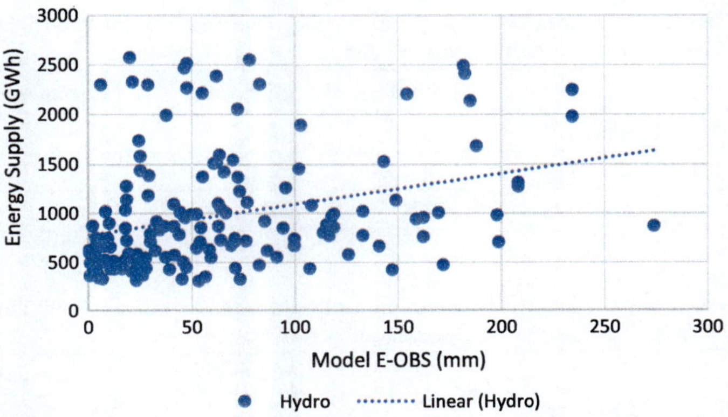

Fig. 3 Regression between the Energy Supply (GWh) and the E-OBS data of accumulated precipitation (mm) between the years 2008–2020 with monthly resolution

Using the regression equation in Fig. 3 ($Y = 3.1449x + 776.11$) and the data from the climate model for future accumulated precipitation (RCP8. 5), we can characterize the electricity production for the period 2041–2070, with a maximum production of 1534 (GWh), a minimum production of 714 (GWh) and an average of 978 (GWh), and for the period 2071–2100, with a maximum production of 1558 (GWh), a minimum production of 681 (GWh) and an average of 958 (GWh). A comparison with the data for the period 2008–2020 shows a maximum production of 2574 (GWh), a minimum production of 301 (GWh) and an average of 977 (GWh).

For the best analysis of the results, the elements have been divided into classes, for the period 2008–2020, the mean of the water production and its respective standard deviation have been calculated and for class 1 the mean plus 25% of the standard deviation has been used, giving a value of 828 (GWh) and for class 3 the mean plus 25% of the standard deviation has been used, giving a value of 1127 (GWh). Class 2 is characterized by the range between classes 1 and 3, i.e. between 828 (GWh) and 1127 (GWh), as shown in Table 1.

After defining the classes for the period 2008–2020, the same classes were used to evaluate the future period, thus defining the number of elements in each class. As can be seen in Table 2 for the future period 2041–2070, we have a reduction of up to 11% in class 3, which represents the maximum production of hydro energy.

Table 1 Class division of the electric energy production of the water variable (GWh) for the period 2008–2020 with its respective elements and percentage

		(GWh)	Elements	Percentage (%)
2008–2020	Class 1 <	828	81	52
	Class 2 =	828–1127	33	21
	Class 3 >	1127	42	27

Table 2 Class division of the electric energy production of the water variable (GWh) for the future period 2041–2070 with its respective elements and percentage

		(GWh)	Elements	Percentage (%)
2041–2070	Class 1 <	828	75	21
	Class 2 =	828–1127	220	61
	Class 3 >	1127	65	18

Table 3 Class division of the electric energy production of the water variable (GWh) for the future period 2071–2100 with its respective elements and percentage

		(GWh)	Elements	Percentage (%)
2071–2100	Class 1 <	828	89	25
	Class 2 =	828–1127	217	60
	Class 3 >	1127	54	15

By comparison, Table 3 shows the future period 2071–2100, where a reduction of up to 12% in the maximum hydro energy production can be observed. In addition, there is an increase in the percentage of class 2, which represents the average production of hydro energy. This situation is because the data are not highly correlated (because hydro energy depends on variables other than precipitation) and therefore the statistical model converges to the mean.

There is a large concentration of data between classes 1 and 3, which would be the class 2 elements. There is a small concentration of elements higher than class 3 and lower than class 1, confirming the reduction of the future supply of hydropower to its maximum potential.

5 Conclusions

Considering this study, it could be verified that there is an influence of climate change on renewable energy production. Renewable energies from solar radiation and wind power are studied, but for this article, considering only hydro energy, it was found that in the scenario considered, there is a trend of decrease in precipitation accumulated over the years until 2100. For the period 2071–2100, there was a 12% decrease in the maximum supply of hydroelectric energy production.

There are more detailed ways to study the future of climate change and its influence on the energy matrix of renewable sources, but there are no detailed data on energy production at a daily and regional level in mainland Portugal, although this study has provided a very relevant initial conclusion.

Greater investment in the integration of renewable energy sources may be symbolic, but it is crucial to draw up contingency plans for possible crises in the future. This critical thinking is essential to mitigate the impact felt in many other sectors.

Finally, incorporating the predictability of future weather events into the planning of actions taken today in order to optimize financing processes is a major but very necessary challenge, and without more ambitious policies, for the use of renewable sources to integrate electricity supply systems, a false perception will be created and the potential of renewable resources may be obscured. The analysis of climate projections is therefore a tool to provide answers that are increasingly close to reality.

References

1. Performance Indexes Analysis of the Reservoir-Hydropower Plant System Affected by Climate Change- Author Parvin Golfam1.
2. IPCC, Homepage, https://www.ipcc.ch/, last accessed 11.12.2022.
3. J Nuno Fidalgo, Débora de São José.: Impact of Climate Changes on the Portuguese Energy Generation Mix (2022).
4. Maurício T. Tolmasquim.: Renewable energy hydraulic, biomass, wind, solar, oceanic (2016).
5. Maria La Salete Ferreira Sousa.: The problem of strategy in the renewable energy sector in Portugal.
6. Assessing the Future wind Energy Potential in Portugal Using a CMIP6Model Ensemble andWRF High-Resolution Simulations – Author André Claro.
7. Hydropower Contribution to the Renewable Energy Transition Under Climate Change – Author José Pedro Ramião.
8. Express, https://expresso.pt/economia/economia_energia/2023-01-19-Producao-hidroeletrica-em-Portugal-alcancou-novo-maximo-historico-a-4-de-janeiro-92952dc5, last accessed 02/18/2023.
9. José Pedro Ramião.: Hydropower Contribution to the Renewable Energy Transition Under Climate Change Water Resources Management.
10. Edivaldo Afonso de Oliveira Serão.: Climate and land use change: future impacts on hydropower and revenue for the amazon.
11. COPERNICUS, Homepage, https://surfobs.climate.copernicus.eu/dataaccess/access_eobs_chunks.php, last accessed 11.15.2022.
12. Homepage Climate Portal, http://portaldoclima.pt/pt/, last accessed 01/15/2023.

Potential for Hydrogen Production Associated to Water and Food in Off-Grid Communities of Southern Africa

Luiz Rodrigues, Luis Araujo, A. J. Gano, P. J. R. Pinto, S. G. Simões, Paulo Brito, Armindo Monjane, and C. M. Rangel

Abstract Solar energy is called to meet electricity demands for isolated, off-grid communities in Africa. However, solar electric energy is intermittent and can be stored, for a limited amount of time, in batteries, which are expensive and cause serious environmental impacts at the end of their lifetime. Conversion of the surplus electric energy to green hydrogen through water electrolysis and back to electricity, when needed, using electrolyser-fuel cells systems, is examined as a potential solution to meet the water-energy-food nexus in Southern Africa. In the framework of the Agrivoltaics concept, the main constrains, opportunities and parameters to consider its applicability are presented and discussed, in terms of its technical, economic, environmental and social impacts. In the second phase of this work the developing of a PEM electrolyser for connection to a solar PV power source and a fuel cell device, for a stand-alone application, is proposed, ensuring high reliability and energy conversion efficiencies, as well as adequate transient response and a competitive cost. It is intended as a low-carbon energy system, realising the potential for synergy in the Agrivoltaic concept, aligned with global and regional sustainability goals.

Keywords Solar hydrogen production · Off-grid autonomous hydrogen systems · PEM electrolysis · Fuel cells · Agrivoltaic

---

L. Rodrigues
Portalegre Polytechnic Institute, Portalegre, Portugal
e-mail: luiz.rodrigues@ipportalegre.pt

L. Araujo
Universidade Agostinho Neto, Luanda, Angola

A. J. Gano · P. J. R. Pinto · S. G. Simões · C. M. Rangel (✉)
National Laboratory for Energy and Geology, Lisbon, Portugal
e-mail: carmen.rangel@lneg.pt

P. Brito
VALORIZA—Research Center for Endogenous Resource Valorization, Portalegre, Portugal

A. Monjane
Universidade Pedagógica de Maputo, Maputo, Mozambique

J. R. da. C. S. Galvão et al. (eds.), *Proceedings of the 3rd International Conference on Water Energy Food and Sustainability (ICoWEFS 2023)*, Springer Proceedings in Earth and Environmental Sciences, https://doi.org/10.1007/978-3-031-48532-9_31

1 Context and Potential for Hydrogen Production

Subsistence agriculture and pastoralism are within the main economic activities in Southern Africa, particularly in off-grid communities. However, both activities are quite endangered by the growing desertification that the region is facing because of climate change, that leads to the scarcity of water and rainfall, essential for both activities. A virtuous combination of photovoltaic solar energy production and agriculture, Agrivoltaic, appears with an enormous potential to address these issues, including the use of photovoltaic in water irrigation systems [1, 2]. In fact, although photovoltaic electricity generation is seen as a source of clean energy, it demands extensive soil occupation, which might be in competition with arable soil [1]. The use of soils underneath the solar panel to perform activities related to agriculture and/or pastoralism, allows a better use of both resources, soil and sun, and synergistic gains: soil productivity increases due to the reduction of water evaporation, and energy conversion efficiencies increase due to refreshment of solar panels [2].

The relationship between climate change and increasing desertification has been recognized for a long time in vast regions of southern Africa [3–5]. The predictions of the United Nations (UN) Intergovernmental Panel for Climate Changes (IPCC) points out, for the southern Africa region, a significant drought in the dry season months, but no significant differences are projected in the rainfall for the wet season and no increased risk of drought events are reported [3]. However, despite the remaining uncertainty, combined evidence of current trends (increased rainfall variability and intensity of extreme events, including drought) with global climate model predictions of warming and drying across the region, add significant concerns that climate change will exacerbate land degradation and desertification [3]. Most of southern Africa is dry with assessments of the desertification potential pointing that all Namibia and Botswana, and in fact more than half of South Africa is rated as potential desert [5]. Moreover, according do United Nations Environmental Program (UNEP), one-third of Africa is affected by desertification and 73% of the agriculturally used lands are degraded. In most southern Africa drylands, the failure of any seasonal rain in a year result in social and economic difficulties because most population is rural, depending on subsistence farming and/or direct exploitation of vegetative resources [4]. Indeed, two or three successive years of drought conduct to severe environmental stress that often culminates in serious food shortages, famine, and loss of both human and animal life. Behind this problem, the Southern Africa Development Community (SADC), that includes as member countries Tanzania, Zambia, Zimbabwe, Namibia, Angola, Botswana, Lesotho, Eswatini, Malawi, Mozambique, South Africa, Madagascar, Mauritius, and Democratic Republic of Congo, formulated and worked out mechanisms of implementing Regional and National Action Programs to correspond to the 1994 Convention to combat desertification [4].

On the other hand, it is well known that access to clean water, healthy food, cooking facilities and energy, particularly in the form of electricity, are still important aspects that need immediate attention throughout the continent [6].

According to the International Energy Agency (IEA), in 2019 only around 56% of the population of Africa had access to electricity, and the situation is even worse in rural areas, with only 37% compared with 81% in urban areas [7]. In Sub-Saharan Africa the average rural electricity access rate is approximately 25% [8], and in South Africa, for example, around 1.3 million households have no access to grid connection [9]. Remote off-grid localities may benefit from electrification of the different activities, namely those related to agriculture, pastoralism, and small-scale industrial tasks, or in schools and clinics [9, 10], since the extension of the electricity grid may be simply prohibitive.

Besides, it is recognized that Africa has an abundance of renewable energy sources that can help the continent to meet its needs for electricity, to promote the needed economic growth while meeting the global targets for carbon dioxide emissions reduction [11]. Therefore, Africa is also a suitable place to produce green hydrogen, i.e., hydrogen produced through water electrolysis using electricity from renewable sources, since the needed electricity may be generated at low cost from abundant renewable sources, in particular solar energy, see Fig. 1. Due to the relatively high solar radiation potential of large extensions of the African territory, solar energy is seen as the main source for the electrification of the entire continent and to produce green hydrogen for auto-consumption and exportation [6]. Hydrogen may play an important role in this context of renewable energy harnessing [11, 12] and in the decentralized generation of electricity, particularly in remote off-grid African communities, since it can be produced and consumed locally, compensating for the fluctuations related to the availability of renewable sources [10, 12].

To ensure a sustainable hydrogen economy in Africa it is also vital to address water sources for hydrogen production via electrolysis, considering the whole dimensions of the Water-Energy-Food Nexus. Furthermore, a coherent policy and regulatory framework is necessary to harvest synergies between the goals of energy poverty alleviation, food provision, decarbonisation and safeguarding water resources and ecosystems.

2 Projects and Strategies

Despite some major green hydrogen projects foreseen for some African regions, mainly confined to the North, West, Namibia, and South Africa [6, 11, 13–15], especially pointed to the exportation to Europe [16], only a reduced number of African countries have designed, till now, a comprehensive strategy to combine renewable energy harnessing with green hydrogen and synthetic gases production. In fact, South Africa established its HySA (Hydrogen South Africa) strategy in 2008 to take better advantage of the country's huge platinum group metal (PGM) resources, providing a clear direction for effective deployment of hydrogen and fuel cell technology, and to produce green hydrogen from clean and renewable resources for meeting the electricity and transportation gap in the country, especially in the rural settings [9,

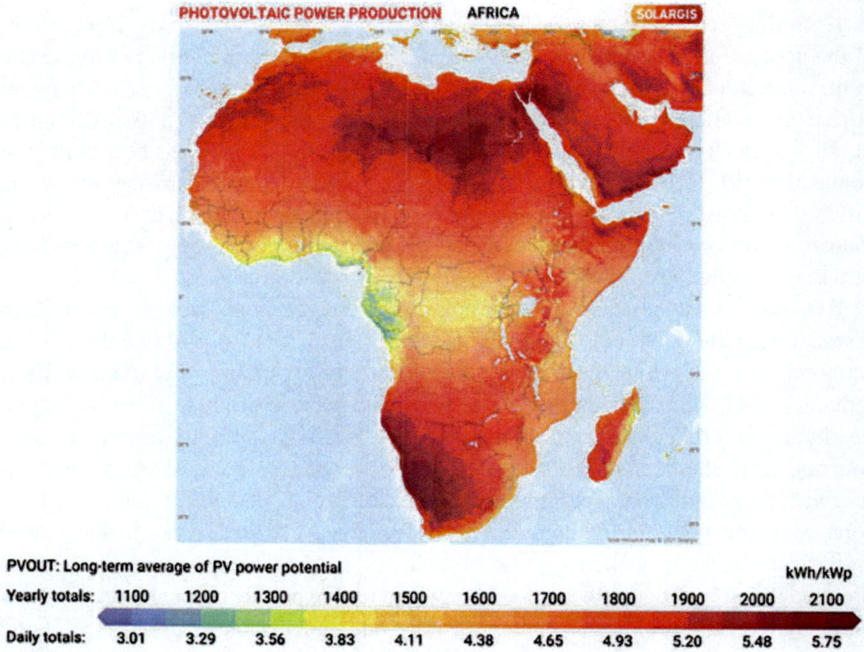

Fig. 1 Africa long-term average of PV power potential, period 1994–2020. Map © 2021 Solargis

11, 12]. South Africa has indeed announced plans to support a pipeline of green H_2 worth about US$17.8 billion, over the next decade [17].

In Egypt, an agreement between the Egypt Sovereign Fund and two foreigner companies was reached for the development of 50–100 MW electrolysers to produce green hydrogen, and lately, green hydrogen was even introduced in the Egypt's 2035 Energy Strategy [11]. In 2020, a partnership between Moroccan and Germany governments was launched to improve the production of green hydrogen through power-to-X and power-to-power type projects [11]. In Nigeria, short-, medium- and long-term strategies were designed to facilitate the implications of green hydrogen energy in the country [11]. In Uganda a project was implemented to store overabundant energy generated by solar panels through the production of green hydrogen, to warrant constant supply of electricity at night [11].

The lack of strong hydrogen energy policies and strategy in Southern Africa certainly hinders the region vision for a green hydrogen economy. The recent project regarding the production of a H_2 Atlas-Africa, a joint initiative between the German Federal Ministry of Education and Research and African partners in the Sub-Saharan region, namely, SADC and ECOWAS, provides a decision support tool for the concept and design of pilot hydrogen plants. It focusses on detailed technological, environmental, economic, and social feasibility assessment, taking present and future local energy demands into consideration and assessed the availability and suitability of land and water resources, while considering land use for agriculture and local demand for water across a total of 31 sub-Saharan countries [9, 11], among them 12 from the SADC region [18]. H_2 Atlas Africa intends to map out these locations on an interactive atlas to show hydrogen production hotspots, and demonstrate through pilot projects how the production, distribution and export of green hydrogen can be performed in an economically efficient manner [9].

A recent study into Africa's green hydrogen potential, commissioned by the European Investment Bank, International Solar Alliance and the African Union highlights the benefits of harnessing solar power to generate green H_2 in four African hubs: Mauritania, Morocco, Southern Africa, and Egypt [19]. The study concludes that Africa can secure access to clean and sustainable energy, accelerate low-carbon economic growth across the continent, reduce greenhouse gas emissions by 40%, and become a global energy player through green H_2 exports. According to this study, Africa may produce 50 million tons of green H_2 per year around 2035 through solar energy, supplying about 7 EJ of energy, more than 1/3 of the 19.9 EJ consumed across the continent in 2021.

Imasiku et al. [18] reviewed the policy for green H_2 in the southern Africa. Besides the already referred South African HySA strategy for H_2, the review mentions an additional restricted action plan in Namibia that includes the production of green H_2 and ammonia. Some green and grey H_2 productions were identified in Zimbabwe and South Africa related to the production of fertilizers and fossil fuels that may contribute to foster the green H_2 industry development in both countries. The authors concluded that most of the green H_2 projects implemented in southern Africa are either privately founded or funded by international players [18].

Within the Hydrogen energy policies and strategy in southern Africa apart from support to projects, investment in R&D and capacity building is required, aligned with the demonstration of commercial viability of hydrogen-agriculture-photovoltaic systems as a potential solution to meet the water-energy-food nexus in southern Africa, in order to stimulate innovation and achieve a cost-competitive hydrogen value chain, in the path towards a future carbon–neutral economy.

3 The Agrivoltaics Concept

It is well known that photovoltaic solar energy capturing demands for a large share of land occupation [2], and with the mentioned foreseen megaprojects for green hydrogen production in Africa, mainly based on electricity generated from sun energy, the land occupation problem will be intensified, certainly conflicting with food production [1]. Land is, indeed, a vital asset for any economy based on agriculture such as most of the southern Africa countries [1]. Agrivoltaic involves a compromise between agriculture and photovoltaic energy generation that contributes to the increase of the Land Equivalency Ratio (LER). The system, known also as "agrophotovoltaics" (APV), "solar sharing", or "PV agriculture", provides mutual benefits across the food-energy-water nexus [1]. Rainwater harvesting from PV systems may bring improved access to useable water, along with the diversion of the same water for PV surface cleaning and irrigation. Reduction in water extracted for irrigation can also support improvements in environmental outputs and input cost management. On the other hand, this approach provides off-grid farms with energy for productive uses such as pumping water, operating small-scale machinery, raising poultry, and producing fodder, as well as for value adding processes, such as grinding, drying, packing, etc. [20].

Despite these advantages and its rapid increase in popularity in many countries, Agrivoltaic systems have not yet gained the need traction in Africa [20]. In fact, only few projects and studies have been till now undertaken to explore or assess the benefits of Agrivoltaic systems in Africa. A research project is underway in Ghana to develop an Agrivoltaic system to bolster the national resilience of renewable energy and food production security to a changing climate [21]. The economic potential of agrophotovoltaics in South Africa was assessed through the case study of a grape farm in West Cape [22]. The results show an optimistic potential of implementing APVs in South Africa to help to target the 6000 MW solar energy increase by 2030 without giving up land for agriculture and minimizing other land conflicts related to the acquisition of land for energy production.

The economic feasibility of Agrivoltaic systems in Food-Energy-Nexus context was modelled and assessed through a case study in Niger [6, 23]. Croplands, grasslands, and wetlands in Southern Africa are considered among the best places for capturing solar energy using photovoltaic panels (PVP) since they combine coolness with high irradiance, and these are places where typical crops would normally grow [24]. But even in arid and semiarid regions Agrivoltaic may be applied to cultivate certain crops that would not normally grow there, PVPs promotes shading and moisture preservation.

It is shown that significant quantities of food and jobs may be created in Africa if Agrivoltaic systems can increase their surface area by securing large energy offtakers, whether in the form of off-grid industrial clients, via on-grid feed-in tariff schemes, or large-scale PV-to-fuel intercontinental export operations expected for the next years [24].

Even though Agrivoltaic may substantially has the potential to increase land use efficiency, technical and economic viabilities of the integrated concept need to be demonstrated in rural communities, moreover with added storage capability using hydrogen.

4 Hydrogen in Stand-Alone Energy Systems

Hydrogen stand-alone energy systems will be discussed in the context of the proposed application (Fig. 2). The interest on the concept of integrating renewable energy sources with hydrogen storage systems, for these applications, has increased in the past decade. Results have been reported in the literature focusing on system performance and viability [25, 26], simulation studies centring on system energy management, control strategies, sizing and modelling [27–32].

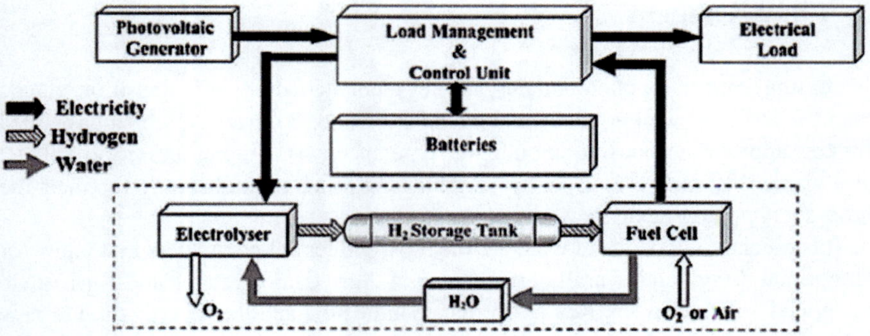

Fig. 2 Schematic diagram showing stand-alone system with flows of electricity, water, and hydrogen [35]

With the increasing share of renewable energy sources, new requirements to access less costly electrical energy feedstock are emerging, and the markets for electrolytic hydrogen require more flexible electrolysis plants, which call for electrolysers to be operated dynamically. With the inherent variability and intermittency of the sources, flexibility, and reactivity to changes in the input conditions and loading effects are paramount. Within those with high readiness level, the PEM electrolysis technology is considered the one that better supports transient part load transitions, when connected with renewable energy [33, 34].

Regarding the numerous end uses of hydrogen, the re-electrification pathway by employing a fuel cell will be discussed as well as the advantages that support the use of green hydrogen for isolated locations. Hydrogen-based power systems will be compared against main competitors such as batteries and diesel generators, which, presently, are the conventional options in off-grid scenarios. The system will be optimized for sizing and power control strategy, with the assistant of smart management systems that will consider not only the production variability but also consumption estimations from the stand-alone profiles. The influence of the location and the impact of different load energy consumption profiles will be explored. Optimal configurations will be also ascertained, also from an economic viewpoint.

The main barrier to implementation of hydrogen-based technologies is the high cost of the system and the scarce availability of commercially competitive off-grid electrolysers, especially incorporating designs that work under variable renewable power inputs.

System characteristics and advantages for the envisaged application will be discussed.

5 Final Remarks

Increasing demand for photovoltaics, as a key element in the energy transition, entails land use issues as well as concerns on landscape and biodiversity, which has called for new approaches, considering integrated perspectives. Among these, Agrivoltaic, as a viable dual-land use, appear very promising as a potential solution to meet the food-energy-water nexus.

It is crucial to understand what are the most influential parameters that allow for an optimal design and overall performance of Agrivoltaic systems and Agrivoltaic integrated with hydrogen. Key performance indicators and the investigation of new innovative models, able to include additional activities, community functions and tasks, that may be developed in the defined context, are being addressed.

Inherent variability and intermittency of the renewable power sources require electrolyser's flexibility and reactivity to changes in the input conditions and loading effects. There is a lack of systematic studies to quantify the impact of operating parameters linked to the durability of critical components in the hydrogen production system. Understanding of degradation mechanisms, namely under dynamic operation, will be tackled.

Apart from the technical challenges, the economic feasibility of the proposed systems and expected cost-competitiveness along the value chain are to be demonstrated. Capacity building is required and is identified to be implemented at various levels, in order to establish the basis for educational and training strategies to ensure specialist skills and knowledge can be acquired in a timely manner. Also, to be considered, is the fact that implementing a new technology on agricultural land, which is highly valued, involves law and planning challenges.

The integration of hydrogen and its storage capability with Agrivoltaic in stand-alone energy systems have yet to demonstrate their significant potential to enable the transition to clean energy in off-grid locations, ensuring climate change mitigation and energy supply reliability.

Acknowledgements Acknowledgements are due to the HYLANTIC Project_Atlantic Network for Renewable Generation and Supply of Hydrogen to promote High Energy Efficiency EAPA_204/2016, co-financed by the European Regional Development Fund in the framework of the Interreg Atlantic Programme.

References

1. M.A. Al Mamun, P. Dargush, D. Wadley, N.A. Zulkarnain, A.A. Aziz, "A review of research on Agrivoltaic systems", Renewable and Sustainable Energy Reviews, 161, 112351 (2022)
2. A. Agostini, M. Colauzzi, S. Amaducci, "Innovative Agrivoltaic systems to produce sustainable energy; An economic and environmental assessment", Applied Energy, 281, 116102 (2021)
3. L.C. Stringer, J.C. Dyer, M.S. Reed, A.J. Dougill, C. Twyman, D. Mkwambisi, "Adaption to climate change drought and desertification: local insights to enhance policy in southern Africa", Environmental Science Policy (2009). https://doi.org/10.1016/j.ensci.2009.04.002.

4. J.P. Msangi, "Drought hazards and desertification management in the drylands of Southern Africa", Environmental Monitoring and Assessment, 99, 75–87 (2004)
5. P.D. Tyson, "Climate and desertification in Southern Africa", GeoJournal Suplementary Issue, 2, 3–10 (1981)
6. R. Bandhari, "Green hydrogen production potential in West Africa – Case of Niger", Renewable Energy, 196, 800–811 (2022)
7. IEA (2020). World Energy Outlook 2020 – Analysis. https://www.iea.org/reports/world-energy-outlook-2020.
8. J. Macdonald, L. Probst, J.R. Cladera, AIP Conference Proceedings 2635, 070002 (2022); https://doi.org/10.1063/5.0105526 Published Online: 06 December 2022
9. D. Bessarabov, G. Human, A. J. Kruger, S. Chiuta, P. M. Modisha, S. P. du Preez, S. P. Oelofse, I. Vincent, J. Van Der Merwe, H. W. Langmi, J. Ren, N. M. Musyoka, "South African hydrogen infrastructure (HySA infrastructure) for fuel cells and energy storage: Overview of a projects portfolio", International Journal of Hydrogen Energy, 42, 13568–13588 (2017)
10. E.K.P. Augusto, "Projecto de Pilha de Combustível de 30 kW para abastecer uma zona remota", Tese de Mestrado e Catálise e Energias Renováveis, Universidade Agostinho Neto, Luanda, Angola, 2020
11. N. AbouSeada, T.M. Hatem, "Climate Action: Prospects of green hydrogen in Africa", Energy Reports, 8, 3873–3890 (2022)
12. T.R. Ayodele, J.L. Munda, "Potential and economic viability of green hydrogen production by water electrolysis using wind energy resources in South Africa", International Journal of Hydrogen Energy, 44, 17669–17687 (2019)
13. Reuters, "Africa´s first hydrogen power plant seen producing electricity in 2024", September 12, 2022 (accessed through https://www.reuters.com/business/energy/africas-first-hydrogen-power-plant-seen-producing-electricity-2024-2022-09-12/ at February 12, 2023)
14. S. Timmerberg, M. Kaltschmitt, "Hydrogen from renewables: Supply from North Africa to Central Europe as blend in existing pipelines – Potential and costs", Applied Energy, 237, 795–809 (2019)
15. M.D. Mukelabai, U.K.G. Wijayantha, R. E. Blanchard, "Renewable hydrogen economy outlook in Africa", Renewable and Sustainable Energy Reviews, 167, 112705 (2022)
16. C. Chance, "Focus on hydrogen: a new energy frontier for Africa", January 2021, 1–10 (accessed through https://www.cliffordchance.com/content/dam/cliffordchance/briefings/2021 /01/focus-on-hydrogen-a-new-energy-frontier-for-africa.pdf at February 12, 2023)
17. P. Vecchiatto, "South Africa in talks with investors on green hydrogen projects", Bloomberg, February 11, 2022 (accessed through https://www.bloomberg.com/news/articles/2022-02-10/south-africa-in-talks-with-investors-on-green-hydrogen-projects?leadSource=uverify%20wall on February 22, 2023)
18. K. Imasiku, F. Farirai, J. Olwoch, S.N. Agbo, "A policy review of green hydrogen economy in Southern Africa", Sustainability 13, 13240 (2021).
19. European Investment Bank, "New study confirms €1 trillion Africa's extraordinary green hydrogen potential", 21 December 2022, 1–4
20. A.E. Cheo, N. Adelhardt, T. Kriger, J. Berneiser, F.A.S. Santillano, B. Bingwa, N. Suleiman, P. Thiele, A. Royes, D. Gudopp. A. Sidibe, K.M.T.M. Fahmy, E.G. Tambo, Y. Diallo, B. Sogouba, "Agrivoltaic across the Water-Energy-Food-Nexus in Africa: Opportunities and challenges for rural communities in Mali", *Research Square*, 1–15, https://doi.org/10.21203/rs.3.rs-1503422/v1.
21. F. Kemausour, W. Amponsah, "Agrivoltaic technology in drylands of West Africa: Strengthening National Innovation System for diffusion and market development at the Water-Energy-Food Nexus", Presentation at the 7th. Ghana Renewable Energy Fair, October 13, 2021
22. M.B. Barantandikiye, "Exploring the economic potential of agrophotovoltaics in South Africa: Case of a grape farm in Worcester, Wester Cape", International MSc. Degree in Science in Rural Development Thesis at Ghent University, a.o., 2022

23. S.N. Bhandari, S. Schlüter, W. Kuckshinrichs, H. Schlör, R. Adamou, R. Bhandari, "Economic feasibility of agrivoltaic systems in Food-Energy-Water Nexus context: Modelling and case study in Niger", Agronomy 11, 1906 (2021). https://doi.org/10.3390/agronomy11101906.
24. R.J. Randle-Boggis, E. Lara, J. Onyango, E.J. Temu, S.E. Hartley, "Agrivoltaics in East Africa: Opportunities and Challenges", AIP Conference Proceedings 2361, 090001 (2021); https://doi.org/10.1063/5.0055470 Published Online: 28 June 2021
25. Bergen, L. Pitt, A. Rowe, P. Wild, N. Djilali, "Experimental assessment of a residential scale renewable–regenerative energy system", Journal of Power Sources, 186, 158–166 (2009)
26. L. Gracia , P. Casero, C. Bourasseau and A. Chabert, "Use of hydrogen in off-grid locations, a techno-economic sssessment", Energies, 11, 3141 (2016)
27. Y. Kalinci, A. Hepbasli, I. Dincer, "Techno-economic analysis of a stand-alone hybrid renewable energy system with hydrogen production and storage options", International Journal of Hydrogen Energy, 40, 7652–7664 (2015)
28. F. Khalid, M. Aydin, I. Dincer, M.A. Rosen, "Comparative assessment of two integrated hydrogen energy systems using electrolyzers and fuel cells". International Journal of Hydrogen Energy, 41, 19836–19846 (2016)
29. D. Ipsakis, S. Voutetakis, P. Seferlis, F. Stergiopoulos, C. Elmasides, "Power management strategies for a stand-alone power system using renewable energy sources and hydrogen storage", International Journal of Hydrogen Energy 34, 7081–7095 (2009)
30. C. Darras, S. Sailler, C. Thibault, M. Muselli, P. Poggi, J. C. Hoguet, S. Melscoet, E. Pinton, S. Grehant, F.Gailly, C. Turpin, S. Astier, G. Fontès, "Sizing a photovoltaic system with hydrogen/oxygen storage based on the ORIENTE model", International Journal of Hydrogen Energy, 35, 3322–3332 (2010)
31. S.S. Deshmukh, R.F. Boehm, "Review of modeling details related to renewably powered hydrogen systems", Renewable and Sustainable Energy Reviews, 12, 2301–2330 (2008)
32. S. Pedrazzi, G. Zini, P. Tartarini, "Complete modeling and software implementation of a virtual solar hydrogen hybrid system", Energy Conversion and Management, 51, 122–129 (2010)
33. Buttler and H. Spliethoff, "Current status of water electrolysis for energy storage, grid balancing and sector coupling via power-to-gas and power-to-liquids: A review", Reneweable and Sustainable Energy Review, 82, 2440–2454 (2017)
34. PEM Water Electrolysis - Volume 2, D. Bessarabov, P. Millet, Academic Press (Elsevier), 2018, ISBN: 978–0–08–102830–8
35. Djafour, M.S. Aida, and B. Azoui. "Photovoltaic assisted fuel cell power systems", Energy Procedia, 50, 306–313 (2014)

Sustainability

The following part contains papers related to "Sustainability".

The editors had assistance from the following conference organizing committee members in the editing of this part:

- Flávio Gabriel da Silva Craveiro
- Joel Oliveira Correia Vasco

Analysing Vegetation Cover Changes in Protected Areas: An Example from Lithuania

Miguel Inácio and Paulo Pereira

Abstract Protected areas are implemented to safeguard natural and cultural values. Their level of protection largely dictates which socio-economic activities are allowed. Protected areas where nature protection objectives co-exist with socio-economic activities are exposed to their potential anthropogenic impacts, such as land use changes. Land use changes are an important driver of land degradation in protected areas. This study analyses the potential impact of land use changes in Lithuanian Regional and National Parks by analysing changes in vegetation cover. In a multi-temporal analysis, this was done by calculating the Normalised Difference Vegetation Index (NDVI), a proxy commonly applied to infer vegetation change. NDVI values above 0.66 represent dense vegetation, which we selected to be analysed. NDVI was calculated using Landsat Imagery in a fiver year interval from 1990 to 2020. The analysis was done in Google Earth Engine and ArcGIS 10.8 platforms. The results show a general increase in dense vegetation cover from 1990 to 2020, although at different magnitudes in each park. Changes in comparison to 2020 were analysed, and the results suggest that until 2010/2015, there was an increase in dense vegetation cover. After this, most parks show stagnation or even a decrease. Principal analysis showed that most parks were clustered except Kauno Marios, Pajuris, Kursiu Nerija and Dzūkija parks.

Keywords NDVI · Dense vegetation · Landsat · Remote sensing · Google earth engine · Protected areas

M. Inácio (✉) · P. Pereira
Environmental Management Laboratory, Mykolas Romeris University, Ateities G. 20, 08303 Vilnius, Lithuania
e-mail: miguel.inacio@mruni.eu

J. R. da. C. S. Galvão et al. (eds.), *Proceedings of the 3rd International Conference on Water Energy Food and Sustainability (ICoWEFS 2023)*, Springer Proceedings in Earth and Environmental Sciences, https://doi.org/10.1007/978-3-031-48532-9_32

1 Introduction

Land use changes are recognised as one of the most critical anthropogenic drivers of environmental degradation and biodiversity loss [1]. More than 32% of the world's terrestrial ecosystems have experienced land use alteration to a certain degree from 1960 to 2019 [2]. Protected areas have been established to safeguard natural and cultural values worldwide [3]. It is estimated that around 16.64% of the earth's surface (terrestrial and marine) is protected by one or several protection schemes [4]. The advantages of establishing protected areas are widely recognised, reinforcing the narrative that more protection is needed, especially in areas exposed to multiple anthropogenic drivers of change [4].

Global agendas, like the Convention for Biological Diversity, define nature protection targets to be achieved. Other agendas, like the United Nations Sustainable Development Goals, aim at ensuring nature's sustainability in the long run. United Nations Educational, Scientific and Cultural Organization (UNESCO) identifies areas of cultural heritage that must be preserved. In Europe, e.g., the Biodiversity 2030 or Green Deal Strategies aim to restore the ecosystems and their services and ensure the sustainable use of natural resources. European Member states have a long tradition of establishing protected areas, e.g., under the Natura 2000, Habitats Directive, Ramsar frameworks, and UNESCO. According to the European Environmental Agency, 26% of European ecosystems are under protection. However, despite all efforts to avoid the influence of anthropogenic drivers, protected areas are also subject to environmental degradation [5]. Protected areas have different conservation objectives, in which the level of protection is defined. Areas with stricter protection tend to allow very few or no anthropogenic activities, while others allow for the development of socio-economic activities. Hence, the allowed socio-economic activities can negatively impact the environment in the latter's case. Several authors [6, 7] highlight the role of anthropogenic activities like deforestation and agriculture expansion in compromising the environmental integrity of protected areas. Thus, assessing the impacts of these socio-economic activities within the protected areas is necessary to ensure its long-term protection objectives.

Lithuania's network of protected areas includes areas where socio-economic interactions are allowed. However, to the best of our knowledge, a study has yet to be conducted in Lithuania to analyse the impacts of socio-economic activities within protected areas. Thus, this study's main objective is to assess changes in forest cover, using the Normalised Difference Vegetation Index as a proxy.

2 Materials and Methods

2.1 Study Area

Lithuania is one of the three Baltic States located in north-eastern Europe. It covers an area of 63,500 km^2 and has a population of around 2.8 million inhabitants [8]. About 68% (1,913,385) of the population lives in urban areas, while 32% (892,613) live in rural areas. Thus, the population density is relatively low, with 44 inhabitants per km^2 [8]. The predominant land uses in Lithuania are *croplands* (around 50% of the territory), woodland and forest (35%), and grasslands (7%) [9]. Agriculture plays an important role in the country's economy.

Lithuania has a diversity of landscapes, ecosystems, and species. Nature is intrinsically embedded in the country's cultural heritage, traditions, and recreational activities. The importance of preserving natural habitats is of utmost importance, not only from an environmental perspective but also from a socio-economic standpoint. This has been ensured by establishing a network of protected areas that preserve cultural and natural heritage [10]. Currently, the network of protected areas is defined following an integrated approach. It is composed of *Territories of conservational preservation priority* (e.g., strict reserves and reserves), *Integrated protected areas* (e.g., regional and national parks), *Territories of ecological protection priority* (ecological protection zones), and *Territories of restorative protection priority* (e.g., recuperation plots). This network covers an area of approximately 11,837 km^2, corresponding to 18% of the country's area [10].

National and Regional Parks occupy circa 50% of Lithuania's total protection area and belong to the *Integrated protected areas* category, which according to the State Protected Areas Service [10], fuse nature protection and conservation with delineated areas for socio-economic activities. This means that these areas, although protected, are exposed to potential land-use changes and anthropogenic impacts such as deforestation and agricultural expansion. In total, five National Parks occupy an area of 1440.2 km^2, and 35 Regional Parks an area of 4489.5 km^2 (Fig. 1).

2.2 Methodology

To assess potential environmental degradation within Regional and National Parks, we analysed the dense forest cover per park from a multi-temporal perspective. For this, we used the proxy Normalized Difference Vegetation Index (NDVI), a commonly used indicator for identifying vegetation cover (e.g., [11]). The range of NDVI values, from −1 to 1, vegetation is represented by values above 0 [12, 13]. NDVI values from 0 to 1 allow the identification of the coverage of different vegetation densities. In this study, we are interested in analysing dense vegetation cover to represent the dense forest. According to Zaitunah et al. [14] NDVI values above 0.60 represent dense, healthy vegetation.

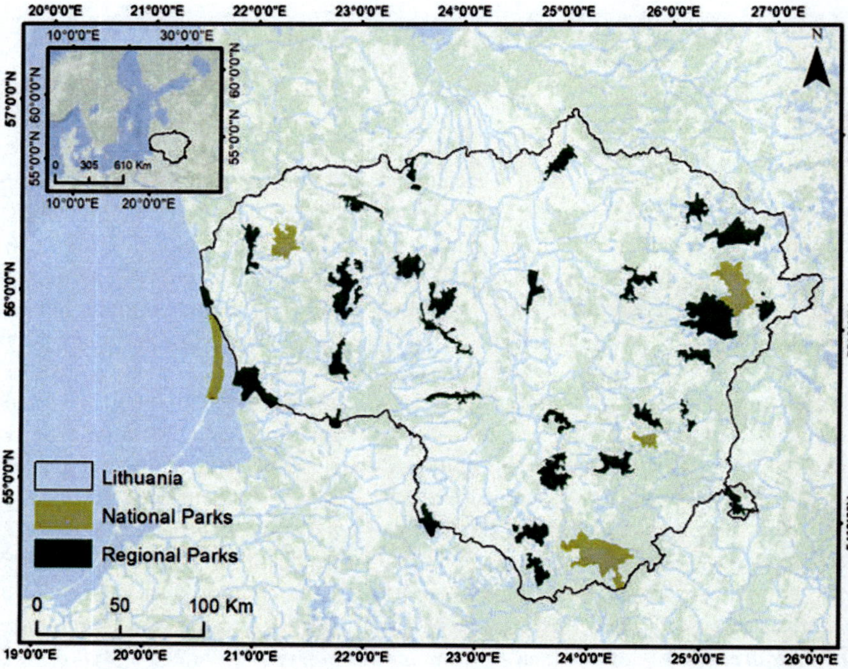

Fig. 1 Location of national and regional parks in Lithuania

We calculated NDVI for all Regional and National parks using Remote Sense Imagery to analyse changes in dense vegetation cover over time. A time span of 30 years was defined, from 1990 to 2020. The analysis was done following a 5-year gap; thus, the analysed years were: 1990, 1995, 2000, 2005, 2010, 2015, and 2020. To calculate NDVI from remote sensing, multispectral images are necessary. Based on the defined years of analysis, Landsat emerges as a suitable satellite to acquire images. Several Landsat missions were launched over the years. Our study utilises Landsat 5 Thematic Mapper (TM) and Landsat 8 Operational Land Imager (OLI). These satellites have a spatial resolution of 30 m. NDVI can be calculated using the Near Infrared (NIR) and Red (R) bands, based on formula (1):

$$\frac{(NIR - Red)}{(NIR + Red)} \tag{1}$$

For Landsat 5, NDVI is calculated using Band 4 (NIR) and Band 3 (R). For Landsat 8, NDVI is calculated based on Band 5 (NIR) and Band 4 (R). NDVI is a normalize index, therefore, it is unitless.

The acquisition and processing of Landsat imagery and calculation of NDVI were made using the Google Earth Engine (GEE) Platform [15]. GEE is a cloud-computing platform in which the whole repository of all Landsat mission imagery is available and allows the processing of large amounts of data. The platform works based on

JavaScript language and allows to import of shapefiles as assets and export files in multiple extensions. There are already pre-defined scripts within the GEE platform that can be freely used, for example, to mask clouds and apply scale factors to satellite images.

The multi-temporal calculation of NDVI for all parks was done based on several steps. The first step was to import the parks' polygons (shapefile), provided by the State Protected Areas Service, an asset to GEE. This asset was added to the script to represent our Region of Interest (ROI). The second step was to select the Landsat imagery from the GEE database. We chose Landsat Level 2 data (L2) dataset, which contains atmospherically corrected surface reflectance data. The third step was to import the cloud mask script available in the library of GEE (different Landsat 5 and 8). This script removes clouds and applies scaling factors to Landsat images. The fourth step was to import L2 data and filter it according to the defined criteria. We filtered images based on date to account for one whole year (e.g., from 01-01-1990 to 01–01-1991), months (from May to August), and cloud coverage (less than 10% over the ROI). The fifth step was to make a median composite of images per year and clip the image using the parks' polygons. The sixth step was calculating NDVI by applying the formula to the median composite and creating a threshold for values above 0.66. The seventh step was to generate pixel area values (originally in m^2), transform them to hectares, and, using zonal statistics, calculate the total area (sum) of NDVI values above 0.66 per analysed park. The last step in GEE was to export zonal statistics information as a shapefile. These steps were performed per analysed year. Due to the unavailability of satellite images to cover all parks, 2000 was replaced by 1999. For other years, the filtered months were extended to include September and the percentage of cloud cover.

In ArcGIS 10.8 [16], the zonal statistics shapefile information was extracted and processed in Microsoft Excel. The percentage of dense vegetation coverage per park (dividing NDVI area by the total area of each park) and its change over time were calculated. A principal component analysis (PCA) was applied based on the correlation matrix, and a varimax method was used to rotate the factors. The statistical analyses were carried out using Statistica 10.0 and the graphics using Plotly (https://chart-studio.plotly.com).

3 Results and Discussion

3.1 Dense Vegetation Cover Within National and Regional Parks

Regarding the percentage of dense vegetation cover per park, the results show an overall increase for all parks (Table 1). Hence, the magnitude of change is different among parks. Five parks (e.g., *Dubysos Regional Park* and *Tytuvėnai Regional Park*) show an increase of less than 20% regarding dense vegetation cover area between

1990 and 2020). More than half of the parks (18) show a 20% to 50% increase in dense vegetation cover area between 1990 and 2020) (e.g., *Aukštadvaris Regional Park* and Meteliai Regional Park). Eleven parks (e.g., *Neris Regional Park* and *Sirvėta Regional Park*) show a 50–70% increase. Finally, only one park (*Dzūkija National Park*) shows an increase above 80%. The park that registered the lowest increase (8%) was *Panemuniai Regional Park*. It is also important to highlight that despite an overall increase, several parks (e.g., *Sirvėta Regional Park*) show fluctuations between the years.

The general trends observed (the increase in dense vegetation cover) align with the timeline of establishing regional parks and implementing conservation priorities in national parks. Prior to the independence of Lithuania from the Soviet Union, important work on delineating and protecting territories was already a concern [10]. In 1992 the newly established Lithuanian Government approved the establishment of 30 Regional Parks [10, 17]. Thus, almost all parks show an increase in dense vegetation cover in the subsequent years, results in line with previous studies (e.g., [18, 19]). The fluctuations over the years can be explained by land use changes such as forest cuts by agroforestry industries or agricultural expansion [20]. Nevertheless, the positive trend in dense forest cover demonstrates the protection efforts imposed by protected areas.

3.2 Changes in Dense Vegetation Cover Within Regional and National Parks

We analysed changes in vegetation cover for all years in comparison with 2020. The results show that up to 2005/2010, in general, there has been an increase in forest cover. After this period, there has been a stagnation or decrease in forest cover compared to 2020. Only a few parks show an increase (e.g., *Kauno Marios Regional Park*) or stagnation (e.g., *Sirvėta Regional Park)* from 2015 to 2020.

From the PCA three main factors were identified. Factor 1 explained the majority of the variance (79.36%). Factors 2 and 3 explained 9.43% and 6.04%, respectively. Kauno Marios, Pajuris, Kursiu Nerija and Dzūkija parks are separated from all the other parks (Fig. 2). The first three parks had the lowest percentage of dense vegetation, while Dzūkija had the highest.

4 Conclusions

Protected areas are established to safeguard natural and cultural values. However, if not properly managed, these areas are also prone to the effects of anthropogenically driven land use changes. This study analysed vegetation cover changes within Regional and National Parks from 1990 to 2020. The results show a positive trend

Table 1 Percentage of dense vegetation cover in Lithuanian Regional and National parks over time

R & N Parks	1990	1995	1999	2005	2010	2015	2020	*Diff*
Anykščiai (R)	31	44	49	51	64	68	65	*53*
Asveja (R)	34	64	59	61	67	69	67	*49*
Aukštadvaris (R)	39	51	49	67	70	69	62	*36*
Aukštaitija (N)	19	59	52	59	35	67	67	*72*
Biržai (R)	38	30	42	43	52	64	51	*25*
Dieveniškės His. (R)	23	56	47	62	71	67	61	*61*
Dubysos (R)	61	64	65	66	73	72	69	*12*
Dzūkija (N)	10	24	24	66	62	65	54	*82*
Gražutė (R)	26	60	50	59	40	65	65	*60*
Kauno Marios (R)	25	29	31	35	35	36	37	*33*
Krekenava (R)	47	42	48	46	56	72	58	*20*
Kursiu Nerija (N)	8	18	15	17	17	18	17	*52*
Kurtuvenai (R)	65	68	69	67	73	75	76	*14*
Labanoras (R)	15	62	52	59	55	69	66	*77*
Meteliai (R)	35	44	40	50	49	46	51	*30*
Nemuno Delta (R)	45	59	57	54	57	63	61	*26*
Nemuno Kilpos (R)	43	51	52	67	65	64	66	*35*
Neris (R)	35	60	61	70	73	78	74	*52*
Pagramantis (R)	60	71	68	67	73	76	77	*22*
Pajuris (R)	18	27	26	26	29	31	31	*43*
Panemuniai (R)	48	50	54	57	62	66	53	*8*
Pavilniai (R)	42	73	71	69	76	76	77	*46*
Rambynas (R)	54	58	65	60	66	74	72	*25*
Salantai (R)	45	60	59	58	62	65	59	*23*
Sartai (R)	31	58	44	58	52	63	68	*54*
Sirvėta (R)	30	66	41	52	54	69	69	*56*
Trakai Historical (N)	31	35	40	52	59	59	56	*45*
Tytuvėnai (R)	53	61	63	60	65	68	65	*18*
Varniai (R)	48	63	62	60	71	73	71	*33*
Veisiejai (R)	25	32	43	58	55	60	59	*57*
Venta (R)	43	58	54	49	62	69	56	*24*
Verkiai (R)	36	70	67	69	73	75	74	*51*
Vištytis (R)	57	63	66	72	72	76	73	*22*
Žagarė (R)	50	61	58	53	61	70	62	*19*
Žemaitija (N)	47	68	64	58	72	74	74	*37*

R—Regional. N—National. Diff.—the difference in dense vegetation cover between 1990 and 2020

Table 2 Change in percentage of dense vegetation cover in Lithuanian Regional and National parks, in comparison to 2020

R & N Parks	1990	1995	1999	2005	2010	2015
Anykščiai (R)	34	21	16	14	0	−3
Asveja (R)	33	3	8	7	0	−2
Aukštadvaris (R)	22	11	12	−5	−8	−7
Aukštaitija (N)	48	8	15	8	32	0
Biržai (R)	13	21	9	8	−1	−13
Dieveniškės His. (R)	37	5	14	−1	−10	−6
Dubysos (R)	8	5	4	3	−3	−3
Dzūkija (N)	44	30	30	−12	−7	−11
Gražutė (R)	40	5	16	7	25	0
Kauno Marios (R)	12	7	6	1	2	1
Krekenava (R)	11	16	10	12	2	−14
Kursiu Nerija (N)	9	−1	2	0	0	−1
Kurtuvenai (R)	10	8	7	9	2	0
Labanoras (R)	51	4	14	8	11	−2
Meteliai (R)	15	7	11	0	1	5
Nemuno Delta (R)	16	2	4	8	4	−2
Nemuno Kilpos (R)	23	15	14	0	1	2
Neris (R)	38	14	13	3	1	−5
Pagramantis (R)	17	7	10	11	4	2
Pajuris (R)	13	4	5	5	2	−1
Panemuniai (R)	4	3	−1	−4	−9	−14
Pavilniai (R)	35	4	6	8	1	1
Rambynas (R)	18	14	7	13	6	−1
Salantai (R)	13	−1	−1	1	−3	−7
Sartai (R)	36	9	24	10	16	4
Sirvėta (R)	39	3	27	17	15	0
Trakai Historical (N)	25	21	16	3	−3	−3
Tytuvėnai (R)	12	4	2	5	0	−3
Varniai (R)	23	8	9	10	0	−3
Veisiejai (R)	34	27	16	1	4	−1
Venta (R)	13	−3	2	7	−6	−14
Verkiai (R)	37	4	7	5	1	−1
Vištytis (R)	16	10	7	1	1	−3
Žagarė (R)	12	1	4	9	0	−8
Žemaitija (N)	28	6	10	16	2	0

R—Regional. N—National

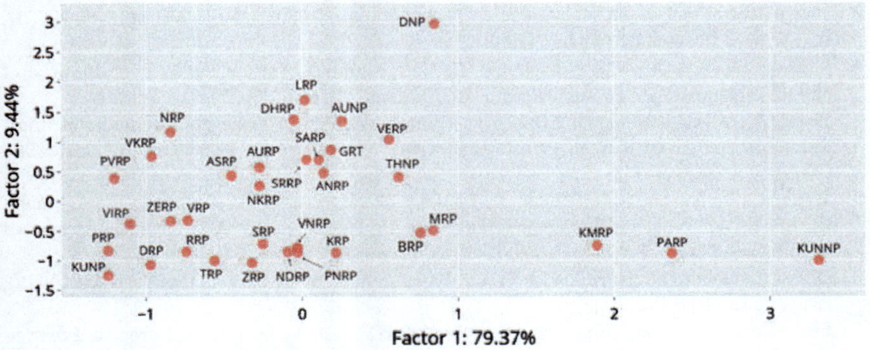

Fig. 2 Relation between Factor 1 and 2. Anykščiai (ANRP); Asveja (ASRP); Aukštadvaris (AUNP); Biržai (BRP); Dieveniškės (DHRP); Dubysos (DRP); Dzūkija (DNP); Gražutė (GRT); Kauno Marios (KMRP); Krekenava (KRP); Kursiu Nerija (KUNNP); Kurtuvenai (KUNP); Labanoras (LRP); Meteliai (MRO); Nemuno Delta (NDRP); Nemuno Kilpos (MKRP); Neris (NRP); Pagramantis (PRP); Pajuris (PARP); Panemuniai (PNRP); Pavilniai (PVRP); Rambynas (RRP); Salantai (SRP); Sartai (SARP); Sirvėta (SRRP); Trakai Historical (THNP); Tytuvėnai (TRP); Varniai (VRP); Veisiejai (VERP); Venta (VNRP); Verkiai (VKRP); Vištytis (VIRP); Žagarė (ZRP); Žemaitija (ZERP)

with vegetation cover increasing from 1990 to 2005–2010, likely the result of conservation efforts both from national and international frameworks. However, a shift occurred from 2010 to 2020 when most parks showed a stagnation or decrease in dense vegetation cover. From our analysis, three parks emerged with the lowest dense forest cover and one with the highest. This information is essential for stakeholders to understand and identify potential causes for the dense vegetation decrease and align their management with conservation efforts and priorities. This will be addressed in future works by identifying potential causes for the dense vegetation cover decrease.

Acknowledgements This work was funded by the project Study on the landscape of structural changes of the state parks and description of the landscape aesthetic's value and financed by the State Service for the Protected Areas under the Ministry of Environment. Nr. 2022-06-21 Nr. F4-173/ 2022-06-22 Nr. 2ST-190.

References

1. Pereira, P., Gomes, E., Rocha, J.: Preface. In: Pereira, P., Gomes, E., and Rocha, J. (eds.) Mapping and Forecasting Land Use: The Present and Future of Planning. pp. xvii–xviii. Elsevier (2022)
2. Winkler, K., Fuchs, R., Rounsevell, M., Herold, M.: Global land use changes are four times greater than previously estimated. Nat. Commun. 12, 2501 (2021).
3. Watson, J.E.M., Dudley, N., Segan, D.B., Hockings, M.: The performance and potential of protected areas. Nature. 515, 67–73 (2014).

4. UNEP-WCMC and IUCN. Protected Planet Report 2020. UNEP-WCMC and IUCN: Cambridge UK; Gland, Switzerland (2021).
5. Shrestha, N., Xu, X., Meng, J., Wang, Z.: Vulnerabilities of protected lands in the face of climate and human footprint changes. Nat. Commun. 12, 1632 (2021).
6. Veach, V., Moilanen, A., Di Minin, E.: Threats from urban expansion, agricultural transformation and forest loss on global conservation priority areas. PLoS One. 12, e0188397 (2017).
7. Wade, C.M., Austin, K.G., Cajka, J., Lapidus, D., Everett, K.H., Galperin, D., Maynard, R., Sobel, A.: What Is Threatening Forests in Protected Areas? A Global Assessment of Deforestation in Protected Areas, 2001–2018. Forests. 11, (2020).
8. OSP. Lithuanian Official Statistics Portal. Population as of 1 of July. https://osp.stat.gov.lt/), last accessed 10th Dec 2022
9. EEA: Copernicus Land Monitoring Service: CORINE Land Cover 2018. European Environment Agency (EEA) (2018).
10. State Protected Areas Service: Protected Areas in Lithuania. Publishing House Lututė, Vilnius (2019).
11. Jiang, S., Chen, X., Smettem, K., Wang, T.: Climate and land use influences on changing spatiotemporal patterns of mountain vegetation cover in southwest China. Ecol. Indic. 121, 107193 (2021).
12. Kriegler, F.J., Malila, W.A., Nalepka, R.F., Richardson, W.: Preprocessing Transformations and Their Effects on Multispectral Recognition. In: Remote Sensing of Environment, VI. p. 97 (1969).
13. Jones, H.G., Vaughan, R.A.: Remote sensing of vegetation: principles, techniques, and applications. Oxford university press (2010).
14. Zaitunah, A., Samsuri, Sahara, F.: Mapping and assessment of vegetation cover change and species variation in Medan, North Sumatra. Heliyon. 7, e07637 (2021).
15. Gorelick, N., Hancher, M., Dixon, M., Ilyushchenko, S., Thau, D., Moore, R.: Google Earth Engine: Planetary-scale geospatial analysis for everyone. Remote Sens. Environ. 202, 18–27 (2017).
16. ESRI: ArcGIS Desktop: Release 10.8. Redlands, CA: Environmental Systems Research Institute, (2020).
17. Kavaliauskas, P.: Land Management in Lithuania: Past and Future. GeoJournal, (33)97–106 (1994).
18. Gomes, E., Inácio, M., Bogdzevič, K., Kalinauskas, M., Karnauskaitė, D., Pereira, P.: Future land-use changes and its impacts on terrestrial ecosystem services: A review. Sci. Total Environ. 781, 146716 (2021).
19. Juknelienė, D., Česonienė, L., Jonikavičius, D., Šileikienė, D., Tiškutė-Memgaudienė, D., Valčiukienė, J., Mozgeris, G.: Development of Land Cover Naturalness in Lithuania on the Edge of the 21st Century: Trends and Driving Factors, (2022).
20. Jõgiste, K., Frelich, L.E., Laarmann, D., Vodde, F., Baders, E., Donis, J., Jansons, A., Kangur, A., Korjus, H., Köster, K., Kusmin, J., Kuuluvainen, T., Marozas, V., Metslaid, M., Metslaid, S., Polyachenko, O., Poska, A., Rebane, S., Stanturf, J.A.: Imprints of management history on hemiboreal forest ecosystems in the Baltic States. Ecosphere. 9, e02503 (2018).

Wet Cleaning Technologies for Biomass-Derived Syngas: A Review

J. R. C. Rey, P. S. D. Brito, and C. Nobre

Abstract Biomass gasification and pyrolysis are thermochemical conversion technologies with significant potential for the future development of a sustainable energy system worldwide, which can help to replace several fuels and chemicals currently provided by fossil fuels. Gasification and pyrolysis generate solid, liquid, and gaseous products. The gas is a mixture of gases composed mainly of CO, CO_2, H_2, CH_4, N_2, and small concentrations of impurities that are relevant and can have significant drawbacks in its subsequent use as fuel. Among those impurities are included tars, particulate matter, nitrogen-based compounds (NH_3, HCN), sulfur-based compounds (H_2S, COS), hydrogen halides (HCl, HF), and trace metals (Na, K). Thus, cleaning syngas is an essential process to get a gas with suitable characteristics before use in further applications such as internal combustion engines, gas turbines, hydrogen production, and other chemical applications. In this context, this work presents a review of the main wet cleaning systems that have been developed for syngas cleaning and conditioning. For wet cleaning, wet scrubbers are generally used, and this is a relatively mature and highly effective process. The main drawbacks are the generation of a residual liquid stream that is difficult to treat and eliminate, and the decrease in energy efficiency, by cooling the syngas. Therefore, this review critically examines the significant challenges in syngas wet cleanup and discusses its major recent advances.

Keywords Biomass · Thermochemical conversion · Contaminants · Wet cleaning

J. R. C. Rey (✉) · P. S. D. Brito · C. Nobre
Research Centre for Endogeneous Resource Valorizatiob, VALORIZA, Polytechnic Institute of Portalegre, Portalegre, Portugal
e-mail: jose.rey@ipportalegre.pt

J. R. da. C. S. Galvão et al. (eds.), *Proceedings of the 3rd International Conference on Water Energy Food and Sustainability (ICoWEFS 2023)*, Springer Proceedings in Earth and Environmental Sciences, https://doi.org/10.1007/978-3-031-48532-9_33

1 Introduction

With the rapid growth of the world population and the development of modern society, the global demand for energy is rapidly increasing. Facing this increase, the scientific community has been tasked with finding alternative energy generation sources that are more environmentally friendly and economically sustainable. Several technologies for the generation of energy from renewable sources are close to being competitive with the energy generated from fossil fuels and nuclear energy in a growing number of countries. In this sense, biomass is emerging as one of the most promising alternative sources for the global supply of reliable, clean, and affordable primary energy that can substitute fossil fuels in several applications [1].

According to Jen et al. [2], there are three main methods of thermochemical conversion of biomass, i.e., combustion, pyrolysis, and gasification. Biomass combustion produces heat and energy, which is traditionally used in the process industry and heating homes. However, combustion releases dangerous and potentially harmful substances such as NO_2, CO_2, particulates, etc., which must be considered during its use [3]. Pyrolysis is the thermal decomposition of biomass, into solids, liquids, and gases, by heating at high temperatures (400 °C to 800 °C) in the absence of oxygen [4]. The pyrolysis process conditions affect the physicochemical properties, quality, and yield of the pyrolyzed products. Thus, the pyrolysis process can be modified to enhance the yield and quality of pyrolysis products [5]. Gasification is a promising thermochemical process that converts of biomass into gas under different atmospheres (air, steam, O_2, CO_2, etc.), and the produced syngas could also be used for highly valued chemicals production via Fischer–Tropsch synthesis [2].

One of the main problems limiting the use of gasification or pyrolysis syngas on a large scale is that the gas contains small but significant amounts of impurities, such as different tar compounds, particulate matter, nitrogen compounds like ammonia (NH_3) and hydrogen cyanide (HCN), sulfur-containing inorganic compounds like hydrogen sulfide (H_2S), carbonyl sulfide (COS) and carbon disulfide (CS_2), hydrogen halides and halogens such as hydrogen chloride (HCl) and chlorine (Cl), and trace metals like sodium (Na) and potassium (K). These contaminants are released during the thermochemical conversion of the biomass and are consequently present at widely varying concentrations in syngas [6]. The presence of these contaminants in syngas poses several technical and operational issues including equipment corrosion (H_2S) and fouling (tar), catalyst deactivation (tar, H_2S, NH_3, HCl and trace metals), and environmental pollution (NH_3) [7].

There are different processes of improving syngas quality by decreasing the concentration of contaminants, and those include primary or secondary processes. The primary processes reduce the concentration of contaminants *in-situ*, that is, during the thermochemical conversion process and the secondary processes consist of gas treatment after the thermochemical conversion [8]. This document focuses on the secondary cleaning method, specifically on wet-cleaning systems. It also updates previous revisions to include new advances and challenges of syngas wet-cleaning systems.

2 Gas Wet Cleaning Technology

The wet gas cleaning (WGC) is considered an efficient and economical gas cleaning technology that reduces the concentration of relevant contaminants up to the lowest desirable limits. This method is the most widely used to remove contaminants from the raw syngas, due to its proven reliability and high efficiency [6]. The main distinguishing feature of this approach is that it is carried out at temperatures typically below 100 °C [9]. As gasification and pyrolysis processes are carried out at high temperatures (400 to 900 °C), the main disadvantage of this method is the penalty of the thermal efficiency of the plant due to the cooling of the syngas and the additional costs to treat or eliminate contaminants retained in the product streams resulting from the cleaning process used.

In general, these technologies use wet scrubbers (spray and wash towers, cyclonic and dynamic scrubbers, Venturi scrubbers, and wet electrostatic precipitators (W-EPS) [6]), with water to remove contaminants from the syngas stream [6], although other cleaning fluids such as motor oil and vegetable oils can also be used [10]. The spray scrubber systems are the simplest and use inertial forces to separate impurities from the gas stream. The dynamic and cyclonic systems can be integrated into a single device, as shown in Fig. 1a). In this case, the tangential inlet removes the larger impurities by inertial separation followed by a moistened paddle to capture the smaller impurities. After the dynamic section of the system, the remaining impurities in the gas stream are removed by the final cyclonic motion [8].

The Venturi scrubber is made up of a convergent-divergent channel, in which the cross-sectional area decreases and then increases. The narrowest part is known as the throat area. In the converging section, the decrease in the area increases the velocity and turbulence of the dirty gas. The lavage fluid is injected slightly upstream of the throat or directly into the throat section. The turbulence that occurs in this area causes the atomization of the liquid, improving the gas–liquid contact. The gas–liquid mixture then decelerates as it moves through the diverging section, causing

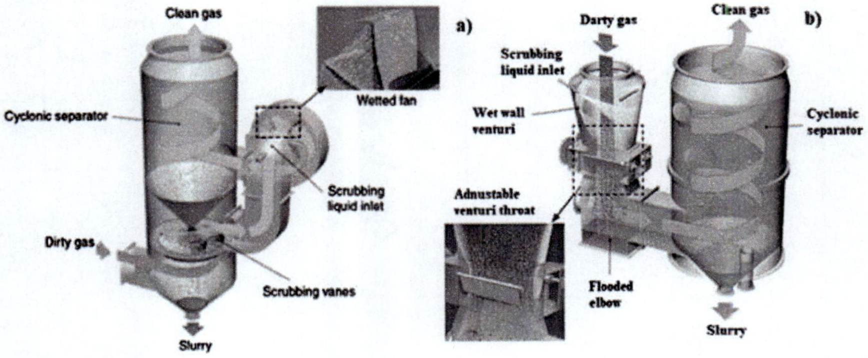

Fig. 1 Wet scrubbers: **a** Integrating wet dynamic and cyclonic scrubbers; **b** Venturi wet scrubber [11]

additional particle and droplet impacts and agglomeration of the droplets [11]. The Venturi scrubber is generally integrated with a cyclone separator, located in the stripping section, as shown in Fig. 1b). In this section by the action of the centrifugal forces of the cyclone, the liquid droplets are separated from the gas stream.

The electrostatic precipitators (ESP) are systems for collecting particles widely used in different industrial applications, due to their low energy consumption, low-pressure drop, and high efficiency of particle removal. ESPs are devices that use electrical forces to move the particles from the gas stream to the collection plates. There are two basic types of ESP, the Dry-ESP (see Fig. 2a and the Wet-ESP Fig. 2b). The Wet-EPS operates in three stages: charge, recollection, and cleaning of the particles from the collecting electrode, these equipment's, consist of a vertical or horizontal chamber with tubes or plates placed in parallel and isolated, through which the electric current circulates and incorporates a wet scrubber before and/or from the chamber. The mixture of water and particles becomes electrically charged as it passes through the chamber and is separated from the gas stream when colliding with the collecting electrode. Cleaning of the collecting electrode is done by washing the collecting surface with water. As Wet-ESPs operate in a humid environment, they can remove, in addition to particles, other pollutants such as tar and acid gases present in the syngas [6].

The WGC process is the most used because it allows the easy removal of multiple contaminants such as particles, tar, NH_3, HCl, and H_2S, some of which are soluble in water. This process works at temperatures close to room temperature, in order to limit

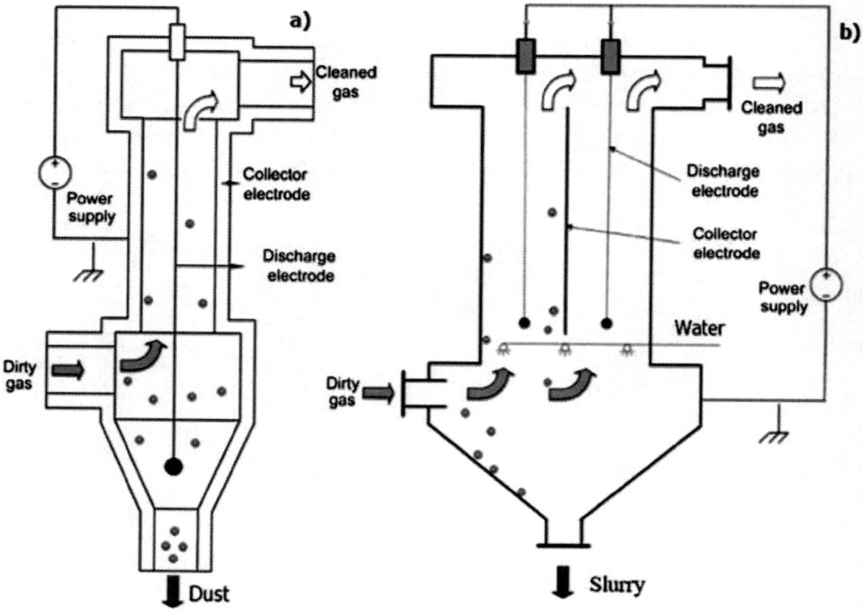

Fig. 2 Electrostatic precipitators: **a** Dry-EPS, **b** Wet-EPS

evaporation losses of the scrubbing liquid used in the cleaning process [8]. However, despite these low operating temperatures energetically penalizing the process, WGC techniques will continue to be important gas treatment technologies in the future due to their high efficiency and proven reliability.

WGC systems have a particle removal efficiency greater than 90% for particle sizes greater than 5 μm (>PM5) and is between 60 and 70% for submicron particles [12], while for alkali it exceeds 95% [13]. The efficiency of particle removal from the syngas stream by means of a tube-type wet electrostatic precipitator was studied by Parihar et al. [14]. The analyzed system has the capacity to process 100 Nm^3/h of gas. This gas is produced in a downdraft gasifier and the maximum concentration of particles determined is 700 mg/Nm^3. The results of the study show that the Wet-EPS managed to reduce the volume of particles to 36 mg/Nm^3, a value below the permitted limits for naturally aspirated reciprocating engines (50 mg/Nm^3). The Wet-EPS achieved a maximum particle removal efficiency of 83%. In another study, Corrêa Jr et al., [12] studied the effect of the use of syngas derived from rice husk as a complement to natural gas in a microturbine. To use the gas in the microturbine, it was first subjected to a cleaning process that combines a dry cyclone filter, responsible for removing the largest particles, and a Venturi-type scrubber to remove the smallest particles. According to the reported results, the system is capable of efficiently removing particles down to levels below 30 mg/Nm^3, a necessary condition to be able to use the syngas in the microturbine without affecting its performance. Overall, wet scrubbers are a mature particle removal technology, available on a commercial scale and of high efficiency, fundamentally for particle sizes greater than > PM5, however, for submicron particles more sophisticated techniques are required that imply greater energy consumption, greater complexity, and high economic costs.

In addition to removing particles, wet scrubbers also remove tar in the same process. As stated by Parihar et al. [14] the Wet-ESP system achieves a tar removal efficiency of 62%, while in Corrêa Jr et al. [12] they report a tar concentration lower than 100 mg/Nm^3 after passing through the Venturi scrubber. Water-based wet scrubbers show high removal efficiency of some light, oxygenated and polar compounds that make up tar, while heavy and non-polar heterocyclic compounds have low removal efficiency [15]. These non-polar compounds are also removed from the gas stream by condensing because the wet scrubber reduces the gas temperature enough for these tars to reach the dew point.

Wet scrubbers produce a highly polluted water stream, requiring complex and expensive treatment and purification systems, which is the main drawback of using water for wet cleaning of syngas [6]. Other scrubbing liquids that have been studied include engine oil and vegetable oils [10]. Among the benefits from the use of oil-based scrubbing liquids are included the high efficiency of contaminants removal, primarily the heavy and heterocyclic tars, the possibility of operating at higher temperatures, thus improving the overall efficiency of the process, and the ability to regenerate the scrubbing oil with hot gas, which allows the contaminants to be concentrated for later recovery or disposal [16]. The main drawbacks of using oils include the increase in the complexity of the process and the high cost of the oils,

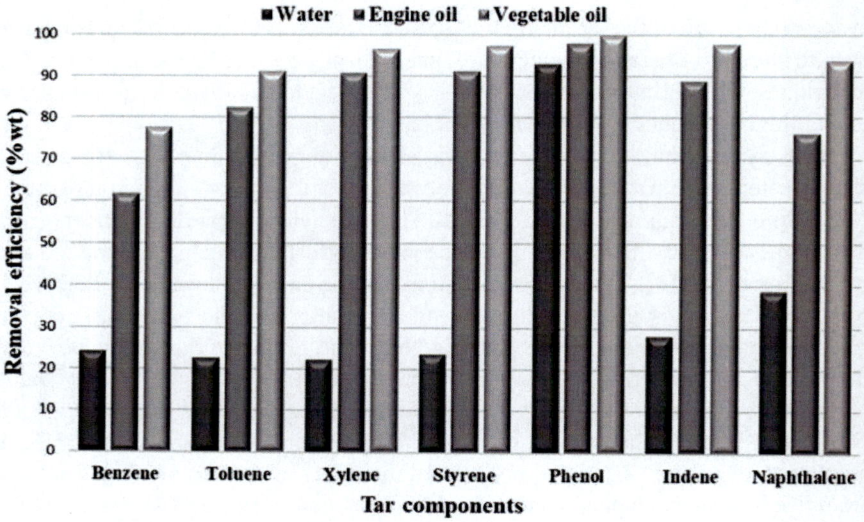

Fig. 3 Tar compounds removal efficiency for different scrubbing liquids [8]

which ultimately results in a general increase in the production costs of the clean syngas [6]. As an example, and because tar is a main contaminant present in the raw syngas, Fig. 3 shows a comparison between the efficiency of removal of tar compounds for the most widely used scrubbing liquids.

The wet removal of sulfur pollutants from syngas at low temperatures is achieved through physical absorption, chemical reaction, or mixed processes. For the absorption of sulfur, alkanolamines, alkaline salts, or aqueous ammonia are commonly used, as well as organic solvents with high affinity for acid gases as chemical and physical solvents [17]. Many of the end applications for syngas require extremely low levels of sulfur contaminants. For example, in the use of syngas as a fuel in a gas turbine, sulfur concentrations must be below 30 mg/Nm^3, to meet the SO_2 emission limits in the stack. The removal of sulfur contaminants from produced gas for gas turbine applications was discussed by Descamps et al. [18]. This study uses chemical solvents based on methyl-diethanolamine amines (MDEA), reporting an H_2S removal efficiency of 95%. Other applications that require very low levels of sulfur contaminants include chemical synthesis. For these applications, the sulfur concentration needs to be less than 1 mg/Nm^3, which is achieved using physical absorption solvents such as Selexol® and Rectisol® are used [19].

The main nitrogen-derived compounds present in the syngas stream are NH_3 and HCN. These compounds are very soluble in water, so this liquid is very suitable for their elimination, even the condensation of the water steam contained in the syngas is capable of substantially reducing the content of NH_3 and HCN when condensing. This was confirmed by Pinto et al., [20] when using a cooled condenser to eliminate the water contained in a stream of syngas derived from sewage sludge, which resulted in an elimination of more than 90% in NH_3 by the condensed water.

In the case of chlorine, the main compounds present in raw syngas are HCl gas and/or NH_4Cl in the form of solid particles, which are formed by the reaction of HCl with NH_3 when the gas is cooled to about 300 °C [16]. These species of chlorine are very soluble in water, so water-based wet scrubbing is appropriate and, to date, it continues to be the most widely used solvent to remove chlorine contaminants to the levels necessary for each application [21]. Nevertheless, the main disadvantage of using water is that an expensive ionic exchanger is required to remove chlorine from the solution [22]. The elimination of the chlorine species can be improved by adding sodium hydroxide (NaOH) to the water. This is a logical solution, since the reaction between HCl and NaOH is potentiated, and sodium chloride (NaCl) is produced. The NaCl formed can be subsequently removed by precipitation, which constitutes one of the benefits of wet caustic scrubbing [21]. Furthermore, another added benefit is that caustic water can remove other pollutants such as COS and CO_2 [16]. A summary of the main technologies for syngas wet scrubbing is shown in Table 1.

3 Demonstration and Commercial Scale of Syngas Wet Cleanup Systems

Although we only address wet cleaning systems in this study, typically in many commercial-scale applications, wet-cleaning and dry-cleaning methods are combined, providing high-efficiency, low-temperature contaminant removal solutions. A relevant example of this combination on a commercial scale is provided by the downdraft gasification plants of the Indiana company ANKUR [26]. This company offers a series of solutions for the generation of electrical and thermal energy with a wide range of biomass gasifiers ranging in size from 5 to 500 kW output. For syngas cleaning, they employ a combination of wet and dry-cleaning methods, including cyclones, wet scrubbers, and various filters. The syngas passes through a cyclone, located at the outlet of the gasifier, to remove the largest particles (ash, coal, and others), then it is sent to a wet scrubber Venturi-type, based on water, which removes particles, tar, and other impurities. The moisture is then removed from the gas and sent to a filtering system, see Fig. 4 [27]. The filter system is composed of three or more filters, the first filters use sand, sawdust, or carbon as filter material, and the last one is a fabric filter. This cleaning system results in a final tar and particulate content below 25 mg/Nm^3, clean enough to be used in internal combustion engines (ICE). But, cleaning with water results in large amounts of tar-contaminated water, about 0.6 kg of wastewater per kilogram of biomass [8]. Therefore, a complex, and expensive wastewater treatment and purification system is required, which is the main drawback of this syngas cleaning technology.

Another technology that combines wet and dry cleaning methods is the OLGA technology, developed by the Center for Energy Research in the Netherlands (ECN), now known as Netherlands Organization for Applied Scientific Research (TNO) [15]. This system comprises a cyclone, oil-based multi-stage wet scrubbers (OLGA), and

Table 1 Wet technologies for syngas cleaning [17, 23–25]

Cleaning system	Contaminant removed and efficiency	TRL	Comments	Application
Washing tower (water)	Particles > 98% Tars > 25% Chlorine > 90% Nitrogen > 99%	8–9	High filtration efficiency; Simple and compact construction but with a high operating cost due to effluent treatment	Combustion, ICE
Bio-oil scrubber	Tars 51.6–64.5%	6–8 6–8	It simultaneously removes a wide range of contaminants; High filtration efficiency; but the use of oils increases the complexity and the operating costs of the system	Combustion, ICE
Venturi scrubber (water + NaOH)	Particles > 99% Tars > 90% Chlorine > 90% Nitrogen > 99% Chlorine > 99%	8–9	It simultaneously removes a wide range of contaminants; Small size, but with a high operating cost due to effluent treatment	Combustion, ICE, GT
W-ESP (water)	Particles > 99% Tars 40–70% Chlorine > 90% Nitrogen > 99%		High filtration efficiency; Simple and compact construction but with a high operating cost due to effluent treatment	
Scrubber-oil + AC	Tars > 98%	6–8	High filtration efficiency; but the use of oils increases the complexity and costs of the cleaning process	Combustion, ICE, GT, Methanol synthesis, FT synthesis
Washing tower (limestone slurry)	Sulfur > 95%	8–9	High filtration efficiency; Simple and compact construction; No mechanical components	
Washing tower (DEPG[a])	Sulfur > 99%	6–8		

[a]Dimethyl ether of polyethylene glycol

water-based wet scrubbers [28]. This technology can operate at temperatures between 60 and 450 °C. With an efficiency of over 99% in removing contaminants from raw syngas, compared to conventional water scrubbers that achieve efficiencies between 49 to 79%, or even when using a Wet-ESP, where maximum removal efficiency reaches 99% for the heaviest contaminants and up to 74% for lighter contaminants [8]. OLGA is divided into four main stages, as shown in Fig. 5.

In the first stage, the gas passes through a cyclone that removes the largest particles by centrifugal force. Then, in the second stage (the collector), the syngas is cooled by the scrubbing oil. The heaviest contaminants (tars with more than three rings and fine particles) condensate and are separated by filtration from the scrubbing oil and introduced into the gasifier as feedstock together with the biomass [8]. In the third stage, the scrubbing oil absorbs the lightest contaminants (mainly one to two ring tar compounds) in the absorber [15]. The oil loaded with light tars is regenerated in

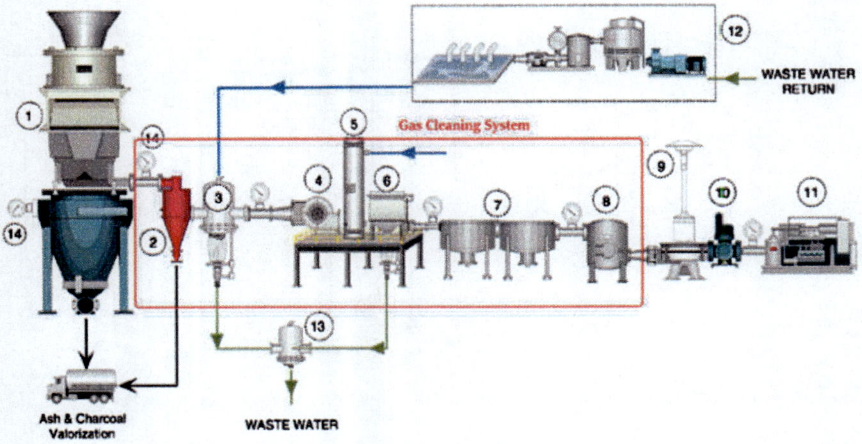

Fig. 4 Scheme for the ANKUR gas cleaning system: 1. gasifier; 2. cyclone and ash collection box; 3. Venturi scrubber and drain box; 4. wet blower; 5. heat exchanger; 6. mist eliminator; 7. fine filters; 8. pleated filter; 9. flare; 10. header box; 11. gas engine; 12. waste water treatment; 13. combo tank; 14. different pressure points [27]

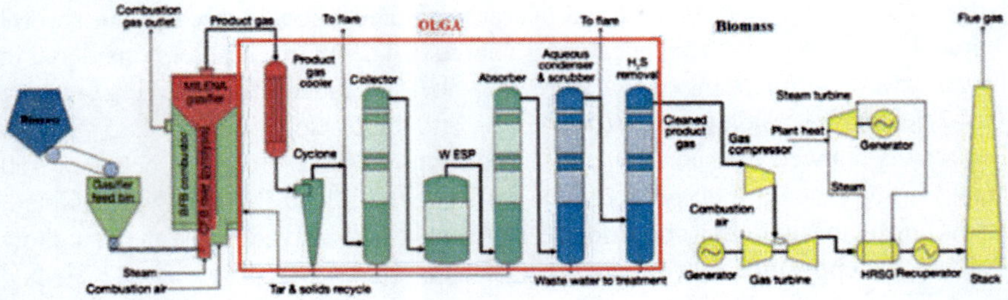

Fig. 5 Syngas cleaning with heat exchanger and OLGA system in the Gussing gasification plant [29]

the stripper which makes it possible to recover approximately 99% of the oil used in cleaning the gas, while the light tars are fed together with the gasifying agent to the reactor [15].

An advantage of the OLGA process is the possibility of recycling light and heavy tars into the gasifier, which contributes to the increase in the energy efficiency of the technology [15]. Another advantage is that the light components, such as CH_4, CO, and H_2, are practically unaffected by this treatment [8]. The main disadvantage is the high costs of oil-based absorbents. It is relevant to state that the OLGA technology is already commercially available [8].

The Polytechnic Institute of Portalegre is working on a hydrogen purification system for the syngas produced in an activated carbon process. The gas generated during the activation process, after cleaning and purification, will be used to produce hydrogen through a membrane separation system. The system is made up of a rotary

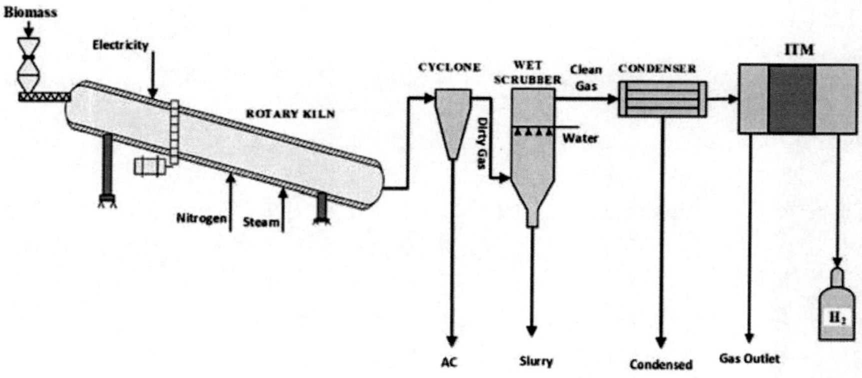

Fig. 6 Schematic diagram of the integrated carbon activation and hydrogen recovery system

kiln for pyrolysis and activation, a cleaning, and purification system for the activation gas, and a hydrogen production system using ceramic membrane separation (see Fig. 6).

At this moment, work is being done on the construction of the gas cleaning and purification system and the hydrogen separation system. The laboratory-scale carbon activation system is built and operational. This allowed to carry out several charcoal activation tests from eucalyptus charcoal. The main variables considered in the study were the operating temperature (800–950 °C), the reactor operating pressure (0.5–2.5 bar), the residence time (5–60 min), and the particle diameter (1.5–5 mm). The results showed that an activated carbon with a surface area of between 560 and 1156 m^2/g and a total pore volume of between 0.2 and 0.7 cm^3/g is obtained, depending on the operating conditions. The results of the activation gas analysis show an average composition by volume of 17.5% CO_2, 13.4% CO, 40.1% H_2, 15.2% N_2, and 8.8% O_2, with a lower calorific value of 7.51 MJ/Nm^3. The activation gas has a high percentage of hydrogen in its composition, which demonstrates its enormous potential to produce hydrogen. Considering the high efficiency of wet gas cleaning systems for this project, it was decided to build a Venturi-type wet-scrubber with water as the cleaning fluid. The system also has a condenser to remove moisture from the gas before it is sent to the separator for hydrogen production.

4 Conclusion

This study presented a review of the main characteristics of the wet gas cleaning technologies applied in the elimination of contaminants generated during the thermal conversion of biomass by pyrolysis and gasification, making special considerations regarding their composition, classification, and elimination processes. Wet cleaning systems necessarily operate at temperatures close to room temperature because the upper limit of operation corresponds to the upper temperature that liquid adsorbents withstand before evaporation. Wet gas cleaning is a mature and efficient pollutant removal technology. It is based on a physical–mechanical treatment, generally uses water as the working fluid, and has pollutant removal efficiencies of up to 99%, but with the disadvantage of reducing energy conversion efficiency and generating highly hazardous waste. An interesting variant that eliminates the problem of hazardous waste is the OLGA technology. This system uses oil as a working fluid. It allows the recycling of the waste to the reactor, providing energy and environmental benefits, its main drawback is the increased costs of the cleaning system.

Finally, our considerations after doing this review and evaluation of the advantages, disadvantages and current trends of the syngas different cleaning and conditioning options, are that there is not yet a sufficiently simple, low-cost, and environmentally friendly technology that guarantees the efficient removal of particles, tar, and contaminants at adequate levels for most of the final applications of the syngas. So, it is still a demanding task to develop efficient cleaning processes, to provide a combination of low cost and minimal environmental impact.

Acknowledgements This work was funded by FCT—Fundação para a Ciência e a Tecnologia within the R&D Units Project Scope UIDP/04077/2020.

References

1. Y. Qian, S. Sun, D. Ju, X. Shan, and X. Lu, "Review of the state-of-the-art of biogas combustion mechanisms and applications in internal combustion engines," *Renew. Sustain. Energy Rev.*, vol. 69, no. November 2016, pp. 50–58, 2017. https://doi.org/10.1016/j.rser.2016.11.059.
2. J. Ren, J. P. Cao, X. Y. Zhao, F. L. Yang, and X. Y. Wei, "Recent advances in syngas production from biomass catalytic gasification: A critical review on reactors, catalysts, catalytic mechanisms and mathematical models," *Renew. Sustain. Energy Rev.*, vol. 116, no. March, p. 109426, 2019, doi: https://doi.org/10.1016/j.rser.2019.109426.
3. M. Plaza Recobert, "Carbón activado: evaluación de nuevos precursores y del proceso de activación con dióxido de carbono." 2015.
4. D. Mohan, C. U. Pittman, and P. H. Steele, "Pyrolysis of Wood/Biomass for Bio-oil: A Critical Review," *Energy Fuels*, vol. 20, no. 3, pp. 848–889, May 2006. https://doi.org/10.1021/ef0502397.
5. S. R. Naqvi *et al.*, "Recent developments on sewage sludge pyrolysis and its kinetics: Resources recovery, thermogravimetric platforms, and innovative prospects," *Comput. Chem. Eng.*, vol. 150, p. 107325, Jul. 2021. https://doi.org/10.1016/j.compchemeng.2021.107325.
6. N. Abdoulmoumine, S. Adhikari, A. Kulkarni, and S. Chattanathan, "A review on biomass gasification syngas cleanup," *Appl. Energy*, vol. 155, pp. 294–307, 2015. https://doi.org/10.1016/j.apenergy.2015.05.095.
7. W. P. Chan *et al.*, "A hot syngas purification system integrated with downdraft gasification of municipal solid waste," *Appl. Energy*, vol. 237, no. January, pp. 227–240, 2019. https://doi.org/10.1016/j.apenergy.2019.01.031.
8. C. Courson and K. Gallucci, "Gas cleaning for waste applications (syngas cleaning for catalytic synthetic natural gas synthesis)," in *Substitute Natural Gas from Waste*, vol. 8, M. Materazzi and P. U. Foscolo, Eds. Elsevier Inc., 2019, pp. 161–220. https://doi.org/10.1016/b978-0-12-815554-7.00008-8.

9. Y. Richardson, M. Drobek, A. Julbe, J. Blin, and F. Pinta, *Biomass Gasification to Produce Syngas*. 2015. https://doi.org/10.1016/B978-0-444-63289-0.00008-9.
10. C. M. Meijden, "Development of the MILENA gasification technology for the production of Bio-SNG. PhD. Thesis.," *ECN Energy Res. Neth.*, no. December, p. 206, 2010.
11. D. Mussatti, "Wet Scrubbers for Particulate Matter," in *Particulate Matter Controls*, 2002, p. 62.
12. P. S. P. Corrêa, J. Zhang, E. E. S. Lora, R. V. Andrade, L. R. de Mello e Pinto, and A. Ratner, "Experimental study on applying biomass-derived syngas in a microturbine," *Appl. Therm. Eng.*, vol. 146, no. September 2018, pp. 328–337, 2019. https://doi.org/10.1016/j.applthermaleng.2018.09.123.
13. O. Hirohata *et al.*, "Release behavior of tar and alkali and alkaline earth metals during biomass steam gasification," *Energy Fuels*, vol. 22, no. 6, pp. 4235–4239, 2008. https://doi.org/10.1021/ef800390n.
14. A. Kumar, S. Parihar, T. Hammer, and G. Sridhar, "Development and testing of tube type wet ESP for the removal of particulate matter and tar from producer gas," *Renew. Energy*, vol. 74, pp. 875–883, 2015. https://doi.org/10.1016/j.renene.2014.09.006.
15. R. W. R. Zwart, "Gas cleaning downstream biomass gasification," *ECN-E*, vol. 1, no. ECN-E--08–078, p. 65, 2009.
16. P. J. Woolcock and R. C. Brown, "A review of cleaning technologies for biomass-derived syngas," *Biomass Bioenergy*, vol. 52, pp. 54–84, 2013. https://doi.org/10.1016/j.biombioe.2013.02.036.
17. Prabhansu, M. K. Karmakar, P. Chandra, and P. K. Chatterjee, "A review on the fuel gas cleaning technologies in gasification process," *J. Environ. Chem. Eng.*, vol. 3, no. 2, pp. 689–702, 2015. https://doi.org/10.1016/j.jece.2015.02.011.
18. C. Descamps, C. Bouallou, and M. Kanniche, "Efficiency of an Integrated Gasification Combined Cycle (IGCC) power plant including CO 2 removal," *Energy*, vol. 33, pp. 874–881, 2008. https://doi.org/10.1016/j.energy.2007.07.013.
19. B. Burr and L. Lyddon, "A COMPARISON OF PHYSICAL SOLVENTS FOR ACID GAS REMOVAL," Texas, U.S.A., 2008.
20. F. Pinto, R. N. Andre, H. Lopes, I. Gulyurtlu, and I. Cabrita, "Effect of Experimental Conditions on Gas Quality and Solids Produced by Sewage Sludge Cogasification . 2 . Sewage Sludge Mixed with Biomass," vol. 39, no. 13, pp. 2314–2325, 2008. https://doi.org/10.1021/ef700767q.
21. SVB. Van Paasen, MK. Cieplik, and NP. Phokawat, "Gasification of Non-woody Biomass," 2006.
22. L. B. Allegue and J. Hinge, "Biogas and bio-syngas upgrading," 2012.
23. M. L. Rios, A. M. González, E. E. S. Lora, lmazán del Olmo, and O. Agustin, "Reduction of tar generated during biomass gasification: A review," *Biomass Bioenergy*, vol. 108, no. November 2017, pp. 345–370, 2018. https://doi.org/10.1016/j.biombioe.2017.12.002.
24. S. Nakamura, S. Kitano, and K. Yoshikawa, "Biomass gasification process with the tar removal technologies utilizing bio-oil scrubber and char bed," *Appl. Energy*, vol. 170, pp. 186–192, May 2016. https://doi.org/10.1016/j.apenergy.2016.02.113.
25. S. Anis and Z. A. Zainal, "Tar reduction in biomass producer gas via mechanical , catalytic and thermal methods : A review," *Renew. Sustain. Energy Rev.*, vol. 15, no. 5, pp. 2355–2377, 2011. https://doi.org/10.1016/j.rser.2011.02.018.
26. ALBIZ, "Products catalog _ Ankur Scientific Energy Technologies Private, Limited _ ALL," *Ankur scientific*, 2021. https://58372-in.all.biz/goods (accessed Feb. 16, 2021).
27. D. Vera, F. Jurado, N. K. Margaritis, and P. Grammelis, "Experimental and economic study of a gasification plant fuelled with olive industry wastes," *Energy Sustain. Dev.*, vol. 23, pp. 247–257, 2014. https://doi.org/10.1016/j.esd.2014.09.011.
28. B. Drift, "Biomass gasification in the Netherlands," no. July. ECN, Netherlands, p. 33, 2013.
29. M. P. Systems, "MILENA and OLGA get together for high efficiency and low tar - Modern Power Systems," 2015. https://www.modernpowersystems.com/features/featuremilena-and-olga-get-together-for-high-efficiency-and-low-tar-4815584/ (accessed Aug. 25, 2021).

Seaweed Harvesting and Aquaculture: An Overview of the Past 70 years

Teresa Mouga

Abstract In the past decades seaweeds consumption changed from traditional uses such as food and fertilizer to other biotechnological applications, boosting a growing economic activity. Therefore, since the end of the Second World War, seaweeds' aquaculture has been replacing the traditional harvest, accounting for 99.5% of the available biomass in 2020. Nevertheless, in different parts of the globe, this change has happened at different pace. On the other hand, commercial aquaculture only covers a little more than fifteen species, the ones that are economically viable. Asia, notably China, dominates the seaweeds cultivation with over 97.5% and 70% of the market share, respectively. Europe and America are still engaged in the harvesting of wild seaweeds, but this traditional activity has some sustainability implications for the resource. As to Portugal, the activity is still emergent, consistent with little interest among the population in consuming seaweeds. The ongoing investment in research and in the development of pilot scale farming projects in Europe, America, and parts of Africa, however, are enhancing commercial seaweed aquaculture. We expect that these innovative approaches may soon raise the attractiveness of seaweeds aquaculture outside Asia.

Keywords Seaweed production · Traditional harvest · Sustainable resources

1 Introduction

The use of seaweeds is ancient, at least since the Neolithic, over 8000 years ago [1]. The use as food and pharmaceuticals is also recorded in writing in third century Chinese documents [2]. Indeed, coastal populations of Europe, America, Asia, Pacific Ocean Islands, among others, have an ancient and close relationship with seaweeds, harvesting attached to rock or beach-cast seaweeds for domestic use [3]. The most widespread use of seaweeds is likely to be in Asian cuisine, with a typical daily

T. Mouga (✉)
MARE / ARNET, ESTM, Polytechnic University of Leiria, Leiria, Portugal
e-mail: mougat@ipleiria.pt

J. R. da. C. S. Galvão et al. (eds.), *Proceedings of the 3rd International Conference on Water Energy Food and Sustainability (ICoWEFS 2023)*, Springer Proceedings in Earth and Environmental Sciences, https://doi.org/10.1007/978-3-031-48532-9_34

portion size consumed of 8 g dry weight [4]. In other parts of the world, however, seaweeds have been traditionally used mostly for fodder, medicine, or fertilizer [5]. In Portugal the harvesting of this biomass is a long-standing activity, being used mainly as fertilizer. This was a commercial activity regulated since the fourteenth century by the royal charter of king D. Dinis [6].

Since the Second World War, the traditional use of seaweeds has now spread from direct human consumption to other processed formations, such as food additives and nutraceuticals, boosting a growing economic activity [2, 7], sustained by the growing awareness of the seaweeds' nutritional and health benefits [8, 9]. Furthermore, the production of phycoccolloids for industry is also a major trend, as carrageenans, agar or alginates are widely used in the food industry to produce processed foods. Undeniably, the demand from the food industry worldwide is increasing due to the effectiveness of these compounds in adding nutritional value and increasing thickness and stability to food products [10]. Besides, research is unraveling many other biotechnological uses of seaweeds. These include, e.g., pharmacological uses as antioxidant, antimicrobial [11], immunostimulant activities [12], cytoprotective [13] and antitumor properties [14], hypocholesterolemic and hypolipidemic responses [15], diabetes and obesity prevention[16]. Other applications are the energy sector [17], the production of biomaterials for packaging and biofilms [18, 19], the agriculture sector as biostimulant and biofertilizer [20] and bioremediation [21]. More, awareness on sustainable production is driving the consumers preferences to consider healthier and environmentally friendly products. All in all, seaweeds are gaining popularity across the blue economy sector driving the market growth in the forthcoming years.

Accordingly, the seaweed market peaked in 2020 (last year with data) to over 35 million tons (Mt) fresh weight (fw), and a value of $US 16.44 billion (15.50 billion €). The market is expected to continue to grow at an annual rate of over 7.5% [22]. To meet demand, the seaweed market is increasingly supported by aquaculture production, with harvesting becoming less and less important.

This paper looks at the growth of seaweed aquaculture production since the 1950s' compared to the harvesting market. It also analyses the world market, by continent, the main species produced and harvested, and the Portuguese market.

2 Methods

The data analyzed were obtained from the Food and Agriculture Organization of the United Nations (FAO) database FishStatJ, version 4.02.08 [23]. Queries were made to include all aquatic plants. Then, all microalgae were excluded (e.g., *Spirulina, Haematococcus, Dunaliella* and *Tetraselmis*) as well as vascular plants (e.g., aquatic plant nei), as these are not under the seaweed *taxa* Rhodophyta (red algae), Chlorophyta (green algae) or Phaeophyceae (brown algae).

The data obtained were analyzed between 1950 to 2020, for the harvesting and the aquaculture production.

The data were then divided by continent, to assess the production by region.

Finally, the data were analyzed by *taxa* Rhodophyta, Chlorophyta, and Phaeophyceae, to assess the production by taxonomic group.

On specific Portuguese and European data, the European Marine Observation and Data Network [24] was accessed. A query was made for "Algae Production", selecting "Macroalgae (seaweeds)", per country, and including all large, medium, and small enterprises. For Portugal, all regions were included, either continental or islands.

Finally, to get data on Portuguese companies, the online business information platform Racius [25] was used.

3 Results and Discussion

3.1 World Seaweed Production

One of the ways of safeguarding wild seaweeds' populations is to guarantee enough biomass through cultivation. In the early 1950s, wild harvesting and aquaculture of seaweed contributed in similar amounts to global production, with around 0.5 Mt fw each (Fig. 1). Yet, while the harvest grew to around one Mt in the late 1960s and has fluctuated around this number ever since, seaweeds' aquaculture increased to 35 Mt in 2020, currently contributing 99.5% of total seaweeds in the market [23]. The demand on red seaweeds phycocolloids fueled aquaculture development because the wild stocks of Irish moss (*Chondrus crispus* and *Mastocarpus stellatus*), the natural source of carrageenan, were unable to meet the needs of the market [26]. Intense research in aquaculture protocols developed in Asian countries allowed a significant increase in world seaweed's production since early 1990'. The growth of aquaculture has shown a steep growth ever since.

The number of cultivated species though is very limited, including little more than fifteen *taxa*. The most produced *taxon* worldwide in 2020 is the cold-water brown algae *Laminaria/Saccharina japonica* (kombu), with over 12.4 Mt of fw (35.6% of worldwide production). In second are the red seaweeds *Kappaphycus/Eucheuma* with 10 Mt produced in 2020 (28%). In third place is the red seaweed *Gracilaria* sp. with 5.2 Mt and a world share of 14.9%, followed by the red *Porphyra/Pyropia* (nori) with more than 3 Mt and a 9% share. In fifth is the brown *Undaria pinnatifida* (wakame) with 8% and, finally, is the brown *Sargassum fusiforme* with about 0.2% (Table 1 and Fig. 2).

The brown seaweeds *Laminaria japonica, Undaria pinnatifida, Sargassum fusiforme,* and *Macrocystis pyrifera* are produced mainly for human consumption, animal feed, are used in agriculture as a bio-fertilizer/bio-stimulant and in the phycocolloid industry, as they produce alginate. The cultivated red seaweeds' *Pyropia/Porphyra* sp. (nori) are produced for food, *Gracilaria* sp. for the agar industry, and *Eucheuma/Kappaphycus alvarezii* for the extraction of carrageenans, all phycocolloids widely used in the food industry.

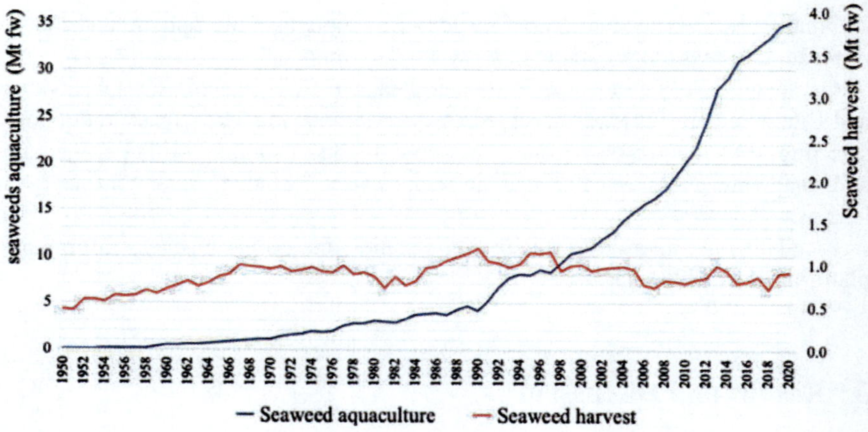

Fig. 1 Seaweed's aquaculture (left, in blue) and harvest (right, in red) in millions of tons of fresh weight, produced between 1950 and 2020 (note the difference in y-axis scales)

Table 1 Main species of seaweeds produced in aquaculture, quantity produced in 2020 (thousand tones fresh weight), mean value per ton, top producing countries, and main market

Taxa	kt (fw)	$US/t	Top producing countries	Purpose
Laminaria japonica	12 469.80	$ 372.47	China, Japan	Food, Feed
Eucheuma sp.	8 283.50	$ 203.74	Indonesia, Tanzania, Philippines, Zanzibar	Carrageenan production
Gracilaria sp.	5 182.10	$ 456.11	China, Indonesia, Philippines	Feed, agar production
Porphyra/Pyropia sp.	3 048.02	$ 925.95	China, Japan, Korea	Food
Undaria pinnatifida	2 810.60	$ 750.40	China, Japan, Korea, Russia	Food, Feed
Kappaphycus alvarezzi	1 609.40	$ 999.77	Indonesia, Malaysia, Philippines, Vietnam	Carrageenan production
Sargassum fusiforme	292.90	$ 800.33	China, Japan, Korea	Food

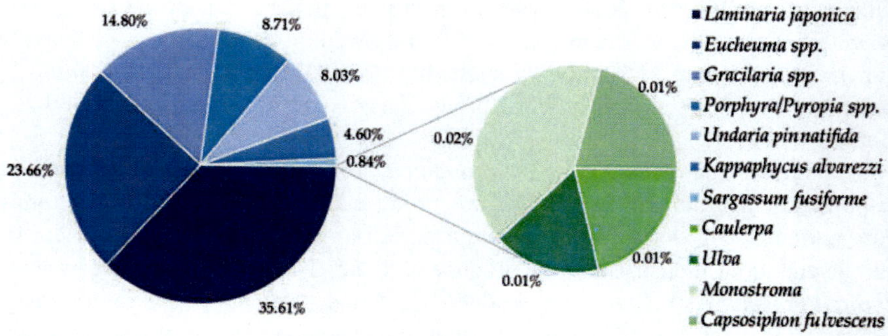

Fig. 2 Perentage of the most produced seaweeds in 2020

Produced in much lower quantities, the most cultivated *taxa* of green seaweeds are *Ulva* sp. (sea-lettuce), *Monostroma nitidum, Capsosiphon fulvescens, Caulerpa* sp. and *Codium* sp., all accounting for only 15.5 mt in 2020 corresponding to less than 0.1% of the total (Fig. 2). These seaweeds are mainly used for food, but have other applications, such as in agriculture, as a biofertilizer/biostimulant, in the pharmaceutical industry and in bioremediation [27].

Seaweed's aquaculture is dominated by Asian countries, which in 2020 have contributed with more than 99% of total seaweed aquaculture biomass (Table 2). The top producing country is China, with 60% of the total market, followed by Indonesia, with 27% of the market, South Korea with 5% and the Philippines with 4%. Less than 1% is shared by the other continents, with Zanzibar, in Africa, being the top producer (0.26%), Chile in South America (0.05%), and Norway in Europe (0.001%).

Although Europe currently has a minor contribution to the world market (0.71%), the European Union aims to develop seaweeds aquaculture in the next years. Research is being undertaken to improve the technology and production protocols, in line with the "Strategic guidelines for more sustainable and competitive EU aquaculture for the period 2021 to 2030" [28]. These guidelines aim at more competitive and resilient aquaculture sector, ensuring the health and sustainability of the seafood, creating economic opportunities, and reducing the seafood imports while decreasing the carbon footprint. So, although in small quantities, other species are being produced in Europe. Examples are the brown seaweeds *Alaria esculenta* (Babberlocks) in Ireland, Faroe Islands and Norway, *Saccharina latissima* (Sea belt) in Denmark, Faroe Islands, France, Norway and Spain, *Laminaria hyperborea* (Tangle) in Faroe Islands and Ireland, and *Undaria pinnatifida* (Wakame) in France and Spain. Besides these species, other red and green seaweeds are reported, such as *Ulva* sp. (Sea-lettuce) and *Gracilaria* sp. [27].

The production of new species in aquaculture demands for significant extra research efforts and funding to select seaweed species and strains with specific properties adapted for the new biotechnological applications. At the same time, these species must be economically viable, optimizing the cost benefit ratio [26, 29]. New ongoing pilot scale farming projects in Europe, America and parts of Africa are underway enhancing commercial seaweed aquaculture outside Asia [30].

Table 2 Seaweeds' harvest and aquaculture, in tones of fresh weight, per continent, in 2020

Continent	Seaweeds harvested	Seaweeds aquaculture	Total seaweeds	% harvested seaweeds	% world aquaculture	% world production
Asia	143 086	34 884 438	35 023 524	0.41%	99.633%	97.49%
America	484 871	23 874	508 745	95.31%	0.068%	1.42%
Europe	254 023	641	254 664	99.75%	0.002%	0.71%
Africa	30 466	103 941	134 407	22.67%	0.297%	0.37%
Oceania	2 647	105	2 752	96.18%	0.000%	0.01%
Total	915 095	35 012 998	35 928 093	2.55%		

3.2 *Seaweed Aquaculture* Versus *Harvest*

Even though the world production already relies almost entirely on aquaculture (97,49%), there are still some species wild harvested. In 2020, 915 thousand tones (kt) of seaweeds were harvested, of which 53% in America, 28% in Europe, 16% in Asia, 3% in Africa, and less than 0,3% in Oceania (Table 2). America is the continent most contributing to harvesting, with 485 kt of seaweeds harvested (95.3% harvested seaweeds, only 4.7% cultivated). Notably *Ascophyllum nodosum* (Rockweed) in Canada and United States of America is used as a fertilizer, and *Macrocystis* sp., *Lessonia* sp. (Chilean kelp), *Gigartina* sp. *Gracilaria* sp., and *Sarcothalia crispata* (Leister) in Chile and Peru [27]. Chilean brown seaweeds are responsible for the supply of about 10% of the raw material for alginates [31].

Following is Europe, with more than 250 kt harvested, corresponding to almost the total of the seaweeds' European market (99.75%) for only 641 t were produced in 2020. Seaweeds harvest includes the species *Alaria esculenta* (Babberlocks), harvested in Iceland, *Laminaria* sp. harvested in Norway, Spain, France, Ireland, and Iceland, *Saccharina latissima* in Denmark, Spain, France, Greece, Ireland, Netherlands, and Norway. *Ascophyllum nodosum* (Rockweed) is harvested in Spain, France, Iceland, Ireland, and Norway. *Palmaria palmata* (Dulse) is harvested in Spain, Ireland, Netherlands, and Norway. *Chondrus crispus* (Irish Moss) is harvested in Ireland. *Porphyra* (Laver) is harvested in Ireland, Iceland, and Norway. *Fucus* sp. is harvested in Denmark, Spain, France, and Ireland. *Gelidium corneum* is harvested in Spain and Portugal. *Himanthalia elongata* (Sea thong) is harvested in Spain, and *Ulva* sp. (Sea-lettuce) is harvested in Spain, Italy, and Ireland [24]. All these species have been harvested for hundreds of years, sharing histories of use by coastal human communities [31].

Africa harvests a little more than 30 kt (22% harvest, 78% cultivation) including mostly red seaweeds (*Gelidium corneum* harvested in Morocco and kelps in South Africa). Asia harvests less than 0.5% of the seaweeds, including *Sargassum* sp. in Indonesia and kelps (*Saccharina/Laminaria*) in Japan [27].

The most sought-after wild seaweeds are kelps (brown seaweeds), mainly from *Lessonia* sp. from Chile, North European kelp *Laminaria hyperborea* from Norway and France, and *Sargassum* sp. from Indonesia, accounting for more than 520 kt. The most harvested red seaweeds are *Gracilaria* sp. and Leister *Sarcothalia crispata,* harvested in Chile, with 78 kt (Fig. 3).

Wild seaweed harvesting has been carried out sustainably for decades, notably in Canada, hence the numbers remain rather stable. The exception is the harvesting of *Gelidium*, required to produce microbiological-grade agar. Gelidiales have experienced over-harvesting in Japan and on the European–North African Atlantic coast [29]. Currently, the demand relies mostly on Morocco *Gelidium corneum* agar, and the local resources are poorly managed [32]. Because there is no farmed substitute for the microbiological-grade agar producing raw material, this over-harvest causes serious concerns on the sustainability of the resource [33]. Likewise, Red, and Brown Chilean seaweeds' have also experienced over-harvesting [30]. Over-harvest of these

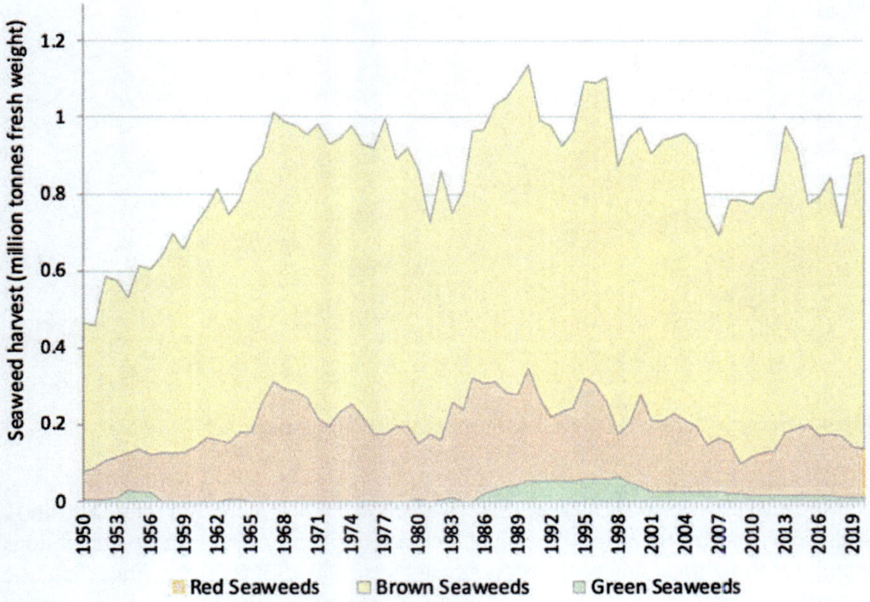

Fig. 3 Harvest of main *taxa* of seaweeds (1950 to 2020)

marine resources has ecological impacts on the marine ecosystems. Being foundational species, these canopy-forming seaweeds provide shelter, food, and nursery habitat to many marine species [34]. The reduction of the seaweeds' communities can have severe impact on the recruitment of other marine species. It may also reduce the contribution to the marine carbon cycle, decrease nutrient removal and the carbon sequestration [35]. Sustainable harvest relies on accurate data of seaweeds standing stocks, but there is a large knowledge gap of long-term biomass data in most countries [31]. These data are mandatory to safeguard robust harvest management plans. Other threats to these seaweeds are the warming of the seawater [36] and the introduction of alien species [37].

3.3 The Portuguese Seaweed Market

The harvesting of seaweeds, to be used as fertilizer, is an ancient activity all along the Portuguese coast. Taking different names (*moliço*, *sargasso* or *limo*), mixtures of seaweeds were beach-cast or harvested in the sea, dried, and then spread on agricultural fields. This activity, though, fell into disuse with the introduction of chemical fertilisers [6]. Around the same time, nonetheless, red macroalgae, especially the genera *Gelidium* and *Pterocladiella*, started being used as a feedstock for agar extraction. Japan dominated the market, but during the Second World War Portugal

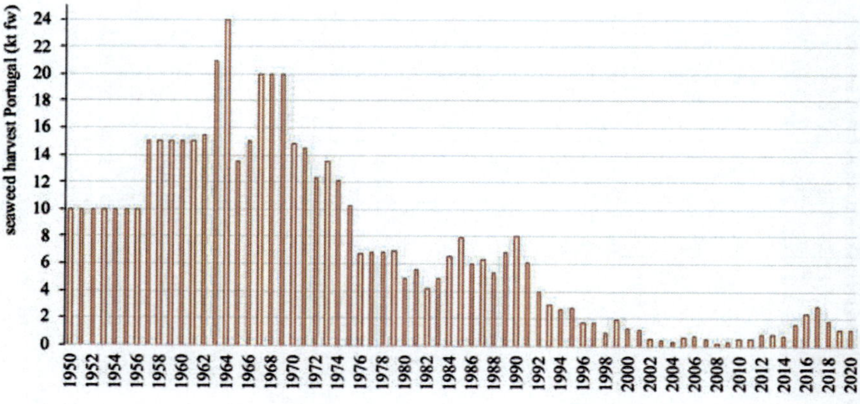

Fig. 4 Seaweeds' harvest in Portugal, in thousand tons of fresh weight (1950 to 2020)

significantly increased its market share. Thus, the harvest increased until the 1960', with a decrease until the 1980', a small increase until early 1990', these fluctuations being due to political and socio-economical constrains [32] (Fig. 4). Since then, the harvest has significantly decreased, except for one area in Portugal where the activity still has important local economic impact [6]. In 2020 around 1750 t of red seaweeds were harvested [38] of which 85% in S. Martinho do Porto and 10% in the Azores archipelago. The harvest is managed by local harvesters' associations and the harvest is purchased by an Iberian company for the agar industry (Iberagar SA.).

As to seaweeds market, there are only five companies registered in Portugal, producing seaweeds in aquaculture systems. Only two operate on an industrial scale: Algaplus, Lda. a land based IMTA industrial semi-intensive aquaculture (https://www.algaplus.pt) and Unidade Bioquímica Lda. (UBQ II), in the Madeira Archipelago, which we were unable to confirm the status. The other three (Aquazor, Lusalgae and Seaforester) are still pilot scale companies. These five companies are producing mostly *Porphyra* sp., *Codium* sp., *Gracilaria* sp., *Ulva* sp., *Laminaria* sp. and *Palmaria palmata.* Since 2013, these companies produced green (up to 35 t a year), red (up to 1.27 t a year) and brown seaweeds (up to 0.16 t a year) [23, 24], directed mainly for export to European countries.

Regarding harvesters, besides the above-mentioned associations, a company in the Azores archipelago harvests several seaweed species for food, cosmetics, and biotechnology (SeaExpert, http://seaexpert-azores.com/). Again, most of the seaweeds are destined to other European countries. Another 23 companies have registered seaweed harvesting but didn't report this economic activity [25]. Therefore, the seaweed market in Portugal is incipient, consistent with little interest among the population in consuming seaweeds. Increasing knowledge of the uses of algae, such as in agriculture and biotechnology, may soon raise the attractiveness and therefore allow the market to expand.

4 Conclusions

Seaweeds market has been expanding in the past decades, due to new innovative applications and it is expected to continue growing in the next years. Despite the dominance of Asia in aquaculture production, Europe is the main world importer of seaweeds, mainly for the food industry [39], driving the European market to continue to grow exponentially. Therefore, ongoing research for innovative, less expensive, more sustainable, and efficient technology, and the development of more productive seaweed's strains are boosting the interest in the cultivation outside Asia.

The ecosystem services paid by seaweeds cultivation besides provision services, including the carbon sequestration, buffer for the ocean acidification, nutrient removal, the habitat, and enrichment provided to other marine organisms, and, in coastal areas, the wave attenuation [40] are other drivers for the growing attractiveness of seaweed cultivation. The European seaweed market is therefore expected to change significantly over the next decade.

Acknowledgements This work was funded by national funds through FCT—Fundação para a Ciência e a Tecnologia for I.P., within the scope of the project MARE (UIDB/04292/2020), and the project LA/P/0069/2020, granted to the Associate Laboratory ARNET.

References

1. Becket, A., Macgregor, G., Duffy, P., Finlay, N., Miller, J., Sheridan, A.: Forest grazing and seaweed foddering: early Neolithic occupation at Maybole, South Ayrshire. Proceedings of the Society of Antiquaries of Scotland. 139, 105–122 (2009)
2. Yang, L.-E., Lu, Q.-Q., Brodie, J.: Rhodophyta) in China: history, culture and taxonomy. Eur J Phycol. 52, 251–263 (2017). https://doi.org/https://doi.org/10.1080/09670262.2017.1309689
3. Pereira, L.: Edible Seaweeds of the World. CRC Press (2016)
4. Macartain, P., Gill, C.I.R., Brooks, M., Campbell, R., Rowland, I.R.: Nutritional Value of Edible Seaweeds. Nutr Rev. 65, 535–543 (2007).
5. Pereira, L., Cotas, J.: Historical Use of Seaweed as an Agricultural Fertilizer in the European Atlantic Area. In: Pereira, L., Bahcevandziev, K., and Joshi, N.H. (eds.) Seaweeds as Plant Fertilizer, Agricultural Biostimulants and Animal Fodder. pp. 1–22. CRC Press, Taylor & Francis Group, Boca Raton (2019)
6. Cavaco, M., Duarte, A., Bernardino, S., Afonso, C., Mouga, T.: Sustainable Use of Seaweeds from S. Martinho do Porto, Portugal - Past, Present, and Future Perspective. In: da Costa Sanches Galvão, J.R., et al. (eds.) Proceedings of the 1st International Conference on Water Energy Food and Sustainability (ICoWEFS 2021). pp. 1–11. Springer International Publishing, Leiria (2021)
7. McHugh, D.J.: A Guide to the Seaweed Industry. Rome, Italy (2003)
8. van der Spiegel, M., Noordam, M.Y., van der Fels-Klerx, H.J.: Safety of novel protein sources (insects, microalgae, seaweed, duckweed, and rapeseed) and legislative aspects for their application in food and feed production. Compr Rev Food Sci Food Saf. 12, 662–678 (2013). https://doi.org/https://doi.org/10.1111/1541-4337.12032
9. Pacheco, D., Souza Araújo, G., Cotas, J., Gaspar, R., et al.: Invasive Seaweeds in the Iberian Peninsula: A Contribution for Food Supply. (2020)

10. Kraan, S.: Algal Polysaccharides, Novel Applications and Outlook. In: Chang, C.F. (ed.) Carbohydrates - Comprehensive Studies on Glycobiology and Glycotechnology. pp. 489–532. InTech Open Science, Croatia, Rijeka (2012)
11. Freitas, M. V., Inácio, L.G., Ruas, A., Silva, I.A., Mouga, T., Pereira, L., Afonso, C.: Antioxidant and Antimicrobial Properties of Selected Red Seaweeds from Central Portugal. Applied Sciences. 13, 157 (2022). https://doi.org/https://doi.org/10.3390/app13010157
12. Passos, R., Correia, A.P., Ferreira, I., Pires, P., Pires, D., Gomes, E., do Carmo, B., Santos, P., Simões, M., Afonso, C., Baptista, T.: Effect on health status and pathogen resistance of gilthead seabream (Sparus aurata) fed with diets supplemented with Gracilaria gracilis. Elsevier B.V. (2021)
13. Pinteus, S., Silva, J., Alves, C., Horta, A., Fino, N., Rodrigues, A.I., Mendes, S., Pedrosa, R.: Cytoprotective effect of seaweeds with high antioxidant activity from the Peniche coast (Portugal). Food Chem. 218, 591–599 (2017). https://doi.org/https://doi.org/10.1016/j.foodchem.2016.09.067
14. Pádua, D., Rocha, E., Gargiulo, D., Ramos, A.A.: Bioactive compounds from brown seaweeds: Phloroglucinol, fucoxanthin and fucoidan as promising therapeutic agents against breast cancer. Phytochem Lett. 14, 91–98 (2015)
15. Cardoso, S.S.M., Carvalho, L.L.G.L., Silva, P.P.J., Rodrigues, M.S.M., Pereira, O., Pereira, L.: Bioproducts From Seaweeds: A Review With Special Focus On The Iberian Peninsula. Curr Org Chem. 18, 896–917 (2014). https://doi.org/https://doi.org/10.2174/138527281807140515154116
16. Cherry, P., O'hara, C., Magee, P.J., Mcsorley, E.M., Allsopp, P.J.: Risks and benefits of consuming edible seaweeds. Nutr Rev. 77, 307–329 (2019)
17. Fasahati, P., Dickson, R., Saffron, C.M., Woo, H.C., Liu, J.J.: Seaweeds as a sustainable source of bioenergy: Techno-economic and life cycle analyses of its biochemical conversion pathways. Renewable and Sustainable Energy Reviews. 157, 112011 (2022)
18. Andrade, M.A., Barbosa, C.H., Souza, V.G.L., Coelhoso, I.M., Reboleira, J., Bernardino, S., Ganhão, R., Mendes, S., Fernando, A.L., Vilarinho, F., Silva, A.S., Ramos, F.: Novel active food packaging films based on whey protein incorporated with seaweed extract: Development, characterization, and application in fresh poultry meat. Coatings. 11, 1–12 (2021)
19. Yu, G., Zhang, Q., Wang, Y., Yang, Q., Yu, H., Li, H., Chen, J., Fu, L.: Sulfated polysaccharides from red seaweed Gelidium amansii: Structural characteristics, anti-oxidant and anti-glycation properties, and development of bioactive films. Food Hydrocoll. 119, (2021)
20. Tuchy, L., Chowańska, J., Chojnacka, K.: Seaweed extracts as biostimulants of plant growth: Review. Chemik. 67, 636–641 (2013)
21. Neveux, N., Bolton, J.J., Bruhn, A., Ras, M.: The Bioremediation Potential of Seaweeds: Recycling Nitrogen, Phosphorus, and Other Waste Products. In: la Barre, Stephane Bates, S.S. (ed.) Blue Biotechnology: Production and Use of Marine Molecules. pp. 217–239. Wiley (2018)
22. Fortune Business Insights: Commercial Seaweed Market, https://www.fortunebusinessinsights.com/industry-reports/commercial-seaweed-market-100077
23. FAO: Fishery and Aquaculture Statistics. Global capture production 1950–2020 (FishStatJ), www.fao.org/fishery/statistics/software/fishstatj/en
24. EU: European Marine Observation and Data Network, https://www.emodnet-humanactivities.eu/download-data.php
25. Racius, online business information platform, https://www.racius.com/apanha-de-algas-e-de-outros-produtos-do-mar/em-atividade/
26. Hafting, J.T., Craigie, J.S., Stengel, D.B., Loureiro, R.R., Buschmann, A.H., Yarish, C., Edwards, M.D., Critchley, A.T.: Prospects and challenges for industrial production of seaweed bioactives. J Phycol. 51, 821–837 (2015). https://doi.org/https://doi.org/10.1111/jpy.12326
27. Cai, J., Lovatelli, A., Aguilar-Manjarrez, J., et al.: Seaweeds and microalgae: an overview for unlocking their potential in global aquaculture development. FAO, Rome, Italy (2021)
28. EU: Strategic guidelines for more sustainable and competitive EU aquaculture for the peri-od 2021 to 2030, https://eur-lex.europa.eu/legal-content/EN/TXT/?uri=CELEX: 52021DC0236

29. Buschmann, A.H., Camus, C., Infante, J., Neori, A., Israel, Á., Hernández-González, M.C., Pereda, S. v., Gomez-Pinchetti, J.L., Golberg, A., Tadmor-Shalev, N., Critchley, A.T.: Seaweed production: overview of the global state of exploitation, farming and emerging research activity. Eur J Phycol. 52, 391–406 (2017). https://doi.org/10.1080/09670262.2017.1365175
30. Buschmann, A.H., Hernández-González, M.D.C., Varela, D.: Seaweed future cultivation in Chile: Perspectives and challenges. Int J Environ Pollut. 33, 432–456 (2008)
31. mac Monagail, M., Cornish, L., Morrison, L., Araújo, R., Critchley, A.T.: Sustainable harvesting of wild seaweed resources. Eur J Phycol. 52, 371–390 (2017)
32. Santos, R., Melo, R.A.: Global shortage of technical agars: back to basics (resource management). J Appl Phycol. 30, 2463–2473 (2018). https://doi.org/https://doi.org/10.1007/s10811-018-1425-2
33. Mouga, T., Fernandes, I.B.: The Red Seaweed Giant Gelidium (Gelidium corneum) for New Bio-Based Materials in a Circular Economy Framework. Earth. 3, 788–813 (2022)
34. Verdura, J., Sales, M., Ballesteros, E., Cefalì, M.E., Cebrian, E.: Restoration of a canopy-forming alga based on recruitment enhancement: Methods and long-term success assessment. Front Plant Sci. 9, 1832 (2018). https://doi.org/https://doi.org/10.3389/FPLS.2018.01832/BIBTEX
35. Duarte, C.M., Gattuso, J., Hancke, K., Gundersen, H., et al: Global estimates of the extent and production of macroalgal forests. Global Ecology and Biogeog. 31, 1422–1439 (2022)
36. Franco, J.N., Tuya, F., Bertocci, I., Rodríguez, L., Brezo Martínez, I., Sousa-Pinto, I., Arenas, F.: The 'golden kelp' Laminaria ochroleuca under global change: Integrating multiple eco-physiological responses with species distribution models. J Ecol. 106, 47–58 (2018).
37. Blanco, A., Larrinaga, A.R., Neto, J.M., Troncoso, J., Méndez, G., Domínguez-Lapido, P., Ovejero, A., Pereira, L., Mouga, T.M., Gaspar, R., Martínez, B., Lemos, M.F.L., Olabarria, C.: Spotting intruders: Species distribution models for managing invasive intertidal macroalgae. J Environ Manage. 281, 111861 (2021). https://doi.org/https://doi.org/10.1016/j.jenvman.2020.111861
38. DGRM: Recursos Pesca, https://www.dgrm.mm.gov.pt/web/guest/recursos-da-pesca-dgrm
39. Kraan, S.: Seaweed and Alcohol. In: Seaweed in Health and Disease Prevention. pp. 169–184. Elsevier (2016)
40. Kim, J.K., Yarish, C., Hwang, E.K., Park, M., Kim, Y.: Seaweed aquaculture: Cultivation technologies, challenges and its ecosystem services. Algae. 32, 1–13 (2017).

Production and Material Types for Pallets: A Review

Alejandro Pulgar and Nelson S. Oliveira

Abstract Pallets are used extensively in the supply chain for the transportation of goods between suppliers, manufacturers, distributors, and retailers. This article reviews pallet production, its mechanical properties related to material options (according to European standards and industrial applications) and management systems. In the global supply chain, polymer, metal, and composites pallets compete in line with the concept of sustainable development. The supply chain requires pallets according to specific characteristics, such as sizing, maximum load capacity, durability (lifespan), cost and more information is needed about their environmental impacts. Some types of pallets (stringer and blocks) can have several configurations, number, width and spacing. Furthermore, the pallets management system is divided in three models; EMP (extensive management of pallets), TPO (transfer pallet's ownership) and PR (pallet rent), where suppliers, distributors and retailers select the most suitable option. The systematic review of the literature identified 15 documents with the key methodological elements, such as, specifications, inventory data, goals, and system boundaries for a life cycle assessment of pallets.

Keywords Pallets production · Pallets materials · Pallets sustainability

A. Pulgar
School of Technology and Management, Polytechnic Institute of Leiria, Leiria, Portugal

N. S. Oliveira (✉)
E-LCM—Laboratory of Separation and Reaction Engineering—Laboratory of Catalysis and Materials, Escola Superior de Tecnologia E Gestão, Instituto Politécnico de Leiria, Leiria, Portugal
e-mail: nelson.oliveira@ipleiria.pt

ALiCE—Associate Laboratory in Chemical Engineering, Faculty of Engineering, University of Porto, Rua Dr. Roberto Frias, Porto, Portugal

J. R. da. C. S. Galvão et al. (eds.), *Proceedings of the 3rd International Conference on Water Energy Food and Sustainability (ICoWEFS 2023)*, Springer Proceedings in Earth and Environmental Sciences, https://doi.org/10.1007/978-3-031-48532-9_35

1 Introduction

Pallets are the small piece in the machine that moves goods in the global industry. The impact of pallets in the today´s supply chain gives a relevant importance to this revision, even from the environmental perspective. In addition, waste recycling in the European Union supports using eco-friendly materials to make a transition towards circular economy [1–2]. According to [3], in Europe, pallets production will have an increase of $6.2 bn between 2022 and 2026 (this represents 65% of European pallet market growth only from western Europe), this data was obtained from principal vendors and pallets producers, depending on materials and geographical landscape. This report involved areas such as sizing, forecast and industry analysis. On the other hand, war in Ukraine impacts in the European pallet market, due to sanctions against Russia and Belarus from EU, and raw materials are no available from those countries and lorry drivers creating a potential delay in the transportation. At the same time an increase in demand for EPAL Euro pallets (new or used) will appear during this period.

This literature review was focused on pallets made of different materials and their mechanical behaviour. Moreover, revision on pallet types is needed to compare between management strategy and industrial demand. The main goal of this research is to identify parameters that affect the selection of the best pallet material and the management process. The result of this review provides information to the development of a Multicriteria analysis (MCA) which will include the results of a Life Cycle Assessment (LCA) to determine the best pallet option (less environmental impact and energy consumption). Optimizing the pallet production process with less waste while increasing product quality and functionality.

2 Review Process and Studies Reviewed

2.1 *Review Process*

This study systematically reviews literature on the pallets manufacturing process and materials, based on the PRISMA Statement guideline. The search for literature was performed using the search strings of "pallets", "manufacturing", "process", "management" and "materials" in keywords, title and abstract of SCOPUS and internet (google academic/scholar and refseek). The result yielded 154 in total, 133 from SCOPUS and 21 from internet. A total of 46 studies were removed for being duplicated. Out of them, two studies could not be retrieved by the authors, namely [4–5]. Thus, 106 articles were analyzed in the review process, out of where 71 were excluded for being outside the scope of this paper. Finally, a total of 15 studies were left for quantitative and qualitative review. Figure 1 presents the PRISMA flow diagram of the review process.

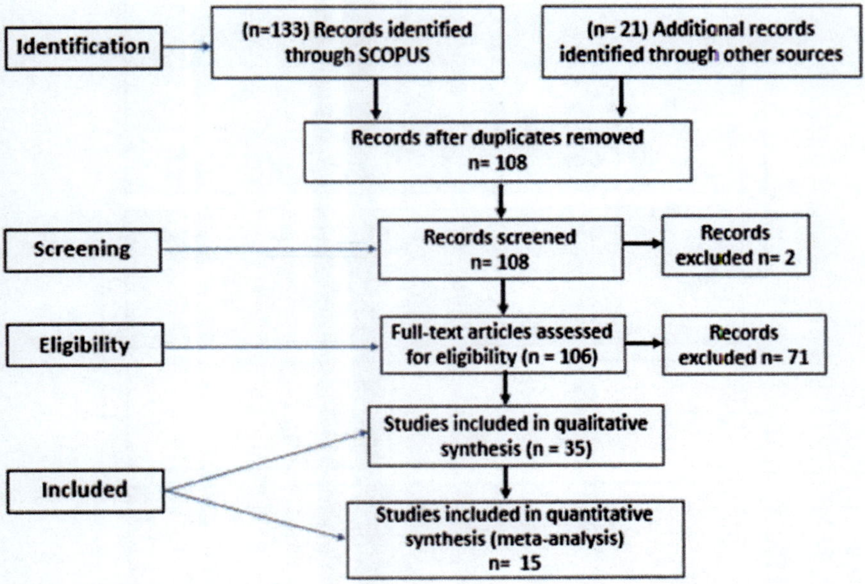

Fig. 1 Process review flow diagram for pallets. Flow diagram adapted from [6]

2.2 *Studies Reviewed*

In the final review process a total of 15 documents were analyzed. A total of 10 were scientific journal articles and focused on types of pallets (properties and the comparison of pallets produced from different materials, such as wood and plastic), models of production, manufacturing and data from their life cycle stages (References: [1, 7–15]). Other documents showed a comparison among different pallet frames and new materials behavior on pallet design (References: [2, 16–19]).

3 Classification of Pallets

Pallets are made of different raw materials, with different dimensions, types of production and management strategy. Materials such as wood, composites, polymers, and metal are used, however, wood (softwood, hardwood, and technical wood), plastic (virgin plastic and recycled) and composites are the most common. Moreover, some sizes are used depending on the region, Europe produces a standard dimension, China, GMA (grocery manufacturers association—USA), Australia and others make pallets with a specific size also (Fig. 2). Pallet types can be differentiated between stringer (a stringer pallet is called that because it uses "stringers", which support the unit load. The stringers are the boards, typically 50.8 × 101.6 mm or 76.2 ×

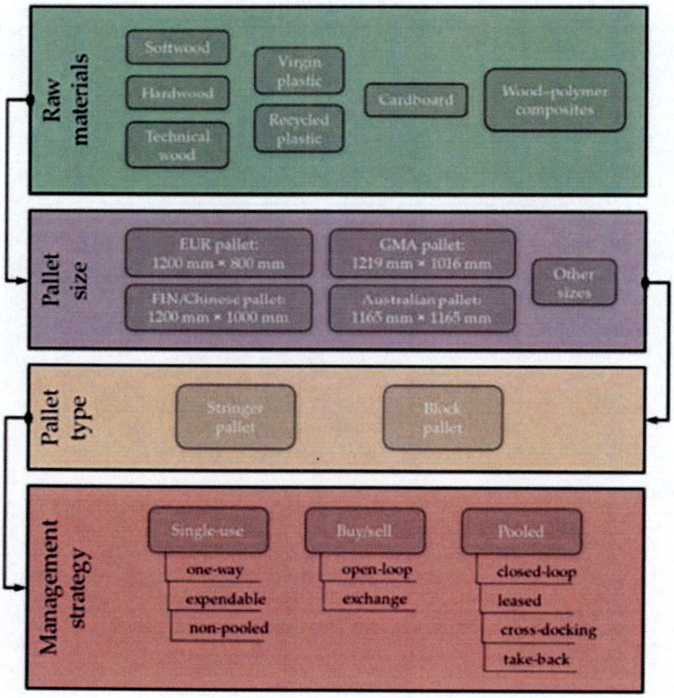

Fig. 2 Classification of pallets [8]

101.6 mm, sandwiched between the top and bottom deck boards) or block configuration (block pallets use 4 to 12 blocks to support the top deckboards). Stringer pallets are the most used in USA [20]. Additionally, there are three common industry strategies for managing pallets [7]; single use expandable pallet system (that involves an open-loop system with no salvage value), buy/sell program with salvage value (open-loop system) and closed-loop system (leased pallet pooling program or rental mode).

4 Material and Production of Pallets

Several pallets are founded in the market; wood and composites are commonly used; however, other materials (metal or polymers) can be obtained. The general pallets manufacturing process can follow a similar journey, this starts receiving a raw material (pallets lumber/cants for example), then are trimmed, ripped and sawn into the proper size (sizing), for whatever type of pallet is needed.

Table 1 Wooden pallet weights in Australia and China [14]

Pallet type	Australia (1165 mm × 1165 mm)	China (1200 mm × 1000 mm)
Softwood pooled	34 kg	34 kg
Hardwood pooled	43 kg	43 kg
Softwood simple/one-way	22.5 kg	22.5 kg

4.1 Wooden Pallets

Wooden Pallets are the most used around the world, however, depending on the location wood can vary, due to the availability of the area. Pallets are basically made of hardwood from Angiosperm trees (oaks, maples, and sycamore) and softwood from gymnosperm trees such as conifers. On the other hand, wooden pallets are obtained from recycling (wood) where their mechanical properties can be different.

During the production of wooden pallet exists several inputs such as, raw material (wood), nails to fasten, electricity for the process equipment, thermal energy, or fuel (for phytosanitary treatment) where conventional heat treating is usually used. Pallets dimensions and mass can vary, due to density and moisture of wooden materials or sizing specifications. According to the European Pallet Association, the Euro pallet (1200 × 800 mm) has an average weight of 25 kg, however, the SFN-EN 13,698–1 standard estimate the total weight of the Euro pallet around 21.82 kg. Moreover, the electricity demand has a range from 0.43 to 7.9 MJ and a thermal energy demand of 4.1 ± 2.5 MJ (average) per EUR pallet. Fuel demand has a range from 0.017 to 0.16 dm^3 per EUR pallet [8].

Pallets made of wood in USA are produced in different sizes, depending on the industry, a 1219 × 1016 mm pallet (GMA) has a weight range between 16.58 and 18.57 kg (with dry and moisture content), and to produce a single pallet is necessary 0.11 MJ of electricity, 0.189 cm^3 of diesel and 9.32 kg of water [9]. In Australia and China [14], wooden pallets have a weight according to the wood type (Table 1), with an electricity consumption from 1.7 to 3.6 MJ in Australia and 1.8 to 3.6 MJ in China. The consumption of fuel oil is higher in Australia (20 kg oil eq.) than China (0.6 kg oil eq.).

4.2 Plastic Pallets

Plastic pallets production requires polymers itself (virgin or recycled), commonly polypropylene and polyethylene [10]. Electricity for thermoforming and fuel for equipment is necessary and no phytosanitary treatment is needed. On the other hand, there is not a standard for sizing on pallets, then, size varies significantly, also is common to see sizing equals to wooden pallets, and usually plastic pallets have a maximum weight of 34 kg (with an input polymer raw material of 21 ± 7 kg/m^2). This type of pallets requires more electricity than a wooden pallet (between 24.5

to 1292 MJ per 1 m^2 of the pallet surface). Small fuel consumption (diesel) was reported by some researchers from 0.04 to 0.63 dm^3 [8].

4.3 Composite Pallets

Wood-plastic composite (WPC) is made by mixture of wood (in shape of sawdust) and polymeric materials (PE or PP). Moreover, WPC pallets are heavy weighted (density: 650–800 kg/m^3) and do not require any fixing elements. Extrusion process (from a 3D profile) and moulding (final product) permit redesign, optimize and produce pallets with a high shear rate [11]. Also, PP can be mixed with glass fibers or natural fibers, and another biocomposites with vary polymers matrix types, including PVC (polyvinyl chloride), PE or PP, in addition with several fillers, can be used to manufacture pallets [21].

4.4 Metal Pallets

Metal pallets design process is challenging and do not follow a specific standard, those can be made of steel or aluminum. Aluminum is more expensive than steel, and both represent an advantage in companies that take a long-term approach to their assets. Also, metals pallets represent about 1.5 times the cost of wood or plastic pallets, however, durability is the higher selling point of this product [19]. Main characteristics of those pallets such as weight and durability, also determine their cost and their impact on the life cycle. The value for electricity consumption is about 11.05 MJ (to produce one pallet); nevertheless, metal pallets make up less than 1% of the market [12].

4.5 General Properties of Pallets

For a given application of pallets, it has been necessary to define at least five performance indicators; Strength, Stiffness, Durability, Functionality and Purchase price. It was not found in the literature performance indicators relating to its environmental performance, considering its life cycle. Optimizing one has an effect on the others. Then, it must be important to find a balance to select the best pallet option. Pallets have different characteristics and behavior depending on their material. For instance, wooden pallets are easy to prototype (applying Pallet Design System), and their properties are referenced from ASME MH1 to ensure conformance to industry best practices [18]. Additionally, wood pallets can support a load of 2500 kg (high strength), also, are cheaper and more stable. However, water and environmental conditions affect their mechanical behaviour [17].

Table 2 Comparative analysis of different pallets materials [12]

Factor	Wood	Plastic	Paper	Metal
Cost	3	1	2	1
Durability/lifespan	1	3	1	3
Mechanical resistance	1	3	1	3
Stiffness	N/A	1	1	3
Sustainability	2	1	3	2
Use	3	3	1	1
Weight	1	3	3	1

Several materials have been analyzed from a literature [12], based on a set of factors (used to determine the pallet's suitability for a given application). According to this, it was mentioned (Table 2) different pallets materials and their value assigned, using a qualitative scale from 1 to 3 (1 = weak, 2 = satisfactory or 3 = good). Metal pallets possess the best mechanical performance and paper pallets the worst mechanical performance. In relation to cost and use wooden pallets perform better, however, can be heavier (such as metal pallets).

4.6 *Types of Pallets*

Regardless of pallet material, there are two basic main types of pallets design: block and stringer. Nevertheless, there are seven models in the market (Table 3) based on those main design (custom or standard size); from those, roto-molded pallets are the most economical and is made using rotational molding method [16].

4.7 *Production Phases of Pallets*

The model of a pallet production made of wood, plastic, metal, or recycled material is presented in four stages (Fig. 3). First, designing pallets is necessary raw material or recycled material depending on the requirement. Later, electricity, oil or fuel can be used during the first stage of production. In the second stage (maintenance), only pallets made of wood are repaired. At the end of the process (third stage—end of life) can vary, for instance, metal pallet can be reused or melted and transformed into another functional product, however, a wood pallet could use it in a composting process to produce heat, also can be incinerated to generate heat and electricity. Plastic and WPC pallet at the end of their life also can produce energy from incineration.

Table 3 Comparative pallets design [16]

Pallet	Characteristics
Roto Molded	Economical and made using rotational molding method, those pallets are used in industries where hygiene is a priority
Injection Molded	Different sizes and colours. The main raw material is high-density polyethylene (HDPE)
Block/ Deck Board	High handling efficiency and can be made of different materials (wood, plastic, or metal)
Stringer	Can be lifted only from two sides (that are open), provides additional support over block pallets
Double Face	For strong and heavy-duty tasks, top and bottom sides can be used
Double Wing	Top and bottom side have wings (both sides deck extends beyond the stringers to offer additional surface area to transport more goods all at once)
Solid Deck	For small goods, it offers a flat surface (with no spacing), and can be made of wood, plastic, or metal

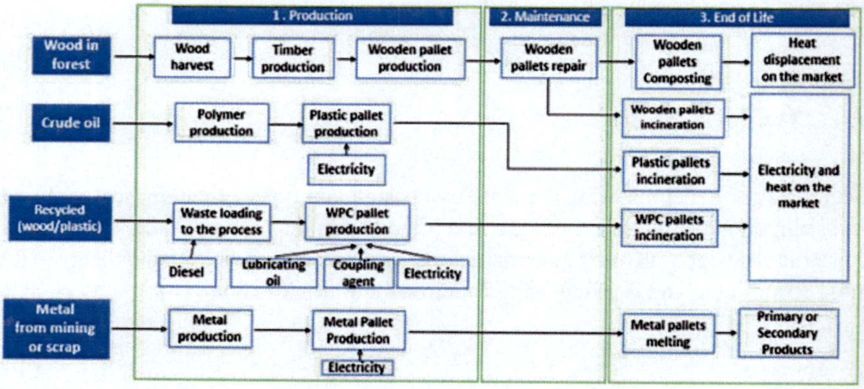

Fig. 3 Pallet's production stages [1, 15]

5 Management Strategies

Due to the diversity in the structure of pallets, the management pallet is becoming a challenging strategy. Researchers have developed different pallets management strategies (PMS), and those are included in this section; EMP (extensive management of pallets), TPO (transfer pallet's ownership) and PR (pallet rent). Pallets can be used for a single utilization (one-way or no-pooled pallets) and represent an open-loop system with no salvage value. In this system, pallets do not return to their manufacturer or distributor (usually they are inexpensive and expendable). On the other hand, pallets can be sold to customers and later resold, for instance, to a local recycling facility (this model represents an open-loop system with salvage value). Finally, there is a closed-loop rental system, where a customer leases pallets from a pooling provider. And several agreements are offered according to the customer strategy (some pallets are collected from any stage in the supply chain), and the return depots can be managed either by the local pallets recycled or the provider [7, 8, 13].

The most popular pallets management system in the world is the EMP (costumers can select important characteristics such as size, material, and quality). However, EMP system represents excessive costs and low logistic efficiency. Moreover, this model is no eco-friendly, due to companies under this system prefer wooden pallets. In EMP systems, pallets are not moving with cargos, for instance, in a basic pallet supply chain (supplier, distributor and retailer), a supplier buys pallets to move their own goods and later, those pallets are unloaded (Fig. 4). In the TPO system (Fig. 5), retailers buy pallets from distributors and those are resold to customers (with cargoes) or recyclers. Damage pallets can be put in a maintenance station before being used by retailers. Finally, in the PR system, pallets are rented from a pallet provider, which offers logistics, maintenance, and recycling services. If any pallet is damaged, the supplier (distributor or retailer) pays a fee to the provider (Fig. 6).

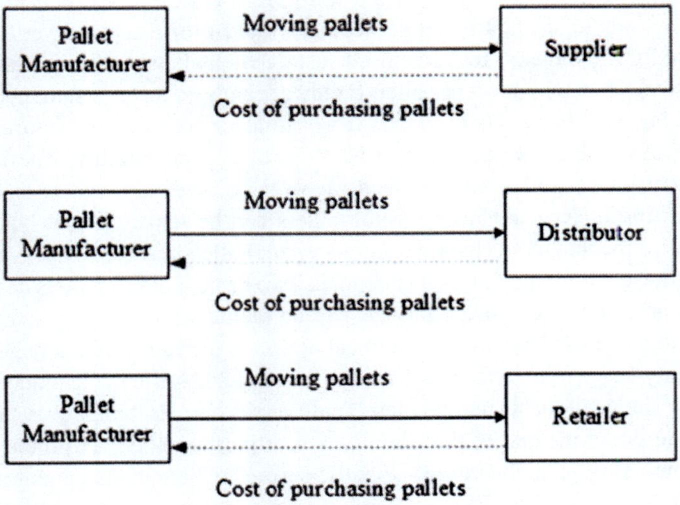

Fig. 4 Operations of EMP [13]

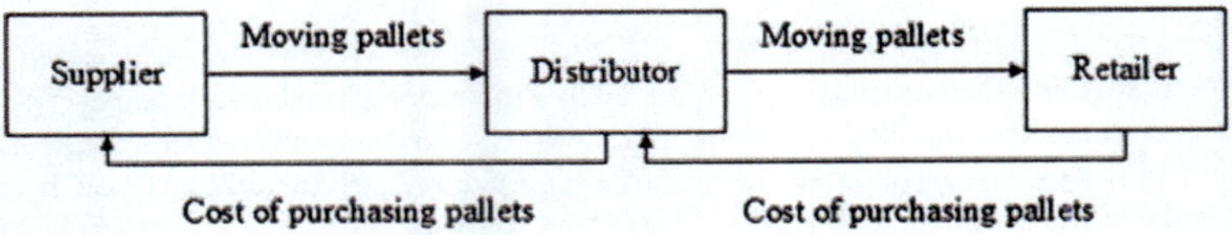

Fig. 5 Operations of TPO [13]

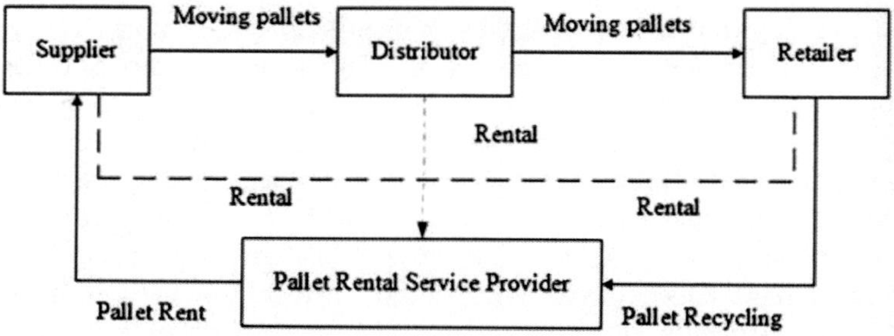

Fig. 6 Operations of PR [13]

6 Conclusion

The present article presented different pallet options for the global supply chain. The review process was conducted analyzing 15 documents (journals, data from articles and specialized web pages). From this analysis, around the world some regions use pallets according to their demand. North America (USA and Canada) prefers pallets size conforming to the GMA; however, European countries select pallets with other dimensions (for example 1200 × 1000 mm). This research will help to evaluate different options to select a pallet according to their properties and verify the best alternative for the industry towards more sustainable pallets. It is important to note that the environmental impact of pallets highly depends on materials, manufacturing, and handling processes. Therefore, it is essential to consider for future research the entire lifecycle of pallets when evaluating their sustainability. The impact of the industry in the global environment involves eco-friendly materials for pallet design, for instance, wooden and metal pallets can be recycled, and both present non-identical mechanical behaviour and lifespan. Pallets made of plastic and paper are the lightest comparing to wood and metal. Two basic types of pallets design were reviewed, and from these seven models can be manufactured (using a standard or costume dimension). The most economical option is the roto-molded pallet and for a heavy-duty task a double face pallet is the best option. A life cycle model of pallet production showed four stages and their inputs versus outputs, where vary depending on the material, at the end of the life of wood or plastic pallet can produce energy (from composting or incineration). Finally, several pallets management strategies have been developed. The EMP system, mostly used around the world, is not an eco-friendly model, is expensive and has a low logistic efficiency, and this strategy can be competitive if pallets are reused to be competitive in terms of environmental impact.

Acknowledgements The work has been supported by Professor Nelson Oliveira, supervisor of this dissertation and member of LSRE-LCM—Laboratory of Separation and Reaction Engineering—Laboratory of Catalysis and Materials, financially supported by LA/P/0045/2020 (ALiCE), UIDB/50020/2020 and UIDP/50020/2020 (LSRE-LCM), funded by national funds through FCT/MCTES (PIDDAC).

References

1. Khan, M. H., Deviatkin, I., Havukainen, J., Horttanainen, M.: Environmental impacts of wooden, plastic, and wood-polymer composite pallet: A life cycle assessment approach. The International Journal of Life Cycle Assessment, 26(8), 1607–1622 (2021).
2. Mazeika, A.: Environmental impact analysis of alternative pallet management systems. Thesis Master of Science in Sustainable Engineering, Rochester Institute of Technology (2001).
3. Technavio. https://www.technavio.com/report/pallet-market-in-europe-industry-analysis, last accessed 11/06/2022.
4. Alanya-Rosenbaum, S., Bergman, R., Gething, B.: Assessing the life-cycle environmental impacts of the wood pallet sector in the United States. Journal of Cleaner Production, 320, 128726, (2021).
5. Zacchei, E., Tadeu, A., Almeida, J., Esteves, M., Santos, M.I., Silva, S.: Development of a Modular Metal Pallet for Transportation and Stationary Conditions: Numerical Analyses and Experimental Characterizations. In: Dimitrovová, Z., Biswas, P., Gonçalves, R., Silva, T. (eds) Recent Trends in Wave Mechanics and Vibrations. WMVC 2022. Mechanisms and Machine Science, vol 125. Springer, Cham, (2023).
6. Moher, D., Liberati, A., Tetzlaff, J., Altman, D. G.: Preferred reporting items for systematic reviews and meta-analyses: The PRISMA statement. Journal of Clinical Epidemiology, 62(10), 1006–1012, (2009).
7. Roy, D., Carrano, A., Pazour, J., Gupta, A.: Cost-effective pallet management strategies. Transportation Research Part E (93). 358–371 (2016).
8. Deviatkin, I., Khan, M., Ernst, E., Horttanainen, M.: Wooden and plastic pallets: A review of life cycle assessment (LCA) studies. Sustainability, 11 (20). 5750 (2019).
9. Anil, S.K., Ma, J., Kremer, G.E., Ray, C.D., Shahidi, S.M.: Life cycle assessment comparison of wooden and plastic pallets in the grocery industry. Journal of Industrial Ecology, 24 (4). 871-886 (2020).
10. Tepic, J., Todic, V., Tanackov, I., Lukic, D., Stojic, G., Sremac, S.: Modular system design for plastic euro pallets. Metalurgija, 51 (2). 241 – 244 (2012).
11. Soury, E., Behravesh, A., Rouhani, E., Zolfaghari, A.: Design, optimization and manufacturing of wood and plastic composite pallet. Materials & Design, 30 (10). 4183 – 4191 (2009).
12. Zacchei, E., Tadeu, A., Almeida, J., Esteves, M., Santos, M.I., Silva, S.: Design of new modular metal pallets: Experimental validation and life cycle analysis. Materials & Design 214. 110425 (2022).
13. Ren, J., Zhao, Q., Liu, B., Chen, C.: Selection of pallet management strategies from the perspective of supply chain cost with analogic software. PLOS ONE, 14 (6). e0217995 (2019).
14. Bengtsson, J., Logie, J.: Life cycle assessment of one-way and pooled pallet alternatives. In: The 22nd CIRP conference on Life Cycle Engineering, 29, pp. 414–419, Elsevier, (2015).
15. Liu, G., Müller, D.A.: Addressing sustainability in the aluminum industry: A critical review of life cycle assessments. Journal of Cleaner Production, 35, 108-117 (2012).
16. Bizongo. https://www.bizongo.com/blog/types-of-pallets. Last accessed 11/15/2022.
17. Baig, M.: Designing novel grooved pallets for industrial application. Thesis Master of Science in Civil Engineering. Cleveland State University (2018).
18. Clarke, J.: Industry overview and wood, plastic, paper & metal options, grate pallet. https://abe.psu.edu/research/areas/bio-based-products/wood-packaging/publications/articles-from-non-penn-state-authors/pallets-101-industry-overview-and-wood-plastic-paper-metal-options/view (2004). Last accessed 11/25/2022.
19. Trebilcock, B.: Pallets 101: Iron man, or what you need to know about metal pallets. https://www.mmh.com/article/pallets_101_iron_man_or_what_you_need_to_know_about_metal_pallets (2010). Last accessed 11/14/2022.
20. Buck, C.: Pallets: Block vs stringer. https://www.linkedin.com/pulse/pallets-block-vs-stringer-chip-buck/ (2020). Last accessed 11/07/2022.

21. Korol, J., Burchart-Korol, D., Pichlak, M.: Expansion of environmental impact assessment for eco-efficiency evaluation of biocomposites for industrial application. Journal of Cleaner Production, 113, 144-1552 (2016).
22. Alanya-Rosenbaum, S., Bergman, R., Gething, B., Mousavi-Avval, S.H.: Life cycle assessment of the wood pallet repair and remanufacturing sector in the United States. Biofuels, Bioproducts and Biorefining, 16 (5), 1342-1352 (2022).

Combined and Separately Effect of Tert-Butylhydroquinone and Propyl Gallate on a Biolubricant Based on Cardoon Oil

Sergio Nogales-Delgado, Pedro Jesús Gómez Farrona, José María Encinar Martín, and Juan Félix González González

Abstract The use of biolubricants is becoming important, as the incipient industrialization is requiring alternatives for petrol-based products. Thus, a double transesterification from vegetable oils could provide biolubricants for these purposes, apart from other interesting by-products like fatty acid methyl esters, glycerol or methanol (which can be re-used in the system). Consequently, this synthesis could adjust to the concept of biorefinery, which is highly appreciated due to its implication with green chemistry or circular economy, contributing to sustainability. However, one challenge related to these biolubricants is their low oxidative stability, which can alter the properties of these products during storage. To avoid it, some interesting antioxidants like tert-butylhydroquinone (TBHQ) or propyl gallate (PG) could be added to biolubricants to keep their main properties during oxidation. The aim of this work was to assess the effectiveness of TBHQ and PG in a biolubricant based on cardoon oil through double transesterification with methanol and pentaerythritol. It was proved that PG was more effective than TBHQ, obtaining higher oxidative stabilities at lower concentrations (exceeding 11 h at 500 ppm, whereas TBHQ required 2000 ppm to obtain a similar value). In contrast, the combined use of TBHQ and PG did not offer any synergistic effect.

Keywords Transesterification · Fatty acid methyl ester · Antioxidant · Viscosity · Acid number

S. Nogales-Delgado (✉) · J. F. González González
Department of Applied Physics, University of Extremadura, Avda. De Elvas s/n, 06006 Badajoz, Spain
e-mail: senogalesd@unex.es

P. J. G. Farrona · J. M. E. Martín
Department of Chemical Engineering and Physical-Chemistry, University of Extremadura, Avda. De Elvas s/n, 06006 Badajoz, Spain

J. R. da. C. S. Galvão et al. (eds.), *Proceedings of the 3rd International Conference on Water Energy Food and Sustainability (ICoWEFS 2023)*, Springer Proceedings in Earth and Environmental Sciences, https://doi.org/10.1007/978-3-031-48532-9_36

1 Introduction

Due to the current concern about the environment, as well as emerging challenges like geopolitics and high prices of petrol-based products, the search for alternatives is becoming vital, especially in some regions like Europe [1]. Apart from the obvious environmental advantages of the production and use of these alternatives (such as zero-net CO_2 emissions, greenhouse gas reduction, or the production of biodegradable products), many regions are trying to reduce their energy or material dependence, which can promote a sustainable energy growth in developing countries.

In that sense, the implementation of a biorefinery could adjust to these policies related to green chemistry, reusability and circular economy, as the use of natural raw materials or wastes (which can be difficult to manage) are the cornerstone for the implementation of industries based on energy and material production. Besides, many of the intermediate products can be reused in the process, and multiple products with different purposes can be generated [2].

One interesting example of these biorefineries is the production of biodiesel and biolubricant based on vegetable oils (for instance, an interesting waste like cardoon oil) through double transesterification with methanol and a more complex alcohol like trimethylolpropane or pentaerythritol [3, 4]. The parallelisms with a biorefinery are evident: the use of a natural raw material or a waste like frying oil (otherwise, its management would be problematic); generation of multiple interesting products like glycerol, fatty acid methyl esters (biodiesel, the intermediate product) and fatty acid esters (biolubricant, the final product); reuse of intermediate products like methanol (released during the second transesterification and reused in the first one). Consequently, the atom efficiency of the process is high, contributing to environmental protection [5].

However, one of the main challenges related to biodiesel and biolubricants is their low oxidative stability, especially if the content of some fatty acids (such as linoleic or linolenic acids) is high. In order to avoid this inconvenience, the use of antioxidants (natural or synthetic) is advisable, considerably increasing oxidative stability [6, 7].

Considering the above, the aim of this work was to obtain biodiesel and biolubricant from cardoon oil through a double transesterification (with methanol to obtain biodiesel and with pentaerythritol to obtain biolubricant), adding PG and TBHQ as antioxidants. Thus, the final biolubricant was characterized, and the antioxidant addition was optimized according to extreme oxidation conditions.

Table 1 Reaction conditions for cardoon biolubricant production

Property	Units	Optimum value
FAME/alcohol ratio	n.a.[a]	1:1/4
Pressure	mmHg	260
Catalyst concentration	% w/w	1.0
Temperature	°C	160

[a]Not applicable

2 Materials and Methods

2.1 Biodiesel and Biolubricant Production

The raw material used in this study was cardoon oil from *Cynara Cardunculus*, provided by by the Research Center "La Orden-Valdesequera" (Badajoz-Spain) Section of Non-Food (CICYTEX). The oil was mechanically extracted and filtered, showing low acidity, and stored in opaque containers for further processing.

For biodiesel production, previous chemical conditions were selected, like the following [8]: reaction temperature, 65 ◦C; reaction time, 60 min; methanol/oil ratio, 6:1; catalyst concentration (sodium methoxide), 1% w/w. Thus, after purification, the FAME content of biodiesel was over 96.5%, complying with the standard and making it a suitable reagent for the next step [9].

Regarding biolubricant production, optimum chemical conditions were selected according to previous studies (see Table 1) [3]:

Afterwards, the biolubricant was filtered through vacuum filtration and stored for further treatments.

2.2 Biodiesel and Biolubricant Characterization

Once cardoon biodiesel and biolubricant were obtained and purified, a typical characterization including FAME content through gas chromatography (for biodiesel), moisture (by using a titrator 870 KF Trinitro Plus, Metrohm, Herisau, Switzerland), density, viscosity by using a viscosimeter (at 40 °C for biodiesel and also at 100 °C for biolubricant), acid number, iodine number, cold filter plugging point (CFPP), flash and combustion points (following the Cleveland open cup method), and oxidative stability.

All these parameters were obtained according to UNE-EN 14,214 standard [9] as explained in previous studies [8]. Concerning oxidative stability, 3 g of sample was placed in a tube, at 110 °C, bubbling air (15 L·min^{-1}) which was trapped in a container with ultrapure water and registering the subsequent increase in conductivity when degradation products due to oxidation are dissolved in water. The induction point

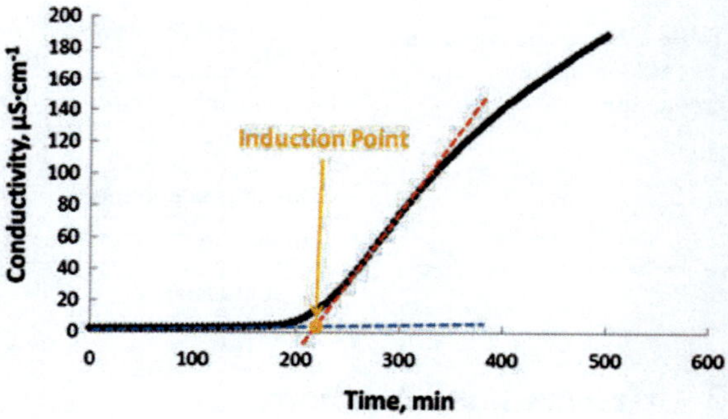

Fig. 1 Induction point (oxidative stability) determination. It is the point where the stable line (in blue) and the unstable line (in red) cross. The black line is the experimental data. In this example, the induction point is 210 min

(expressed in minutes or hours) was defined as the time when stable (with constant conductivity) and unstable stages (with a considerable increase in viscosity) cross, as observed in Fig. 1.

2.3 *PG and TBHQ Addition*

Two antioxidants, propyl gallate (PG) and tert-Butylhydroquinone (TBHQ) were purchased from Sigma-Aldrich, Saint Louis, MO, United States, with a purity exceeding 97%. They were added to cardoon biolubricant at different concentrations, depending on the initial oxidative stability of the control sample (from 0 to 2000 ppm).

Thus, the suitable amount of PG or TBHQ was added to a certain volume of biodiesel (20 mL), using ultrasound for 5 min to assure the dissolution of the antioxidant, storing them in opaque flasks for further analysis or treatments.

2.4 *Biolubricant Oxidation*

The extreme oxidation conditions were based on the Rancimat method [10], adapted for higher amounts of sample: The sample started the oxidation once it reached the working temperature (110 °C), the air used for bubbling was synthetic air (Linde, Barcelona, Spain) and the experiment stopped after 8 h, cooling down the sample as soon as possible, storing it in opaque flasks at room temperature. Afterwards, the evolution of two parameters like viscosity and acid number, which usually present changes during storage or oxidation, was determined. To sum up, the experimental design of this research work is included in Fig. 2.

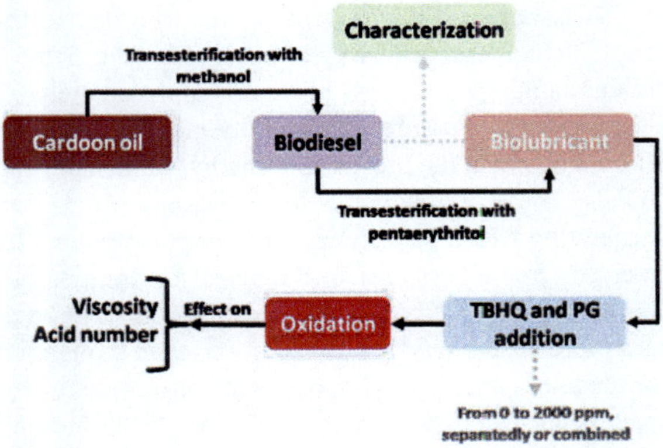

Fig. 2 Experimental design for this research work

3 Results and Discussion

3.1 *Biodiesel and Biolubricant Characterization*

Concerning cardoon fatty acid methyl esters (FAMEs, biodiesel), the main results obtained in this work are included in Table 2. As observed, cardoon biodiesel presented a high conversion, exceeding the lower limit for FAME content, and most parameters were within the limits of UNE-EN 14,214 standard (especially interesting was the high flash and combustion points, clearly exceeding 170 °C, which implies safety during storage) [9]. However, the oxidative stability of cardoon FAMEs was extremely low, around 2 h, which makes the use of antioxidants like PG or TBHQ necessary.

Table 2 Main properties of cardoon biodiesel

Property	Units	Results	UNE-EN 14,214 limits
FAME content	%	97.39	> 96.5
Moisture	$mg \cdot kg^{-1}$	300	< 500
Density	$kg \cdot m^{-3}$	881	860–900
Viscosity (40 °C)	$mm^2 \cdot s^{-1}$	4.5	3.5–5.0
Acid number	$mg_{KOH} \cdot g^{-1}$	0.2	< 0.5
Oxidative stability	h	2.08	> 8
Iodine number	$g_{I2} \cdot 100\ g^{-1}$	118	< 120
C. F. P. P.[a]	°C	−1	−20 - + 5
Flash point	°C	176	> 120
Combustion point	°C	192	n.a.[b]

[a] Cold Filter Plugging Point
[b] Not applicable

This short oxidative stability could be due to the FAME profile obtained for cardoon biodiesel. This way, as observed in Fig. 3, the majority fatty acid methyl ester in this sample was methyl linoleate (approximately 60%), with two double bonds in its molecular structure, followed by methyl oleate (25%), with one double bond. It should be noted the presence of methyl linolenate, exceeding 2%, with three double bonds. These double bonds act as starting points during auto-oxidation processes, generating free radicals that, after propagation and termination, can generate polymers or free fatty acids in biodiesel. Consequently, the high presence of methyl linoleate, compared to methyl oleate, could explain the low oxidative stability, as it is a more unstable FAME. In addition, even though it is at low concentration, methyl linolenate could play an important role in this parameter, with a further decrease in this parameter. Thus, the oxidative stability and the amount of antioxidant required for the improvement of this parameter could depend on fatty acid methyl ester composition.

Regarding cardoon biolubricant, its main characteristics are included in Table 3. As observed, the conversion of biodiesel was high, exceeding 90%. Viscosity at 40 °C was considerable (exceeding 55 $mm^2 \cdot s^{-1}$), as the esters derived from pentaerythritol are very complex molecules, promoting intermolecular interactions like Van der Waals forces or hydrogen bonds. These interactions reinforce the resistance to flow of cardoon biolubricant, increasing its viscosity at 40 and 100 °C compared to biodiesel.

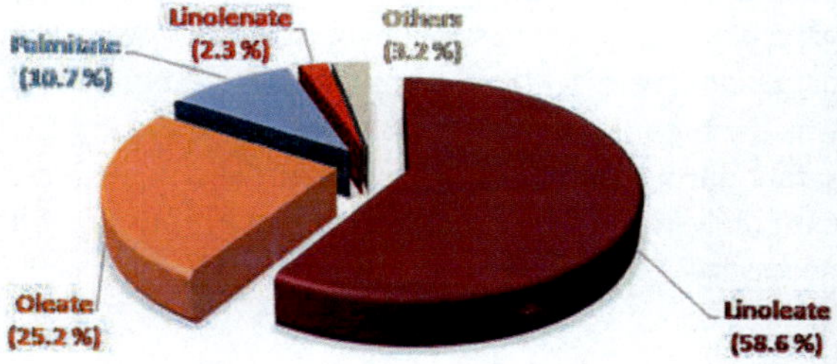

Fig. 3 Fatty acid methyl ester profile of cardoon biodiesel

Table 3 Main properties of cardoon biolubricant

Property	Units	Results
Conversion	%	92.8
Moisture	$mg \cdot kg^{-1}$	200
Density	$kg \cdot m^{-3}$	928
Viscosity (40 °C)	$mm^2 \cdot s^{-1}$	57.38
Viscosity (100 °C)	$mm^2 \cdot s^{-1}$	11.19
Viscosity index	n.a.[a]	209
Acid number	$mg_{KOH} \cdot g^{-1}$	0.3
Oxidative stability	h	2.03
Flash point	°C	185
Combustion point	°C	199

[a] Not applicable

Another interesting aspect of this biolubricant is its high viscosity index, which implies less noticeable changes in viscosity with temperature. However, the oxidative stability was low, due to the instability of cardoon oil and biodiesel (on account of the presence of the subsequent linoleate esters), requiring the addition of antioxidants.

3.2 *Effect of PG and TBHQ Addition on Cardoon Biolubricant*

As previously explained, the use of PG and TBHQ is vital to improve the oxidative stability of biodiesel and, especially, biolubricant, as the auto-oxidation process could imply an increase in viscosity (mainly due to polymerization of esters) and acidity (due to the generation of free fatty acids).

Figure 4 shows the effect of PG and TBHQ on oxidative stability of cardoon biolubricant, pointing out the effectiveness of both antioxidants to increase the oxidative stability of this product. Thus, less than 2000 ppm were required to obtain an oxidative stability of at least 8 h.

Specifically, PG was more efficient, as it reached 8 h of oxidative stability at lower concentrations (345 ppm) than TBHQ (1555 ppm). These data are similar to those found in the literature, where the effectiveness of these antioxidants was proved, whereas there is controversy about the efficiency of this kind of products [11].

Furthermore, the combined effect of both antioxidants was studied. Thus, a mixture of PG and TBHQ (at the corresponding concentrations to assure 4 h of oxidative stability separatedly) was added to cardoon biolubricant, in order to compare

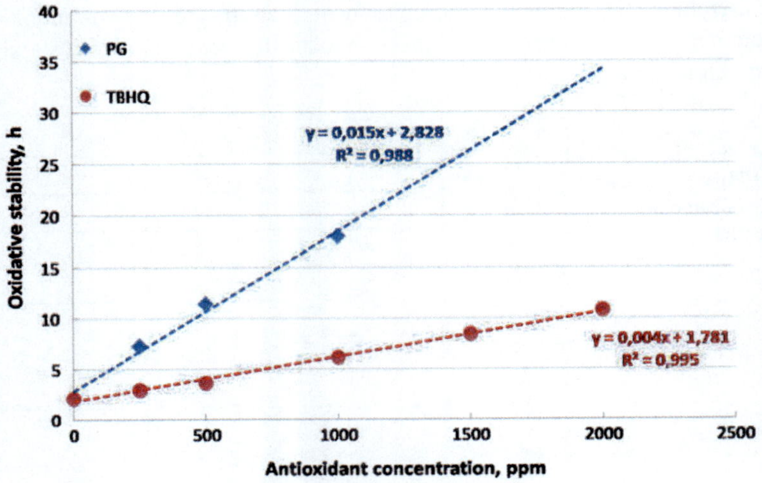

Fig. 4 Effect of PG and TBHQ on oxidative stability of cardoon biolubricant

with the addition of each antioxidant to obtain 8 h of oxidative stability. In order to assess the possible synergy of this mixture, according to previous studies [12], Eq. (1) was used:

$$Synergy = \frac{(IP_{Mixture} - IP_0) - \left[W_{TBHQ} \cdot \left(IP_{TBHQ} - IP_0\right) + W_{PG}(IP_{PG} - IP_0)\right]}{IP_{Mixture} - IP_0} \quad (1)$$

where

- $IP_{Mixture}$ is the oxidative stability (induction point) of the mixture of TBHQ and PG, at the quantities established to obtain an induction point of 4 h each.
- IP_0 is the oxidative stability of control sample (that is, biolubricant without any antioxidant addition).
- IP_{TBHQ} is the oxidative stability for TBHQ at 555 ppm (concentration calculated to obtain 4.0 h of oxidative stability).
- IP_{PG} is the oxidative stability for PG at 78 ppm (concentration calculated to obtain 4.0 h of oxidative stability).
- W_{TBHQ} is the weight fraction of TBHQ in the mixture.
- W_{PG} is the weight fraction of PG in the mixture.

This way, Table 4 shows the results obtained. It should be noted that the mixture of antioxidants (with the corresponding concentration to obtain 4 h of oxidative stability separatedly) did not reach 8 h of induction point. Additionally, the weight fraction of PG was low, due to its high efficiency at reaching 4 h of oxidative stability. As a result, synergy was below 1, which shows that there is no synergistic effect. On the contrary, an inhibitory effect was found. In that sense, this effect (along with additive and synergistic effects) are typical when antioxidant mixtures are used, as explained in the literature [11]. In this specific experiment, the separately use of antioxidants is recommended in order to obtain higher efficiency to increase oxidative stability of cardoon biolubricant.

Table 4 Combined effect of PG and TBHQ on oxidative stability of cardoon biolubricant

$IP_{Mixture}$, h	6.40
IP_{TBHQ}, h	4.00
IP_{PG}, h	4.00
IP_0, h	2.03
W_{TBHQ}	0.87
W_{PG}	0.13
Synergy	0.55

Table 5 Changes in viscosity and acid number of cardoon biolubricant with different antioxidant additions

Property	Units	B_0		B_{PG345}	$B_{TBHQ1555}$	$B_{PG78+TBHQ}$
		0 h	8 h	8 h	8 h	8 h
Viscosity (40 °C)	$mm^2 \cdot s^{-1}$	57.38	188.98	57.43	57.98	59.73
Acid number	$mg_{KOH} \cdot g^{-1}$	0.3	6.42	0.66	0.71	0.78

Finally, different cardoon biolubricant samples underwent extreme oxidation conditions:

- B_0: control sample without any antioxidant addition.
- B_{PG345}: biolubricant with the optimum PG concentration (345 ppm) to achieve 8 h of oxidative stability.
- $B_{TBHQ1555}$: biolubricant with the optimum TBHQ concentration (1555 ppm) to achieve 8 h of oxidative stability.
- $B_{PG78+TBHQ554}$: biolubricant with PG and TBHQ concentrations expected to achieve 8 h of oxidative stability (78 and 554 ppm, respectively). In reality, it achieved 6.4 h, as seen in Table 4.

Table 5 shows the main results obtained, comparing the starting point (0 h of oxidation) and the final results after 8 h of extreme oxidation. As expected, there was a considerable increase in viscosity and acid number in control sample, due to the abovementioned reasons. Nevertheless, the use of antioxidants reduced this effect, as observed in previous studies [3], and PG was more effective at retaining the original properties of cardoon biolubricant. Again, the combined use of PG and TBHQ did not show any benefits, and there was a slight increase in viscosity and acid number compared to the separate use of PG and TBHQ. In conclusion, the use of PG is recommended in this case, due to its effectiveness and efficiency.

4 Conclusions

The main findings of this research work were the following:

- Cardoon biodiesel and biolubricant presented interesting properties to be used as fuel and lubricant, respectively, except for their low oxidative stability, which can alter their main properties (like viscosity or acid number) during storage or oxidation processes.
- Thus, the use of antioxidants was advisable. PG and TBHQ addition showed efficiency at increasing oxidative stability in biolubricant. Especially, PG had a high efficiency, with a considerably increase in oxidative stability at low concentrations.

However, the combined use of these antioxidants did not offer any advantage, not observing any synergistic effect.

- Consequently, TBHQ and especially PG kept the main properties of cardoon biolubricant (that is, viscosity and acid number) during oxidation, whereas the combination of both had a poorer performance. In that sense, it is advisable the separate use of these antioxidants to assure the main quality parameters during storage or oxidation processes.

References

1. Vakulchuk, R.; Overland, I.; Scholten, D. Renewable energy and geopolitics : A review. *Renew. Sustain. Energy Rev.*, *122*, 109547 (2020), doi:https://doi.org/10.1016/j.rser.2019.109547.
2. Liberato, V.; Benevenuti, C.; Coelho, F.; Botelho, A.; Amaral, P.; Pereira, N.; Ferreira, T. Chemicals Production in a Biorefinery Context. *Catalysts*, 1–37 (2019).
3. Encinar, J.M.; Nogales-Delgado, S.; Álvez-Medina, C.M. High oleic safflower biolubricant through double transesterification with methanol and pentaerythritol: Production, characterization, and antioxidant addition. *Arab. J. Chem.*, *15*, 103796 (2022), doi:https://doi.org/10.1016/j.arabjc.2022.103796.
4. Ma, X.; Zhang, Y.; Song, Z.; Yu, K.; He, C.; Zhang, X. Enzyme-catalyzed synthesis and properties of polyol ester biolubricant produced from Rhodotorula glutinis lipid. *Biochem. Eng. J.*, *173*, 108101 (2021), doi:https://doi.org/10.1016/j.bej.2021.108101.
5. Martinez-Guerra, E.; Gude, V.G. Assessment of sustainability indicators for biodiesel production. *Appl. Sci.*, *7* (2017), doi:https://doi.org/10.3390/app7090869.
6. Dunn, R.O. Antioxidants for improving storage stability of biodiesel. *Biofuels, Bioprod. Biorefining* (2008).
7. Quinchia, L.A.; Delgado, M.A.; Valencia, C.; Franco, J.M.; Gallegos, C. Natural and synthetic antioxidant additives for improving the performance of new biolubricant formulations. *J. Agric. Food Chem.*, *59*, 12917–12924 (2011), doi:https://doi.org/10.1021/jf2035737.
8. Encinar, J.M.; Nogales-Delgado, S.; Sánchez, N.; González, J.F. Biolubricants from rapeseed and castor oil transesterification by using titanium isopropoxide as a catalyst: Production and characterization. *Catalysts*, *10* (2020), doi:https://doi.org/10.3390/catal10040366.
9. UNE-EN 14214:2013 V2+A1:2018 Liquid petroleum products - Fatty acid methyl esters (FAME) for use in diesel engines and heating applications - Requirements and test methods. (2018).
10. Nogales-Delgado, S.; Encinar, J.M.; González, J.F. Safflower Biodiesel : Improvement of its Oxidative Stability by using BHA and TBHQ. *Energies*, 19–22 (2019).
11. Jemima Romola, C.V.; Meganaharshini, M.; Rigby, S.P.; Ganesh Moorthy, I.; Shyam Kumar, R.; Karthikumar, S. A comprehensive review of the selection of natural and synthetic antioxidants to enhance the oxidative stability of biodiesel. *Renew. Sustain. Energy Rev.*, *145*, 111109 (2021), doi:https://doi.org/10.1016/j.rser.2021.111109.
12. Rokosik, E.; Siger, A.; Rudzińska, M.; Dwiecki, K. Antioxidant activity and synergism of canolol and α-tocopherol in rapeseed oil is affected by the presence of phospholipid association colloids. *LWT*, *133*, 110095 (2020), doi:https://doi.org/10.1016/j.lwt.2020.110095.

Impact of Air Quality on Mental Health of Higher Education Students

Ana Ferreira, Ana Ferreira, João Paulo Figueiredo, António Loureiro, Sílvia Seco, and Lúcia Simões Costa

Abstract The decrease in air quality in cities is now an object of concern, as a large part of the population lives in these areas. Problems related to mental health are on the rise. Around 20% of children and adolescents worldwide have some mental health problem, with suicide being the second leading cause of death among young people aged 15 to 29. This study aimed to evaluate the air quality in two distinct areas in Portugal and to analyze its impact on the mental health of higher education students. Data collection consisted in assessment of air quality, in two different areas of one municipality of Portugal, a green one, with no traffic and characterized by being a forested area and one with the presence of road traffic. The second moment consisted of the application of a questionnaire addressed to all university students who participated in the study. We found that in both areas, $PM_{2.5}$ concentrations exceeded the exposure limit value and PM_{10} only exceeded the limit value in the area with traffic. It was found that students had higher values of negative emotional symptoms in the zone with road traffic, when compared to the green zone. The monitoring of atmospheric pollutants is a fundamental tool in the identification, assessment and

A. Ferreira (✉) · A. Ferreira · J. P. Figueiredo · L. S. Costa
Polytechnic Institute of Coimbra, Coimbra Health School, Rua 5 de Outubro—S. Martinho do Bispo, Apartado 7006, 3046-854 Coimbra, Portugal
e-mail: anabaferreira@estesc.ipc.pt

A. Ferreira
e-mail: anaferreira@estesc.ipc.pt

J. P. Figueiredo
e-mail: jpfigueiredo@estesc.ipc.pt

L. S. Costa
e-mail: lucias@estesc.ipc.pt

A. Loureiro · S. Seco
Polytechnic Institute of Coimbra, Occupational and Environmental Health Service, Rua da Misericórdia, Lagar dos Cortiços—S. Martinho do Bispo, 3045-093 Coimbra, Portugal
e-mail: antonio.loureiro@ipc.pt

S. Seco
e-mail: silvia.seco@ipc.pt

J. R. da. C. S. Galvão et al. (eds.), *Proceedings of the 3rd International Conference on Water Energy Food and Sustainability (ICoWEFS 2023)*, Springer Proceedings in Earth and Environmental Sciences, https://doi.org/10.1007/978-3-031-48532-9_37

resolution of air quality problems, quantifying the levels of atmospheric pollutants and ensuring that they are within the limits tolerated by human health and ecosystems, through the implementation of measures within the scope of air quality management.

Keywords Air quality · Mental health · Students

1 Introduction

Today, more than ever, it is important to know the quality of the air we breathe, since it is increasingly contaminated due to atmospheric pollution, which ends up influencing, directly or indirectly, all the spaces we occupy in our daily lives. [1]. The air, considered pure, consists mainly of nitrogen (78.0%), oxygen (21.0%) and 1.0% of other gases such as argon, water vapor and carbon dioxide [2].

The concept of outdoor air quality is related to the level of concentration of atmospheric pollutants in the lower layer of the atmosphere, the troposphere. Over the years, man has verified that this change has negative repercussions on public health and on the well-being of populations, also exerting a harmful influence on fauna and flora [3].

Air pollution is one of the main environmental problems, of which road transport is one of the main sources of pollution in urban areas as a result of the emission of pollutants into the atmosphere [4, 5].

The World Health Organization (WHO) lists a series of toxic components for health related to air pollution, highlighting Suspended Particulates (PM), carbon monoxide (CO), nitrogen dioxide (NO_2), sulfur dioxide (SO_2) and ozone (O_3) [5, 6]. Particles or aerosols suspended in the air are pollutants constituted by a complex mixture of solid and liquid particles present in a gas and vary in size and composition, depending on their source and formation, being classified according to size and formation mechanism [7].

According to data provided by the WHO, in 2018, air pollution causes about 7 million deaths per year worldwide [8]. At the same time, mental illness is an important and rapidly growing cause of morbidity worldwide. Both problems disproportionately affect disadvantaged groups and can show marked urban–rural differences [9].

In Europe, air pollution is considered the main risk of an environmental nature to human health [10]. It carries considerable risks, with an emphasis on stroke, coronary heart disease, lung cancer and acute and chronic respiratory diseases [11].

Several studies on air quality and its possible impact on mental health have been published recently. Exposure to environmental pollution may be a risk factor for mental health problems among higher education students, with general well-being and sleep quality also being associated with mental health [12]. There is an association between $PM_{2,5}$ exposure and high anxiety symptoms, and it is possible to reduce exposure to this particulate matter through individual behavior and regulatory changes [13].

In view of the above, the present study aimed to evaluate the air quality in two distinct areas, separated by 5 km (green area and area with road traffic), in the one municipality of Portugal and to analyze its impact on the mental health of higher education students.

2 Material and Methods

The study sample was non-probabilistic and consisted of 20 higher education students. The inclusion criterion for participation of students in the study was the non-use of a protective mask during measurements.

Data collection consisted of two investigation moments, the first being the assessment of air quality, by measuring the environmental parameters CO, CO_2, volatile organic compounds (VOC), formaldehyde (CH_2O), hydrogen sulfide (H_2S), NO_2, SO_2, $PM_{1.0}$, $PM_{2.5}$, PM_{10}, respirable particles and PM_{Totals} and meteorological variables temperature (T°), relative humidity (Hr) and air velocity (V_{ar}) in 2 zones, in a total of 642 measurements carried out at the green area and 833 measurements at the area with road traffic, the second moment consisted of the application of a questionnaire addressed to all higher education students who participated in the study.

All measurements were carried out in a period of 60 min, per day, with minute-by-minute sampling in the identified areas, being carried out between 17:30 h and 18:30 h and the equipment was placed at least 1,0 m from any obstacle.

Air quality measurements were also carried out to evaluate the parameters $PM_{2.5}$ and PM_{10} by dosimetry. These devices were placed on the students who participated in the study, next to the upper respiratory tract. Each of the participants performed three 15-min walks per measurement day, making a total of 45 min with minute-by-minute sampling. The walks were carried out in the surroundings of the evaluated areas, implying the passage next to the point where the remaining equipment was located to measure the quality of the ambient air.

To carry out the analytical collection of the parameters evaluated, specific portable real-time reading equipment were used.

It was considered as a reference for the exposure limit values for $PM_{2.5}$, 22 $\mu g/m^3$ (0.022 mg/m^3), for PM_{10}, 50 $\mu g/m^3$ (0.05 mg/m^3) and for CO, 8.7 ppm, as mentioned in Portuguese legislation, Decree-Law No. 102/2010, of September 23 [14].

The parameters H_2S, NO_2 and SO_2 were excluded as they did not contribute significantly to the study.

In the second moment of investigation, a questionnaire was applied to obtain data on habits, lifestyles, health condition, the student's perception of air quality and mental health (with Depression and Stress Anxiety Scale-EADS-21).

3 Results

Of the 20 students who participated in the study, 8 were male (40.0%), with an approximate mean age of 21.0 ± 0.93 years and 12 were female (60.0%), and the age mean was approximately 20.4 ± 1.16 years. Of the 20 students, 13 usually lived in a rural area and 7 in an urban area. During the school year, 14 lived in an urban area, 7 in a rural area and only 1 in an industrial area.

The presence of particulate matter in relation to the existence or not of car traffic in the place where the measurements were carried out is in Table 1.

According to Table 1, there were statistically significant differences (p-value < 0.05) between the suspended particles evaluated as a function of the whether or not there is road traffic. These differences showed that in places with traffic, the particles evaluated showed significantly higher variations compared to places without traffic, and the maximum reached by $PM_{2.5}$ and PM_{10} exceeded the legally established exposure limit value of 0.022 mg/m^3 and 0.05 mg/m^3, respectively.

Table 1 Mean concentrations of particulate matter depending on whether or not there is road traffic in the assessed areas

Parameter	Road traffic	M ± SD	Min	Max	N	M_W; *p-value*
$PM_{1.0}$ (mg/m^3)	Doesn't exist	0.029 ± 0.026	0.007	0.089	193	9029.500; < 0.0001
	Exists	**0.049 ± 0.015**	0.024	0.092	184	
	Total	0.039 ± 0.023	0.007	0.092	377	
$PM_{2.5}$ (mg/m^3)	Doesn't exist	0.031 ± 0.026	0.007	0.090	193	8932.500; < 0.0001
	Exists	**0.050 ± 0.016**	0.026	**0.096**	184	
	Total	0.040 ± 0.024	0.007	0.096	377	
PM_{10} (mg/m^3)	Doesn't exist	0.035 ± 0.027	0.009	0.096	193	8304.500; < 0.0001
	Exists	**0.061 ± 0.025**	0.030	**0.173**	184	
	Total	0.048 ± 0.029	0.009	0.173	377	
Breathable Particles (mg/m^3)	Doesn't exist	0.032 ± 0.026	0.008	0.091	193	8881.000; < 0.0001
	Exists	**0.052 ± 0.017**	0.027	0.106	184	
	Total	0.042 ± 0.024	0.008	0.106	377	
PM_{Total} (mg/m^3)	Doesn't exist	0.043 ± 0.028	0.011	0.108	193	5737.500; < 0.0001
	Exists	**0.094 ± 0.054**	0.035	0.388	184	
	Total	0.068 ± 0.050	0.011	0.388	377	

N: Number of measurements; M: Mean; SD: Standard Deviation; Min: Minimum; Max: Maximum
Test: Wilcoxon-Mann–Whitney (M_W)

Through the analysis of Table 2, it was found that the spaces with the presence of road traffic showed significantly higher mean values of $PM_{2.5}$ compared to the exposure limit value established in the legislation ($p < 0.05$). The mean exceedance in places with traffic compared to the exposure limit value was 0.028 mg/m^3. Regarding the PM_{10} values, it was found that in spaces with no traffic, the PM_{10} values obtained were below the exposure limit value, while in spaces with traffic showed values of PM_{10} significantly higher than the exposure limit value ($p < 0.05$). The mean exceedance in places with traffic compared to the exposure limit value was 0.011 mg/m^3. As for the CO, it was found that, either in spaces with no traffic and in spaces with the presence of traffic, the values obtained were below the exposure limit value.

In Table 3, is the mean concentration values of the parameters $PM_{1.0}$, breathable particles, PM_{Total}, CH_2O, CO_2 and VOC at the different measurement locations. The parameters $PM_{1.0}$, breathable particles, PMT_{otals}, CH_2O and CO_2 showed higher values at the location with car traffic compared to the location without traffic, however the VOC values were higher at the green area measurement location.

The evaluation of the meteorological variables, temperature, relative humidity and air velocity in the different measurement locations is in Table 4.

It was verified the presence of statistically significant differences ($p < 0.05$) between the evaluated meteorological variables and the measurement location/zone. At green area, T° and Hr showed higher mean values compared to those recorded at traffic road, however V_{ar} obtained higher mean values at traffic road.

Table 5 refer to $PM_{2.5}$ and PM_{10} assessments performed using portable dosimeters that were placed on each student to perform the 3 15—minute walks. The presence of statistically significant differences ($p < 0.05$) was verified, with the mean concentration values of $PM_{2.5}$ and PM_{10} recorded being higher in area with road traffic.

The student's perception of the conditions of each evaluated area and the respective symptoms associated with exposure to certain pollutants and atmospheric particles in the two areas was evaluated. It was intended to understand if the students felt affected by the poor air quality and in what type of area. It was found that of the 20

Table 2 Mean concentrations of $PM_{2.5}$, PM_{10} and CO in the two areas evaluated

Parameter		Green area	Traffic road
$PM_{2.5}$ (ELV = 0.022 mg/m^3)	N	193	184
	M ± SD	0.031 ± 0.026	**0.050 ± 0.016**
	p-value	< 0.0001	< 0.0001
PM_{10} (ELV = 0.05 mg/m^3)	N	193	184
	M ± SD	0.035 ± 0.027	**0.061 ± 0.025**
	p-value	< 0.0001	< 0.0001
CO (ELV = 8.7 ppm)	N	193	184
	M ± SD	0.721 ± 1.104	**1.110 ± 1.068**
	p-value	< 0.0001	< 0.0001

N: Number of measurements; M: Mean; SD: Standard Deviation; VLE: Exposure Limit Value; Test: t-Student for 1 sample

Table 3 Mean concentrations of $PM_{1.0}$, respirable particles, PM_{Total}, CH_2O, CO_2 and VOC per zone evaluated

Parameter	Measurement location	N	M ± SD	*p-value*
$PM_{1.0}$ (mg/m^3)	Green area	193	0.030 ± 0.026	< 0.0001
	Traffic road	184	**0.049 ± 0.015**	
	Total	377	0.039 ± 0.023	
Breathable Particles (mg/m^3)	Green area	193	0.032 ± 0.026	< 0.0001
	Traffic road	184	**0.052 ± 0.017**	
	Total	377	0.042 ± 0.024	
PM_{Total} (mg/m^3)	Green area	193	0.043 ± 0.028	< 0.0001
	Traffic road	184	**0.094 ± 0.054**	
	Total	377	0.068 ± 0.050	
CH_2O (ppm)	Green area	193	0.000 ± 0.000	< 0.0001
	Traffic road	184	**0.035 ± 0.548**	
	Total	377	0.017 ± 0.421	
CO_2 (ppm)	Green area	193	543.650 ± 29.624	0.002
	Traffic road	184	**555.970 ± 24.094**	
	Total	377	549.670 ± 27.725	
VOC ($\mu g/m^3$)	Green area	193	**0.638 ± 0.409**	< 0.0001
	Traffic road	184	0.487 ± 0.296	
	Total	377	0.564 ± 0.366	

N: Number of measurements; M: Mean; SD: Standard Deviation; Test: Wilcoxon-Mann–Whitney

Table 4 Weather variables Var, T° and Hr per assessed location

Parameter	Measurement location	Mean	N	*p-value*
Air speed (m/s)	Green area	0.72 ± 0.255	193	< 0.0001
	Traffic road	**0.379 ± 0.294**	184	
	Total	0.324 ± 0.280	377	
Temperature (°C)	Green area	**15.770 ± 5.439**	193	< 0.0001
	Traffic road	11.089 ± 1.072	184	
	Total	13.485 ± 4.600	377	
Relative humidity (%)	Green area	**78.415 ± 22.388**	193	< 0.0001
	Traffic road	76.087 ± 5.362	184	
	Total	77.279 ± 16.471	377	

N: Number of measurements; Test: Wilcoxon Mann–Whitney

Table 5 Mean concentrations of $PM_{2,5}$ and PM_{10} assessed using the dosimeter, depending on whether or not there is road traffic in the assessed areas

Parameter	Place	M ± SD	Min	Max	N	M_W; p-value
$PM_{2.5}$ mg/m^3	Green area	0.025 ± 0.024	0.001	0.150	233	
	Traffic road	**0.038 ± 0.037**	0.000	**0.408**	290	23,292.000; < 0.0001
	Total	0.032 ± 0.0323	0.000	0.408	523	
PM_{10} mg/m^3	Green area	0.066 ± 0.050	0.000	0.180	179	
	Traffic road	**0.169 ± 0.128**	0.000	**0.393**	345	17,088.000; < 0.0001
	Total	0.134 ± 0.118	0.000	0.393	524	

N: Number of measurements; M: Mean; SD: Standard Deviation; Min: Minimum; Max: Maximum; Test: Wilcoxon-Mann–Whitney (M_W)

students who participated in the study, only 6 (30.0%) reported having felt affected by poor air quality, being the area where they felt most affected the one with road traffic.

Next, it was assessed the existence of symptoms and/or pathologies frequently felt by the students (Table 6). Of the 20 students, only 11 claimed to have at least one of the symptoms/diseases. The symptoms/diseases that the students most indicated they had were "sneezing crisis" (45.5%), "wheezing/whistling" (36.4%) and "allergic rhinitis" (36.4%).

Regarding the assessment of their health, 70.0% of the students rated it as "very good" or "good", 25.0% as "satisfactory" and only 5.0% (1 student) as "bad".

Table 6 Symptoms/diseases frequently experienced by students who participated in the study

Symptoms/diseases (N = 11)	N	%	% of cases
Asthma	3	11.5	27.3%
Wheezing/whistling	4	15.4	**36.4%**
Sneezing crisis	5	19.2	**45.5%**
Allergic rhinitis	4	15.4	**36.4%**
Dry cough	3	11.5	27.3%
Headaches	3	11.5	27.3%
Dizziness	1	3.8	9.1%
Itching, burning or irritation in the eyes	2	7.7	18.2%
Breathing difficulties	1	3.8	9.1%
Total	26	100.0	236.4%

N: Number of cases

The association of different types of negative emotional symptoms (anxiety, depression and stress) with the different areas under study is in Table 7. Although there were no statistically significant differences ($p > 0.05$), the participants showed higher levels of depression, anxiety and stress in the area with road traffic.

Table 8, shows the relationship between the gender variable with the existence of negative emotional symptoms (anxiety, depression and stress). It was found that there were no statistically significant differences between staying in the green zone or in road traffic and the presence of negative emotional symptoms ($p > 0.05$), with the exception of anxiety. Anxiety showed statistically significant differences, with females showing more symptoms than males in an area with road traffic ($p < 0.05$). Regardless of the existence or not of statistically significant results, it was observed that symptoms of depression are higher for males in any of the areas, green or with road traffic, and symptoms of anxiety and stress are higher for females, in any of the zones.

Table 7 Expression of the different scales evaluated depending on the area where the measurements took place

N = 20	Mean	SD	*p-value*
Depression scale—green zone	0.650	1.785	0.906
Depression scale—zone with road traffic	**0.700**	2.003	
Anxiety scale—green zone	0.850	1.424	0.292
Anxiety scale—zone with road traffic	**1.350**	1.899	
Stress scale—green zone	0.650	1.599	0.353
Stress scale—zone with road traffic	**1.150**	1.872	

N: Number of cases; SD: Standard deviation; Test: Wilcoxon-Mann–Whitney

Table 8 Expression of the negative emotional symptoms evaluated according to gender

Gender		Depression green zone	Depression traffic zone	Anxiety green zone	Anxiety traffic zone	Stress green zone	Stress traffic zone
Male	M	**0.875**	**0.750**	0.500	0.250	0.125	–
	SD	2.475	2.121	0.756	0.463	0.354	–
	N	8	8	8	8	8	8
Female	M	0.500	0.667	**1.083**	**2.083**	**1.000**	**1.917**
	SD	1.243	2.015	1.730	2.151	2.000	2.109
	N	12	12	12	12	12	12
Total	M	0.650	0.700	0.850	1.350	0.650	1.150
	SD	1.785	2.003	1.424	1.899	1.599	1.872
	N	20	20	20	20	20	20
p-value		0.901	0.804	0.495	0.020	0.408	–

N: Number of cases; M: Mean; SD: Standard Deviation; Test: Wilcoxon-Mann–Whitney

4 Discussion

Through the results obtained, it was possible to conclude that although the measurements were all carried out in the same season of the year, the cloudy days were the ones that presented mostly higher concentrations of suspended particles, that is, the meteorological conditions have an influence on the concentrations of atmospheric pollutants.

It was verified that in places with the presence of road traffic, the suspended particles evaluated showed significantly higher variations compared to places without traffic, and the maximum reached by $PM_{2.5}$ and PM_{10} exceeded the established exposure limit value. Road transport is the main source of NO_2 and PM in urban areas in Europe, and, consequently, is the main responsible for the exceedances of the limit values legislated by the EU for these pollutants [4]. One a study carried out in Southwest Europe reports that traffic contributes about 35.0% and 23.0% to the concentrations of $PM_{2.5}$ and PM_{10}, respectively in urban areas [15], but pollutants such as CO, CO_2, VOC, nitrogen oxides (NOx) are also anthropogenic sources of motor vehicle emissions [16].

With regard to $PM_{2,5}$ concentrations, we can see that, both at green area and at traffic road, they showed significantly higher mean values compared to the legally established exposure limit value, ie 0.022 mg/m^3. When analyzing each area evaluated, it was found that it was the area with road traffic, the one that presented, on mean, concentrations of higher $PM_{2.5}$. Atmospheric pollutants emitted in one place can contribute to poor air quality elsewhere, which means that it is not just in cities that pollution levels exceed legislated limits. Some rural areas also have high levels of pollution, it is the case that part of the pollutants found in Europe are emitted in Asia and North America, as well as part of the pollutants emitted in Europe are transported to other continents, thus explaining the concentrations of $PM_{2.5}$ obtained in the green zone [17].

Regarding the mean concentrations of PM_{10}, it was found that traffic road presented mean concentration values above the exposure limit value. From the analysis of the mean values of CO, it was found that both zones had, on mean, values lower than the legally established value, that is, lower than 8.7 ppm, however the area with road traffic was the one with the highest mean concentrations.

With regard to the observed values of $PM_{1,0}$, Breathable Particles, PM_{Totals}, CH_2O, CO_2 and VOC per zone, it was found that the $PM_{1.0}$, Breathable Particles, PM_{Totals}, CH_2O and CO_2 showed higher values in the place with traffic, compared to the place without traffic, however the values of registered VOC were higher in the green zone. As indicated by some studies, it was found that the mean concentration of CO_2, VOC and CH_2O presented higher values in urban areas when compared to rural areas [18]. The main source of CO_2 comes from road transport, so the values of this pollutant are generally higher in urban areas where traffic is higher [17]. The results found in another study show that it is in urban areas (predominantly) that the mean concentrations of CO_2 in the environment are higher, compared to the mean concentrations in rural areas [19]. Hence, the area with road traffic (urban) is the place where this parameter was higher, since it is in this area that there is a greater number of cars in circulation.

Exposure to air pollution is associated with increased occurrence of respiratory and cardiovascular disease, cancer, and mortality, but recent studies also suggest associations with mental health [20]. In this study, the most prevalent symptom/disease indicated by the students was "sneezing attacks", followed by "wheezing/whistling" and "allergic rhinitis". Students who reported having symptoms/diseases reported that these tend to worsen, with greater emphasis, with the spring season. The short-term effects are temporary and range from a simple discomfort, such as irritation of the eyes, nose, skin, throat, coughing and chest tightness, and breathing difficulties, to more serious conditions such as asthma, pneumonia, bronchitis, and lung and heart problems. Short-term exposure to air pollution can also cause headaches, nausea, and dizziness. These problems can be exacerbated by long periods of exposure to pollutants, which is harmful to the neurological system [21, 22].

The levels of depression, anxiety and stress expressed by the students, were similar regardless of whether they remained in the green area or in the area with road traffic. Although in urban areas, these symptoms have greater expression. Regarding this, a study developed in China, in 2021, shows that students residing in cities with the highest air pollution were 6.67 times more likely to experience severe mental stress compared to those residing in cities with less air pollution [5]. Zhang et al. found that normal air quality was not strongly associated with lower levels of negative mental health, but poor air quality had strong links with higher levels of negative mental health [23]. Another study confirmed a significant association between higher levels of $PM_{2.5}$ and PM_{10} and increased mental health problems. Even in cases of relatively low levels of PM_{10} air pollution, exposure to PM_{10} has significant negative impacts on subjective satisfaction with life and well-being [24].

Finally, it was found that regardless of the existence or not of statistically significant results, symptoms of depression are higher for males in any of the areas, green or with road traffic, and symptoms of anxiety and stress are higher for females, in any of the zones. A study developed by Zue et al. in China, in 2020, contrasts the results obtained in our study, where the "gender" factor stands out and where men were more than 1.6 times more likely to increase health problems compared to women [24].

5 Conclusion

The evaluated environmental parameters presented, on mean, higher concentration values in the area with road traffic compared to the green area. Specifically, the mean concentrations of $PM_{2.5}$ and PM_{10} showed exposure values above those legally established. The impact of air quality on student's mental health was not very expressive, but negative emotional symptoms are more evident in areas with road traffic and especially evidenced by females. On the other hand, it is women who show more negative symptoms of anxiety and stress and men who show more symptoms of depression.

Knowing that pollutants present in the atmospheric air can be transported over long distances, crossing regional and national borders and that they can have a negative impact on human health, it is increasingly important to encourage and promote pedestrian circulation, the circulation of bicycles, the use of public transport and shared mobility. The reduction of road traffic, the creation of cycle paths and the increase of green spaces in urban areas, will contribute to the reduction of atmospheric pollutants and noise and will lead to numerous benefits for physical and mental health and for the reduction of mortality. It also seems to be relevant to carry out studies that monitor the atmospheric pollutants responsible for the change in air quality and that relate their presence to changes in mental health, namely with the existence of symptoms of anxiety, stress and depression.

References

1. de Carvalho, J. M., Ferreira, A., Figueiredo, J. P., Loureiro, A. (2020). Indoor Air Quality in Classroom Labs. In: Arezes, P., et al. Occupational and Environmental Safety and Health II. Studies in Systems, Decision and Control, vol 277. Springer, Cham. https://doi.org/10.1007/978-3-030-41486-3_31
2. Faria, T. F. Poluição do ar por partículas em Portugal: Gestão da Qualidade do Ar ambiente, das emissões de poluentes para a atmosfera e da exposição ocupacional. Dissertação de Mestrado, Universidade Lusófona de Humanidades e Tecnologias. (2016).
3. Oliveira, K. Quantificação da relação entre fontes de emissão e a qualidade do ar na região de Aveiro. Dissertação de Mestrado, Universidade de Aveiro. (2018).
4. Gomes, R. Estudo do Impacte de uma Zona de Emissões Reduzidas em Coimbra. Dissertação de Mestrado, Faculdade de Ciências e Tecnologia da Universidade de Coimbra. (2016).
5. Zhang, W., Peng, S., Fu, J., Xu, K., Wang, H., Jin, Y., Yang, T., & Cottrell, R. Urban Air Pollution and Mental Stress: A Nationwide Study of University Students in China. Public Health 9:685431 https://doi.org/https://doi.org/10.3389/fpubh.2021.685431 (2021).
6. Rezende, C. Poluição atmosférica como fator de risco à saúde humana e seu impacto na economia. Caderno saúde e desenvolvimento, 10 (18) (2021).
7. Filipe, T. S., Vasconcelos Pinto, M., Almeida, J., Alcobia Gomes, C., Figueiredo, J. P., Ferreira, 331 A. Indoor air quality in sports halls. CRC Press Taylor & Francis Group, A Balkema Book, 332 Occupational Safety and Hygiene. 2013: 175–179. ISBN: 978–1–138–00047–6 (Hbk); ISBN: 333 978–0–203–72965–6 (2013)
8. Organização Mundial de Saúde. Air pollution, https://www.who.int/health-topics/air-pollution#tab=tab_1, last acecessed 2022/04/18.
9. Lane, C. Depression and suicide risk linked to air pollution, https://www.ucl.ac.uk/news/2019/dec/depression-and-suicide-risk-linked-air-pollution, last acecessed 2022/03/25.
10. European Environment Agency. Air quality in Europe 2019. https://www.eea.europa.eu/publications/air-quality-in-europe-2019, 2019.
11. Fernandes, A. Análise da qualidade do ar e preocupações com a saúde. Dissertação de Mestrado, Faculdade de Engenharia da Universidade do Porto. (2017).
12. Sui, G., Liu, G., Jia, L., Wang, L., & Yan, G. The association between ambient air pollution exposure and mental health status in Chinese female college students: a cross-sectional study. Pub Med, 25(28), 28517-28524 (2018).

13. Power, M., & Kioumourtzoglo, M. The relation between past exposure to fine particulate air pollution and prevalent anxiety: observational cohort study. BMJ 2015;350:h1111 https://doi.org/10.1136/bmj.h1111 (2015).
14. Decreto-Lei n.º 102/2010, de 23 de setembro. Diário da República n.º 186/2010, Série I, Ministério do Ambiente e do Ordenamento do Território. (2010)
15. Correia, A. M., Costa, C., Gonçalves, D., Henriques, G., & Almeida, S. M. Estudo da exposição da população a poluentes do ar nos transportes em ambiente urbano: artigo de revisão. Instituto Politécnico de Lisboa. http://hdl.handle.net/10400.21/14119. 2021
16. Zhang, K., & Batterman, S. Air pollution and health risks due to vehicle traffic. Science of The Total Environment, 450-451, 307-316 (2013).
17. Rocha, S. Previsão de Indicadores de Qualidade de Ar Exterior em Ambiente Urbano. Dissertação de Mestrado, Faculdade de Economia do Porto (2014).
18. Apascaritei, M., Popescu, F., & Ionel, H. Air Pollution level in urban region of Bucharest and in rural region 11th International Conference on Sustainability in Science Engineering, 330–335. 2008.
19. Ferreira, A., Cardoso, S. M. Effects of indoor air quality on respiratory function of children in the 1st cycle of basic education of Coimbra, Portugal. CRC Press Taylor e Francis Group, A Balkema Book, Occupational Safety and Hygiene II. 2014: 347–350. ISBN: 978–1–138–00144–2. 2014.
20. Lynch, K. M., & Mirabelli, M. C. Outdoor Air Quality Awareness, Perceptions, and Behaviors Among U.S. Children Aged 12–17 Years, 2015–2018. J Adolesc Health 68(5):882–887 (2021).
21. Manisalidis, I., Stavropoulou, E., Stavropoulos, A., & Bezirtzoglou, E. Environmental and Health Impacts of Air Pollution: A Review. Frontiers in Public Health 8(14), 11-13 (2020).
22. Ferreira, A., & Barros, N. COVID-19 and Lockdown: The Potential Impact of Residential Indoor Air Quality on the Health of Teleworkers. Int. J. Environ. Res. Public Health, 19(10), 607 (2022).
23. Zhang, X., Zhang, X., & Chen, X. Happiness in the air: How does a dirty sky affect mental health and subjective well-being? Journal of Environmental Economics and Management, 85, 81-94. https://doi.org/https://doi.org/10.1016/j.jeem.2017.04.001 (2017).
24. Zu, D., Zhai, K., Qiu, Y., Pei, P., Zhu, X., & Han, D. The Impacts of Air Pollution on Mental Health: Evidence from the Chinese University Students. Int J Environ Res Public Health 17(18): 6734 (2020)

Indoor Air Quality in Dwellings Located in an Industrial Zone and Its Effects on the Health of the Population

Inês Gregório, Ana Ferreira, João Paulo Figueiredo, António Loureiro, Sílvia Seco, and Fernando Moreira

Abstract In recent decades there has been a significant increase in atmospheric concentrations of hazardous substances, resulting from urban and industrial activities. The good quality of the air we breathe is considered a basic requirement for human health and well-being. The possible effects on human health, derived from these emissions from large industrial sources, have been a long-standing concern, especially when these activities are located in the vicinity of urban areas, where housing is present. This study aimed to evaluate the Indoor Air Quality (IAQ) in the homes of individuals residing near the industrial zone of Alpiarça, as well as analyzing individual's perceptions of AQ and the prevalence of symptoms/diseases. The sample consisted of 8 dwellings and the 14 occupants who lived in them. Data collection consisted of two investigation moments: evaluation of environmental parameters (CO, CO_2, $PM_{0,5}$, $PM_{1,0}$, $PM_{2,5}$, $PM_{5,0}$, PM_{10}, PM_{Totals} and ultrafine particles) and meteorological variables (temperature and relative humidity); application of a questionnaire to the inhabitants of the evaluated residences. It can be concluded that it's necessary to take measures to improve the AQ in some of the evaluated dwellings

I. Gregório · A. Ferreira (✉) · J. P. Figueiredo · F. Moreira
Polytechnic Institute of Coimbra, Coimbra Health School, Rua 5 de Outubro—S. Martinho do Bispo, Apartado 7006, 3046-854 Coimbra, Portugal
e-mail: anaferreira@estesc.ipc.pt

I. Gregório
e-mail: inesgregorio0@gmail.com

J. P. Figueiredo
e-mail: jpfigueiredo@estesc.ipc.pt

F. Moreira
e-mail: fernando.moreira@estesc.ipc.pt

A. Loureiro · S. Seco
Polytechnic Institute of Coimbra, Occupational and Environmental Health Service, Rua da Misericórdia, Lagar Dos Cortiços, S. Martinho Do Bispo, 3045-093 Coimbra, Portugal
e-mail: antonio.loureiro@ipc.pt

S. Seco
e-mail: silvia.seco@ipc.pt

J. R. da. C. S. Galvão et al. (eds.), *Proceedings of the 3rd International Conference on Water Energy Food and Sustainability (ICoWEFS 2023)*, Springer Proceedings in Earth and Environmental Sciences, https://doi.org/10.1007/978-3-031-48532-9_38

as improve air renewal systems in order to make this renewal more efficient and effective, opting whenever possible for natural ventilation.

Keywords Indoor air quality · Industrial zone · Public health

1 Introduction

The interest given to the issue of air quality has expanded considerably. In recent decades there has been a significant increase in atmospheric concentrations of hazardous substances, resulting from urban and industrial activities [1].

Industrial Parks or Industrial Zones constitute a territorial space, in which a series of industrial or business activities that may be related to each other are grouped [2]. Atmospheric emissions from industrial processes or combustion in industrial units pose risks and cause serious impacts on human health and the environment [3].

One of the top five environmental risks to public health is indoor air pollution, with pollutant levels inside buildings being two to five times higher than outdoors, and in extreme cases, being 100 times higher. Currently, society spends much of its day indoors, making these levels of contamination of great importance [4].

The World Health Organization defines Sick Building Syndrome as the set of illnesses and discomfort suffered by people who live in a building [5]. This is caused by inadequate ventilation, decompensation of temperature and relative humidity, lighting conditions, among others [6].

Indoor Air Quality (IAQ) can have a significant impact on human health. Recently, there has been growing concern about the adverse health effects of various chemicals released from building materials and appliances indoors [7]. The pollutants commonly analyzed in the characterization of the IAQ are carbon dioxide (CO_2), carbon monoxide (CO), some hydrocarbons, with special attention to Volatile Organic Compounds (VOC) and particulate matter (PM) [8].

The health problems caused by poor air quality are similar to the symptoms we experience when we have the flu or a cold: headaches, problems with the frontal sinuses, congestion, nausea, tiredness, eye, nose and throat irritation. The indoor environment is almost never suspected of being the cause of symptoms exhibited by occupants, unless the symptoms are shared by multiple occupants, have an unreasonable persistence, or the air quality is abnormal and suspect [9].

Studies have shown that various actions carried out inside homes (for example, cooking, cleaning, various jobs, heating, among others) can result in the emission and resuspension of particulate matter, especially $PM_{2,5}$ [10]. In Portugal, a study carried out in homes where pollutants such as PM_{10}, CO and VOC were measured observed that in 60.0% of the homes studied, the concentrations of these pollutants exceeded, at some point, the legal limits, alerting to the need to develop more studies [11].

The monitoring of air quality in industrial areas is essential to assess the exposure of populations, and to propose solutions that aim to improve the air quality of places affected by emissions from activities carried out in these areas [2].

In view of the above, the main objectives of this study were to evaluate the air quality in homes located next to the industrial area in the municipality of Alpiarça, district of Santarém, Portugal, as well as to analyze the perception of individuals about air quality and the prevalence of symptoms/diseases.

2 Material and Methods

The present study was an observational (analytical) level II study, with a transversal timeline. The study consisted of evaluating the quality of indoor in 8 houses in the municipality of Alpiarça, district of Santarém, Portugal located next to an industrial area, as well as the respective residents (14).

Data collection consisted of two investigation moments, the first being the evaluation of the environmental parameters CO, CO_2, $PM_{0,5}$, $PM_{1,0}$, $PM_{2,5}$, $PM_{5,0}$, PM_{10}, PM_{Totals} and the ultrafine particles and variables meteorological temperature (T°) and relative humidity (Hr); the second, the application of a questionnaire addressed to all residents of the dwellings under study.

About the first moment of investigation, measurements were carried out over a period of 30 min, with minute-by-minute sampling. The measurements took place, in each household, in the morning and in the afternoon in two spaces of the house, which were the kitchen/living room and bedroom, which is where the individuals spent most of their time. The equipment's was placed in the most central position possible in each evaluated space and at the height of the inhabitant's airways, in the seated position. It is also important to mention that the samples were collected at a height of 1.0 m from the ground and at least 3.0 m from the walls. To carry out the analytical collection of the analyzed parameters, portable real-time reading equipment was used.

It was considered as a reference for the maximum concentration of CO, 10.0 mg/m^3 (9 ppm), for CO_2, 2250 mg/m^3 (1250 ppm), for $PM_{2.5}$, 25 μg/m3 (0.025 mg/m^3), for PM_{10} 50 μg/m3 (0.05 mg/m^3), as referred to in Ordinance No. 138-G/2021 (Issuer: Health and Environment and Climate Action), of 1 July [12]. Given the lack of reference values for the concentration of ultrafine particles, the analytical average values recorded in the measurements of outdoor air were taken as a reference point. According to Decree-Law No. 243/86 (Issuer: Ministry of Labor and Social Security), of August 20, the reference comfort environmental conditions for T° must oscillate between 18 and 22 °C, while Hr must oscillate between 50 and 70% [13].

With regard to the second moment of investigation, the questionnaire was divided into four fundamental parts: the first proposed to obtain personal data from individuals; the second addressed their perception of air quality; the third part was intended to obtain data on the health condition of individuals, that is, if there was a history of chronic disease and symptoms of respiratory or cardiovascular disease, or smoking habits in the family; the fourth part aimed to obtain information on the housing conditions of the residences under study.

The collected data were statistically treated using the IBM SPSS Statistics software, version 27.0. Regarding the hypothesis tests applied, statistical tests were used: t-Student for a sample, Thurs Independence Square and Fisher's exact test. Data interpretation was based on a significance level of *p-value* ≤ 0.05 for a confidence level of 95%.

3 Results

Of the 14 individuals who participated in the study, 10 were female (71.4%), and 4 were male (28.6%). In terms of educational qualifications, 50.0% of the residents had secondary education, with 4 female residents and 3 male. Regarding the years in which they lived in the residence, 85.7% of the respondents had lived in their homes for more than 20 years and the rest (14.3%) between 10 and 15 years. Regarding the distance from the house to the industrial park, 10 respondents were living less than 500 m (71,4%) from the industrial area and the rest lived in a house located between 500 m and 1 km (28.6%) of the referred area.

In Table 1, the estimated analytical values of the average concentrations of CO, CO_2, $PM_{2.5}$ and PM_{10} recorded inside the different evaluated dwellings were compared with the legally established values (protection threshold).

According to the results obtained, it was observed that there were no statistically significant differences between the analytical mean values of CO concentration and the protection threshold in House 7 (*p-value* > 0.05). All other dwellings showed significant differences. Differences were recorded in the average concentration of CO, with House 7 having the highest average values, while House 2 had the lowest average values. It is also important to mention that in none of the evaluated dwellings the legally established limit value of protection was exceeded. With regard to the analytical average values of CO_2 concentration, it was found that there were statistically significant differences between the estimated average values and the protection threshold in all the dwellings under study. It was House 7 that presented the highest average CO_2 values and, consequently, was the only house that recorded concentration values above the protection threshold value (1515.49 ppm). On the other hand, House 5 was the one with the lowest average values. Regarding the analytical average concentrations of $PM_{2,5}$, it was found that there were statistically significant differences in relation to the protection threshold in all the residences, with only House 7 being the one that recorded the highest average value and, consequently, exceeded the value protection threshold when recording the average concentration of 29.22 $\mu g/m^3$. On the other hand, it was House 3 that registered the lowest average values. With regard to the average concentrations of PM_{10}, there were statistically significant differences in Houses 1, 2, 3, 4, 5 and 6, while in House 7 there were no significant differences. The average values of this pollutant were above the protection

Table 1 Average concentrations of CO, CO_2, $PM_{2,5}$ and PM_{10}, inside the different houses evaluated

Housing									
		House 1	House 2	House 3	House 4	House 5	House 6	House 7	House 8
CO (ppm)	N	120	120	120	120	120	121	120	120
	M	0.158	0.110	0.305	0.471	0.665	0.174	8,855	0.338
	PD	0.075	0.033	0.178	0.266	0.558	0.083	5.104	0.196
	LP	<	<	<	<	<	<	<	<
	DM	−8,572	−8,619	−8,424	−8,258	−8,064	−8,555	0.126	−8,391
	p-value	< 0.001	< 0.001	< 0.001	< 0.001	< 0.001	< 0.001	0.787	< 0.001
CO_2 (ppm)	N	120	120	120	120	120	121	120	120
	M	756.64	642.53	770.51	773.10	567.85	620.83	1515.49	533.88
	PD	154.108	62,794	58,046	60,699	82,578	88,912	65,603	59,547
	LP	<	<	<	<	<	<	>	<
	DM	−493,358	−607,467	−479,492	−476,900	−682,150	−629,174	265,492	−716,125
	p-value	< 0.001	< 0.001	< 0.001	< 0.001	< 0.001	< 0.001	< 0.001	< 0.001
$PM_{2,5}$ ($\mu g/m^3$)	N	120	120	120	120	120	121	120	120
	M	3,514	3,835	3,373	7,587	6,228	5,150	29,223	6,726
	PD	1,598	2,079	0.493	2,392	3,140	1.703	11,474	1.815
	LP	<	<	<	<	<	<	>	<
	DM	−21,486	−21,165	−21,627	−17,413	−18,772	−19,850	4,223	−18,274
	p-value	< 0.001	< 0.001	< 0.001	< 0.001	< 0.001	< 0.001	< 0.001	< 0.001

(continued)

Table 1 (continued)

Housing		House 1	House 2	House 3	House 4	House 5	House 6	House 7	House 8
PM_{10} (μg/m^3)	N	120	120	120	120	120	121	120	120
	M	18,334	16,873	28,283	42,549	44,507	22,885	63,908	30,983
	PD	8,434	8,169	14,834	15,687	21,248	5.059	20,973	5,117
	LP	<	<	<	<	<	<	>	<
	DM	−31,666	−33,127	−21,717	−7,451	−5,493	−27,115	13,908	-19.017
	p-value	< 0.001	< 0.001	< 0.001	< 0.001	0.005	< 0.001	0.787	< 0.001

N = Number of measurements performed; M = Average; DP = Standard deviation; LP = Protection Limit; DM = Mean Difference. LPCO = 9 ppm; LP CO_2 = 1250 ppm; LP $PM_{2,5}$ = 25 μg/m^3; LP PM_{10} = 50 μg/m^3. Test: t-Student for one sample threshold value in House 7, while in the others it was below the protection threshold value. House 1 was the one with the lowest average values.

Afterwards, we tried to confirm whether there were any exceedances of the reference value imposed by the legislation for the meteorological variables (T° and Hr) in the indoor air of the different evaluated dwellings (Table 2).

Analyzing the data related to T°, it was found that of the measurements carried out in all the dwellings under study, 78.0% were within the legislated reference range [18 °C 22 °C], 13.8% recorded values below 18 °C and 8.2% (78 measurements) were above 22 °C. House 7 recorded the highest number of measurements outside the reference range and, consequently, above this value (>22 °C). With regard to the meteorological variable Hr, of the 961 measurements carried out inside the dwellings under study, it was found that 67.5% of the measurements were within the legislated reference range [50–70%] and 32.5% above this range. It should be noted that there was no record of Hr values below the reference range.

Next, we intended to assess the presence or absence of symptoms/diseases depending on the sex of the study participants (Table 3).

It was found that all the participants in the study (14) revealed to have some of the symptoms/diseases under investigation. Analyzing the results obtained in Table 3, it

Table 2 Meteorological variables T° and Hr in the different dwellings under study

Housing		Temperature			Relative humidity	
		< 18.00 °C	[18.00–22.00 °C]	> 22.00 °C	[50.0–70.0%]	> 70.0%
House 1	N	0	120	0	0	120
	% column	0.0%	16.0%	0.0%	0.0%	38.5%
House 2	N	28	92	0	59	61
	% column	21.1%	12.3%	0.0%	9.1%	19.6%
House 3	N	4	116	0	71	49
	% column	3.0%	15.5%	0.0%	10.9%	15.7%
House 4	N	0	119	1	71	49
	% column	0.0%	15.9%	1.3%	10.9%	15.7%
House 5	N	101	19	0	88	32
	% column	75.9%	2.5%	0.0%	13.6%	10.3%
House 6	N	0	121	0	120	1
	% column	0.0%	16.1%	0.0%	18.5%	0.3%
House 7	N	0	43	77	120	0
	% column	0.0%	5.7%	98.7%	18.5%	0.0%
House 8	N	0	120	0	120	0
	% column	0.0%	16.0%	0.0%	18.5%	0.0%
Total	N	133	750	78	649	312
	% column	100.0%	100.0%	100.0%	100.0%	100.0%

N = Number of measurements performed; Test: Qui Square of Independence

Table 3 Presence or absence of symptoms/diseases depending on the sex of the participants

Symptoms/Diseases		Sex		
		Feminine	Male	Total
		n (%)	n (%)	n (%)
Wheezing / Whistling	Yes	1 (10.0%)	0 (0.0%)	1 (7.1%)
	No	9 (90.0%)	4 (100.0%)	13 (92.9%)
Allergies (rhinitis)	Yes	2 (20.0%)	3 (75.0%)	5 (35.7%)
	No	8 (80.0%)	1 (25.0%)	9 (64.3%)
Headaches	Yes	8 (80.0%)	1 (25.0%)	9 (64.3%)
	No	2 (20.0%)	3 (75.0%)	5 (35.7%)
Pruritus (itching), irritation, burning or dryness in the eyes	Yes	4 (40.0%)	1 (25.0%)	5 (35.7%)
	No	6 (60.0%)	3 (75.0%)	9 (64.3%)
Pruritus (itching), irritation or dryness of the skin	Yes	2 (20.0%)	0 (0.0%)	2 (14.3%)
	No	8 (80.0%)	4 (100.0%)	12 (85.7%)
Dizziness	Yes	4 (40.0%)	0 (0.0%)	4 (28.6%)
	No	6 (60.0%)	4 (100.0%)	10 (71.4%)
Odor sensitivity	Yes	6 (60.0%)	1 (25.0%)	7 (50.0%)
	No	4 (40.0%)	3 (75.0%)	7 (50.0%)
Breathing difficulties	Yes	2 (20.0%)	0 (0.0%)	2 (14.3%)
	No	8 (80.0%)	4 (100.0%)	12 (85.7%)
Dry cough	Yes	3 (30.0%)	2 (50.0%)	5 (35.7%)
	No	7 (70.0%)	2 (50.0%)	9 (64.3%)
Sneezing, runny nose or stuffy nose (when not constipated)	Yes	3 (30.0%)	2 (50.0%)	5 (35.7%)
	No	7 (70.0%)	2 (50.0%)	9 (64.3%)

was observed that female subjects manifested more symptoms/diseases than males. 100% of the inhabitants who reported wheezing/whistling, itching or skin irritation, dizziness and breathing difficulties were female. It was also found that 60.0% of people who reported having allergies were male and 88.9% of people who reported having headaches were female. It was proved that 80.0% of people who reported feeling itchy or irritated eyes were female and 85.7% of people who said they were sensitive to odors as well. Finally, and in relation to the symptoms of dry cough and sneezing crisis, 60.0% of the people who indicated feeling these symptoms were female.

It is also important to mention that when asked if they stopped having the symptoms/diseases when they moved outside the house, all respondents answered no.

In Table 4, an attempt was made to assess the presence or absence of different symptoms/diseases depending on the distance from the evaluated dwellings to the industrial park that was located nearby.

Table 4 Presence or absence of symptoms/diseases depending on the distance from the house to the industrial park

Symptoms/Diseases			Distance from housing to industrial park		
			Less than 500 m	Between 500 m and 1 km	*p-value*
Wheezing/Whistling	Yes	n (%)	0 (0%)	1 (25%)	0.286
	No	n (%)	10 (100%)	3 (75%)	
Allergies (rhinitis)	Yes	n (%)	4 (40%)	1 (25%)	0.999
	No	n (%)	6 (60%)	3 (75%)	
Headaches	Yes	n (%)	6 (60%)	3 (75%)	0.999
	No	n (%)	4 (40%)	1 (25%)	
Pruritus (itching), irritation, burning or dryness in the eyes	Yes	n (%)	2 (20%)	3 (75%)	0.095
	No	n (%)	8 (80%)	1 (25%)	
Pruritus (itching), irritation or dryness of the skin	Yes	n (%)	0 (0%)	2 (50%)	0.066
	No	n (%)	10 (100%)	2 (50%)	
Dizziness	Yes	n (%)	2 (20%)	2 (50%)	0.520
	No	n (%)	8 (80%)	2 (50%)	
Odor sensitivity	Yes	n (%)	4 (40%)	3 (75%)	0.559
	No	n (%)	6 (60%)	1 (25%)	
Breathing difficulties	Yes	n (%)	1 (10%)	1 (25%)	0.505
	No	n (%)	9 (90%)	3 (75%)	
Dry cough	Yes	n (%)	4 (40%)	1 (25%)	0.999
	No	n (%)	6 (60%)	3 (75%)	
Sneezing, runny nose or stuffy nose (when not constipated)	Yes	n (%)	3 (30%)	2 (50%)	0.508
	No	n (%)	7 (70%)	2 (50%)	
Test: Fisher's Exact					

It was observed that 10 of the study participants lived less than 500 m from the industrial park and the remaining 4 lived at a distance between 500 m and 1 km from it. Through the analysis of the data obtained, it was found that there was no association between the distance from the dwellings to the industrial park and the presence or absence of symptoms/diseases (*p-value* > 0.05). Regarding the symptoms/diseases with the highest number of cases, of the 14 inhabitants in the study, 9 indicated having the symptom headaches, with 66.7% living less than 500 m from the industrial park and 33.3% between 500 m 1 km; 7 reported being sensitive to odors, with 57.1% living less than 500 m from the industrial park and 42.9% between 500 m and 1 km.

4 Discussion

After analyzing the results obtained, it was found that the respondents lived in houses located less than 500 m and between 500 m and 1 km from the industrial area, which is why it is important to treat pollution control and mitigation as an immediate need and not as an option. Industrial zones have been facing major challenges in terms of the environment, and for this reason they have come to be considered as targets in the fight against excessive atmospheric emissions [14].

Regarding the average concentrations of CO, it was found that, on average, the concentrations of this pollutant were below the protection threshold value in all dwellings. When the average values obtained in each evaluated space were analyzed, it was observed that they were higher in the afternoon compared to the morning, with the highest value of CO concentration being verified in the "living room" space, which is where individuals manifested spending most of their time. Studies reveal that domestic combustion appliances and industrial installations are common sources of air pollution and are therefore of great concern for the health of populations [15].

Regarding the observed values of CO_2 concentration in the dwellings, it was found that House 7 exceeded, on average, the legally established protection threshold, this may be due to the fact that, despite the existence of natural ventilation in the dwelling, the inhabitants mentioned that they never opened the windows. According to a study developed by Wong et al., the concentration of CO_2 in indoor air is used as an indicator of air quality and ventilation efficiency [16]. Of the 3 spaces evaluated, the average values obtained for the CO_2 parameter were higher in the morning, with the exception of the evaluated space "Sala de vida". The main source of CO_2 in indoor air results from the biological metabolism of its occupants, so the values of this air pollutant are generally higher in building areas and at times when occupants stay longer, being directly related with the number of occupants per square meter. As mentioned by Lu et al. (2011), the increase in CO_2 concentration inside buildings shows poor ventilation of the space and, consequently, poor IAQ [17].

With regard to the average concentration values of $PM_{2,5}$ and PM_{10}, it was found that in practically all the evaluated dwellings, the values obtained did not exceed the protection threshold value. However, in House 7, the mean values obtained exceeded the protection threshold value established for $PM_{2,5}$ (25 μg/m^3) and for PM_{10} (50 μg/m^3). According to a study developed by Ferreira (2007) in interior spaces, particulate matter may originate from the materials present in the building, but also from the occupants themselves, either by the agitation caused in the air or by the release of particles [6].

From the analysis of the average values of T° found inside the studied dwellings, it was found that, although these, in general, did not present a reason for great concern, since they were, for the most part, within the reference range. [18–22°C], it should be noted that House 7 obtained average values above this reference value. Regarding the average levels of relative humidity, it was found that most of the evaluated dwellings had humidity levels within the reference range [50–70%], except for the results obtained in House 1 and 2 in which the mean values were higher than the reference range. According to the study developed by Amorim, T° and Hr are relevant factors to consider, as thermal comfort enables greater productivity in leisure and work activities, so it is necessary to periodically evaluate these parameters in the dwellings, safeguarding the comfort and well-being of the inhabitants [18].

According to the study carried out, it was found that the difference obtained in the results of the parameters evaluated inside House 7, when compared with the other dwellings, was noticeable. The analytical average values estimated in its interior were always higher than those of the other dwellings, with the exception of the meteorological variable relative humidity. This difference between House 7 and the other dwellings may be due, as mentioned above, to the fact that despite the fact that there is natural ventilation in the dwelling, the inhabitants indicated that they are not in the habit of opening windows. Ventilation, natural or mechanical, is the second most effective way to provide acceptable indoor air conditions [19]. The lack of ventilation worsens the air quality, namely where the thermal and acoustic insulation of the buildings means that the windows remain closed, without a sufficient air renewal system [20].

Regarding the symptoms perceived by the inhabitants in the study, it was found that females manifested more diseases than males. According to Smith, an estimated 4% to 6% of the total disease burden can be attributed to women, as they spend most of their time indoors [21]. In this study, the most prevalent symptom/disease indicated by the inhabitants were headaches, sensitivity to odors, allergies, pruritus (itching), irritation, burning or dry eyes, dry cough and sneezing crisis. The existence of the aforementioned symptoms may be related to exposure to atmospheric pollutants in indoor air, such as CO, which can cause symptoms such as headaches and dizziness; exposure/inhalation of particulate matter associated with breathing difficulties, upper airway irritation, decreased lung function, among others [20]. Several studies claim that many illnesses and deaths (due to long-term exposure) are related to inhaling indoor air pollutants. Most air pollutants contribute to these negative effects on human health due to their particulate nature, particularly when the particle diameter is very small, as these particles can lodge in the pulmonary alveoli and quickly pass into the bloodstream [22–24].

5 Conclusions

With the results obtained in this study, it can be concluded that it is necessary to take measures to improve the IAQ in some homes, since there were atmospheric pollutants, such as CO_2, $PM_{2,5}$ and PM_{10} which were higher than legally established values, namely in House 7. The importance of improving air renewal systems is highlighted, to make this renewal more efficient and effective, opting whenever possible for natural ventilation, such as the simple habit of opening windows daily. The composition of the cleaning products and other chemicals used should also be analyzed to see if there are any compounds that could cause risks to the air quality and health of the residents.

In the present study, it was not possible to verify that the houses closer to the industrial area had a worse air quality, because comparing the dwellings, it was found, for example, that the values of CO_2, $PM_{2,5}$ and PM_{10} were higher than the threshold value. protection in House 7, which was located at a distance of 500 m from the industrial area, while House 5 was located 100 m from it and did not behave in the same way. However, to have greater control and reduction of atmospheric emissions from industries, it is advisable to take measures, namely the use of pollutant control equipment and the optimization of combustion processes. In this sense, actions must be taken to protect and improve air quality, in order to make it impossible for atmospheric pollutants to reach concentrations that are capable of jeopardizing the health of the inhabitants. In addition, it is also advisable to carry out regular assessments of the air quality in industrial areas with the main objective of maintaining adequate levels, so as not to harm the quality of indoor and outdoor air in the vicinity of these areas.

References

1. Fernandes, A. Analysis of Air Quality and Health Concerns. Faculty of Engineering, University of Porto. https://repositorio-aberto.up.pt/bitstream/10216/102619/2/180885.pdf (2017).
2. Alves, R. $PM_{2.5}$ levels in the Industrial Zone of Northeast Portugal. Escola Superior Agrária de Bragança. https://bibliotecadigital.ipb.pt/bitstream/10198/19233/1/Alves_Rafaela.pdf (2018).
3. Gomes, P. Air Quality in Industrial Areas - Barreiro's case study. Faculty of Science and Technology - Universidade Nova de Lisboa. (2008).
4. Ferreira, J. Relationship between Air Quality and Human Exposure to Atmospheric Pollutants. University of Aveiro]. https://ria.ua.pt/handle/10773/697 (2007).
5. WHO. Development of Who Guidelines for Indoor Air Quality - Report on a Working Group Meeting. In World Health Organization Regional Office for Europe. (2006).
6. Ferraz, M. (2016). The concept of health. Pharmacovigilance: Critique and Ways Forward, 31(5), 161–172. https://doi.org/10.1007/978-3-319-40400-4_15 (2016).
7. Suzuki, N., Nakaoka, H., Hanazato, M., Nakayama, Y., Tsumura, K., Takaya, K., Todaka, E., & Mori, C. Indoor air quality analysis of newly built houses. International Journal of Environmental Research and Public Health, 16, 11. https://doi.org/10.3390/ijerph16214142 (2019).
8. Silva, S. Modeling indoor air quality. University of Aveiro. (2016).
9. de Carvalho, J.M., Ferreira, A., Figueiredo, J.P., Loureiro, A. Indoor Air Quality in Classroom Labs. In: Arezes, P.M. et al (eds) Occupational and Environmental Safety and Health II, Studies in Systems, Decision and Control, vol 277. Springer, Cham. https://doi.org/10.1007/978-3-030-41486-3_31 (2020).
10. Abreu, C. Indoor air quality. In The Indoor Environment and the Health of Residential Building Occupants. http://hdl.handle.net/10400.6/3579 (2010).
11. Coriander, S. Indoor Air Quality in Homes: Pollutant Emitting Sources. Universidade Nova de Lisboa. https://run.unl.pt/bitstream/10362/17472/1/Coentro_2015.pdf (2015).
12. Ordinance No. 138-G/2021 of July 1st. Ordinance No. 138-G/2021 of July 1st. Diário da República, 1st Series. (2021).
13. Decree-Law No. 243/86 of August 20th. Decree-Law No. 243/86 of August 20 (General regulation of hygiene and safety at work in commercial, office and service establishments). Republic Diary. (1986).

14. Gomes, L. Air Quality in Industrial Areas: Analysis of Temporal and Spatial Patterns of Ozone and Nitrogen Oxides in the Urban Industrial Area of Mirandela (Portugal). Escola Superior Agrária de Bragança]. In Instituto Politécnico de Bragança (2020).
15. Maia, A. Assessment of Exposure to Atmospheric Pollutants in the City of Aveiro. University of Aveiro (2014).
16. Wong, L., Mui, K., Shi, K., & Hui, P. An energy impact assessment of indoor air quality acceptance for air-conditioned offices. Energy Conversion and Management 49(10), 2815-2819. (2008).
17. Ferreira, A., Figueiredo, J.P., Pêgo, F., Loureiro, A., Seco, S. Air Quality Assessment in Kitchens and Effects on Workers Health. In: Arezes, P., et al (eds) Occupational and Environmental Safety and Health II. Studies in Systems, Decision and Control, vol 277. Springer, Cham. https://doi.org/10.1007/978-3-030-41486-3_25 (2020).
18. Amorim, D. Optimization of natural ventilation conditions in service buildings to ensure indoor air quality without excessive energy consumption. Universidade do Minho. https://repositorium.sdum.uminho.pt/handle/1822/29576 (2012).
19. Carmo, A., & Prado, R. Indoor Air Quality [USP Polytechnic School]. In Control of Atmospheric Pollution. http://www.casoi.com.br/hjr/pdfs/Qualidade Interna do Ar.pdf (1999).
20. Smith, K., Samet, J., Romieu, I., & Bruce, N. Indoor air pollution in developing countries and acute lower respiratory infections in children. 15. Thorax 2000;55:518–532 ((2000).
21. Ferreira, A., Cardoso, S. M. Effects of indoor air quality on respiratory function of children in the 1st cycle of basic education of Coimbra, Portugal. CRC Press Taylor e Francis Group, A Balkema Book, Occupational Safety and Hygiene II. 2014: 347–350. ISBN: 978–1–138–00144–2. (2014).
22. Ferreira, A., & Barros, N. COVID-19 and Lockdown: The Potential Impact of Residential Indoor Air Quality on the Health of Teleworkers. International Journal of Environmental Research and Public Health, 19, 23. https://doi.org/https://doi.org/10.3390/ijerph19106079 (2022).
23. Filipe, T.S., Vasconcelos Pinto, M., Almeida, J., Alcobia Gomes, C., Figueiredo, J.P., Ferreira, A. Indoor air quality in sports halls. CRC Press Taylor & Francis Group, A Balkema Book, Occupational Safety and Hygiene. 2013: 175–179. ISBN: 978–1–138–00047–6 (Hbk); ISBN: 978–0–203–72965–6 (2013).
24. Marques. C., Ferreira, A., Figueiredo, J. P. Evaluation of indoor air quality in kindergartens – case study. CRC Press Taylor e Francis Group, A Balkema Book, Occupational Safety and Hygiene IV. 2016: 375–380. ISBN: 978–1–138–02942–2 (Hbk), ISBN: 978–1–315–62896–7 (2016).

Air Quality Evaluation in Urban and Rural Areas

João Soares, Ana Ferreira, João Paulo Figueiredo, António Loureiro, Sílvia Seco, and Fernando Moreira

Abstract Deteriorating Air Quality (AQ) is one of the most severe health problems for people living in cities. The increasing use of road transport, the expansion of industry and domestic heating contribute significantly to the worsening of this problem. The AQ indicates the level of pollution in the air we breathe, which is caused by various chemical substances present in the air, which, in turn, change the natural composition of the atmosphere. This study aimed to: evaluate the concentration of atmospheric pollutants and particles—carbon monoxide, carbon dioxide, ultrafine particles, $PM_{2.5}$, PM_{10} and PM_{Totals}—and the meteorological variables—temperature and relative humidity—in two distinct areas, one urban and one rural; analyze the ambient AQ in these areas in order to investigate whether there are significant differences between them. Assessments were performed twice a day (morning and afternoon) at each site for 1 week (7 days). They were carried out in periods of thirty minutes, with sampling each minute. We found that the concentrations of pollutants vary depending on the area and that PM_{10} exceeded the legally established exposure limit value. Since pollutants act globally and not just in the place where

J. Soares (✉) · A. Ferreira · J. P. Figueiredo · F. Moreira
Polytechnic Institute of Coimbra, Coimbra Health School, Rua 5 de Outubro—S. Martinho do Bispo, Apartado 7006, 3046-854 Coimbra, Portugal
e-mail: j_cancela_@hotmail.com

A. Ferreira
e-mail: anaferreira@estesc.ipc.pt

J. P. Figueiredo
e-mail: jpfigueiredo@estesc.ipc.pt

F. Moreira
e-mail: fernando.moreira@estesc.ipc.pt

A. Loureiro · S. Seco
Polytechnic Institute of Coimbra, Occupational and Environmental Health Service, Rua da Misericórdia, Lagar Dos Cortiços, S. Martinho Do Bispo, 3045-093 Coimbra, Portugal
e-mail: antonio.loureiro@ipc.pt

S. Seco
e-mail: silvia.seco@ipc.pt

J. R. da. C. S. Galvão et al. (eds.), *Proceedings of the 3rd International Conference on Water Energy Food and Sustainability (ICoWEFS 2023)*, Springer Proceedings in Earth and Environmental Sciences, https://doi.org/10.1007/978-3-031-48532-9_39

they originate and can cause harmful effects on health and climate, creation of low-emission zones for road traffic, reinforcement of green infrastructure consolidation of the cycling network, among others can be strategies that gradually reduce the values reached.

Keywords Air quality · Air pollutants · Public health

1 Introduction

Air is an indispensable resource for human being and its quality affects the health and quality of life of populations. Air pollution is a leading cause of premature death and illness and is the biggest environmental health risk. There are studies that show that nine out of ten people breathe air with high levels of pollutants. The risk of exposure to pollution is higher in cities, mainly due to road traffic and mobility patterns based on the private vehicle. Other important additional sources are residential and commercial biomass burning, industrial activity and even natural sources such as forest fires [1]. The deterioration of air quality is one of the most severe problems for the health of those who live in these spaces. The increasing use of road transport, the expansion of industry and domestic heating contribute significantly to the worsening of this problem [2]. The effects of air pollution on public health are very serious. One third of deaths from stroke, lung cancer and heart disease are due to air pollution. Most cities in the world live with higher values of air pollution than the maximum values recommended by the World Health Organization (WHO) [3].

From a physical–chemical point of view, atmospheric pollutants belong to two broad categories: gaseous and particulate matter. The first correspond to gases, such as carbon monoxide (CO) and carbon dioxide (CO_2) emitted by different sources or generated in the atmosphere. Particulate matter (ultrafine particles, $PM_{2,5}$, PM_{10} e PM_{Totals}) consists of a complex mixture of particles of varying sizes, solid or liquid, suspended in the air [4].

According to the WHO, air pollution is the biggest environmental risk to health in the European Union (EU), causing around 400,000 premature deaths every year and generating hundreds of billions of euros of external health-related costs. People in urban areas are particularly exposed. Airborne particulates are the air pollutants responsible for most of these premature deaths [5].

Air quality tells us the level of pollution in the air we breathe, which is caused by various chemical substances present in the air, which, in turn, change the natural composition of the atmosphere [6]. This is a very important environmental component for public health and for the quality of life of citizens. Air pollution problems with repercussions on human health and ecosystems, mainly in terms of pollutants and fine particles, exist mainly in densely populated urban areas [7]. The assessment of ambient air quality is therefore essential so that the most appropriate measures can be implemented with a view to improving the levels of the various pollutants [8].

According to some studies, the main effects of air pollutants on health are reflected in the respiratory and cardiovascular system. These effects can vary and depend on the exposure time, concentration and vulnerability of each person [9–13]. The solution to these problems increasingly involves the design and implementation of new models of space organization, in which green spaces must emerge as fundamental elements [2]. According to a study by Martins et al., the natural ability of green vegetation to capture airborne contaminants resulted in air quality improvement of up to 16% in an area with dense vegetation in the United States of America [14]. The transpiration process of trees affects air temperature, radiation absorption, heat accumulation, wind speed, relative humidity, among other variables. Green areas have the potential to mitigate some of the negative effects of urbanization [2].

In this sense, the objective of this study was to evaluate the concentration of atmospheric pollutants and particles and the meteorological variables temperature (T°) and relative humidity (Hr) existing in two different areas, an urban area and a rural area.

2 Material and Methods

The study was carried out in the central region of Portugal. This study was of the analytical observational type, level II, of timeline, of the prospective type, where the atmospheric pollutants defined in the places under study were evaluated. Two different points were selected to carry out the data collection, one located in an urban area and the other in a rural area. The collection points are approximately 12.7 km apart.

Data collection consisted of evaluating the atmospheric pollutants CO, CO_2, Ultrafine Particles, $PM_{2,5}$, PM_{10} e PM_{Totals} and the meteorological variables T° and Hr in the two locations under study. Assessments were performed twice a day (morning and afternoon) at each location for 7 days. They were carried out in periods of thirty minutes, with sampling from minute to minute. The periods of the day were selected since it is in the early morning and late afternoon period that there is a greater amount of car traffic in the evaluated areas. For the analytical collection of the data, portable real time reading equipment was used.

As reference values, there were considered 10 mg/m^3 (8.7 ppm) for the limit value for CO, 25 $\mu g/m^3$ for $PM_{2,5}$ and 50 $\mu g/m^3$ for PM_{10}, as referred to in Decree-Law No. 102/2010, of September 23 [15].

The collected data were statistically treated using the IBM SPSS Statistics software version 27.0. In the analysis of the performed evaluations, the Wilcoxon Mann–Whitney non-parametric test was applied to compare 2 independent samples and the t-Student parametric test for 1 sample. For the estimation of statistical inference, we considered a confidence level of 95% and a random error of less than or equal to 5%.

3 Results

Data collection was carried out at two different sampling points, in an urban area, a very urbanized area where there is a large circulation of cars, and in a rural area with a large area of surrounding forest.

We then proposed to evaluate the behavior of the CO as a function of the periods defined by "morning" and "afternoon" between the different areas, urban and rural, adjusted to the days of the week.

According to the results obtained, we found that on the days of the week: "Tuesday", "Wednesday", "Thursday", "Friday" and "Saturday", the CO values were always significantly higher in the area designated as urban compared to the rural area (*p-value* < 0.05). There were also significant differences on the days of the week about "Sunday" and "Monday" where CO values were higher in urban areas than in rural areas in the afternoon (*p-value* < 0.05). However, in the morning periods on "Sunday" and "Monday" there were no statistically significant differences between the average concentrations of CO between urban and rural areas.

We proposed to evaluate the behavior of CO_2 in terms of the periods defined by "morning" and "afternoon" between urban and rural areas, adjusted to the days of the week. We found that there were statistically significant differences regarding the levels of CO_2 in the "morning" and "afternoon" periods on "Sunday", "Monday", "Tuesday" and "Wednesday", where CO_2 values were always significantly higher in rural areas (*p-value* < 0.05). There were also significant differences on "Friday" and "Saturday", where in the morning the CO_2 values were significantly higher in the urban area and higher in the afternoon in the rural area (*p-value* < 0.05).

Then, we evaluated the mean values of ultrafine particles as a function of the periods defined by "morning" and "afternoon" between the different areas, urban and rural, adjusted to the days of the week (Table 1).

We found significant differences between the levels of ultrafine particles detected in urban and rural areas, where these values were always significantly higher in the urban area, both in the "morning" and in the "afternoon" period (*p-value* < 0.05).

In Table 2 we seek to evaluate the behavior of the mean concentrations of $PM_{2,5}$ depending on the periods defined by "morning" and "afternoon" between the different areas, urban and rural, adjusted to the days of the week. We found that on "Monday", "Tuesday" and "Friday", $PM_{2,5}$ concentration values were always significantly higher in urban areas than in rural areas. There were also significant differences on "Wednesday" and "Thursday", where the average values of $PM_{2,5}$ were higher in the urban area in the morning.

The mean $PM_{2,5}$ concentration values were higher in the rural area on "Sunday" and "Wednesday" in the afternoon. These values were also higher in the rural area on "Saturday" during the morning and afternoon periods (*p-value* < 0.05). However, there were no significant differences between the study sites on "Sunday" during the "morning" period and "Thursday" during the "afternoon" period (*p-value* > 0.05).

We propose to evaluate the differences in the mean concentration of PM_{10} as a function of the periods defined by "morning" and "afternoon" between the different

Table 1 Mean concentrations of Ultrafine Particles as a function of the "morning" and "afternoon" periods between urban and rural areas adjusted for the days of the week

Ultrafine particles (Particles/cm^3)		Zone		
Day of the week	Period	Urban (M ± SD)	Rural (M ± SD)	*p-value*
Sunday	Morning	6531 ± 3555	3087 ± 351	≤ 0.0001
	Afternoon	10,324 ± 5178	2157 ± 372	≤ 0.0001
Monday	Morning	20,535 ± 19,225	9787 ± 4999	≤ 0.0001
	Afternoon	1120 ± 53	2808 ± 521	≤ 0.0001
Tuesday	Morning	25,787 ± 13,167	1479 ± 161	≤ 0.0001
	Afternoon	8580 ± 3020	2053 ± 498	≤ 0.0001
Wednesday	Morning	17,601 ± 3079	3574 ± 179	≤ 0.0001
	Afternoon	8032 ± 4469	2151 ± 323	≤ 0.0001
Thursday	Morning	20,716 ± 7854	2223 ± 433	≤ 0.0001
	Afternoon	21,240 ± 10,007	3234 ± 99	≤ 0.0001
Friday	Morning	46,500 ± 20,301	3961 ± 392	≤ 0.0001
	Afternoon	16,472 ± 6032	8053 ± 453	≤ 0.0001
Saturday	Morning	10,652 ± 7936	3661 ± 665	≤ 0.0001
	Afternoon	13,616 ± 2705	5323 ± 1454	≤ 0.0001

Table 2 Mean concentrations of $PM_{2,5}$ as a function of the "morning" and "afternoon" periods between urban and rural areas adjusted to the days of the week

$PM_{2,5}$ (µg/m^3)		Zone		
Day of the week	Period	urban (M ± SD)	Rural (M ± SD)	*p-value*
Sunday	Morning	2.901 ± 0.683	2.746 ± 0.532	0.647
	Afternoon	1.043 ± 0.313	1.137 ± 0.258	0.036
Monday	Morning	4.722 ± 1.163	2.632 ± 0.312	≤ 0.0001
	Afternoon	3.244 ± 1.346	2.646 ± 0.477	0.029
Tuesday	Morning	8.164 ± 1.487	5.457 ± 1.467	≤ 0.0001
	Afternoon	4.783 ± 1.136	3.317 ± 1.068	≤ 0.0001
Wednesday	Morning	10.656 ± 3.380	7.670 ± 0.501	≤ 0.0001
	Afternoon	1.814 ± 0.423	2.180 ± 0.716	0.002
Thursday	Morning	6.069 ± 2.459	3.718 ± 1.348	≤ 0.0001
	Afternoon	6.034 ± 3.710	5.468 ± 0.456	0.062
Friday	Morning	18.272 ± 2.642	10.361 ± 0.679	≤ 0.0001
	Afternoon	5.342 ± 1.143	4.551 ± 0.237	≤ 0.0001
Saturday	Morning	2.111 ± 0.483	2.481 ± 0.142	≤ 0.0001
	Afternoon	1.773 ± 0.425	2.642 ± 0.520	≤ 0.0001

M: Mean; SD: Standard Deviation; Test: Wilcoxon Mann–Whitney areas, urban and rural, adjusted for the days of the week. PM_{10} values were significantly higher in the urban area in both periods, "morning" and "afternoon", on "Wednesday". On the other hand, they were significantly higher in rural areas in both periods on "Sunday" and "Saturday". As for the days "Tuesday", "Thursday" and "Friday", significantly higher values were recorded in the urban area in the morning and in the rural area in the afternoon. Regarding "Monday", there were no statistically significant differences between the two zones in the "morning" period. However, in the afternoon, the rural area recorded significantly higher values when compared to the urban area.

Then, we proposed to evaluate the behavior of the mean concentrations of PM_{Totals} as a function of the periods defined by "morning" and "afternoon" between the urban and rural areas, adjusted to the days of the week.

We found the average values of PM_{Totals} recorded are significantly higher in the rural area on "Sunday" and "Saturday" in the "morning" and "afternoon" periods.

The average values of PM_{Totals} were also significantly higher in the rural area on "Wednesday", during the morning period. On the other hand, "Monday", "Tuesday", "Thursday" and "Friday", these values were higher in rural areas during the afternoon (*p-value* < 0.05). As for the urban area, it obtained higher values on "Wednesday" in the afternoon and on "Tuesday", "Thursday" and "Friday" during the "morning" period (*p-value* < 0.05). During the morning period of "Monday" there was no significant difference between the mean concentration values of PM_{Totals}, with the *p-value* being > 0.05.

Finally, we tried to compare the values of concentrations of CO, $PM_{2,5}$ and PM_{10} with the exposure limit values legally established in Decree-Law No. 102/2010, of September 23 (Table 3).

Through the analysis of Table 3, we concluded that there were statistically significant differences between the analytical average values of concentration of CO, $PM_{2,5}$ and PM_{10} and the legally established exposure limit value, both in urban and rural areas, so that the mean values of the evaluated atmospheric pollutants are below the exposure limit value in both zones. However, the maximum values of $PM_{2.5}$ and PM_{10} exceeded the exposure limit value defined by Decree-Law No. 102/2010 in the urban area. As for the limit value of $PM_{2,5}$, it is 25 $\mu g/m^3$ and the maximum value

Table 3 Comparison of the values of CO, $PM_{2,5}$ and PM_{10} obtained in the areas under study with the exposure limit values legally established in Decree-Law No. 102/2010

Parameter	ELV	Mean of the urban area ± SD	Max; Min in urban area	Mean of the rural area ± SD	Max; Min in the countryside
CO	8.729 ppm	0.273 ± 0.522*	4.1; < 0.001	0.038 ± 0.075*	0.7; < 0.001
$PM_{2,5}$	25 $\mu g/m^3$	5.495 ± 4.749*	27.78; 0.80	4.072 ± 2.499*	11.99; 0.86
PM_{10}	50 $\mu g/m^3$	22.199 ± 19.367*	98.44; 2.67	19.540 ± 10.605*	76.03; 3.97

ELV: Exposure Limit Value; SD: Standard Deviation; Max: Maximum; Min: Minimum: Test: t-Student for 1 sample; **p-value* ≤ 0.0001

reached in the urban area was 27.78 $\mu g/m^3$. For PM_{10}, the exposure limit value is 50 $\mu g/m^3$, the maximum value reached in this area being 98.44 $\mu g/m^3$. It should also be noted that in the rural area, in the case of PM_{10}, the maximum value of 76.03 $\mu g/m^3$ was reached, that is, above the established exposure limit value. It is important to mention that the maximum value reached by $PM_{2.5}$, in the urban area, was recorded on "Wednesday" during the "morning" period. About the maximum values reached by PM_{10}, in the urban area it was recorded during the "morning" of "Thursday" and in the rural area in the period of the "afternoon" of "Tuesday".

4 Discussion

As for the results obtained, regarding the CO, except for the mornings of "Sunday" and "Monday", where there were no significant differences between the study sites, the values were higher in the urban area compared to the rural area. This behavior on the part of CO was to be expected, since this is a colorless and odorless gas, poisonous and a product, mainly, of the combustion of motor vehicles. For this reason, carbon monoxide pollution is found, at high levels, in areas with intense vehicle traffic [16].

Regarding the CO_2 parameter, it showed a different behavior to that observed in the CO, since on most days higher values were recorded in the rural area compared to urban area, with this phenomenon not occurring only on the mornings of "Friday" and "Saturday" and the afternoon of "Thursday", when the average concentration of this pollutant was higher in the urban area. Several studies indicate that the increase in CO_2, in most cases, is a consequence of human activities based on the daily consumption of energy, water, paper, fossil fuels and the large amount of waste produced [17].

For all these reasons, it was expected that the highest values would be detected in urban areas, however this did not happen. According to a study developed by Salustiano et al., carried out in Brazil, where the differences between CO_2 values during the daytime compared to the nighttime period were analyzed, the following conclusions were drawn: the daytime period is the main responsible for CO_2 consumption due to the performance of photosynthesis by plants, while at night the respiration of plants and living microorganisms responsible for the decomposition of dead matter, release CO_2 into the atmosphere [18].

In this study, the minimum value of the atmospheric pollutant was detected at 15:00 h and the maximum value at 02:00 h. Bearing in mind that measurements in rural areas were carried out between 6:30 am and 7:30 am during the month of April, it was expected that CO_2 levels would be significantly higher in rural areas, since during this period there was still no sunlight, which is necessary for photosynthesis, contrary to what happened during data collection in the urban area. This happens because during the night there is no CO_2 consumption by the plants, which influences the amount of this atmospheric pollutant present in the environment. In addition to this factor, it is important to consider that during the period of the year in which the measurements were carried out, it is common to use fireplaces for residential heating and burning, mostly in rural areas. These two processes imply the burning of organic matter, which is one of the main contributors to the production of CO and CO_2. In addition to gases (CO and CO_2), there are also other types of pollutants, such as particulate matter. This results from a mixture of solid and liquid particles suspended in the atmosphere. Its origin can be natural (sea salts, volcanic ash, soil erosion and pollen) or anthropogenic, resulting mainly from combustion processes [19].

Studies carried out indicate that the vast majority of particles are generated mainly by industrial activities and the burning of fuels, however, they also come from the natural suspension of dust in the soil or due to anthropic activities [20].

According to the results obtained, the average values of ultrafine particles were significantly higher in urban areas, on all days of the week and periods, when compared with data obtained in rural areas. According to a study carried out by Lelieveld, fine dust particles are formed chemically, mainly in the atmosphere from emissions from transport, industry and agriculture [21]. This type of particles can travel directly from the nose to the brain via the olfactory nerve. Some studies even indicate that they affect the central nervous system of children. Inhaling these types of particles can also accelerate the rate of cognitive decline in adults and increase the risk of developing Alzheimer's [22].

Regarding $PM_{2,5}$, the differences recorded between urban and rural areas were quite different, since higher values, were only recorded in rural areas on 3 days: in the afternoons of "Sunday" and "Wednesday", and the day of "Saturday". On the remaining days and periods, the average concentration values were higher in the urban area. The particulate matter—PM_{10}—registered higher average concentration values in the urban area only on the mornings of "Tuesday", "Thursday" and "Friday". About PM_{Totals}, they behaved as follows: the average concentration values were higher in the rural area on "Sunday", afternoon of "Monday", "Tuesday", "Thursday" and "Friday", morning of "Wednesday" and throughout the day of "Saturday". While in the urban area they were only higher in the morning of "Tuesday", "Thursday" and "Friday" and in the afternoon of "Wednesday".

According to the Portuguese Environment Agency (APA), once released or formed in the atmosphere, particles are transported by the wind over long distances, and may be responsible for high concentrations even in places far from their source [23]. For this reason, even if it is expected that they originate in urban areas due to the burning of fossil fuels, a greater number of cars in circulation, a greater number of people in these areas and a greater number of companies/industries, these particles are very easy to transport, which causes them to be found at great distances from their emission sources, for example in rural areas. Physiologically, the effects of these pollutants are related to the respiratory system with consequences on lung function, increased inflammation, producing respiratory symptoms and developing lung diseases and infections, mainly affecting children and the elderly [24].

When we compared the values obtained between the two study areas we found that for CO, ultrafine particles and $PM_{2,5}$, there were higher concentrations of these pollutants in urban areas. Regarding the parameters CO_2, PM_{10} and PM_{Totals}, a greater number of periods of the day with higher levels were recorded in the rural area. This happened because the number of particles of a biological nature is usually higher in rural or semi-rural areas [25]. Regarding the legally established exposure limit values for CO, $PM_{2.5}$ and PM_{10}, compared to the average concentrations obtained, the variables $PM_{2,5}$ and PM_{10} exceeded the legally established value. This value is 25 $\mu g/m^3$ for $PM_{2.5}$ and the maximum value recorded in the urban area was 27.78 $\mu g/m^3$ during the "Wednesday" morning. As for PM_{10}, the legislated value is 50 $\mu g/m^3$ and was exceeded in the urban area with a value of 98.44 $\mu g/m^3$ during the morning of "Thursday". This value was also exceeded in the rural area with a maximum value reached of 76.03 $\mu g/m^3$ in the afternoon of "Tuesday". These values were exceeded, mostly, during periods of the day when the highest traffic was registered, and since road traffic is one of the main contributors to pollution in large cities and with worldwide repercussions, this fact may be related to higher concentrations of atmospheric pollutants in these periods. Another issue to consider is that particles are a product, not only of the emission of exhaust gases, but also of the wear of tires and brakes and the resuspension of the particles deposited [26].

5 Conclusions

By conducting this study, it was confirmed that there are some differences between the air quality in urban areas compared to rural areas. However, these differences did not make it possible to classify any of the areas under study as better compared to the other, since some of the parameters presented the best values in the rural area, while others present the lowest results in the urban area, for the achieved tests, under the described test conditions. In this study, some concentration values above the legally established exposure limit value were also detected, as was the case of $PM_{2,5}$ and PM_{10}. With the application of the present study, it was possible to understand that although poor air quality is often associated with large urban centers, in some types of pollutants, it can be worse in rural areas.

Thus, since pollutants act globally and not just in the place where they originate and Portugal must reduce emissions by 17% by 2030, strategies must be implemented to gradually reduce the values reached. Some of these measures may include the creation of low-emission zones (for road traffic), reinforcement of the green infrastructure, consolidation of the cycling network, expanding its extension and ensuring its continuity, installing free parking areas for bicycles, investing in the network of public transport, renewal of the municipal fleet through the acquisition of electric vehicles, the requalification of pedestrian paths and the reduction/greater control of burning and burning in rural areas.

In conclusion, it is important to carry out air quality assessments regularly, to find effective solutions to reduce the high levels of atmospheric pollutants and, consequently, prevent the effects on the health of populations, climate change, as well as on the ecosystem in general.

References

1. Lopes, M. Air quality - right to health. Industry and Environment, 61. (2020)
2. Ferreira, A., Figueiredo, J.P., Pêgo, F., Loureiro, A., Seco, S. Air Quality Assessment in Kitchens and Effects on Workers Health. In: Arezes, P., et al (eds) Occupational and Environmental Safety and Health II. Studies in Systems, Decision and Control, vol 277. Springer, Cham. https://doi.org/10.1007/978-3-030-41486-3_25. (2020).
3. Amaral, L. Ir a Ares. Air Quality - Right to Health. Industry and Environment. 15(120), 2–3 (2020)
4. Fernandes, J., Carvalho, A., Campos, J., Costa, L., & Filho, G. Air pollution and respiratory, cardiovascular and reproductive effects on human health. 2009
5. World Health Organization. Atmospheric pollution: our health is still not sufficiently protected. European Court of Auditors. https://op.europa.eu/webpub/eca/special-reports/air-quality-23-2018/en/, last accessed 2022/05/15.
6. Algarve Regional Coordination and Development Commission. Air quality. https://www.ccdr-alg.pt/site/info/qualidade-do-ar. Last accessed 2022/05/15
7. Borrego, C. (2020). Qualidade do Ar. Air Quality - Right to Health. Industry and Environment. 15(120), 2–3 (2020)
8. Portuguese Environment Agency. Air Quality Assessment. https://apambiente.pt/ar-e-ruido/avaliacao-da-qualidade-do-ar, last accessed 2022/04/26. (2021a)
9. Silva, LT, & Mendes, JFG. Determination of the Air Quality Index in a Medium Size City. Universidade do Minho. 2006
10. Ferreira, A., & Barros, N. COVID-19 and Lockdown: The Potential Impact of Residential Indoor Air Quality on the Health of Teleworkers. Int. J. Environ. Res. Public Health, 19(10), 607 (2022)
11. Filipe, T.S., Vasconcelos Pinto, M., Almeida, J., Alcobia Gomes, C., Figueiredo, J.P., Ferreira, A. Indoor air quality in sports halls. CRC Press Taylor & Francis Group, A Balkema Book, Occupational Safety and Hygiene. 2013: 175–179. ISBN: 978–1–138–00047–6 (Hbk); ISBN: 978–0–203–72965–6 (2013)
12. de Carvalho, J.M., Ferreira, A., Figueiredo, J.P., Loureiro, A. Indoor Air Quality in Classroom Labs. In: Arezes, P.M. et al (eds) Occupational and Environmental Safety and Health II, Studies in Systems, Decision and Control, vol 277. SSDC. Springer, Cham, pp 285–292. ISBN 978–3–030–41485–6 https://doi.org/10.1007/978-3-030-41486-3_31
13. Ferreira, A., Cardoso, S. M. Effects of indoor air quality on respiratory function of children in the 1st cycle of basic education of Coimbra, Portugal. CRC Press Taylor e Francis Group, A Balkema Book, Occupational Safety and Hygiene II. 2014: 347–350. ISBN: 978–1–138–00144–2 (2014).
14. Martins, APG, Ribeiro, AP, Ferreira, ML, Martins, MAG, Negri, EM, Scapin, MA, De Oliveira, A., Saiki, M., Saldiva, PHN, & Lafortezza, R. Infraestrutura verde para monitorar e minimizar os impactos da poluição atmosférica. Energia e ambiente 35(102), 31-57 (2021)
15. Decree-Law No. 102/2010. Ministry of the Environment and Territorial Planning. https://dre.pt/dre/detalhe/decreto-lei/102-2010-342123. (2010)
16. Fernandes, AR, Rosa, C., & Ferreira De Castro, S. Analysis of air quality and health concerns. University of Porto. 2017
17. Paes, C., Silva, L., & Lopes, M. Carbon Dioxide Emission: Report of an Experience with the Use of the Senac Environmental Calculator. Experiências em Ensino de Ciências 12(4), 154-165. 2017
18. Salustiano, R., Silva J. D., Antônio, M., Moura, L., Meixner, FX, Kormann, R., Fernando Da, R., Lyra, F., Ferreira, M., & Filho, N. Study of the concentration of atmospheric CO_2 in a pasture area in the Amazon region. Revista Brasileira de Geofísica 22(3):259-270 (2005)
19. Maria, A., & Silva, E. Fugitive emissions of particles into the atmosphere: characterization, impacts and mitigation. University of Aveiro. 2017
20. Magalhães, LC. Study of Atmospheric Particulate Material and Metals Associated with Total Suspended Particles in the City of Ouro Preto. Universidade Federal de Ouro Preto. 2005

21. Lelieveld, J. (2018). Ultrafine particles in the atmosphere can cause cardiovascular disease. Max-Planck Institute of Chemistry. https://www.dn.pt/ciencia/poluicao-provoca-mais-de-tres-milhoes-de-mortes-prematuras-por-ano-4782299.html, last accessed 2022/04/24.
22. Brugge, D., & Lane, K. Fine particles of pollution are the new enemies of our health. 2018.
23. Portuguese Environment Agency. Suspended Particles (PM). https://apambiente.pt/ar-e-ruido/particulas-em-suspensao-pm, last accessed 2022/04/26. (2021b)
24. Carneiro, C., Ramos, C., Alves, A., Uriona, K., Koga, G., Marinho, L., Junior, R., & Cardoso, M. Effects of air pollution on respiratory diseases: a narrative review. Ciências da Saúde: Desafios, perspectivas e possibilidades 2(23), 230-251(2021)
25. Tobias, L. Characterization of aerosols from Évora. University of Évora. 2013
26. Faria, T. Particulate air pollution in Portugal: Management of ambient air quality, pollutant emissions into the atmosphere and occupational exposure. 2016

Consumers' Acceptance and Willingness to Pay for Innovative and Sustainable Food – an Exploratory Study

C. João Maurício and Maria Eduarda Fernandes

Abstract The agricultural sector is currently facing several challenges, and needs to evolve to a more sustainable system. Consumers have a major role to play in this transformation. In this sense, this study aims to explore Portuguese consumers' receptiveness and willingness to pay (WTP) for innovative and more sustainable food. Particularly, for apples produced in an innovative and sustainable way, making using of algae from the Portuguese coast. Additionally, the impact of information provided to consumers about the consequences of consuming this new product, on their receptiveness and willingness to pay, is tested: focusing on health benefits of the new product vs. on the environmental benefits. To this end, an online questionnaire was conducted in Portugal in July 2022. From a sample of 411 valid answers and 396 apple consumers, a wide majority (86.9%) were receptive to this innovative and sustainable apples and was willing to pay a price premium of 0.35 €/kg compared to the conventional apples (average price of 1.95 €/kg for sustainable apples, compared to 1.60 €/kg for conventional apples). This WTP was not influenced by the type of information provided in terms of benefits of the new product (health vs. environment) but the acceptance for the innovative apple increased with the provision of information. Additionally, some respondent's ecological consumption habits and socioeconomic characteristics demonstrated to be statistically significant for the WTP. These results indicate that consumers are aware of the need to move towards a more sustainable food sector and value this innovative (and more sustainable) apples.

Keywords Sustainable food · Consumers' receptiveness · Willingness to pay

C. João Maurício (✉)
ESTG, CARME, Instituto Politécnico de Leiria, Leiria, Portugal
e-mail: joao.miguel.costa@ipleiria.pt

M. E. Fernandes
ESTG, CARME, Instituto Politécnico de Leiria, Leiria, Portugal
e-mail: eduarda.fernandes@ipleiria.pt

J. R. da. C. S. Galvão et al. (eds.), *Proceedings of the 3rd International Conference on Water Energy Food and Sustainability (ICoWEFS 2023)*, Springer Proceedings in Earth and Environmental Sciences, https://doi.org/10.1007/978-3-031-48532-9_40

1 Introduction

The agri-food sector is one of the pillars of society and currently faces difficult challenges, such as the need for increased productivity, reduced environmental impact and food waste, while ensuring safe access to food to the world's growing population. Agriculture is currently responsible for emitting 26% of greenhouse gases, using about 50% of habitable territory and 70% of drinking water [1]. Also, this economic activity uses a large quantities of plastics, for irrigation, crop protection and product packaging [2] and excessively rely on the use of fertilizers [1] that infiltrate soils and are carried through the water cycle to the oceans harming their balance [1, 3, 4]. Currently, the agricultural sector is at the top of the list of the ones responsible for the highest production of waste, with about one third of all food produced in Europe for human consumption being lost or wasted before reaching the consumer [4, 5], reaching 46% in the case of fruits and vegetables.

The agricultural sector therefore assumes enormous importance in the transition to a sustainable development [6–8], namely, to achieve some of the United Nations (UN) Sustainable Development Goals (SDGs) [9, 10]: SDG #2 (Eradicate hunger) and SDG #12 (Sustainable production and consumption). Healthy and sustainable food is necessary to achieve sustainable development and ensuring food security [11]. Sustainable diets are those with small environmental impact, contributing to food and nutrition security, a healthy life for present and future generations and protect and respect biodiversity and ecosystems. Additionally, sustainable diets are culturally acceptable and accessible, economically just and achievable, nutritionally adequate, safe and healthy while optimizing natural and human resources [12]. However, access to sustainable food is determined by a complex and dynamic set of social practices [13]. Consumers are generally unaware of the entire food supply chain, and ethical implications of their purchasing behaviors [7]. Still, certain choices and consumption habits make it easier to move towards more sustainable food. Namely, preferences for plant-based [7, 14], organic [15, 16], local and seasonal[17] food, reduction of meat consumption [18, 19], buying food that is traceable [17], with sustainable packaging [17, 20], and, finally, reducing food waste [4, 11, 12].

To face the challenges necessary to build a more sustainable agri-food sector, all stakeholders must be involved (governments, consumers and producers, among others) [7]. One of the challenges is to change consumer's perceptions about products that are labelled as more sustainable [15]. It is thus necessary to invest in education and communication directed to sustainable product choices by consumers [15], leading them to accept more sustainable methods in the food sector [18, 20]. Access to information is relevant since consumers with access to information or more informed about the sustainable attributes of the product are more receptive to it [21] and even more willing to pay a price premium for those attributes [22]. However, the type of information provided to consumers is also relevant for their behavior: consumers value information on health benefits more [23–25] compared to other types of information, such as information on environmental benefits [23, 25] associated with the consumption of sustainable food. Other factors and characteristics explain their willingness to pay (WTP) for sustainable food: gender [26, 27]; higher levels of education [28] upper-middle incomes [26]; household size [26]; ethnocentrism [29, 30]; environmental concerns [26]; food safety [31]; familiarity with sustainable food [26, 32]; habit of consuming organic products [33]; awareness of sustainable attributes [26, 34]; widespread knowledge about sustainability [22, 32], health benefits [22], environmental benefits [22, 35] and attitude towards innovation[36].

Considering the challenges of agriculture and sustainable development, projects like ORCHESTRA (POCI-01-0247-FEDER-070155) arise, aiming to develop new products, based on algae extracts. These algae are from invasive species of the Portuguese coast and the extracts obtained have potential for: pest control, induction of resistance and plant fortification, allowing the use of synthetic chemicals/additives to combat pests and nutrient deficiencies to be reduced. The final residues are also used for soil quality improvement (as fertilizer), maximizing the whole life cycle of the product. Additionally, the project aims to develop bio products to improve fruit shelf life during storage and during the distribution chain, by maximizing the phytonutrients, optimizing and making long-term preservation more sustainable [37].

The main objective of this work is to understand whether Portuguese consumers value a food product, in particular apple, that is produced in a more sustainable way, based on an innovative algae-based product, compared to a conventionally produced apple. Additionally, we also intend to test the impact of the type of information given to the consumer for this same valuation: highlighting the advantages of the innovative product for the consumer's health or the advantages for the environment.

2 Methodology

To achieve the goals of our research, an exploratory study was implemented using a quantitative empirical methodology, with primary data collected through an online survey. The questionnaire was shared on social media (Instagram and Facebook) and by email during the months of June and July 2022 and the sample was constituted by personal and professional contacts of the researchers. Therefore, the study is based on a non-random convenience sample, with a self-selection sampling technique.

The questionnaire was constituted by three main parts. The first section included questions to characterize respondents in terms of their environmental attitudes, green consumption habits, level of agreement with innovations in the food sector, sustainable eating behaviors, apple consumption frequency and most valued attributes. The second part of the questionnaire assessed consumer's acceptance and willingness to pay for a more sustainably produced and preserved apple, and the impact of information on environmental versus health benefits of its consumption. The last section aimed to analyze the sociodemographic characteristics of the consumers. 411 valid answers to this questionnaire were obtained. Respondents were mainly women (69.8%), ageing between 15 and 34 years old (58.2%), with small households (65.9% with 1 to 3 people in the household), without children or elderly in the household (62.8%) and a higher level of education (75.0%).

3 Results and Discussion

Results reveal that the majority of respondents consider themselves subjects with ecological concerns, which reflects on their Ecological Consumption (median value of 4[1] in all questions used to measure this variable). Our sample is also receptive to innovations in the food sector, particularly: the use of new raw materials; the reinforcement of nutritional characteristics; and packaging in a protective atmosphere. Seasonal and local consumption of fruits and vegetables, and avoidance of over-packaged fruits and vegetables were the most frequent sustainable behaviors respondents consider to adopt (all with a median of 4[2]). The majority of the individuals in our sample consume apples (396 out of 411) and the characteristics they consider to be the most important is taste [38, 39] followed by health benefits, freshness, food safety, texture, and aroma.

When asked about their acceptance for a more sustainably produced apple, the majority of respondents answered positively (86.9%). These results are in line with previous studies that conclude consumers are more receptive when the product demonstrates a sustainable attribute [40, 41].

Our results show that both information about environmental or health benefits of this more sustainably produced apples cause a positive impact on the receptivity for this product (more than 70% of respondents). This result is in line with previous findings that pointed out a worse perception and receptivity by consumers without access to information [26, 42], and that consumers with access to information or more informed about the sustainable attributes of the product are more receptive to it [21]. However, the type of information given, either about the benefits on the environment or on consumer's own health, is not determinant for the change in consumer behavior (there is no statistically significant difference according to the Mann–Whitney test: p-value = 0,232, z = −1,196, for those who initially answered "maybe" would buy the sustainable apple).

3.1 Willingness to Pay (WTP) for Sustainable Apples

When asked about the specific amount they would be willing to pay for a sustainable apple, responses ranged from a minimum of €0.10 to a maximum of €3 (given a price

[1] On a 5-point agreement Likert scale where: 1-Strongly disagree, 2-Disagree, 3-Neither agree nor disagree, 4-Agree, 5-Strongly agree.

[2] On a 5-point frequency Likert scale where: 1-Never, 2-Rarely, 3-Sometimes, 4-Often, 5-Always.

of €1.60 / kg for a conventional apple). A total of 305 consumers (out of 396 apple consumers of our sample) would be willing to pay a price premium to buy apples produced in a more sustainable way (median price of €1.95/kg, which corresponds to a €0.35 price premium, considering the price of €1.60/kg for a conventional apple). That is, consumers would be willing to pay a premium of 21.9% for a sustainable apple, which corroborates the results of previous studies [22, 28, 31].

On the other hand, for 17.4% of the subjects, a discount of about 37.5% would be necessary to accept the more sustainable apple (median value of 1.00 €/kg for a more sustainable apple). Furthermore, different types of information provided (on health benefits vs. on environmental benefits) had no impact on the WTP for sustainable apples, according to the non-parametric Mann–Whitney test (p-value = 0.163; Z = −0.976), contrarily to previous results [23, 25].

3.2 Influence of Consumer Characteristics and Behaviors on WTP for Sustainable Apples

Consumers who are more concerned with the consequences of their consumption habits on the environment are more receptive to sustainable food, confirming previous findings [43]. Based on Spearman's correlation test, we conclude that consumers whose consumption habits are affected by their concern about the impact on the environment and are willing to change consumption habits and lose quality of life, in favor of the environment, are more willing to pay a price premium for a sustainable apple ($\rho = 0.129$, $\alpha = 0.05$ and $\rho = 0.201$, $\alpha = 0.01$, respectively) than those who do not. The variable Ecological Consumption, is positively correlated with the WTP ($\rho = 0.175$, $\alpha = 0.01$) for apples produced more sustainably. Also, similarly to other studies [36], our results show that there is a positive and statistically significant correlation between respondents WTP for the sustainable apples and their acceptance of innovative food products, such as vegetable proteins and synthetic meat ($\rho = 0.105$, $\alpha = 0.05$), the use of new raw materials such as algae, vegetable drinks, Stevia etc. ($\rho = 0.120$, $\alpha = 0.05$). Thus, consumers in our sample who are more concerned about the consequences of their consumption on the environment and who already have sustainable eating behaviors are more receptive to buying a more sustainable apple, which is in line with previous literature [17, 44]. As for the potential impact of respondents' socio-demographic characteristics on their WTP for sustainable apples, we found that higher levels of education and financial comfort lead to higher WTP, also in line with the findings of previous authors [22, 27, 31]. It was also observed, a negative relationship with the number of household members and WTP, consistent with previous findings [31, 41] but contrary to others [26].

4 Conclusion

The main objective of this exploratory study was to assess the receptivity and willingness to pay of Portuguese consumers for a more sustainably produced and preserved apple. Our results show that among the respondents in our sample who consume apples, there is a good receptivity towards apples produced in an innovative and sustainable way, as about 87% admitted the possibility of buying it, if it was available in the supermarket. Besides, the majority of the subjects in this study stated their willingness to pay a price premium for an apple produced in a more sustainable way—an average of 35 cents/kilo, against a reference value of 1.60€/kg for a conventional apple. This WTP did not changed with the provision of information about the environmental benefits of this new product nor about the health benefits. However, for those subjects that were unsure about buying this more sustainable apples if they were available in the supermarket, the provision of information proved to be relevant, since about 70% of respondents changed their answer and stated to be receptive to this new product. Therefore, Portuguese consumers seem to value firms' efforts to innovate and produce food in more circular and sustainable way, as it is the case of ORCHESTRA project, that aims to develop an innovative algae-based product to be applied to fruit (apples and pears), with the objective of increasing productivity in a sustainable way. Being receptive to this type of innovation and signaling their willingness to pay a premium for apples produced and preserved in a more sustainable way, for example through the use of algae extracts, is good news for the food sector and society in general. This exploratory research indicates the consumer interest in sustainable food, which encourages firms efforts towards sustainability and is a positive sign in favor of a more sustainable society.

Being an exploratory study, and particularly because of the sampling method used (self-selection and convenience), the current study has obvious limitations in terms of statistical inferences for the Portuguese population. Therefore, as a future research avenue, it would be pertinent to pursue a new study with a probability, representative sample.

Funding This work was supported by National Funds of the FCT – Portuguese Foundation for Science and Technology within the project [UIDB/04928/2020] and POCI-01–0247-FEDER-070155.

References

1. FAO: The State of Food and Agriculture. (2019)
2. Vox, G., Loisi, R.V., Blanco, I., Mugnozza, G.S., Schettini, E.: Mapping of Agriculture Plastic Waste. Agric. Agric. Sci. Procedia. 8, 583–591 (2016). https://doi.org/https://doi.org/10.1016/j.aaspro.2016.02.080
3. Withers, P., Neal, C., Jarvie, H., Doody, D.: Agriculture and Eutrophication: Where Do We Go from Here. Sustainability. 6, 5853–5875 (2014). https://doi.org/https://doi.org/10.3390/su6095853

4. Ellen MacArthur Foundation: Growth within: a circular economy vision for a competitive europe. (2015)
5. Toop, T., Kirby, M., Theodorou, M., Hull, M., Ward, S., Oldfiel, T.: AgroCycle - developing a circular circular economy in agriculture. Energy Procedia. 123, 76–80 (2017). https://doi.org/https://doi.org/10.1016/j.egypro.2017.07.269
6. Avillez, F.: A Agricultura Portuguesa: As últimas décadas e perspectivas para o futuro. Fundação Francisco Manuel dos Santos (2015)
7. Jurgilevich, A., Birge, T., Kentala-Lehtonen, J., Korhonen-kurki, K., Pietikäinen, J., Saikku, L., Schösler, H.: Transition towards Circular Economy in the Food System. Sustainability. 8, 1–14 (2016). https://doi.org/https://doi.org/10.3390/su8010069
8. Henriksson, P.J.G., Tran, N., Mohan, C.V., Chan, C.Y., Rodriguez, U.P., Suri, S., Mateos, L.D., Utomo, N.B.P., Hall, S., Phillips, M.J.: Indonesian aquaculture futures – Evaluating environmental and socioeconomic potentials and limitations. J. Clean. Prod. 162, 1482–1490 (2017). https://doi.org/https://doi.org/10.1016/j.jclepro.2017.06.133
9. Farming First: The Story of Agriculture and the Sustainable Development Goals, https://farmingfirst.org/sdg-toolkit#section_2
10. BCSD: Objetivos de Desenvolvimento Sustentável (ODS) - BCSD Portugal, https://www.ods.pt/
11. Capone, R., Bilali, H. El, Debs, P., Cardone, G., Driouech, N.: Food System Sustainability and Food Security: Connecting the Dots. J. Food Secur. 2, 13–22 (2014). https://doi.org/10.12691/jfs-2-1-2
12. FAO: Sustainable Diets And Biodiversity Directions And Solutions For Policy, Research And Action. (2012)
13. Brons, A., Oosterveer, P.: Making Sense of Sustainability: A Practice Theories Approach to Buying Food. Sustainability. 9, (2017). https://doi.org/10.3390/su9030467
14. Goodland, R.: Environmental sustainability in agriculture: diet matters. Ecol. Econ. 23, 189–200 (1997). https://doi.org/https://doi.org/10.1016/S0921-8009(97)00579-X
15. Aschemann-witzel, J., Peschel, A.: How circular will you eat ? The sustainability challenge in food and consumer reaction to either waste-to-value or yet underused novel ingredients in food. Food Qual. Prefer. 77, 15–20 (2019). https://doi.org/https://doi.org/10.1016/j.foodqual.2019.04.012
16. Azzurra, A., Massimiliano, A., Angela, M.: Measuring sustainable food consumption : A case study on organic food. Sustain. Prod. Consum. 17, 95–107 (2019). https://doi.org/https://doi.org/10.1016/j.spc.2018.09.007
17. Mancini, P., Marchini, A., Simeone, M.: Which are the sustainable attributes affecting real consumption behaviour? Consumer understanding and choices. Br. Food J. 119, 1839–1853 (2017). https://doi.org/10.1108/BFJ-11-2016-0574
18. García-Oliveira, P., Fraga-Corral, M., Pereira, A., Prieto, M., Simal-Gandara, J.: Solutions for the sustainability of the food production and consumption system. Crit. Rev. Food Sci. Nutr. 1–17 (2020). https://doi.org/10.1080/10408398.2020.1847028
19. Verain, M.C.D., Sijtsema, S.J., Dagevos, H., Antonides, G.: Attribute segmentation and communication effects on healthy and sustainable consumer diet intentions. Sustain. 9, 1–19 (2017). https://doi.org/https://doi.org/10.3390/su9050743
20. Ketelsen, M., Janssen, M., Hamm, U.: Consumers' response to environmentally-friendly food packaging - A systematic review. J. Clean. Prod. 254, (2020). https://doi.org/10.1016/j.jclepro.2020.120123
21. Schaefer, F., Blanke, M.: Opportunities and Challenges of Carbon Footprint, Climate or CO2 Labelling for Horticultural Products. Erwerbs-Obstbau. 56, 73–80 (2014). https://doi.org/https://doi.org/10.1007/s10341-014-0206-6
22. Cagalj, M., Haas, R., Morawetz, U.: Effects of quality claims on willingness to pay for organic food Evidence from experimental auctions in Croatia. Br. Food J. 118, 2218–2233 (2016). https://doi.org/https://doi.org/10.1108/BFJ-11-2015-0453
23. Castellari, E., Ricci, E.C., Stranieri, S., Marette, S., Sarnataro, M., Soregaroli, C.: Relationships Between Health and Environmental Information on the Willingness to Pay for Functional

Foods: The Case of a New Aloe Vera Based Product. Nutrients. 11, 1–16 (2019). https://doi.org/https://doi.org/10.3390/nu11112781
24. Ran, Y., Lewis, A.N., Dawkins, E., Grah, R., Vanhuyse, F., Engström, E., Lambe, F.: Information as an enabler of sustainable food choices: A behavioural approach to understanding consumer decision-making. Sustain. Prod. Consum. 31, 642–656 (2022). https://doi.org/https://doi.org/10.1016/j.spc.2022.03.026
25. Wang, H., Chen, Q., Zhu, C., Bao, J.: Paying for the Greater Good ?—What Information Matters for Beijing Consumers' Willingness to Pay for Plant-Based Meat? Foods. 11, 1–17 (2022). https://doi.org/https://doi.org/10.3390/foods11162460
26. Haghjou, M., Pishbahar, E., Mohammadrezaei, R., Hayati, B., Dashti, G., Pishbahar, E., Mohammadrezaei, R., Dashti, G.: Factors Affecting Consumers' Potential Willingness to Pay for Organic Food Products in Iran: Case Study of Tabriz. J. Agric. Sci. Technol. 15, 191–202 (2013)
27. Rodríguez-bermúdez, R., Miranda, M., Orjales, I., Ginzo-villamayor, M., Al-soufi, W., López-Alonso, M.: Consumers' perception of and attitudes towards organic food in Galicia (Northern Spain). Int. J. Consum. Stud. 44, 206–219 (2020). https://doi.org/https://doi.org/10.1111/ijcs.12557
28. Minh, N., Masahiro, M., Susumu, F.: Willingness to Pay for Organic Vegetables in Vietnam: An Empirical Analysis in Hanoi capital. Fac. Agric. Kyushu Univ. 58, 449–458 (2019). https://doi.org/https://doi.org/10.5109/27378
29. Darby, K., Batte, M., Ernst, S., Roe, B.: Willingness to pay for locally produced foods: A customer intercept study of direct market and grocery store shoppers. (2006)
30. Krystallis, A., Chryssohoidis, G.: Consumers' willingness to pay for organic food: Factors that affect it and variation per organic product type. Br. Food J. 107, 320–343 (2005). https://doi.org/https://doi.org/10.1108/00070700510596901
31. Aye, M.S., Takahashi, Y., Yabe, M.: Effects of Consumer Preferences on Environmentally Friendly Tomatoes in Myanmar. J. Agric. Sci. 11, 29–41 (2019). https://doi.org/https://doi.org/10.5539/jas.v11n13p29
32. Díaz, F., Pleite, F., Paz, J., García, P.: Consumer knowledge, consumption, and willingness to pay for organic tomatoes. Br. Food J. 114, 318–334 (2012). https://doi.org/https://doi.org/10.1108/00070701211213447
33. Denver, S., Jensen, J.D.: Consumer preferences for organically and locally produced apples. Food Qual. Prefer. 31, 129–134 (2014). https://doi.org/https://doi.org/10.1016/j.foodqual.2013.08.014
34. Akgüngör, S., Miran, B., Abay, C.: Consumer Willingness to Pay for Organic Food in Urban Turkey. J. Int. Food Agribus. Mark. 22, 299–313 (2010). https://doi.org/https://doi.org/10.1080/08974431003641455
35. Cicia, G., Giudice, T. Del, Ramunno, I., Tagliafierro, C.: Splitting consumer's willingnes s to pay premium price for organic products over main purchase motivations. In: 98 th EAAE Seminar 'Marketing Dynamics within the Global Trading System: New Perspectives'ynamics within the Global Trading System: New Perspectives.' pp. 1–13 (2006)
36. Marette, S., Disdier, A.C., Beghin, J.C.: A comparison of EU and US consumers' willingness to pay for gene-edited food: Evidence from apples. Appetite. 159, (2021). https://doi.org/10.1016/j.appet.2020.105064
37. MARE: add-value to ORCHards through thE full valoriSaTion of macRoalgAe, https://www.mare-centre.pt/pt/proj/orchestra
38. Botelho, A., Dinis, I., Lourenço-Gomes, L., Moreira, J., Costa Pinto, L., Simões, O.: The role of consumers in agrobiodiversity conservation: the case of traditional varieties of apples in Portugal. Agroecol. Sustain. Food Syst. 42, 796–811 (2018). https://doi.org/https://doi.org/10.1080/21683565.2018.1426669
39. Santeramo, F.G., Carlucci, D., Devitiis, B. De, Seccia, A., Stasi, A., Viscecchia, R., Nardone, G.: Emerging trends in European food, diets and food industry. Food Res. Int. 104, 39–47 (2018). https://doi.org/https://doi.org/10.1016/j.foodres.2017.10.039

40. Güney, O., Giraldo, L.: Consumers' attitudes and willingness to pay for organic eggs A discrete choice experiment study in Turkey. Br. Food J. 122, 678–692 (2019). https://doi.org/https://doi.org/10.1108/BFJ-04-2019-0297
41. Tonsor, G.T., Shupp, R.: Valuations of Sustainably Produced Labels on Beef, Tomato, and Apple Products. Agric. Resour. Econ. Rev. 38, 371–383 (2009)
42. Yin, S., Wu, L., Du, L., Chen, M.: Consumers' purchase intention of organic food in China. J. Sci. Food Agric. 90, 1361–1367 (2010). https://doi.org/https://doi.org/10.1002/jsfa.3936
43. Siegrist, M., Visschers, V., Hartmann, C.: Factors influencing changes in sustainability perception of various food behaviors: Results of a longitudinal study. Food Qual. Prefer. 46, 33–39 (2015). https://doi.org/https://doi.org/10.1016/j.foodqual.2015.07.006
44. Vermeir, I., Verbeke, W.I.M.: Sustainable food consumption: exploring the consumer '"attitude – behavioral intention"' gap. 169–194 (2006). https://doi.org/10.1007/s10806-005-5485-3

Educational Experience: Active Carbon as a “Bacteria Hotel” and Its Comparison with Insect Hotels to a Better Understanding of Sustainable Biogas Production

Sergio Nogales-Delgado, Beatriz Ledesma Cano, Silvia Román Suero, and Juan Félix González González

Abstract Sustainability in education is becoming more and more important nowadays, as future professionals (especially in STEM disciplines) should consider a sustainable approach when developing their tasks. Indeed, many steps such as the so-called Sustainable Development Goals have been taken by institutions and governments. On the other hand, engagement of students in STEM subjects is equally important, in order to improve their knowledge and skills and to promote these disciplines at higher educational levels. In that sense, the role of universities is vital in order to offer a wide and varied educational offer at different levels, such as university or master degrees, but also in exhibitions where the engagement of students in STEM disciplines applied to sustainability is equally important. The aim of this work was to propose a pedagogical guide to explain (in scientific exhibitions) how active carbons work to improve the yield of biogas production in sewage sludge biodigestion. Thus, a comparison between active carbon as “bacteria hotel” and typical insect hotels could be interesting to attract the interest of a wide range of audience (between primary education to undergraduate students or public in general). As a result, the methodology and the main resources required to carry out this exhibition were exposed, and students showed a general interest and understanding of the subject.

Keywords Primary education · Secondary education · STEM · Biodigestion · Sewage sludge

---

S. Nogales-Delgado (✉) · B. L. Cano · S. R. Suero · J. F. González González
Department of Applied Physics, University of Extremadura, Avda. De Elvas S/N, 06006 Badajoz, Spain
e-mail: senogalesd@unex.es

J. R. da. C. S. Galvão et al. (eds.), *Proceedings of the 3rd International Conference on Water Energy Food and Sustainability (ICoWEFS 2023)*, Springer Proceedings in Earth and Environmental Sciences, https://doi.org/10.1007/978-3-031-48532-9_41

1 Introduction

1.1 Increasing Motivation and Engagement in STEM Education

STEM (Science, Technology, Engineering and Mathematics) is currently becoming more and more present in society, especially in education, mass media, business and industry. On the other hand, scientific and technical disciplines, as other subjects, require high levels of motivation and engagement, especially when further steps towards higher education and university degrees are considered. Thus, fostering students' motivation can have a positive impact on their tendency to choose STEM disciplines [1]. Apart from that, educational institutions (especially in higher education and universities) should consider the interrelation among different elements (such as research centers or other education centers), integrate sustainable development in policies and strategies especially in long terms, implementing it at all educational levels and thanks to interdisciplinary STEM. Apart from that, specific STEM activities, where STEM concepts are put in practice, can enhance student's confidence and engagement in these subjects for the future [2, 3].

1.2 Current Situation in Spain and Extremadura

Regarding sustainability, many aspects can be applied to STEM education. Thus, educational curricula are becoming more adapted to sustainability applied to scientific and technical subjects, which is vital to attract students from primary and secondary education to STEM disciplines, in order to promote their equivalent university degrees. This way, and according to Fig. 1a, there are opposite trends in the number of students enrolled in Science and Technology courses at secondary education. On the one hand, there was a steady increase in student at a national level, whereas in the case of Extremadura a constant decrease was found.

This trend is worrying in Extremadura, due to the fact that, as observed in Fig. 1b, the population in Extremadura and Spain similarly increased until 2010, decreasing Extremadura's population from then on. Thus, students in all disciplines in this region can plummet in future years.

Nevertheless, there seems to be an increasing interest in disciplines related to Science and Technology in secondary education, as it can be seen in Fig. 2, where a similar trend can be found in Spain and Extremadura between 2014 and 2020. Thus, the ratio of students interested in Science and Technology increased from 45 to 48% in Extremadura, with a similar increase in the case of Spain.

Consequently, and although there is a decrease in students in Extremadura, they seem to be more interested in STEM disciplines, which can present a series of opportunities, as the lack of students can promote a more personalized and focused education, offering a wider range of educational offer, with the subsequent improvement

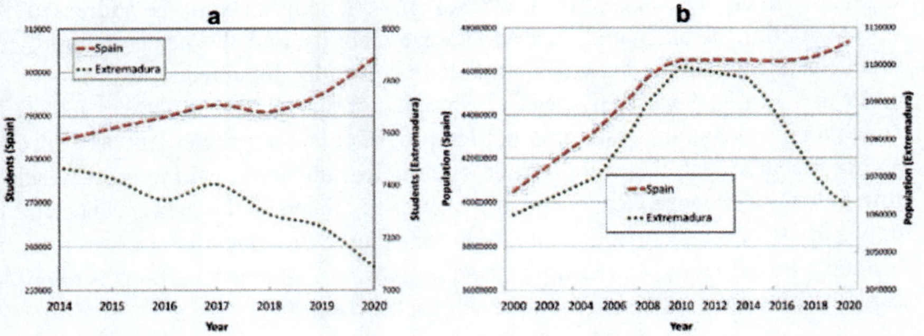

Fig. 1 Students evolution in Spain and Extremadura for Science and Technology in secondary education **a** and population evolution **b** [4, 5]

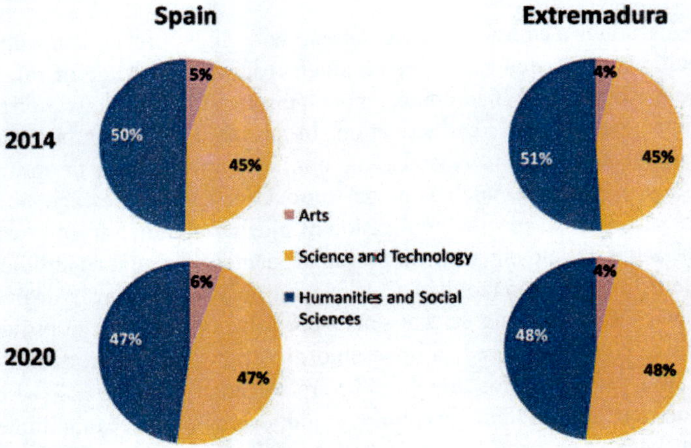

Fig. 2 Science and Technology ratio of students in Spain and Extremadura in 2014 and 2020

in quality. Thus, the possible application of aspects such as sustainability to STEM education at different educational levels can be feasible under these circumstances.

1.3 *Education to Foster Sustainability*

Sustainability has attracted special attention recently at all educational levels, possibly due to the encouragement perceived from international institutions and agencies like UN and their Sustainable Development Goals, highly recommended in high education institutions [6]. Thus, the education for sustainability or education for sustainable development is an interesting subject to be included in educational programs, where the roles of universities and research centers can be important, especially in higher education [7], where many challenges should be addressed like deconstruction of old paradigms about sustainability and the need of improving practices in educational activities rather than curricular documents [8].

In that sense, the use of specific examples of sustainable processes in STEM education can be an interesting point to improve student's involvement, as a specific and practical aspect of general subjects can help students to gain interest in scientific fields. As a consequence, specific examples related to the area of expertise of university departments could be an important source for educational exhibition, for instance. In this study, an example based on biochar and active carbons is exposed, explaining its foundations in the following subsection.

1.4 Active Carbon

An active carbon is a carbonaceous adsorbent with high porosity and, subsequently, high specific surface area. It can be obtained from a wide range of raw materials, including wastes with a difficult management, through multiple activation techniques such as chemical and physical activation. In any case, active carbons are used in different applications, such as metal extraction, water purification or optimization of microbiological processes such as biogas production through sewage sludge anaerobic digestion. Thus, a specific application of biochar obtained from sewage sludge in order to act as a yield enhancer in biogas production through anaerobic digestion can be interesting, as it can facilitate CH_4 generation by promoting hydrolysis, acidogenesis and methanogenesis. In that sense, pore size of biochar can provide microhabitats to microbes involved in anaerobic digestion to proliferate, and the most representative pore sizes for this purpose are included in Table 1, where microbe distribution depends on their morphology and size. Micropores facilitates biofilm formation, which can protect microbes during the process, especially under acid stress conditions.

Additionally, the presence of surface function groups could promote nutrient retention and contaminant removal, which can be advantageous in methane generation. To sum up, the use of sewage sludge to produce biochar in order to improve biogas production is an excellent example of sustainability, green chemistry and circular economy, which can be important as a subject in STEM disciplines.

Table 1 Most suitable pore sizes for microbes involved in anaerobic digestion

Pore size, μm	Microbe
0.3–13	Bacteria/archaea
0.8–7	Methanosaeta
1.2–120	Methanobacterium
2–80	Fungus
7–30	Protozoa

Table 2 Equivalences between an active carbon and an insect hotel

Active carbon in biogas production	Insect hotel
Pores as microhabitat for microbes	Holes to host insects like wasps and solitary bees
Pore size determines microbe distribution	Hole size determines insect selection
These microbes report benefits like better biogas production	Insects act as pollinators and pest controllers
This process is sustainable and further chemical treatments are not required	Their function is a good replacement for pesticides, enhancing sustainability
This process enhances the positive role of microbes, which are normally despised	The use of insect hotels places value on insects, usually undervalued

However, active carbons, where many concepts such as pore size are related to the microscopic world, which can make difficult the explanation of some concepts. Thus, a comparison with an equivalent macro level could help in that sense. For instance, a comparison of biochar with insect hotels could be feasible, as they present similarities like the ones explained in Table 2.

1.5 Objective of This Work

Considering the above, the aim of this work was to propose an educational methodology adapted to different kinds of audience in order to promote sustainability through a specific example like the use of active carbon to improve the production yield of biogas during sewage anaerobic digestion. Thus, a specific example was used in scientific exhibitions for heterogeneous audience in order to foster sustainability in science and technology.

2 Context and Methodology

The context of this work is similar to previous studies, that is, there is a heterogeneous audience (from primary to university level) and also teacher's profile is heterogeneous (from PhD students to university professors). In that sense, as proved in our previous work, the use of a brief and adaptive teaching programme for such specific subjects is vital to offer a high-quality educational experience [9]. In general, the time required to complete the scientific exhibition is around 20 min, according to the requirements of the events where these exhibitions take place. According to these circumstances, the content of this teaching unit (adapted to each kind of audience according to their level) is explained in Table 3.

As observed, at Primary Education levels, the more complex levels are skipped, whereas the Secondary Education level and University Level share most of the

Table 3 Teaching unit content adapted to different audiences

Concept	P. E	S. E	U. L
Role of insects and microorganisms	Yes	Yes	Yes
Insects and microorganisms for specific uses	No	Yes	Yes
Active carbon: definition and main uses	Yes	Yes	Yes
Active carbon: how it is made and characteristics	No	Yes	Yes
Water treatment and sewage sludge generation	No	Yes	Yes
Active carbon and its use in biogas production	No	Yes	Yes
Insect hotel: definition and uses	Yes	Yes	Yes
Insect hotel: hole size vs insect size	Yes	Yes	Yes
Initial comparison between active carbon and insect hotels	Yes	Yes	Yes
Further comparison: pore and hole size distribution	Yes	Yes	Yes
Explanation of SEM and porosimetry	No	No	Yes

* P.E., Primary Education; S. E., Secondary Education; U. L., University level

concepts, except for porosimetry and scanning electron microscopy (SEM), the most difficult concept explained in this unit. As explained in the following section, these concepts (when explained for all the educational levels) are differently explained, increasing its complexity with the educational level of the audience. Also, the exercises are adapted to each situation. The teaching materials required for this work are the following:

- Educational poster, where the main concepts about active carbon and insect hotels are explained and compared.
- Two insect hotels, two samples of active carbon and two images from the electronic microscope.
- A vernier caliper, preferably digital, to determine the diameter of holes in insect hotels.

3 Results

In this context, the main characteristics of each concept are included, adapted to each educational level, as follows:

Role of insects and microorganisms: It is explained at all educational levels, breaking the myth that insects and microorganisms are bad to our health (bite, sting, diseases, etc.). On the contrary, they both present many advantages for human beings, such as pollination or pest control in the case of insects and food processing in the case of microorganisms. For further educational levels, specific examples of these advantages can be added, depending on the time available.

Insects and microorganisms for specific uses: For S. E. and U. L., the specific use of insects as pollinators and pest controllers, along with the role of microorganisms in biogas production during sewage sludge biodigestion, are explained. In that sense, the use of insect hotel and active carbon is introduced for the following explanations.

Active carbon, definition and main uses: In that sense, a brief definition about active carbon is introduced, comparing its porosity with other similar concepts at macroscopic scale such as the holes in a gruyere cheese or the typical insect hotels. In that sense, it is vital to explain how active carbons can be used in a wide range of applications such as microorganism proliferation. For further educational levels, the use of active carbon as adsorbents or catalyst support can be explained.

Active carbon, how it is made and characteristics: For S. E. and U. L., the production of active carbon (through pyrolysis, activation with steam or CO_2 or through hydrocarbonization, for instance) is explained, pointing out the main characteristics of these products such as porosity or surface area, which are vital to understand their performance.

Water treatment and sewage sludge generation: For S. E. and U. L., the context of sewage sludge generation in a wastewater treatment plant is explained, pointing out biogas production through biodigestion and the importance of high yields in CH_4 in biogas.

Active carbon and its use in biogas production: For S. E. and U. L., as explained in the previous point the role of microorganisms is vital during biodigestion, and active carbon can enhance their performance during this process by promoting their proliferation. Thus, pore size distribution is important to determine the main prevailing kind of microorganism in sewage sludge, as it is related to microorganism size.

Insect hotel, definition and uses: For all educational levels, the concept of insect hotel is explained, as a host of solitary bees, wasp or ladybirds, for instance, in order to promote pollination and pest control (such as cottony cushion or aphid).

Insect hotel, hole size vs insect size: As all the audience can imagine, the holes included in an insect hotel will determine the presence of specific kinds of insect, as a bee, wasp or ladybird have different sizes. Consequently, depending on the average or majority hole size in an insect hotel, there will be more or less insects of a specific size.

Initial comparison between active carbon and insect hotels: As previously explained, for all educational levels, a comparison between active carbon and insect hotels is possible, as we can compare pores and microorganisms in active carbon with holes and bees, wasps or ladybirds in an insect hotel.

Further comparison, pore and hole size distribution: For higher educational levels, a pore size distribution obtained through porosimetry for different active carbons and a hole size distribution obtained manually for different insect hotels is shown, explaining the similarities between each kind of histogram.

Explanation of SEM and porosimetry: Equally, for higher educational levels, there might be an explanation about the way how microscopy images through scanning electron microscopy (SEM) and porosimetry histograms are obtained.

The proposed exercises or activities related to this exhibition are the following:

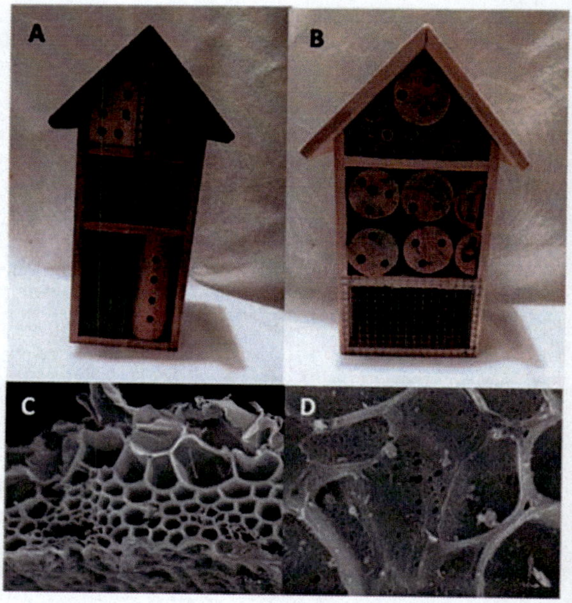

Fig. 3 Comparison between insect hotels (A and B) and active carbons (C and D)

- **Visual comparison between insect hotels and active carbons:** In this case, different insect hotels can be used, paying attention to the differences in their structure (see Fig. 3a and b). It should be noted that these changes in hole sizes are important to determine the ratio of insects that are going to visit the hotel. In contrast, different active carbons can be shown, in order to see visual differences or, especially, differences under the microscope. For that purpose, SEM figures can be used, to point out the differences in pore sizes in different active carbons (see Fig. 3c and d). As in the case of insect hotels, these differences in pore distribution are important to determine the kind of microorganisms that are going to proliferate during biodigestion of sewage sludge.
- **Hole size distribution in an insect hotel and comparison with pore size distribution in active carbon:** For this exercise, a vernier caliper is used to measure the inner diameter of each hole in an insect hotel. Afterwards, the hole size distribution (from the minimum to the maximum hole size) is represented, by using percentages of holes that share the same size. In that sense, concepts like percentage, average, standard deviation, histogram, etc., can be introduced depending on the academic level of the audience. Thus, Fig. 4 shows hole size distributions for two different insect hotels. As observed, not all the insect hotels share the same hole distribution, which makes them useful to host, preferably, a specific group of insects depending on their size. For instance, Hotel 1 can admit a wide range of insects (from different kinds of ladybirds to wasps), whereas Hotel 2 is specially suitable for bees.

On the other hand, a similar comparison between two different active carbons can be made according to their pore size distribution (see Fig. 5) Similarly to the

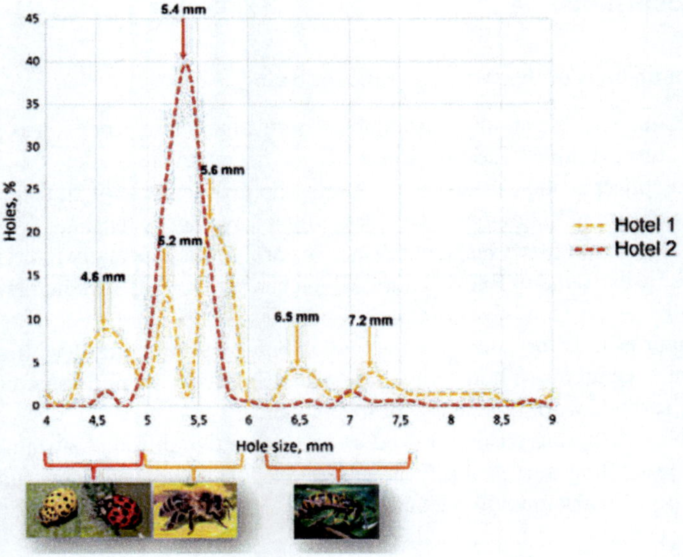

Fig. 4 Hole size distribution in different insect hotels

previous comparison with insect hotels, a different profile in pore size can be done in this case, which will make each kind of active carbon suitable for a certain kind of microorganisms. In that sense, the equivalence between insect hotels and active carbons can be feasible.

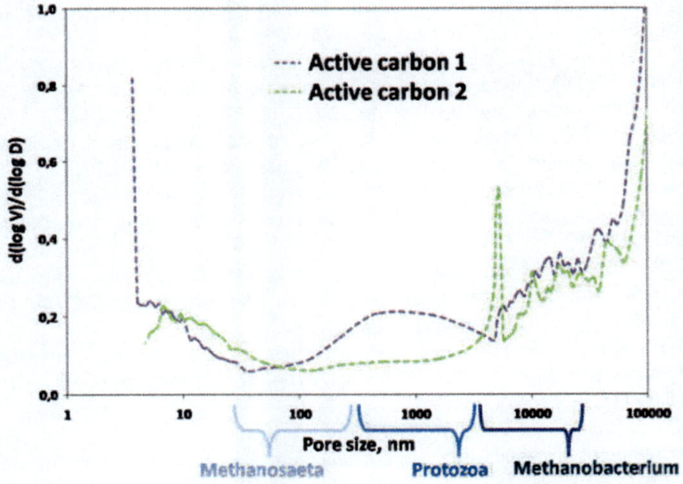

Fig. 5 Comparison between two different active carbons and the ideal microorganism distribution depending on pore size

4 Conclusions

The main findings of this work were the following:

- A specific example about a sustainable improvement of a process was presented.
- With this example, various aspects have been covered, like the change of biased concepts about bacteria or insects, the practical use of statistics (like the interpretation of histograms, the calculation of average values, etc.).
- The valorization of wastes to obtain active carbon was especially interesting, as it can be easily found in Extremadura, which can promote its sustainable economic growth.
- The approach to the concept of active carbon by comparing it with something tangible like an insect hotel was successful, as it made the understanding of how active carbons work more easy for the audience.
- This way, this comparison allowed us to explain two different sustainable practices, promoting sustainability at different educational levels by adapting the educational curriculum to the audience.

References

1. Aeschlimann, B.; Herzog, W.; Makarova, E. How to foster students' motivation in mathematics and science classes and promote students' STEM career choice. A study in Swiss high schools. *Int. J. Educ. Res.* **2016**, *79*, 31–41, doi:https://doi.org/10.1016/j.ijer.2016.06.004.
2. Lozano, R.; Ceulemans, K.; Alonso-Almeida, M.; Huisingh, D.; Lozano, F.J.; Waas, T.; Lambrechts, W.; Lukman, R.; Hugé, J. A review of commitment and implementation of sustainable development in higher education: Results from a worldwide survey. *J. Clean. Prod.* **2015**, *108*, 1–18, doi:https://doi.org/10.1016/j.jclepro.2014.09.048.
3. Craig, C.A.; Petrun Sayers, E.L.; Gilbertz, S.; Karabas, I. The development and evaluation of interdisciplinary STEM, sustainability, and management curriculum. *Int. J. Manag. Educ.* **2022**, *20*, 100652, doi:https://doi.org/10.1016/j.ijme.2022.100652.
4. Government, S. Web del Ministerio de Educación y Formación Profesional Available online: https://www.educacionyfp.gob.es/portada.html (accessed on Jan 23, 2023).
5. Government, S. Instituto Nacional de Estadística Available online: https://www.ine.es/dyngs/INEbase/es/operacion.htm?c=Estadistica_C&cid=1254736176951&menu=ultiDatos&idp=1254735572981.
6. Serafini, P.G.; Moura, J.M. de; Almeida, M.R. de; Rezende, J.F.D. de Sustainable Development Goals in Higher Education Institutions: A systematic literature review. *J. Clean. Prod.* **2022**, *370*, doi:https://doi.org/10.1016/j.jclepro.2022.133473.
7. Figueiró, P.S.; Raufflet, E. Sustainability in higher education: A systematic review with focus on management education. *J. Clean. Prod.* **2015**, *106*, 22–33, doi:https://doi.org/10.1016/j.jclepro.2015.04.118.
8. Viegas, C. V.; Bond, A.J.; Vaz, C.R.; Borchardt, M.; Pereira, G.M.; Selig, P.M.; Varvakis, G. Critical attributes of Sustainability in Higher Education: A categorisation from literature review. *J. Clean. Prod.* **2016**, *126*, 260–276, doi:https://doi.org/10.1016/j.jclepro.2016.02.106.
9. Nogales-Delgado, S.; Encinar Martín, J. M.; Román Suero, S. Environmental Education for Students from School to University: Case Study on Biorefineries. *Educ. Sci.* **2019**, *9*, 202, doi:https://doi.org/10.3390/educsci9030202.

Sustainable Cities by Collaborative Last Mile Parcel Delivery - Live Example Insight

Csilla Bartucz and Edit Süle

Abstract The restrictive measures during the Covid pandemic resulted in an unexpected growth in e-commerce, which growth has continued since the epidemic. The growth of e-commerce also increased the number of last mile parcel delivery transactions. The growth of last mile parcel logistics also resulted in an increasing number of cars, emissions and traffic problems. City governments introduce restrictive measures against increased environmental burdens, which imposes an additional burden on parcel logistics providers. Collaboration among last-mile service providers can assist boost productivity while at the same time lowering harmful environmental effects. However, there is little service provider cooperation in this industry. This study analyses the development of cooperation between service providers based on an examination of a Central-Eastern European Courier, Express, Parcel (CEP) market, which has market characteristics that can be referred to as typical for this area such as: a small number of dominant service providers; a large number of small service providers who are usually subcontractors of the dominant ones; individualistic operation. The market structure of the example country being presented is oligopoly, which is typical of the CEP market of the countries in the region. The analysis of the Hungarian market is supplemented by the creation of a model that simulates cooperation, which shows that the quantity of delivery vehicles, the distance travelled by the trucks, and the degree of environmental impact brought on by all of this can be decreased.

Keywords Sustainability · Last mile delivery · CEP market · Collaboration · City logistics · Sustainable city

C. Bartucz (✉) · E. Süle
Széchenyi István University, Egyetem Tér 1, 9026 Győr, Hungary
e-mail: bartucz.csilla@sze.hu

E. Süle
e-mail: sule.edit@ga.sze.hu

J. R. da. C. S. Galvão et al. (eds.), *Proceedings of the 3rd International Conference on Water Energy Food and Sustainability (ICoWEFS 2023)*, Springer Proceedings in Earth and Environmental Sciences, https://doi.org/10.1007/978-3-031-48532-9_42

1 Introduction

In the last few years, the scientific community has also become more concerned about the detrimental impacts of human activity on the environment. World organizations, together with the press, exert pressure on companies to reduce their environmentally harmful activities [20]. According to [15], significant production growth has been accompanied by a significant increase in consumption, which has resulted in additional environmental pollution. These factors include rapid population growth, higher levels of technological advancement, and rising purchasing power.

E-commerce has been showing an upward trend for years, which began to grow in an unpredictable manner during the covid epidemic that began in 2020. According to a study by [11], the sector grew by 15% in 2010, but this growth rate was already 35% in 2020. The pandemic situation, which has opened a new chapter in people's lives through restrictive measures, has affected a special branch of the logistics sector, which is called parcel delivery. On the Courier, Express, Parcel (CEP) market, during parcel delivery, the product ordered by the customer is delivered to the recipient by the parcel delivery service provider. The recipient can request parcel delivery to a delivery point, to a parcel locker or to their home. During the parcel delivery process, the service provider collects the parcels to be delivered, then sorts them according to areas, and finally delivers the parcels to the customers using smaller trucks. The operation of the service providers is characterized by the use of their own collecting and distributing vehicles, the involvement of their own human resources, and the use of their own sorting machine. The most critical process in parcel delivery logistics is the delivery to the recipient after sorting, that is called last mile delivery. For the service provider, the last mile section is the most inefficient and, according to the research of [18], the section accounts for the largest part of the total logistics cost. In addition, [12] believe that the last mile section is also the most energy consuming operation. When vehicles transport the ordered items from the depot to distribution centers or straight to the customer, this is the stage of last mile delivery that generates the most pollution. Another characteristic of parcel delivery is the presence on the market of several service providers, who often deliver parcels to the same area or even to the same customer. Since service providers mostly use traditional fuel-powered vans to carry out deliveries, last mile parcel delivery is also responsible for the increased truck traffic on the roads, emissions, and traffic jams [12, 18].

During the pandemic, the internet shopping habits of people who were condemned to stay at home increased, and while alternative types of delivery were on the rise, during the epidemic they returned to home delivery [17]. The delivery demands caused by the growing number of orders force more and more trucks onto the roads generating increased traffic and more pollution. Environmental regulations are putting increasing pressure on cities and thus on last mile providers. The city administration is introducing restrictive measures in order to improve the situation, which imposes additional burdens on last mile service providers. Based on this, the service providers can adhere to the rules using eco-friendly solutions, but they must also consider cost and operational effectiveness. In recent decades, countless solutions and models have been created to solve the last mile problem outlined earlier. Models have been substantiated by the introduction of alternative vehicles, through the optimization of deliveries to the establishment of service centers. In the vast majority of instances, scientists have provided figures to substantiate the costs and environmental effects of each strategy. Nevertheless, the creation of cooperation between the service providers to alleviate environmental burdens or optimize the use of resources was only a solution proposed in very few cases.

Based on the above, the most important question of this study is that if the cooperation is economically beneficial for the parties and is also useful for the city management, what is the reason why the parties only become partners in delivery in certain cases. In order to answer this question, this study examines the current CEP market trends through the presentation of the last mile parcel delivery market in Hungary, and interprets the possibilities of the development of cooperation in this market. In addition to the analysis of the Hungarian CEP market, a model simulating cooperation is presented, which aims to prove the economic benefits of cooperation. In accordance with the above, the structure of this study is based on the following chapters: The Literature Review chapter provides an overview of the literature on the subject; in the Discussion chapter, the Hungarian CEP market and the existing collaborations are presented, as well as the closely related simulation model. Finally, the Conclusion chapter closes the study and suggests further research opportunities.

2 Literature Review

Since the outbreak of the coronavirus in 2020, which also had an impact on last mile deliveries, e-commerce has experienced an unprecedented growth. The prevalence of the epidemic, according to [4], also altered consumer behavior; the proportion of home deliveries, which had earlier started to fall, started to rise again. Human activity has led to an increase in poisonous gas emissions and dangerous material emissions into the atmosphere. In terms of carbon monoxide (CO), sulfur dioxide (SO2), nitrogen oxides (NOx), and volatile organic compounds, the transportation industry is one of the top emitters (VOC). By 2050, emissions should be decreased by 80 to 95 percent compared to the 2011 European Commission projections [8]. To meet these goals, a reduction in the transportation sector of at least 60% must be accomplished [12].

Global urbanization is rising, which results in more transportation activities for distributing goods and providing services. The presence of more trucks as a result of expanding e-commerce and urbanization increased traffic and, at the same time, deteriorated city air pollution data [19]. According to [18, 12] last mile delivery is one of the most energy-intensive operations, which is contributing to increased traffic and air pollution. In addition, according to [17], city logistics are troublesome because of traffic, parking fees, street closures, limitations, and large distances from the parking van to the customer's door. Additionally, [5] contend that city logistics, or growing urban logistics, presents significant challenges due to congestion, high emissions, and a detrimental effect on quality of life.

In order to discover the best solution to the expanding issues, the scientific literature has been addressing the breadth of the last mile delivery challenge in terms of sustainability, environmentally, socially, and economically for years. Although a significant proportion of the last mile delivery related articles published in the last 25 years contained solutions created to protect the environment, only a few articles have already been written to emphasize cooperation. The collaborations emphasized in last mile delivery were strengthened with the development of city logistics [13, 14, 22] and Urban Delivery Centers (UDC) [1, 16, 24], while some scholars also emphasized the importance of cooperation between service providers [2, 25]. The cooperation created in the last mile delivery operation can basically be vertical and horizontal. When the participants are subordinate to each other and each serves a distinct area of the market, the collaboration is vertical. Vertical partnerships can develop during the supply chain process. According to [10], complementary players of the transport and mobility service value chain engage in vertical collaboration. The typical form of vertical cooperation is subcontracting. However, [6] state that horizontal cooperation is a practice among businesses operating at the same level of the supply chain, where the businesses are either independent or in competition with one another.

[4, 21] highlight that service provider cooperation is a desirable alternative to lower overall costs and boost utilization. [23] emphasize that the application of the cooperation theory based on resource sharing can reduce the number of vehicles participating in the operation, the number of kilometers travelled, and in addition to the economic benefit, they can maximize their service level. The aim of the collaborations is to reduce costs and increase efficiency from an economic point of view. The authors also emphasize that if the goal is only to reduce costs, cooperation can result in a cost reduction of up to 45%. Furthermore, [3] analyzed the fuel consumption by simulating the cooperation, based on which they proved that they were able to achieve a 19% reduction in fuel consumption in the case of cooperation. Although the various simulations primarily examined the economic effectiveness of collaborations, each simulation contained additional value that further proved the economic benefits of collaboration models.

3 Discussion

The purpose of this chapter is to present the region's CEP market structure, which it intends to do through a detailed description of the Hungarian CEP market. The following chapter describes in detail the structure of the Hungarian CEP market, presents the key actors of the sector, as well as the characteristic features of the market. After the Hungarian-related insight, a model simulating cooperation presents the economic and environmental advantages of cooperation.

3.1 Hungarian CEP Market Overview

Hungary is one of the member states of the Visegrád Four, a country in Central and Eastern Europe, which lies in the middle of the Carpathian Basin. Its population is 9.7 million, its capital is Budapest. The population of Budapest is 1.7 million. Hungary has 7 other large cities with a population of more than 100,000. The country is basically centered on the capital city, although in the last few years the big cities in the countryside have also achieved significant economic development mainly due to international investments. Logistics parcel delivery can be understood as part of the Courier, Express, Parcel (CEP) market. The CEP service means the delivery of non-palletized parcels under 50 kg. CEP service providers can also be interpreted as players in the international freight forwarding and distribution market. Couriers deal with same-day and next-day delivery of light parcels. The CEP service basically covers a door-to-door service. In terms of domestic conditions, nearly a thousand companies are active in the CEP area in Hungary. The industry's revenue is over sixty billion HUF, but the largest players account for more than two-thirds of this [7].

GKI Economic Research Co. (GKI) is one of Hungary's market research companies with significant references and experience. They assess the Hungarian logistics last mile market every two years and analyze the changes experienced there, as well as provide forecasts for the future. Based on the 2021 study by GKI, from the point of view of parcel logistics, there are 7 significant service providers operating in the country: General Logistics Systems B.V. (GLS), National Hungarian Post (MPL), DPDgroup (DPD), Post 207 Beteiligungs (Express One), Sprinter, Foxpost and Packeta s.r.o (Packeta) (see Table 1).

The sales revenue of the 7 largest domestic CEP service providers almost completely covers the entire sales revenue of the domestic CEP market, which is due to the fact that the nearly 1000 smaller players in the market are typically subcontractors to the 'dominant ones' (see Table 2). It can also be observed that even in the

Table 1 The 7 large CEP players in Hungary in 2021

Rank	Name of company	Revenue in HUE	Revenue in EUR
1	GLS	57 609 882 000	147 717 646
2	MPL	35 925 200 000	92 115 897
3	DPD	13 722 437 000	35 185 736
4	Express one	9 973 005 000	25 571 808
5	Sprinter	4 320 718 000	11 078 764
6	Foxpost	1 718 004 000	4 405 138
7	Packeta	1 644 050 000	4 215 513

Source own elaboration based on GKI data

* Exchange rate 1 EUR = 390 HUF Based on the central exchange rate of the Hungarian National Bank on January 25, 2023

sales revenue of the 7 most significant companies, differences of magnitude can be discovered. In fact, the most dominant players in the market are the first three companies, GLS, MPL and DPD, and a gap can already be observed in the sales revenue of the companies following them. From the point of view of the investigation, vertical subcontracting does not qualify as cooperation. In the case of cooperation, a subordination relationship arises between the parties, which is also called a partnership. Table 2 shows that the dominant players in the Hungarian market do not operate on a partnership basis. At the same time, Foxpost and Packeta stand out among the actors, as they solve the home delivery activity in cooperation and partnership with other dominant actors.

According to the 2021 study of [9], home delivery accounted for 74% of all transactions, and delivery of parcels lockers and parcel points accounted for 10.3% of all transactions. In 2020, the delivery chain related to online trade became much more complex and chaotic due to the pandemic, and the long-term effects of the shutdowns and transformational processes that occurred at that time were felt in 2021. In terms of transaction numbers, the Hungarian online retail market grew by more than 30% in 2021 compared to 2020. Although home delivery continues to enjoy unbroken popularity, parcel locker machines significantly strengthened their market position in 2021. The expansion of parcel points was not so spectacular in 2021, however, the number of delivery points increased here as well. In the long term, the goal of most service providers is certainly diversion—fewer parcels delivered to homes and dropped off at more parcel points/parcel locker machines.

Table 2 Service-related general data of the 7 big players in 2021

Company	Coverage	Parcel point	Parcel locker	Nr of depos	Mode of home delivery	Collaborative partners
GLS	International	1060 +	100 +	86	With subcontractors	–
MPL	International	2900 +	51	23	Subcos + own resources	–
DPD	International	300 +	0	23	With subcontractors	–
Express one	International	160 +	0	30	With subcontractors	Foxpost, Viddl
Sprinter	International	700 +	0	N/A	With subcontractors	–
Foxpost	Nation-wide	0	414	4	Together with Express One	Express One, Gebrüder Weiss
Packeta	International	800 +	40	6	With external partners	MPL, Express One, DPD, Fáma, Sprinter

Source own elaboration based on GKI data

Due to the pandemic, purchases aimed at meeting daily needs have been diverted to the online space, which were previously linked to everyday purchases due to speed. Expectations regarding the speed of delivery have changed: the need for faster delivery has appeared. In the case of non-food products, 1–3 day delivery is the basic expectation, in the case of food and FMCG products, much longer the option of same-day delivery is more common, which is typically available as a premium or surcharged service. Technology is developing to an unprecedented extent, direct telephone access to couriers, communication of delivery time windows, bank card payment available at the courier is already an expectation on the part of customers.

According to [2], the need for cooperation is primarily driven by economic interest, which can also be seen in the Hungarian CEP market. The two market-leading companies (GLS and MPL), presumably because of their capital strength, do not cooperate with other service providers. They solve all further portfolio expansions themselves. However, the other 5 significant market players provide the complementary services required for the expansion, if applicable, on the basis of agreements (cooperation) with other service providers.

Although the Hungarian CEP market was dominated by the 7 players presented earlier, in the last few years the market has diversified in a special way. Two large online stores (Alza = alzaBox, eMAG = Easybox) have started to provide their own customers with automatic delivery of their own parcels. The appearance of alzaBox and Easybox on the market increased the number of parcel locker machines on the one hand (alzaBox = 200 + units; Easybox = 449 units), but on the other hand, these two players also appeared as competitors on the CEP market (Table 3).

GKI supplemented its 2021 Hungarian logistics survey with a primary public survey of 24,000 people, in which it wanted to gain insight into the last mile logistics service from the customer's side. One of the important groups of questions of the questionnaire survey was to learn about the environmental awareness and expectations of customers. According to the result of the survey, for a significant part of Hungarian customers, environmental protection aspects are not important when it comes to online shopping and delivery. In addition, it can also be said that in a certain sense, 37% of customers see online retailing as a more environmentally friendly solution than traditional in-store shopping. However, according to online customers who consider online trade to be less environmentally friendly, modern vehicles are not always used for home delivery. Customers consider the bicycle courier to be the most environmentally friendly solution, while the parcel locker machine of the external service provider comes second. In total, nearly a third of the respondents consider

Table 3 Revenue of the 2 significant web shop players in Hungary in 2021

Name of company	Revenue in HUF	Revenue in EUR
Alza	3 180 299 010	8 154 613
eMAG	90 618 287 000	232 354 582

Source own elaboration based on GKI data
* Exchange rate 1 EUR = 390 HUF Based on the central exchange rate of the Hungarian National Bank on January 25, 2023

the parcel locker machine to be the most environmentally friendly pick-up solution. However, 35% of customers would not pay a surcharge for the delivery of an environmentally friendly parcel.

Environmental responsibility is expected from the majority of companies; thus, most companies are taking action in the field of sustainability. GLS delivers parcels in downtown Budapest with bicycles, electric tricycles and electric vans. MPL also uses 84 electrically powered vehicles in its fleet to carry out deliveries. In addition, it also buys three-wheeled electric mopeds and e-bikes in the name of sustainability. DPD uses environmentally friendly cargo bikes for deliveries in downtown Budapest. Foxpost also uses electric vehicles to deliver the parcel to the parcel locker machines. The use of alternative fuel vehicles and electric cargo bikes is one of the options for reducing the ecological footprint. Another aspect of sustainability is the use of common resources, i.e., the development of cooperation. In large cities, in countless cases, all service providers are present in one part of the city and often make deliveries to the same address or to nearby addresses. Even if deliveries are made with environmentally friendly ways, service providers in the same area will still be present on the roads. The shared use of resources, i.e., the shared use of delivery vehicles, can reduce the number of vehicles on the roads and thus the intensity of traffic and the amount of traffic jams. The authors of this article carried out a simulation, through which they investigated the economic (and possibly environmental) benefits of cooperation for companies and, where applicable, the city administration.

3.2 *Simulation of Cooperative Last Mile Delivery*

According to [12], last mile delivery is environmentally destructive in terms of the number of used trucks and their emissions. A possible alternative to increase the efficiency of last mile delivery is the development of cooperation between service providers. In order to prove that the cooperation is economically beneficial for the participating parties, the authors of this article prepared a simulation that describes with data in which area and in what way savings can be achieved, and how the environmental effects can be reduced while maintaining the same service level. For creating the simulations, the Aptean Paragon vehicle routing and scheduling system has been used which is the product of Paragon Software System Ltd. based in UK. In order to demonstrate the simulation model, two artificial parcel delivery companies have been created with depos in Budapest. To run the simulation, 1 day of limited parcel delivery was modelled in the inner districts of Budapest (Hungary). The delivery addresses were selected and assigned using a random function from the Budapest address database. The examined delivery points (addresses) have been evenly distributed across the area. In order to carry out the simulation efficiently, it was decided that the drivers work 8 h a day, and the number of trucks required for delivery is specified by the simulator. Parcels are delivered on the given day between 10:00 and 14:00. Regarding the size of the parcels, the modelling was done with average-sized parcels. The authors examined 4 scenarios to create the simulation models. The first scenario demonstrated the independent operation of the two service providers. In the second and third scenarios, rudimentary cooperation of service providers was planned, in the framework of which one service provider transported the parcels to the depot of the other service provider (and vice versa). In the fourth scenario, the cooperation was supplemented with a jointly operated cross-dock point, where the cross-dock point was located at almost the same distance from the service providers, but the location was closer to the city. Before running the simulation, the cost factors required for the delivery, such as: the operating cost of the depot and the cross dock, the costs of loading and transportation, as well as the cost and time of the vehicles, were determined in advance.

As a result of running the four scenarios, it became visible how many vehicles are needed to deliver the number of parcels, how many kilometers each vehicle travels during the delivery process, and how much time each vehicle spends in completing the delivery.

Figure 1 shows that without cooperation, 16 vehicles are needed for deliveries, while in the case of cooperation (even in the case of rudimentary cooperation), only 14 vehicles carry out the deliveries. In terms of the number of kilometers travelled, a significant reduction can be seen primarily in the case of the inclusion of the jointly operated cross-dock point. Although the number of kilometers traveled was the lowest in the case of the fourth simulation (operation of a common cross-dock point), it resulted in a lower number in the case of Simul 2 and Simul 3, thus proving the advantage of the cooperation in this regard.

In terms of costs, a decrease can be detected both in terms of the cost of trucks, kilometers travelled, and time (see Fig. 2). Nevertheless, in the case of the cost per kilometer traveled, the greatest cost reduction could also be realized in Simul 4, although even rudimentary cooperation could result in cost reduction.

The simulation presented above showed the delivery of a small amount of artificially created parcel stock in the case of independent service provider operation, as well as during cooperation. Although the small number of parcels does not reflect reality, the positive effect of cooperation can be demonstrated even in the case of such a small data set. The resource sharing realized during the cooperation required 2 fewer vehicles from the suppliers, which result in savings in both time and costs for the service providers. The city also benefits from the establishment of cooperation,

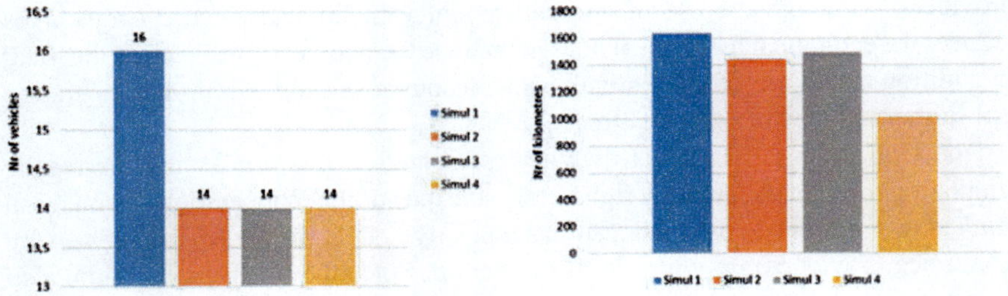

Fig. 1 Number of used vehicle and number of kilometers travelled by simulation model

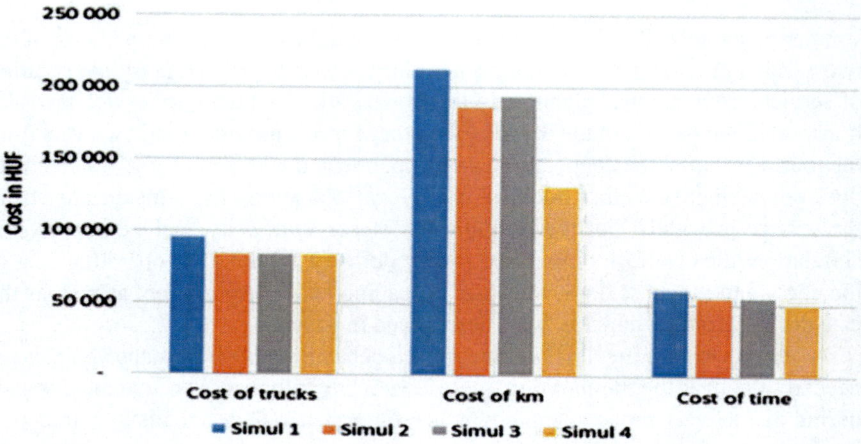

Fig. 2 The evolution of variable cost elements

since there are fewer vehicles on the roads, which results in a reduction in the level of harmful emissions. In order to be able to prove the benefits of cooperation based on realistic results, the authors recommend running a simulation that includes the actual daily parcel volume as well as the emission level of the used vehicles.

4 Conclusion

The coronavirus epidemic that started in 2020 resulted in a significant increase in e-commerce and at the same time a change in customer habits. A significant number of customers switched to online shopping, a habit that remained even after the end of the epidemic. As a result, online ordering has challenged CEP market providers. The increased delivery of parcels also resulted in more and more vehicles appearing on the roads making deliveries, which naturally increased road traffic and air pollution. Reducing harmful emissions and improving air quality have been the objectives of the European Union and other organizations for many years, and the fulfilment of these guidelines and expectations also affects the last mile parcel delivery service providers. Although the use of alternative fuel vehicles and electric cargo bikes reduces the ecological footprint, it does not reduce the number of vehicles on the road. In addition to the use of alternative means, resource sharing based on cooperation can further reduce the negative effects.

The Hungarian CEP market overview presented in this article and the results simulating relevant cooperation rightly raise the question: if cooperation can increase efficiency and reduce costs and also has a positive effect on the environment, why do service providers rarely turn in this direction. Since the scientific literature also reports few realized collaborations, it can be assumed that the realization of collaborations is not only based on the economic interests of the parties. In case of favorable economic condition, if a crisis of confidence between the parties or a low willingness to take risks can be assumed, the joint work will not take place. Examining the Hungarian market, however, it can be stated that a change can be seen in the last few years. Complementary type collaborations promoting portfolio expansion were established. Foxpost, the market leader in the parcel locker market, cooperates with the company Express One to be able to offer a home delivery service. In order to expand its portfolio, Packeta also entered into cooperation with well-known CEP market players. Since consumer needs change quickly, the service provider often cannot solve the need overnight with its own resources, so it looks for a cooperative partner. From a sustainable city's point of view, these types of collaborations are very favorable, since Foxpost, for example, does not build its own fleet for home delivery, but uses others. As a result, it does not increase the number of vehicles on the roads, and does not contribute to the increase of harmful emissions. In addition to the Hungarian market analysis, the authors prepared a model simulating cooperation with a small number of samples. The simulation proved that the number of delivery vehicles, the kilometers travelled and the amount of time used can be reduced in the case of cooperation in logistics last mile parcel delivery.

References

1. Aljohani, K. and Thompson, R. G.: Profitability of freight consolidation facilities: A detailed cost analysis based on theoretical modelling. Research in Transportation Economics, 90, 101122 (2021).
2. Bartucz, Cs. and Süle, E.: Can the trust in Uber-like platform use be translated into parcel logistics? ENTRENOVA – Enterprise Research Innovation Conference. https://www.econstor.eu/escollectionhome/10419/262228 (2021).
3. Bhasker, A. and Sarmah, S. P.:Collaborative last-mile delivery and pick-up in city logistics. International Journal of Logistics Systems and Management, Vol. 34, No. 4 (2019).
4. Bhati, A., Akram, H., Basit, H., M., Khan, A., U., Naqvi, S., M., R. and Bilal, M.: E-commerce trends during COVID-19 Pandemic. International Journal of Future Generation Communication and Networking, Vol. 13, No. 2, pp. 1449–1452 (2020).
5. Cleophas, C., Cottrill, C., Ehmke, J. F. and Tierney, K.: Collaborative urban transportation: Recent advances in theory and practice. European Journal of Operational Research, 273, pp: 801-816 (2019).
6. Cruijssen, F., Dullaert, W. and Fleuren, H.: Horizontal cooperation in transport and logistics: A literature review. Transportation Journal, 46 (3), pp. 22-39 (2007).
7. Diófási-Kovács, O. and Szilágyi, Sz.: A Magyarországon működő CEP szolgáltatók széndioxid kibocsátásának komparatív elemzése. 176. sz. Műhelytanulmány, HU ISSN 1786–3031 (2019).
8. European Commission (2011a). A roadmap for moving to a competitive low carbon economy in 2050. Communication. Belgium: Brussels. European Commission (2011b). White Paper: Roadmap to a single European transport area – Towards a competitive and resource efficient transport system. Communication. Belgium: Brussels.
9. GKI Gazdaságkutató Zrt – Onine csomaglogisztikai körkép 2022 – gkid Research & Consulting Kft. (2022).
10. Gonzalez-Feliu, F., Pronello, C. and Salanova Grau, J. M.: Multi-stakeholder collaboration in urban transport: State-of-the-art and research opportunities. Transport, vol. 33, no. 4, pp. 1079–1094 (2018).

11. Hagen, T. and Scheel-Kopeinig, S.:Would customer be willing to use an alternative (chargeable) delivery concept for the last mile? Research in Transportation Business & Management, doi: https://doi.org/10.1016/j.rtbm.2021.100626 (2021).
12. Halldórsson, Á. and Wehner, J.: Last-mile logistics fulfilment: A framework for energy efficiency. Research in Transportation Business & Management, Vol. 37, 100481. https://doi.org/https://doi.org/10.1016/j.rtbm.2021.100626 (2020).
13. Hribenik, M., Zero, K., Kummer, S. and Herold, D. M.: City logistics: Towards a blockchain decision framework for collaborative parcel deliveries in micro-hubs, Transportation Research Interdisciplinary Perspectives, 8, 100274 (2020).
14. Jamshidi, A., Jamshidi, F., Ait-Kadi, D. and Ramudhin, A.: A review of priority criteria and decision-making methods applied in selection of sustainable city logistics initiatives and collaboration partners. International Journal of Production Research, DOI: https://doi.org/10.1080/00207543.2018.1540892 (2018).
15. Kazancoglu, Y, Ozbiltekin-Pala, M and Ozkan-Ozen, Y. D.: Prediction and evaluation of greenhouse gas emissions for sustainable road transport within Europe. Sustainable Cities and Society, Vol. 70, 102924 (2021).
16. Letnik, T., Mencinger, M. and Perus, I.: Flexible Assignment of Loading Bays for Efficient Vehicle Routing in Urban Last Mile Delivery. Sustainability, 12, 7500 (2020).
17. McLeod, F., Cherrett, T., Bates, O., Bektas, T., Lamas-Fernandez, C., Allen, J., Piotrowska, M., Piecyk, M. and Oakey A.: Collaborative parcels logistics via the carrier's carrier operating model. Transportation Research Record, Vol. 8, pp. 384-393. https://doi.org/https://doi.org/10.1177/0361198120920636 (2020).
18. Olsson, J., Hellström, D. and Palsson, H.: Framework of last mile logistics research: a systematic review of the literature. Sustainability, vol. 11, 7131 (2019).
19. Paddeu, D., Parkhurst, G., Fancello, G., Fadda, P. and Ricci, M.: Multi-stakeholder collaboration in urban freight consolidation schemes: drivers and barriers to implementation. Transport, vol. 33, issue. 4, pp. 913-929 (2016).
20. Seroka-Stolka, O.: The development of green logistics for implementation sustainable development strategy in companies. Procedia – Social and Behavioral Sciences, Vol. 151, pp.302–309 (2014).
21. Ukko, J., Saunila, M., Nasiri, M. and Rantala, T.: The importance of sustainability engagement in small businesses supplier collaboration. Sustainable Development, 30:1-9 (2021).
22. Taniguchi, E., Thompson, R. G. and Qureshi, A. G.: Modelling city logistics using recent innovative technologies. Transportation Research Procedia 46, pp. 3-12 (2019).
23. Wang, Y., Zhang, J., Assogba, K., Liu, Y, Xu, M. and Wang, Y.: Collaboration and transportation resource sharing in multiple centers vehicle routing optimization with delivery and pickup. Knowledge-Based Systems (160), pp. 296–310 (2018).
24. Zelenska, I. and Svadlenka, L.: The Possibilities of Using a UCC in the Czech Republic Cities and Important Criteria Evaluation. 23rd international scientific conference TRANSPORT MEANS (2019)
25. Zhang, M., Fu, Y., Zhao, Z., Pratap, S. and Huang, G. Q.: Game theoretic analysis of horizontal carrier coordination with revenue sharing in E-commerce logistics. International Journal of Production Research, https://doi.org/https://doi.org/10.1080/00207543.2018.1492754 (2018).

Numerical Modelling of Beach Morphodynamics in Post-nourishment Conditions—Case Study

Filipa S. B. F. Oliveira

Abstract The planning of effective and optimized beach nourishment interventions needs to be supported by numerical modelling predictions of the spatio-temporal evolution of the nourishment and the surrounding native beach. The topo-bathymetry of an urban oceanic beach was monitored immediately after a 1 million m^3 sand nourishment and two months later. The coastal stretch has two types of backshore alongshore: dune and seawall. A process-based numerical model, XBeach—surfbeat mode, was applied to simulate the beach morphdynamics during the two months between surveys. The model was settled on the post-nourishment topo-bathymetry and forced with synoptic hydrodynamic data. Its performance was evaluated through indicators of impact and error. According to a frequently applied classification for numerical morphodynamic models, the model performance was evaluated as reasonable in the coastal stretch extreme sectors and good in the inner sector. The results suggest that the morphological evolution observed was not only due to the erosive incident waves but also to a readjustment to the beach natural profile following the sand deposition.

Keywords Beach evolution · Coastal protection · Coastal risks

1 Introduction

Like in many countries worldwide, an increasing investment in beach nourishment operations, as result of change in coastal protection paradigm, is being observed in the Portuguese coastal zone. Over the last few years, the complement of the structural approach (groynes, jetties, seawalls, revetments and detached breakwaters) with non-structural solutions (dune restoration and rehabilitation, geotextile encapsulated-sand systems, beach nourishment and dredging operations) has been prioritized for

F. S. B. F. Oliveira (✉)
Laboratório Nacional de Engenharia Civil, Lisboa, Portugal
e-mail: foliveira@lnec.pt

J. R. da. C. S. Galvão et al. (eds.), *Proceedings of the 3rd International Conference on Water Energy Food and Sustainability (ICoWEFS 2023)*, Springer Proceedings in Earth and Environmental Sciences, https://doi.org/10.1007/978-3-031-48532-9_43

providing an integrated and more efficient protection against climate change impacts in coastal zones, particularly erosion and inundation.

This study is part of a set of morphodynamic studies, supported by numerical modelling, developed by Laboratório Nacional de Engenharia Civil (LNEC) for the Environmental Portuguese Agency (Agência Portuguesa do Ambiente—APA). It concerns the validation of the morphodynamic model XBeach, in its surfbeat mode, for the beaches of Costa da Caparica, Almada, Portugal, for the two-month period September–October/2014, after a sand nourishment intervention of 1 million m^3 performed between 26/June-25/August of 2014. The objective of the study was to simulate numerically the beach morphological response in post-nourishment conditions and, thus, to contribute with guidance for actions to be developed in the short-to-medium term.

The coastal stretch under analysis has a coastline with about 3.4 km and main direction varying between N140° and N150° (Fig. 1). The limits are: at northwest, the groyne designated has EV1 (Fig. 1); and at southeast, the southeast limit of the alongshore seawall which is located at about 260 m from further southeast groyne, designated as EC1 (Fig. 1). The stretch has seven sandy cells: the first cell, the one further northwest, is the longest, with 1.4 km length alongshore (between EV1 and EC7, in Fig. 1); the length of the other six cells, between EC1 and EC7, varies between 250–400 m.

Fig. 1 Study zone: coastal stretch of Costa da Caparica (between groyne EV1 and the southeast extreme of the seawall, 260 m further southeast of groyne EC1)

The groynes EV1 and EC7, which limit the first cell, have 360 m and 140 m, respectively. Within this cell: the northwest and central sectors have a backshore dune system, which frontal dune crest reaches an elevation of about 11 m above the nautical reference level, designated as Zero Hidrográfico (ZH), 2.26 m below the mean sea level; the southeast sector of this first cell has a seawall at the backshore with 460 m of alongshore length (until the groyne EC7) that limits the expansion of the natural beach profile.

The seawall and the groynes have an elevation of approximately 8 m and 6 m above ZH, respectively. The seawall extends about 460 m further northwest of EC7 and 260 m further southeast of EC1. The length of the groynes EC1 to EC7 (in the cross-shore direction) varies between 100–250 m.

The coastal stretch of Costa da Caparica has been under intense human pressure since 1929, when relevant morphologic changes started to be observed [1]. The majority of the coastal defense structures were built between 1959 and 1971. The stretch remained stable until the winter 2000/2001, when an erosion and overtopping process started. In response, several interventions were performed: transference of sand from the beach face to reinforce the frontal dune, and rehabilitation and consolidation of the defense structures—as emergency intervention, in the winters 2002/2003, 2003/2004 and 2006/2007; and emerged beach nourishment, with 3.5×10^6 m^3 of sand, deposited in 2007, 2008, 2009 and 2014 (Table 1), with sediments incoming from dredging the inlet channel, located north of the stretch [2].

The impacts of the maritime storms occurred during the maritime winter season (October–March), are the following: (i) qualitatively, the general retreat of the coastline, the lowering of the beach face, and the overwash and inundation of the frontal dune, sometimes followed by dune breaching; (ii) quantitatively, in the case of the storm of 19/January/2013 and in the first cell, the erosion sand volume between the elevations mean sea level and high tide at spring water level was 50×10^3 m^3, and the average and maximum retreat of the coastline at mean sea level was 20 m and 41 m, respectively.

Table 1 Beach nourishment interventions performed in Costa da Caparica between 2007 and 2014

Date of the operation	Volume of sand deposited in the emerged beach (m^3)	Location of deposition
Aug/2007	500,000	Between EV1 and EC4 (extension 2.4 km)
Aug/2008	1,000,000	Between EV1 and about 500 m south of EC1 (extension 3.8 km)
Aug/2009	1,000,000	Between EV1 and about 500 m south of EC1 (extension 3.8 km)
26/Jun/2014–25/Aug/2014	1,000,000	Between EV1 and about 500 m south of EC1 (extension 3.8 km)

2 Data and Method

Following the execution of the Digital Terrain Models (DTMs) from the topo-bathymetric surveys of August/2014 and October/2014, it was defined the maximum common area between the DTMs, about 2962 km^2. Through the comparison of the DTMs with this common area, the morphological evolution of the study zone during the study period, between August/2014 and October/2014, was calculated.

The morphodynamics of the active beach (includes the surf zone, the beach face, the berm, and the dune—the last not always existing) during the study period was simulated through the model XBeach [3], version 1.22.4867, designated by Kingsday [4], in the non-stationary or surfbeat mode. The resolution of the coastal processes was done along the six shore-normal profiles, PCC3, PCC5, PCC7, PCC9, PCC11 e PCC14, with regular horizontal grid spacing of 5 m. Based on results of a sedimentological analysis performed in the scope of a monitoring program, the sediments were considered sand with median grain size diameter D50 equal to 0.3 mm [5].

The model was forced with synoptic hydrodynamic data, sea level and wave conditions (parameters: significant height, Hs; peak period, Tp; and mean direction, Dir), at the offshore boundary points 2, 3, 4, 5, 7 and 10 (Fig. 2), determined through hydrodynamic modelling, applied at larger geospatial scales, regional and intermediate [6] (not described here).

The morphodynamic model performance was evaluated through indicators of impact and error. The indicators of impact used were the volume of erosion per alongshore beach length and the retreat at the horizontal levels − 1 m, 0 m and + 2 m ZH. As the error indicator, it was applied the Brier Skill Score (BSS) [7].

3 Results

The morphological evolution of the study zone observed between August/2014 and October/2014 (Fig. 3) reveals that the northern and central sectors suffered larger morphological changes than the southern sector. The larger lowering of the topo-bathymetry was observed at the southeast extreme of Santo António beach (Fig. 3).

The hydrodynamics forcing parameters at the offshore boundary points indicate that the tide has a large influence in the wave direction at the northern points and that this influence decreases towards south (Fig. 2).

The visual comparison of the numerical and observed results for the six validation shore-normal profiles can be seen in Fig. 4. Overall, the model either reproduced approximately or overestimated the retreat of the beach berm above the beach face. The phenomenon of sediment extraction from the upper beach face and berm and its transport towards offshore was generally well predicted by the model, even though the maximum Hs during the study period was only 3 m and the study period was two months, a much larger period than in most of the applications of the model reported in the literature.

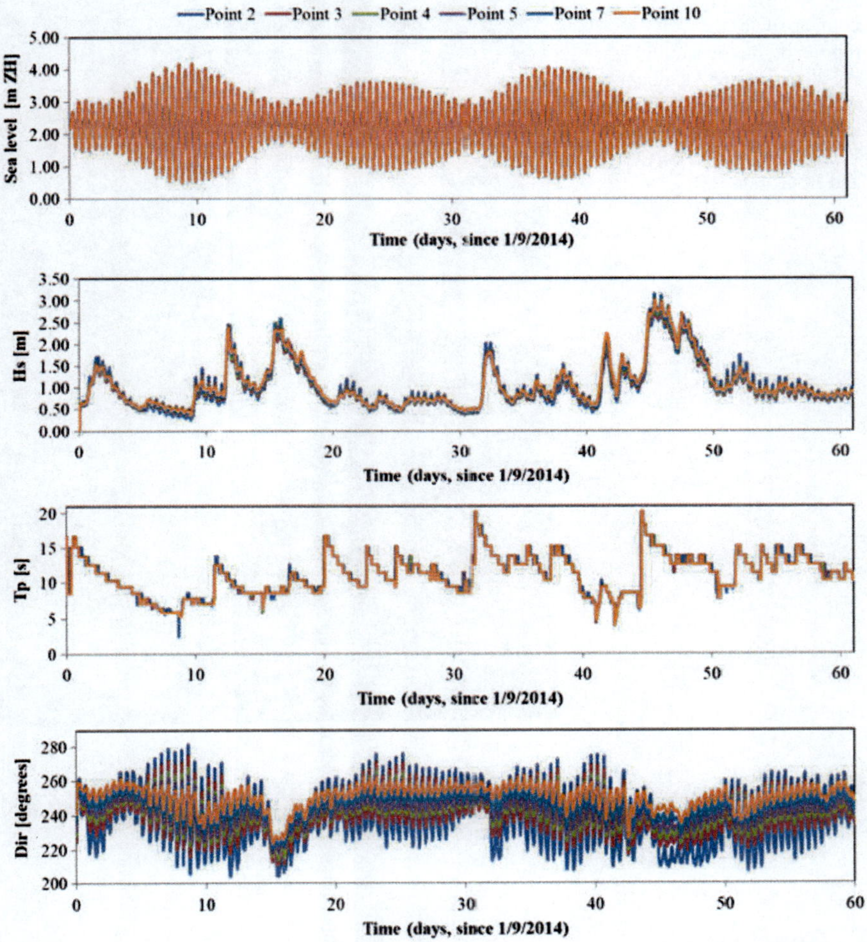

Fig. 2 Hydrodynamic forcing conditions: sea level and wave parameters Hs, Tp and Dir, at the offshore points 2, 3, 4, 5, 7 and 10, of the respective shore-normal profiles PCC3, PCC5, PCC7, PCC9, PCC11 and PCC14

The retreat indicator shows that the numerical results at the vertical reference levels are qualitatively correct, with the exception of the profiles PCC9 e PCC14 (at the extreme of the beaches with seawall) at the reference level + 2 m ZH (Table 2). Regarding the erosion volume indicator, the model only underestimated this metric for the northern profile, PCC3. It overestimated this indicator for the profiles PCC7 and PCC14, and calculated approximated values for the three other profiles (Table 2).

According to a classification for the performance of morphodynamic numerical models proposed in the literature [6] (0.8 $\leq$ excellent < 1.0; 0.6 $\leq$ good < 0.8; 0.3 $\leq$ reasonable < 0.6; 0 $\leq$ weak < 0.3; bad < 0), and for an observation error (δ) of

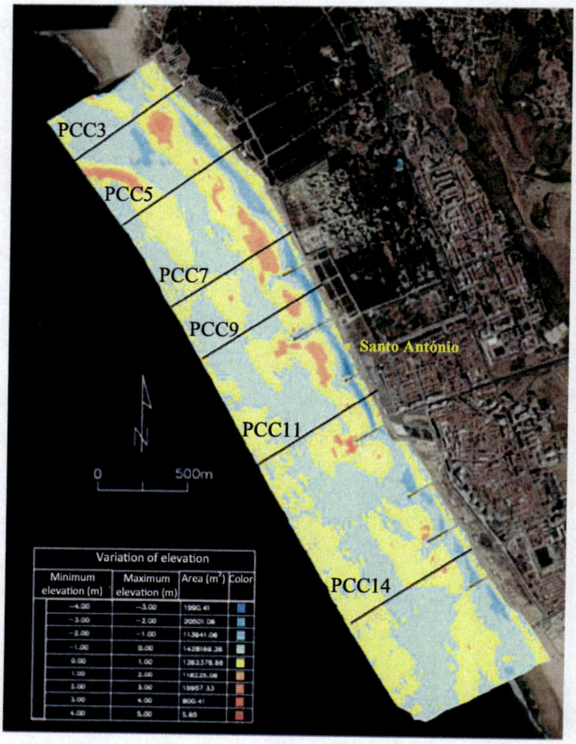

Fig. 3 Result of the comparison of the topo-bathymetric surveys of August/2014 and October/2014, and location of the shore-normal profiles PCC3, PCC5, PCC7, PCC9, PCC11 and PCC14

0.15 m, the model varies between reasonable for the profiles PCC3 and PCC14, and good for the other four profiles (Table 2).

4 Conclusions

The performance of the morphodynamics model XBeach (version 1.22.4867, designated by Kingsday [4], in the non-stationary or surfbeat mode, profile approach) developed for the beaches of Costa da Caparica was reasonable for the extreme parts of the coastal stretch and good for the inner part. The weaker performance in the extremes might be due to the influence of the alongshore processes neglected in a profile approach. Since the study period starts immediately after a nourishment intervention, it is likely that the morphological evolution process did not occur exclusively due to the wave action but also due to the natural readjustment of the beach profile. The results of the study suggest that for this particular case study, a better validation of the model might be achieved using a 2D model approach.

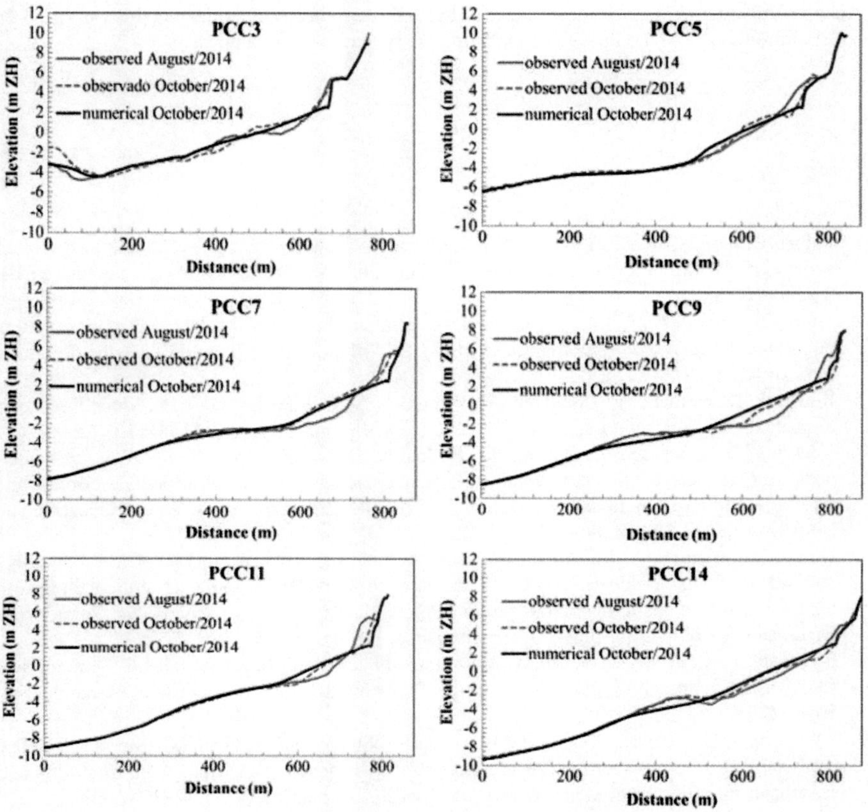

Fig. 4 Numerical and observed evolution of the shore-normal profiles PCC3, PCC5, PCC7, PCC9, PCC11 and PCC14, between August/2014 and October/2014

Table 2 Impact and error performance indicators of the morphodynamics model

Profile	Performance indicators				
	Impact (numerical/observed)			Error	
	Retreat (m) at the vertical levels			Erosion volume (m^3/m)	BSS ($\delta = 0.15/0.3$)*
	− 1 m ZH	0 m ZH	+ 2 m ZH		
PCC3	20/40	− 55/− 85	0/0	77/141	0.5/0.6
PCC5	− 30/− 15	− 25/− 35	25/30	100/113	0.7/0.8
PCC7	− 70/− 75	− 45/− 65	15/10	157/77	0.7/0.8
PCC9	− 95/− 60	− 70/− 40	− 10/25	199/215	0.7/0.8
PCC11	− 65/− 35	− 45/− 30	20/15	143/153	0.6/0.6
PCC14	− 45/− 30	− 30/− 35	− 10/25	135/86	0.4/0.6

* based on [2]

Acknowledgements The author thanks to APA for the permission of publishing this work, and the collaboration of Dr. André Fortunato, Dr. Paula Freire and Mr. Fernando Brito.

References

1. Veloso-Gomes, F., Taveira-Pinto, F., Pais-Barbosa, J., Costa, J., Rodrigues, A.: Estudo das intervenções na Costa da Caparica. 1as Jornadas de Hidráulica Recursos Hídricos e Ambiente, FEUP, 27–35 (2066).
2. Pinto, C., Taborda, R., Silveira, T.: Alimentação artificial das praias da Costa da Caparica. Síntese dos resultados da monitorização (2007 a 2014). Contributo para o Grupo de Trabalho do Litoral (Anexo VI). pp 213–237 (2014).
3. Roelvink, D., Reniers, A., Dongeren, A., Vries, J.T., Mccall, R., Lescinski, J.: Modelling storm impacts on beaches, dunes and barrier islands. Costal Engineering, 56, 1133–1152 (2009).
4. Roelvink, D., Dongeren, A., Mccall, R., Hoonhout, B., Rooijen, A., Geer, P., Vet, L., Nederhoff, K., Quartaert, E.: XBeach Technical Reference: Kingsday Release. Model description and reference guide to functionalities. Deltares, UNESCO-IHE Insti-tute of Water Education and Delft University of Technology, 141 p (2015).
5. Diogo, Z.S., Silveira, T.M., Sousa, H., Carapuço, A.M., Silva, A.N., Lira, C., Taborda, R., Andrade, C.F.: Estudo de caso da Costa da Caparica. Caracterização da variabilidade morfodinâmica sazonal e pós-temporal das praias da Costa da Caparica. Entregável 2.2.c. Relatório da Faculdade de Ciências da Universidade de Lisboa, 119 p (2013).
6. Fortunato, A.B., Freire, P., Bertin, X., Rodrigues, M., Ferreira, J., Liberato, M.L.: A numerical study of the February 15, 1941 storm in the Tagus estuary. Continental Shelf Research, 144, 50–64 (2017).
7. Van Rijn, L.C., Walstra, D.J.R., Grasmeijer, B., Sutherland, J., Pan, S., Sierra, J.P.: The predictability of cross-shore bed evolution of sandy beaches at the time scale of storms and seasons using process-based profile models. Coastal Engineering, 47, 295–327 (2003).

Reactive Materials and Solutions Towards Treatment and Reuse of Waters with Contaminants of Emerging Concern

Adriano Santos Silva, Fernanda F. Roman, Ana P. F. da Silva, Jose L. Diaz de Tuesta, Marzhan Kalmakhanova, D. Snow, and Helder T. Gomes

Abstract Intense population growth has significantly impacted the quality and access to water, resulting in increased production and release of contaminants of emerging concern (CECs), such as pharmaceutical compounds. Catalytic wet peroxide oxidation (CWPO) is a promising technology for the removal of CECs that relies on the use of solid catalysts to accelerate the reaction, its reactivity and stability depending greatly on the catalyst used. This work aims to compare three typically studied catalysts: a clay-based, a carbon-based and a hybrid material, consisting in carbon-shell metal nanoparticle structure. Hybrid catalysts combine the benefits of metal-based catalysts (high activity) and carbon-based catalysts (low leaching), indicating to be a suitable choice. However, it is highlighted that the development of proper solutions for treatment and reuse of waters must pass through detailed identification and quantification of CECs, allowing better catalyst evaluations under real scenario conditions.

Keywords Water pollution · Remediation technologies · Heterogeneous catalysis · Fenton-like systems · Advanced oxidation processes

A. Santos Silva · F. F. Roman · A. P. F. da Silva · J. L. Diaz de Tuesta · H. T. Gomes (✉)
Centro de Investigação de Montanha (CIMO), Instituto Politécnico de Bragança, Bragança, Portugal
e-mail: htgomes@ipb.pt

Laboratório Associado Para a Sustentabilidade E Tecnologia Em Regiões de Montanha, (SusTEC), Instituto Politécnico de Bragança, Bragança, Portugal

J. L. Diaz de Tuesta
Department of Chemical and Environmental Technology, ESCET, Rey Juan Carlos University, Madrid, Spain

M. Kalmakhanova
Department of Chemistry and Chemical Engineering, M.Kh. Dulaty Taraz Regional University, Taraz, Kazakhstan

D. Snow
School of Natural Resources and Nebraska Water Center, part of the Robert B. Daugherty Water for Food Global Institute, University of Nebraska, Lincoln, USA

J. R. da. C. S. Galvão et al. (eds.), *Proceedings of the 3rd International Conference on Water Energy Food and Sustainability (ICoWEFS 2023)*, Springer Proceedings in Earth and Environmental Sciences, https://doi.org/10.1007/978-3-031-48532-9_44

1 Introduction

Intense population growth brought the need of solutions to various environmental problems directly related to the fast economic development over the years. The growing population also increased the demand for freshwater, a resource deeply affected by human activity [1]. Water is the most important resource for human life, but a significant population share still has no access to drinking water worldwide (*ca.* 34%) [2]. The number of people with access to drinking water has registered an increase since 2015, but it is not enough to achieve universal access to this resource by 2030, as targeted by the 17 goals of the UN agenda [3]. In this scenario, water reservoirs are under pressure due to contamination problems, soil erosion and inadequate sanitation [4]. Pathogens, detergents, chemicals, fertilizers, heavy metals and pharmaceuticals are examples of pollutants commonly found in water bodies, being a matter of public concern. In recent years, organic pollutants at trace levels have been identified in the environment, including in effluent streams of wastewater treatment plants, presenting new challenges for public authorities and environmental agencies [5]. This class of pollutants, known as contaminants of emerging concern (CECs), has been the focus of many studies [6].

The biggest problem related to the ubiquitous presence of CECs in water bodies is the lack of information on the effects caused by the presence of these compounds in water environments. Despite not being completely understood, some works reported that CECs presence in water directly affects human health and well-being. Damage to aquatic ecosystems and habitats were also reported in some studies [7]. This specific class of pollutants raises concern due to the expectation that pollution is likely to increase in the following years, caused by the predicted economic development. Aligned with the increasing pollution, there is a lack of evolution of mechanisms and regulations to address this environmental problem [8]. In this context, CECs pollution has become a matter of global concern for the water and wastewater treatment industry, public authorities, and society in general. Among CECs, pharmaceuticals have been a matter of special attention in the last years due to the proven negative effect these compounds have on aquatic life [9].

Several studies reported treatment options for water and wastewater contaminated with CECs, including adsorption, membrane filtration and electrochemical oxidation [10]. In addition, advanced oxidation processes (AOPs) have received great attention as suitable tertiary treatment alternative processes [11]. In recent years, studies exploring AOPs to treat water are focused on developing proper catalysts to promote robust oxidation power processes [12]. The mechanism behind AOPs action is related to the formation of highly oxidizing reactive oxygen species, such as hydroxyl radicals ($HO^{\cdot}$), capable of breaking down the contaminants into non-toxic simpler molecules. Among AOPs, catalytic wet peroxide oxidation (CWPO) has the potential to transform and remove organic pollutants from wastewater [13]. CWPO typically operates at mild conditions (25–130 °C and 1–4 bar), and a solid catalyst is used to promote decomposition of hydrogen peroxide into hydroxyl and hydroperoxyl radicals, which are highly oxidizing species [14]. The solid catalyst can be recovered and reused, which has been the main challenge for the validation of the feasibility of this technology. For instance, early studies demonstrated that a suitable catalyst for this process should be comprised of a support material (high surface area and mesoporous nature) and an active phase (transition metal) [12]. However, the need to solve iron leaching issues has led to new findings, showing that carbonaceous materials have a high potential to be used as catalysts due to their redox properties, allowing them to interact with hydrogen peroxide to form the oxidative radicals [15, 16].

In this work, a comparison between the performance of different catalysts for the CWPO of paracetamol (PCM), chosen as a model pharmaceutical pollutant, will be shown. The catalysts used for this purpose are examples of typical materials (pillared clays), carbon materials (carbon nanotubes), and hybrid magnetic carbon-based materials (multi-core shell ferrite nanoparticles). The discussion will be presented on the performance of each catalyst, showing weaknesses of the traditional materials compared to the most advanced and recent catalysts used in CWPO. Last, based on the conclusions gathered, future studies will be briefly discussed in the quest to develop reactive materials and alternative solutions for the treatment and reuse of waters containing CECs, to contextualize the importance of the work that has been done so far.

2 Methodology

2.1 Reactants and Materials

Natural clay was supplied from Kazakhstan. Iron (III) chloride hexahydrate (99%, Aldrich) and cobalt (II) chloride hexahydrate (99%, Fischer Chemical) were used for the preparation of pillared clay. Alumina (99%, BASF), ethanol absolute (96%, Fischer Chemical), ethylene glycol (99%, Fischer Chemical), iron (II) chloride tetrahydrate (99%, Acros organics), iron (III) chloride hexahydrate (99%, VWR chemicals) and polypropylene (average Mw ~ 250,000, average Mn ~ 67,000, Sigma-Aldrich) were used to synthesize carbon nanotubes. Iron (III) chloride hexahydrate (99%, Aldrich), cobalt (II) chloride hexahydrate (99%, Fischer Chemical), ethanediol (99.8%, Fischer Chemical), formaldehyde (38% w/w), resorcinol (99%, Alfa Aesar), tetraethyl orthosilicate (TEOS) (98%, Fluka Chemika), sodium hydroxide (98%, Labkem) and ammonia solution (28–30%, Merck) were used to prepare the carbon-coated magnetic nanoparticles. Paracetamol (98%, Alfa Aesar), hydrogen peroxide (30% m/v, Fischer Chemical), titanium (IV) oxysulfate (99.99%, Aldrich), sulfuric acid (98%, Labkem), sodium sulfite anhydrous (98%, Panreac), acetonitrile (99.9%, Fischer Chemical), and ortho-phosphoric acid (85%, Riedel-de-Haen) were used for oxidation experiments.

2.2 Preparation of Catalysts and Characterization

Pillared clay (PCLAY) was prepared by mixing a pillaring solution with natural clay (2 wt%) previously washed with phosphate buffer solution to remove impurities. The pillaring solution was prepared by dropwise addition of 0.5 M NaOH solution, at room temperature, to an aqueous solution of 0.5 M $FeCl_3$ and 0.25 M $CoCl_2$ to obtain a final solution with a molar ratio OH/(Fe + Co) = 2:1. The suspension was stirred at room temperature for 3 h, and left aging for another 72 h (no stirring). In the end, the final solid was recovered by filtration and washed until rinsing waters reached pH 7, dried overnight at 60 °C, and calcined at 600 °C for 5 h, leading to PCLAY sample.

Carbon nanotubes (CNTs) were prepared by chemical vapor deposition using polypropylene (PP) as a carbon source, and iron oxide supported on alumina (IO/Al_2O_3) as substrate. The metal substrate was prepared by adapting a sol–gel procedure described elsewhere [17, 18]. In brief, a $FeCl_2{\cdot}4H_2O$ solution in ethanol and a $FeCl_3{\cdot}6H_2O$ solution in ethanediol were mixed (molar ratio 1:2) with 6.6 g of alumina. The amount of reactants was adjusted to obtain 20% iron oxide supported on alumina. The final powder was calcined at 300 °C for 12 h and 600 °C for 24 h. The CVD process was carried out in a one-chamber reactor under N_2 flow (50 mL min^{-1}) considering 1 g of catalyst to 5 g of carbon source. The synthesis was conducted by heating at 850 °C for 1 h. In the end, the CNTs were recovered and washed with a 50% vol./vol. H_2SO_4 solution at 140 °C for 3 h to remove metal impurities, leading to a PP-derived CNT (CNT@PP sample).

Hybrid carbon-coated magnetic nanoparticles were prepared in a procedure divided into 4 steps: (I) synthesis of magnetic core, (II) carbon coating, (III) annealing, and (IV) etching. The synthesis of cobalt ferrite was performed by sol–gel, as described in other works [19]. Briefly, $CoCl_2.6H_2O$ solution in ethanol and $FeCl_3.6H_2O$ solution in ethanediol were mixed (molar ratio = 1:2) and stirred at 60 °C for 2 h. The resulting solution was stirred and the temperature was raised to 189 °C to remove all the solvent from the solid. The solid obtained from this step went through heat treatment at 300 °C for 24 h and 600 °C for 12 h in an oxidative atmosphere. The recovered material was washed to remove impurities and dried overnight in oven at 60 °C, leading to CoFe. For the carbon coating, the procedure involved the hydrolysis reaction of TEOS in the presence of ammonia and the polymerization reaction between resorcinol and formaldehyde. In brief, 0.25 g of nanoparticles were sonicated in 50 mL of water and transferred to a round bottom flask using 150 mL of ethanol. The flask was previously loaded with 0.1 g of resorcinol and 1.2 mL of ammonia solution. The mixture was stirred at 30 °C for 1 h when 0.15 mL of formaldehyde and 0.21 mL of TEOS were added to the flask. After 6 h of mixing, the temperature was raised to 80 °C, and the mixture was kept stirring for another 8 h. In the end, the nanoparticles coated with resin were washed several times until rinsing waters reached neutral pH. The last 2 steps involved the thermal treatment in N_2 atmosphere (1 h at 120 and 400 °C, followed by 3 h at 600 °C) and etching with NaOH 10 M solution for 16 h to remove the silica formed by TEOS hydrolysis, leading to production of CoFe@C.

Physisorption with liquid nitrogen and Fourier Transform Infra-Red (FT-IR) spectroscopy were performed to characterize the materials, as described in other works [12]. In brief, textural properties were gathered by analysis of N_2 isotherms at 77 K obtained in QuantaChrome NOVA TOUCH LX[4], equipped with long cells with a bulb and outer diameter of 9 mm. Prior to run N_2 adsorption analysis, samples were degasified at 120 °C during 16 h at vacuum. Specific surface area was determined by BET method and total pore volume at 0.98 of relative pressure. FT-IR analysis was carried out using a Perkin Elmer Spectrum Two FT-IR spectrometer at a resolution of 4 cm^{-1} and a scan range of 4000 to 450 cm^{-1}.

2.3 Liquid-Phase Experiments

Batch oxidation experiments were conducted in a well-stirred 250 mL round bottom flask equipped with a condenser and temperature controller. The initial concentration of PCM of 100 mg L^{-1} was used to evaluate treatment capacity of an organic pollutant in water. For the reaction, 100 mL of the model solution with pH adjusted to 3.5 using 1 M H_2SO_4 was loaded in the reactor, followed by the addition of the stoichiometric amount of H_2O_2 for complete mineralization of PCM. Once the temperature reached 80 °C, the catalyst was loaded, marking the beginning of the reaction (t_0). Non-catalytic and adsorption runs were performed under the same operating conditions in the absence of catalyst and H_2O_2, respectively.

PCM and H_2O_2 concentrations were monitored by periodically withdrawing aliquots from the reactor. H_2O_2 was determined based on a colorimetric method described in previous works [20], and PCM concentration was measured by a Jasco HPLC system at a wavelength of 277 nm (UV-2075 Plus detector) coupled with a Nucleosil 100-5C18 (15 cm × 2.1 mm id) column. The separation of the reaction products was obtained under an eluent flow of 0.300 mL min^{-1}, consisting of a mixture of ultrapure water with 0.15 wt%. of phosphoric acid (A) and acetonitrile (B) (A:B volume ratio = 90:10 v/v) for 6 min, followed by a gradient elution to reach A:B volume ratio of 65:35 up to 12 min (hold for 6 min). Leached iron was determined by atomic absorption spectroscopy (Varian SpectrAA 220) at the end of reaction.

3 Results and Discussion

3.1 Characterization

FT-IR spectra is shown in Fig. 1. The band observed at 780 cm^{-1} for PCLAY is related to quartz impurity in the material, commonly observed in materials originating from natural clays [12]. The band observed in 1025 cm^{-1} for PCLAY is ascribed to stretching vibrations of the Si–O group. For CNT, bands at 1636 and 1385 cm^{-1} can be attributed to C = C and carboxylate groups, respectively [21]. The band found in 3440 cm^{-1} is due to the presence of -OH groups of the water adsorbed in the CNT. For the CoFe@C nanoparticles, the band in 1632 cm^{-1} can be attributed to the bending vibration of the OH groups in the carbon-based shell. At last, both CoFe@C and PCLAY materials present a response around 450–500 cm^{-1}, ascribed to the presence of transition metals in their structure. These results confirm the majority carbon composition of CNTs, the metal-based composition for PCLAY and the hybrid nature of carbon-coated magnetic nanoparticles.

Regarding the textural properties (Table 1), it is possible to observe that PCLAY is the catalyst with the lowest BET-surface area (S_{BET}), which is related to the nature of the raw material used to prepare PCLAY. The CoFe@C catalyst has a lower surface area compared to CNT@PP, however, the hybrid nanoparticle possesses a higher surface area compared to the bare core (result not shown here). The same tendency observed for S_{BET} is repeated for the total pore volume (V_T) of the catalysts. The N_2 isotherms classification for the materials is typical of catalysts composed of mesopores for CNTs and CoFe@C, and macropores for PCLAY.

Fig. 1 FTIR results

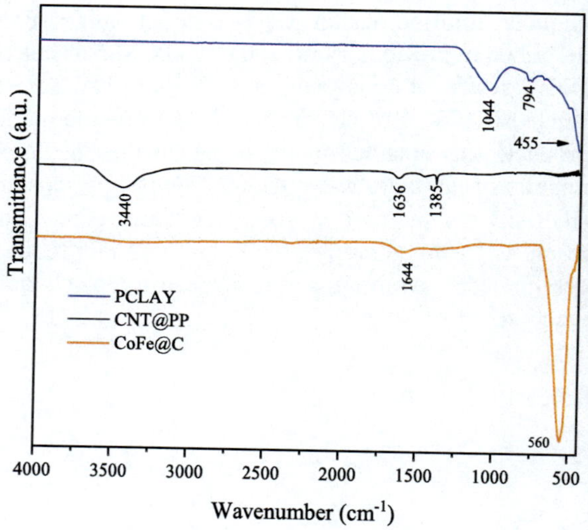

Table 1 Textural properties

Catalyst	S_{BET} (m^2 g^{-1})	V_T (mm^3 g^{-1})
PCLAY	19	72
CoFe@C	57	60
CNT@PP	94	247

3.2 Liquid-Phase Reactions and Insights

The results obtained for PCM and H_2O_2 concentrations upon reaction time in the CWPO of PCM are shown in Fig. 2. All materials display high catalytic activity, allowing to achieve more than 95% of PCM removal in 8 h of reaction. The catalytic activity towards H_2O_2 decomposition and PCM oxidation follow the same order: CoFe@C, followed by PCLAY and then by CNT@PP. PCM removal profile for CoFe@C is sharper compared to the other catalyst. For CoFe@C, the fast H_2O_2 decomposition observed in the beginning of the reaction slowed the performance, likely by allowing radical recombination, resulting into non-active species [22]. On the other hand, CNT@PP did not result in complete removal of PCM due to its inability to fully decompose H_2O_2.

Characterization results showed the presence of iron for CoFe@C and PCLAY, whereas CNT@PP did not display the presence of metal-based species, according to FT-IR spectra. This indicates that transition metals play a significant role in this process. Most likely, the iron present in those materials is the main characteristic in influencing the catalytic activity in a CWPO process. However, carbon-coated materials are preferred compared to PCLAY due to easy recovery and higher stability of the active phase arising from the protection of the core due to the carbon layer. Textural properties revealed a behavior previously reported in literature: higher surface areas do not necessarily lead to increased activity of catalyst in CWPO, as other parameters may be more relevant, such as the presence of metallic phases or carbon content. Tuning the metallic content may affect the stability of the catalysts, as increased leaching is expected, resulting in water pollution and loss of catalytic activity. On the

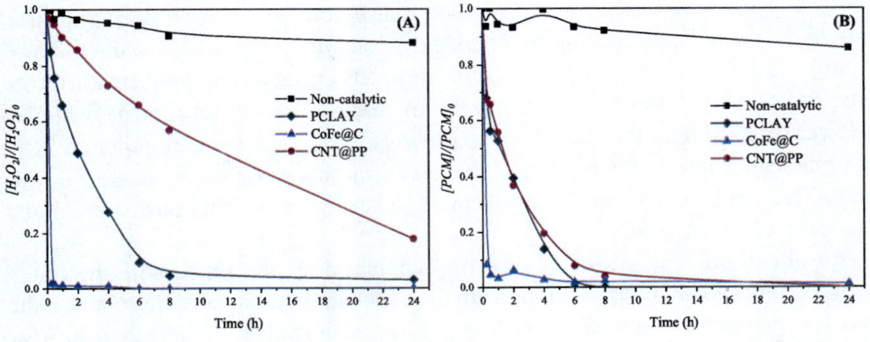

Fig. 2 Results of H_2O_2 **a** and PCM **b** concentrations through time for oxidation reactions

other hand, carbon-based materials do not have such high catalytic activity compared to metal-based ones, so surface modifications (e.g., doping) are required to increase the activity [16, 20].

The activity of PCLAY in the process is well-known and studied in the literature [12]. In brief, introducing transition metals in the clay structure during the pillaring process is the key feature ascribing activity for this material in H_2O_2 decomposition into hydroxyl radicals. The advantage of this material is related to its low-cost nature, which includes the use of a widely available resource (clay) with a low-cost modification process that can be performed at room temperature, not requiring energy to be carried out. However, the use of such material for liquid-phase reactions is complicated due to its hard recovery, which hinders the reutilization of the catalyst for other reactions. The reutilization feature is an important characteristic in CWPO studies because it is considered one advantage of this process compared to homogeneous Fenton reactions.

Carbon materials were already studied as catalysts for removing other organic pollutants by CWPO [23]. The catalytic activity of these materials for CWPO process is due to the electron donor–acceptor properties of the carbon structure, allowing the electron-transfer mechanism between the catalyst and H_2O_2. The study of non-supported carbon materials for CWPO is more recent than traditional metal-supported catalysts, and its application brings considerable advantages. For instance, non-supported carbon materials do not face iron leaching problems. Therefore, these catalysts do not face inactivation issues. It is important to highlight that carbon materials are stable for this application due to the mild temperature in which the process is carried out, which is the main issue related to the application of carbon catalysts in other fields. Herein, the CNT used to promote PCM degradation has yet one advantage compared to traditional carbon materials: the low iron content that cannot be removed from CNT (< 1%) confers magnetic properties to the nanostructured carbon-based material. This characteristic is useful for removing the magnetic CNT catalyst from the reaction medium, which increases the feasibility of using the same material several times as a catalyst for oxidation reactions.

Hybrid materials comprised of carbon and metal, such as carbon-coated magnetic nanoparticles, represent another material with potential applicability in CWPO [24]. These catalysts have several advantages, since the coating is responsible for protecting the magnetic core and promote the hydrogen peroxide decomposition into hydroxyl radicals, leading to increase the catalytic activity toward the removal of organic pollutants by CWPO. In this work, for example, the bare cobalt ferrite core can remove 43% of pollutants in 8 h of reaction (result not shown), which is a lower removal than the coated nanoparticle (> 95%). For the case reported here, the innovation lies in the synthesis method of the magnetic core, sol–gel, in this case. This method is known to result in nanoparticles with high purity, yield and monodispersibility of sizes.

Results of adsorption tests after 6 h of contact time performed with the tested materials are shown in Fig. 3, along with the concentration of iron determined at the end of experiments. Removal of PCM by adsorption did not reach more than 30% for all catalysts, with more expressive adsorption observed for CNT@PP due to its higher

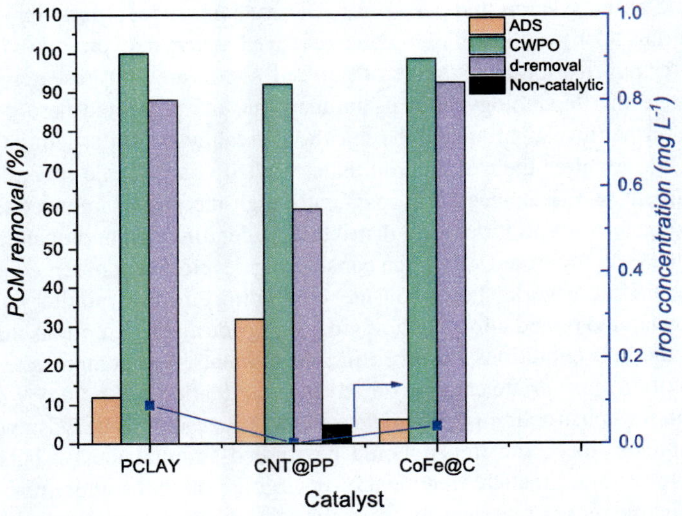

Fig. 3 PCM removal by CWPO and adsorption (left axis) and iron concentration at the end of the CWPO reaction (right axis)

surface area. The parameter d-removal calculated for all tests represents the difference between removal by CWPO and by adsorption, being useful to assess the catalytic efficiency of each material. For instance, according to this parameter, CoFe@C and PCLAY remove PCM with the highest efficiency compared to CNT@PP. However, CNT@PP is the catalyst with the lowest iron leached at the end of the reaction, which is related to the lower amount of iron in this material. Iron leaching for PCLAY was higher than CoFe@C, which could be expected due to the nature of the CoFe@C catalyst, where the carbon layer protects the magnetic core against leaching.

4 Final Remarks and Prospects

All materials used in this work showed catalytic activity towards the degradation of PCM, chosen as model CEC, by CWPO. Each catalyst was assessed in equal operating conditions and chosen to illustrate a specific case from literature. The work reported here shows the increased efficiency and feasibility of carbon-based magnetic nanoparticles over traditional metal-based catalysts (represented by PCLAY) and non-supported carbon catalysts (represented by CNT@PP). Nevertheless, carbon catalysts are still preferred over traditional metal-based materials due to iron leaching problems faced by the second class. All reactions carried out in this study were made in batch experiments, which is a useful tool for evaluating catalysts' performance. Still, continuous experiments should be considered in future works to evaluate the feasibility of the wastewater treatment studied here in real systems, including membrane-based systems and permeable reactive barriers incorporating novel reactive materials. Construction of permeable reactive barriers may provide a large-scale process for purification and storage of purified wastewater, but research needed to demonstrate this technology is in its infancy. Results presented here provide the basis for further investigation and the selection of catalysts for upscaling CWPO (or other AOPs) requires their evaluation under realistic, large-scale scenarios. Effective treatment of CECs necessarily passes through the proper knowledge of types of compounds, concentrations, and distribution. Identification and quantification of those CECs is an important step when considering development of remediation technologies, and future works should consider combining efforts to monitor pollutants in real wastewater to remediation technologies, aiming designing solutions that approximate to the real conditions observed in real effluents. We demonstrate here that selection of an appropriate catalyst for advanced oxidation technology will provide the means for construction of a complete treatment system for CECs and a scientific symbiosis since both fields should be studied simultaneously. This research show promise for a practical treatment of emerging pollutants under real scenarios and sustainable reuse of wastewater through both membrane and permeable reactive barrier systems.

Acknowledgements This work was financially supported by CIMO (UIDB/00690/2020 through FEDER under Program PT2020, by the Science Committee of the Ministry of Education and Science of the Republic of Kazakhstan (Grant No. AP13067715), project PLASTIC_TO_FUEL&MAT (PTDC/EQU-EQU/31439/2017) and project Strengthening Portuguese Research and Academic Water Sustainability Programs (2022/0148-G-2022-0239) by FLAD/UP Ensino Superior 2022-2023. Fernanda F. Roman acknowledges the Foundation for Science and Technology (FCT) and the European Social Fund (FSE) for the individual research grant with reference SFRH/BD/143224/2019. Adriano S. Silva and Ana Paula Silva were supported by the doctoral Grant SFRH/BD/151346/2021and PRT/BD/153090/2021 financed by FCT with funds from NORTE2020, under MIT Portugal Program. Jose L. Diaz De Tuesta acknowledges the financial support through the program of *Atracción al Talento* of *Comunidad de Madrid* (Spain) for the individual research grants 2020-T2/AMB-19836 and 2022-T1/AMB-23946.

References

1. M. J. M. De Vidales, E. Palomo De La Fuente, E. Atanes-Sánchez, and F. Fernández-Martínez, "New compact multi option photo reactor for the removal of contaminants of emerging concern from wastewater," *J. Environ. Chem. Eng.*, vol. 10, no. 3, 2022.
2. I. M. Alexe and S. Toma, "Learning about the Importance of Water for Life through Informatics Products," *Procedia - Soc. Behav. Sci.*, vol. 76, pp. 66–70, 2013.
3. H. Ritchie and M. Roser, "Clean Water and Sanitation," *Our World Data*, 2021.
4. I. V Muralikrishna and V. Manickam, "Chapter One - Introduction," in *Env. Man.* I. V Muralikrishna and V. Manickam, Eds. Butterworth-Heinemann, 2017, pp. 1–4.
5. P. Chakraborty, N. W. Shappell, M. Mukhopadhyay, S. Onanong, K. R. Rex, and D. Snow, "Surveillance of plasticizers, bisphenol A, steroids and caffeine in surface water of River Ganga and Sundarban wetland along the Bay of Bengal: occurrence, sources, estrogenicity screening and ecotoxicological risk assessment," *Water Res.*, vol. 190, p. 116668, 2021.

6. Al-Mashaqbeh, Alsafadi, Dalahmeh, Bartelt-Hunt, and Snow, "Removal of Selected Pharmaceuticals and Personal Care Products in Wastewater Treatment Plant in Jordan," *Water*, vol. 11, no. 10, p. 2004, Sep. 2019.
7. K. N. Kim, H. R. Jung, and W. J. Lee, "Hollow cobalt ferrite-polyaniline nanofibers as magnetically separable visible-light photocatalyst for photodegradation of methyl orange," *J. Photochem. Photobiol. A Chem.*, vol. 321, pp. 257–265, 2016.
8. J. C. Espíndola and V. J. P. Vilar, "Innovative light-driven chemical/catalytic reactors towards contaminants of emerging concern mitigation: A review," *Chem. Eng. J.*, vol. 394, no. August 2019, p. 124865, 2020.
9. C. Chokejaroenrat *et al.*, "Enrofloxacin and Sulfamethoxazole Sorption on Carbonized Leonardite: Kinetics, Isotherms, Influential Effects, and Antibacterial Activity toward S. aureus ATCC 25923," *Antibiotics*, vol. 11, no. 9, pp. 1–20, 2022.
10. R. Sivaranjanee and P. S. Kumar, "A review on remedial measures for effective separation of emerging contaminants from wastewater," *Environ. Technol. Innov.*, 23, p. 101741, 2021.
11. S. Ryskie, C. M. Neculita, E. Rosa, L. Coudert, and P. Couture, "Active treatment of contaminants of emerging concern in cold mine water using advanced oxidation and membrane-related processes: A review," *Minerals*, vol. 11, no. 3, pp. 1–23, 2021.
12. A. S. Silva, M. S. Kalmakhanova, B. K. Massalimova, J. L. D. de Tuesta, and H. T. Gomes, "Wet peroxide oxidation of paracetamol using acid activated and Fe/Co-pillared clay catalysts prepared from natural clays," *Catalysts*, vol. 9, p. 705, 2019.
13. R. S. Ribeiro, J. Gallo, M. Bañobre-López, A. M. T. Silva, J. L. Faria, and H. T. Gomes, "Enhanced performance of cobalt ferrite encapsulated in graphitic shell by means of AC magnetically activated catalytic wet peroxide oxidation of 4-nitrophenol," *Chem. Eng. J.*, vol. 376, no. 2019, p. 120012.
14. J. L. Diaz de Tuesta, A. Quintanilla, D. Moreno, V. R. Ferro, and J. A. Casas, "Simulation and optimization of the CWPO process by combination of aspen plus and 6-factor doehlert matrix: Towards autothermal operation," *Catalysts*, vol. 10, no. 5, pp. 548, 2020.
15. M. T. Pinho, H. T. Gomes, R. S. Ribeiro, J. L. Faria, and A. M. T. Silva, "Carbon nanotubes as catalysts for catalytic wet peroxide oxidation of highly concentrated phenol solutions: Towards process intensification," *Appl. Catal. B Environ.*, vol. 165, pp. 706–714, 2015.
16. J. L. Diaz de Tuesta, A. Quintanilla, J. A. Casas, and J.J. Rodriguez, "P-, B- and N-doped carbon black for the catalytic wet peroxide oxidation of phenol: Activity, stability and kinetic studies" *Catalysis Communications*, vol. 102, no. 5, pp. 131–135, 2017.
17. F. Roman *et al.*"Selective denitrification of simulated oily wastewater by oxidation using Janus-structured carbon nanotubes" *Catalysis Today, 2023.*
18. R.S. Riberio, O. Vieira, R. Fernandes, F. Roman, J. L. Diaz de Tuesta, A.M.T.Silva, H.T. Gomes, "Synthesis of low-density polyethylene derived carbon nanotubes for activation of persulfate and degradation of water organic micropollutants in continuous mode" *Journal of Environmental Management*, vol. 308, pp. 114622, 2022.
19. N. M. C. Guari, A. S. Silva, J. L. Diaz de Tuesta, W. E. Pottker, P. Y. Cordeiro, and H. T. Gomes, "Magnetic $CoFe_2O_4$@carbon yolk-shell nanoparticles as catalysts for the catalytic wet peroxide oxidation of paracetamol: kinetic insights," 2022.
20. J. L. Diaz de Tuesta *et al.*, "The pH effect on the kinetics of 4-nitrophenol removal by CWPO with doped carbon black catalysts," *Catal. Today*, vol. 356, pp. 216–225, 2020.
21. Y. Huaccallo-Aguilar *et al.*, "Optimization Parameters, Kinetics and Mechanism of Naproxen Removal by Catalytic Wet Peroxide Oxidation with a Hybrid Iron-Based Magnetic Catalyst," *Catalysts*, vol. 9, no. 3, p. 287, 2019.
22. R. S. Ribeiro, A. M. T. Silva, J. L. Figueiredo, J. L. Faria, and H. T. Gomes, "Catalytic wet peroxide oxidation: a route towards the application of hybrid magnetic carbon nanocomposites for the degradation of organic pollutants. A review," *Appl. Catal. B Environ.*, vol. 187, pp. 428–460, 2016.
23. M. Martin-Martinez *et al.*, "Carbon nanotubes as catalysts for wet peroxide oxidation: The effect of surface chemistry," *Catal. Today*, vol. 357, no. October 2018, pp. 332–340, 2020.

24. R. S. Ribeiro, A. M. T. Silva, P. B. Tavares, J. L. Figueiredo, J. L. Faria, and H. T. Gomes, "Hybrid magnetic graphitic nanocomposites for catalytic wet peroxide oxidation applications," *Catal. Today*, vol. 280, pp. 184–191, 2017.

Strategic Environmental Assessment and Environmental Report Proposal for Detailed Plans: The Case Study of Vale Da Cabrita in Leiria

Ana Margarida Morais, Luísa M. S. Gonçalves, and Lizete Heleno

Abstract Human activity has a direct or indirect impact on the environment and, therefore, a protection policy focused on prevention is crucial and implies a prior assessment of environmental impacts, before taking decisions. Thus, minimizing damage to the environment by weighing ecological consequences/impacts, for example, of master plans, or large-scale urban planning projects or considering environmental values within the scope of the decision-making process, is fundamental. This procedure constitutes a strategic instrument for sustainability which was given the designation of Strategic Environmental Assessment (SEA). In the present work, a methodology developed to obtain the Strategic Environmental Assessment and the Environmental Report of a detailed urban plan is presented. The goal is to contribute to the integration of environmental sustainability considerations in the development of a detailed urban plan, to detect and enhance opportunities, minimize strategic risks, encourage stakeholders' involvement and public discussion in decision-making processes, and contribute to sustainable strategic planning. The methodology was applied to the Detailed Urban Plan of Vale da Cabrita in Leiria used as a case study. In conclusion, the detailed urban plan developed induces, in general, significant positive impacts in the study area. Planning and management guidelines were proposed that will allow a reduction/mitigation of the potential negative impacts and associated risks, as well as a governance framework for this purpose. The work also reveals the need for deeper integration of certain aspects of sustainable development, in order to positively influence local territorial strategic development.

Keywords Strategic environmental assessment · Territory management instruments · Environmental report

A. M. Morais (✉)
Municipality of Leiria, Leiria, Portugal
e-mail: anamargaridamorais@gmail.com

L. M. S. Gonçalves
INESC Coimbra, ESTG, Polytechnic Institute of Leiria, Nova IMS University, Lisbon, Portugal

L. Heleno
School of Technology and Management, Polytechnic Institute of Leiria, Leiria, Portugal

J. R. da. C. S. Galvão et al. (eds.), *Proceedings of the 3rd International Conference on Water Energy Food and Sustainability (ICoWEFS 2023)*, Springer Proceedings in Earth and Environmental Sciences, https://doi.org/10.1007/978-3-031-48532-9_45

1 Introduction

The Strategic Environmental Assessment (SEA) is mentioned for the first time by Wood and Djeddour, in 1989, in the context of a European research project as an "environmental assessment adequate to policies, plans, and programs with a more strategic nature [1]. This evolution intended to answer to the need to integrate sustainability considerations and strategic decisions at a planning level, since only the Environmental Impact Assessment (EIA) figure existed, formally developed in the US National Environmental Policy Act of 1969.

In the case of urban plans, SEA should contribute: (i) to the integration of environmental and sustainable considerations (e.g. including economic, social, and governance issues); (ii) in their development and approval, in order to detect and enhance opportunities and minimize strategic risks; (iii) to encourage public participation and discussion in decision-making processes; (iv) to encourage change in mentalities and (v) contribute to sustainable strategic planning [2].

There is no universal approach to SEA, and SEA itself has been the subject of diverse interpretations [3, 4]. Currently, it is understood as an additional instrument of a holistic response to the main challenges of development.

In accordance with Portuguese regulation, Decree-Law No. 232/2007, of 15 June, which transposes Directive 2001/42/EC of the European Parliament, as well as the European Council of 27 June 2001, related to the assessment of the effects of certain plans and programs on the environment, SEA consists on the "…identification, description, and assessment of any significant impacts on the environment resulting from a plan or program, carried out during the procedure for their development and drawing and before it is approval or submission to legislative procedure…".

In the specific case of a detailed urban plan, for the purposes of assessing submission to SEA, their impacts must be considered, cumulatively, in accordance with the legal provisions of the Legal Regime of Territorial Management Instruments (RJIGT), approved by Decree-Law No. 80/2015, of 14 May, amended.

The present work consists in developing a proposal for an Environmental Report (ER), carried out within the scope of the SEA process of a Detailed Urban Plan (DUP) of Vale da Cabrita, hereinafter referred to as PPVC, located in the City and Municipality of Leiria, according to Portuguese regulation.

2 Methodological Analysis of SEA

The SEA approach applied to a DUP follows the Portuguese methodological reference guides, which the main stages are shown in Fig. 1. In the work developed only the first two stages (A and B) were applied. For the purposes of a better definition of the methodology to be applied and a better definition of Critical Decision Factors (CDF) and respective criteria and indicators, a careful reading of the specialized bibliography and a comparative examination of the SEA of eight Portuguese master plans, were carried out. Were also analyzed four DUP, as they were more adequate to the scale of this work [2, 5].

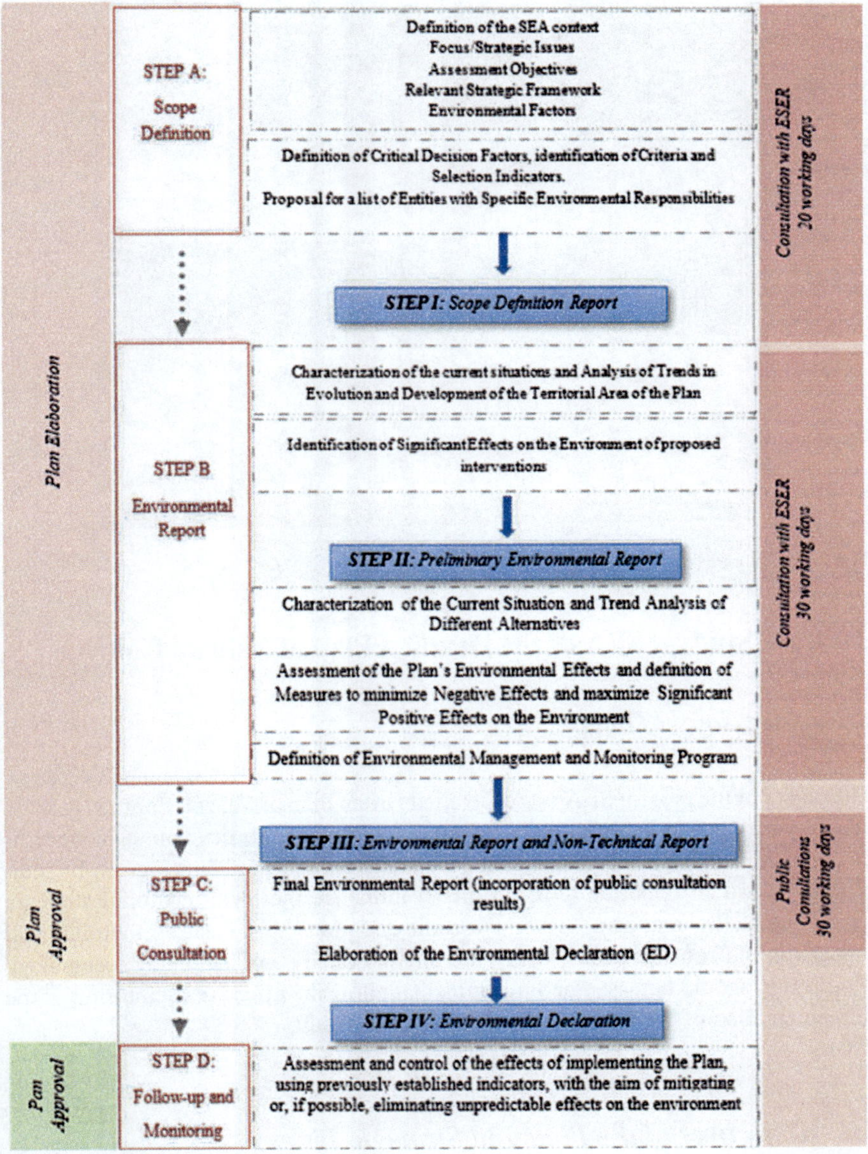

Fig. 1 DUP SEA methodology [2, 5]

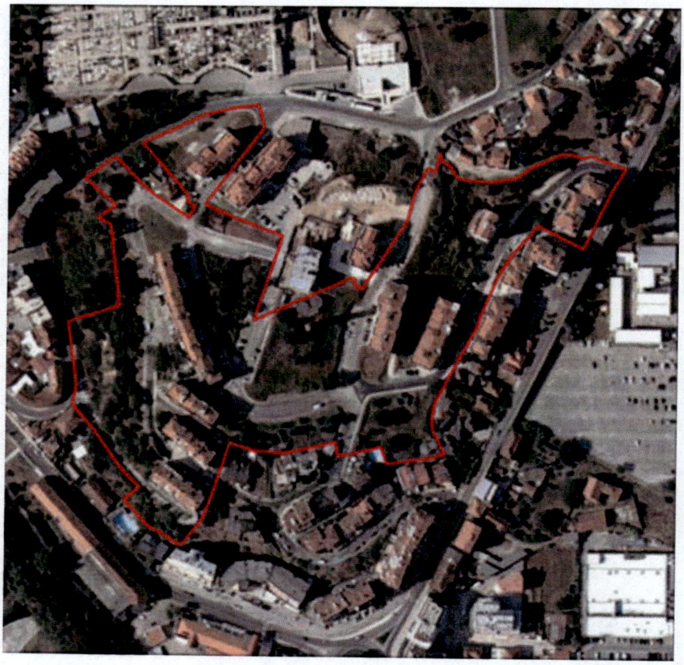

Fig. 2 Study area [6]

3 Case Study—SEA of the Detailed Plan of Vale da Cabrita

3.1 Assessment Object

The object of the present SEA is the Detailed Urban Plan of Vale da Cabrita (PPVC), which is an opportunity for a deeper integration of sustainable development concerns in the options that were established for the study area of the Plan (Fig. 2). In addition, it is also an opportunity to positively influence the local territorial strategic development, by proposing several recommendations that ensure environmental objectives and considerations, which are internationally and nationally recognized. Monitoring should be based on indicators that allow the effective monitoring of the implementation of the PPVC in the field of sustainability.

3.2 Objectives of the Plan and Strategic Lines

The objectives of the PPVC include the urban design and requalification of a built area, aiming do obtain a building framework for the rehabilitation of existing buildings or the construction of new ones in the existing and newly construction plots, on the legal premises of the Master plan of the city. It encompasses, infrastructure to be built or rehabilitated, the definition of building alignments, and the establishment of continuity with the surrounding public space allowing the requalification of the public space for further development of this particular urban area.

3.3 *Critical Decision Factors*

Within the scope of SEA, it is crucial to recognize the key issues that contribute to the sustainability of the strategic decision that will be taken. These key issues are called Critical Decision Factors (CDF).

The identification of CDF results from an integrated analysis of the following elements:

- Strategic Reference Framework (SRF), which establishes the framework for international, national, regional, and local urban plans and programs with an impact on the DUP.
- Strategic issues (SI) of the DUP proposal, which determine the strategic objectives and respective lines of action to which the plan must respond (Table 1).
- Environmental factors (EF), which define the relevant environmental scope, guided by the definition of legally established environmental factors in Decree-Law No. 232/2007, of 15 June.

The CDF considered within the scope of this SEA of the PPCV, intend to focus on the relevant aspects, namely those related to the current situation and those that are relevant within the scope of the SEA for the options of the PPVC: Environmental efficiency and human health (green city); Social and territorial cohesion (fair city); Sustainable economic development (productive city) For each CDF, the respective sustainability objectives and evaluation criteria were defined, which are explained in Table 2.

The rationale for the proposed CDF is further supported by two vectors its articulation with the PPVC strategic objectives and its pertinence to the specificities of the territory. The integrated analysis of these aspects allows the assessment of how achieving the objectives of the PPVC contributes to the valuation of strengths/opportunities and mitigation of the weaknesses/threats identified in the territory, according to each of the CDFs considered.

3.4 *Strategic Assessment and Evaluation of the Effects*

Within the scope of this point, an assessment is made for each of the CDF considering the existing situation and analysis of the main trends, based on the identified criteria and indicators. A SWOT (Strengths, Weaknesses, Opportunities, and Threats) analysis was carried out before the execution of the Plan. The opportunities

Table 1 Strategic issues to be considered and strategic objectives of the PPVC

Strategic issues (SI)	Strategic objectives PPVC
Friendly and attractive Leiria	SO1-Promoting urban morphological balance and coherence, in the sense of promoting the city image, quality of life, and enhancing sociability
	SO2-Improve the urban environment by increasing the municipality's attractiveness and improving the well-being of the population
	SO3-Create urban public space in an integrated way, enhancing its enjoyment in a safe, accessible, and inclusive way
Sustainability and environment	SO4-Promote environmental quality, ensuring efficiency/effectiveness in the use of natural resources, in line with the principles and objectives of sustainable development and minimizing environmental risks
	SO5-Promoting urban quality and land qualification, consolidating a predominantly housing occupation, and improving the public space to support mobility systems, and basic infrastructure, through solutions with low environmental impact
	SO6-Adopt solutions to improve bioclimatic and acoustic comfort, and to adapt urban space to climate change through energy efficiency conditions
Urbanity	SO7-Promoting urbanism of proximity, ensuring availability for the location of different uses: housing, economic activities, equipment, and infrastructure
	SO8-Ensuring a balanced occupation of the territory, aiming at the rehabilitation and consolidation of existing and proposed buildings, in order to respond to the intervention area
	SO9-Ensure that the area in question has spaces with various vocational uses, namely accommodation, connecting spaces, built spaces, open spaces, green spaces, preservation of views, good accessibility, and architectural quality

and risks arising from the implementation of the PPVC are evaluated, and guidelines are subsequently defined for planning and management follow-up.

After the diagnosis made in the previous points, the PPVC proposals were analyzed, considering the CDF1, CDF2 and CDF3 developed.

3.5 *Trend Analysis*

Table 3, summarize the evolution trend of the defined evaluation criteria, respectively for CDF1, CDF2 and, CDF3, considering the implementation or not of the PPVC proposal.

Table 2 CDF: description of the evaluation criteria for the PPVC SEA

CDF1—environmental efficiency and human health (green city)	
Evaluation criteria	General objectives/Scope Description/
C1.1-infrastructural services	Promote the provision of structural public services, essential to the general well-being and public health, with the aim of responding to the needs and expectations of users, ensuring the effective use of natural resources in line with the principles and objectives of sustainable development. Includes urban waste management services, water supply services, urban wastewater management services, and rainwater management
C1.2-climate changes (adaptation)	Minimize the negative effects of the impacts of climate change on biophysical and socioeconomic systems and enhance the positive effects (make the city more resilient and adapted to the impacts of climate change C
C1.3-quality of the sound environment	Promoting the physical and mental well-being of populations by minimizing exposure to noise
C1.4-air quality	Promote environmental air quality and protect human health
CDF2—social and territorial cohesion (fair city)	
C2.1-urban morphological balance and coherence	Restore the urban values that have been built and those that are yet to be built, respecting the parameters and zoning defined by the Municipal Master Plan/PPVC in force
C2.2-qualification of the public space	To give urban coherence to the territory of extended occupation for the promotion of the image, the quality of life, and the potentiation of sociability
C2.3-climate change (mitigation) and sustainable mobility	Reduce the emission of Greenhouse Gases (GHG) into the atmosphere through sustainable mobility
C2.4-accessibility	Foster the construction of inclusive, participatory societies that promote collective life and that promote connectivity, and social inclusion
CDF3—sustainable economic development (productive city)	
C3.1-quality of life of the population and urban living	Take advantage of urbanization for structural transformation, high productivity, value-added activities, leveraging local economies, and improving urban metabolism
C3.2-energy efficiency	Ensuring the energy transition and promoting the decarbonization of society with a view to carbon neutrality

3.6 *Guidelines Related to Opportunities and Risks*

Considering the strategic objectives (SO) and lines of action of the PPVC, an analysis was made of the effects of the plan's proposals, both positive/opportunities and negative/threats, in line with Decree-Law no. 232/2007, namely, taking into account its nature (positive, negative), occurrence (short, medium and long term), effect (secondary, cumulative and synergistic) and duration (temporary or permanent).

Table 3 Analysis of evolution trends for CDF1, CDF2, and CDF3

Rating Criteria		Current Situation	Evolution trend	
			Without PPVC	With PPVC
CDF1	Infrastructure Services		↓	↑
	Climate Change (Adaptation)		↓	↑
	Sound Ambient Quality		↔	↑↑
	Air quality		↓	↑
CDF2	Urban Morphological Balance and Coherence		↓	↑
	Public Space Qualification		↓	↑↑
	Climate Change (Mitigation)		↔	↑
	Accessibility		↓	↑
CDF3	Quality of Life of the Population and Urban Living		↓	↑
	Energy Efficiency		↓	↔

Legend:

Distance to the desirable situation (SRF guidelines)

Very far away	Far	Close	Very close

Evolution trend

Very negative ↓↓	Negative ↓	No change ↔	Positive ↑	Very positive ↑↑

3.7 *Follow-Up Guidelines/Recommendations*

From the analysis carried out and considering that the PPVC proposal represents an improvement in terms of most of the defined criteria, it is advised, however, to consider some recommendations (planning and management guidelines) for each CDF, to stimulate potential opportunities, arising from its implementation.

CDF1—Environmental Efficiency and Human Health

- Incorporate climate change adaptation measures into the PPVC Regulation and into the occupation solutions provided for the Plan, to make the territory more resilient to the impacts associated with their effects.
- Include ecological solutions in the infrastructure to be built or rehabilitated.

CDF2—Social and Territorial Cohesion

- Evaluation of the possibility of increasing the green space area ("green zones" in the PPVC proposal) and parking, through the transfer of existing municipal domain areas, without jeopardizing urban commitments assumed by the owners of the private lots.
- Improvement of the urban design to express a better definition of public spaces for road and pedestrian circulation, parking, layout, dimensioning, and implementation of urban furniture.

- Improvement of sidewalks, which promotes pedestrian movements, in order to create a safe, attractive, comfortable, legible, continuous network, respecting the principles of universal accessibility.
- Analysis and evaluation of the public transport supply network in the intervention area.
- Detail the road network in the designed parts of the PPVC and integrate aspects related to road traffic, accessibility, signaling, and safety.
- Elaborate a landscape architecture project to join the pieces of the PPVC proposal, which would materialize all the concerns related to the framework and landscape enhancement of the site, as well as those related to the need to increase the resilience of the urban space to the impacts of climate change contribute to revert loss of biodiversity and degradation of ecosystems and restoration of water cycles.

CDF3—Sustainable Economic Development (Productive City)

- Monitor the occupation of lots regarding the installation of economic activities to verify trends and adapt public space equipment and infrastructure to them.
- Consideration of energy-environmental indicators in the Proposal Regulation.
- Integrate energy and climate sustainability measures for buildings and intervention in public spaces, in the plan regulation proposal, whether in the form of recommendations/incentives or obligations.
- Consideration of measures relating to the sustainable construction of mixed buildings in the Plan Proposal Regulation.

4 Governance

The governance framework identifies who has the institutional obligations in implementing the Plan, and which stakeholders should be involved in the process. To ensure the successful implementation of the planned actions it is important to guarantee who shares the responsibility, ensuring the principles of good governance of the Public Administration, especially those of decision, good faith, collaboration, participation, responsibility, and open administration.

In the case of the PPVC, the Portuguese Environment Agency, the Coordination and Regional Development Commission of the Center, the Municipality of Leiria, in addition to the Non-Governmental Organizations for the Environment, the population in general, representatives of academic institutions and representatives of the economic activities assume a preponderant role.

5 Follow-Up and Monitoring

It should be noted that the Plan's SEA procedure does not end with the preparation and subsequent approval of the ER after the institutional and public consultation phase. After the ER approval, the Environmental Declaration is issued, and its publication is followed by the follow-up phase.

Follow-up presupposes developing or, the follow-up guidelines previously listed for each of the identified CDFs and a set of other tasks that should be reflected in Evaluation and Control Reports [2] namely, evaluate the efficiency of the staff governance and institutional changes; check SRF changes and additional conditions or guidance; investigate uncertainties and unexpected events.

Furthermore, the follow-up of the Plan's SEA is based on the monitoring and evaluation of the performance of the environmental effects of the recommended actions, for the three identified CDFs, which in the PPVC SEA are based on the indicators [7], and respective metrics which may be readjusted at this stage if warranted. Thus, for the 10 criteria assigned to the CDF, 25 measurable indicators were established, with well-defined goals.

6 Discussion and Analysis

The main objective of the work was to develop an ER proposal, within the scope of the SEA procedure for a DUP of a specific area in the City of Leiria—PPVC, following the methodology considered in the reference methodological Portuguese guides. The goal is to contribute to the integration of environmental sustainability considerations in the preparation of a Plan, to detect and enhance opportunities and minimize strategic risks, encourage stakeholders' involvement and public discussion in decision-making processes, and contribute to sustainable strategic planning.

From the analysis and evaluation carried out, it is considered that the PPVC proposal will induce significant positive effects in the intervention zone if compared to the absence of the Plan.

In view of the weaknesses identified, it is considered that this report constitutes an opportunity for improvement for the deeper integration of aspects of sustainable development in the options established for the Plan's intervention area.

After evaluating the strategic options of the territorial model, where it was intended to assess the risks and opportunities resulting from the proposal, follow-up guidelines/recommendations were listed, which ensure the consideration of international and national recognized environmental objectives.

The guidelines for action were defined according to the Critical Decision Factors considered, namely: CDF1—Environmental Efficiency and Human Health (Green City), CDF2—Social and Territorial Cohesion (Fair City), and CDF3—Sustainable Economic Development (Productive City).

The ER concludes that the PPVC proposal represents an improvement in terms of infrastructure, adaptation to climate change, quality of the sound environment and air quality, as well as in terms of urban morphological balance and coherence, qualification of public space, climate change, and accessibility. Also contributes positively to sustainable urban development and represents an increase in the population's quality of life and urban experience.

The proposal could benefit, however, if, at the plan regulation level, adaptation measures were incorporated to increase the resilience of the territory to climate change. Also, energy-environmental indicators and integrated energy sustainability measures for buildings and public space, whether in the form of recommendations/incentives or obligations could be considered, as well as the consideration of measures relating to sustainable construction of mixed buildings.

At the same time, the SEA points to the need for a greater improvement of urban design. There is, also a need for detail, regarding the road network to integrate aspects related to road traffic, accessibility, signaling, and safety. It is also suggested the possibility of analysis and evaluation of the public transport network in the area.

The proposal would also benefit if a landscape architecture project were drawn up, considering all the concerns related to the framework and landscape enhancement of the site, namely those related to the need to increase the resilience of the urban space, contributing to reverse the loss of biodiversity, degradation of ecosystems and reestablishment of water cycles.

Likewise, the inclusion of ecological solutions in the infrastructures to be built or rehabilitated would constitute an opportunity for innovation and progress toward sustainability. Within the scope of the development of the territorial model, the possibility of increasing the area of green spaces and parking should be evaluated, through the transfer of areas of the existing municipal domain, without jeopardizing urban commitments assumed by the owners of private lots. On the other hand, monitoring the occupation of lots with regard to the installation of economic activities in order to verify trends and adapt public space equipment and infrastructure to those, would also be beneficial for the adaptive management of the territory.

7 Conclusion

SEA assumes a relevant role in planning and land management, representing a strategic instrument of environmental policy, preventing the environmental impacts of Plans and Programs options entail, and contributing to enriching them, through the improvement of the planning methodology.

SEA is still seen by decision-makers with some distrust [8], being considered more a formal administrative procedure, to be complied with by law and its strategic relevance for sustainable development is undervalued. In this sense, it is crucial for the success of the SEA, that communication between the political decision makers/top management, the stakeholders, and the team that carries out the assessment of the plan option 'impacts, must be established. Thus, without joint analysis and effective consideration of governance issues, the role of SEA will be reduced.

In this sequence, its role could be more decisive if the Environmental Declaration were binding, similar to what is established in the Environmental Impact Assessment regime for the Environmental Impact Declaration.

Ideally, there will come a time when SEA is no longer distinguishable from planning as sustainability issues will be intrinsically considered in the decision-making process and the negative environmental externalities of actions proposed in the plan and/or program will be considered and taken into account.

Acknowledgements This work is supported by the Fundação para a Ciência e a Tecnologia (FCT) and Centro2020 through the following Projects: UIDB/04044/2020, UIDP/04044/2020, UIDB/00308/2020 and PAMI—ROTEIRO/0328/2013 (N° 022158)

References

1. M. F. Tetlow e M. Hanusch, "Strategic environmental assessment: the state of the art," Impact Assessment and Project Appraisal, vol. 30, n° 1, pp. 15–24, 2012.
2. M. R. Partidário, "Guia de melhores práticas para Avaliação Ambiental Estratégica - Orientações metodológicas para um pensamento estratégico em AAE," APA, I.P – Agência Portuguesa do Ambiente, Lisboa, 2012.
3. L. White e B. F. Noble, "Strategic environmental assessment for sustainability: A review of a decade of academic research," Environmental Impact Assessment Review, vol. 42, pp. 60–66, 2013.
4. G. Vicente e M. R. Partidário, "SEA – Enhancing communication for better environmental decisions," Environmental Impact Assessment Review, vol. 26, n° 8, pp. 696–706, 2006.
5. DGOTDU, "Guia de Avaliação Ambiental dos Planos Municipais de Ordenamento do Território," Direção-Geral do Ordenamento do Território e Desenvolvimento Urbano, Lisboa, 2010.
6. CML e MPT, "Plano Estratégico de Mobilidade e Transportes de Leiria," CML (Câmara Municipal de Leiria) & MPT, Leiria, 2016.
7. DGT, "Formação dos planos territoriais. Questões Estratégicas e Indica-dores para Avaliação Ambiental. ISBN 978-989-8785-24-4.," Lisboa, 2021.
8. B. Dalal-Clayton e B. Sadler, "STRATEGIC ENVIRONMENTAL ASSESSMENT: An International Review - With a special focus on developing countries and countries in transition," England, 2004.

Sustainability Performance of Buildings in the Project Stage—Residence Students

Lizete Heleno, Pedro Baptista, Marta Gregório, Nelson S. Oliveira, and Sílvia Monteiro

Abstract In recent years, the concept of sustainable construction has been promoted by demand and by regulations. Several methodologies to assess the sustainability of the construction have been spread across the world, by several countries for almost all continents. This study focuses on the most used construction sustainability assessment methodologies, which were compared with the Portuguese methodology, LiderA. It was found that the LiderA is the sustainability system with the greatest number of relevant criteria, considering the three dimensions of sustainability, environment, economy and social. Considering the 5 phases of the construction life cycle, the importance to apply these methodologies in construction is crucial in the early phases. It can help to draft the sustainability classification of the building. To apply the LiderA was used a residence of students, a building composed of apartments, in the phase of planning. All of the 40 criteria from the LiderA were classified from F to A++, according to the project of the residence of students. The final classification, as it was projected, was A, meaning that the building has an environmental performance 50% higher than usual practice. The category of criteria with the highest contribution was the category of quality of service and resilience, with a percentage of achievement greater than 50%. The local dynamics and the resources category were the categories with the lowest contribution, and they have the potential to be reevaluated by the building promoter, since it has the potential to improve the sustainability classification of the residence of students.

L. Heleno · P. Baptista
School of Technology and Management, Polytechnic Institute of Leiria, Leiria, Portugal

M. Gregório
Fractus Construção Modular, Lda, Avenida Cidade Maringá, Edifício D. João III, 34, Leiria, Portugal

N. S. Oliveira · S. Monteiro (✉)
LSRE-LCM—Laboratory of Separation and Reaction Engineering—Laboratory of Catalysis and Materials, School of Technology and Management, Polytechnic Institute of Leiria, Leiria, Portugal
e-mail: silvia.monteiro@ipleiria.pt

ALiCE—Associate Laboratory in Chemical Engineering, Faculty of Engineering, University of Porto, Rua Dr. Roberto Frias, Porto, Portugal

J. R. da. C. S. Galvão et al. (eds.), *Proceedings of the 3rd International Conference on Water Energy Food and Sustainability (ICoWEFS 2023)*, Springer Proceedings in Earth and Environmental Sciences, https://doi.org/10.1007/978-3-031-48532-9_46

Keywords Green building classification systems · LiderA · Sustainability performance

1 Introduction

1.1 Sustainable Buildings

The development of sustainable buildings is crucial for the sustainable use of natural resources and it is also relevant to achieve the 17 Sustainable Development Goals (SDG). World Green Building Council (WGBC) prioritized eleven SDGs for the building and construction sector to enable a profound shift towards sustainable development. The SDGs are the following: SDG 3 "Good Health and Well-Being", SDG 6 "Clean Water and Sanitation", SDG 7 "Affordable and Clean Energy", SDG 8 "Decent Work and Economic Growth", SDG 9 "Industry, Innovation and Infrastructure", SDG 10 "Reduced Inequalities", SDG11 "Sustainable Cities and Communities", SDG 12 Responsible Consumption and Production", SDG 13 "Climate Action", SDG 15 "Life on Land" and SDG 17 "Partnership for the Goals" [1].

In this sense, the construction sector has played a role in world development, the means that in recent years, sustainable construction practices have been adopted in the sense that sustainable construction can contribute to the economy of natural resources [2]. For the analysis of sustainable construction solutions, several Construction Sustainability Assessment Methodologies were developed, which have been used in recent years [3]. These tools make it possible to assess and recognize the effort and commitment of companies/organizations in adopting sustainable solutions in the medium and long term, based on environmental certification systems [4]. It should be noted that the certification associated with a high level of environmental performance reflects the concern for the environment on the part of the promoting companies/organizations, and their desire for their brands to be associated with "environmentally responsible" buildings/constructions, this being an asset to your success. On the other hand, these entities also contribute to a more sustainable society by verifying the feasibility and advantages of the option for sustainable construction, presenting real projects that take into account the future of humanity and the planet [5].

Sustainability in construction is a theme that involves the creation of projects that guarantee levels of excellence and quality in search of continuous improvement, the optimization of consumption, the reduction of waste and the commitment to durability. For this reason, it is important to choose labour and partners, as well as to choose and use sustainable and environmentally friendly materials. In this context, the use of more recent and innovative technologies can lead to resource optimization and waste generated minimization [6].

It should be noted that the principles of sustainable construction enhance self-sufficiency and even self-sustainability, which means considering the ability to manage and recycle the site's resources, minimizing the environmental impact [7]. Sustainable construction has several advantages over others; since it focuses on saving energy, water, materials, use of recyclable and reusable materials, reducing pollutant emissions and increasing the lifespan of a building due to the choice of materials with high durability. These advantages may not be visible in the short term, but in the long term, they represent an advantageous solution, improving the constructive and housing quality of buildings. Thus, and applying the principles of sustainable construction, it is possible to carry out an analysis by adopting an evaluation and certification system in order to improve the quality in terms of the performance of buildings [8]. A building is considered sustainable when it meets the minimum requirements for functionality, guarantees a good performance in terms of its sustainability and when the construction is adequate to the needs of the users and its purpose [9].

The negative impacts caused by built environments are of high importance and the Construction Industry is considered one of the main causes [6]. With the growing environmental concern in this area, methods and tools began to emerge to study the environmental performance of buildings, based on the following vectors: consumption of materials, water consumption, polluting emissions, energy consumption, waste production and interaction with the surrounding environment and its location [10].

Different methodologies are currently recognized for Green Building Classification Systems (GBRS), developed in different countries. According to Bernardi et al. (2017) the two most representative GBRS at international level are the systems *Leadership in Energy and Environment Design* (LEED) and *Building Research Establishment's Environmental Assessment Method* (BREEAM), and the two GBRS with more representation in Portugal are *Sustainable Building* Tool (SBTool) and *Leading for the Environment for sustainable construction* (LiderA) [11].

LEED is a very comprehensive methodology with adaptability, these facts being an advantage in its use. It is also a highly flexible system, so it can be applied to any type of building, whether residential or in the tertiary sector. For these reasons it is one of the most recognized systems worldwide, being the target of several updates over time [12]. In BREEAM measures are used to assess the performance of the building in the design, construction and operation phases. These measures are represented in a range of areas with pre-defined requirements, and according to the fulfilment of that [13]. SBTool is a framework for evaluating the environmental performance of a building, constituting an instrument for management and urban planning with the objective of improving the organization of space, helping to preserve and improve the quality of the environment and the quality of life in urban areas [11]. The LiderA system assesses the environmental performance of buildings, based on an integrated approach to ecological principles and resource efficiency, promoting the planning and execution of projects to minimize environmental impacts, contributing to the environmental, social and economic balance [14]. Reis (2022) assessed several methodologies and verified that LiderA is the methodology with the greatest number of relevant criteria for the assessment of building sustainability [15].

1.2 Construction Sustainability Assessment Method—LiderA

LiderA was developed in Portugal at Instituto Superior Técnico, with the aim to be used to support the search for solutions in the design phase, during construction, use and end of life, accompanying the entire life cycle of a building. This methodology, which was presented and made available in 2005 in the V1.02 version, was mainly intended for the building and its immediate surroundings. A second version of the methodology, V2.0, extended the application of the methodology to spaces outside the construction lot, the area where it is located, as well as its users from a perspective of sustainable communities. Subsequently, version 3.0 was created, which added the concept of the life cycle of buildings to the parameters analysed in previous versions. Currently, the LiderA methodology has version V4.00a [16]. It is an evaluation system that covers constructive and environmental issues, as well as enhances the framework of the options taken in the different areas analysed in the context of the use of the building, as well as social. The LiderA methodology is based on six principles that translate into six categories, each of which was assigned different weights, representing their proportionality of importance in this system (Table 1) [14].

The LiderA methodology classifies the performance of each building in levels from A^{+++} to G, according to the following [14]:

- Class A^{+++} Regenerative Situation;
- Class A^{++} (90% higher than usual practice);
- Class A^{+} (75% higher than usual practice);
- Class A (50% higher than usual practice);
- Class B (37.5% higher than usual practice);
- Class C (25% higher than usual practice).
- Class E (usual practice).

Table 1 Relationship between principles, categories and ponderation [14]

Principle	Category	Ponderation (%)
1. Improve local dynamics and promoting appropriate integration	Local integration (habitat)	12
2. Promote the efficient use of resources	Resources (flows)	30
3. Reduce the impact of environmental loads (both in value and in toxicity)	Management of environmental loads (emissions)	10
4. Ensure the environments' quality, by focusing on environmental comfort	Quality of service and resilience	15
5. Promote sustainable socio-economic experiences	Socioeconomic experiences	22
6. Ensure the sustainable use of the built environment, through environmental management and innovation	Sustainable uses	11

- Class F and G (below usual practice).

The class A^{+++} (Regenerative Situation) is considered when a building produces more resources than it needs, for example, it produces more energy than its energy needs, it produces better water than what it receives.

Based on the information described, LiderA is considered to have the following advantages [14]:

- Attribution of an environmental classification to the building without losing focus on the individual evaluations of the analysed criteria.
- Application to new buildings as well as existing buildings or buildings undergoing rehabilitation.
- Portuguese origin, which was developed taking into account the specific contingencies of the country and national and community regulations.

2 Methodology

This study aims to apply LiderA system, for the support and development of sustainable solutions regarding a building project, proposed by the partner company *Fractus Construção Modular, Lda.* In this study, the technical information provided was related to a student apartment building, which should be attractive and viable in its specific market. The LiderA version V4.00a was applied [16], which is based on a set of six principles, translated into six categories that include areas of intervention operationalised by 40 criteria, which allow the guidance and evaluation of the level of sustainability.

Thus, the methodology of this study followed the official recommendations for the application of this tool, using the LiderA's thresholds table, and the weighting defined for each. The performance levels are numeric, but from the point of view of communication are transformed into class (from A^{+++} to G). This happens with the performance level of the criteria, as well as with the final performance level of the project. The final value of the performance class of the project is obtained by the sum of the values obtained for each criterion. It is also possible to determine and analyse the level of performance by category or by area allowing us to analyse the project proposed by the partner entity in more detail, and to identify more easily the main gaps and the main opportunities for improvement.

3 Analysis of the Results

As already mentioned, the LiderA methodology used in this study is based on a set of six principles of good environmental performance (6 category), which are specified into 20 areas and 40 programmatic criteria, each with different weights (Wi) representing their proportionality of importance in this system. To assess the sustainability performance of student residence, each criteria was analysed and classified. The results of the analysis and corresponding classification is presented by each category.

The "Local Integration" category is subdivided into 3 areas and includes the first 6 criteria. For the analysis of this aspect, it was important to consider the place foreseen for the implementation of the building, the architectural project provided by the promoting company *Fractus Construção Modular*, as well as the municipal plan of Leiria, namely *Plano Diretor Municipal de Leiria* (PDM) [17]. The results obtained for the first 6 criteria are systematized in Table 2, verifying that 4 criteria presented a performance class superior to the usual practice (class E). Regarding this category, stands out the fact that the architectural project (1) defines the height of the building by the average of the existing buildings in the vicinity, (2) guarantees a palette of colours within the existing ones in the place, as well as (3) ensures the landscaping enhancement, provided for the existence of green spaces with native species.

The "Resources" category under analysis include energy, water, materials and food supply (Table 3). The results obtained for the energy area are quite satisfactory, highlighting factors such as the orientation of the building to the south and consequent favouring of natural lighting, the high thermal insulation guaranteed by the proprietary Fractus SIP panels used in the walls, as well as the expected use of energy from renewable sources. In the water area, it was considered that the promoter in the project phase will carry out an analysis to reduce drinking water consumption levels. It is also assumed that rainwater will be collected and used for secondary consumption, namely for watering green spaces. The promoting entity informed that it will use innovative materials with good environmental and healthy performance for its future residents in building construction. These data, together with those from the technical sheets, allowed assigning ratings well above the usual practice for criteria in the area of materials. The criteria P14 was assigned by F classification since the existence of any type of vegetable and/or animal food production in the areas belonging to the building envelope or in the building itself was not considered.

The first two criteria of the "Management of Environmental Loads" category are related to waste and wastewater management (Table 4). Criteria P15 was assigned an F classification since the promoter did not consider measures that enhance the

Table 2 Results obtained for "local integration" category

Area	Number	Wi (%)	Criteria	Class
Soil	P1	2	Territorial organization	A
	P2	2	Enhance soil functions	C
Natural ecosystems	P3	2	Ecological valorisation	E
	P4	2	Ecosystems services	A
Landscape and heritage	P5	2	Landscape valorisation	A+
	P6	2	Built heritage valorisation	E

Table 3 Results obtained for the "Resources" category

Area	Number	Wi (%)	Criteria	Class
Energy	P7	5	Passive performance	A
	P8	5	Energy systems	A
	P9	5	Carbon management	C
Water	P10	5	Moderate water use	A
	P11	2	Local water management	B
Materials	P12	1	Products and materials of responsible origin	A+
	P13	6	Durability of built environments	C
Food supply	P14	1	Contribution to local food production and access	F

reduction and reuse of wastewater. On the other hand, criterion P16 was assigned an A+ rating, as it was considered that the promoter will favourably intervene in the production of waste at all stages of the life cycle of the building, such as providing equipment that allows the separation of waste, as well as making users aware of good waste management practices. The last three criteria in this category are related to noise management, management of atmospheric emissions and other loads, which stands out the A++ classification attributed to criteria P18, since it is considered that combustion equipment will not be installed.

The four criteria for "Quality of Service and Resilience" category present excellence ratings (Table 5). Examples of some of the factors that contributed to this classification have highlighted the layout of the interior spaces, which will allow for a good rate of natural ventilation and natural lighting; installation of protection and intervention systems in case of emergency; the dimensioning of the building and surrounding spaces in the design phase for potential natural risks, including the building's safety and resilience in the face of external difficulties.

The results obtained for the "Socioeconomic Experiences" category were based on the analysis of architectural plans, technical data sheets provided by the promoter and maps of the planned location for the implementation of the building (Table 6). In these documents it is possible to identify particularities, such as the guarantee of the existence of paths next to the building with adequate dimensions for the flow of users; the distance of the building to public transport; the guarantee of accessible and safe streets and public spaces, including the existence of cycle paths for all, not forgetting

Table 4 Results obtained for the "Management of Environmental Loads" category

Area	Number	Wi (%)	Criteria	Class
Wastewater	P15	2	Wastewater management	F
Waste	P16	3	Waste management	A+
Other emissions	P17	3	Noise management	E
	P18	1	Management of atmospheric emissions	A++
	P19	1	Other loads	C

Table 5 Results obtained for "Quality of Service and Resilience" category

Area	Number	Wi (%)	Criteria	Class
Service quality	P20	7	Environmental quality and other aspects	A+
	P21	2	Safety and control of risks (human)	A++
Structural Adaptation	P22	3	Climate adaptation and other natural hazards	A++
	P23	3	Resilience and adaptation	A+

the importance of lowered sidewalks and adequate night lighting. Also, the location of the building in an urban space, the promoter's care for the building's aesthetic harmony, the high recyclability of building materials, as well as the commitment to prioritize equipment with low operating costs, allows it to obtain a good classification for the criteria P31, P32 and P33.

The implementation of a new constructive solution, based on an easy-to-assemble industrialized and prefabricated structure, associated with this project, significantly contributes to the A classification of criteria P28. Regarding criteria P29 and P34, as it is a building in the design phase, the classification of F was assigned as it is not possible to predict the type of interaction with the local community. However, knowing that the building is intended for students, the P30 criteria considered a high probability of them interacting with other students residing nearby, and consequently the occurrence of social and cultural activities.

Table 6 Results obtained for the "Socioeconomic Experiences" category

Area	Number	Wi (%)	Criteria	Class
Accessibility	P24	3	Active mobility	A
	P25	1	Efficient transport systems	D
Space for all	P26	3	Inclusive built areas	A
	P27	1	Inclusive spaces—accessible and safe streets and public spaces	A++
Social vitality	P28	2	Flexibility and complementarity of uses	A
	P29	2	Contribution to community well-being	F
	P30	1	Social responsibility	A
Amenities and culture	P31	2	Friendly amenities	B
	P32	1	Contribution to culture and identity	B
Green economy	P33	5	Low life cycle costs	A+
	P34	1	Contribution to circular economy	F
	P35	1	Contribution to green jobs	E

In the analysis of the "Sustainable Use" category, it is highlighted that the building is intended for young higher education students (Table 7). In this sense, a good marketing strategy will be essential, encompassing a message of concern for environmental sustainability on the part of the promoting entity, without forgetting the technological and digital demands that young people consider essential. This fact is evidenced by the request for an environmental performance assessment by the promoter, which resulted in the current document.

In addition to the analysis by category and by criterion, it is important to jointly assess all the parameters and verify their impact on the final class of environmental performance of the building. This study makes it possible to more effectively direct the focus of attention to the weakest points, which may represent sources of origin for negative or less positive environmental impacts.

Thus, applying LiderA's methodology a performance level of 29.61% was obtained, which corresponds to a class A. This means that the building has an environmental performance 50% higher than usual practice.

Figure 1 compares the contribution of the six categories to the final sustainability performance level of the studied building (case of study), either among themselves or with the maximum possible predicted by this tool (base values). The results show that despite the category "Resources" having the greatest impact on the LiderA methodology, the contribution to the case of study is only the second with the most impact, with 6.61% in 30% possible, compared with the contribution of 9% in the category "Quality of Service and Resilience" in 15% possible. This means that there was a percentage of achievement of 60% in the "Quality of Service and Resilience" category and only 22% achievement in the "Resources" category. It should be noted that, except for the "Quality of Service and Resilience" category, all other categories have percentages of achievement below 30%.

The "Local Integration" category has the worst performing with only 2.3% of the possible 12%, representing 19.3% of achievement. The main factors that contributed to this result were the limitation of available permeable soil, inherent to the building's design, the biodiversity of the implementation area, and the penalty inherent to the LiderA methodology in criterion P6, taking into account that this study does not involve the preservation or requalification of pre-existing buildings.

Table 7 Results obtained for the "Sustainable Use" category

Area	Number	Wi (%)	Criteria	Class
Connectivity	P36	3	Connectivity and interaction (digital systems)	A+
Sustainable management	P37	3	Information management for sustainable performance	A+
	P38	1	Maintenance and management for sustainability	E
	P39	1	Monitoring and governance	E
Marketing and innovation	P40	3	Marketing and innovation	A

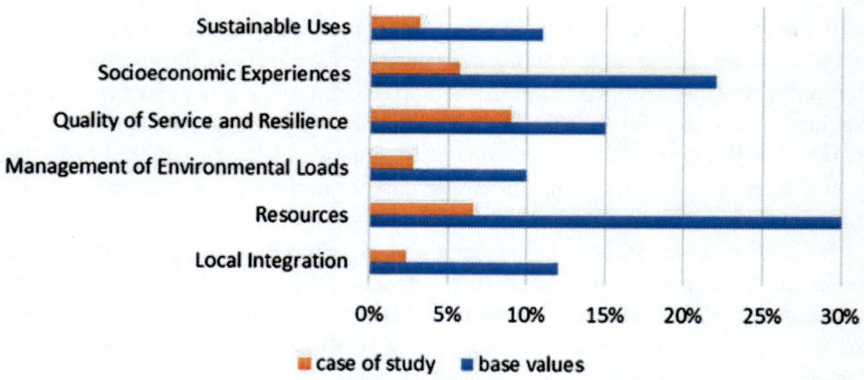

Fig. 1 Comparison between the contribution of the classification of the student residence for each category and the maximum contribution of each category

4 Conclusions

In the global analysis of the building's environmental performance, it is verified that thirty of the forty criteria present a performance class superior to the usual practice (class E), representing 75% of the total criteria. In this sense, it can be considered that the project proposed by the promoter has a philosophy favourable to the minimization of negative impacts in relation to the three pillars of sustainability. It should be noted that twenty-two criteria are accounted for with a performance class equal to or greater than A, that is, more than half of the criteria have an efficiency level 50% higher than usual practice. With a final percentage of 29.61%, the building under study obtained a classification of A according to the LiderA methodology.

The analysis by category of the results shows that the quality of service and resilience category is the most valued with a percentage of achievement greater than 50%, which means that the main approach of the proposed project was to ensure the service quality and environmental resilience, focused on environmental comfort. The principle of sustainability that appears less valued in the basic design of the building under study was the "1. Improve local dynamics and promoting appropriated integration", followed by the "2. Promote the efficient use of resources" principle.

In future work, hypotheses for improvement will be analysed jointly with the promoting entity, based on the feasibility of implementation without changing the main objective of the initial project, and simultaneously valuing the principles of sustainability implicit in the LiderA methodology. There will be a focus on the Resources and Socio-Economic Experiences categories, as they are the most relevant categories in this methodology. Examples of suggestions for improvement could include creating areas for food production, installation of systems that allow 75% of the energy consumed to come from renewable sources, and creating conditions for sharing equipment among users, such as washing machines.

Acknowledgements This work was financially supported by LA/P/0045/2020 (ALiCE), UIDB/50020/2020 and UIDP/50020/2020 (LSRE-LCM), funded by national funds through FCT/MCTES (PIDDAC).

References

1. WGBC, "World Green Building Council (WorldGBC)," 02 01 2023. [Online]. Available: https://worldgbc.org/sustainable-development-goals/.
2. X. Xu, Y. Wang and L. Tao, "Comprehensive evaluation of sustainable development of regional construction industry in China," Journal of Cleaner Production, pp. 1078–1087, 2019.
3. M. Keeler and P. Vaidya, Fundamentos de Projeto de Edificações Sustentáveis, vol. 2ª Edição, Porto Alegre: Bookman Editora LTDA, 2018.
4. V. G. d. Silva, "Habitação mais Sustentável," Projeto Finep, São Paulo, 2007.
5. J. Cidell, "Building Green: The Emerging Geography of LEED-Certified Buildings and Professionals," The Professional Geographer, vol. 61, no. 2, pp. 200–215, 2009.
6. M. Yılmaz and A. Bakış, "Sustainability in Construction Sector," Procedia - Social and Behavioral Sciences, vol. 195, pp. 2253–2262, 2015.
7. J. Saparauskas and Z. Turskis, Evaluation of Construction Sustainability by Multiple Criteria Methods, vol. XII, Lithuania: Ūkio Technologinis ir Ekonominis Vystymas, 2006, pp. 321–326.
8. P. Sousa and M. Amado, "Construção Sustentável," Faculdade de Ciências e Tecnologia, Caparica, 2012.
9. L. Bragaça, R. Mateus and H. Koukkari, "Building Sustainability Assessment," MDPI, BASEL, 2010.
10. L. M. Campos, A. C. Trierweiller, D. N. d. Carvalho and J. Šelih, "Environmental Management Systems in the Construction Industry," Environmental Engineering and Management Journal, vol. 16, no. 2, pp. 453–460, 2016.
11. E. Bernardi, S. Carlucci, C. Cornaro and R. Bohne, "An Analysis of the Most Adopted Rating Systems for Assessing the Environmental Impact of Building," Sustainability, vol. 7, 2017.
12. S. Chance, "Planning for Environmental Sustainability," Planning for Higher Education, vol. 41, no. 1, p. 194, 2012.
13. BRE Global Limited, "BREEAM New Construction 2018," BREGlobal Limited, UK, 2019.
14. M. D. Pinheiro, "LiderA," Lisboa, 2019.
15. A. Reis, "Avaliação de soluções de reabilitação de edifícios no âmbito da sustentabilidade ambiental," Instituto Politécnico de Leiria, Leiria, Portugal, 2022.
16. LiderA, "Sustainable construction leadership and certification," 2019. [Online]. Available: https://en.lidera4all.com/publicacoes. [Accessed 01 2023].
17. C.M.Leiria, "Portal de Informação Geográfica," 10 Janeiro 2022. [Online]. Available: https://geoportal.cm-leiria.pt/MuniSIG/Html5Viewer/index.html?viewer=MapaFisico.MapaFisico.

Exposure to Infrasound and Low Frequency Noise—A Case Study of Guarda City Population (Portugal)

João Almeida, Luís Marques, João Paulo Figueiredo, and Ana Ferreira

Abstract Wind energy plays an unavoidable role in the Portuguese electricity sector. Despite the several environmental advantages associated to this infrastructures and technological advances to make wind turbines quieter, they still produce sounds resulting from their mechanics and aerodynamics, in particular infrasound and low frequency noise. The aim of this study was to access the levels of infrasound and low frequency noise that the Guarda city population was exposed to. A total of 134 measurements were carried out in 67 points of urban areas. CESVA SC420 sound level meter (with "FFT Analysis" module) was used. Based on the measurements taken, infrasound and low frequency noise were calculated. The statistical analysis of the results was carried out using the IBM SPSS software. Statistically significant differences were found between the northern and north-western zones of the city compared to the western zone, regarding infrasound and infrasound exposure and low frequency noise. Regarding the low frequency noise there were no statistically significant differences between the various areas of the city. It was also possible to see that with the increase of air speed, there was also an increase in sound pressure levels. It was possible to conclude that the infrasound results are higher than the low frequency noise results. Although the infrasound frequency is below the audible threshold, it can bring problems to human health.

Keywords Wind turbines · Noise · Infrasound · Low frequency noise

J. Almeida (✉) · L. Marques · J. P. Figueiredo · A. Ferreira
Instituto Politécnico de Coimbra, Escola Superior de Tecnologia da Saúde de Coimbra, Rua 5 de Outubro—Fala, S. Martinho Do Bispo, 3045-043 Coimbra, Portugal
e-mail: joaoalmeida@estesc.ipc.pt

L. Marques
e-mail: luis.manuel.marques@hotmail.com

J. P. Figueiredo
e-mail: jpfigueiredo@estesc.ipc.pt

A. Ferreira
e-mail: anaferreira@estesc.ipc.pt

J. R. da. C. S. Galvão et al. (eds.), *Proceedings of the 3rd International Conference on Water Energy Food and Sustainability (ICoWEFS 2023)*, Springer Proceedings in Earth and Environmental Sciences, https://doi.org/10.1007/978-3-031-48532-9_47

1 Introduction

The wind sector in Portugal has been growing in recent years, constituting an expanding market with potential at various levels: commercial energy exploitation, equipment production, construction and maintenance of wind turbines, services and research. However, with the increasing expansion and development of the construction of wind turbines, and consequent territorial dissemination, it is expected a significant attention to this issue by the public entities with responsibilities in the area of land planning and planning [1, 2].

With the evolution in the use of wind energy, in the number and size, and taking into account the specific characteristics of their operating components, it was found that current legislation did not keep up with the needs regarding the impact of noise, making it essential to standardize the criteria for assessing the noise caused by the installation and operation of wind turbines [3]. Despite technological advances to make wind turbines quieter, they still produce sounds resulting from their mechanics and aerodynamics, as well as infrasound and low frequency noise, which can affect the quality of life of human and animal populations [4]. Aerodynamic noise is directly related to the airflow around the blades, which results from forces applied on the blades. Aerodynamic sound generally increases with rotor speed [5–7].

The physical proximity between the wind farm and the community maximizes the conflict of interests between those responsible for installing the wind farm and the residents, accentuating the socio-environmental impacts to which the residents are subjected from the installation of the wind turbines [8, 9].

The propagation of wind turbine noise differs considerably from the propagation of the ideal spherical wave, since a series of factors, called attenuations, influence it, such as sound absorption by the air, the existence of meteorological phenomena such as air turbulence and refraction, the existence of natural barriers such as mountains or forests, absorption by the ground itself, among others [10, 11].

In terms of direct effects on the hearing system, the noise levels measured in the vicinity of wind turbines in operation, do not have sufficient intensity to compromise the hearing health of the population. However, the "non-auditory" effects are significant, but tend to be ignored as they are not necessarily disturbing to human perception [5]. The consequences identified in more recent studies are classified into two distinct categories: Vibroacoustic Disease and Wind Turbine Syndrome [12, 13].

There is physiological evidence from experimental and field studies showing that the mechanisms underlying the cardiovascular effects of noise are responses to biological stress. In a crisis situation, of biological stress, the division of the sympathetic autonomic nervous system, the region of the brain responsible for sending nerve signals that stimulate sympathetic activation (central and peripheral) is the hypothalamus, is activated immediately. Sympathetic autonomic nervous system activation causes vasoconstriction in the circulatory system, raising blood pressure and heart rate, dilating bronchi and pupils, inhibiting digestion, among other physiological regulations. In a situation of noise, deregulation may occur in the promotion of the atherosclerotic process which, in the long term, may cause hypertension and cardiovascular events harmful to health [14].

The aim of this study was to understand the levels of infrasound and infrasound and low frequency noise to which the population of Guarda city is exposed. It was also intended to perform a characterization of the city in relation to the level of exposure to these types of noise.

2 Material and Methods

This was a descriptive observational study, analytical in nature and cross-sectional in time. The study consisted of data collection regarding the exposure of the population of the city of Guarda to infrasound and low frequency noise. After data collection, data processing was carried out.

The sample was of the non-probabilistic type and, as to technique, by convenience. 67 measurement points were selected (Fig. 1) and the sample composed by 134 evaluations of sound pressure levels.

In a first stage, to define the measuring points, a matrix was prepared, which was obtained by defining a representative area of the site under study (6 km long and 3 km wide) [15]. The points were approximately 500 m apart from each other (Fig. 1). Subsequently, and for better data treatment, lines were drawn following the cardinal and collateral points, where the measurement points were inserted. The software Google Earth pro (version 7.3.3.7786) was used to identify the measurement sites.

For data collection, regarding sound pressure levels, it was used a CESVA SC420 sound level with the module "FFT Analysis" (FF420) for Fast Fourier Transform

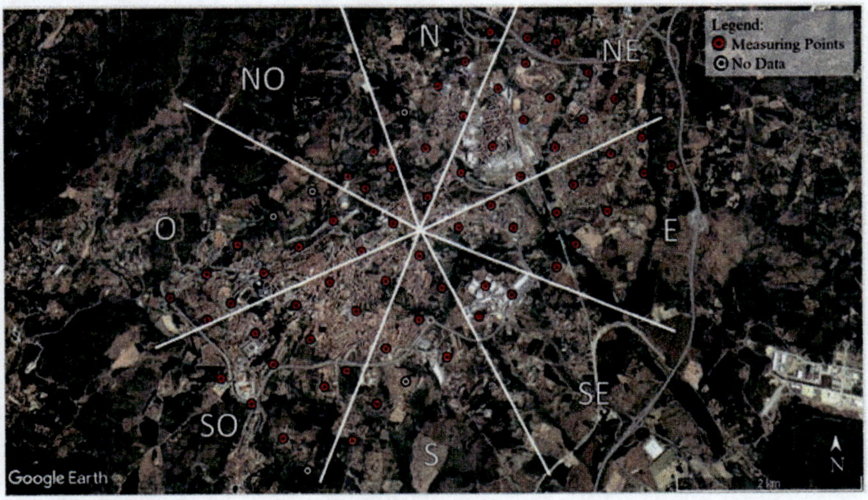

Fig. 1 Measuring points and division by cardinal points

(FFT) analysis of 10,000 lines from 2 to 20,000 Hz (2 Hz/line). The equipment was calibrated before and after the measurements using the CESVA CB006 acoustic calibrator.

The analysis was carried out per frequency band of constant bandwidth from 2 Hz to 20,000 Hz, in real time and over the whole dynamic range of measurement. Each point was evaluated twice to obtain a longer sampling time. Based on the collected frequencies the following noise types were calculated: Infrasound (IS) from 2 to 20 Hz and Low Frequency Noise (LFN) from 20 to 500 Hz.

To carry out the measurements, the following methodological conditions were assumed: the sound level meter was aimed at the nearest wind turbine; the microphone was positioned 1.5 m from the ground and at a distance of 2 m from building and other infrastructures; the sampling time was 1 min (maximum time possible for the sound level meter with the FFT analysis module); the amplitude of the frequencies measured was between 2 and 20,000 Hz [4]. CESVA Lab software (version 3.2) was used to obtain the measurement data from the sound level meter.

The climatological data were collected from the website of the Portuguese Institute of the Sea and Atmosphere (IPMA) regarding the meteorological station of Guarda and relative to the measurement periods. Values of air direction and speed were collected.

The data obtained were statistically treated using IBM SPSS Statistics software. For that purpose, measures of central tendency and dispersion were used. To compare the mean of the results obtained with the various measuring zones, the parametric Analysis of Variance (ANOVA) test was used, as the p-value was less than 0.05, and equality of variance was not guaranteed. The Games-Howell (G-H) multiple comparisons test was used, which guarantees the robustness of the result even though there was no equality of variance. Pearson's Linear Correlation test was also used. For this test, the reference values for interpreting the correlation were used: $r < 0.2$ indicates a very low linear correlation; r between 0.2 and 0.39, a low linear correlation; r between 0.4 and 0.69 moderate linear correlation; r between 0.7 and 0.89 high linear correlation; and r between 0.9 and 1, a very high correlation. For the estimation of statistical inference, a confidence level of 95% and a random error less than or equal to 5% were taken into consideration.

3 Results

For a better characterization of the site under study, wind turbines were identified within a radius of 30 km. In this area, 148 wind turbines were identified.

During the measurements, the wind flow had a southerly direction and speeds ranged between 5.278 m/s (19 km/h) and 9.722 m/s (35 km/h).

The calculation of the IS value was obtained with the recorded values of the frequencies between 2 and 20 Hz (Fig. 2). The highest sound pressure level was 98.9 dB, located in the most south-western part of the city. The measurement with the lowest result was registered in the Northwest zone and had the result of 69.4 dB.

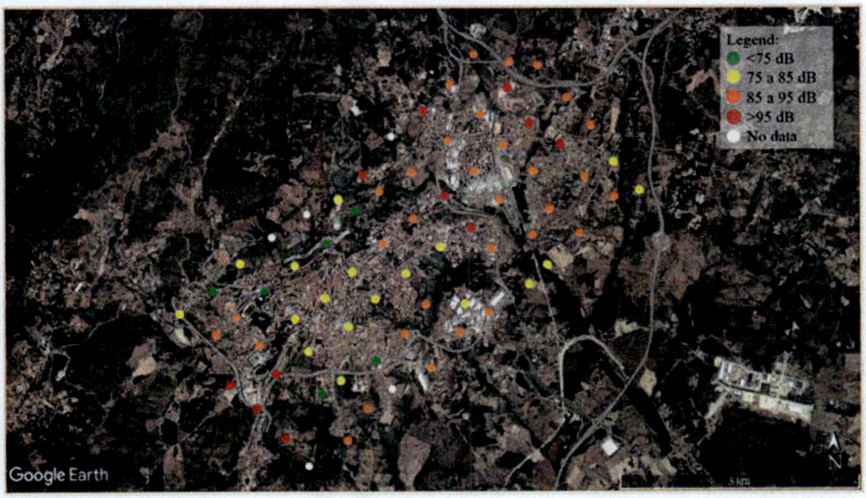

Fig. 2 Result of the IS measurements

42 points were obtained with sound pressure levels of Infrasound higher than 85 dB, 11 of which had levels higher than 95 dB.

When analyzed the mean values by geographical location (Table 1), it was possible to verify statistically significant differences in the levels of IS (dB) according to the cardinal points identified in the city of Guarda ($p < 0.05$). According to the Games-Howell (G-H) multiple comparisons test, the cardinal point "North" showed a noise level of IS significantly higher than the cardinal point "West" in the same city, i.e., the latter geographical point revealed lower noise levels compared to the northern location of the city under study ($d_{\bar{x}} = 11.45$; $p < 0.0001$). Similar profile occurred between the "North-East" cardinal point versus the "West" cardinal point ($d_{\bar{x}} = 9.69$; $p = 0.002$). However, the remaining cardinal points showed a relatively similar pattern of exposure to IS ($p > 0.05$).

By plotting a linear correlation between infrasound variables with air velocity, it was possible to see that there is a positive correlation between air velocity and infrasound sound pressure levels. This association was possible to verify through the value of the slope of the trend line different from zero, existing a statistically significant and positive correlation because the value of the slope is positive ($r = 0.398$). That is, the higher the air speed, the higher the sound pressure level. Based on the linear r^2 we can assume that this positive pattern is effective in 16% of the measurements ($r^2 = 0.159$).

The LFN sound pressure level calculation was obtained with the recorded values of the frequencies between 20 and 500 Hz. Unlike IS, LFN is already picked up by the human ear as it is already within the hearing threshold.

Table 1 Results of infrasound sound pressure levels, in dB, by location

Location	n	M	DP	LI	LS	Min	Max	F; (gl_1;gl_2); p
North	13	**92.18**	3.72	89.93	94.43	86.96	97.74	3557; (7; 38.339); 0.005
North East	27	**90.43**	5.49	88.26	92.61	79.59	101.74	
East	22	87.67	5.46	85.25	90.09	78.50	96.96	
South East	6	89.37	5.80	83.28	95.46	79.06	96.59	
South	14	85.79	7.79	81.29	90.29	68.72	92.98	
South West	24	86.82	8.72	83.14	90.50	70.06	100.47	
West	20	**80.74**	8.51	76.76	84.72	69.22	95.18	
Northwest	8	83.99	13.59	72.62	95.35	66.72	100.42	
Total	134	87.14	8.01	85.77	88.51	66.72	101.74	

Test: Brown-Forsythe; Games-Howell Multiple Comparisons Test
Dependent variable: IS levels (dB)

In Fig. 3, it can be observed the results obtained from LFN measurements. The highest sound pressure level measured was 81.6 dB, located in the most Northeast area of the city. The measurement with the lowest result was registered in the Northwest area with 54.4 dB. Ten points with low frequency noise sound pressure level higher than 75 dB were obtained.

Regarding the results of LFN sound pressure levels, by location (Table 2), and considering the Games-Howell (G-H) multiple comparisons test, it was possible to verify that there is not the presence of statistically significant differences of LFN levels (dB) according to the cardinal points identified in Guarda city. This means that

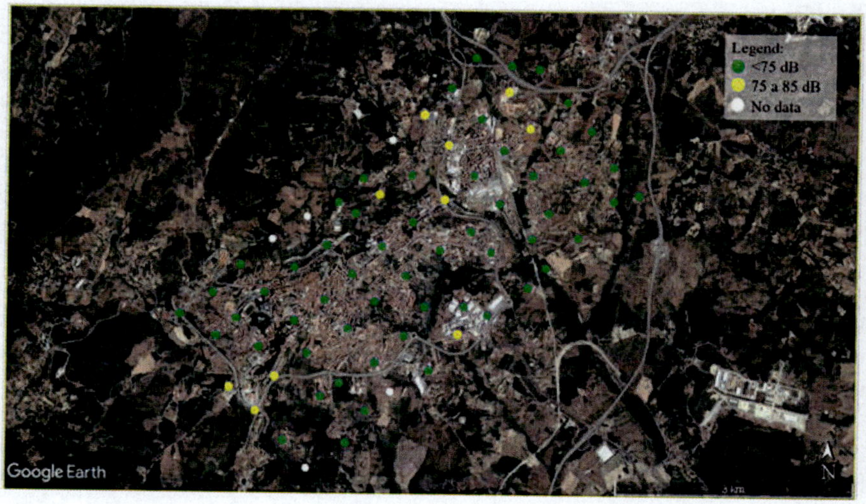

Fig. 3 Result of LFN measurements

Table 2 Results of LFN sound pressure levels, in dB, by location

Location	n	M	DP	LI	LS	Min	Max	F; (gl_1;gl_2); p
North	13	70.95	5.85	67.41	74.49	61.94	79.06	3498; (7; 45.764); 0.004
North East	27	69.22	7.31	66.33	72.11	52.45	85.77	
East	22	65.60	4.53	63.58	67.61	58.64	72.81	
South East	6	70.46	5.35	64.85	76.08	65.56	80.21	
South	14	64.56	3.82	62.35	66.76	56.21	70.82	
South West	24	67.33	7.90	63.99	70.66	53.17	79.77	
West	20	66.57	6.31	63.61	69.52	52.82	74.50	
Northwest	8	73.44	6.22	68.24	78.65	66.41	83.23	
Total	134	67.88	6.60	66.75	69.01	52.45	85.77	

Test: Brown-Forsythe; Games-Howell Multiple Comparisons Test
Dependent variable: LFN levels (dB)

there is a similar statistical exposure in the different cardinal points regarding LFN ($p > 0.05$).

Concerning the linear correlation between the LFN variable with air speed, it was possible to see that there is a positive correlation between air speed and LFN sound pressure levels. This association was possible to verify through the value of the slope of the trend line different from zero, existing a statistically significant and positive correlation because the value of the slope is positive ($r = 0.195$). That is, the higher the air speed, the higher the sound pressure level. Based on the linear r^2 we can assume that this positive pattern is effective in 4% of the measurements ($r^2 = 0.038$).

4 Discussion

The environment is defined as that which surrounds or surrounds living beings, also a set of conditions that sustain them in the biosphere as a whole or part of it. No living being lives without the relationships it establishes with the environment; in fact, living beings only survive thanks to the constant exchanges they establish with the environment. The human being itself is considered a product of its environment, however, unlike other living beings, man is the only species that has the ability to modify, organize, degrade or even destroy the environment [16].

The Portuguese territory counted more than 5500 MW of installed power. The district of Guarda had 517.3 MW of installed power, making it the fourth Portuguese district with more installed power [17].

The results obtained showed that there is a problem regarding infrasound and low frequency noise, which is accentuated when analyzing the frequencies in the non-audible range (infrasound), frequencies where the highest values were obtained. It was possible to verify 42 points where the sound pressure level was higher than

85 dB, 11 of which presented levels higher than 95 dB. The difference in the results obtained, can be due to several factors, among them: wind direction, air speed, atmospheric attenuation coefficient, sound physics, geomorphological factors, surrounding environment, among others.

With regard to wind direction, in this study, the wind was only from the south. Although it is not the dominant wind in this region, it becomes worrying since it comes from the location of wind turbines, being potentiator of a higher level of sound pressure [4, 10]. Regarding air speed, it was found that there is a moderate linear correlation, i.e., there is an increase in sound pressure when air speed is higher. This increase is more relevant in the IS values. In fact, it was possible to verify that the higher the air speed, the more tendency there is for the sound pressure level to be also higher. With the increase of the air speed, the more the wind turbine rotor turns, the more times the blade passes through the tower and consequently, more low frequency noise (including infrasound) is produced [18].

Regarding atmospheric absorption, it varies with the sound frequency and distance. The atmospheric absorption varies with temperature and relative humidity [10]. The lowest frequencies have a very low attenuation coefficient per km, revealing a very small loss during propagation. This small reduction in noise levels (particularly low frequency noise), in relation to distance, is also due to a very particular characteristic of this type of noise: extremely long wavelengths. This characteristic means that noise can be transmitted over long distances without any loss of energy [19–21]. That is why it is so important to identify existing wind turbines at great distances.

Regarding the topography of the terrain, these seem to influence the propagation and the levels of noise, namely by reducing the noise due to the mountains and valleys on the route between the wind farm and the measurement points [4]. Although not significant, there was a tendency for the noise levels in the south and southeast to be lower, which may be due to the irregular relief (mountains and valleys) and the fact that the measurement points are not in direct line with the wind turbines. The southwestern points recorded slightly higher sound pressure levels due to the orientation of the wind farm located southwest of the study area.

The sound attenuation by the forest cover, concerning low frequency acoustic propagation is not affected by the vegetation, but only depends on the "ground" effect, namely at the sound impedance level [22]. The author also mentions that the transmission of low frequency sound waves are less susceptible to absorption by intersecting elements, such as tree trunks and leaves [23, 24].

When comparing the IS and ILFN results they are similar, a fact that is explained by the frequencies that compose them, namely from 2 to 20 Hz and that are used for the calculations of these noise variables. In LFN there is a low linear correlation. This result can be explained since this variable, composed of audible noise, is influenced by various sources existing in urban space (road traffic, commerce, industry, among others) that have a behavior, from the acoustic point of view, different from infrasound.

The main symptoms diagnosed in Wind Turbine Syndrome are problems in concentration and learning, dizziness, instability and nausea, headaches that increase with frequency and sleep disturbance, in which the individual has difficulty falling asleep due to the audible noise, however, not all residents near wind turbines manifest the symptoms, because everyone has a different perception. The noise coming from wind turbines directly influences the quality of sleep of the population, which can interfere with the restorative power that sleep promotes in the human body, through stress loads, which causes the individual to have an incomplete night's sleep, as they wake up several times during the night, causing a restriction of sleep that can lead the individual to have less psychomotor capacity, lack of creativity, lack of memory and difficulties to communicate. The lack of balance and psychomotor difficulty are due to the fact that noise interferes directly with our ears, more specifically the vestibular apparatus, responsible for maintaining balance [4, 13, 25].

Vibroacoustic Disease is considered a systemic pathology that occurs due to excessive exposure to LFN. Exposure to LFN causes a thickening of the middle layers of the walls, interfering with blood flow in the human body [26, 27]. LFN mainly affects the individual's respiratory system and cardiovascular system, and may also affect the nervous system, tissues and organs. The damage to organs and tissues happens through the propagation of sound vibration in the tissues, known as the mechanotransduction method, which promotes alterations in the receipt of information from the cells, causing the excessive appearance of collagen and elastin. The most diagnosed effects in Vibroacoustic Disease, are related to problems in the heart, such as alteration in its vascular structures, mitral valve prolapse, and thickening without apparent inflammation in the pericardium [25, 28, 29].

5 Conclusion

The results of the study show that there are high levels of sound pressure present in Guarda city, mainly at the level of infrasound and infrasound and low frequency noise, which although not perceptible are present. It was possible to conclude that there are significant differences of exposure to IS, from the Western zone to the Northern and North-eastern zones. The North and Northeast zones are the zones where there is more exposure to this type of noise. We can also conclude that in the exposure to LFN there are no statistically significant differences between the various zones of measurement. It was also possible to observe that air speed is directly linked with the production of infrasound and infrasound and low frequency noise, since with an increase in air speed there is also an increase in the rotation speed of the wind turbines.

As study limitation, it is identified the impossibility of studying the wind direction, to analyze whether the wind direction influences the sound pressure levels, since in this study the measurements always occur with a southerly wind. Air speeds and the influence of other climatic factors such as temperature and relative humidity were also not studied, as well as the analysis in other seasons of the year and in other periods of the day (day and night).

References

1. Almeida, B.: Localização de parques eólicos e ordenamento do território. MSc, Universidade do Porto (2013).
2. Portal Energia - Energias renováveis: Vantagens e desvantagens da energia eólica. Available from: https://www.portal-energia.com/vantagens-desvantagens-da-energia-eolica/ (2019).
3. Maia D.: Ruído de parques eólicos - Análise e caracterização. MSc, Universidade do Porto (2010).
4. Almeida J.: Análise geográfica do risco de exposição a infrassons e ruído de baixa frequência nas turbinas eólicas. PhD, Universidade de Coimbra (2018).
5. Simões S.: Caracterização do ruído produzido por um parque eólico. Efeito sobre a população. MSc, Instituto Politécnico de Setúbal (2015).
6. Salgado C.: Base de dados do potencial eólico em Portugal Continental. MSc, Universidade Nova de Lisboa (2014).
7. Gomes L.: Avaliação de ruídos em aerogeradores situados no complexo eólico Serra Azul - Ba. MSc, Universidade Católica do Salvador (2017).
8. Silva L. Paisagem sonora e análise dos impactos causados por ruídos em parques eólicos na comunidade Xavier, Camocim, Litoral Oeste do Ceará. MSc, Universidade Federal do Ceará (2019).
9. Turunen, A. W., Tiittanen, P., Yli-Tuomi, T., Taimisto, P., & Lanki, T.: Symptoms intuitively associated with wind turbine infrasound. Environmental Research, 192. (2021).
10. Luz E.: Simulação do ruído de aerogeradores em parques eólicos. Universidade Federal de Santa Catarina (2018).
11. Instituto do Ambiente: O ruído e a cidade. Lisboa (2004).
12. Pardal, T.: Ruído ocupacional - Baixa frequência: Doença vibroacústica vs. Síndrome da turbina eólica. MSc, Instituto Politécnico de Setúbal, 2013.
13. Magalhães A.: Doença vibroacústica: Haverá "ruído" além do "ouvido"? MSc, Universidade de Coimbra (2010).
14. Carizio, B.; Esteves, T.; Carreira, P.; Faria, J.; Rodrigues, S.: Risco de patologias cardiovasculares induzidas por ruídos ocupacionais: Uma revisão integrativa. Hispeci & Lema On-Line 9(1), 36–55 (2018).
15. Almeida, J.; Santos, S.; Figueiredo, J.; Ferreira, A.: Evaluation of sound pressure levels of infrasounds and low frequency noise in Coentral Grande. European Journal of Public Health, 29(Supplement 1), 8 (2019).
16. Gonçalves, M.: A influência do ambiente na saúde uma reflexão sobre o ambiente na saúde infantil. MSc, Universidade do Porto (2016).
17. e2p. (2022). Parques eólicos em Portugal - Dezembro de 2021. Energias Endógenas de Portugal. http://e2p.inegi.up.pt
18. van den Berg, G.: The beat is getting stronger: The effect of atmospheric stability on low frequency modulated sound of wind turbines. Journal of Low Frequency Noise Vibration and Active Control 24(1), 1–24 (2005).
19. Berglund, B.; Hassmén, P.; Job, S.: Sources and effects of low-frequency noise. Journal of the Acoustical Society of America, 99, 2985–3002 (1996).
20. Murphy, E.; King, E.: Environmental noise pollution: Noise mapping, public health, and policy. Elsevier. USA (2014).
21. Goelzer, B.; Hansen, C.; Sehrndt, G.: Occupational exposure to noise: Evaluation, prevention and control. World Health Organisation, Geneva (2001).
22. Gronberg, A.: Sound propagation through Australian forest land with special regards to noise generated by wind turbines. Faculty of Engineering LTH of Lund University (2015).
23. Almeida, J.; Videira, A.; Figueiredo, J.; Ferreira, A.: Infrasound and low-frequency noise produced by wind turbines in housings - Case studies. In: Galvão, J.; Brito, P.; Neves, F.; Craveiro, F.; Almeida, H.; Vasco, J.; et al., editors. Proceedings of the 1st International Conference on Water Energy Food and Sustainability (ICoWEFS 2021). Gewerbestrasse: Springer, p. 400–410 (2021).

24. Almeida, J.; Santos, M.; Stefuryn, R.; Figueiredo, J.; Ferreira, A.: Variation of infrasound and low frequency noise produced bywind turbines, regarding distance and vegetation - Case study. In: Galvão, J.; Brito, P.; Neves, F.; Craveiro, F.; Almeida, H.; Vasco, J.; et al., editors. Proceedings of the 1st International Conference on Water Energy Food and Sustainability (ICoWEFS 2021). Gewerbestrasse: Springer, p. 400–410 (2021).
25. Liu, A.; Ferraretto, R.; Sousa, B.: Impactos negativos para a saúde humana ocasionados pelo ruído dos geradores eólicos. 9ª Jornada Científica e Tecnológica da Faculdade de Tecnologia de Botucatu, Brasil (2020).
26. Alves-Pereira, M., Castelo Branco, N.: Sobre o impacto de infrassons e ruído de baixa frequência na saúde pública - Dois casos de exposição residencial. Revista Lusófona de Ciências e Tecnologias da Saúde, 4(2), 186–200 (2009).
27. Almeida, J.; Alves-Pereira, M.; Nossa, P.: The impact of wind turbine noise on health. European Journal of Public Health, 29(Supplement 1), 20 (2019).
28. Costa, A.: O ruído ambiental de aerogeradores de pequena dimensão. MSc, Universidade do Porto (2014).
29. Almeida, J.; Simões, B.; Simões, H.; Figueiredo, J.; Ferreira, A.: Ruído ambiental versus infrassom e ruído de baixa frequência - Caso de estudo de população próxima de turbinas eólicas. Millenium - Journal of Education, Technologies and Health, 2(1 special edition), 397–405 (2016).

Carbon Footprint and Energy Assessment of Plastic Products Produced at Neutroplast

Carlos Hernandez, Carla Rodrigues, Beatriz Penha, Tânia Simões, João Redol, and Fausto Freire

Abstract This article presents a carbon footprint (greenhouse gas emissions) and cumulative energy demand assessment of plastic products produced at Neutroplast in 2020 and 2021. Neutroplast produces nutraceutical, pharmaceutical and personal care products. A cradle-to-gate assessment was performed, aiming at identifying hotspots and opportunities for environmental improvement. The results show that polymers are the main contributors to environmental impacts (mainly high-density polyethylene, largest input at Neutroplast). Therefore, using alternative materials with lower carbon footprint and cumulative energy demand has the greatest potential to reduce impacts. Production at Neutroplast can also reduce impacts with increase in renewable electricity production.

Keywords Life cycle assessment · Cradle-to-gate · Packaging · Plastic

1 Introduction

In 2021, plastic production reached 50.3 million tonnes in the European Union, in which the packaging industry accounted for 40.5% of plastic consumption [1]. Plastics have become essential materials to ensure a sustainable future as they deliver numerous applications and benefits to society. However, the extensive use of plastic products is increasing concerns about their environmental impacts [1, 2]. The Life Cycle Assessment (LCA) methodology can be employed to assess the environmental impacts of plastic products and to identify hotspots for reducing impacts.

This article presents a carbon footprint and energy demand assessment of plastic products produced at Neutroplast, based on production data in 2020 and 2021. A

C. Hernandez (✉) · C. Rodrigues · F. Freire
Univ Coimbra, ADAI, Department of Mechanical Engineering, Rua Luís Reis Santos, Pólo II, 3030-788 Coimbra, Portugal
e-mail: carlos.hernandez@dem.uc.pt

B. Penha · T. Simões · J. Redol
Neutroplast, Zona Industrial Casal Espinheira, Lote 10, 2590-057 Sobral Monte Agraço, Portugal

J. R. da. C. S. Galvão et al. (eds.), *Proceedings of the 3rd International Conference on Water Energy Food and Sustainability (ICoWEFS 2023)*, Springer Proceedings in Earth and Environmental Sciences, https://doi.org/10.1007/978-3-031-48532-9_48

cradle-to-gate assessment was performed aiming at identifying hotspots and opportunities for environmental improvement. The findings include recommendations to improve the environmental performance of plastic products produced at Neutroplast.

2 Materials and Methods

This chapter is divided into two sections. Section 2.1 describes and analyses production at Neutroplast. Section 2.2 presents the life cycle model and inventory developed to assess the carbon footprint and energy demand assessment of plastic products produced at Neutroplast.

2.1 *Production at Neutroplast*

Neutroplast is a company, located in Sobral de Monte Agraço, Portugal, that develops and produces plastic products for the nutraceutical, pharmaceutical and personal care sectors. As shown in Fig. 1, the company has five production lines with a wide range of products, including different types of bottles (pills bottles, spray bottles, dropper bottles, and beauty bottles), bottle tops and dosing devices. The products are manufactured using extrusion blowing, injection blowing or injection molding. Some of the products are also labelled at Neutroplast using screen and pad printing.

Figure 2 shows the global production data for Neutroplast (2020 and 2021, excluding August, since production was partially closed). In 2020 and 2021, Neutroplast had a total electricity consumption of 2,022,749 kWh. A photovoltaic (PV) flat-roof system satisfied 16% of the electricity demand, replacing 314,114 kWh of electricity from the national grid. Regarding materials, there was an input of 951,778 kg of polymers, mainly: (i) high-density polyethylene "HDPE" (817,003 kg), (ii) polypropylene "PP" (100,162 kg) and (iii) low-density polyethylene "LDPE" (34,613 kg). Production losses represent 0.9% (8513 kg) of polymer inputs, rejected

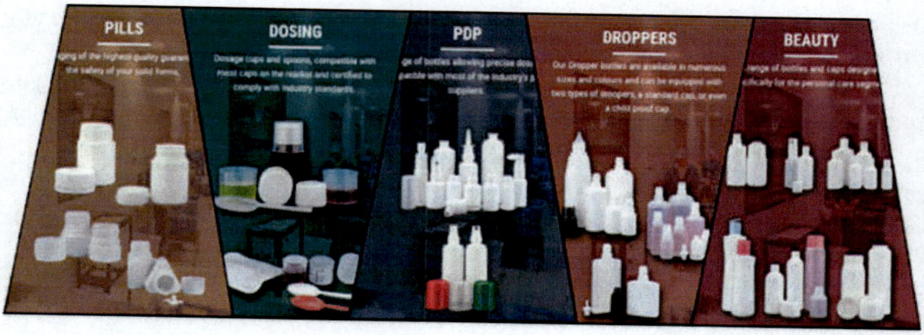

Fig. 1 Product lines of Neutroplast

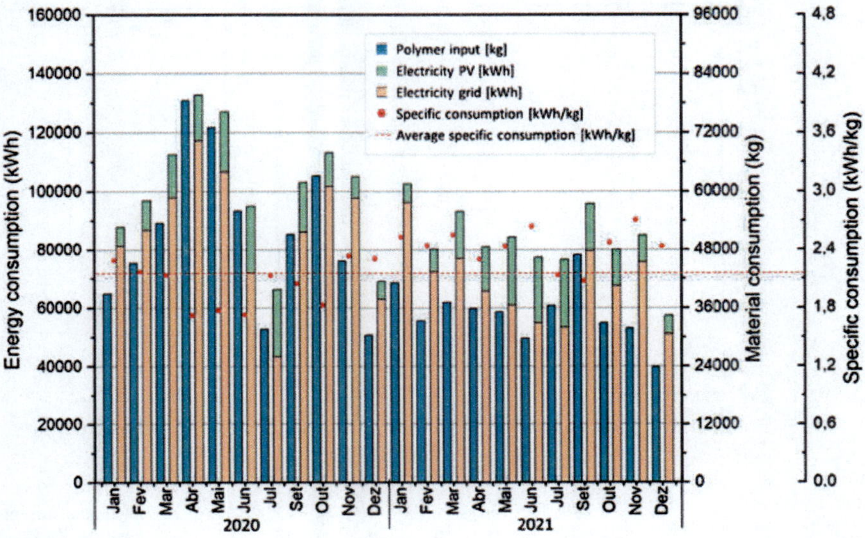

Fig. 2 Energy (kWh), material (kg), and specific electricity consumption per kg of plastic consumed (kWh/kg) in 2020 and 2021 (excluding August, production was partially closed)

material from purges and used in quality tests. Other rejected material is internally recycled. Neutroplast presented an average bi-annual specific energy consumption of 2.13 kWh per kg of polymer input, with monthly averages between 1.69–2.67 kWh/kg.

Figure 3 shows the electricity consumption of an injection equipment of HDPE bottles. The amount of electricity and HDPE consumed per production batch was measured in an injection equipment, for various batches of HDPE bottles produced, including bottle sizes of 20 ml (measurements in 15 production batches); 30 ml (3); 50 ml (6); 75 ml (38); 200 ml (14); and 500 ml (6). It can be noted that electricity consumption per kg of HDPE decreases with increase in bottle volume, ranging from $1.36^{+0.25}_{-0.24}$ kWh/kg (20 ml) to $0.45^{+0.05}_{-0.03}$ kWh/kg (500 ml).

2.2 *Life Cycle Model and Inventory*

A life cycle model was developed for plastic products produced at Neutroplast to assess greenhouse gas (GHG) emissions (based on IPCC 2013, for 100-year time horizon) [3] and cumulative energy demand (CED, for non-renewable and renewable sources [4]), aiming at identifying hotspots and opportunities for environmental improvement. The functional unit is defined as an average 10 g plastic product, to be representative of the various product lines produced at Neutroplast in 2020 and 2021. Figure 4 shows the system boundaries, which include polymers and production at Neutroplast. The life cycle model was developed based on primary data from

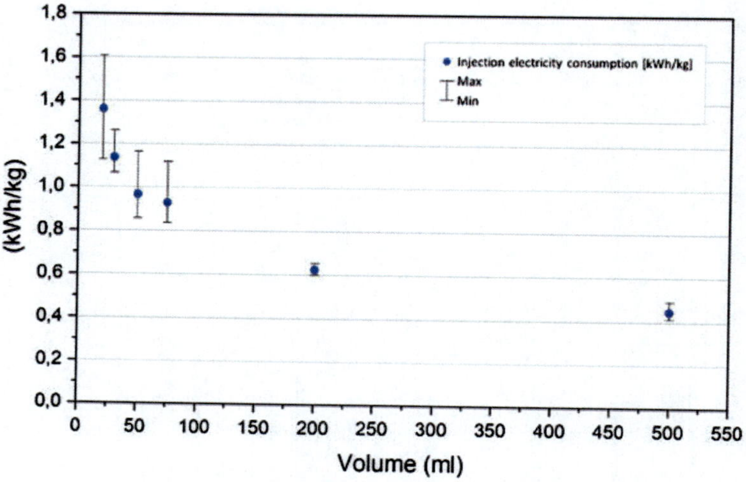

Fig. 3 Electricity consumption per kg of injected HDPE for bottle volumes of 20 ml (measurements in 15 production batches in selected injection equipment); 30 ml (3); 50 ml (6); 75 ml (38); 200 ml (14); and 500 ml (6)

Neutroplast and secondary data from ecoinvent 3.6 database [5] and PlasticsEurope [6, 7]. The transportation of raw materials is included based on generic data. GHG emissions and CED per kg of polymers are shown in Table 1. Regarding electricity, the PV flat-roof system (peak power: 173 kWp, 630 multi-crystalline silicon panels, 25 years lifetime, 1030 m^2) was modelled based on [8]. Electricity from the Portuguese grid was based on [9, 10]. SimaPro 9.1.1 was used for system modelling and calculation of environmental impacts, adopting the system model Allocation at the Point of Substitution (APOS) (Fig. 4).

Table 1 GHG emissions and CED per kg of polymer. *Source* Ecoinvent 3.6 database [5] and PlasticsEurope [6, 7]

	GHG emissions	CED non-renewable	CED renewable
	kg CO2 eq/kg	MJ/kg	MJ/kg
HDPE	2.33	79.45	1.04
PP	2.29	80.91	0.90
LDPE	2.49	81.51	1.52

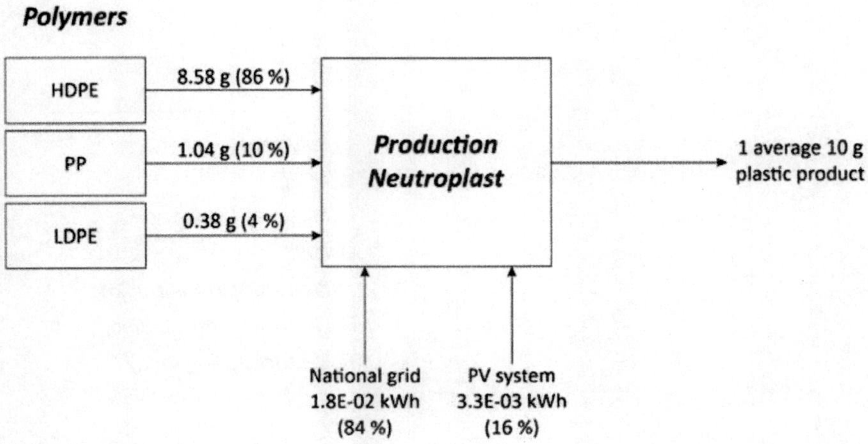

Fig. 4 System boundaries (cradle-to-gate), considering energy and material consumption of an average 10 g plastic product produced by Neutroplast (2020–2021)

3 Results

Results show that GHG emissions and CED non-renewable of plastic products produced at Neutroplast are mainly associated with polymers, as shown in Fig. 5. Polymers account for 78% of GHG emissions and 91% of CED non-renewable (most from HDPE, which represents 86% of total polymers inputs) and electricity 22% of GHG emissions. The CED renewable accounts for 6% of CED, mainly associated with electricity.

4 Conclusions

This paper presents a carbon footprint (GHG emissions) and energy demand (CED) assessment of plastic products produced at Neutroplast. A cradle-to-gate assessment was performed aiming at identifying hotspots and opportunities for environmental improvement. Results show that polymers are the highest contributors to GHG emissions and CED (particularly HDPE, the largest input of polymer at Neutroplast). These results emphasize the importance of future assessment of ecodesign strategies for HDPE products using alternative materials (e.g. biopolymer or recycled plastic) with a lower carbon footprint. It was also concluded that production at Neutroplast has room for environmental improvement with increase in renewable electricity production.

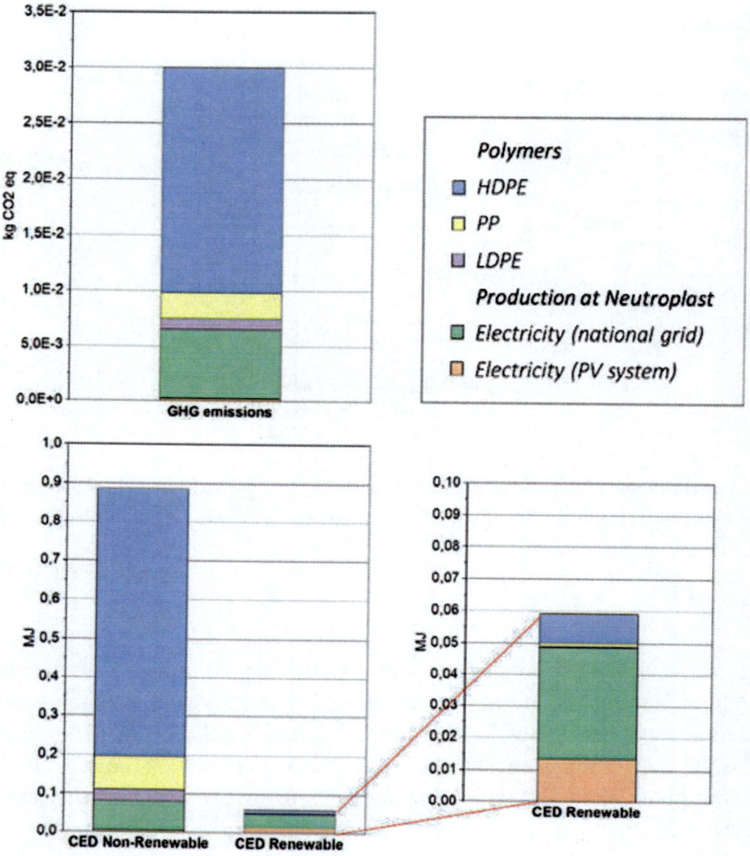

Fig. 5 GHG emissions and CED (non-renewable and renewable) of an average 10 g plastic product produced by Neutroplast (2020–2021)

Acknowledgements This research was supported by European Regional Development Fund (FEDER) through Compete2020 (Operational Program Competitiveness and Internationalization) under the S4Plast project POCI-01-0247-FEDER-046089, as well as, the SET-LCA project (CENTRO-01-0145-FEDER-030570), co-funded by European Regional Development Fund (FEDER) through POCentro (PT2020) and the Portuguese Foundation for Science and Technology (FCT). The first author is grateful for the financial support from the S4Plast project and FCT through the MIT-Portugal doctoral grant PRT/BD/154263/2022.

References

1. PlasticsEurope, "Plastics – the Facts 2022," no. October, 2022, [Online]. Available: https://plasticseurope.org/knowledge-hub/plastics-the-facts-2022/.

2. European Technology Platform for Sustainable Chemistry (SUSCHEM), "Sustainable Plastics Strategy," 2021, [Online]. Available: http://suschem.org/publications.
3. I. P. on C. Change, *Climate Change 2013 – The Physical Science Basis*. Cambridge University Press, 2014.
4. R. Hischier *et al.*, "Implementation of Life Cycle Impact Assessment Methods," St. Gallen, 2010.
5. R. Hischier, "Life Cycle Inventories of Packaging and Graphical Papers (part II): Plastics," Dubendorf, 2007.
6. PlasticsEurope, "Eco-profiles and Environmental Product Declarations of the European Plastics Manufacturers: High-density Polyethylene (HDPE), Low-density Polyethylene (LDPE), Linear Low-density Polyethylene (LLDPE)," 2016. [Online]. Available: https://www.plasticseurope.org/en/resources/eco-profiles.
7. PlasticsEurope, "Eco-profiles and Environmental Product Declarations of the European Plastics Manufacturers: Polypropylene (PP)," 2016. [Online]. Available: https://www.plasticseurope.org/en/resources/eco-profiles.
8. N. Jungbluth, M. Stucki, and R. Frischknecht, "Photovoltaics Part XII," Dübendorf, Switzerland, 2009.
9. R. Garcia, P. Marques, and F. Freire, "Life-cycle assessment of electricity in Portugal," *Appl. Energy*, vol. 134, pp. 563–572, 2014, https://doi.org/10.1016/j.apenergy.2014.08.067.
10. J. Kabayo, P. Marques, R. Garcia, and F. Freire, "Life-cycle sustainability assessment of key electricity generation systems in Portugal," *Energy*, vol. 176, pp. 131–142, 2019, https://doi.org/10.1016/j.energy.2019.03.166.

Raising Awareness for Sustainable Development Goals Through Hands-On Experiments

Maria C. Costa, Dina M. R. Mateus, and Henrique J. O. Pinho

Abstract With the aim of protecting our planet, the United Nations defined 17 Sustainable Development Goals (SDG), which requires the involvement of all countries to make an endeavour to achieve better living conditions. In this regard, stakeholders such as governments, regional and international organizations, and civil society need to work together to make efforts to meet these objectives. In addition, Higher Education Institutions have a crucial role, not only in developing research and disseminating SDG practices but also in intervening in society. This paper presents a collaborative project led by a Higher Education Institute and targeted to the school community. With the objective of strengthening consciousness for the development of good practices related to SDG, a workshop was prepared and implemented with primary school teachers. Besides providing knowledge about SDG, several hands-on experiments were presented to teachers to be reproduced with their students. Based on participant observation and enquiry forms applied to teachers and students, it was verified that the project was effective in enhancing recognition of teachers and students about the importance of providing good practices related to SDG. It is concluded that collaboration between higher education institutions and schools can trigger the implementation of sustainable development practices in the community.

Keywords Sustainable development goals · Hands-on · Sustainability

M. C. Costa (✉) · H. J. O. Pinho
Ci2—Smart Cities Research Center, Instituto Politécnico de Tomar, Tomar, Portugal
e-mail: ccosta@ipt.pt

D. M. R. Mateus
Techn&Art—Technology, Restoration and Arts Enhancement Center, Instituto Politécnico de Tomar, Tomar, Portugal

J. R. da. C. S. Galvão et al. (eds.), *Proceedings of the 3rd International Conference on Water Energy Food and Sustainability (ICoWEFS 2023)*, Springer Proceedings in Earth and Environmental Sciences, https://doi.org/10.1007/978-3-031-48532-9_49

1 Introduction

The 2030 Sustainable Development Goals (SDG) [1] were outlined to provide better living conditions for all and ensure the sustainability of our planet. Achieving these objectives requires cooperation and collaboration between various stakeholders from civil society to government, among others [2, 3]. Therefore, efforts and actions are needed among all countries to correspond to the referred goals. In this regard, Higher Education Institutions have a crucial role not only in developing research on this topic, as well as disseminating and promoting SDG practices; but also in intervening in society, where it can be an example and a driving force for the implementation of good practices [4–7]. This paper presents a collaborative project led by a Higher Education Institute (HEI) and targeted to the school community, with the objective of strengthening consciousness for the development of actions related to SDG. The project includes one workshop targeted at teachers from different levels of primary school (First to sixth grade, 6–11 years old) and 20 workshops targeted at students from the first cycle of six primary schools (about 400 students, from first to fourth grade, 6–9 years old). Because it is related to education for sustainability and promotes partnerships in this sense, the project is related to Goal 4, Quality Education, and goal 17, Partnerships for the Goals.

In a preliminary work, the authors of this paper described several hands-on experiments related to SDG that were developed by researchers of the HEI, in the context of funded projects in Portugal [8]. In particular, the experiments were focused mainly in goals 6, 7, 11, 12, 13, 14, and 15: Clean Water and Sanitation, Affordable and Clean Energy, and Responsible Consumption and Production, among others. These experiments were introduced in a workshop targeted to teachers with the aim of increasing their knowledge about SDG and strengthening consciousness of their role and the role of schools in promoting actions and good practices related to this thematic.

The present study extends the previous research by providing a more comprehensive intervention project that, in addition to workshops for teachers, also involves workshops for primary school students regarding the need to promote initiatives and practices related to measures necessary to protect our planet and improve our quality of life. Thus, two dimensions are approached. On the one hand, teachers' perceptions about the SDG, and on the other hand the students' perceptions about the same topic. Moreover, it is provided the impact of the SDG workshops on teachers and students. In this regard, the following research questions are raised: Does a collaborative project between Higher Education and the school community can change teachers' and students' knowledge and perceptions about SDG? What are the insights regarding attitudes and the promotion of good practices in the community?

The empirical study was developed with teachers and students who participated in the workshops and voluntarily answered enquiry forms before and after the workshops. Also, it gives insights about the hands-on experiments developed at the workshops. This paper adds an important contribution to the literature on community intervention projects and their role in contributing to the implementation of good practices related to the theme under study.

The paper is organized as follows. Section two describes the methodology and the hands-on experiments implemented in the workshops, section three the analysis and discussion of enquiry forms responses. Finally, in section four conclusions and future work are presented.

2 Methodology and Description of the Hands-On Experiments

This study employs a mixed methodology approach, combining both quantitative and qualitative methods [9]. The participants in this research consist of teachers who willingly enrolled in a workshop related to the SDG, as well as the students who engaged in the workshop led by facilitators at the request of their teachers. The qualitative component of the study involves participant observation during the workshops attended by both teachers and students. For the quantitative aspect, the research team designed questionnaires using Google Forms, and these questionnaires were administered during the teacher's workshop and at one of the workshops conducted for students. The collected questionnaire data was analysed using the SPSS® software package, version 27. A total of 63 teachers, who actively participated in the teacher's workshop, voluntarily completed the questionnaires both before and after the workshop.

Approximately 90% of the teachers possess over 40 years, with 65% of them dedicated to teaching students in the primary cycle (comprising the first cycle), and a significant majority, 92.9%, are female.

Table 1 summarizes the topics provided in some questions of the enquiry forms applied to teachers, where questions numbered with 1 were applied before the workshop and numbered with 2 were applied after the workshop. The questions with a Likert scale were classified with 1 for "Not at all" to 5 for "Very high".

In addition, 27 students who participated in one of the 20 student-oriented workshops completed additional questionnaires during the workshop held in their classroom. Of these students, 15 were in the 4th year of elementary school, while 12 were in the 3rd year, almost 67% of whom were male. Table 2 shows the questions included in the students' enquiry forms.

Table 1 Topics of the questions applied to teachers before and after the workshop

Code	Question topic
1_a/2_a	Confidence and knowledge to introduce SDG to students
1_b/2_b	Need of professional development in SDG topics
1_c/2_c	Role of schools in promoting SD practices in the community
2_d	Relevance of hands-on experiments in encouraging students to carry out SDG practices
2_e	Role of teachers in sensitizing students to develop SD practices

Table 2 Questions applied to students

Code	Question	Proposed answers
3_a	Why our planet needs help?	Climate changes
		Forest fires
		People do not understand each other
		High consumption of natural resources
		Greenhouse effect
		Too much waste
		Air pollution
		Oceans and forests pollution
		Effects of pollution on animals and plants
		Too much pollution
		Earth temperature increase
		Scarcity of potable water
3_b	How can we help the planet?	Reducing water consumption
		Reducing energy consumption
		Reducing all waste and recycling
		Separating the garbage
		Avoiding food waste
		Having a responsible consumption
		Make use of public mobility
		Teaching people to protect the Earth
		Planting more trees
		Doing studies and experiments
		Make use of renewable energy
		Actively monitor threats to the planet

The hands-on experiments carried out during these workshops are detailed in Table 3.

More details of the hands-on experiments are available in the referred previous work [8].

Table 3 Short description of the hands-on experiments implemented in the workshops

Experiment	Description	Covered SDG	Message
A filtration system for greywater made from greener and reused materials	Filtration of wastewater through filters constructed from recycled plastic bottles that are filled with various natural or repurposed granular materials	6, 11, 12, 13, 14, 15	Implementing eco-friendly and sustainable technologies in wastewater treatment plays a crucial role in achieving cost-effective and universal water sanitation solutions, while also safeguarding freshwater resources. The repurposing of solid waste aids in reducing resource usage and waste disposal, serving as a proactive measure against climate change
Reuse of plastic lids	Creation of diverse ornamental objects crafted from plastic caps	11, 12, 13, 14, 15	This experience aimed to raise awareness and encourage positive attitudes towards the responsible use of plastics, with particular emphasis on their reuse and recycling as a way of combating pollution and tackling climate change
Sustainable energy sources	The peel of citrus fruits is pressed close to the flame of a candle, producing a radiant effect. In addition, citrus fruits were used as batteries, using the electricity generated to light an LED lamp	7, 11, 12, 13	Citrus peel was selected as an example of food waste with the potential for generating energy. The objective of this experiment is to show children, engagingly and straightforwardly, that food waste can serve as a sustainable energy source

3 Analysis and Discussion of the Answers to the Enquiry Forms

This section is divided into three subsections. The first and second subsections give an account of teachers' and students' answers to the enquiry forms. The third subsection presents an outline of the discussion providing more information based on the qualitative approach.

3.1 Analysis of Teachers' Responses to the Enquiry Forms

The following figures provide results regarding teachers' responses to the questions whose topics can be found in Table 1. In Fig. 1, it can be observed that teachers' confidence and knowledge regarding the introduction of SDG topics to students has

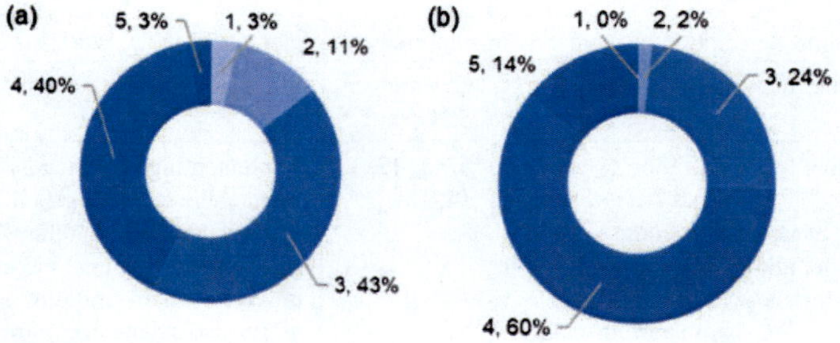

Fig. 1 Results from questions 1_a (**a**) and 2_a (**b**)

increased after the workshop. For example, from 40 to 60% in the "High confidence (4)" option, and from 3 to 14% in the "Very high confidence (5)" option. It should be noticed that before the workshop most teachers had responded they have "Enough (3)" (43%) or "Some (2)" (11%) confidence, which revels lack of confidence in this matter. This is very important because without confidence and knowledge teachers will not be able to develop this approach in class.

Regarding recognition of necessity for professional development related to SDG, all teachers acknowledged they need training in this matter before and also after the workshop (Questions 1_b e 2_b, Fig. 2). In fact, after the workshop there was an increasement in this recognition; for example, from 29 to 75% in the "Very high (5)" option. This result shows that besides needing to improve knowledge about SDG, they recognize the importance of this subject matter.

This result is reinforced by the teachers' realisation of their role and that of the schools in promoting SD practices in the community (Questions 1_c e 2_c, Fig. 3). Indeed, about 90% find this role "Very high" or "High", which means they are very aware of the importance of their mission in this approach.

In addition, teachers find the relevance of the hands-on experiments (Table 3) in encouraging students to carry out SDG practices, and also that they have a relevant role in sensitizing students to develop SD practices (Questions 2_d e 2_e, Fig. 4).

Indeed, more than 90% responded "Very high" or "High", which reveals that they find the experiments developed by the facilitators very relevant to students and also that they can sensitize students to develop SD practices.

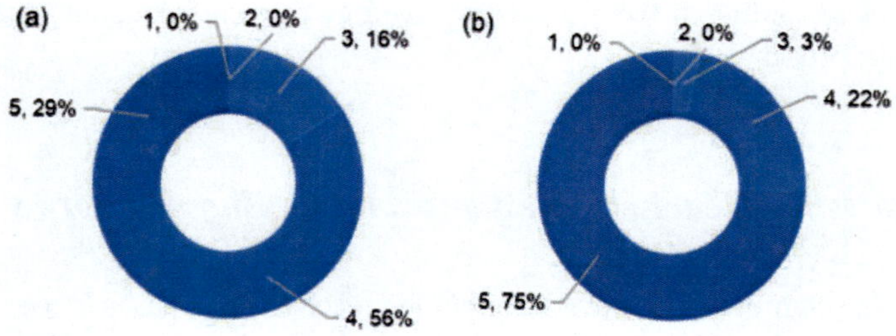

Fig. 2 Results from questions 1_b (**a**) and 2_b (**b**)

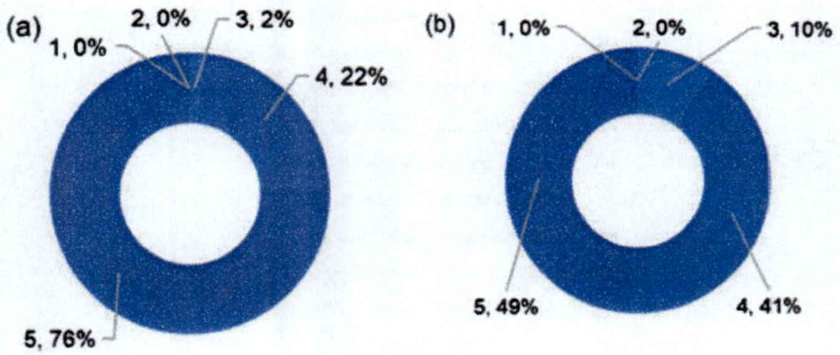

Fig. 3 Distribution of answers to questions 1_c (**a**), and 2_c (**b**)

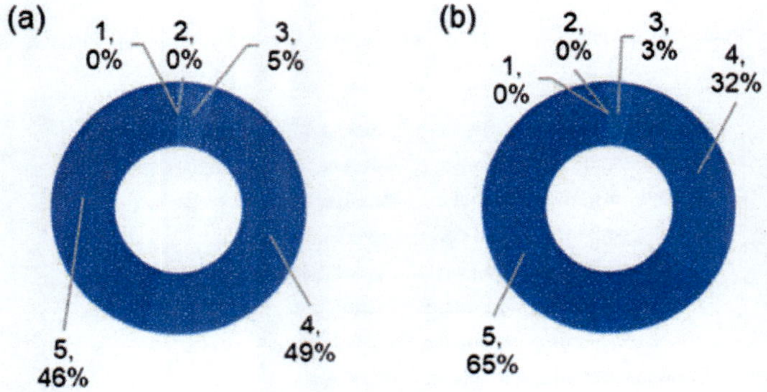

Fig. 4 Distribution of answers to questions 2_d (**a**), and 2_e (**b**)

Based on the results of teachers' responses to the enquiry forms, it was verified that there was an increasement in their knowledge regarding SDG and also in the recognition of their mission and of schools in promoting actions related to this matter.

3.2 Analysis of Students' Answers to the Enquiry Forms

Figure 5 presents the results of the answers of students to the questions applied after the workshop implemented in their class.

As can be observed in Fig. 5, most students selected "Oceans and forests pollution" as the main reason why our planet needs help, followed by "Air pollution", "Forest fires" and too much waste.

Regarding measures to help the planet, most students selected "Reducing all waste and recycling", followed by "Separating the garbage" and "Reducing water consumption" (Fig. 6).

These answers show that the hands-on experiments were effective in raising awareness of the ODS addressed, namely: The need to reuse and recycle materials (11 and

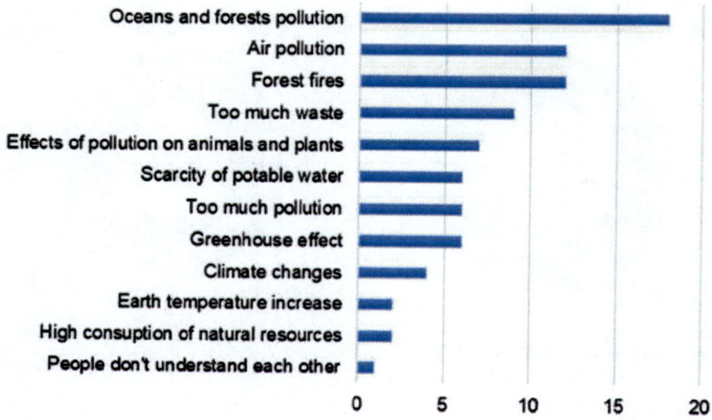

Fig. 5 Distribution of student´s answers to the question "Why do you think the Planet needs help?"

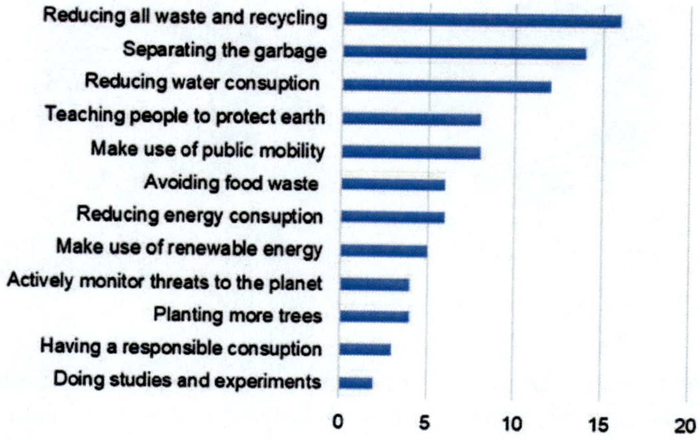

Fig. 6 Distribution of student's answers to the question "What can we do to protect the environment and help the planet?"

12) to prevent pollution of forests, oceans, and air (14 and 15) as well as climate change (13); The problems associated with water scarcity (6, 13, 14); The need to look for greener energy sources (7, 13).

3.3 Outline and More Insights About the Discussion

Besides the enquiry forms, participant observation was performed during the workshops with teachers and students. Regarding the focus group that was performed with the teachers during the workshop, we highlight the appreciation of teachers for the training that made them be able to develop experiments with students. Most of them expressed interest in more training in this area and said they wanted to implement this approach in their school, namely the experiments presented at the workshop. Moreover, they refer to the hands-on experiments as adequate to be implemented in schools and useful to better help students to understand this subject matter.

Therefore, the workshop had impact on teachers namely by improving their knowledge about the subject matter approached and also motivated them to develop this approach in their schools. Indeed, they revealed consciousness of their role and of schools regarding the development of actions related to SDG with students and in the community.

In what concerns students, before the workshop an open question related to measures about how to help our planet was posed to them. Several responses were similar to usual ones such as "Pollution", "Reducing waste", and "Recycling", among others. After the workshop, the same open question was placed and other answers, related to the examples approached in the workshop, were observed. For example: "Watering our plants with the water we already use", "Collect rainwater to water the plants and wash the floor"; "We should start reusing bottles", "Riding in hydrogen cars", "We need to walk and cycle" and "Building hydrogen cars", Don't buy more stuff than we need, among others. Therefore, the workshop provided knowledge to students, and hopefully, they will be motivated to promote SD practices that will help to protect our planet.

4 Final Considerations and Future Work

The 17 SDG were defined to face the need for measures to protect our planet and provide better living conditions for all [1]. This paper presents an empirical study that resulted from a collaborative project led by a HEI and targeted to the school community. In this context, theoretical contents and hands-on experiments related to SDG (Table 3) was presented to teachers in a workshop, with the objective of strengthening consciousness for the development of good practices related to SDG. Also, 20 workshops were implemented with about 400 students in six schools to help teachers in their mission to provide knowledge and raise awareness of SDG with students.

It was verified that the workshop improved teachers' knowledge, motivation, and confidence to introduce SDG content to students. Besides acknowledging the necessity of professional development, teachers strengthened consciousness of their role and of schools in promoting actions and good practices related to SDG.

Furthermore, the students also extended their knowledge about SDG and understood the need to protect our planet, namely by measures such as reutilizing water among other actions.

Therefore, both workshops had impact in teachers and students by improving their knowledge in this matter and also about the need to be active in taking measures to improve our quality of life.

Although it has already been demonstrated by previous work that higher education students' participation in sustainability-related research projects improves their awareness of good practices [10], the present work extends this approach to lower levels of education and the community. So, it was verified that the project was effective in raising awareness of teachers and students about the importance of implementing good practices related to SDG. It is concluded that collaboration between HEI and schools can trigger the implementation of SDG practices in the community.

Acknowledgements This work has been partially funded by national funds through FCT—Fundação para a Ciência e a Tecnologia, I.P., under the Project UIDB/05567/2020 and "Verão com Ciência" action. The authors also acknowledge the support of: AcademiaCAP.IPT, BIOTEC.IPT and LAB.IPT; and Comunidade Intermunicipal do Médio Tejo, under the Project PEDIME—Plano estratégico de Desenvolvimento Intermunicipal da Educação no Médio Tejo (funded by Centro 2020).

References

1. United Nations: The 17 Goals, https://sdgs.un.org/goals last accessed 2023/01/25.
2. Caiado, R., Leal Filho, W., Quelhas, O., Nascimento, D., Ávila, L.: A literature-based review on potentials and constraints in the implementation of the sustainable development goals. J. Clean. Prod. 198, 1276–1288 (2018).
3. Riviera, F., Hermosilla, P., Delgadillo, J., Echeverría, D.: The Sustainable Development Goals (SDGs) as a Basis for Innovation Skills for Engineers in the Industry 4.0 Context. Sustainability, 12(16), 6622, 1–14 (2020).
4. Aleixo A., Leal, S., Azeiteiro, U.: Conceptualization of sustainable higher education institutions, roles, barriers, and challenges for sustainability: an exploratory study in Portugal. J. Clean. Prod. 172, 1664–1673 (2018).
5. de Amorim, W., da Silva Neiva, S., Castro, B., Deggau, A., Jonck, A., de Albuquerque Junior, C., de Andrade, J.: Higher Education Institutions as Drivers of Sustainable Communities: A Case Study of the University of Southern Santa Catarina Empowering the Community. In: Leal Filho W., Tortato U., Frankenberger F. (eds.) Universities and Sustainable Communities: Meeting the Goals of the Agenda 2030. World Sustainability Series. Springer, Cham (2020).
6. Leal Filho, W., Eustachio, J., Caldana, A., Will, M., Salvia, A., Rampasso, I., Anholon, R., Platje, J., Kovaleva, M.: Sustainability Leadership in Higher Education Institutions: An Overview of Challenges. Sustainability, 12(9), 3761, 1–19. (2020).
7. Ma, Y., Men, J., Cui, W.: Does Environmental Education Matter? Evidence from Provincial Higher Education Institutions in China. Sustainability, 12(16), 6338, 1–12 (2020).
8. Mateus, D., Costa, M., Gomes, M., Pinho, H.: Promoting education for sustainable development: a collaborative project between a higher education institution and the surrounding school community. In Leal Filho, W., Portela de Vasconcelos, C. (eds). Handbook of Best Practices in Sustainable Development at University Level. Springer, Cham (2022).
9. Cohen, L., Lawrence, M., Keith, M.: Research Methods in Education. 6th edn. Taylor and Francis, London (2007).
10. Mateus, D., Pinho, H., Nogueira, I., Rosa, M., Cartaxo, M., Nunes, V.: Participation of students in the project Valorbio: A case study to accelerate the implementation of sustainability principles in the curriculum, Int. J. Sustain. High. Educ., 21(2), 244–263 (2020).

Towards Sustainable Mobility: The Roadway Bridges Contribution

José C. Almeida

Abstract Bridges, due to their unique nature, are fundamental components of the road network. They have an important role in sustainable economic development as they are structures that allows to overcome obstacles created by nature, making circulation faster and safer. Climate action is one of the Sustainable Development Goals that The United Nations Development Programme defined, in 2015. Climate change is increasing the severity and frequency of climate-related extreme weather events around the globe. One of the parameters that enhances the global warming is the greenhouse gas emissions. For the past century, almost all internal combustion engines have been powered on petroleum fuels. Transportation is one of the most important contributors of greenhouse gases. In this paper it´s presented a methodology that allows the quantification of the life cycle costs and enhances the emission mitigation comparing different material alternatives of reinforced concrete roadway bridges. For the different material solutions, utilizing a probabilistic deterioration model, the lifetime of the different materials is computed, and the repair is scheduled. The methodology allows to define, and rank, the alternatives considering the user costs (vehicle operating costs; time costs; accidents costs, and toll costs) quantifying the correspondent vehicle greenhouse gases emission.

Keywords Sustainability · Life cycle analysis · Roadway bridges

1 Introduction

To solve the world's environmental and social challenges like poverty, inequality, conflict, and injustice as well as environmental degradation and climate change it's mandatory to promote sustainability [1]. Actual society is facing enormous challenges to meet the population´s basic need while protecting the environment. The

J. C. Almeida (✉)
CISE—Electromechatronic Systems Research Centre, Instituto Politécnico da Guarda, Guarda, Portugal
e-mail: jcalmeida@ipg.pt

J. R. da. C. S. Galvão et al. (eds.), *Proceedings of the 3rd International Conference on Water Energy Food and Sustainability (ICoWEFS 2023)*, Springer Proceedings in Earth and Environmental Sciences, https://doi.org/10.1007/978-3-031-48532-9_50

concept of sustainable development was defined by the World Commission on Environment and Development through its Brundtland report, a document entitled Our Common Future. According to this report, sustainable development should "meet the necessities of the present generation without harming the future generation´s capacity to meet their own" [2]. A sustainable approach must deal with the integration of social, environmental, and economic performance of human lives to benefit both current and future generations [3]. Related with the sustainable levels of resources consumption, presently, there is no country that meets the basic needs of its population [4]. Society is facing a multiplicity of challenges, to stimulate action in areas of critical importance for humanity and the planet. With the aim of overcome actual and future sustainability development the United Nations defined 17 Sustainable Development Goals and 169 targets [5].

According to [6], Sustainable Development Goals are evolving in an increasingly complex global environment in the context of climate change, population growth, economic transformation, and policy change. To identify the pillars underpinning overall sustainability and more holistically predict the response of Sustainable Development Goals to global change, it is crucial to identify the key axes of Sustainable Development Goals progress as a representation of global sustainable development.

Sustainability deals with the balanced integration of the three sustainable pillars [7, 8]. Then, to promote organizations and societies capable of addressing the world´s sustainability challenges is essential to have effective sustainable management education [9].

The sustainability consciousness is increasing, but if the goal is to make advances on pressing environmental concerns such as global climate change and ecosystem degradation it's fundamental to change the way that society produces energy and uses natural resources [10].

1.1 The Roadway Bridges Impact

It's known that one of the major contributors to greenhouse gas emission is the loss of forest [11]. But, based on contamination and the material´s function in each stage of its life cycle production, from cradle to grave, the sustainable impact of construction materials differs [12, 13]. The carbon emissions of transportation systems impact the safety of bridges, which are crucial components of transportation systems.

The emissions related to the transport systems usually comes from two different causes: vehicle emissions and construction materials. The emissions that result from vehicle directly discharge greenhouse gases to the atmosphere. The construction typically used infrastructures are materials like concrete and steel. Those produce greenhouse gases emissions due to the production and transport processes. To mitigate the effects of global warming the goal for 2050 is to reduce the net global CO_2 emissions below zero [14]. The global consumption of construction materials for highway infrastructure improvement will still increase over the next three decades [15]. Subsequently its mandatory to reduce the greenhouse emission for the highway infrastructures systems. A sustainable transportation system that aims to diminish greenhouse gases emissions can help to accomplish two of the sustainable development goals as climate change and sustainable cities and communities [16–18].

One of the largest consumers of natural resources is the construction industry [19]. It consumes over 40% of all raw stone, gravel, and sand, 25% of all raw timber, 40% of energy, and 16% of the annually water used in the world. As stated, bridges, due to their nature, are critical infrastructures of the transport system, that permits the connection of communities allowing the reduction of the travelling time and distance. The sustainability goals can be reached utilizing bridge management systems that allows to implement a cost-effective construction and maintenance [20, 21].

There are many factors that contribute to the degradation of bridges as part of the communication infrastructure, including negative environmental impact, increased loads, material fatigue, design, and construction errors as well as inaccurate repairs [22]. An important and crucial aspect of managing bridges is to establish some criteria to show the current condition of the bridge elements based on inspection results [23]. To address and have structured decision making and planning/scheduling for bridge infrastructure inspection, maintenance and repair or retrofit it´s necessary the utilization of Bridge Management Systems. The data collected from inspections, however, constitutes a sound database that can be used to understand the deterioration of managed bridges under certain conditions related to the environment, traffic, or durability [24]. The focus of this paper is to define criteria to determine the best material options taking user costs and environmental impact into account.

2 Methodology

For this work different models of deterioration of reinforced concrete structures, with special focus on damages resulting from carbonation and chloride ingress in bridge decks was studied. An applicable model is identified that allows modelling of reinforced concrete behaviour, to estimate time for corrosion onset and progression. Information is presented on construction practices which, depending on the level of exposure to the elements, can reduce deterioration and extend structure longevity. Several alternative materials to the common steel utilization—A0—were considered: A1—epoxy coated reinforcement; A2—galvanized steel reinforcement; A3—solid stainless-steel reinforcement; A4—coated stainless steel reinforcement; A5—use of corrosion inhibitors; and A6—protection/cathodic prevention.

This model defines the users costs through the application of existing international cost models. It also specifies a model to determine travel times based on distance and queueing possibilities. For the travelled distance the vehicle operating costs, time costs, accident costs, and toll costs are computed. Then, considering the different type of vehicles that move across the roads, the carbon emissions to the atmosphere are quantified.

2.1 Deterioration Process

The most widely used construction material in the world is reinforced concrete [25]. Among the benefits of concrete is that it provides a protective layer for the encased steel, provided by is alkaline nature, as well as being capable of holding up against the effects of time and stress in adverse situations. The durability and safety of reinforced concrete structures are directly affected by corrosion [26].

Corrosion of steel rebars is mainly caused by damage to their passive films caused by carbonate and chloride ions that destroy reinforced concrete structures [27, 28]. As a result of carbonation, calcium hydroxide in concrete reacts with carbon dioxide, destroying the alkaline passive film. The free chloride ions produced by chloride ion corrosion can corrode the passive film and increase conductivity, thus accelerating corrosion of the reinforcement. In this process, electrochemical reactions are associated with the corrosion of steel rebars in concrete [29].

Based on Tuutti´s model [30], the service life of a concrete structure can be divided into an incubation period and a propagation period.

The incubation period refers to the penetration of chloride into porous materials and the accumulation of chloride near the rebar. The duration of the corrosion process is determined by the concrete cover thickness and the concentration of chloride needed to initiate the corrosion process.

The propagation period corresponds to the stage during which the chloride ions depassivate the reinforcing steel surface, causing corrosion and, ultimately, the local destruction of the structure.

For the initiation time it was adopted probabilistic approach formulation based on the Fick's 2nd Law, or diffusion law, presented in the Eq. (1).

$$C(x,t) = C_0 + (C_s - C_0) \cdot \left[1 - erf\left(\frac{x}{2\sqrt{D_0 \cdot \left(\frac{t}{t_0}\right)^{\alpha} \cdot t}}\right)\right] \tag{1}$$

where: $C(x,t)$—chloride ion concentration at distance x after a t period of exposure [kg/m^3]; C_0—initial concentration of chloride ion inside the concrete [kg/m^3]; C_s—chloride ion concentration on the concrete surface [kg/m^3]; *erf*—error function; D(t)—diffusion coefficient for time t; D_0 –initial diffusion coefficient; t—time; t_0—reference initial time; α—exponent relative to the evolution of the diffusion coefficient, which depends on the type of material.

In the adopted model, the propagation time is estimated using the determination of the reduction of the cross section of the rebars that causes the beginning of the cracking of the concrete foreseen in the LNEC E-465 Specification [31]. Thus, the parameter is defined that estimates the relative reduction of the diameter of the reinforcements that originates the beginning of the concrete cracking. This parameter is quantified using the Eq. (2).

$$k = \frac{7.45 + 0.73 \cdot \frac{R}{\emptyset_0} - 1.74 \cdot f_{cd}}{\frac{\emptyset_0}{2}} \quad (2)$$

where: k—relative reduction of the reinforcement diameter that initiates concrete cracking; R—concrete cover; $\emptyset_0$—initial rebar diameter; f_{cd}—design compressive strength of concrete.

2.2 User Costs

To maintain bridges, work must be performed on the structure, which often causes a greater or lesser degree of disturbance, depending on the type and the scope of the work. Aside from the construction costs themselves, known as direct costs, these works also result in costs for the users of the structure. Carrying out works can often lead to a decrease in the speed of traffic, resulting in disruption to the normal flow. This could come with additional costs for users if they experience delays while passing through a construction site, or lengthy queues, or if their fuel consumption increases, as well as any extra distance resulting from diverted routes and an increase in accidents caused by these works [32].

According to [33], user costs in work zones are all those resulting from the increase in vehicle operating costs and time due to delays resulting from construction, maintenance, or rehabilitation activities. It can also be added the costs resulting from the increase in the accident rate. User costs depend on the opportunity, duration, frequency, scope and characteristics of the work area; the volume and characteristics of the operation.

In this study it was considered the user costs represented in Eq. (3).

$$C = VOC + TC + AC + ToC \quad (3)$$

where: VOC—vehicle operating costs; TC—time costs; AC—accidents costs; ToC—toll costs (when applicable).

To determine vehicle operating costs, fuel, tire, maintenance, and engine oil costs were quantified, as well as the devaluation and idling costs. The quantification was carried out considering the different types of vehicles. The adopted values, in relation to the cost of time, were computed according to distribution proportion of the purpose of the trips along the days of the week; average monthly earnings of workers in general and workers in the transport sector; defining an hourly cost of business travel; and defining a percentage of the cost of non-work trips based on the cost of work trips. The accident costs depend on the type of victims that can result from the road accidents. In this model it was quantified the costs for damage only, light injuries, serious injuries and for fatalities. Considering the accident data of Portugal, the number of accidents was calculated considering the formulation stated in [34]. This formulation for the quantification of highways number of accidents, considers the traveled length and the annual average daily traffic. To consider all the traffic circulation situations it was introduced some correction factors like type and road location. Since the traffic flow can occur in highways with tolls, this factor was also considered.

All the costs were quantified considering four classes of vehicles: passenger cars; commercial cars; buses; and trucks. This study considers, for all the material alternatives, the costs that results from crossing the road without works being carried on comparing them with the flow crossing work zones or taking a detour road. The speed was quantified taking in consideration the level of service of the roads [35].

2.3 *Traffic Environmental Impact*

The carbon emissions resulting due to the reconstructions of bridges are around 336 kg/m^3 [36, 37]. According to the authors the carbon emission resulting from the preventive repairs depends on the type of applied technique and from the repair materials. Beside the user cost that result from the disturbance of the normal traffic flow it was also considered the vehicle emissions. According to [38] the gasoline consumption of light traffic in the US is about 0.098 L/km, and the diesel consumption of heavy traffic is 0.326 L/km [39]. The US Environmental Protection Agency states that the carbon emission of gasoline is 2.35 kg/L and the carbon emission of diesel is 2.69 kg/L [39].

3 Case Study

This methodology was then applied in the bridge over the Vila Chã stream. The bridge, that belongs to the E80 Highway located in the Guarda district between Celorico da Beira and Fornos de Algodres (Fig. 1). The original bridge was duplicated in 2003. Alongside of the original bridge it was built a new one with four spans, and a total length of 74.00 m. The total length between the highway nodes is 12.100 kms. Just along with the highway E80 is located the detour road N16 with a total length between nodes of 13.300 kms.

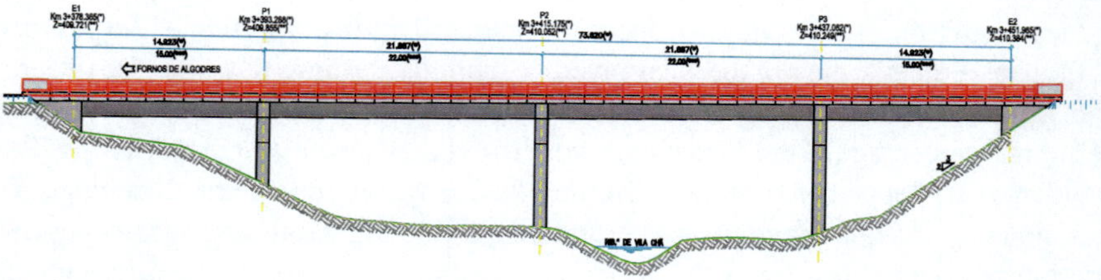

Fig. 1 Bridge under study

For the materials employed in the bridge construction, considering the type of concrete and the respective cover, employing the deterioration models, the lifetime of the alternative A0 was defined. This period corresponds to the end of the initiation time. Considering the existent values, available in the literature, it was quantified the lifetime extension of each alternative. Those values are presented in the Table 1.

It can be observed in Table 1 that the different materials behave differentially. Considering the bridge design lifetime, it can be noted that it will be necessary to perform three interventions for alternative A0 and the alternative A3 has a service life bigger that the lifespan of the bridge. Admitting the life span of each intervention it was calculated the number of interventions in the 100 years period.

When repair operations are carried out, and due to restrictions on the free flow circulation, additional costs for users are created. Here, a comparative study is carried out between the costs of users without considering any disruption due to the existence of works and the costs that occur due to the existence of works. The differential between these costs will be the assigned value to each intervention.

The user costs related to distance, time, accident, and toll, were computed for the different categories of vehicles: passenger, commercial, trucks and buses. The traffic was considered admitted the available data of the Portuguese Institute for Mobility and Transport.

The vehicle operating costs were quantified for the different vehicle classes the fuel costs, tire costs, maintenance costs, and depreciation costs. The adopted values, regarding the cost of time, were calculated due to statistical distribution of trips according to the day of the week; number of occupants for vehicle type; average monthly earnings of workers in the general sector and workers in the transportation sector; definition of an hourly cost for business travel; and definition of a percentage of non-work travel costs based on work travel costs. All the costs were quantified considering the density of vehicles in the roads and the correspondent level of service. For each level of service, it was defined the travelling crossing speed and the maximum capacity of the road. When the number of vehicles is greater than the road capacity queues are formed. Considering that the free flow will be disturbed with the construction works, idling costs were also computed. For all the scenarios it was defined a maximum number of queued vehicles and the correspondent number of vehicles that were detoured. The accident costs were defined considering the accident data of the Portuguese National Road Safety Authority. The number of accidents were computed considering the formulation presented in [34] utilizing correction factors related to the type and location of the road. Considering the accident unitary costs in the international literature the total accident costs were computed utilizing the human capital cost method and the global cost method [40]. The total accident costs

Table 1 Service life and number of interventions of different repair options

Alternative	A0	A1	A2	A3	A4	A5	A6
Service life	30	A0 + 20	A0 + 5	A0 + 80	A0 + 50	A0 + 20	A0 + 35
Interventions	3	2	2	0	1	2	1

Table 2 Service life and number of interventions of different repair options

Alternative	A0	A1	A2	A3	A4	A5	A6
User costs [k€]	18,622	10,391	14,709	0	4984	10,391	5869
CO_2 emissions [ton]	221	171	125	0	53	171	76

were found according to the statistical and unitary costs for fatal victims, serious injuries, light injuries, and damage only.

It was found that in the unrestricted flow scenario the biggest user costs are related with the vehicle operating costs, approximately 45% of the total amount, the toll costs represent around 30%, the time costs about 20%. The remaining 5% are linked with the accident costs. In this unrestricted flow, even considering a 2% increasing rate, there will be no queue formation in the entire studied period.

For the circulation with the construction works it was found that the biggest share of the costs is transferred to the time costs with the queue formation (50%) and the second biggest share (20%) is related with the operating vehicle costs in queued conditions. The total toll cost drops to less than 15%. All the other costs are related to unforced flow and detour costs.

For all the circulation scenarios, and for all the material alternatives, the tail pipe emissions were computed. The total user costs and the CO_2 emissions are plotted in Table 2.

It can be observed that the alternative A3 is the best solution regarding not only the user costs but also the amount of CO_2 emissions. This is related to the lifetime of the stainless-steel material. The lifetime of this material solution leads to a situation that the bridge will never suffer any intervention in the design expected life. In the other end, as expected, is the common reinforced concrete utilization. Due to is shorter life span, this cheaper material solution, leads to a superior number of interventions causing bigger user costs and higher CO_2 emissions.

4 Conclusions

This study demonstrates that the best, and more sustainable, strategy is to utilize materials that causes the minimum users disruption. Putting only direct costs in perspective, stainless steel reinforcement is the most expensive solution. Based on the immediate cost comparison, this option is nearly 2.5 times more expensive than common steel reinforcement. The time costs are proved to be the main contributor for the total costs. The total costs, direct and user costs, for alternative A0 is more ten times the costs of alternative A3. Since the main cost are the user costs, reducing the total number of intervention days reduces total costs almost proportionally. Considering the next generations, environmental and economic sustainability, the policy makers should never make their decisions considering only in the direct costs. Comparing the results is should be stated that life cycle analysis and life cycle costs should be mandatory for bridge construction projects.

Acknowledgements This work was supported by the Portuguese Foundation for Science and Technology (FCT) under Projects UIDB/04131/2020 and UIDP/04131/2020.

References

1. United Nations: Sustainable development goals: Communications materials. https://www.un.org/sustainabledevelopment/news/communications-material/, last accessed 2020/07/30.
2. WCED: Our Common Future: a report from the United Nations World Commission on Environment and Development. Oxford University Press, Oxford (1987). https://sustainabledevelopment.un.org/content/documents/5987our-common-future.pdf., last accessed 2022/08/19.
3. Ranjbari, M., Morales-Alonso, G., Shams Esfandabadi, Z., Carrasco-Gallego, R.: Sustainability and the Sharing Economy: Modelling the Interconnections. Dirección y Organización, 68, pp. 33–40 (2019).
4. O'Neill, D.W., Fanning, A.L., Lamb, W.F., Steinberger, J.K.: A good life for all within planetary boundaries. Nature Sustainability, 1(2), pp. 88–95 (2018).
5. UN: Transforming Our World: the 2030 Agenda for Sustainable Development. https://sdgs.un.org/2030agenda, last accessed 2022/10/16.
6. Yin, C., Zhao, W., Fu, B., Meadows, M.E., Pereira, P.: Key axes of global progress towards the Sustainable Development Goals. Journal of Cleaner Production, 385 (2023).
7. Geissdoerfer, M., Savaget, P., Bocken, N.M.P., Hultink, E.J.: The Circular Economy – A new sustainability paradigm? Journal of Cleaner Production, 143, pp. 757–768 (2017).
8. Hansmann, R., Mieg, H.A., Frischknecht, P.: Principal sustainability components: Empirical analysis of synergies between the three pillars of sustainability. The International Journal of Sustainable Development and World Ecology, 19(5), pp. 451–459 (2012).
9. Greenland, S., Saleem, M., Misra, R., Mason, J.: Sustainable management education and an empirical five-pillar model of sustainability. The International Journal of Management Education, 20(3) (2022).
10. Hall, J.K., Daneke, G.A., Lenox, M.J.: Sustainable development and entrepreneurship: Past contributions and future directions. Journal of Business Venturing, 25, pp. 439–448 (2010).
11. Bustamante, M.M.C., Roitman, I., Aide, T.M., Alencar, A., Anderson, L.O., Aragão, L., Asner, G.P., Barlow, J., Berenguer, E., Chambers, J., Costa, M.H., Fanin, T., Ferreira, L.G., Ferreira, J., Keller, M., Magnusson, W.E., Morales-Barquero, L., Morton, D., Ometto, J.P.H.B., Palace, M., Peres, C.A., Silvério, D., Trumbore, S., Vieira, I.C.G.: Toward an integrated monitoring framework to assess the effects of tropical forest degradation and recovery on carbon stocks and biodiversity. Global Change Biology, 22(1), pp. 92–109 (2016).
12. Balaras, C.A., Droutsa, K., Dascalaki, E., Kontoyiannidis, S.: Heating energy consumption and resulting environmental impact of European apartment buildings. Energy and Buildings 37(5), pp. 429–442 (2005).
13. Moncaster, A.M., Symons, K.E.: A method and tool for 'cradle to grave' embodied carbon and energy impacts of UK buildings in compliance with the new TC350 standards. Energy and Buildings, 66, pp. 514–523 (2013).
14. Miller, S.A., Horvath, A., Monteiro, P.J.: Readily implementable techniques can cut annual CO2 emissions from the production of concrete by over 20%. Environmental Research Letters, 11(7), Article 074029, (2016).
15. Gursel, A.P., Masanet, E., Horvath, A., Stadel, A.: Life-cycle inventory analysis of concrete production: A critical review. Cement & Concrete Composites, 51, pp. 38–48 (2014).

16. Bibri, S.E., Krogstie, J.: On the social shaping dimensions of smart sustainable cities: A study in science, technology, and society. Sustainable Cities and Society, 29, pp. 219–246 (2017).
17. Ibrahim, M., El-Zaart, A., Adams, C.: Smart sustainable cities roadmap: Readiness for transformation towards urban sustainability. Sustainable Cities and Society, 37, pp. 530–540 (2018).
18. Silva, B.N., Khan, M., Han, K.: Towards sustainable smart cities: A review of trends, architectures, components, and open challenges in smart cities. Sustainable Cities and Society, 38, pp. 697–713 (2018).
19. Martí, J.V., García-Segura, T., Yepes, V.: Structural design of precast-prestressed concrete U-beam road bridges based on embodied energy. Journal of Cleaner Production, 120, pp. 231–240 (2016).
20. Mohammadi, A., El-Diraby, T.: Toward user-oriented asset management for urban railway systems. Sustainable Cities and Society, 70, Article 102903 (2021).
21. Yang, Y., Ng, S.T., Xu, F.J., Skitmore, M.: Towards sustainable and resilient high-density cities through better integration of infrastructure networks. Sustainable Cities and Society, 42, pp. 407–422 (2018).
22. Fabianowski, D., Jakiel, P., Stemplewski, S.: Development of artificial neural network for condition assessment of bridges based on hybrid decision making method – Feasibility study. Expert Systems with Applications, 168, article 114271 (2021).
23. Tarighat, A., Miyamoto, A.: Fuzzy concrete bridge deck condition rating method for practical bridge management system. Expert Systems with Applications, 36(10), pp. 12077–12085 (2009).
24. Frangopol, D.M., Dong, Y., Sabatino S.: Bridge life-cycle performance and cost: analysis, prediction, optimisation and decision-making. Structure and Infrastructure Engineering, 13(10), pp. 1239–1257 (2017).
25. Nisbet, M.A., VanGeem, M.G., Gajda, J., Marceau, M.: Environmental life cycle inventory of portland cement concrete. World cement, 28, pp. 100–103 (1997).
26. James, A., Bazarchi, E., Chiniforush, A.A., Aghdam, P.P., Hosseini, M.R., Akbarnezhad, A., Martek, I., Ghodoosi, F.: Rebar corrosion detection, protection, and rehabilitation of reinforced concrete structures in coastal environments: A review. Construction and Building Materials, 224, pp. 1026–1039 (2019).
27. Sassine, E., Laurens, S., François, R., Ringot, E.: A critical discussion on rebar electrical continuity and usual interpretation thresholds in the field of half-cell potential measurements in steel reinforced concrete. Materials and Structures, 51, article 93 (2018).
28. Kaur, K., Goyal, S., Bhattacharjee, B., Kumar M.: Electrochemical impedance spectroscopy to study the carbonation behavior of concrete treated with corrosion inhibitors. Journal of Advanced Concrete Technology, 15(12), pp. 738–748, (2017).
29. Jiang, L., Wang, J., Yang, M., Lin, N., Liu, J.: Effect of Ca_2+ and Na+ on the curing performance of chloride ions in cement under carbonization. Case Studies in Construction Materials, 13 (2020).
30. Tuutti, K.: Corrosion of steel in concrete. CBI Research Report n°. 4.82. Swedish Cement and Concrete Research Institute, Stockholm, Sweden (1982).
31. LNEC E-465: Concrete – Methodology for estimating concrete performance properties that allows to satisfy the design lifetime of reinforced or prestressed concrete structures under environmental exposures XC and XS. (in portuguese), Laboratório Nacional de Engenharia Civil, Lisbon, Portugal (2007).
32. Branco, F.A., de Brito, J.: Handbook of Concrete Bridge Management. American Society of Civil Engineers, ASCE Press Series, (2004).
33. NJDOT: Road user cost manual. New Jersey Department of Transportation, New Jersey, USA (2001).
34. Lopes, S.A., Cardoso, J.L.: Accident prediction models for Portuguese motorways. Laboratório Nacional de Engenharia Civil, Lisbon, Portugal, (2007).
35. National Academies of Sciences, Engineering, and Medicine. Highway Capacity Manual 7th Edition: A Guide for Multimodal Mobility Analysis. Washington, DC: The National Academies Press (2022).

36. U.S. Environmental Protection Agency: Greenhouse Gas Emissions from a Typical Passenger Vehicle. Ann Arbor (2018).
37. Worrell, E., Price, L., Martin, N., Hendriks, C., Meida, L.O.: Carbon dioxide emissions from the global cement industry. Annual Review of Environment and Resources, 26(1), pp. 303–329 (2001).
38. FHWA: Highway statistics. Federal Highway Administration, Washington, D.C., USA, https://www.fhwa.dot.gov/policyinformation/statistics/2020/, last accessed 2021/09/07.
39. Schoettle, B., Sivak, M., Tunnell, M.: A Survey of Fuel Economy and Fuel Usage by Heavy-Duty Truck Fleets. Arlington, Virginia: The University of Michigan, USA (2016).
40. Blincoe, L.J., Seay, A.G., Zaloshnja, E., Miller, T.R., Romano, E.O., Luchter, S., Spicer, R.S.: The economic impact of motor vehicle crashes 2000, DOT HS 809 446. National Highway Traffic Safety Administration, U.S. Department of Transportation, Washington D.C., USA (2002).

Sustainability Assessment of Building Rehabilitation Solutions: The Mid 70s Portuguese Building

Alexandre Reis, Lizete Heleno, Silvia Monteiro, and Nelson S. Oliveira

Abstract The Energy Performance of Buildings Directive (EPBD) establishes improvements to buildings' comfort conditions and energy performance, bringing them closer to buildings with almost zero energy needs, namely the Nearly Zero Energy Building (NZEBs). The assessment tools for sustainable construction, more recognized internationally, are LEED and BREEAM, and in Portugal is the LiderA. The application of those tools promotes energy efficiency and decarbonization of buildings, as referred to in the Portuguese Long-Term Strategy for Building Renovation (LTRS-PT) until 2050. In this case, a house representative of the construction of the period between 1971 and 1980 in Portugal was chosen to evaluate environmental sustainability using LiderA. In the current conditions, the LiderA rating class of the house was D. To improve the performance of the building were identified balanced rehabilitation solutions without excessive automation. The presented solutions should avoid transforming the house into a complicated device with operational difficulties or specialized operation. In this way, the proposed interventions in the building and systems are based on renewable sources, prioritizing energy and water efficiency. Furthermore, as a contribution to the health and well-being of the occupants, we considered measures related to indoor air quality (IAQ) and noise minimization. In addition, the encouragement of native species has contributed to ecological enhancement. With that proposals, the house rating class using LiderA improved from class D to class A+.

Keywords EPBD · Green building rating systems · LiderA · Environmental sustainability

A. Reis · L. Heleno
School of Technology and Management, Polytechnic Institute of Leiria, Leiria, Portugal

S. Monteiro · N. S. Oliveira (✉)
LSRE-LCM—Laboratory of Separation and Reaction Engineering-Laboratory of Catalysis and Materials, School of Technology and Management, Polytechnic Institute of Leiria, Leiria, Portugal
e-mail: nelson.oliveira@ipleiria.pt

ALiCE—Associate Laboratory in Chemical Engineering, Faculty of Engineering, University of Porto, Rua Dr. Roberto Frias, Porto, Portugal

J. R. da. C. S. Galvão et al. (eds.), *Proceedings of the 3rd International Conference on Water Energy Food and Sustainability (ICoWEFS 2023)*, Springer Proceedings in Earth and Environmental Sciences, https://doi.org/10.1007/978-3-031-48532-9_51

1 Introduction

The Energy Performance of Buildings Directive (EPBD) establishes that improvements must be introduced in the comfort conditions and the energy performance of existing buildings, bringing them Nearly Zero Energy Buildings (NZEBs). To this end, financing instruments should be available, especially for people experiencing energy poverty. The EPBD also guarantees full respect for the national differences and specificities of each Member State—different countries, different cultures, and different people.

In 1987, the World Commission on Environment and Development (WCED), created in 1983, published a report entitled "Our common future". The document became known as the Brundtland Report in honor of Commission President Gro Harlem Brundtland, having launched the guiding principles for sustainable development as it is currently known. The Brundtland Report established that the critical global environmental problems were primarily the result of massive poverty in the South and unsustainable consumption and production patterns in the North. Therefore, it required a strategy that united development and the environment, now known as sustainable development. It is defined as follows: "Sustainable development is a development that meets the needs of the present without compromising the ability of future generations to meet their own needs". In 1989, the United Nations (UN) General Assembly debated the report and decided to organize the UN Conference on Environment and Development [1].

From the 1980s onwards, many buildings began to be considered sustainable based on energy efficiency, thus ensuring a lower environmental impact and less dependence on air conditioning equipment. The preliminary study phase began to be considered decisive, with joint action by teams of architects and engineers to minimize the environmental impacts of the construction process itself, starting to value bioclimatic architecture more and more [2]. The successive concern with sustainable construction eventually led to the emergence of Green Building Rating Systems (GBRS) in the early 1990s.

The Long-Term Strategy for Building Renovation makes explicit reference to the most internationally recognized GBRS, the Leadership in Energy and Environment Design (LEED), and the Building Research Establishment's Environmental Assessment Method (BREEAM), as well as to the Portuguese LiderA, acronym of Leading for the Environment for sustainable construction. Renovated and sustainable buildings in the European Union (EU) will help pave the way for a clean and decarbonized energy system, as buildings are one of Europe's most significant sources of energy consumption, responsible for over a third of EU emissions. But, per year, only 1% of buildings are subject to rehabilitation actions, so effective action is crucial to make Europe climate neutral (net zero emissions) by 2050. Currently, around 75% of buildings in the EU are not energy efficient, but 85% to 95% of current buildings will still be in use in 2050. Therefore, the renovation of public and private buildings is a necessary step and was identified in the European Green Deal [3] as a critical initiative to drive energy efficiency in the sector and meet targets.

Commission's recovery plan aims to continue to support the rehabilitation of EU buildings. To achieve this dual ambition of greater energy efficiency and economic growth, in 2020, the European Commission published a new strategy to boost renovation called "A Renovation Wave for Europe—Greening our buildings, creating jobs, improving lives" [4]. In addition to reducing Greenhouse gas (GHGs) emissions, these building rehabilitation actions will improve the occupants' quality of life, with the prospect of creating many additional "green" jobs in the construction sector. Furthermore, with almost 34 million Europeans unable to adequately heat their homes, the Renovation Wave can also help fight energy poverty, providing better comfort and well-being conditions to the most vulnerable [5]. In Portugal, the existing building stock cannot provide families with adequate living conditions and well-being. Current issues are bad indoor air quality and thermal and acoustic discomfort, giving rise to the occurrence of pathologies in the construction elements that can cause health problems to its occupants [6], especially the elderly, because of the ageing of the Portuguese population [7].

The main objective of this study is to assess the sustainability of rehabilitation solutions to a dwelling using the Portuguese GBRS LiderA. It was based on a three bedroom house [8] that was designed with constructive solutions from the period corresponding to the construction boom between 1971 and 1980 [9]. We attempted to recreate the practice in force at the time: a reinforced concrete structure, masonry with plate [9], and spans glazed with single glass [10].

2 Literature Review

Many organizations and academia have widely developed building Sustainability Assessment Systems (BSAS), emphasizing GBRS within the concept of BSAS [11]. Environmental Impact Assessment (EIA) tools link the environmental context of a building to the design decision-making framework. Life Cycle Assessment (LCA) and GBRS are the two commonly used approaches to analyze the environmental performance of the entire building holistically. While the GBRS rely primarily on a checklist with many qualitative criteria, the LCA obliges decision-makers to base their analyzes on numerical evidence, making it easier to compare design options. Some GBRS, such as LEED, BREEAM, and Green Star, have incorporated the LCA into their assessment system. This practice brings transparency to the design process and increases designers' awareness of the building's environmental impact [12].

The first GBRS appeared in the United Kingdom in 1990, BREEAM, through the Building Research Establishment (BRE) Group [13]. In Europe, the Code for Sustainable Homes (CSH) was identified, also from the United Kingdom [14], similar to BREEAM; the Deutsche Gesellschaft für Nachhaltiges Bauen (DGNB), from Germany [15]; EcoEffect, from Sweden [16]; the Haute Qualité Environnementale (HQE), from France [17]; LiderA, from Portugal; PromisE, from Finland [18]; the Protocollo ITACA, from Italy [19] and the Total Quality Building Assessment (TQB), from Austria [20]. In 1994, LEED [21] emerged in the United States of

America, which has increasingly become one of the GBRS of reference worldwide. In the American continent, in addition to LEED, there is Green Globes from Canada [22]. The International Initiative for a Sustainable Built Environment (iiSBE) [23] is registered in Canada, and it developed the Sustainable Building Tool (SBTool), which has been adapted to several countries, including Portugal. In Oceania, were identified the National Australian Built Environment Rating System (NABERS) [24] and the Green Star [25].

Asia was the continent where it was possible to identify the most significant number of GBRS, 11 in total. With distribution across ten countries, namely China, with the Assessment Standard for Green Building (ASGB) [26], the Comprehensive Assessment System for Building Environmental Efficiency (CASBEE) from Japan [27], the Green Building Index (GBI) from Malaysia [28], the Green Mark from Singapore [29], the Global Sustainability Assessment System (GSAS) from Qatar [30], the Indian Green Building Council (IGBC) from India [31], the LOTUS from Vietnam [32], the Pearl Building Rating System (PBRS) from the United Arab Emirates [33] and finally Hong Kong with two GBRS, the BEAM [34] and the CEPAS [35]. We did not identified any GBRS originating in Africa. Some GBRS have local representation, such as the SBTool, with a specific development for each country. In addition, BREEAM, DGNB, Green Globes, Green Star, HQE, and LEED are also represented in countries other than the location where the initial GBRS were developed. However, even within the same country, there can be considerable variations in climate and culture, as is the case, for example, in Brazil and China [11].

Numerous researchers have analyzed and compared different GBRS from different perspectives. Shan and Hwang [36] concluded that the most mentioned GBRS in the researched scientific articles are LEED, BREEAM, CASBEE, Green Star, ASGB, and BEAM [36]. Elena Bernardi et al. [37] concluded that the most relevant GBRS are LEED, BREEAM, CASBEE, SBTool, and HQE [37]. Considering the number of citations in Scopus of a work carried out in April 2017 according to specific keywords, Bernardi et al. [37] prioritized several GBRS. Based on the study of these authors [37], we made an update for the year 2021, and the two most representative GBRS internationally (LEED and BREEAM) and the two GBRS with representativeness in Portugal (SBTool and LiderA) were selected. Table 1 summarizes the result of this hierarchy, showing an increase in citations from 2017 to 2021.

The LiderA system, which evaluates the environmental performance of buildings, recommends that they are planned and executed with the concern to minimize environmental impacts, enabling the achievement of a level of service compatible with the desired. On the other hand, the fact that a project is thought out from the beginning in order to integrate sustainable measures may contribute to a balanced solution of the costs in the life cycle of the construction, thus bringing not only environmental but also economic benefits. The system was developed by Manuel Duarte Pinheiro, a lecturer at the Department of Civil Engineering and Architecture of the Instituto Superior Técnico (IST).

In 2005 the first version was made available, which was mainly intended for buildings and their surroundings. As a consequence of the applications made, version

Table 1 Citations in Scopus of the two most representative GBRS internationally and the two GBRS with representativeness in Portugal

GBRS	Keywords	Citations in Scopus	
		2017	2021
Leed	Leed OR "leadership in energy and environmental design" AND sustainable AND building AND (assessment OR evaluation)	256	480
Breeam	Breeam OR ("bre environmental assessment method" OR "building research establishment environmental assessment methodology") AND sustainable AND building AND (assessment OR evaluation)	132	219
SBTool	sbtool AND sustainable AND building AND (assessment OR evaluation)	28	42
LiderA	Lidera OR "liderar pelo ambiente para a construção sustentável" AND sustainable AND building AND (assessment OR evaluation)	5	6

2.0 was presented in 2010 in a perspective of sustainable communities, extending the possibility of application of the system, not only to the built environment, but also to the built environment, including the search for buildings, outdoor spaces, blocks, neighbourhoods, zones and their users. The system has already been implemented in different types of projects and by different agents, having achieved the certification of projects from the planning and design phase to the operation phase. There are 6 aspects, 20 areas and 40 criteria, based on the principle that the legal requirements are met and adopted as minimum requirements in the different areas considered, with the search for sustainability based on the improvement in relation to the common practice [39].

3 Case Study

This work considers a single house [8] designed with building solutions corresponding to the construction boom between 1971 and 1980 [9]. It attempted to recreate the practice in force at the time: a reinforced concrete structure, masonry walls [9], and windows glazed with single glass [10].

The assumptions considered in this case study for the base scenario were the following:

- The house is vacant and on a plot without any trees or fences. It has shrubs in the north and south boundaries;
- The lot is in the process of being abandoned. It has a lawn next to the entrance and flower beds, but abandoned;
- House framed in the surrounding area in terms of colors and materials;

- It does not have any system for storing or treating rainwater. However, the water from the roof is conducted through gutters to downspout tubes that, in turn, connect to boxes that are connected to the municipality's rainwater network;
- There is a connection of wastewater to the municipal sewage network;
- The supply of drinking water is ensured through the public network;
- It additionally has a water well with an independent net that serves the lawn, the flowerbeds and the washing of the impervious areas;
- It has two floors and is located in the municipality of Oliveira de Azeméis at an altitude of 188 m. The ground floor consists of a lobby, bathroom 1, living room, kitchen, laundry, pantry, and access stairs to the 1st floor. On the 1st floor, there are the following compartments: hall, bedroom 1, bedroom 2, bedroom 3, and two bathrooms;
- The outer walls have a superficial thermal transmission coefficient of 1.09 W/(m^2 °C);
- As for the roof, it consists of a mat slab in contact with a highly ventilated attic, with an overall heat transfer coefficient of 2.55 W/(m^2 °C);
- Regarding the floors, the ground floor is in contact with the ground and has an overall heat transfer coefficient of 0.64 W/(m^2 °C);
- The floor is in contact with non-useful, highly ventilated spaces, the laundry room, and the pantry through the floor slab, which has an overall heat transfer coefficient of 1.57 W/(m^2 °C);
- The exterior glazed spans whose values of the overall heat transfer coefficient without night occlusion device and the overall heat transfer coefficient day-nigth are 6.50 W/(m^2°C) and 4.10 W/(m^2 °C), respectively;
- No air conditioning systems or domestic hot water (DHW) production exists.

In October 2020, in a first approach, the provisional Energy Performance Certificate (EPC) was issued using the PTnZEB calculation tool [38]. According to the EPC, there is no risk of overheating in summer. On the other hand, the house will be uncomfortable in winter, 115% less efficient than the reference building. Concerning DHW, it is 11% less efficient than the reference. There is no contribution from renewable energy systems (RES), and the EPC rating is D.

To turn the house into an NZEB we considered the following improvement measures:

- Application of thermal insulation (10 cm of mineral wool) on the roof mat slab;
- Application of thermal insulation (6 cm of expanded polystyrene) on the outer walls;
- Replacement of the rolling shutter box with new ones in expanded polystyrene;
- Replacement of windows with new ones in aluminium with thermal break and low-emissivity double glazing, including self-regulating ventilation grilles in the living room, kitchen, and bedrooms, as well as a compensation grille for the kitchen hood;
- Installation of a solar thermal system for DHW with backup through electrical resistance;
- Installation of radiators in the main compartments, living room and bedrooms;

- Installation of a pellet stove in the living room, connected to the radiators.

With the implementation of the improvement measures, there is still no risk of overheating in the summer. On the other hand, given the interventions in the building and the adoption of a heating system based on a renewable source (biomass), conditions in winter are substantially improved, 100% more efficient than the reference building and 100% renewable. As for the DHW, the improvement is also significant due to the solar thermal system, 69% more efficient than the reference building and 70% renewable. Finally, the renewable energy contribution is 93%, and the EPC rating is A+.

To assess the sustainability of rehabilitation solutions to the base scenario, we used the LiderA methodology. The performance level starts with a lower performance from the G class to the highest performance of A++ with the best sustainable practices. It comprises 40 criteria divided into six evaluation categories (site and integration, resources, environmental loadings, environmental comfort, socioeconomic experience, and sustainable use), covering environment, economy, and social aspects. The application of the methodology to the base scenario reached the classification of D.

The LiderA methodology was applied to the new scenario with rehabilitation solutions, considering the new improvement measures for the EPC and the improvement measures for each criterion. The improvement measures took care not to turn the building into a mechanism, seeking to minimize the excessive use of systems and controls that could result in complex handling by the occupants. The proposed rehabilitation solutions could increase the classification from D (base scenario) to A+. It achieved a four-class upgrade with improvements to the surroundings, including indications regarding the structural elements and their maintenance, such as lubricating the hardware of the window frames, cleaning the self-regulating ventilation grilles, and cleaning the roofs, gutters, and drop tubes. In addition to the interventions in the surroundings, we proposed systems based on renewable sources. Pellet stove for heating and solar thermal system for the production of hot sanitary water, transforming the building into a NZEB, without neglecting the awareness information for energy efficiency, such as, identification of the equipment with the highest consumption, the stand-by of specific equipment, shower times and watering timer in periods of the day that minimize evaporation, the availability of operating manuals for the pellet stove (including information on the use, profitability and maintenance of the boiler thermostat—lower temperature avoiding energetically inefficient overheating), the solar thermal system (optimized for the periods of the day with sunlight, limiting the functioning of the auxiliary electrical resistance in these periods) and the appliances (which must be class Energy A or higher, as well as lighting systems), the reduction of drinking water consumption levels, the installation of mixer taps and flow reducers with water efficiency class A, the use of double-flush cisterns with efficiency class hydric A and automated and efficient irrigation system.

In addition to the concern with the water-energy nexus, we emphasized the placement of native species. Thus, it was encouraged to plant orange and lemon trees, grow aromatic herbs, reduce the production of urban solid waste (including composting of organic waste), and implement solutions to minimize noise emissions. Furthermore, we ensured adequate natural ventilation of the spaces and the minimization of potential emissions of indoor pollutants.

4 Conclusions

Buildings represent an essential component of energy consumption and consequent GHG emissions. Therefore, the commitment to sustainable renovation of buildings with the introduction of improvements in comfort conditions and energy performance, bringing them closer to NZEB, could contribute to the decarbonization of the current Portuguese building stock, in line with the assumptions of the Long-Term Strategy for Building Renovation.

The literature review identified numerous green building classification systems, GBRS, originated from several countries from the continents of Europe, America, Asia, and Australia/Oceania. We concluded that the most cited globally in Scopus is LEED, launched in the USA by the USGBC, and BREEAM, developed by BRE from the United Kingdom. In Portugal, we identified the SBTool and LiderA. Except for BREEAM, which does not address the economic aspect, all systems take into account the three dimensions of sustainability: economic, social, and environmental. Concerning the parameters analyzed in each of the systems, it is possible to verify that the Internal Environment, Energy, Resources (soil, water, and materials), and Health and Well-Being are part of the methodologies provided by LEED, BREEAM, SBTool, and LiderA.

With the application of LiderA to a single-family house, we reached class D in the base scenario, as it was built in the mid 70s in Portugal. Then, with the possibility of integrating sustainable measures, we identified opportunities for improvement in all aspects without the ambition of transforming the building into a complicated device. This option could make operating systems and controls more difficult, especially with an increasingly aged population. Instead, interventions were proposed in the surroundings, complemented by systems based on renewable sources, a pellet stove and a solar thermal system, with particular emphasis on the water-energy nexus, using measures to monitor and minimize consumption. Furthermore, to ensure ecological enhancement, we proposed the use of plantations of native species, as well as orange and lemon trees. In addition, we suggested measures related to indoor air quality and noise minimization for health and well-being. Even giving up excessive monitoring and automation, the proposed solutions add environmental and economic benefits to the building, with the classification according to LiderA methodology improving from D to A+.

Acknowledgements This work was financially supported by LA/P/0045/2020 (ALiCE), UIDB/50020/2020, and UIDP/50020/2020 (LSRE-LCM), funded by national funds through FCT/MCTES (PIDDAC).

References

1. World Commission on Environment and Development, «1987: Brundtland Report», 1987. https://www.are.admin.ch/are/en/home/medien-und-publikationen/publikationen/nachhaltige-entwicklung/brundtland-report.html (acedido 16 de novembro de 2021).
2. P. A. da P. Nunes, «Afinal o que é a sustentabilidade na construção?», masterThesis, Universidade de Coimbra, 2015. Acedido: 16 de novembro de 2021. [Em linha]. Disponível em: https://estudogeral.sib.uc.pt/handle/10316/38700
3. European Commission, «Delivering the European Green Deal», *European Commission - European Commission*, 2019. https://ec.europa.eu/info/strategy/priorities-2019-2024/european-green-deal/delivering-european-green-deal_en (acedido 17 de novembro de 2021).
4. European Commission, «Renovation wave», *Energy - European Commission*, 2020. https://ec.europa.eu/energy/topics/energy-efficiency/energy-efficient-buildings/renovation-wave_en (acedido 17 de novembro de 2021).
5. European Commission, «Commission Recommendation (EU) on energy poverty», 32020H1563, 2020. Acedido: 17 de novembro de 2021. [Em linha]. Disponível em: http://data.europa.eu/eli/reco/2020/1563/oj/eng
6. Presidência do Conselho de Ministros, «Resolução do Conselho de Ministros n.º 8-A/2021 | DRE», *Diário da República Eletrónico*, 2021. https://dre.pt/ (acedido 19 de março de 2022).
7. PORDATA, «Indicadores de envelhecimento», 2020. https://www.pordata.pt/Portugal/Indicadores+de+envelhecimento-526 (acedido 30 de março de 2022).
8. Instituto Nacional de Estatística, «Portal do INE», 2021. https://www.ine.pt/xportal/xmain?xpid=INE&xpgid=ine_publicacoes&PUBLICACOESpub_boui=437127876&PUBLICACOESmodo=2 (acedido 19 de março de 2022).
9. S. A. de M. Barbosa, «Comparação do índice de desconforto passivo com a classe energética de edifícios de habitação reabilitados do sul da europa», 2020. Acedido: 16 de maio de 2021. [Em linha]. Disponível em: https://repositorio-aberto.up.pt/handle/10216/128487
10. Instituto Nacional de Estatística, I.P. (INE) e Direção-Geral de Energia e Geologia (DGEG), «Inquérito ao Consumo de Energia no Sector Doméstico - 2010». 2011. [Em linha]. Disponível em: https://www.ine.pt/xportal/xmain?xpid=INE&xpgid=ine_publicacoes&PUBLICACOESpub_boui=127226704&PUBLICACOESmodo=2
11. N. Lazar e K. Chithra, «A comprehensive literature review on development of Building Sustainability Assessment Systems», *Journal of Building Engineering*, vol. 32, p. 101450, nov. 2020, https://doi.org/10.1016/j.jobe.2020.101450.
12. T. Sartori, R. Drogemuller, S. Omrani, e F. Lamari, «A schematic framework for Life Cycle Assessment (LCA) and Green Building Rating System (GBRS)», *Journal of Building Engineering*, vol. 38, 2021, https://doi.org/10.1016/j.jobe.2021.102180.
13. BREEAM BRE, «BREEAM: the world's leading sustainability assessment method for masterplanning projects, infrastructure and buildings», *BREEAM*, 1990. https://www.breeam.com/ (acedido 3 de janeiro de 2021).
14. Ministry of Housing, Communities & Local Government, «Code for sustainable homes: technical guidance», *GOV.UK*, 2006. https://www.gov.uk/government/publications/code-for-sustainable-homes-technical-guidance (acedido 28 de março de 2021).
15. DGNB Germany, «DGNB – German Sustainable Building Council», 2008. https://www.dgnb.de/en/index.php (acedido 10 de janeiro de 2021).
16. Ecoeffect, «EcoEffect», 2006. http://www.ecoeffect.se/English/Environmental%20assessment/summary/Summary%20040817.htm (acedido 10 de janeiro de 2021).
17. beHQE, «HQE definition - What does HQE stands for ?», 1997. https://www.behqe.com/presentation-hqe/what-is-hqe (acedido 10 de janeiro de 2021).
18. PromisE, «Sustainable Building - VTT Materials and Construction», 2006. http://virtual.vtt.fi/virtual/proj6/environ/ympluok_e.html (acedido 10 de janeiro de 2021).
19. Protocollo ITACA, «Itaca», 2004. http://www.itaca.org/valutazione_sostenibilita.asp (acedido 10 de janeiro de 2021).

20. Austrian Sustainable Building Council, «TQB Assessment General Procedure», 2009. https://www.oegnb.net/en/ablauf.htm (acedido 6 de março de 2021).
21. LEED United States GBC, «LEED rating system | U.S. Green Building Council», 1998. https://www.usgbc.org/leed (acedido 3 de janeiro de 2021).
22. Green Globes ECD Canada, «Green Globes - About Green Globes», 2000. https://www.greenglobes.com/about.asp (acedido 3 de janeiro de 2021).
23. SBTool iiSBE, «SBTool and SNTool | International Initiative for a Sustainable Built Environment», 2002. http://www.iisbe.org/sbmethod (acedido 3 de janeiro de 2021).
24. NSW Department of Planning, Industry and Environment, «NABERS», 2001. https://www.nabers.gov.au/ (acedido 3 de janeiro de 2021).
25. Australian Green Building Council, «What is Green Star? | Green Building Council of Australia», 2003. https://new.gbca.org.au/rate/green-star/ (acedido 28 de março de 2021).
26. Y.-W. Ding e S. T. and S. Roh, «Comparison of green building standards in the United States and China», pp. 176–182, set. 2019, https://doi.org/10.22712/susb.20190018.
27. CASBEE, «Welcome to CASBEE website!!», 2004. https://www.ibec.or.jp/CASBEE/english/ (acedido 10 de janeiro de 2021).
28. Green Building Index SDN BHD, «Green Building Index», 2009. https://www.greenbuildingindex.org/ (acedido 28 de março de 2021).
29. Singapore Building & Construction Authority, «Green Mark Certification Scheme | Building and Construction Authority (BCA)», *Singapore Building & Construction Authority*, 2005. https://www1.bca.gov.sg/buildsg/sustainability/green-mark-certification-scheme (acedido 10 de janeiro de 2021).
30. GSAS, «GSAS OVERVIEW | GORD», 2007. https://www.gord.qa/trust-gsas-resource-center-overview (acedido 10 de janeiro de 2021).
31. Indian Green Building Council, «IGBC Rating System - Green Building Rating System in India | IGBC», 2013. https://igbc.in/igbc/redirectHtml.htm?redVal=showratingSysnosign (acedido 28 de março de 2021).
32. Vietnam GBC, «LOTUS Rating Systems», *VGBC*, 2007. https://vgbc.vn/en/lotus-en/rating-systems/ (acedido 10 de janeiro de 2021).
33. PBRS, «Pearl Building Rating System», 2010. https://www.dmt.gov.ae/en/Urban-Planning/Pearl-Building-Rating-System (acedido 10 de janeiro de 2021).
34. BEAM HK, «BEAM Hong Kong», 1996. https://www.beamsociety.org.hk/en_about_us_0.php (acedido 10 de janeiro de 2021).
35. CEPAS, «Notices and reports - Buildings Department», 2002. https://www.bd.gov.hk/en/resources/codes-and-references/notices-and-reports/index_CEPAS.html (acedido 10 de janeiro de 2021).
36. M. Shan e B. Hwang, «Green building rating systems: Global reviews of practices and research efforts», *Sustainable Cities and Society*, vol. 39, pp. 172–180, mai. 2018, https://doi.org/10.1016/j.scs.2018.02.034.
37. E. Bernardi, S. Carlucci, C. Cornaro, e R. A. Bohne, «An Analysis of the Most Adopted Rating Systems for Assessing the Environmental Impact of Buildings», *Sustainability*, vol. 9, n.º 7, Art. n.º 7, jul. 2017, https://doi.org/10.3390/su9071226.
38. PTnZEB, «PTnZEB – Plataforma para a promoção dos nZEB em Portugal», 2020. https://pt-nzeb.pt/ (acedido 29 de novembro de 2021).
39. Pinheiro, M. D. (2019). LiderA. Instituto Superior Técnico, Universidade Técnica de Lisboa, Lisboa.

Wood and Mineral Byproducts Soft Activated as Adsorbents to Liquid-Phase Application

Paulo Mourão and Claúdio Cassavela

Abstract Human activities generate byproducts and wastes that need to be valorised. This paper presents a comparative study on the valorisation of different natural byproducts, of vegetal and mineral origin, through their transformation into adsorbent materials. A series of byproducts (European wood pellets, African natural woods, natural charcoal, lignocellulosic ash, perlite and vermiculite minerals and expanded clay), as received and after smooth activation processes, were tested for adsorption from liquid phase. The aqueous phase tests were carried out with representative molecules from several classes of hazardous compounds such as dyes, pesticides and pharmaceuticals. The molecules selected were methylene blue (MB), acid yellow 17 (AY 17), (4-chloro-2-methyl-phenoxy) acetic acid (MCPA), 4-nitrophenol (4N) and sodium diclofenac (SD). The results obtained with this set of natural, alternative and non-conventional low-cost adsorbents were interesting, particularly in the case of adsorbents of vegetal origin, natural woods and also in the case of lignocellulosic ash. Regarding the soft chemical modification, it has been shown to be a process that allows the preparation of materials with improved performance for some adsorptive/adsorbent systems, namely natural woods/(AY 17; MB; 4N) and natural charcoal and lignocellulosic ash/all the adsorptives. Note that the adsorption processes were relatively fast, with an equilibrium time of around 24 h. In conclusion, the byproducts analysed, some of them after soft activation, present a potential for valorisation through their use as adsorbents for hazardous polluting compounds from the aqueous medium at a temperature of 25 °C.

Keywords Wood · Mineral · Byproducts · Adsorption · Dye · Pesticide · Drug

P. Mourão (✉)
Departamento de Química e Bioquímica, MED & CHANGE, IIFA, ECT, Universidade de Évora, Évora, Portugal
e-mail: pamm@uevora.pt

C. Cassavela
Universidade de Évora, LIFE BIOAs Project, Évora, Portugal

J. R. da. C. S. Galvão et al. (eds.), *Proceedings of the 3rd International Conference on Water Energy Food and Sustainability (ICoWEFS 2023)*, Springer Proceedings in Earth and Environmental Sciences, https://doi.org/10.1007/978-3-031-48532-9_52

1 Introduction

The perspective of a linear economy, which focuses on the product and forgets about the extraction process and all the byproducts, residues and waste generated, is outdated and the perspective of a circular economy, which focuses on sustainability, is increasingly proving to be the right way forward [1]. This focus on circularity, and consequently on sustainability, is the approach that will allow current generations not to compromise future generations [1, 2]. One of the strategies that can contribute to the increase of sustainability is the valorisation of byproducts and wastes through their direct use or after their transformation into new products with high application potential [3].

The transformation of byproducts and wastes into materials that can be used in the field of environmental remediation, in particular for the removal of pesticides, dyes, pharmaceuticals and other hazardous compounds, is increasingly attracting the attention of the scientific and technological communities. The natural byproducts of vegetal origin, i.e. lignocellulosic materials, and those of mineral origin, i.e. stone rubble, have the potential to be used as adsorptive materials, i.e. with the ability to retain substances that are in the liquid or gaseous phase [4]. The literature shows that this retention capacity, the adsorption capacity, can be increased by modifying these materials through physical and chemical processes [5, 6].

From the point of view of sustainability, research into modification processes that minimise the negative effects of the modifying agents or activating agents used, either in terms of quantity, hazard or other parameters, is becoming increasingly relevant. The strategy is to use soft activation processes in which the conditions used are minimised in terms of activating agent (e.g. type, quantity, mass ratio of activating agent to precursor material) and thermal conditions (e.g. activation temperature, heating rate, activation atmosphere) [7–9]. The aim of this soft activation is to improve the structural properties, namely, to achieve the development of porosity, or to modify the chemical nature of the material, in particular the surface chemistry, by introducing specific chemical groups. Alternatively, both objectives can be achieved with this activation step.

In the field of environmental remediation, challenges continue to arise, not so much in terms of the processes used, which are well known, such as adsorption, but rather in terms of the cost of materials and the procedures to be implemented to achieve a positive cost/efficiency balance. The choice of adsorption as one of the most efficient processes is unanimous, but the challenge is to find materials that can be used in specific situations (e.g. contamination of water systems with dyes, pharmaceuticals, pesticides, metals and other toxic pollutants) and whose cost is not prohibitive. Therefore, the focus nowadays is on using and valorising byproducts of human activities such as agriculture (wood residues), construction (e.g. clay, basalt debris), energy production (e.g. ash), etc. by applying them in environmental remediation processes.

Fig. 1 Representative samples of the byproducts tested on adsorption studies from liquid-phase

2 Materials and Methods

2.1 Materials

The byproducts used as natural adsorbents were materials of vegetal and mineral origin, namely Angolan Nuati and Njilite woods wastes, wood pellets, the minerals perlite, vermiculite and expanded clay wastes, natural charcoal and lignocellulosic ash. On Fig. 1 are presented representative images of these materials.

All the adsorptive molecules evaluated in this work are of analytical grade available commercially.

2.2 Methods

Activation and Modification of Byproducts. In this work the byproducts were submitted to two modifications of distinct nature, one acid with HCl and the other basic with NaOH. From the commercial HCl solution (37%) a solution of 5 M concentration was prepared to be used in the byproducts impregnation phase. Then, 10 g of each original material was placed in a 250 mL erlenmeyer to which 100 mL of 5 M HCL solution was added. The original material was kept in contact with this solution for a period of 48 h and then separated from the impregnating solution by vacuum filtration. Then, in order to remove the excess of modifying agent, the modified material was washed with distilled water until the pH of the lavage water was equal to that of the distilled water. Finally, the modified material was dried and stabilized by drying in an oven, above 100 °C, for a period of 72 h. An analogous procedure was carried out for the modification of basic nature, using a 5 M NaOH solution.

Dosage and Quantification of Adsorptive Molecules. All the kinetics and equilibrium adsorption studies of the selected adsorptive hazardous molecules were made with a fixed amount, 0.2 g, of byproduct adsorbent added to 20 mL of aqueous solutions of the different hazardous molecules with known concentrations, in unchanged pH. Relevant physical and chemical characteristics of those molecules are shown in Table 1. The suspensions were maintained in stoppered erlenmeyers of 50 mL, under agitation in a thermostatic shaker bath, at 25 °C, in the dark, for the diseased time. The residual amount, not adsorbed, on solution was evaluated, in triplicate, by UV/Vis spectrophotometry, using a Nicolet Evolution300 UV–Vis Spectrophotometer, at appropriate wavelength.

3 Results and Discussion

3.1 Calibration Curve

Before the kinetics and equilibrium adsorption essays and for all the adsorptive probe molecules the characteristics wavelengths were determined by UV–Vis scanning spectrum, between 200 and 700 nm. After, for the chosen characteristic wavelength the calibration curve was established (Fig. 2).

3.2 Adsorbent/Adsorptive Mass Ratio, Kinetic and Equilibrium Isothermic Tests

In order to reproduce in our experiments scenarios close to real aqueous systems, we did not adjust the pH of the adsorptive/adsorbent system. The first experiments also allowed us to determine that a higher adsorption capacity is guaranteed with an adsorbent:adsorptive mass ratio greater than 0.2:15. As an example, Fig. 3 shows the results for MB in two byproducts of different nature, vegetal (Nuati wood) and mineral (Perlite). It can be seen that increasing the mass of the samples does not lead to an increase in Qads, but rather to a tendency towards stabilisation. Under the conditions used, Qads is maximum for an adsorbent:adsorptive mass ratio of ~ 17% and 13% respectively for Nuati wood and Perlite.

Figure 4 shows examples of the kinetic profile of natural byproducts of both vegetal and mineral origin, namely Nuati wood and Perlite mineral, for the methyl

Table 1 Physical and chemical characteristics of the adsorptive molecules [10]

Adsorptive	Structure	M (g/mol)	WS (g/L)	λ (nm)	pKa
MB	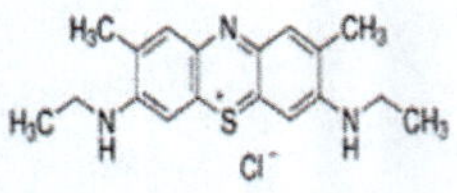	319.85	43.6	668	3.14
AY 17	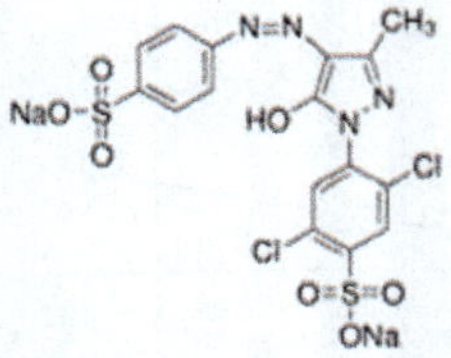	551.29	70.0	404	5.5
MCPA	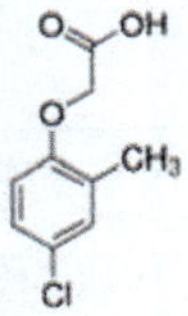	200.62	270.0	279	3.07
4N	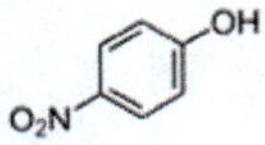	139.11	16.0	402	7.15
SD	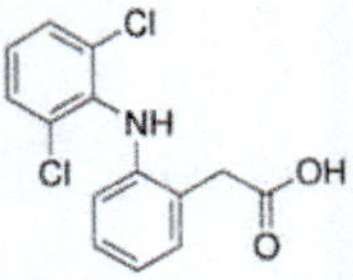	318.13	15.9	276	4.15

M—Molar mass, WS—Water solubility, λ—Characteristic wavelength

blue adsorbent. The results show that the adsorption of the vegetal byproduct is slower in the initial phase, the first 6 h, compared to the mineral. With time, the amount adsorbed (Qads) continues to increase gradually in the case of the mineral byproduct, while there is a significant increase in the case of the vegetal byproduct (between 10 and 24 h). However, at longer times, between 24 and 48 h, it can be seen that Qads tends to stabilise, with reductions in the order of 46% and 14% for Nuati wood and Perlite respectively. It should also be noted that the maximum adsorption value for the mineral byproduct occurs around 8 h (~ 30%).

An equilibrium time of 24 h was chosen to ensure that all systems were tested under similar conditions.

The adsorptive/adsorbent systems studied also show distinct behaviour when the adsorbed amounts are analysed as a function of the equilibrium concentration, for

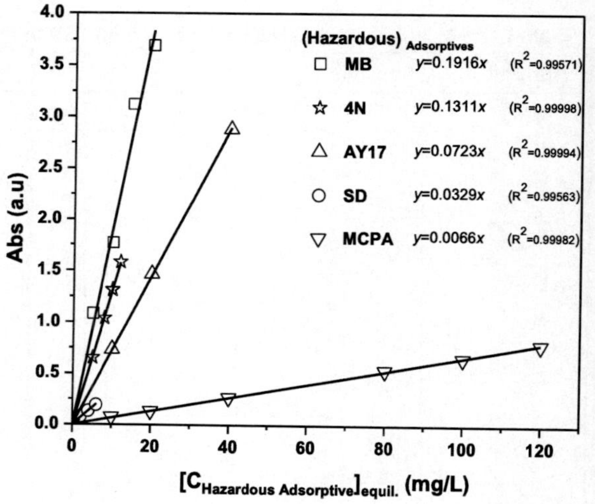

Fig. 2 Calibration curves for the different adsorptive hazardous molecules

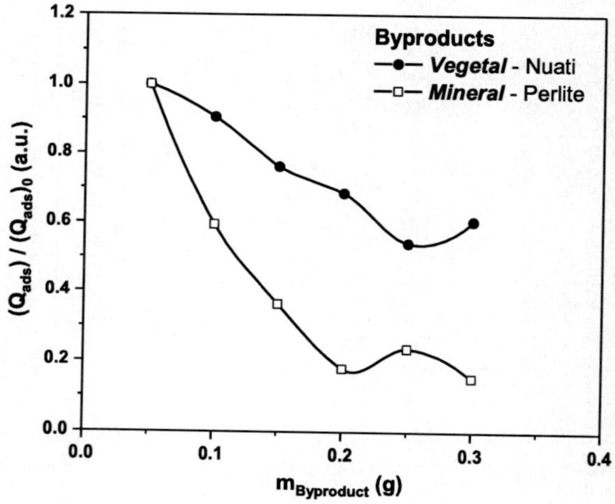

Fig. 3 Influence of Adsorbent/Adsorptive mass ratio on the adsorption capacity to MB by Nuati wood and Perlite ($[C_{MB}]_{initial} = 15$ mg/L, $t_{equilibrium} =$ 24 h)

a relatively wide range of initial concentrations. Figure 5 shows the experimental isotherms obtained with MB for the different byproducts. It can be seen that for the wood byproducts, the maximum adsorption capacity is obtained in the lower concentration range, while for the samples of mineral origin, the adsorbed amounts are lower and tend to stabilise with increasing concentration.

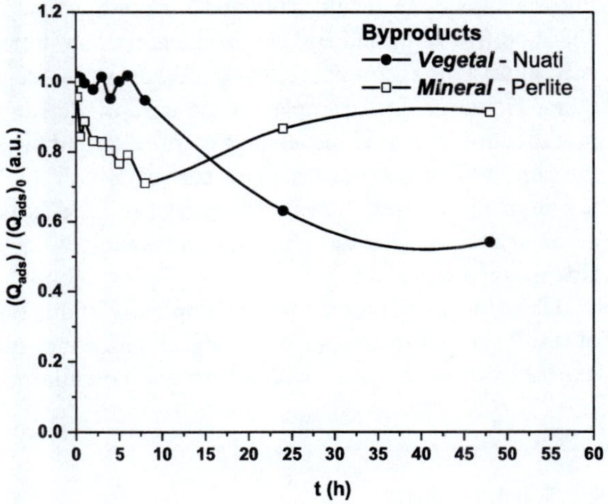

Fig. 4 Kinetic studies for the adsorption capacity to MB by Nuati wood and Perlite ($[C_{MB}]_{initial} = 15$ mg/L)

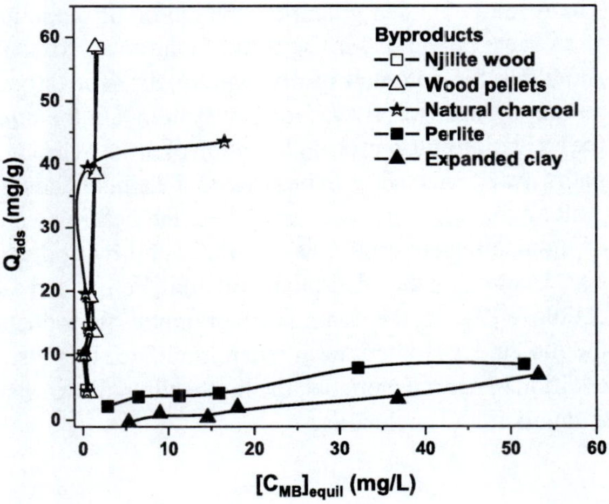

Fig. 5 Adsorption isotherms of MB, obtained at 25 °C, on representative vegetal and mineral unmodified byproducts

3.3 *Unmodified and Soft Modified Byproducts*

In order to evaluate the effect of the modification of the natural byproducts, of vegetal and mineral origin, by soft acid and basic activation, their adsorptive capacity was determined for specific initial concentrations of the studied hazardous compounds. The results show that the effect of acidic or basic modification of the byproducts on their adsorptive capacity is quite heterogeneous. As an example, we present the results for two different molecules, MB and AY 17. These molecules have different size, molar mass and chemical nature, which means that the results obtained reflect different interactions. Consequently, the adsorptive capacity is also highly dependent on the nature of the adsorbent/adsorptive system [9, 11]. Again, as an example,

Fig. 6a shows the results obtained for methyl blue (MB) for unmodified (UNM) and modified (AM and BM) byproducts. For an initial concentration of 40 mg/L of MB, almost all adsorbents show good adsorption capacity and are similar between them. However, for example, in the case of Perlite and Expanded Clay, the acid modification inhibited the adsorption process and the basic modification potentiated the same adsorption process. In the case of dye AY 17, Fig. 6b, the adsorption process is practically non-existent in the case of mineral byproducts. However, in the case of natural wood, natural charcoal and lignocellulosic ash, significant amounts of adsorption are present.

These two examples show the complexity of this type of process, but the diversity of results allows us to state that for each hazardous molecule there is always one or more adsorbent materials with adsorption potential.

4 Conclusions

The results obtained guarantee that some of the natural byproducts studied in this work have potential for use in the removal of toxic pollutants. However, they also show that the potential of the byproducts as adsorbents is highly dependent on the nature of the adsorptive/adsorbent system. On the other hand, it has been found that the activation and gentle modification of these byproducts, either by the acidic or basic pathway, is reflected in an increase in the amount adsorbed, depending on the system. Among the different systems studied, the best results were obtained with adsorbents of lignocellulosic origin, i.e. natural wood byproducts (Nuati and Njilite). This can be explained by the adsorptive potential, before and after modification, of lignin and cellulose [9]. On the contrary, the mineral byproducts show a low performance for the molecules studied, even when modified. It is also important to mention that, in addition to natural charcoal, the lignocellulosic ash also reached interesting adsorbed amounts for all molecules.

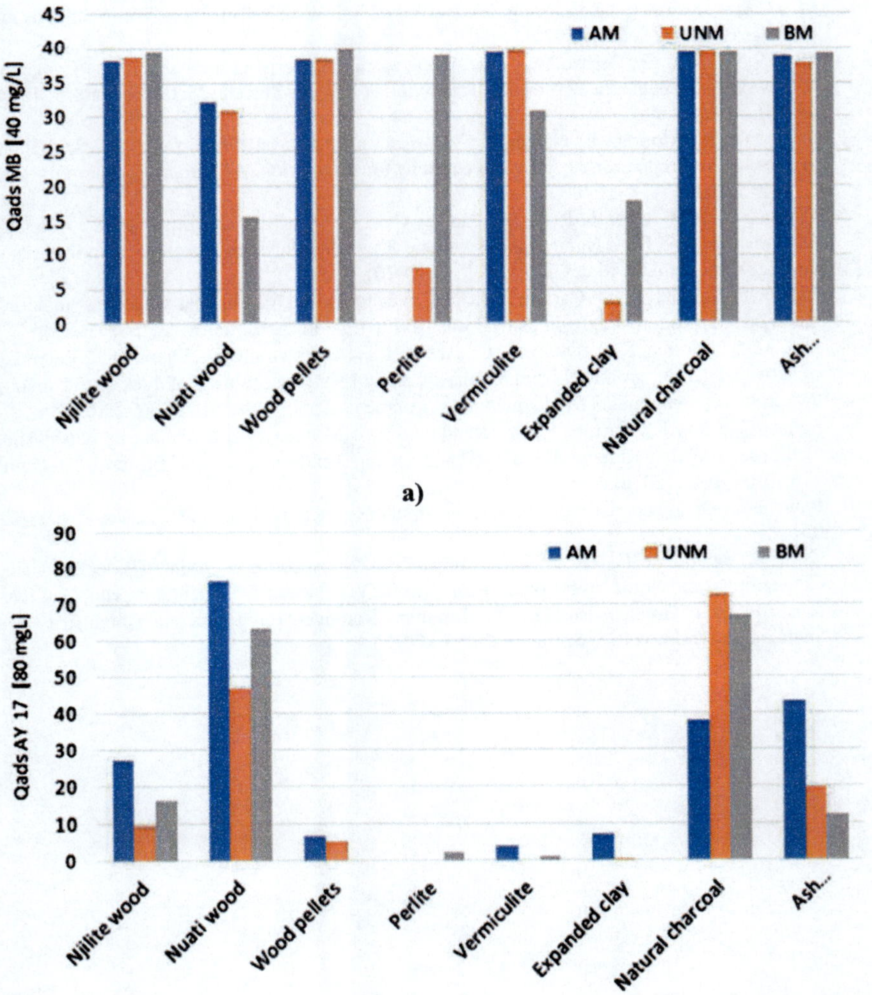

Fig. 6 Adsorbed amounts of MB (**a**) and AY 17 (**b**) obtained at 25 °C, on the Unmodified (UNM) and Acid or Basic soft Modified (AM and BM) byproducts. Initial concentrations of 40 and 80 mg/L for MB an AY 17, respectively

References

1. Nygaard, A.: From Linear to Circular Economy: a Transaction Cost Approach to the Ecological Transformation of the Firm. Circ. Econ. Sust. 2, 1127–1142 (2022).
2. Brizga, J., Khadraoui, S. E.: The Circular Economy and Green Jobs in the EU and Beyond ([edition unavailable]). London Publishing Partnership. Retrieved from https://www.perlego.com/book/3739220/the-circular-economy-and-green-jobs-in-the-eu-and-beyond-pdf (Original work published 2022).

3. Lin, C.S.K., Kaur, G., Li, C., Yang, X., Stevens, C.V.: Waste Valorisation: Waste Streams in a Circular Economy. Wiley (2020).
4. Gisi, S.; Lofrano G.; Grassi, M., Notarnicola, M.: Characteristics and Adsorption Capacities of Low-Cost Sorbents for Wastewater Treatment: A Review. Sustain. Mater. Technol. 9, 10–40 (2016).
5. Tchikuala, E., Mourão, P., Nabais, J.: Valorisation of Natural Fibres from African Baobab Wastes by the Production of Activated Carbons for Adsorption of Diuron. Procedia Eng. 200, 399–407 (2017).
6. Cansado, I. P. P., Mourão, P. A. M., Ribeiro Carrott, M. M. L., Carrott, P. J. M.: Activated Carbons Prepared from Natural and Synthetic Raw Materials with Potential Applications in Gas Separations. Adv. Mater. Res. 107, 1–7 (2010).
7. Ensuncho-Muñoz, A.E., Carriazo, J.G.: Characterization of the carbonaceous materials obtained from different agro-industrial wastes. Environ. Technol. Innov. 36, 5–8 (2015).
8. Ani, J.U., Akpomie, K.G., Okoro, U.C., Aneke, L.E., Onukwuli, O.D., Ujam, O.T.: Potentials of activated carbon produced from biomass materials for sequestration of dyes, heavy metals, and crude oil components from aqueous environment. Appl. Water Sci. 10, 69 (2020).
9. Suhas, Gupta, V.K., Carrott, P.J.M., Randhir Singh, Monika Chaudhary, Sarita Kushwaha.: Cellulose: A review as natural, modified and activated carbon adsorbent, Bioresour. Technol. 216, 1066–1076 (2016).
10. PubChem - Open chemistry database (https://pubchem.ncbi.nlm.nih.gov/), accessed 2023/02/05
11. Crini, G., Lichtfouse, E., Wilson, L.D., Morin-Crini, N.: Adsorption-Oriented Processes Using Conventional and Non-conventional Adsorbents for Wastewater Treatment. In: Crini, G., Lichtfouse, E. (eds) Green Adsorbents for Pollutant Removal. Environmental Chemistry for a Sustainable World, vol 18. Springer, Cham. (2018).

Online Sport Infrastructure Management for the Sustainability: Football, Futsal, and Beach Football

Ricardo Gomes, Ricardo Martinho, Alexandrino Gonçalves, Vitor Pereira, Sandro Brito, Luis Monteiro, Júlio Vieira, and Manuel Nunes

Abstract In Portugal there are 22 Football Associations—which correspond to 18 districts of the mainland, one in Madeira island and three in the Azores. One of the main tasks of the Football Associations is certifying the sport infrastructures to its official competitions: futsal, football, and beach football. Over the last years, the Technical Council of Leiria Football Association has developed an innovative web platform for Sport Infrastructure Management, used in official competitions, named (in Portuguese) "GIDO—Gestão de Infraestruturas Desportivas Online". Based on a user-friendly interface, this platform enables the observation of the sport infrastructure characteristics over time, the infrastructures management and promotes benchmarking between them. Additionally, this platform is the best way to optimize the travel time between any geographic point and a sport infrastructure and to know which clubs use that sports infrastructure.

Keywords Benchmarking · Sustainability · Online sport infrastructure management · "GIDO—Gestão de Infraestruturas Desportivas Online" · Football · Futsal · Beach football

R. Gomes (✉)
Department of Civil Engineering, Polytechnic of Leiria, Leiria, Portugal
e-mail: ricardo.gomes@ipleiria.pt

R. Martinho · A. Gonçalves
Department of Computer Engineering, Polytechnic of Leiria, Leiria, Portugal

A. Gonçalves
CIIC—Polytechnic Institute of Leiria, Leiria, Portugal

R. Gomes
INESC Coimbra—Institute of Systems and Computer Engineering of Coimbra, Coimbra, Portugal

R. Gomes · V. Pereira · S. Brito · L. Monteiro · M. Nunes
Leiria Football Associations, Leiria, Portugal

J. Vieira
Portuguese Football Federation, Cruz Quebrada—Dafundo, Lisboa, Portugal

J. R. da. C. S. Galvão et al. (eds.), *Proceedings of the 3rd International Conference on Water Energy Food and Sustainability (ICoWEFS 2023)*, Springer Proceedings in Earth and Environmental Sciences, https://doi.org/10.1007/978-3-031-48532-9_53

1 Introduction

In Portugal, Football Associations are the figures, at the district level, whose responsible for all official competitions in football, futsal, and beach football. The competitions involve senior levels and training—both male and female teams. In order to fulfill their functions, Football Associations have different management bodies with different tasks, such as, the General Assembly, the Association President, and the Disciplinary, Justice, Arbitration, Fiscal and Technical Councils, composed of a President, Vice-President and Vowels.

An important task of Football Associations is certifying the sport infrastructures to its official competitions, training coaches and referees and granting licenses to teams and managers and other sports agents. Several courses and training are given annually, which later give recognition to professionals and, therefore, the right to be in the area reserved for game elements. Additionally, each Football Association has a collective insurance, which is then responsible for reimbursing athletes and other players in case of accidents or injuries.

On the other hand, given the proximity to the local population, it is also important to highlight here the enormous role of Football Associations to promote sport practice and physical activity in all parts of the country, which involves publicizing sports and everything that involves them, to attract more and more athletes to football, futsal, and beach football. The Leiria Football Association (https://afleiria.fpf.pt) was founded on May 20, 1929, with the objective of promoting, regulating, and directing the practice of non-professional football, futsal, and beach football in the district of Leiria. It currently has 120 affiliated clubs, 11,000 players (of which around 8500 in training), 2000 directors, 700 coaches and 181 referees, organizing around 9000 games per sports season.

The conditions of infrastructures are one of the main factors to encourage sports. In this context it is important to highlight the commitment and determination of clubs' leaders, which, in close articulation with all local authorities, have made a huge effort to improve their sports infrastructures, thus contributing to increase the number of teams, players, coaches and managers, work on training and retaining talent, and achieving sporting successes. Given the importance of this factor, over the last years, Leiria Football Association (by its Technical Council) has developed an innovative web platform for the Online Sport Infrastructure Management used in their official competitions, named (in Portuguese) "GIDO—Gestão de Infraestruturas Desportivas Online". Based on a user-friendly interface, this platform permits the observation of the sport infrastructure characteristics over time, the infrastructures management and promotes benchmarking between them. Additionally, the platform is the best way to optimize the travel time between any geographic point and a sport infrastructure and to know which clubs use that sports infrastructure. In the next sections, this innovative platform for Online Sport Infrastructure Management is described, and then the main conclusions are presented.

2 Online Sport Infrastructure Management Description

Sports infrastructures used in official competitions of the Leiria Football Association are subject to a regular visit by the Technical Council, which checks their suitability for sporting activities. In 2015, the Leiria Football Association launched a challenge to the Polytechnic of Leiria to propose the development of a digital repository for sports infrastructure related to official futsal, football, and beach football competitions, as well as a web platform for supporting the activity of the Technical Council during their regular visits at sports infrastructures. The Departments of Computer Engineering and Civil Engineering of the Polytechnic of Leiria presented a prototype for this platform, based on a standard form defined by the Technical Council of Leiria Football Association, considering the recommendations of the Portuguese Football Federation (www.fpf.pt) for each sport modality [1]. Afterwards, the Janela Digital software company (www.janeladigital.com), developed in 2018 the web platform named in Portuguese "GIDO—Gestão de Infraestruturas Desportivas Online", which is now available from (Figs. 1 and 2): http://infraestruturas.afleiria.com

The GIDO platform is composed of a web application that permits anyone to know the clubs affiliated to the Leiria Football Association, and which clubs use a specific sports infrastructure. To access the web application, it can be used from a computer, mobile phone or tablet [2]. There is also a private area for clubs to consult the "status" of their associated infrastructures, check deadlines for the inspections of each of the infrastructures and which of them require corrective actions. The digital platform contains a back-office module that allows the registration, in real-time, of inspections carried out by the Technical Council of Leiria Football Association.

A user-friendly interface allows users to have access to the geographic location of each sport infrastructure and optimizes the travel time until each sport infrastructure

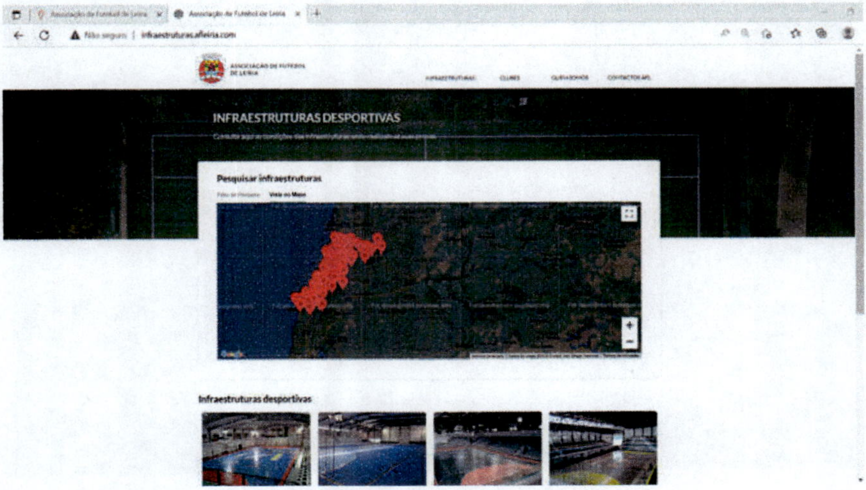

Fig. 1 Online sport infrastructure management front office [2]

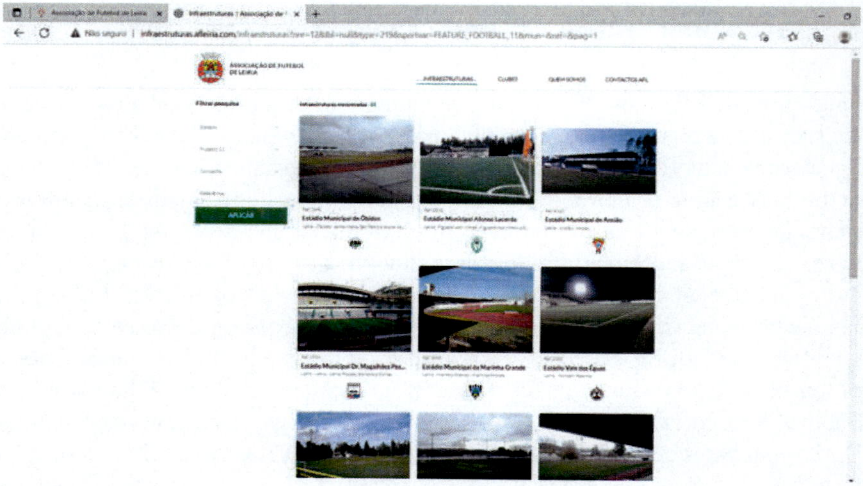

Fig. 2 Sports infrastructures using a search filter (search by stadiums) [2]

(Figs. 3 and 4). For each sport infrastructure, the main characteristics are listed, including photos (Fig. 5). In this way, it is possible to compare the evolution of sport infrastructures over time and promote benchmarking between them.

The GIDO platform intends to evolve in order to provide other relevant information for clubs, directors, supporters and all people involved with the sports modalities, assuming itself as an important working tool for the development of district football. As an example, in the next updates, an infrastructure occupancy calendar (which is the responsibility of the Leiria Football Association Technical Council)

Fig. 3 Geographic location of each sport infrastructure [2]

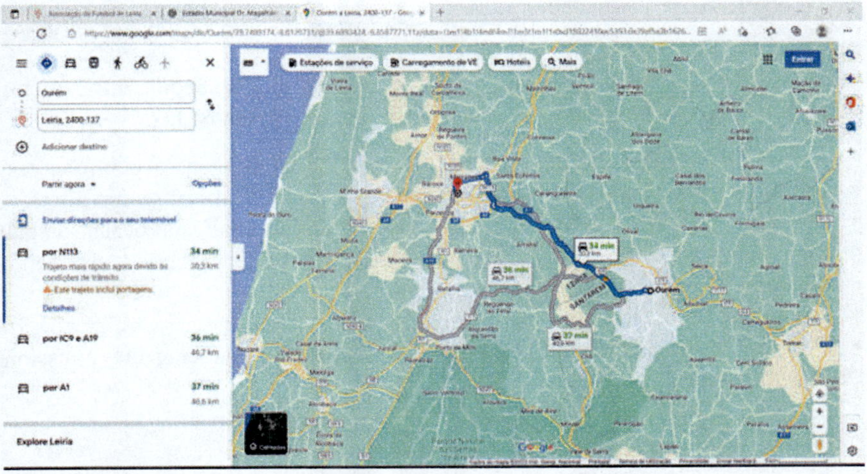

Fig. 4 Optimal travel time between two different geographic points [2]

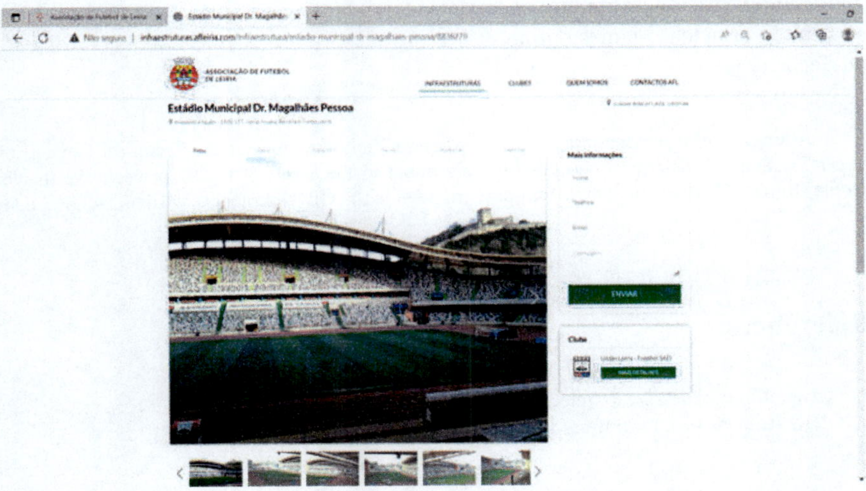

Fig. 5 Main characteristic of sport infrastructure (Leiria stadium) [2]

and some parameters for calculating analysis and statistics performance indicators will be considered.

3 Conclusion

GIDO is the first web platform available worldwide for sports infrastructure management in real-time and some advantages can be presented here:

- Allows for three user profiles (Football Association, Clubs, general public access);
- Quick search for sports infrastructure from geographic maps or search filters;
- Digital repository for sports infrastructure of futsal, football, and beach football;
- Registration in real time of inspections carried out by the Technical Council of the Football Association;
- Promotes benchmarking;
- Optimize the travel time until each sports infrastructure, reducing CO_2 emissions and others costs.

Next steps:

- To create an exclusive interaction channel with clubs;
- Insert performance indicators about each sports facility (population, restaurants, car parks, …);
- Insert a sports infrastructure occupancy calendar;
- Link to broadcast games online;
- Security plan.

Acknowledgements In Memoriam of José Adelino Vigia, member of the Technical Council of Leiria Football Association, who greatly contributed to the result of GIDO project. We thanks Janela Digital software company for support provided in the GIDO project: www.janeladigital.com

References

1. "Laws of the Game 22/23". International Football Association Board.
2. GIDO Homepage, http://infraestruturas.afleiria.com, last accessed 2023/01/31.

Building a Sustainable Future: Examining the Impact of Earth-Based Materials on Civil Engineering Development

Yeşim Tarhan, İsmail Hakkı Tarhan, Flavio Craveiro, and Helena Bartolo

Abstract Earth-based construction materials, such as adobe, rammed earth, and compressed stabilized earth blocks, have been utilized for centuries, and are now experiencing renewed interest due to the pressing environmental threat posed by greenhouse gas emissions. Their economic attractiveness, widespread availability, recyclability, and superior engineering characteristics, including thermal conductivity, insulation, and moisture absorption, make them vital to sustainable development in civil engineering. By combining these time-honored materials with cutting-edge construction methods developed thanks to rapidly advancing technology, civil engineering is on the verge of a paradigm shift. This study provides a comprehensive exploration of the history of earth-based construction materials from the past to the present, delving into the challenges and potential solutions associated with their use in modern construction.

Keywords Earth-based materials · Sustainability · Modern construction techniques

---

Y. Tarhan (✉)
Ardahan University, Ardahan, Türkiye
e-mail: yesimtarhan@ardahan.edu.tr

İ. H. Tarhan
Tokat Gaziosmanpaşa University, Tokat, Türkiye

F. Craveiro · H. Bartolo
CIAUD, Universidade de Lisboa, Lisboa, Portugal

School of Technology and Management, Polytechnic of Leiria, Leiria, Portugal

J. R. da. C. S. Galvão et al. (eds.), *Proceedings of the 3rd International Conference on Water Energy Food and Sustainability (ICoWEFS 2023)*, Springer Proceedings in Earth and Environmental Sciences, https://doi.org/10.1007/978-3-031-48532-9_54

1 Introduction

1.1 Evolution of Earth-Based Materials

Earth-based construction materials have a rich history that spans thousands of years, and continue to be used today in many parts of the world. Over two billion people currently reside in earthen homes, while roughly 10% of the world's cultural heritage structures are earthen structures [1]. Earth is a readily available material, commonly found in urban environments where construction sites produce large volumes of excavated materials. The advantages of using earth as a building material include low cost, good insulation properties, recyclability, and low environmental impact [2]. However, the development of earth construction has been limited by factors such as high labor costs, poor mechanical properties, water sensitivity, and low durability [3].

There is a clear need for an improved earth-based mix that combines fast texturing with early strength. Adapting scientific knowledge and experience from the concrete industry to soil construction can provide significant environmental and economic benefits [4]. By overcoming the current limitations of soil-based construction materials, it is possible to enhance the viability of these materials in modern construction, while simultaneously reducing the environmental impact of the construction industry.

Numerous earth-based construction techniques are employed in the construction industry, as depicted in Fig. 1. Examples include adobe (also known as mudbrick or cob), bulk earth technique, rammed earth technique, compressed earth brick technique (also called compressed earth block masonry), and keeled adobe technique (also known as wattle and daub) [2, 5].

In 2015, the United Nations established the Sustainable Development Goals (SDGs), a comprehensive plan aimed at addressing some of the most critical global challenges (Anonymous 2023). The SDGs are a universal call to action that involves all countries, regardless of their income level, to achieve a sustainable future for all. Several of the SDGs are particularly relevant to the construction industry, such as promoting sustainable industrialisation, building resilient infrastructure, and making cities and communities safe, sustainable, and resilient. In addition, the efficient use of natural resources, including reducing material use per capita, is an important goal.

The construction industry is a major contributor to global greenhouse gas emissions, with concrete being a particularly high emitter. Therefore, the widespread use of earthen construction as an alternative building material to concrete is critical to achieving sustainable development goals. Earthen construction has the potential to enhance economic, social, and environmental performance in the construction sector. By promoting the use of earthen construction, we can significantly reduce the construction industry's carbon footprint while contributing to the achievement of the SDGs.

Fig. 1 Several construction techniques use earth-based materials **a** adobe, **b** rammed cast earth technique, **c** rammed earth brick technique, **d** keeled adobe technique

2 Advantages of Earth-Based Materials

Earth construction has several important environmental advantages. Not only is it generally non-toxic and non-polluting, but raw earth has a high adsorption capacity for volatile compounds due to its clay content, making it an excellent choice for promoting indoor air quality [6]. Additionally, compared to other building materials, earth-based materials have superior hygrothermal performance. They can absorb and desorb moisture to a greater extent and stabilise indoor climate due to their low thermal conductivity and strong breathability. This characteristic prevents condensation from forming on the walls and ceilings because the vapour formed in the earthen building can easily pass through the walls [7].

Earth is a unique building material that offers several advantages, including its ability to balance humidity levels in the air. Unlike other materials, earth absorbs and desorbs moisture quickly and to a greater extent, which helps to stabilize the indoor climate. Research has shown that unbaked earth bricks can absorb up to 30 times more moisture than baked bricks when the relative humidity in a room suddenly rises from 50 to 80% over two days [8].

Another significant advantage of the earth is its ability to store heating energy within its structure, making it an excellent energy-storing material after water. Earth can balance the temperature by gradually releasing the energy it stores to the space, even after the heating source is removed. Additionally, the nature of earth construction allows vapor to pass through walls, which prevents condensation from forming on exterior walls and ceilings [9].

Moreover, earthen building materials do not require embodied energy during production and are usually supplied from the construction site, which saves transportation costs and is, therefore, more economical. Raw earth has an exceptionally low environmental impact when applied without industrial additives and is a fully recyclable material [10].

3 Stabilization of Earth-Based Materials

The development of various low environmental impact binders for earth stabilization can promote the use of local resources with minimal CO_2 impact. In recent years, the use of Portland Cement for stabilization has become more common, slightly increasing earth-based materials' environmental impact [11]. The addition of Portland Cement to earth is a common method of improving its strength, durability, and resistance to water erosion. This technique can also transform some earth materials that are initially unsuitable for construction into materials that can be used effectively [2]. Efforts are also being made to adapt modern concrete technologies used in reinforced concrete structures for use with earth-based materials.

Various stabilizers have been employed to enhance the strength, disaggregation, and erosion resistance of earth-based construction materials, including clay, tree resins, natural bitumen, gum, agave juice, cactus juice, cowpats, and casein. Additionally, certain additives used in conventional Portland cement-based concrete can be added to correct the mixture and produce self-compacting earth concrete [12].

The use of mineral admixtures such as calcined gypsum, lime, and pozzolana as binders has been a common practice in many local constructions. The ancient Romans, for example, used volcanic ash from the town of Pozzuoli, a material rich in alumina, silica, and alkali, along with lime and aggregates, to produce their concrete [13].

Crushed ceramics such as fired bricks, pottery, and tiles have also been historically utilized as pozzolanic materials. These materials were mixed with lime or unfired clay materials to create stabilized earth bricks [14]. Crushed brick or tile is a chemically active and stable material with a glassy structure rich in silica (SiO_2), which also possesses weak acidic properties. When clay is fired at 600–800 °C, its kaolinite component loses water and decomposes into silica and alumina. Lime, on the other hand, is a strong alkali. The amorphous structures in crushed brick react with lime to form calcium silicate hydrates (C-S-H) and calcium aluminate hydrates (C-A-H), which contribute to the strength, water hardenability, and hydraulic properties of earth-based mortars and bricks [15].

The use of local materials such as volcanic tuff, lime, and pozzolans as binders has a long history in earth-based construction. Tras, a volcanic tuff from the Vulkaneifel region in Germany, has been used in mortar and concrete production since Roman times [15]. In ancient Jerash (Jordan), crushed pozzolans, lime, and gypsum were found in mortars [16], and at the Kom el-Dikka archaeological site in

Egypt, pozzolanic material (Roman brick fragments) was used as aggregate and a lime-gypsum mixture as a binder phase [17].

Today, ready-made mixtures of trass and lime are commercially available and used as plaster and masonry mortars in earth-based construction. In Turkey, a mixture called Alker has been used for adobe production since the 1980s, which stabilizes the earth materials with fast-setting gypsum (8–10 wt% of dry soil) and slow-setting lime (2.5–5 wt%) [18]. Lime-containing soil-based mixtures act as an activator of pozzolanic reactions, and the presence of a sulfate source (hemihydrate, phosphogypsum) in the mixture allows the formation of additional products such as ettringite [19–22]. This not only increases the strength and water hardenability of earth-based mortars and bricks, but also helps to reduce the environmental impact of the binders used in earth-based construction.

To fully leverage the advantages of earth-based materials in the construction industry, it is necessary to reconsider the traditional mineral stabilizers used in local earth-based mixtures and develop a new binder formulation strategy for soil-based mixtures. This new strategy should take into account recent advancements in mixtures containing gypsum, lime, and pozzolans, as well as the latest knowledge of cement chemistry. Doing so can make earth-based construction more sustainable, eco-friendly, and cost-effective. This would also help to promote the use of locally available resources, minimize environmental impact, and reduce the carbon footprint of the construction industry. Therefore, developing a new binder formulation strategy is crucial for the wider adoption of earth-based materials in the construction industry.

4 Building Techniques with Earth Materials

Soil is a fundamental component of earth-based construction and can be sourced locally or engineered by mixing soils from different origins. The properties of the soil vary depending on the location, including factors such as type, particle size distribution, and amount of fine materials. These factors can significantly impact the soil's characteristics and performance as a building material.

To ensure the suitability of the soil for construction, tests can be performed to determine its strength, plasticity, and other properties. Soil stabilization techniques can be employed to enhance engineering properties such as increasing strength, reducing compressibility, and improving durability. When selecting soil for earth construction, the texture, plasticity, compactibility, and cohesion of the soil are critical factors to consider. Soils with a balanced mix of particle sizes, high plasticity, cohesion, and the ability to be compacted are preferred for construction purposes. Plasticity is the soil's ability to deform without cracking or disintegrating and is important for adobe and rammed earth construction. Compactibility, or the potential for the soil to be compacted and reduce its porosity, affects its stability and strength. Cohesion, or the soil's ability to sustain tensile stress, is critical for structural stability. By evaluating and optimizing these properties, it is possible to use soil as a durable and sustainable building material in earth construction [23, 24].

The unified Soil Classification System from ASTM D2487, a standard for geotechnical engineering, considers particle size distribution as a controlling factor for all other soil properties. It identifies the main types of subsoils and their suitability for earth construction. According to this system, the most suitable soils for earth construction are silty or clayey gravels, silty or clayey sands, and to a lesser extent, low-plasticity silts and clays.

Throughout history, vernacular builders have adapted their construction techniques to the properties of the natural soils available to them. This has led to the development of a rich diversity of constructive cultures influenced by local environmental and climatic conditions and social habits. While there is a wide variety of traditional earth-building techniques, from a technical perspective, they can be broadly categorised into four main variants: wattle and daub [25], cob [26], masonry with adobe (mud bricks) [27], and rammed earth [28]. Wattle and daub involve weaving thin branches together and plastering them with a mixture of clay and straw, while cob involves shaping a mixture of clay, sand, and straw into large monolithic blocks. Adobe bricks are made by hand, often with the addition of straw, and left to dry in the sun before being laid with a clay mortar. Rammed earth involves compacting damp soil into wooden layers to create a solid, monolithic wall. These techniques have been used for centuries to create durable, sustainable buildings that blend harmoniously with their natural surroundings. In recent years, compressed earth blocks have been introduced as an innovative addition to this repertoire of techniques, offering a more standardized and efficient approach to earth construction [29].

5 Challenges and Solutions for Earth-Based Materials

Earth-based materials exhibit both cohesive and frictional properties, meaning their mechanical behaviour is determined by the combined effects of these forces [30, 31]. When used in earthen buildings, raw earth is typically an unsaturated porous medium containing adsorbed water and liquid water bridges (capillary water) in equilibrium with the internal atmospheric relative humidity. Cohesion in raw earth arises from various inter-particle surface forces, including dispersion (van der Waals), capillary, and ionic correlation forces, the relative strengths of which depend mainly on factors such as the clay content, nature and hydration state of the material [32]. Despite technological advancements, earth mixtures as a construction material, compared to conventional construction materials like concrete, steel, and wood still generally exhibit lower mechanical properties such as lower tensile strength and greater susceptibility to moisture and weathering [2].

Experimental studies have provided insights into the relationship between the strength and stiffness of compacted earth and its water content and matric suction [33]. This data can be used to guide the selection of appropriate moisture content and compaction levels for achieving the desired mechanical properties of the earth-based material. Similar to other cohesive-frictional materials such as Portland Cement concrete, the strength and stiffness of raw earth also increase as its density, or more accurately, its solid volume fraction, increases [34].

If the earth-based is suitable for the chosen construction method, and proper architectural guidelines are followed, then unstabilized earthen buildings can endure over time, even in moderately wet climatic conditions. Nonetheless, it should be noted that unstabilized earthen walls may not satisfy the durability standards set for industrial materials, as they may not pass certain tests like the spray test, drip test, or wet-to-dry strength ratio [35].

To improve the mechanical and durability properties of earth-based mixtures, various stabilization techniques can be employ, such as adding cement or other binders, applying compressive forces during construction, or using specialized construction methods like rammed earth or compressed earth block construction [36]. Stabilization techniques such as the addition of cement, lime, or other binders can enhance the mechanical strength and durability of earth-based mixtures [37]. The use of reinforcing elements such as bamboo, straw, or other fibers can also improve the mechanical properties of the material and increase its resistance to cracking or deformation [38, 39]. In addition, the design and construction of earth-based structures can be optimized to minimize the stresses on the material and ensure its long-term durability and stability. Overall, the mechanical properties of earth-based mixtures in construction depend on various factors, and careful consideration and proper design are required to ensure their suitability for different applications.

6 Conclusions

The use of earth as a building material offers several advantages, including low cost, good insulation properties, recyclability, and low environmental impact. However, the development of earth construction has been hindered by various factors, such as high labor costs, poor mechanical properties, water sensitivity, and low durability. Despite these challenges, ongoing research and innovation in earthen construction techniques and material science are providing solutions to some of these issues, thereby expanding the use of earth-based materials in the construction industry.

With the growing awareness of sustainable building practices and the need for environmentally friendly construction materials, it is likely that the use of earth as a building material will continue to gain popularity in the future.

Alternative approaches to earthen construction that involve adding organic or inorganic admixtures to enhance crude earth's compressive strength and durability may offer a more environmentally acceptable solution.

It is important to consider socio-psychological aspects and the negative perception that earthen construction often faces. Addressing these issues may require better technical education and communication to build trust and promote the benefits of earthen construction.

Ultimately, the decision to use earthen construction materials depends on a range of factors, including the suitability of the local subsoil, available resources, and the environmental impact of the chosen approach. While unstabilized earthen construction can be a durable and affordable solution for many housing challenges, alternative approaches that enhance the properties of raw earth may offer a more sustainable and environmentally sound option for the future.

References

1. Eloundou, L., & Joffroy, T. (2013). Earthen Architecture in today's World. In Proceedings of the UNESCO International Colloquium on the Conservation of World Heritage Earthen Architecture; UNESCO Publishing: Paris, France. Available online: www. whc. unesco. org/en/series/36 (accessed on 14 June 2021).
2. Van Damme, H., & Houben, H. (2018). Earth concrete. Stabilization revisited. Cement and Concrete Research, 114, 90-102.
3. Perrot, A., Rangeard, D., & Courteille, E. (2018). 3D printing of earth-based materials: Processing aspects. Construction and Building Materials, 172, 670-676.
4. Landrou, G., Brumaud, C., & Habert, G. (2015). Development of a self-compacted clay based concrete-rheological, mechanical and environmental investigations. Academic Journal of Civil Engineering, 33(2), 80-84.
5. Pacheco-Torgal, F., & Jalali, S. (2012). Earth construction: Lessons from the past for future eco-efficient construction. Construction and building materials, 29, 512-519.
6. Melià, P., Ruggieri, G., Sabbadini, S., & Dotelli, G. (2014). Environmental impacts of natural and conventional building materials: a case study on earth plasters. Journal of cleaner production, 80, 179-186.
7. Hall, M., & Allinson, D. (2009). Analysis of the hygrothermal functional properties of stabilised rammed earth materials. Building and Environment, 44(9), 1935-1942.
8. Minke, G. (2021). Building with Earth: Design and Technology of a Sustainable Architecture Fourth and revised edition. Birkhäuser.
9. Cuce, E., Harjunowibowo, D., & Cuce, P. M. (2016). Renewable and sustainable energy saving strategies for greenhouse systems: A comprehensive review. Renewable and Sustainable Energy Reviews, 64, 34-59.
10. Calkins, M. (2008). Materials for sustainable sites: a complete guide to the evaluation, selection, and use of sustainable construction materials. John Wiley & Sons.
11. Van Damme, H., & Houben, H. (2020). Reply to the discussion of the paper "Earth concrete. Stabilization revisited" by ATM Marsh, A. Heath, P. Walker, BV Venkatarama Reddy, and G. Habert. Cement and Concrete Research, 130, 105992.
12. Tarhan, Y., Tarhan, İ. H., Craveiro, F., & Bártolo, H. (2021, May). Sustainable Materials for Additive Manufacturing: Earth-Based Concrete. In Proceedings of the 1st International Conference on Water Energy Food and Sustainability (ICoWEFS 2021) (pp. 708–716). Cham: Springer International Publishing.
13. Tosun, Y., Oltulu, M., Polat, R., Şahin, R., & Bingöl, A. F. (2015). Hidrolik Kireçli Harçlarda Uçucu Kül Kullanımı, 5. Tarihi Eserlerin Güçlendirilmesi ve Geleceğe Güvenle Devredilmesi Sempozyumu, 1, 103–112, Turkish paper with English abstract.
14. G. Artioli, M. Secco, A. Addis, (2019). The vitruvian legacy: mortars and binders before and after the Roman world, Eur. Mineral. Union Notes Mineral., 20, pp. 151-202, https://doi.org/10.1180/EMU-notes.20.4
15. Aprea, S. (2018). The Search for Affordable Cement and Concrete in the 18 th and 19 th Century. In High Tech Concrete: Where Technology and Engineering Meet (pp. 2777–2784). Springer, Cham.
16. Yaseen, I. A. B., Al-Amoush, H., Al-Farajat, M., & Mayyas, A. (2013). Petrography and mineralogy of Roman mortars from buildings of the ancient city of Jerash, Jordan. Construction and building materials, 38, 465-471.

17. Fort, R., Ergenç, D., Aly, N., de Buergo, M. A., & Hemeda, S. (2021). Implications of new mineral phases in the isotopic composition of Roman lime mortars at the Kom el-Dikka archaeological site in Egypt. Construction and Building Materials, 268, 121085.
18. Pekmezci, B. Y., Kafesçioğlu, R., & Agahzadeh, E. (2012). Improved performance of earth structures by lime and gypsum addition. METU J. Fac. Archit. (2012), https://doi.org/10.4305/METU.JFA.2012.2.9
19. Shen, W., Zhou, M., & Zhao, Q. (2007). Study on lime–fly ash–phosphogypsum binder. Construction and Building Materials, 21(7), 1480-1485.
20. Carlos, M. L., Jose, L. R. A., ro, M. R., Jorge, P. P. O., Jose, D. J. P. B., & Angel, M. D. J. (2011). Effect of fly ash and hemihydrate gypsum on the properties of unfired compressed clay bricks. International Journal of Physical Sciences, 6(24), 5766–5773.
21. Morsy, M. S., Alsayed, S. H., & Salloum, Y. A. (2012). Development of eco-friendly binder using metakaolin-fly ash–lime-anhydrous gypsum. Construction and Building Materials, 35, 772-777.
22. Morsy, M. S., Rashad, A. M., Shoukry, H., Mokhtar, M. M., & El-Khodary, S. A. (2020). Development of lime-pozzolan green binder: The influence of anhydrous gypsum and high ambient temperature curing. Journal of Building Engineering, 28, 101026.
23. Houben, H., Guillaud, H., & Hall, B. B. (1994). Earth construction: a comprehensive guide (p. 362). London: Intermediate Technology Publications.
24. Schroeder, H. (2016). Sustainable building with earth (pp. 1-576). New York, NY, USA:: Springer International Publishing.
25. Niroumand, H., Zain, M. F. M., & Jamil, M. (2013). Various types of earth buildings. Procedia-Social and Behavioral Sciences, 89, 226-230.
26. Hamard, E., Cazacliu, B., Razakamanantsoa, A., & Morel, J. C. (2016). Cob, a vernacular earth construction process in the context of modern sustainable building. Building and Environment, 106, 103-119.
27. Delgado, M. C. J., & Guerrero, I. C. (2006). Earth building in Spain. Construction and building materials, 20(9), 679-690.
28. Walker, P., Keable, R., Martin, J., & Maniatidis, V. (2005). Rammed earth: design and construction guidelines.
29. Turco, C., Junior, A. C. P., Teixeira, E. R., & Mateus, R. (2021). Optimisation of Compressed Earth Blocks (CEBs) using natural origin materials: A systematic literature review. Construction and Building Materials, 309, 125140.
30. Schöpfer, M. P., Abe, S., Childs, C., & Walsh, J. J. (2009). The impact of porosity and crack density on the elasticity, strength and friction of cohesive granular materials: Insights from DEM modelling. International Journal of Rock Mechanics and Mining Sciences, 46(2), 250-261.
31. Lu, N., Wu, B., & Tan, C. P. (2007). Tensile strength characteristics of unsaturated sands. Journal of Geotechnical and Geoenvironmental Engineering, 133(2), 144-154.
32. Morel, J. C., Charef, R., Hamard, E., Fabbri, A., Beckett, C., & Bui, Q. B. (2021). Earth as construction material in the circular economy context: practitioner perspectives on barriers to overcome. Philosophical Transactions of the Royal Society B, 376(1834), 20200182.
33. Pandya, S., & Sachan, A. (2020). Variation of collapse potential and stiffness degradation with matric suction of compacted unsaturated cohesive soil. International Journal of Geotechnical Engineering, 14(1), 35-48.
34. Morel, J. C., Pkla, A., & Walker, P. (2007). Compressive strength testing of compressed earth blocks. Construction and Building materials, 21(2), 303-309.
35. Heathcote, K. A. (1995). Durability of earthwall buildings. Construction and building materials, 9(3), 185-189.
36. Gomes, M. I., Faria, P., & Gonçalves, T. D. (2018). Earth-based mortars for repair and protection of rammed earth walls. Stabilization with mineral binders and fibers. Journal of Cleaner Production, 172, 2401-2414.
37. Galán-Marín, C., Rivera-Gómez, C., & Petric, J. (2010). Clay-based composite stabilized with natural polymer and fibre. Construction and Building Materials, 24(8), 1462-1468.

38. Brandt, A. M. (2008). Fibre reinforced cement-based (FRC) composites after over 40 years of development in building and civil engineering. Composite structures, 86(1-3), 3-9.
39. Ardanuy, M., Claramunt, J., & Toledo Filho, R. D. (2015). Cellulosic fiber reinforced cement-based composites: A review of recent research. Construction and building materials, 79, 115-128.
40. United Nations Sustainable Development Goals (SDGs) Homepage, https://www.iso.org/sdgs.html, last accessed 2023/18/02

Approaching Green Building Concept by Eco-Efficient Construction and Demolition Waste (CDW) in Portugal

Burhan Kubilay Kaptan and José Luís Barroso Aguiar

Abstract Recent years have seen widespread building activity, which has resulted in widespread environmental damage, significant CO_2 emissions, and the waste of valuable natural resources throughout the world. Construction is one of the most energy-intensive industries in the European Union (EU), using about 40% of the union's total energy and producing 36% of its GHG emissions. According to the International Green Building Council, over 35% of all landfill waste is made up of CDW, while the construction sector is responsible for more than half of all materials removed worldwide. To avoid these problems, authorities have drafted laws and developed plans to lessen the ecological toll of building supplies through Green Building (GB) concept, Circular Economy (CE), life cycle assessment (LCA) and life cycle costing (LCC) methods. Portuguese law on the harmonization of regulations for the sale, manufacture, and disposal of construction materials is mostly lifted verbatim from its European Union counterpart. This research contributes an eco-efficiency framework for CDW management and CDW use in green building and circular economy concept in Portugal, which is consistent with the principle of Eco-efficiency.

Keywords C&D waste · Green buildings · Circular economy

1 Introduction

Raw materials are consumed at a high rate in modern society, and waste is produced at an even higher rate, especially in the building industry. Around 32% of all resources are used in the construction industry, including up to 40% of all energy and 12% of all available water [1].

B. K. Kaptan (✉) · J. L. B. Aguiar
School of Engineering of University of Minho (UMinho), The Centre for Territory, Environment and Construction (CTAC), Campus de Azurém, 4800-058 Guimarães, Portugal
e-mail: 10631@alunos.uminho.pt

J. L. B. Aguiar
e-mail: aguiar@civil.uminho.pt

J. R. da. C. S. Galvão et al. (eds.), *Proceedings of the 3rd International Conference on Water Energy Food and Sustainability (ICoWEFS 2023)*, Springer Proceedings in Earth and Environmental Sciences, https://doi.org/10.1007/978-3-031-48532-9_55

Water and soil pollution, air pollution, climate change, and the resulting effects on flora and fauna, as well as losses in primary resources, a damaged international reputation, decreased tourism, and increased fuel consumption in transportation are just some of the economic, environmental, and social consequences of CDW [2–4].

Sustainable building, often known as green construction, is the process of designing, constructing, and maintaining a building in a way that minimizes its impact on the natural environment and maximizes its efficiency in terms of both materials and energy use. site selection, design, build, run, maintain, repair, and teardown [5, 6].

Sustainable development, in which the demands of the present are met without sacrificing those of future generations, includes sustainable building as one of its many facets [7]. Hence, environmentally friendly building practices should conserve natural resources without sacrificing functionality. One aspect of green or sustainable architecture is the recycling and use of CDW [8].

The goal of the circular economy is to create an economic system in which raw materials are used as seldom as possible while maintaining their full worth [9]. This means that the kind of recycling and the avoidance of downcycling are now as important as the quantity of recycling or reused materials. To make the change to a circular economy, all phases of a product's lifespan need to be engaged, not only waste management and better recycling [10].

Urban and architectural design in Portugal have come a long way in the previous decade [11]. There were observable successes in terms of both knowledge and execution. Even if the objectives for sustainable construction and buildings are incomplete, sustainable development is expected at the national and regional levels. Water, soil, noise, natural conservation, and urban or construction and demolition waste are just few of the environmental factors that are getting a lot of attention [12].

Promoters of new construction in Portugal have shown an increasing interest in Green Building in recent years as they see the commercial potential of this trend. This is mostly because of the high level of attention that this issue has received. Despite its importance, however, CDW has received very little attention in terms of research into GB's efficacy [13, 14].

Hence, this study aims to evaluate the effects of CDW on GB in Portugal, and then understand the CDW generation and the amount of waste which could be reused for GB purposes. This article also aims to show how sustainable construction materials may be utilized in GB to create healthy, environmentally and economically sound structures.

2 Construction and Demolition Waste

CDW is the largest waste stream in the EU in terms of weight, with mineral content being the highest proportion [15]. CDW generated around 374 million tonnes in 2016 [15], making it the biggest waste source in the EU by weight. The Circular Economy Action Plan [16] identifies C&D as a key sector in the EU for completing the loop, while the updated Waste Framework Directive [17] sets a recovery target of 70% by 2020.

Although the recycling potential of CDW is significant in terms of quantity, it is currently underutilized. The mineral component of CDW, for example, is mostly utilized for road foundations or backfilling at now [18].

The recovery results, although being good, vary greatly amongst EU Member States, ranging from 54 to 100% in 2016 [15]. Unfortunately, present CDW statistics are insufficiently robust; for instance, European Commission research on CDW management found that Member States had varying understandings and accounting procedures for recovery activities of backfilling [19].

The EPA defines CDW as "waste generated during the building, repair, or deconstruction of structures. Structures include both residential and non-residential structures, as well as highways and bridges. Concrete, asphalt, wood, metals, gypsum wallboard, and roofing are common CDW components" [20].

CDW materials are amassed during infrastructure building, modernization, restoration, retrofitting, and deconstruction. Around one million tons of CDW are generated yearly, primarily for construction reasons. It is time to pay close attention to the harm caused by CDW. Most nations lack processing facilities for these CDW materials; hence, waste materials are discarded as opposed to being employed and repurposed in the construction of new buildings [21, 22].

CDW is mostly caused by design faults, faulty procurement and planning, inefficient material handling, natural disasters, raw material residues, and unanticipated changes in building design. Instead, advances in site procedures and building design may contribute significantly to waste reduction [23].

3 Circular Economy Action

In a circular economy, raw resources are not extracted from their cycles; instead, they remain in the economy for as long as feasible via the application of efficiency and intelligence. Also, their worth is retained by maximizing their reuse or recycling. To have the lowest environmental effect, the circular economy needs a fundamental rethinking of value chains and business models, product design, and the economic systems to which they are applied as a whole [24].

This means that in the built environment, buildings and construction materials are designed to be adaptable, demountable, and rarely destroyed. Construction materials or building elements would be recovered fast and efficiently, resulting in the maximum recovery of high-quality materials in a closed loop, with nearly no material going to waste. Dangerous substances, such as asbestos and asphalt, would be eliminated from the material cycle [25].

To achieve this, a new holistic strategy incorporating all actors in the value chain is required, as well as collaboration between diverse industries [26]. The circular economy may have substantial implications on waste management and the achievement of applicable waste targets and policy objectives from the standpoint of CDW

Table 1 Population, total area, population density, GDP per capita, CDW generation and share of CDW values for Portugal

Population in 2021	Total area (km^2)	Population density (per km^2)	GDP per capita	CDW generation in 2017 (10^6 tons)	Share of CDW in 2018
10,269,417	92,090	111.52	$23,145.04	11.40	8.8%

management. In a circular economy, the management of CDW is not considered as a sector in isolation, but rather from a systems approach, where the analysis of interventions takes into consideration all system components.

In addition, it is essential to expand the area of actions that might impact waste management to include other phases of the lifecycles of buildings and other structures. Actions inspired by the circular economy taken in the early phases of a building's lifetime may have a significant impact on its waste management.

4 Construction and Demolition Waste in Portugal

In Resource Effective Utilization of Mixed Wastes—Enhancing Construction and Demolition Waste Management, the dependability of CDW statistics have been evaluated [19]. The quality of the CDW data in Europe is rated as 2.3 on a scale from 1.5 to 4.3 by the research. According to the report, Portugal's CDW statistics are of high quality.

Population, total area, population density, GDP, CDW generation for Portugal are presented in Table 1 [27, 28].

Table 2 demonstrates that the principal constituents forming CDW are mineral waste (ranging from 35 to 73.5%), followed by dredging spoils (40%), asphalt (6%) and metal (5%) in all studies.

5 Green Building Design

The idea of GB originated in the 1960s, when the over-exploitation and overuse of fossil fuels caused an energy crisis and environmental pollution issues that were becoming increasingly anabatic [32]. The construction industry's appeal for sustainability has become stronger during the 1980s as the awareness of the building industry's role as a major source of energy consumption and environmental degradation has grown [33].

Modern GB encompasses the activity of designing, constructing, operating, maintaining, and deconstructing a building in a way that minimizes its negative impact on the environment and maximizes its potential for energy efficiency [34].

Table 2 CDW Composition studies in Portugal

Type of CDW	[29] (%)	[30] (%)	[31] (%)
Concrete	35[a]	20.00	35[b]
Brick		53.50[c]	
Soil/sand/gravel	40[d]		40[e]
Asphalt	6		6
Metal	5	0.47	5
Plastic	1		1
Wood	5	3.25	5
Glass	2	0.08	0.5
Paper/cardboard	1		1
Misc. waste			5[f]
Others	5	22.70	1.5
Total	100	100.00	100

The data refers to [a] Concrete, masonry and mortar; [b] Concrete and brick; [c] Brick includes block, mortar and mixed materials; [d] Excavated soils and restoration of pavement; [e] Excavation soil; [f] Mineral CDW.

To define GB standards and label certificates, several governmental agencies, professional institutions, and independent groups have established and released green building rating systems (GBRSs) [35].

In order to mitigate the destructive effects of the built environment on both people and nature, green buildings are being constructed [34]. Several different governments, professional groups, and independent organizations have issued GBRSs in the last few decades to pave the way for a more sustainable building sector [36].

Through increasing insulation and energy efficiency, the EU green buildings accord hopes to reduce the negative impact of construction on the environment. The target date for achieving this goal is 2030.

In its broadest sense, the term "green building" refers to any structure that has been created, managed, or recycled in a way that prioritizes the health and well-being of its occupants, the efficiency of its workers, the conservation of scarce resources, and the preservation of the natural environment [37].

Considerations for lot design and development efficiency, energy and water efficiency, resource efficiency, indoor environmental quality, building-owner maintenance, and the building's overall impact on the environment are all part of this process, which spans the design, construction, operation, and maintenance phases. Hence, environmentally friendly materials with the same or improved performance characteristics as conventional construction materials should be used as the starting point for green building design [38]. The selection of construction materials is often guided by practical, technical, and economic considerations.

However, with sustainability emerging as a critical issue in recent decades, the construction industry, which is responsible for a sizable portion of annual environmental deterioration, can shoulder the responsibility of contributing to sustainable development by discovering more environmentally benign methods of construction and building. Careful selection of eco-friendly sustainable building materials may be the quickest way for builders to begin integrating sustainable design concepts in buildings. Other potential solutions include new material applications, recycling and reuse, sustainable production of products, and the use of green resources [39].

Due to the multidisciplinary nature of integrated design process (IDP) and the complexity of the data sets it generates data management expertise is essential from the outset of the design process [40]. When it comes to climate, orientation, shadow, sun, energy, acoustics, fire and smoke, fluid dynamics, and life cycle assessment (LCA) and life cycle costing (LCC), GB incorporates building simulation analysis that provides virtual realism.

6 Eco-Efficiency CDW Management

Waste is produced during the entirety of the building process, from the mining and processing of raw materials through the actual building, its subsequent destruction, and the final disposal of garbage in landfills. CDW is any waste left behind after a building has been put through some sort of renovation, deconstruction, or destruction [41].

Deconstruction is the inverse of demolition, when the pieces of a building are demolished to reuse them or improve recycling, whereas demolition is the entire destruction of an existing building, structure, or area that results in mixed debris [32].

Reducing the amount of waste generated on-site and assigning responsibility for the waste management plan to various actors are two solutions proposed by Merino et al. [41]. Reduced onsite fabrication with offsite production of exact quantity needed, buy-back strategy for unused materials, onsite synergy of purpose among contractors and subcontractors to avoid reworking, material auditing prior to deconstruction, designated areas for storage, fabrication, etc., and the use of concrete, bricks, and concrete masonry units (CMU) for subbase in construction are all suggested by Kibert [32]. If CDW is well-managed, it can produce cost savings that help soften the blow of the upfront cash needed for GB projects [42].

When compared to a regular building, a green one needs distinctive materials and frameworks to modify its supportability. Throughout the planning and execution of a construction project, a wide range of methods is combined into a unified whole by practical development. Planned activities and development systems that make use of environmentally friendly materials are one option [43]. Because of their positive impact on the environment, green materials are considered ecologically sound (Greenomics). The practical building materials should prioritize asset efficiency, indoor air quality, energy efficiency, and cost-effectiveness [44].

7 Future Challenges

GB is a new school of thought in the construction industry with the goal of reducing the environmental impact of structures throughout their lifetimes. GB promotes the use of eco-friendlier materials, resource- and waste-saving techniques, and better indoor air quality, among other goals. Potentially positive ecological, monetary, economic, and social outcomes may result from this. Yet, most builders still associate green features with costly technology that drive up cost, therefore GBs are rarely seen as appealing projects despite their proven benefits (e.g., photovoltaic panels, grey water reuse systems). Yet, ideal environmental goals may be achieved at a lesser cost through a rigorous design approach and a comprehensive material selection procedure as opposed to an increased investment in technology [45].

The average investment cost per square foot for some GBs, such as academic buildings, laboratories, community centers, and ambulatory care facilities, is not significantly different from the average investment cost per square foot for non-green buildings with the same characteristics, according to some research.

As a result, the market needs a unified method of distinguishing GBs from conventional buildings by employing transparent, objective, and verifiable green criteria to guarantee that these standards have been met [46]. Construction sites can face a wide range of material difficulties. Grey energy and emissions must be taken into account, and more high-value energy and resources are used on material construction than on building activities.

Building materials include byproducts that can be harmful to the environment, and there are constraints on the availability of many key chemicals. The underlying infrastructure of the built environment should also be taken into account. Resource depletion, corrosion, pollution, durability, longevity, etc. are all issues that may be mitigated with a few technical breakthroughs in the realm of construction materials.

The first step toward a more sustainable built environment is ensuring that newly constructed buildings have a longer lifespan, which may be accomplished by doing away with design characteristics that will become obsolete in a short period of time. It is also important that all necessary elements with short lifespans be created with recycling or raw-material recovery in mind. In order to reach this goal, the complexity of the structure must be dissected into its constituent parts, and practically all trade-offs among linked systems must be understood. A better understanding of the ever-expanding selection of construction materials can help with this.

Finally, the processing of materials at the conclusion of a building's life cycle requires meticulous planning. This should be taken into account early on in the design process, when difficult-to-manage composites may be eliminated. Materials that don't need to be remanufactured after being recycled should be prioritized. If they can't be used again right away, they can be recycled into something new. They should be put to the same usage if they are to be reused.

8 Conclusions

Green buildings are constructed with the goals of reducing energy and material usage, recycling and reusing resources to save the environment and boost economic growth and quality of life for all residents.

The study's primary objective was to determine the capacity development needs towards GB adoption in order to bring developing nations up to speed with industrialized countries in adopting and progressing GB structures. This is significant since there is still a healthy GB even though green elements are clearly being used alongside traditional construction methods in Portugal. This analysis pinpoints the primary domains in which additional capacity is required in order to successfully deploy GB.

The environmental and social costs of construction waste have gotten people's attention across the world. In order to fuel the country's economic expansion, Portugal has engaged in extensive building projects, which has resulted in a tremendous increase in construction waste.

Green building advocacy is a viable approach to these problems. While some research has been done on the efficacy of GBRSs in combating CDW, even less has explored the efficacy of LCA with a focus on Portugal. This dire situation necessitates greater research into the impact of CDW on GB problems in Portugal.

References

1. OECD, Waste Management and the Circular Economy in Selected OECD Countries: Evidence from Environmental Performance Reviews, OECD Environmental Performance Reviews, OECD Publishing, Paris (2019).
2. EEA: Environmental taxation and EU environmental policies, Office of the European Union. European Environment Agency, Luxembourg (2016).
3. World Bank: Solid Waste Management 2018. https://www.worldbank.org/, last accessed 2023/22/03
4. Abubakar, I.R., Maniruzzaman, K.M., Dano, U.L., AlShihri, F.S., AlShammari, M.S., Ahmed, S.M.S., Al-Gehlani, W.A.G., and Alrawaf, T.I.: Environmental Sustainability Impacts of Solid Waste Management Practices in the Global South. International Journal of Environmental Research and Public Health, 19, 12717 (2022).
5. Arum, C.: 2011. Recycling of Construction and Demolition Concrete for Sustainable Buildings. European Journal of Scientific Research. 52, 265–270 (2011).
6. Cole, L.B.: Green building literacy: a framework for advancing green building education. IJ STEM Ed 6, 18 (2019).
7. UNEP: Sustainable Consumption and Production Global edition: A Handbook for Policymakers (2015).
8. Iodice, S.: Garbarino, E.: Cerreta, M.: and Tonini, D.: Sustainability assessment of Construction and Demolition Waste management applied to an Italian case. Waste management (New York, N.Y.), 128, 83–98 (2021).
9. Mandpe, A.: Paliya S.: Gedam, V.V.: Patel, S.: Tyagi, L.: Kumar, S.: Circular economy approach for sustainable solid waste management: A developing economy perspective. Waste Management & Research. 41(3), 499–511 (2023).
10. Moustakas, K.: Loizidou, M.: Effective waste management with emphasis on circular economy. Environ Sci Pollut Res 30, 8540–8547 (2023).

11. Pinheiro, M.: Correia, F.: LiderA - Portuguese voluntary sustainable assessment building system - main lines. The 2005 Sustainable Building Conference in Tokyo 27–29 September (SB05 Tokyo), Tokyo. www.lidera.info. (2005).
12. Guedes, M.C.: Pinheiro, M.: Alves, L.M.: Sustainable architecture and urban design in Portugal: An overview. Renewable Energy, 34 (9), 1999–2006 (2009).
13. Amado, M.P.: Barroso, L.M.: Sustainable Construction: Water Use in Residential Buildings in Portugal. International Journal of Sustainable Construction Engineering and Technology, 4(2), 14–22 (2013).
14. Graça, M.: Alves, P.: Gonçalves, J.: Nowak, D.J.: Hoehn, R.: Farinha-Marques, P.: Cunha, M.: Assessing how green space types affect ecosystem services delivery in Porto, Portugal. Landscape and Urban Planning. 170, 195–208 (2018).
15. Eurostat: Eurostat statistics for waste flow generation 2016. European Commission, https://ec.europa.eu/eurostat, last accessed 2023/22/03.
16. European Commission: Closing the loop-An EU action plan for the Circular Economy. Communication from the Commission to the European Parliament, the Council, the European Economic and Social Committee and the Committee of the Regions COM 614(2), (2015).
17. European Directive 2008/98/EC. Waste Framework Directive targeting construction and demolition waste and waste from households (2008).
18. European Commission: Guidance on the interpretation of the term backfilling 2016, https://ec.europa.eu/eurostat, last accessed 2023/22/03.
19. Deloitte.: Resource efficient use of mixed waste. Improving management of construction and demolition waste. European Commission Brussels (2017).
20. Environmental Protection Agency: Advancing sustainable materials management: 2015 Figures and tables, (2018).
21. Akhtar, A.: Ajit, K.S.: Construction and Demolition Waste Generation and Properties of Recycled Aggregate Concrete: A Global Perspective. J. Clean. Prod. 186, 262–81, (2018).
22. Bao, Z.: Lu, W.: Developing efficient circularity for construction and demolition waste management in fast emerging economies: Lessons learned from Shenzhen, China. Sci. Total Environ. 724, 1–9, (2020).
23. Bossink B.A.G.: Brouwers, H.J.H.: Construction waste: quantification and source evaluation. J Constr Eng Manag 122:55–60, (1996).
24. Bilal, M.: Khan, K.I.A.: Thaheem, M.J.: Nasir, A.R.: 2020. Current state and barriers to the circular economy in the building sector: Towards a mitigation framework. J. Clean. Prod. 276, 123250 (2020).
25. Thelen, D.: van Acoleyen, M.: Huurman, W.: van Brunschot, C.: Edgerton, B.: Kubbing, B.: Scaling the circular built environment, pathways for business and government. World Business Council for Sustainable Development, Geneva, Switzerland and Circle Economy, Amsterdam, Netherlands (2018).
26. Rizos V.: Tuokko K.: Behrens A.: The Circular Economy, A review of definitions, processes and impacts. CEPS Research Report. CEPS, Brussels, Belgium (2017).
27. UNPD: World Population Prospects: 2022 Revision. https://data.worldbank.org last accessed 2023/22/03.
28. European Commission: Eurostat—your key to European statistics, 2019. https://ec.europa.eu/eurostat last accessed 2023/22/03.
29. Passos, J.: Alves, O.: Brito, P.: Management of municipal and construction and demolition wastes in Portugal: future perspectives through gasification for energetic valorisation. Int. J. Environ. Sci. Technol. 17, 2907–2926 (2020).
30. Coelho, A: Brito, J.: Economic viability analysis of a construction and demolition waste recycling plant in Portugal—part I: location, materials, technology and economic analysis. J Clean Prod 39:338–352 (2013).
31. Mália, M.: de Brito, J.: Pinheiro, M.D.: Bravo, M.: Construction and demolition waste indicators. Waste Management & Research. 31(3) (2013).
32. Kibert, C.J.: Green buildings: an overview of progress. Journal of Land Use & Environmental Law, 19(2), 491–502 (2004).

33. Lu, W.: Tam, V.W.: Construction waste management policies and their effectiveness in Hong Kong: A longitudinal review. Renewable and sustainable energy reviews, 23, 214–223 (2013).
34. EPA: Green Building 2016. https://www.epa.gov/, last accessed 2023/22/03.
35. Gou, Z.: Lau, S.S.-Y.: Contextualizing green building rating systems: Case study of Hong Kong. Habitat International, 44(C), 282–289 (2014).
36. Chi, B.: Lu, W.: Ye, M.: Bao, Z.: & Zhang, X.: Construction waste minimization in green building: A comparative analysis of LEEDNC 2009 certified projects in the US and China. Journal of Cleaner Production, 120749 (2020).
37. Singh, C.S.: Green Construction: Analysis on Green and Sustainable Building Techniques. Civil Eng Res J. 2018;002 4(3): 555638 (2018).
38. Darko, A.: Chan, A.P.C.: Strategies to promote green building technologies adoption in developing countries: the case of Ghana, Build. Environ. 130, 74–84 (2018).
39. Qian, Q.K.: Chan, E.H.W.: Khalid, A.G.: Challenges in delivering green building projects: unearthing the transaction costs (TCs), Sustainability 7, 3615–3636 (2015).
40. Balramdas, P.: Meher, S.: Behera, B.: Rath, S.: Dash, P.: Choudhary, A comparison between Normal buildings and Green buildings: a case study approach, Int. Res. J. Eng. Technol. 3, 2372–2395 (2016).
41. Merino, M.R.: Gracia, P.I.: Azevedo, I.S.W.: Sustainable construction: construction and demolition waste reconsidered, Waste Manag. Res. 28, 118–129 (2010).
42. Ma, U.: No waste: managing sustainability in construction, Gower Publishing Ltd., UK (2011).
43. Jadhav, N.Y.: Green and Smart Building Trends. In: Green and Smart Buildings. Green Energy and Technology. Springer, Singapore, (2016).
44. Seyfang, G.: Sustainable Housing: Building a Greener Future In: The New Economics of Sustainable Consumption. Energy, Climate and the Environment Series. Palgrave Macmillan, London (2009).
45. Chan, A.P.C.: Darko, A.: Ameyaw, E.E.: Owusu-Manu, D.: Barriers affecting the adoption of green building technologies, J. Manag. Eng (2016).
46. Pintér, L.: Huppé, G.A.: Capacity-Building Courses for Integrating Sustainable Development into Policy Planning and Implementation, Sustain. Technolog. Build. Constr. Ind. (2014).

A Review on Strategies, Policies and Mechanisms Supporting Bioeconomy and Sustainable Development

Susana Filipe

Abstract Bioeconomy development has established a new paradigm by which fossil-based products and associated services are replaced by sustainable, renewable, circular biobased materials, which will ensure the preservation of natural resources and ecosystems, while ensuring both an economical and a societal balance. A successful implementation of Bioeconomy requires the availability of appropriate public policies and governance mechanisms. Given the multisectorial nature of Bioeconomy, the adequacy of public policies and governance mechanisms will require a framework exploiting synergetic and complementary aspects from agriculture, energy, environment and social areas. Due to its inherent complexity, a successful implementation of Bioeconomy strategies shall consider interministerial cooperation, policy coherence, coordination between local, regional, national, EU, international stakeholders and integration of scientific knowledge provided by Academia and Research driven projects, involving multiple stakeholders from industry, private and public sectors, policy makers, etc. This work reviews the current status of European and National policies supporting Bioeconomy and reflects on some of the gaps that need to be addressed to ensure a successful implementation of the Bioeconomy sector.

Keywords Bioeconomy · Public policies · Sustainability development goals · European Green Deal

S. Filipe (✉)
MED – Mediterranean Institute for Agriculture, Environment and Development & CHANGE – Global Change and Sustainability Institute, Institute for Advanced Studies and Research, Universidade de Évora, Pólo da Mitra, Ap. 94, 7006-554 Évora, Portugal
e-mail: susana.filipe@uevora.pt

J. R. da. C. S. Galvão et al. (eds.), *Proceedings of the 3rd International Conference on Water Energy Food and Sustainability (ICoWEFS 2023)*, Springer Proceedings in Earth and Environmental Sciences, https://doi.org/10.1007/978-3-031-48532-9_56

1 Introduction

1.1 Bioeconomy: Definition and Relevance in the Current Global Situation

Bioeconomy includes all sectors and services that make use of biological resources for the production and processing of products and services. Bioeconomy can be defined as the economical sector that covers agriculture, forestry, fishery, aquaculture and other industries, namely food, textile, pharma, chemistry, etc. for which the knowledge of biological resources, biological processes and biological principles is used together with key enabling and converging technologies, such as bio-, nano-technologies, digitalization and information technology. The Bioeconomy sector covers therefore, all areas that rely on biological processes and resources, namely micro-organisms, plants, animals, as well as organic waste and agroindustrial derived biomass [1, 2]. Due to the importance of Bioeconomy for proper management of land and biological resources within ecologic boundaries, the assurance of sustainability across the whole value chain, as well as the assurance of a social and economic stability and balance, the European Union has established an EU Bioeconomy Strategy [3]. This is aimed at identifying, defining and evaluating appropriate goals, that will ultimately lead to the development of measures, instruments and adequate policies, assuring appropriate trade-offs and maximizing synergies between, economical, social and environmental aspects of the utilization of land, sea and biomass. In essence, Bioeconomy is based on the development of a paradigm by which fossil-based products and associated services shall be replaced by sustainable, renewable, circular biobased materials, which will ensure the preservation of natural resources and ecosystems, while assuring both an economical and a societal balance.

The 2030 United Nations Sustainable Development Agenda [4] and the European Green Deal [5] are aligned and set the ambition for the definition of specific targets which shall ensure a climate neutral, resource efficient, competitive and sustainable development, with models and principles enabling prosperity, equality, social balance, health and wellbeing. The approaches, policies, instruments and governance models to be followed shall be developed while tackling global challenges such as climate change, social, demographic and migration and socio-economic challenges.

The main objectives of the Green Deal Strategy [5] include the assurance of food and nutrition security, the management of natural resources and ecosystem services, the reduction of non-renewable resources and replacement of these by renewable, non-fossil based resources (wind, solar, bioenergy, etc.) and the improvement of European competitiveness, through creation of sustainable business and additional jobs. The current socio-economical European instability set by multiple political and health related issues, which come on top of several other global challenges, has set an even higher relevance to both the United Nations Sustainable Development Goals and the European Green Deal Development Goals. An holistic approach combining an adequate use of knowledge, networks and requirements from academia, industry, government, and civil society, will enable a proper and fast achievement of the EGDDGs and UN-SDGs targets.

2 Bioeconomy: Strategy, Policy Instruments and Future Approaches

2.1 Bioeconomy: European and Portuguese Context

The importance of Bioeconomy does not limit itself to the enhancement of more sustainable, circular and environmentally friendly business models within different industry areas. The introduction of biological concepts, processes and know-how into the economy can work as a catalyst to regional and rural development through the growth of new business models. These have an obvious benefit in what concerns economic growth and reduction of territorial disparities. Examples are the use of agroindustrial wastes as sources of energy and nutrient/biobased components, which might on one hand support the energetic transition and decarbonization and on the other hand exploit the potential of secondary products and further valorization on areas such as the organic fertilizer and the pharmaceutical industry. In addition, processes leading to reuse and reutilization of organic waste and wastewater, retrieved through the application of appropriate technologies, can, while combined with the use of biotechnology and digitalization, enable a step change in areas such as crop and animal productivity. Additional benefits of the development of Bioeconomy related with well-being, nutrition and health improvement, can arise from the application of a more sustainable and efficient use of natural resources and services and decarbonization across the entire value chain. This has been the driving force for the development of European and National Bioeconomy Strategic Programs.

The first European Bioeconomy Strategy dates from 2012 [6] and was mainly focused on the enhancement of R&D activities targeting the development of practices leading to more efficient use of renewable resources. The first revision of this strategy occurred in 2018 [7] and kept the five original objectives defined within the initial strategy, which included: the assurance of food and nutrition security, the sustainable management of natural resources, the dependency decrease of fossil-based, non-renewable resources, the mitigation and adaptation towards climate change and the strengthening of European competitiveness through development of more competitive business and the creation of additional jobs. These objectives, are in line with the targets of the European Green Deal and the United Nations Sustainability Development Goals.

An important target defined within the European Bioeconomy Strategy deals with the assignment of dedicated land and aquatic areas for the preservation and restoration of ecosystems, the implementation of sustainable management practices and the use of nature-based solutions within terrestrial and aquatic ecosystems. This approach is expected to lead to benefits in what concerns climate change mitigation and adaptation, biomass management enhancement and improvement of ecosystem/ecosystem services preservation and resilience. Another key focus of the European Bioeconomy strategy includes appropriate management of agriculture, forestry, land, sea, water resources and biomass. These should ensure primarily food security and secondly compliance with environmental targets, energy transition and other important objectives set by the 18 United Nations SDGs and the European Green Deal. Development of innovative processes, products and new business models, driving an improved use of biomass resources, biomass waste and wastewater valorization, enhanced circularity and energy/resource efficiency, constitute another important pillar from the European Bioeconomy Strategy.

The Portuguese Bioeconomy Sustainability Strategy for 2030 is integrated into the 2030 Strategic Agenda for Portugal [8, 9], which was recently developed and is mainly based on three important pillars: economy, society and territorial development. This strategy implementation, aims to (i) reinforce the use of innovation, technology and digitalization; (ii) ensure a demographic equilibrium, territorial cohesion and social fairness; (iii) enable the climate transition and ensure the sustainable use of resources and (iv) assure a sustainable and competitive economic development for Portugal. The 2030 Bioeconomy Portuguese Strategy [9] action pillars are aligned with the European Biodiversity Strategy and focus mainly on (i) protection and reinforcement of biodiversity; ii) valorization of natural resources and reduction of dependency from non-renewable resources; iii) sustainable management of natural resources; iv) achievement of carbon neutrality; v) valorization of innovative processes, innovation and digitalization as enablers to boost bioeconomy and vi) promotion of rural development.

The achievement of the objectives outlined by the 2030 Portuguese Bioeconomy Strategy [9] shall be enabled by different initiatives, which are linked to knowledge build-up, knowledge transfer, policy and regulation assessment and adaptation, funding initiatives to support R&D and Business development and Market driven initiatives.

Fundamental knowledge build-up shall be supported by research and development initiatives and projects involving multiple actors of the society, namely public and private sectors, research institutes, universities, etc. This shall boost and lead to a faster implementation of ideas into concrete products and services, accelerate knowledge transfer and speed up the development of startups within the areas of Bioeconomy.

The first step for a proper implementation of an adequate Bioeconomy strategy is the assessment of residues and subproducts resulting from both primary and secondary sectors. This includes the materials arising from primary sectors, such as agriculture and animal production, forestry, fishery and aquaculture and the associated secondary transformation sectors (e.g. olive oil industry, wine industry, dairy industry etc.). Once this knowledge is obtained it will be essential to make it available to the general public, via for example the development of specific online platforms/databases.

An essential aspect for the implementation of the Bioeconomy strategy is the evaluation of current legislation, policies and the adjustment of these to enable and facilitate the growth of Bioeconomy. This work shall be done with the support of experts from multisectorial areas, policymakers, agriculture, soil, water, resource valorization specialists, economists, etc. These working groups shall map any existent legislation, policy and governance roadblocks to the development of Bioeconomy, and identify necessary adjustments and changes to these legislation instruments that will ultimately enable the valorisation of byproducts and waste from both primary and secondary industries.

The adaptation of current policies and for example, the change of classification of some primary and secondary industry residues into subproducts, will minimize the barriers for the entry and use (in its total or in specific parts) of this type of subproducts into the Bioeconomy market.

Another important pillar of the strategy is focused on the development of intersectorial R&D activities, which will be the basis for the development of alternative business models and resource valorization within the entire value chain. An outcome of these initiatives shall be the development and implementation of bioeconomy-based technologies ranging from TRL4 to TRL9. The engagement of multisectorial partners in these R&D activities will facilitate startup development, technology transfer and development of commercial products, creating in this way an additional economic value.

Another crucial aspect will be to develop communication and dissemination strategies targeting the acceptance and the increase of public confidence towards biobased products integrating subproducts on their content. Besides this, there will be the need to establish the tools (e.g. legal, regulatory, governance, etc.) enabling an improvement of the competitiveness and valorization of biobased products.

2.2 *Public Policy Adjustment and Development for Bieconomy–The Portuguese Approach*

The adjustment and development of public policies capable of addressing the targets outlined by European and National Bioeconomy Strategies requires an holistic approach, integrating the inputs from different sectors, namely research, technology development, economic infrastructures, regional, national, European actors and civil society. This is particularly important if one considers the different areas of relevance for the Bioeconomy sector, which include but are not limited to energy, agro-food, agroecology and health. In this sense, in order to create adequate platforms to boost Bioeconomy, most European countries are developing holistic national Bioeconomy strategies rather than strategies related to specific policy areas. Furthermore, and even more essential: Bioeconomy-related policies shall be built upon three dimensions—Environment, which addresses mainly the management of land and biological resources within ecological boundaries; Economy, which shall consider the development of sustainable value chains and overall consumption; and Society, that relates with social fairness. The development of adequate consortiums and the strategic alignment of different relevant actors, financially supported by several national and European funding schemes will ultimately result on the achievement of essential knowledge (e.g. technical, business, etc.) which will be of essence to support the development of technically based and adequate public policies.

The appropriate implementation of the Bioeconomy Strategy will require the development of different types of policies, which include direct regulations, economic instruments and voluntary systems [10]. Implementation of specific regulations for waste management, separation, classification, the change of category of given subproducts from primary and secondary industries from waste to secondary by-products, the implementation of restricted areas (ecological zoning) to protect and improve specific habitats and restore important ecosystems and their services, restrictions on the use of chemical fertilizers, regulations to reinforce proper management of manure, wastewater and agro-industry effluents are examples of direct regulations. Direct regulations are designed in order to create obligatory standards and rules that need to be followed by key players. Fulfilment of these regulations is subject to control and non-compliance leads to the implementation of a set of "punishment" measures.

Economic instruments are a type of policy instrument that typically supports key stakeholders to follow the standards and rules defined within direct regulations. Examples of economic instruments are financial incentives supporting the creation of associate and collaborative groups, rural development incentives for the development of more sustainable farming, development and implementation of systems/technologies/practices leading to decarbonization, waste fees, etc. Economic instruments can include specific loans, grants, subsidies, feed-in tariffs, tax incentives, premiums and general R&D funds (national, European).

Voluntary approaches typically enable the development of good codes of practice and positive change of behaviour. Examples include voluntary certification schemes, labelling (e.g. organic farming), development of platforms for infosharing and communication, capacity building and knowledge build-up initiatives are also part of these voluntary instruments.

Considering the Portuguese context only, there are a couple of policy instruments that will support Bioeconomy development [9]. These include (i) the Strategic Plan for the Common Agriculture Policy (PEPAC 2021–2027) [11]; (ii) the Portuguese Agenda for Agriculture Innovation [12]; (iii) the National Strategy for Farm and Agroindustrial effluents (ENEAPAI) 2030 [13]; (iv) the National Strategy for the promotion of Cereal production [14]; (v) the National Plan for the promotion of Biorefineries [15]; (vi) the National Plan for Carbon Neutrality 2050 (RNC2050) [16]; (vii) the National Plan for Energy and Climate 2021–2030 (PNEC2030) [17]; (viii) the European Strategy for Reduction of Methane emissions [18] and (ix) the Fertilizer Regulations (Regulamento (CE) nº. 2019/1009 de 25 de Junho de 2019) [19].

Important measures outlined within the policy instruments referred above address different important pillars: Food, Energy, Valorization of Subproducts, Water and Soil Management, Emissions, Knowledge Buildup and Dissemination, R&D and Economical growth.

Some of the most relevant actions focusing on the Food topic include:

- the establishment of a specific target for the 20% increase of Mediterranean diet being followed
- the promotion of innovation initiatives leading to a more sustainable, efficient and economically balanced agriculture, such as the implementation of precision agriculture and digitalisation, use of biotechnology as a tool to improve species (e.g. adaptation to increase resilience towards climate change), etc.
- the improvement of soil quality and overall crop yield by the application of organic matter as soil fertilizer.

Areas of action to be highlighted within the Energy and Carbon footprint reduction pillar include:

- fostering the increase of energetic efficiency and use of renewable energies within the agriculture sector
- reduction of greenhouse gas emissions by 85% in comparison with 2005, and the capture of 13 million tons of carbon within the agriculture and forestry sectors
- an overall liquid reduction of emissions in the agriculture sector ranging from 21 to 49%
- valorisation in biorefineries of residual agro-industry, forestry and agriculture biomass with low or now value for other sectors (e.g. animal food, other), aiming at the generation of a sustainable business model for residues not used for other applications. This includes use of rice straw, residues from corn crops, pruning biomass from olive orchards, vineyards and fruit orchards, olive oil pomace, wine pomace, manure and other biomass from agro-industries
- increasing annual renewable energy consumption by 47% till 2030
- expanding practices such as conservation agriculture and precision agriculture, reduction of emissions associated with farming and fertilizer usage
- applying agriculture management practices promoting carbon capture, such as biodiverse pasture and regenerative agriculture.

Some examples of measures to be implemented in the areas of Sustainability development include:

- to reach an increase of more than 50% for certified sustainable production land area with respect to the total agriculture area
- to reduce emissions and increase carbon capture within the agriculture sector
- to use additives within animal feed aiming at the reduction of methane emissions
- to reduce the use of nitrogen based fertilizers and replace mineral based by organic based fertilizers
- to develop management systems for manure aiming at a decrease of the emissions.

Approaches for appropriate management of wastewater and other wastes arising from the primary and secondary industry sectors include:

- establishment of valorization schemes for agropecuary and agro-industrial wastes, with special emphasis on the recovery of these and valorization, aiming at the development of a closed cycle for nutrients and water, following the principles of circular economy
- development of suitable management schemes, clear definition of appropriate systems for recovery, treatment, reuse and valorization of agro-industry effluents.

Measures to boost Research and Development, Knowledge Build-up and transfer in the areas of Bioeconomy include:

- increase of the investment in R&D by 60% and general reinforcement from R&D on the areas of relevance for Bioeconomy
- promotion of initiatives to boost training and overall communication to the civil society on topics related with sustainable agroindustry practices, bioeconomy, resource valorization, decarbonization and other relevant topics.

2.3 Future Approaches to Support and Boost Bioeconomy

The United Nations Sustainability Development Goals, the European Green Deal, the Farm to Fork Strategies and the European Strategy for Biodiversity set and define the basis for the creation of an adequate Bioeconomy Strategy. There are two important resources that play a very crucial role and are essential for the achievement of the targets outlined by the strategies above: these are the soil and the water. Climate change and other global challenges developed scenarios that have placed the availability and health of these two crucial resources at risk. Water scarcity is exacerbated by climate change—this is particularly critical in countries of the Mediterranean area, which have been subject to a significant increase in temperatures and drought events. Risks associated with soil quality are also an important challenge worldwide and in particular in the Mediterranean basin. In Portugal only, it is known that 25% of the soil has elevated risk of degradation and very low levels of organic matter [20, 21].

Portuguese governance instruments, mechanisms and processes regulating water management are mainly defined within the Portuguese Directive for Water (Diretiva 60/2000) and other mechanisms for water monitoring, such as qualitative and quantitative evaluation of the waters, which include both water mass and quality evaluation. However, as a consequence of climate change and further reduction of water availability, it is more and more critical to reevaluate, adapt and optimize current legislation and consider a multisectorial approach involving different sectors, agriculture, tourism, industry, etc. Current issues such as water scarcity, pollution and competition of water resources across different sectors, constitute a new challenge for which proper action is required.

As for the soil, there is a need to define, at an European and National scale, the adequate procedures to monitor quality and define adequate measures, policies and instruments to ensure an improvement of soil quality and overall soil restoration.

An essential path towards the improvement of soil quality shall arise from the work done at an European and National level (Missão Solo). Within these platforms, it is aimed to develop systems for monitoring soil health, collection of soil data within Portugal and creation of a network of farms and living-labs devoted to the study of soil and relevant for the Portuguese production systems (e.g. montado, olive orchards, vineyards, pasture, etc.). These initiatives shall be the basis for the creation of adequate national policies for the soil.

There is therefore the need to adapt current governance mechanisms, involving different government areas (agriculture, environment), public and private sectors, central, regional and local administration and scientists. The latter will be ultimately the ones providing technical knowledge, capable of defining solid and adequate measures for improvement of quality and management of these two essential resources—water and soil—which are, as stated before, critical for the development of the Bioeconomy Sector.

3 Concluding Remarks

The Bioeconomy sector constitutes a business with an unique potential, providing means to link multiple sectors, such as the primary and secondary production and other industrial processes focusing on energy, valorization and sustainable development. The growth of the Bioeconomy sectors constitutes an important platform to boost sustainable development, economic stability and social development, being therefore critical for the improvement of circularity, achievement of carbon neutrality and development of new opportunities for jobs and market diversification.

Critical platforms for Bioeconomy development include governance and public policies. Given the links of bioeconomy to many different sectors, the adequacy of public policies and governance mechanisms will require a framework exploiting synergetic and complementary aspects from different areas, ranging from agriculture, energy, environment and social aspects. Due to its inherent complexity, a successful implementation of Bioeconomy strategies will require the fostering of interministerial cooperation, policy coherence and coordination at different levels (local, regional, national, EU, international).

In the current context of global change, it is unavoidable to revisit current knowledge, policies and governance mechanisms. Filling the current gaps and ensuring a sustainable development, will require the acquisition of knowledge, the development of R&D projects focusing on the areas of relevance for Bioeconomy, the involvement of international stakeholders, multisectorial and multidisciplinary approaches. Last but not least, the Bioeconomy sector will only succeed with the acceptance of general society. This will require efficient communication schemes, public society involvement and overall initiatives to engage the society on the development of Bioeconomy sectors [22].

There are however still some gaps that need to be addressed: development of proper mechanisms to regulate the management of land and water across different value chains, solutions to regulate competition between the use of biomass in different sectors and the development of adequate policies for water management and soil quality evaluation/management ensuring the improvement of the quality, health and availability of these two important resources.

References

1. Gould, H., Kelleher, L., O'Neill, E.: Trends and policy in bioeconomy literature: A bibliometric review, EFB Bioeconomy Journal 3, 1-12 (2023).
2. Stephenson, P., Damerell, A.: Bioeconomy and Circular Economy: approaches need to enhance the focus on biodiversity to achieve sustainability, Sustainability,14, 1-20 (2022).
3. European Commission Bioeconomy Homepage, https://research-and-innovation.ec.europa.eu/research-area/environment/bioeconomy_en.
4. United Nations Sustainable Development Goals Homepage, https://sdgs.un.org/goals.
5. European Green Deal Homepage, https://commission.europa.eu/strategy-and-policy/priorities-2019-2024/european-green-deal_en).
6. Directorate-General for Research and Innovation (European Commission), Innovating for sustainable growth - A bioeconomy for Europe, European Commission, Brussels (2012).
7. Directorate-General for Research and Innovation (European Commission), A sustainable bioeconomy for Europe - Strengthening the connection between economy, society and the environment : updated bioeconomy strategy, European Commission, Brussels (2018).
8. Portuguese Strategy 2030 Homepage, https://www.portugal.gov.pt/pt/gc22/comunicacao/documento?i=resolucao-do-conselho-de-ministros-que-aprova-a-estrategia-portugal-2030.
9. Avillez, F., Martinho, S., Campello, F. Serrano, P., Gamito, T., Aires, N., Lopes, R., Brás, S., Linhas Estratégicas dos sectores de produção primária no contexto do desenvolvimento sustentável da estratégia nacional para a bioeconomia sustentável 2030, Gabinete de Planeamento, Políticas e Administração Geral Portugal, Lisboa (2021).
10. Bengtsson, M., Hotta, Y., Hayashi, S. and Akenji, L. *in* Policy Tools for Sustainable Materials Management, Institute for Global Environmental Strategies, Japan (2010).
11. Strategic Plan for the Common Portuguese Agriculture Policy (PEPAC 2021–2027) Homepage: https://www.gpp.pt/index.php/pepac/pepac-plano-estrategico-da-pac-2023-2027.
12. Portuguese Agenda for Agriculture Innovation Homepage - Resolução do Conselho de Ministros n.° 86/2020, https://dre.pt/dre/detalhe/resolucao-conselho-ministros/86-2020-145102353.
13. Estrategia-nacional-para-os-efluentes-agropecuarios-e-agroindustriais (ENEAPAI) 2030 Homepage, https://apambiente.pt/agua/estrategia-nacional-para-os-efluentes-agropecuarios-e-agroindustriais (ENEAPAI) 2030.
14. National Strategy for the promotion of Cereal production (Resolução de Conselho de Ministros n° 101/2018) Homepage, https://dre.pt/dre/detalhe/resolucao-conselho-ministros/101-2018-115777790.
15. National Plan for the promotion of Biorefineries (Resolução do Conselho de Ministros n.° 163/2017 Homepage, https://dre.pt/dre/detalhe/resolucao-conselho-ministros/163-2017-114133883.
16. National Plan for Carbon Neutrality 2050 (RNC2050) Homepage, https://unfccc.int/sites/default/files/resource/RNC2050_EN_PT%20Long%20Term%20Strategy.pdf.

17. National Plan for Energy and Climate 2021–2030 (PNEC2030) Homepage, https://climate-laws.org/geographies/portugal/policies/portugal-s-national-energy-and-climate-plan-for-2021-2030.
18. European Strategy for Reduction of Methane emissions (COM(2020)663 de 14/10/2020) Homepage, https://www.eumonitor.eu/9353000/1/j9vvik7m1c3gyxp/vlcxt8splkzm.
19. Portuguese Fertilizer Regulations (Regulamento (CE) nº. 2019/1009 de 25 de Junho de 2019) Homepage, https://eur-lex.europa.eu/legal-content/en/ALL/?uri=CELEX%3A32019R1009.
20. Pinto Correia, T., Caring for soil is caring for live (2020) *in* https://www.rederural.gov.pt/images/Noticias/2020/Caring_for_soil.pdf.
21. van Delden, H., Fleskens, L., Vanhout, R., Nunes, J.P., Baartman, J., Lesschen, J.P., Verzandvoort, S., Hessel, R. *in* SoilCare Project: Report on the integration and synthesis of Study Site results and their potential for upscaling (2021).
22. Material Economics Sverige AB, EU Biomass Use In A Net-Zero Economy - A Course Correction for EU Biomass, Material Economics, Energy Transitions Commission and the European Forest Institute, Stockolm (2021).

The Role of Construction 3D Printing in Advancing Sustainability and Digitalization in Africa

Karim Fahfouhi, Flávio Craveiro, and Helena Bártolo

Abstract Africa's rising population is increasing pressure on natural resources, resulting in biodiversity loss and affecting the ecosystem. Some African countries are promoting 3D Printing technologies to respond to the rising demand for affordable houses, allowing a more efficient use of natural resources, as well as reducing waste and environmental impacts. This paper gives a brief outline of the digitalization progress in some African countries, with a particular focus on the key role that 3D printing (3DP) can play in revolutionizing construction, converting it into a more sustainable and efficient one. This work also addresses the potential that new printable earthen-based material formulations can bring to foster sustainability and circularity in the built environment. Digitalization can be a major force in advancing Africa's economic development, assisting construction companies in enhancing productivity and efficiency, creating new job opportunities, and accelerating sustainable development goals. Moving forward towards Construction 4.0 will require a multifaceted strategy, including investing in education and training, getting government support, Industry-University collaborations, infrastructure development, establishing public–private partnerships, and fostering IT innovation and sustainability.

Keywords Sustainable development · 3D printing · Digitalization · Sustainability · Africa

---

K. Fahfouhi · F. Craveiro (✉) · H. Bártolo
School of Technology and Management, Polytechnic of Leiria, Leiria, Portugal
e-mail: flavio.craveiro@ipleiria.pt

F. Craveiro · H. Bártolo
CIAUD, Universidade de Lisboa, Lisboa, Portugal

J. R. da. C. S. Galvão et al. (eds.), *Proceedings of the 3rd International Conference on Water Energy Food and Sustainability (ICoWEFS 2023)*, Springer Proceedings in Earth and Environmental Sciences, https://doi.org/10.1007/978-3-031-48532-9_57

1 Introduction

Construction is crucial for a country's economy, and even minor improvements have a significant impact on society and the environment [46]. There is a need to adopt a more circular model to foster further sustainable economic development, accelerating job creation and environmental sustainability in Africa [32]. However, the global construction industry has been hesitant to adopt new technologies compared to other industrial sectors [29], no disruptive changes occurred, and only a few new technological methods were implemented. The construction sector is characterized by many interrelated construction processes affecting the overall productivity at different construction stages. Over the past decades, efficiency improvements have been marginal, and productivity has decreased [31].

On the other hand, demography is rising in Africa, especially for young people under 18s, so it is imperative to solve the continent's housing crisis [18]. According to the World Bank, over 60% of urban residents in Sub-Saharan Africa live in informal settlements, with almost no access to basic services, such as water and sanitation, due to deficient urban infrastructure, increased population growth, and inadequate housing policies [9]. The slow economic development of African countries has impacted the construction sector due to limited access to technology and inadequate infrastructure [39]. Inadequate infrastructure, a shortage of skilled labor, and limited access to funding are significant barriers to growth and development in the industry [13], International Labour Organization, 2023. For instance, inadequate infrastructure in Ethiopia made it difficult to transport cement, leading to delays and increased costs. Additionally, [15] note that on-the-job training leads to poor workmanship and work safety in Tanzania. These challenges, along with political instability, have had a significant impact on the sector's growth and development in Africa [48].

Digital technologies can help African companies optimize supply chains and improve productivity and efficiency. The future of construction can be transformed with the implementation of Industry 4.0 [10, 12], which is a term used to describe the Fourth Industrial Revolution, representing a novel method of setting up industrial processes and resources so they are more adaptable to a constantly changing environment [27]. The United Nations Industrial Development Organization (UNIDO) highlights the potential of Industry 4.0 technologies, such as IoT and AI, to boost industrialization in Africa [52]. According to [4], digitalization will allow increasing visibility across the construction supply chain, as well as boosting automation and efficiency. Additive Manufacturing (AM), also called 3D printing (3DP), can be defined as the process of printing materials or joining composite materials together from a 3D model data, typically with layer-by-layer deposition, as opposed to subtractive manufacturing [16, 41]. 3DP is a growing industry allowing creating 3D objects with complex structural forms, produced with reduced waste. It is commonly used in several industries, such as aerospace, medical devices, and mechanical engineering [44]. 3DP in construction can increase efficiency and improve the quality of projects [38]. The use of 3DP technology can enable construction companies to increase productivity, enhance innovation and achieve building optimization [47]. 3DP can also assist in infrastructure and logistics, allowing on-site production. The construction sector can this way reduce transportation emissions and improve supply chain efficiency. Recent reports show that Africa has made some progress in the adoption of digital technologies for industry and environmental management [33]. Digital technologies allow for enhancing productivity, reducing operational costs, and improving quality control. Digitalization can be a significant driver for a country's economic development, and it might lead to sustainable manufacturing practices across the African continent with the right investment and support.

Recently, the construction sector has started to show positive signs, and African governments have been launching several initiatives aimed at improving infrastructure to provide affordable housing to citizens [1]. In Nigeria, for instance, the government launched the National Housing Fund, to provide low-cost housing loans to low-income families. Similarly, in Kenya, the government has launched a program to deliver affordable housing to citizens, with a target to construct 500,000 new buildings [34]. Some African countries, such as Malawi, Morocco, South Africa, and Kenya, are fostering 3DP technologies to address the housing crises. In Angola, the implementation of 3DP has helped to reduce costs by using 99% locally available materials and improving the quality of construction projects [11].

This paper gives a brief outline of the digitalization progress in some African countries, with a particular focus on the key role that 3DP can play in revolutionizing construction, converting it into a more sustainable and efficient one. This work also addresses the potential that new printable earthen-based material formulations can bring to foster sustainability and circularity in the built environment.

2 Sustainability and Digitalization in Africa

Digitalization has the potential to significantly improve social, economic, and ecological sustainability in Africa [2]. Regarding social sustainability, digital technologies can enhance access to education, healthcare, and other essential services for marginalized populations and reduce poverty. For example, through the use of digital technologies, construction companies can effectively connect with communities and stakeholders, fostering collaboration and engagement in development initiatives. Additionally, digital technologies can facilitate the development of e-learning platforms and other educational resources, enabling workers to enhance their skills and access new opportunities [43].

According to the World Bank [48], ineffective regulations and policies are a major challenge due to a lack of coordination and collaboration between different government agencies. In addition, [24] point out that a lack of industrial competition results in limited innovation and diversity in digital products and services. Furthermore, [5] suggest that a lack of funding for innovation and entrepreneurship leads to inadequate support for startups and SMEs in the digital sector.

According to [25], digitalization can lead to more efficient and cost-effective business practices, as well as increased productivity and economic growth. Digital technologies can help small and medium-sized enterprises (SMEs) to access new markets and customers, increasing their competitiveness and contributing to job creation [26]. In recent decades, the Information, Communication, and Technology (ICT) sector experienced sustained growth and has become a key factor in driving sustained, inclusive growth in Africa. Despite improvements in infrastructure, Africa still confronts problems and will need more investments to attain the required growth [51].

Africa faces severe poverty, inequality, and unemployment across many countries. Rebeca Grynspan, the secretary-general of UNCTAD, stated that "the economic slowdown causes further setbacks in the realization of the 2030 Agenda for Sustainable Development" [50]. ICT accessibility can also support creating new economic opportunities, enhancing information flows that increase efficiency, and assisting citizens in conducting business and generating higher incomes [14].

Digitalization can enable more efficient use of natural resources, reducing waste and environmental impact [35]. Additionally, digital technologies can enhance environmental monitoring and reporting, enabling more effective regulation and management of environmental resources [9].

Reference [28] point out significant challenges and gaps in ecological sustainability in Africa, especially in areas such as natural resources, biodiversity, carbon footprint, and climate change. Reference [6] suggest that rapid population growth exacerbates the pressure on natural resources, leading to biodiversity loss and consequent impacts on ecosystem functioning and services. Additionally, [37] note low technological development and innovation as a challenge, limiting the capacity to adopt sustainable practices and reducing the potential for a low-carbon transition.

3 Digital Fabrication Initiatives in Africa

3.1 South Africa

The potential use of 3DP by the construction sector in South Africa is a promising opportunity to provide affordable and sustainable housing solutions. The country has the competency skills to adopt 3DP, as demonstrated in various successful projects [7, 17, 21]. South Africa has the necessary infrastructure, skilled workforce, and research and development capabilities to support the adoption of 3D printing in the construction sector [17, 21].

South Africa is committed to pursuing the African development agenda and promoting growth and development nationally, continentally, and globally, with a particular focus on science, technology, and innovation (STI) [45].

The "House of the Future" project is an important initiative in South Africa that uses 3D printing technology to address the country's housing crisis. The project is a collaboration between the University of Johannesburg and the Council for Scientific and Industrial Research (CSIR), among other partners. The use of 3D printing technology presents several benefits, including affordability, sustainability, and a reduction in the time taken for construction [30].

3.2 Kenya

African Centre of Technology Studies (ACTS) and Kenyatta University established a 3DP technology center to develop 3DP adoption in industry and economic growth in Africa, while the University of Nairobi has a new research center focused on the development of 3DP in construction [22], therefore, Kenya universities are crucial for the growth of digital fabrication, specifically 3DP [23]. To tackle the housing crisis in Kenya, a project led by the African construction company 14Trees, a joint venture between Holcim Group, British International Investment, and the Green Heart of Kenya Development have started Africa's biggest 3D printed affordable housing project [20]. This project uses a printer manufactured by COBOD [11], a Holcim-patented mix of dry cement, aggregates and chemical admixtures to extrude the composites materials through a nozzle and build a house in 18 h, also requiring handwork from a team of carpenters, plumbers, and electricians to install windows, roofs, doors and connect services [20].

3.3 Morocco

Due to the challenges related to climate change and the saving of natural resources, the construction sector in Morocco is due to change and mitigate the environmental impacts and must drift toward more digitalized construction practices [40]. With the advent global industrial revolution, the Moroccan government aims to adopt innovative technologies such as 3DP to boost productivity in the construction sector [36]. In 2021, the Ministry of Industry and Digital Economy, the Digital Development Agency (DDA) and the Euro-Med University of Fez signed an agreement to join efforts in encouraging SMEs to adopt new technologies and improve manufacturing skills, increase productivity and competitiveness. As part of the initiatives in this agreement, it is also aimed to support scientific research and innovation to make Morocco an attractive and competitive country for advanced manufacturing, product customization, and efficiency in the construction phase. In Morocco, digital technologies can drive economic growth, increase efficiency, and improve competitiveness in various sectors of the Moroccan economy [3]. Morocco has the potential to see substantial advantages from implementing 3DP in its construction sector by utilizing locally available materials, which can reduce dependence on imported goods, boosting the local economy and reducing the carbon footprint of construction projects.

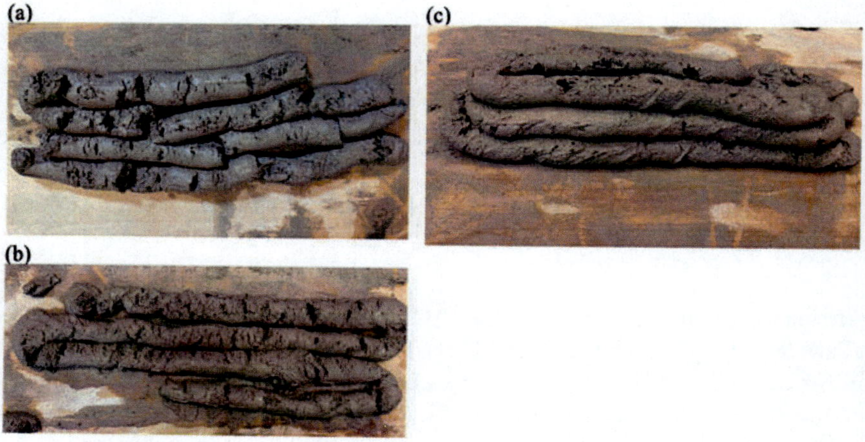

Fig. 1 Printed filaments with different formulations (a, b, and c) based on quantity of water and soil

4 Case Study: Development of Earth-Based Composites

As part of the objectives of this paper, a study on the printing process of earthen materials is ongoing by studying the printing process, fresh properties, and the rheology of printable composites. The use of 3DP with earth-based materials aims to minimize material use and waste, maximizing circularity for a more sustainable and energy-efficient building process [42]. This ongoing experimental work aims to develop different compositions with four components: raw earth, lime-based binder, water and flax fibers. This work is driven by the concept of creating and building new materials which are more environmentally responsible, as more studies on cementitious materials for 3DP are encouraging the use of earthen materials in construction in terms of sustainability [19]. Further experimental work must be carried out to get a better understanding of the behavior of earthen-based composites with other natural fibers.

Figure 1 shows printed filaments using a printing system with a speed of 150 mm/s, a 3 m long hose, and a 25 mm diameter circular nozzle.

5 Conclusions

The built environment must adapt to a low-carbon economy, be more resilient to climate change, increase resource efficiency, and reducing waste, optimizing construction technologies and processes.

Africa's construction sector can be modernized through digitalization and sustainability, advancing sustainable development goals and creating new job opportunities.

Currently, there are positive signs across African countries fostering the adoption of digital technologies for the construction sector. Recent initiatives are focused on skills development and workforce training. The "Future of Construction" program, which aims to build the capacities of construction professionals, companies, and governments in adopting and implementing construction 4.0 technologies, can have a deep impact on Africa's future through the adoption of digital technologies, such as 3DP.

Africa must aim to bridge the talent gap in the digital space to fully leverage the potential of digitalization. This can be achieved by investing in digital skills development programs, promoting public–private partnerships, and establishing the necessary legal and regulatory frameworks to foster innovation. 3DP can be a powerful tool to revolutionize the African industrial landscape [16], reducing environmental impacts and enhancing economic and social well-being.

Fostering Construction 4.0 research and development will involve partnering with universities and research institutions to accelerate the construction industry transformation through greener technologies and sustainable innovation. Thus, moving forward towards Construction 4.0 will require a multifaceted strategy, including investing in education and training, getting government support, Industry-University collaborations, infrastructure development, forming public–private partnerships, and fostering IT innovation and sustainability.

References

1. Adeleye, A., & Winston, S. (2023). *Digitalization of the South African Construction Industry: A Potential for Further Future Research.* 17–27.
2. Akiwumi, P. (2022, September 5). *Digitalization as a driver of structural transformation in African LDCs.* UNCTAD. https://unctad.org/news/blog-digitalization-driver-structural-transformation-african-ldcs
3. Amer, H., & Chafai, M. (2021, September 29). *سبيل المغرب نحو التنمية الاقتصادية ..الرقمنة*. SNRT News. https://snrtnews.com/article/
4. Azmeh, S., Foster, C., & Echavarri, J. (2020). The International Trade Regime and the Quest for Free Digital Trade. *International Studies Review*, *22*(3), 671–692.
5. Baah, B., Bamlford, R., Bayuo, B., Gakuo, C., Mwaya, J., & Tholstrup, S. (2022, February 15). *Supercharging Africa's Startups: The Continent's Path to Tech Excellence.* Tony Blair Institute for Global Change. https://www.institute.global/insights/geopolitics-and-security/supercharging-africas-startups-continents-path-tech-excellence
6. Balogun, A. L., Adebisi, N., Abubakar, I. R., Dano, U. L., & Tella, A. (2022). Digitalization for transformative urbanization, climate change adaptation, and sustainable farming in Africa: trend, opportunities, and challenges, *19*(1), 17–37.
7. Bellamy, C. (2022). How 3D printing can be the solution to the nation's affordable housing crisis. NBC News. https://www.nbcnews.com/news/nbcblk/3d-printing-can-solution-nations-affordable-housing-crisis-rcna10725
8. Bhatia, K., & Manyika, J. (2022, December 13). *Supporting Africa's digital transformation.* https://blog.google/around-the-globe/google-africa/supporting-africas-digital-transformation/
9. Bortolini, R., Rodrigues, R., Alavi, H., Vecchia, L. F. D., & Forcada, N. (2022). Digital Twins’ Applications for Building Energy Efficiency: A Review. *Energies 2022, Vol. 15, Page 7002*, *15*(19), 7002.

10. Boton, C., & Forgues, D. (2023). *Construction 4.0*. Building Transformatinos. https://www.buildingtransformations.org/articles/construction-4-0
11. COBOD. (2022). *Largest 3D printed building in Africa*. https://cobod.com/largest-3d-printed-building-in-africa/
12. Craveiro, F., Duarte, J. P., Bártolo, H. M., Bártolo, J. P. (2019). Additive manufacturing as an enabling technology for digital construction: A perspective on Construction 4.0. *Additive Manufacturing, 103*, 251-267.
13. Desalegn, A., & Solomon, N. (2022). Infrastructure inequities and its effect on poverty reduction across regional states in Ethiopia, *2*(3), 291–309.
14. Economic Commission for Africa. (2018). *Drivers of inclusive growth in Africa*. https://archive.uneca.org/sites/default/files/PublicationFiles/e1701271_0.pdf
15. Evarist, C., Luvara, V. G. M., & Chileshe, N. (2022). Perception on constraining factors impacting recruitment and selection practices of building contractors in Dar Es Salaam, Tanzania.
16. Fahfouhi, K., Freitas, D., Poullain, P., Craveiro, F., & Bártolo, H. (2023). Digital Fabrication in Construction Industry in Africa. In *Proceedings of the 2nd International Conference on Water Energy Food and Sustainability (ICoWEFS 2022)* (pp. 470–479). Springer International Publishing.
17. Fiyaso-Bambi, J. (2023, March 4). *South Africa's first 3D printed low-cost house*. Africanews. https://www.africanews.com/2023/03/04/south-africas-first-3d-printed-low-cost-house-inspire-africa//
18. Ghaedi, M. (2022, November 27). *Africa's population boom may boost economy, global relevance*.
19. Gomaa, M., Jabi, W., Soebarto, V., & Xie, Y. M. (2022). Digital manufacturing for earth construction: A critical review. *Journal of Cleaner Production, 338*, 130630.
20. Holcim. (2023, February 14). Inside the largest 3D-printed affordable housing project to date. https://www.holcim.com/who-we-are/our-stories/largest-3D-printed-affordable-housing
21. Jackson, B. (2019, April 30). *Aeroswift successfully 3D prints 0.5 m tall large scale airframe for UAV* . 3D Printing Industry. https://3dprintingindustry.com/news/aeroswift-successfully-3d-prints-0-5-m-tall-large-scale-airframe-for-uav-154488/
22. KAM. (2018). *Manufacturing Priority Agenda 2018*.
23. Kolade, O., Adegbile, A., & Sarpong, D. (2022). Can university-industry-government collaboration drive a 3-D printing revolution in Africa? A triple helix model of technological leapfrogging in additive manufacturing. *Technology in Society, 69*(March), 101960.
24. Kouladoum, J. C., Wirajing, M. A. K., & Nchofoung, T. N. (2022). Digital technologies and financial inclusion in Sub-Saharan Africa. *Telecommunications Policy, 46*(9), 102387.
25. Kunkel, S., & Matthess, M. (2020). Digital transformation and environmental sustainability in industry: Putting expectations in Asian and African policies into perspective. *Environmental Science & Policy, 112*, 318–329.
26. Kuteyi, D., & Winkler, H. (2022). Logistics Challenges in Sub-Saharan Africa and Opportunities for Digitalization. *Sustainability 2022, Vol. 14, Page 2399, 14*(4), 2399.
27. Lemstra, M. A. M. S., & de Mesquita, M. A. (2023). Industry 4.0: a tertiary literature review. *Technological Forecasting and Social Change, 186*, 122204.
28. Li, D., He, G., Jin, H., & Tsai, F. S. (2021). Sustainable Development of African Countries: Minding Public Life, Education, and Welfare. *Frontiers in Public Health, 9*, 1454.
29. Luo, H., Lin, L., Chen, K., Antwi-Afari, M. F., & Chen, L. (2022). Digital technology for quality management in construction: A review and future research directions. *Developments in the Built Environment, 12*, 100087.
30. Madeleine, P. (2022, March 11). *South Africa Opts for 3D printing to Combat Housing Shortage - 3Dnatives*. https://www.3dnatives.com/en/south-africa-opts-for-3d-printing-to-combat-housing-shortage-110320225/#!
31. Mckinsey. (2019, December 1). The impact and opportunities of automation in construction | McKinsey. *McKinsey*. https://www.mckinsey.com/capabilities/operations/our-insights/the-impact-and-opportunities-of-automation-in-construction

32. Mhlanga, J., Haupt, T. C., & Loggia, C. (2022). Shaping circular economy in the built environment in Africa. A bibliometric analysis. *Journal of Engineering, Design and Technology*.
33. Nchake, M. A., & Shuaibu, M. (2022). Investment in ICT infrastructure and inclusive growth in Africa. *Scientific African, 17*, e01293. https://doi.org/https://doi.org/10.1016/J.SCIAF.2022.E01293
34. NHF. (2023, February 12). *Federal Mortgage*. https://fmbn.gov.ng/National%20Housing%20Fund/nhf.html
35. Ni, Z., Yang, J., & Razzaq, A. (2022). How do natural resources, digitalization, and institutional governance contribute to ecological sustainability through load capacity factors in highly resource-consuming economies?, *Resources Policy, 79*, 103068.
36. North Africa Post. (2021, April 6). *Morocco promotes industry 4.0 for advanced manufacturing & competitiveness – The North Africa Post*. https://northafricapost.com/48813-morocco-promotes-industry-4-0-for-advanced-manufacturing-competitiveness.html
37. Obobisa, E. S., Chen, H., & Mensah, I. A. (2022). The impact of green technological innovation and institutional quality on CO2 emissions in African countries. *Technological Forecasting and Social Change, 180*, 121670.
38. Oguntona, O. A., Akinradewo, O. I., Ramorwalo, D. L., Aigbavboa, C. O., & Thwala, W. D. (2019). Benefits and Drivers of Implementing Green Building Projects in South Africa. *Journal of Physics: Conference Series, 1378*(3), 032038.
39. Ogwu, M. C. (2019). Towards Sustainable Development in Africa: The Challenge of Urbanization and Climate Change Adaptation. *The Geography of Climate Change Adaptation in Urban Africa*, 29–55.
40. Oubrahim, I., Sefiani, N., & Happonen, A. (2023). The Influence of Digital Transformation and Supply Chain Integration on Overall Sustainable Supply Chain Performance: An Empirical Analysis from Manufacturing Companies in Morocco. *Energies 2023, Vol. 16, Page 1004, 16*(2),
41. Pereira, T., Kennedy, J. V., & Potgieter, J. (2019). A comparison of traditional manufacturing vs additive manufacturing, the best method for the job. *Procedia Manufacturing, 30*, 11–18.
42. Perrot, A., Rangeard, D., & Courteille, E. (2018). 3D printing of earth-based materials: Processing aspects. *Construction and Building Materials, 172*, 670–676.
43. Philbin, S., Viswanathan, R., & Telukdarie, A. (2022). Understanding how digital transformation can enable SMEs to achieve sustainable development: A systematic literature review. *Small Business International Review, 6*(1), e473.
44. Rasiya, G., Shukla, A., & Saran, K. (2021). Additive Manufacturing-A Review. *Materials Today: Proceedings, 47*, 6896–6901.
45. Science Business. (2019, May 24). *DST: Harnessing science, technology and innovation for Africa's development*. Science Business. https://sciencebusiness.net/network-news/dst-harnessing-science-technology-and-innovation-africas-development
46. Silvi, J. (2015, January 2). *How can Kenya boost growth?*. World Economic Forum. https://www.weforum.org/agenda/2015/01/how-can-kenya-boost-growth/
47. Srinivasan, D., Meignanamoorthy, M., Ravichandran, M., Mohanavel, V., Alagarsamy, S. v., Chanakyan, C., Sakthivelu, S., Karthick, A., Prabhu, T. R., & Rajkumar, S. (2021). 3D Printing Manufacturing Techniques, Materials, and Applications: An Overview. *Advances in Materials Science and Engineering, 2021*.
48. The World Bank. (2022, December 14). *Africa Overview: Development news, research, data | World Bank*. https://www.worldbank.org/en/region/afr/overview
49. Theconversation. (2022, April 5). *3D printing offers African countries an advantage in manufacturing*. Theconversation. https://theconversation.com/3d-printing-offers-african-countries-an-advantage-in-manufacturing-179777
50. UNCTAD. (2022, October 3). *Africa's economic growth decelerates sharply*. United Natinos Conference on Trade and Development. https://unctad.org/news/africas-economic-growth-decelerates-sharply
51. Unesco. (2022, October 17). *Accelerating the digitalization of vocational training in Africa*. International Institute for Educational Planning. https://dakar.iiep.unesco.org/en/news/accelerating-digitalization-vocational-training-africa

52. UNIDO. (2022, October 17). *Are emerging economies already engaging with Industry 4.0 technologies?* https://www.unido.org/stories/are-emerging-economies-already-engaging-industry-40-technologies

The Potential of Earth as a Construction Material: Review and Perspectives

Lisiane Librelotto, Paulo Ferroli, Karim Fahfouhi, Flávio Craveiro, and Helena Bártolo

Abstract Over centuries, raw earth materials have been used, together with traditional vernacular techniques, to create the first shelters built by humankind. Real vernacular architecture, often partially buried in the ground, was the result of a trial-and-error basis across thousands of years, using surrounding local materials. This paper gives a brief overview of the use of raw earth as a construction material across cities in Europe (Portugal, France and Germany), Africa (Morocco) and America. The investigation was conducted through photographic records, documentary information and research literature. Opportunities for technological innovation are also highlighted, using earth as a base material for a more sustainable construction. The evolution of construction techniques over time allowed pointing out positive and negative aspects from a sustainability view. The main techniques identified were half-timber construction (mixed techniques), adobes and rammed earth, as well as earth coatings and screeds, usually combined wooden structures. Today, adopting sustainable solutions to mitigate climate change is critical, preserving earth resources and the environment by using low-carbon construction materials combined with sustainable construction technologies and practices. Using the earth as a construction material has great potential to address current architectural challenges, and 3D printing stands out as a promising technological innovation towards a more resource-efficient and sustainable built environment.

Keywords Earthen architecture · Raw earth materials · Sustainable construction

L. Librelotto (✉) · P. Ferroli
Federal University of Santa Catarina, Florianópolis, SC, Brazil
e-mail: lisiane.librelotto@ufsc.br

K. Fahfouhi · F. Craveiro · H. Bártolo
School of Technology and Management, Polytechnic of Leiria, Leiria, Portugal

F. Craveiro · H. Bártolo
CIAUD, Universidade de Lisboa, Lisboa, Portugal

J. R. da. C. S. Galvão et al. (eds.), *Proceedings of the 3rd International Conference on Water Energy Food and Sustainability (ICoWEFS 2023)*, Springer Proceedings in Earth and Environmental Sciences, https://doi.org/10.1007/978-3-031-48532-9_58

1 Introduction

Earth architecture represents an important part of the UNESCO World Heritage programme. Over centuries, raw earth materials, a widely available material, have been used to create sustainable housings worldwide, such as the Chinese wall, Portuguese and German earthen architecture, as well as to create sustainable shelters by African and Brazilian indigenous tribes.

A very versatile raw material, earth as a construction material can have several applications. However, its primary usage is mainly focused on adobe, rammed earth blocks, and mixed techniques, such as the half-timber (also called enxaimel in Brazilian Portuguese). Its use differs from country to country, and there is no uniformity of terms and procedures. According to Niroumand et al. (2013) [12], earth is one of the most used construction materials for different architectural and engineering applications, such as construction, moulded hills and buried housings, even if it is most commonly used for vertical wall elements, such as foundations, roofs and floors.

Earth construction can be used in simple or reinforced elements, in the form of adobes, compressed blocks, prefabricated or cast-in-place compacted monolithic panels (rammed earth and hand or hedge), cob elements, bagged earth, earth-straw systems or mixed systems, such as the examples using wood and earth, characteristic of the Half-timber construction, with many variations in composition. Lime, cement, bitumen or even organic elements, such as manure, can be used to enhance the stability of the aggregate, to provide yield stability and reliability. However, the results have not yet been scientifically studied much. Vegetable fibres can also be added to reduce shrinkage.

This research work main aim was to review literature, complementing it with data collection in the field, to examine the use of earth construction in European cities (Portugal, France and Germany), as well as American (Brazil) and African (Morocco) ones, identifying documented vernacular construction techniques and technological innovations for a more sustainable construction. To accomplish this objective, a brief literature review was conducted, together with a photographic record of some earth constructions, as well information on its history, construction techniques and future perspectives.

2 Building with Earth

2.1 *Historical*

Earth construction was a significant part of the first shelters built by humanity, from a mixture of raw materials, erected against earth slopes or partially buried in the ground. Palazzo (2021) [15] reports that the ancestor dwellings were caves, tents and huts, depending on the living conditions or people's mode of subsistence. Modern building techniques are the most widely used for earth construction. Used over centuries, they are emerging today as a more ecological and environmentally friendly alternative for buildings, in a circular economy context based on the use of local and natural materials.

The first archaeological shelter remains date from 12,000 to 10,000 BC, reconstructed by Haklay and Gopher (2015) [7]. These shelters used earth as excavated material to make a support, while the rest of the structure was made from branches and leaves.

Authors such as Minke (2006) [10], Palazzo (2021) [15], and Berge (2009) [2] reported the use of earth construction in 9000 BC at Turkmenistan, at El-Obeid in Mesopotamia (5000–4000 BC) and at Tigris River dated from 7500 BC.

Minke (2006) [10] investigated the writings of Pumpelly (1908) [17], accurately describing the archaeological findings obtained during an excursion in Central Asia, in the region of Turkmenistan. In 2008, Pumpelly (p. 299) details that almost all houses were made of clay, with 30–60 cm wide walls, and roofs made of natural materials, such as straw and wood, or even earth. Pumpelly (1908) [17] also points out that the durability of these constructions' roofs were 30 years.

In the Neolithic settlement of Çatal, in southern Anatolia, a video reconstruction made by ZDF in 2020 showed that dwellings, dating back from 7100 BC, were organized by cells built in adobe. It seems that the cells were built using a mixed load-bearing structure composed of wood and large adobe blocks. The occupation was seasonal, there were few internal courtyards, no streets, and life mostly took place on the dwellings' roofs. Other authors also point out earth constructions in Palestine, China, Peru, Mexico and several African countries, where cities dating back from 8000 BC were erected using locally available earth materials [4, 14].

From the Mesopotamian region, earth constructions would have spread to other regions, such as Africa. Numerous examples can be found across continents, in Africa, Asia and Europe, with some variation in the construction techniques.

In the Iberian Peninsula, earth construction was introduced by the Romans and enriched by the Arabs. In Pre-Columbian America, the techniques were different from the European ones, as the Inca and Aztec civilizations used these raw materials before the arrival of conquerors.

Over centuries traditional techniques and natural materials were used to build settlements, which explains its widespread dissemination throughout the globe. Earth constructions created a vast cultural and historical archive spread across the world. In 2012, UNESCO has made a World heritage inventory of earthen architecture.

However, despite current scientific knowledge in this area, the development of new construction techniques and the addition of innovative materials into earthen compositions, some of these constructions seem not to present the same performance obtained by past earthen constructions. On the other hand, the demand and requisites for the use of vernacular techniques in current construction have also changed in scale, as well as regarding the required performance. An apparent know-how deficit, together with an increase in requirements, is a great challenge for researchers in the field.

In Brazil, Santos, Librelotto & Jacintho (2014) [20] conducted a study and collected data on some of the most common earthen construction techniques used in the Federal District and the State of Santa Catarina (superadobe, hyperadobe), namely bagged earth, rammed earth, cob, stucco, and double stucco filled with plastic bottles, as well through interviews with builders and house dwellers.

Marsh and Kulshreshtha (2021) [9] published a recent study on contemporary construction. They conducted a survey to update information on the amount of existing earthen dwellings in the world. The data was collected and analyzed for 26 countries, representing more than 75% of the world population Information was requested on the material and typology of dwellings through demographic censuses. This study found out that earthen dwellings were predominant in developing countries. The survey showed a decline in the use of earth compared to other materials, mainly due to the negative perceptions of earth materials. The proportion of houses built with earth around the world is estimated to be between 8 and 10%, being the average from 20 and 25%, considering only developing countries. In Brazil, there is less than 5% of dwellings made of earth, including rural and urban areas, which contrasts with other countries, such as Egypt, Bangladesh, Pakistan and India, where more than 60% of dwellings are made with earth in predominantly urban areas [9].

The most common construction techniques are rammed earth, handmade or hedge (mixed technique), cob, adobe and mixed techniques, such as half-timbered ("*enxaimel*"). Rammed earth uses formwork (boxes, shuttering) and the soil is compacted manually using punches or pestles with electric or pneumatic equipment. Usually made in layers with different soil colours that gives it a unique appearance [24].

Cob is a mixture of soil in a plastic state containing clay, water, sand and gravel. In Brazil, it is also common to use "grude" (flour-based glue) in the mixture. The fresh soil in the form of balls is superimposed to make the volumes.

According to Vyncke, Kupers & Denies (2018) [24], earth blocks can be of three types: sun-dried blocks, cut blocks and compressed earth blocks (CEBs or BTCs). Figure 1a shows the blocks cut into the ground.

(a) (b)

Fig. 1 **a** blocks cut out of the ground [24], **b** adobes moulded with shapes in Morocco (Lisiane Librelotto's personal collection)

The half-timbered construction, considered a mixed technique, uses a wooden structure filled with elements, usually of earth in the form of blocks, mixtures with earth and fibres, or even stones and earth mortar as connecting elements and final finishing. The technique is used in several European countries and has its roots in the Alpenina region and in Romanesque construction, in the style that was conventionally called *oppus craticium.* Its use has spread to the Americas since European colonization.

2.2 Earthen Construction in Portugal

The most traditional Portuguese building techniques using earth for construction are rammed earth ("taipal"), adobe and "tabique". The tabique (similar to the Brazilian "taipa de mão", also called "pau a pique" or "taipa de sebe" (wattle and daub), is framed as a mixed construction technique. It consists of a wooden structure covered by a mixture of earth and fibres, also with an earthen covering.

The "taipa" or "taipal" is comparable to the Brazilian "taipa de pilão" and follows similar construction procedures.

In Portugal, it is also noteworthy that the Pombaline cage system is found in historic buildings at Lisbon's Baixa near the river Tagus. The cage consists of an interlocking wooden structure, similar to the balloon frame system, filled with stone masonry or bricks, with lime or earth/"taipa", similar to the half-timbered, whose internal divisions could be made up of partitions. This system was developed after the earthquake in 1775 by Marquês de Pombal's engineers as a more earthquake-resistant solution. Its constituent elements are the props (vertical, 15 × 30 cm in oak or holm oak), the sleepers (horizontal, 10 × 13 cm) and the diagonals, and the wall thickness varies between 15 and 20 cm for the upper floors. Generally, the first floor (ground floor) is made of stone with vaulted ceilings.

From these systems, today, there are hybrid systems, as illustrated in Fig. 2a, with the wooden structure and earth filling, characteristic of the Pombaline cage, and the horizontal slating used in the partitions [23].

Figure 2b shows the chalet of the Countess of Edla in Sintra, whose restoration, after a fire, recreated in detail the decorations of the internal coverings (using cork) and façades, as well the partition elements.

In Brazil, in Maranhão state, its capital called São Luiz, there are several examples influenced by the Portuguese Pombaline construction [21].

2.3 German and French Earth Construction

Germany has a long tradition of earth construction. The most representative techniques are adobe, rammed earth and a mixed technique (half-timbering, also called "fachwerk" or "fachwerkbau"). The half-timbering, resulting from wooden stilts, is

(a) (b) (c) (d)

Fig. 2 **a** Hybrid system [23], **b** Chalet of the Countess D'Edla, in Sintra (Lisiane Librelotto's personal collection), **c** scheme of the structure of the Pombaline Cage [18], and **d** Partition wall [21]

built from wood trusses, fitted into each other (without nails) with structural function; the subsequent openings of the hardwood structure (of high density to provide good fits and rigidity) are filled with adobe, rammed earth, exposed bricks, or covered with a finish, generally earth-based. Figure 3a shows examples of these buildings in the historic centre of Frankfurt, Fig. 3b its origins, and Fig. 3c, d its construction details.

In Europe, half-timbering can be found in several countries, such as Germany, France, the Netherlands, Switzerland, Austria, Italy and England, mainly in places where earthquakes are recorded, as a quite resistant construction to these events [3].

In Brazil, there are many examples of this type of construction, inspired by German and Italian immigrants (from 1815 to 1960), in Rio Grande do Sul, Santa Catarina and Paraná, and a few buildings in São Paulo and Espírito Santo too.

The interfaces between elements were filled with earth and fibres, usually painted with white lime, to visually distinguish them from the wooden elements, which gives them a peculiar architectural appearance, although the whole filling (non-structural), together with the lattice, constituting the structure, forming a single element.

Regarding the constitution of the infill (Fig. 3c), in particular, earth mixed with fibres is deposited on a willow wood batten. One of the first known types of infill, called *opus craticum* (or *craticium*), built by Romans in Herculaneum, Italy, dates

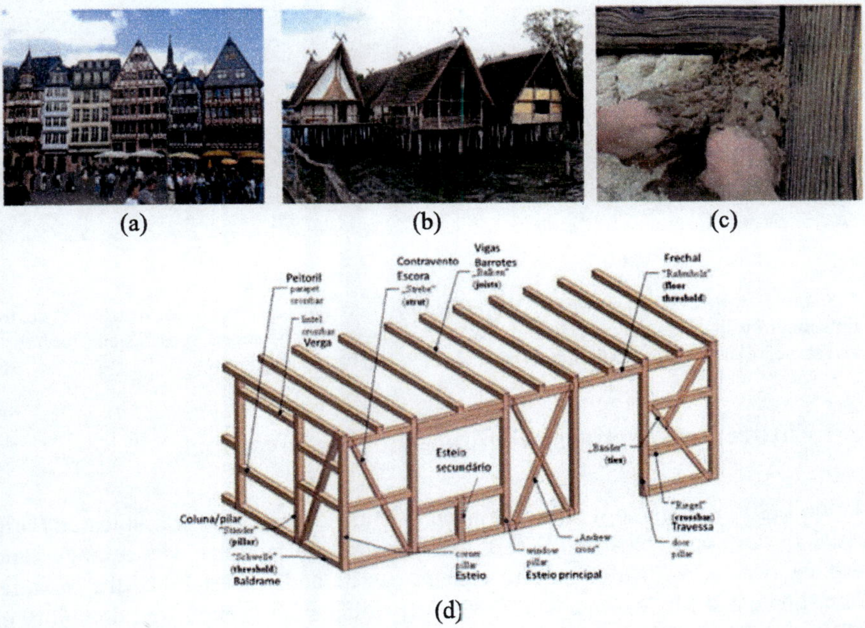

Fig. 3 Half-timbered buildings in Germany: **a** Buildings in Frankfurt (personal collection Lisiane Librelotto), **b** stilts found at 111 archaeological sites from the period 4300 BC to 800 BC in the Alpine region [25], **c** Detail of the ground fill [6], and **d** nomenclature of the components in German, English and Portuguese languages (adapted from Bostenaru and Sassu [3])

back from 79 AD, was a taipa-type construction. Vitruvius mentioned this technique as a wooden frame with a clay filling, but the same description was also used to describe wooden frames with a stone filling laid in mortar, which Romans called *Opus incertum.* [22].

2.4 *Earthen Construction in Africa*

Rodrigues (2021) [19] points out some examples of African earth constructions in different regions: in the North (Libya and Morocco), in the West (Cape Verde archipelago, among others) and the South (Mozambique, Angola). In Mozambique, the most common materials are adobe, mud, and stone, using stalks (plants derived from grasses, such as bamboo). In Morocco, using rammed earth and adobes is one of the main construction techniques. Figure 4 shows some earthen buildings in Morocco.

The rammed earth walls have rustic forms, and the composition of the apparent soil has a high proportion of stones.

(a) (b)

Fig. 4 **a** Hait Ben Haddou—Unesco World Heritage Site, Morocco and **b** Construction style, top finishing of walls and walls in platband, made with transversal and superimposed stalks with earth and stone mixture, Hait Ben Haddou (personal collection Lisiane Librelotto)

3 Future Perspectives: Technological Innovations

In the 1980s, a technology was developed in Brazil called CEBs (Compressed Earth Bricks), developed by COPPE Rio. These bricks are modular with a tongue and groove system, 5–10% of lime or cement can be added to stabilize the mixture. Jayasinghe and Mallawaarachchi (2009) [8] validate the strength and durability of CEBs, while other research group shows the feasibility of building in an industrial scale [13].

Prefabricated and pigmented rammed earth blocks also seem a promising composite building methods for buildings. In Brazil, there are prefabricated mud walls since the 1960s, as reported by Negreiro et al. (2020) [11], as well as wattle and daub, a similar building process.

The compacted and/or pigmented mud, developed from rationalized form systems, also allows industrial production or pre-moulding on site. As a monolithic and self-supporting construction technique, rammed earth shuttering can rely on the resources of rationalized shuttering systems, which are produced with quality and can be automated using advanced manufacturing methods. The use of mechanical compaction systems and the projection of earth at high pressure into the moulds rationalize and speed up the execution of a construction that is more consistent, so with both a better mechanical strength and more resistance to water erosion, as highlighted by Benis, Braga & Salgueiro (2014) [1].

On the other hand, the contemporary "fachwerk" has replaced the traditional timber structures (beams, bars and supports) with glued laminated timber. The cells of the supporting structure are filled with other materials. It was used in medieval Europe in a mixture of straw and limestone or clay and bricks, where the plastered and painted facades were replaced in many cases by wooden and glazed fences, closely resembling the Light Wood Frame constructions.

Finally, one of the most promising technologies to use with earth material compositions is 3D printing (also called Additive Manufacturing). A few groups, mainly from Italy and France, are developing new 3D earth printing formulations. One of the leading groups is the Italian company called WASP. Perrot, Rangeard & Courteille

(2018) [16], a pioneer group from France, point out the importance of finding the right mix of earthen compositions to meet the requisites of the fresh state decisive to allow printability, to obtain the necessary strength for the hardened state. Earth 3D printing can involve different processes, such as extrusion, self-compacting compositions and hyper-compaction. 3D printing presents several advantages over traditional manufacturing processes, it is faster, reduces cost and waste, allows freedom of design and complexity and customization, and improves work safety [5, 16]

4 Final Considerations

This work briefly overviewed the most common earth construction techniques in Europe, Africa and America, identifying some predominant techniques, such as hand and pestle mortar, adobes, compressed earth blocks and half-timbering.

The durability of earth constructions is undeniable, considering the many remaining old examples still stand. However, the conditions of use, the scale of need, the transporting conditions, the availability of materials, energy and labour, or even the mastery of ancient traditional techniques are today completely different from the ancient constructions ones.

There are good examples of historic earth constructions in all surveyed countries. In Portugal, mud-brick, adobes, partition walls and the pigeon cage are techniques that stand out, though there are no specific standards for earth construction. In Germany and France, countries with a long tradition in earth construction, as well in Africa, adobe, rammed earth, and half-timber stand out. In Africa, adobe and rammed earth are the most frequently identified techniques in this survey. Both Germany and France, as well as African countries, have specifications for earth construction.

Many of these techniques are today object of increasing interest, being rescued for their potential to be used in more sustainable constructions, as they have a high thermal inertia, contributing to environmental comfort and are energy efficient, reducing environmental impacts. They are also biologically friendly, can be produced locally due to their wide material availability, and can support self-construction.

However, using earth as a building material has brought associated challenges, such as its durability under adverse climate conditions, requiring performance standards. The large volumes of materials needed for its production is also a limitation, as well as the response to urgent housing demands. It is also difficult to characterize material formulations or determine the resistance of the components produced by the many existing construction techniques.

Different research groups are engaged in the search for sustainable alternatives, which can combine historical knowledge with contemporary needs and the pressure for a more sustainable construction. In this context, 3D printing stands out as a promising technological innovation.

References

1. Benis, K., Braga, J., & Salgueiro, J.: Taipa Contemporânea. Património em Devir. Anuário do Património. Canto Redondo, Lisboa (2014).
2. Berge, B.: The ecology of building materials. 2nd ed., Architectural Press, Elsevier Science (2009).
3. Bostenaru M., Sassu, M.: Half-timbered house in the 'border triangle", Report Number 108, World Housing Encyclopedia (2005).
4. Carvalho, T., Lopes, W.: A Arquitetura de Terra e o Desenvolvimento Sustentável na Construção Civil. In: Congresso Norte Nordeste de Pesquisa e Inovação. IFTO, Palmas (2012).
5. Craveiro, F., Duarte, J. P., Bártolo, H. M., Bártolo, J. P.: Additive manufacturing as an enabling technology for digital construction: A perspective on Construction 4.0. Additive Manufacturing, 103, 251–267 (2019).
6. Fachwerk, https://www.fachwerk.de/fachwerkhaus/sitemap/Lehmbau.html, last accessed 2023/02/01.
7. Haklay, G., Gopher, A.: A New Look at Shelter 131/51 in the Natufian Site of Eynan (Ain-Mallaha), Israel. PLOS ONE 10 (7), e0130121 (2015).
8. Jayasinghe, C., & Mallawaarachchi, R.: Flexural strength of compressed stabilized earth masonry materials. Materials & Design, 30(9), 3859-3868 (2009).
9. Marsh, A., Kulshreshtha, Y.: The state of earthen housing worldwide: how development affects attitudes and adoption. Building Research & Information, 50(5), 485-501 (2021).
10. Minke, G.: Building with earth, design and technology of a sustainable architecture. Birkhäuser Publishers for Architecture (2006).
11. Negreiros, B., Pinto, E., & Pedrini, A.: Experiências Habitacionais com Sistema de Préfabricação em Taipa: Os Casos De Cajueiro Seco-PE e João Câmara-RN. Encontro Nacional de Tecnologia do Ambiente Construído, 18(1), 1–8 (2020).
12. Niroumand, H., Zain, M., Jamil, M.: A guideline for assessing of critical parameters on Earth architecture and Earth buildings as a sustainable architecture in various countries. Renewable and sustainable energy reviews, 28, p. 130-165 (2013).
13. Obonyo, E.: Optimizing the physical, mechanical and hygrothermal performance of compressed earth bricks. Sustainability, 3(4), 596-604 (2011).
14. Olender, M.: A técnica do Pau-a-pique: subsídios para a sua preservação (Master Thesis). Universidade Federal da Bahia, Salvador (2006).
15. Palazzo: Taipas e fundações - TAU 0005 História da Arquitetura e da Arte I (palazzo.arq.br). Material didático, https:// palazzo.arq.br/tau/aula/taipa, last accessed 2021/09/01.
16. Perrot, A., Rangeard, D., Courteille, E.: 3D printing of earth-based materials: Processing aspects. Construction and Building Materials, 172, 670-676 (2018).
17. Pumpelly, R. (Ed.).: Explorations in Turkestan, Expedition of 1904: Ancient Anau and the oasis-world. Carnegie institution of Washington (1908).
18. Ramos, L., Lourenço, P.: Análise das técnicas de construção pombalina e apreciação do estado de conservação estrutural do quarteirão do Martinho da Arcada. Congresso Construção. Universidade do Minho, Guimarães (2000).
19. Rodrigues, J.: Soluções construtivas nas regiões tropicais: caso de Moçambique (Doctoral Dissertation). Universidade Nova de Lisboa, Lisboa (2014).
20. Santos, C., Librelotto, L., Jacintho, C.: Building with Earth - Brazil's Most Popular Raw Earth Building Techniques and the Opinion of Experienced Builders. Key Engineering Materials (Online) v. 600, p. 123-131 (2014).
21. Silva, L.: Arquitetura civil portuguesa: análise das características do sistema construtivo pombalino no centro histórico de São Luís-MA (Master Thesis). Universidade de Brasília, Brasília (2013).
22. Stellacci, S.; Rato, V.: Timber-Framing Construction in Herculaneum Archaeological Site: Characterisation and Main Reasons for its Diffusion. International Journal of Architectural Heritage, p. 1–19 (2019).

23. Torgal, F., Jalali, S.: Earth construction: Lessons from the past for future eco-efficient construction. Construction and building materials, 29, p. 512-519 (2012).
24. Vyncke, J., Kupers, L., Denies, N.: Earth as Building Material – an overview of RILEM activities and recent Innovations in Geotechnics. MATEC Web Conf., 149, 02001 (2018).
25. Wittmann, A.: Fachwerk, a técnica construtiva enxaimel. Vitruvius: projetos (2016).
26. ZDF: erste Großsiedlung der Menschheit (CC BY 4.0), https://www.zdf.de/dokumentation/terra-x/catalhoeyuek-erste-grosssiedlung-der-menschheit-creative-commons-100.html, last accessed 2022/02/01.

Water

The following part contains papers related to "Water".

The editors had assistance from the following conference organizing committee member in the editing of this part:

- Ricardo de Jesus Gomes

The Use of Pervious Pavements to Save Potable Water in Buildings: A Case Study in Universities in Brazil

Igor Catão Martins Vaz, Enedir Ghisi, and Liseane Padilha Thives

Abstract The main objective of this study was to evaluate the use of pervious pavements as an alternative to save potable water in university buildings. A case study was conducted in six university buildings of the Federal University of Santa Catarina (UFSC) in Florianópolis, Brazil. The water consumption of each building was obtained through analysis of water bills and downscaled to daily consumption based on average estimates of the weekly and weekend consumption. Then, based on the water end-use profiles and the rainfall data of the city, a stormwater harvesting system was modelled. The system consists of a pervious pavement parking lot connected to an underground stormwater tank and an upper tank. The stormwater is then provided by gravity towards end-uses such as toilets, cleaning, and gardens. As a main result, potable water savings potentials of 31.39 to 75.14% were obtained, depending on the building and scenario evaluated. In conclusion, pervious pavement is a good alternative for a continuously growing impervious urban context, aligning a broad water dissipation approach and enabling the use of stormwater in buildings.

Keywords Stormwater harvesting · Pervious pavements · Case study · Universities · Sustainability

1 Introduction

Pervious pavements are one of the techniques presented as Sustainable Urban Drainage Systems (SUDS). According to Woods-Ballard et al. [1], SUDS represent an urban water management approach that relies on a more diffuse type of stormwater discharge, copying the natural process of untouched lands. Also, some other features must be considered, resulting in a four-topic design optimisation regarding the water quality, quantity, amenity, and biodiversity as design goals. In an ongoing climate-changing world, such topics become necessary, and SUDS become a tool

I. C. M. Vaz (✉) · E. Ghisi · L. P. Thives
Universidade Federal de Santa Catarina, Florianópolis, SC 88040-900, Brazil
e-mail: igor.catao@posgrad.ufsc.br

J. R. da. C. S. Galvão et al. (eds.), *Proceedings of the 3rd International Conference on Water Energy Food and Sustainability (ICoWEFS 2023)*, Springer Proceedings in Earth and Environmental Sciences, https://doi.org/10.1007/978-3-031-48532-9_59

for better urban centres. Such tools can help create resilience towards more frequent phenomena such as draughts and heavy rainfalls and improve overall sustainability. Overall, SUDS have great potential for sustainable urban centre designs [2, 3].

Regarding pervious pavements, recent studies have shown the potential for many benefits, including the ability to improve water quality [4–6], decrease peak discharge [7], provide better heat island management in urban centres [8], provide traffic noise reduction [9], improve general Life Cycle Assessment (LCA) score [10] and even clean the air [11]. Such benefits can be of great interest to society's public and private sectors, providing answers to many of the problems in the urbanisation and densification scenarios the world has been experiencing.

Another possibility is the optimisation of water management and supply, especially with an ongoing, unpredictable climate and the necessity to improve reliability. With this in mind, many scenarios of reuse and optimisation arise as trends in the design of hydraulic systems. Products such as rainwater or stormwater harvesting schemes, greywater reuse and efficient water fixtures become another great tool for urban centres. These tools have been widely proposed as great environmental [12], financial [13] and technical [14] solutions. The potential of synergy with both types of approaches (SUDS and water optimisation) can also further improve reliability and sustainability in future cities [15].

This paper addresses sustainability within the urban centre by considering the synergy of a stormwater harvesting scheme and the use of pervious pavements. More specifically, we simulated a system within the scope of university buildings in southern Brazil. Similar to other publications on the same topic [16, 17], one can see that the benefits of both systems can be combined, improving general sustainability, and providing non-potable water for use in buildings.

2 Method

This work was performed to provide a general knowledge of the potential for using pervious pavements and stormwater harvesting combined. First, one assessed the systems' design by analysing the standard components and inventory used in the literature. The goal was to understand how both systems could connect to work together. The results of the first part were a complete scheme of the pervious pavement cross-section and the whole system conceptualisation. Figure 1 shows the general concept of the system obtained.

Stormwater that falls onto the pavement is pumped to a stormwater tank and conducted by gravity to non-potable uses. The potable water tank only feeds potable uses and fills the stormwater tank in case of lack of stormwater. In this case, a solenoid valve with a retention function ensures that the water flow only occurs from potable to non-potable tanks, not contaminating the potable source. Also, no water from the roof is harvested, as it was modelled to be drained. The stormwater considered for consumption is only obtained via the pervious pavement, as shown in Fig. 1.

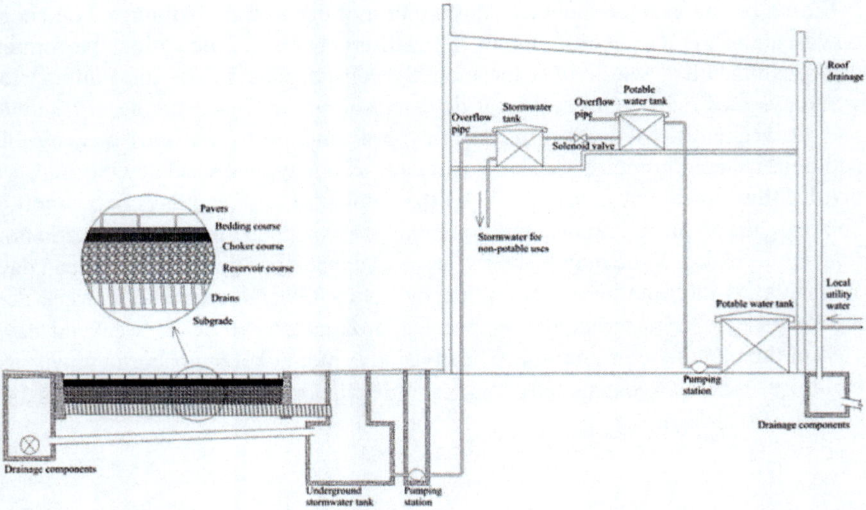

Fig. 1 Stormwater harvesting system scheme using pervious pavements

Secondly, we assessed the buildings' total water consumption. Six buildings in Florianópolis, Brazil, as part of the Federal University of Santa Catarina (UFSC), were considered. They range from one floor to four floors and have different purposes, including classrooms, administration offices, laboratories, and libraries. Data from 2013 to 2018 were obtained, and a profile of daily water consumption was downscaled.

The third part was the assessment of the potable water savings potential of the system. In order to do so, a water balance deterministic approach using the computer programme *Netuno*, version 4, was performed. The main goal was to understand how different the systems would perform under different scenarios. At last, we designed the proposed system via the computer programme Permeable Design Pro (PDP) [18] to obtain the pavements cross-section. Details of the pervious pavement design and local characteristics were also explored to better present the systems proposed and the design process.

2.1 Water Consumption

The water consumption was obtained via data from UFSC [19], measured monthly. It was expected to have differences among the six buildings, and, using such inputs, one could understand how the changes in building realities would influence the outcome. Building consumptions ranged from 300 to 5000 m^3 per year, depending on the building and year of analysis. On average, the six buildings consumed approximately 20,000 m^3 of water annually.

Based on the average monthly consumption of each of the buildings, one could downscale to the daily consumption. Two different sets of calculations performed downscaling. First, we divided the monthly consumption by the total number of weekdays plus one, giving a typical day's consumption. For example, if a month had 22 weekdays, we divided the monthly consumption by 23. This way, by this addition, we could represent all weekend days as one typical weekday. Second, we divided this typical day consumption by the number of weekend days in the month, obtaining the atypical day consumption. Finally, we mapped the typical consumption for each weekday of the month and the atypical consumption for each weekend day. We averaged the monthly consumption by giving a weighted value to week and weekend days, as to weekdays represent approximately 90% and weekend days 10% of the monthly consumption. The division is more likely to occur in university buildings, as at the weekend, only some facilities (laboratories, mostly) are open, but there are no classes.

2.2 *Potable Water Savings Potential*

Potable water savings potential was assessed using the computer programme *Netuno*. *Netuno* is a software developed by Cordova and Ghisi [20] that calculates the daily water balance of the stormwater harvesting systems and returns the overall average values during the time horizon of assessment. This calculation is performed under a deterministic approach based on pre-defined pluviometry data and consumption patterns. Upper and underground stormwater tanks may also be varied during design in an attempt to optimise the system technically.

In this work, we performed eighteen scenarios within *Netuno* regarding three different stormwater demands (as buildings did not have their end-uses defined) and six buildings. In each scenario, simulations were performed by ranging the underground tank size from 1000 to 50,000 L with a step of 1000 L. The optimal tank size was then defined by the threshold indicated within *Netuno*. When the aggregation of 1 m^3 to the underground tank did not provide over 1% of potable water savings, the value of the optimised tank size was reached. Table 1 shows the parameters used.

In order to best address the results, the average 77% of non-potable end-uses was chosen for results comparison. The minimum and maximum non-potable end-uses of 69 and 85% were chosen as the sensitivity range. This way, one can better visualise the influence of the end-uses on the potable water savings potential.

As for the quality of the non-potable water harvested, we considered that the infiltration through the pavement would be enough to retain the significant contaminants, such as the oil from vehicles. Chlorination is also mandatory in Brazilian standards for non-potable water harvesting. Nevertheless, the first flush disposal was considered 0 mm, as no studies were identified for pervious pavement stormwater harvesting systems. In future studies, a specific structure for water quality assessment may be provided, diverting the first flush and assuring the necessary inspections.

Table 1 Stormwater harvesting parameters considered in the simulations using *Netuno*

Parameters	Value
Daily rainfall data	Florianópolis daily data [21]
First flush disposal	0 mm
Threshold of potable water savings per tank size increase for selecting the optimal tank size	1%/m^3
Harvesting area	Six values depending on the building
Water demand	
Stormwater demand (% of water demand)	69, 77 and 85%
Pervious pavement available water coefficient	88.1%
Upper stormwater tank size	Daily average stormwater demand
Underground stormwater tank size	Range between 1000 and 50,000 L (step of 1000 L)

2.3 *Pervious Pavement and Stormwater Harvesting Design*

According to the literature on pervious pavements [1], designers must pay attention to two performance characteristics: the hydraulic and structural features. The first must be addressed to prevent the pavement from presenting runoff, thus creating ponding. The hydraulic feature may be addressed by general expressions of the water input and output or be further simulated by a water balance daily or hourly programme.

The structural design must be performed for the pavement to exhibit minimal deflections and provide safety and drivability throughout its life span. In order to do so, one must address the number of vehicles and their weight, together with variables regarding design choices for the road, soil bed under the pavement and water infiltration characteristics.

In this study, we used Permeable Design Pro (PDP) [18] software to design the pavement's hydraulic and structural features. The hydraulic and structural parameters used were based on the literature. Table 2 shows the parameters used in Permeable Design Pro. Specific literature in the AASHTOs method was used for thicknesses and structural and hydraulic coefficients. For the remaining data, literature and the PDP manual were considered. The California Bearing Ratio (CBR) of the subgrade was considered to be 5%, regarding the soil at UFSC. For other sites assessed, CBR must be checked for local conditions. Finally, we also presented a simplified materials inventory according to stormwater harvesting and pervious pavement designs.

Table 2 Stormwater harvesting parameters considered in the simulations using Permeable Design Pro

Pervious pavement simulation parameters		Value
Structural parameters		
Surface layer	Thickness (mm)	90
	Structural coefficient (dimensionless)	0.3
Choker course	Thickness (mm)	100
	Structural coefficient (dimensionless)	0.11
Reservoir course	Thickness (mm)	Result[a]
	Structural coefficient (dimensionless)	0.08
	Drainage coefficient (dimensionless)	1.0
Traffic	Equivalent Single Axle Load (ESALs)	100,000
Subgrade	California Bearing Ratio (%)	5.0
Hydraulic parameters		
Permeability (mm/s)	Pavers	1.972
	Choker course	0.011
	Reservoir course	1.014
	Subgrade	Impervious
Parking lot premises	Daily rainfall distribution	Type III
	Drainage area	Depends on building
Drain conditions	Distance between drains (m)	6
	Drain diameter (cm)	≤ 20
	Drain slope (%)	≤ 1
	Drain roughness (dimensionless)	0.012
PDP definitions	System's initial water volume (m^3)	0
	Maximum water height in the reservoir (%)	100
	Time step for simulation (hour)	0.1

[a] The main result of the simulation is the thickness of the reservoir course

3 Results and Discussion

First, the water consumption downscaling is presented alongside with characteristics of the object of study. Some of the features of each building are also presented. Second, the potable water savings simulation is presented through the optimisation results. Volumes of potable water savings are also presented to compare the systems. Third, the pervious pavement design and an inventory of materials are provided. Discussions on the topic of design and pavement considerations are also presented.

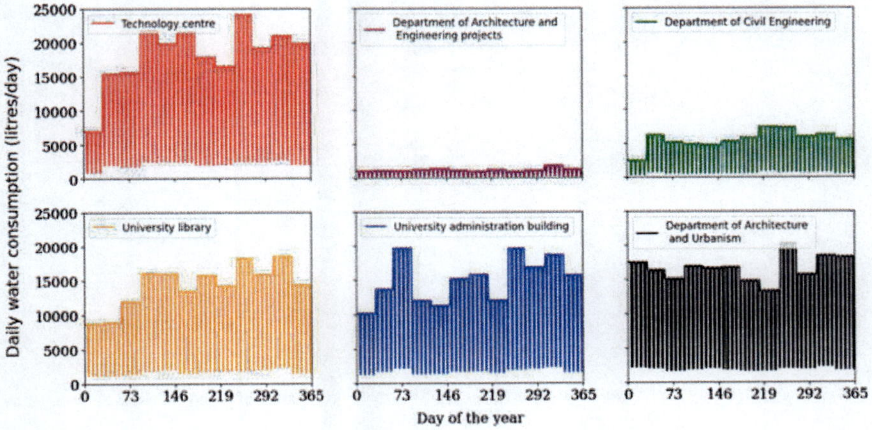

Fig. 2 Water consumption in the six university buildings

3.1 Water Consumption Patterns

The results for each of the six buildings are presented in Fig. 2. One can notice that many different water patterns were involved in the selection, ranging from a simple office building with less than twenty workers to a classroom building involving more than a thousand students. The water consumption over weekdays is represented by the upper parts of each graphic, while consumption over weekends is represented by the lower spikes closer to zero. According to the downscaling method proposed, this trend was expected, where weekdays have a much higher consumption pattern than weekends.

Results affect the potable water savings simulation by the differences in total demanded water and distribution patterns within the year. For example, one can see that the department of Architecture and Urbanism presented a much more homogeneous consumption during the year, while the Technology centre presented a much more seasonal demand. Most of the fluctuations can be explained by the school period, which extends in Brazil from March to December, with a minor winter break in July. The changes in weekday to the weekend are also responsible for changes in the final results of potable water savings potential. Overall, the results show variability within the selected buildings, which is better addressed in the following sections.

3.2 Potable Water Savings Potential

Potable water savings potential was obtained via *Netuno*. The main results obtained are the optimal stormwater tanks and the respective potable water savings. Through the six simulated buildings, one obtained six optional sizes and savings. In order to

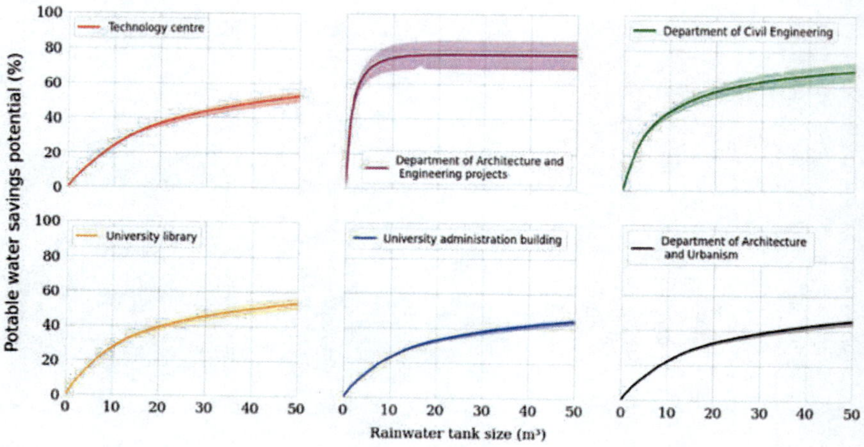

Fig. 3 Potential for potable water savings in the six buildings

best present the results, we show in Fig. 3 the different curves of optimisation of the six buildings, and Table 3 shows the optimal tank size of each of the buildings.

Without the system, the university consumed, on average, a total of 19,344.20 m^3 of water during one year from the local utility. With the proposed system, the university would consume 12,017.31 m^3 of water, i.e. a total reduction of 7326.89 m^3 (37.88%). With an average fee of R$4.50 for the first ten m^3 and R$12.50 for the remaining monthly consumption, we assessed the total return of the system. It was estimated that a total of R$236,042.50 would be spent on potable water with the conventional supply system. By implementing stormwater harvesting, the university would have spent R$144,456.38, granting a reduction of R$91,586.12 in the water bills. The money could be further applied to the university for scholarships, maintenance, and other expenses.

Table 3 Optimal tanks and respective potential for potable water savings

Building	Optimal underground stormwater tank with 77% of non-potable end-uses (m^3)	Potable water savings potential (%)
Technology Centre	21	37.82
Department of Architecture and Engineering Projects	10	75.14
Department of Civil Engineering	21	57.57
University Library	22	40.70
University Administration Building	18	31.39
Department of Architecture and Urbanism	19	32.03

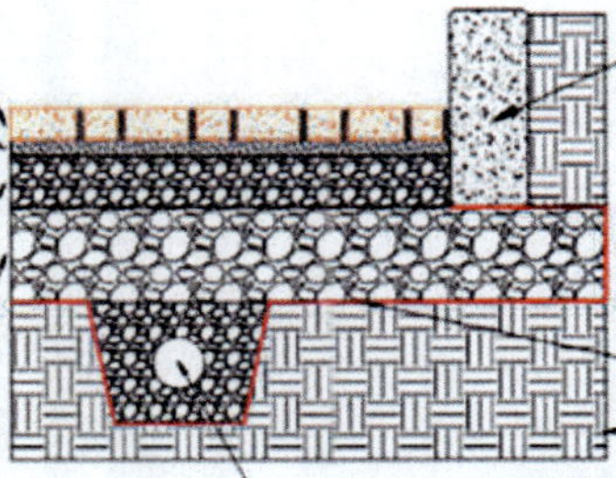

Fig. 4 Cross section of the designed pervious pavement

3.3 *Pervious Pavement and Stormwater Harvesting Design*

The final part of the study was the pervious pavement design. The main goal was to assess how much material would be used for the pavement, which can be used later to assess the economic (Life Cycle Cost Assessment) or environmental (Life Cycle Assessment) aspects. Figure 4 shows the final cross-section obtained for the pavement according to the design parameters presented in Table 2.

The main result was that, for a soil CBR of 5%, the driving aspect of the design is the structural component. The hydraulic design depends on many aspects, such as drains' location, diameter and spacing, slope, and rugosity. Although, the hydraulic aspect was easily fulfilled with the structural design thickness. Furthermore, the selected pervious pavement surface used pavers of Portland cement, and the design would change if an asphaltic coating were used.

Compared to an impervious pavement design, one can notice a much thicker necessity of the reservoir course material, to which the current pavement could be compared. In an impervious pavement, the subbase material (the equivalent of the reservoir course) for this traffic design is usually the minimal value, closer to 100 mm, or even disregarded. With this design in mind, and for constructing the six building systems, the university would need all materials included in Table 4.

Table 4 Inventory of materials for systems' construction

Material	Quantity	Unit
Pervious pavement design		
Pervious pavers (Portland cement)	Units	785,000
Bedding course (Gravel)	m^3	471
Choker course (Gravel)	m^3	1570
Reservoir course (Gravel)	m^3	2826
Impervious underground membrane (HDPE[b])	m^2	16,013
[a] *Other elements such as drains, curbs, culverts and stormwater pipes must also be assessed in detail*		
Stormwater harvesting design		
111 m^3 of an underground tank (HDPE[b])	Kg	2110
41 m^3 of an upper tank (HDPE[b])	Kg	697
Pumping stations (Cast iron)	Units	12

[a] Other elements such as filters, first flush, chlorinator and piping must also be assessed in detail
[b] HDPE stands for high-density polyethylene

4 Conclusion

The most crucial conclusion obtained from the results of this study is that pervious pavements are a possible synergy connection for sustainable buildings. The results show that the university system would demand additional materials for drainage consideration and stormwater harvesting. However, these additional elements would provide potable water savings of almost 40% and annual savings of almost R$100,000. This amount could be redirected towards scientific investments, such as scholarships, laboratory materials and campus maintenance.

Further studies can be initiated regarding the materials inventory by studying the systems proposed herein. The possibility of analysing city-wide resilience and optimisation in a climate-changing world is one of the goals. Other studies, such as LCCA and LCA, derive as possible drivers for decision-making in a much-needed science-based future. Policies and decisions for urbanisation and future cities will decide how we will be able to cope with the future climate. With these topics in mind, technologies such as pervious pavements, SUDS and stormwater harvesting become interesting. In this paper, we provide some data about this exciting cooperation.

References

1. Woods-Ballard, B., Kellagher, R., Martin, P., Jefferies, C., Bray, R., & Shaffer, P. (2011). The SuDS Manual. In Ciria C697.

2. Joshi, P., Leitão, J. P., Maurer, M., & Bach, P. M. (2021). Not all SuDS are created equal: Impact of different approaches on combined sewer overflows. Water Research, 191, 116780. https://doi.org/10.1016/j.watres.2020.116780
3. Cotterill, S., & Bracken, L. J. (2020). Assessing the effectiveness of sustainable drainage systems (SuDS): Interventions, impacts and challenges. Water, 12, 11, 3160. https://doi.org/10.3390/w12113160
4. Martins Vaz, I. C., Antunes, L. N., Ghisi, E., & Thives, L. P. (2021). Permeable Pavements as a Means to Save Water in Buildings: State of the Art in Brazil. Sci, 3(4), 36. https://doi.org/10.3390/sci3040036
5. Winston, R. J., Arend, K., Dorsey, J. D., & Hunt, W. F. (2020). Water quality performance of a permeable pavement and stormwater harvesting treatment train stormwater control measure. Blue-Green Systems, 2(1), 91–111. https://doi.org/10.2166/bgs.2020.914
6. Selbig, W. R., Buer, N., & Danz, M. E. (2019). Stormwater-quality performance of lined permeable pavement systems. Journal of Environmental Management, 251, 109510. https://doi.org/10.1016/j.jenvman.2019.109510
7. Mahmoud, A., Alam, T., Sanchez, A., Guerrero, J., Oraby, T., Ibrahim, E., & Jones, K. D. (2020). Stormwater Runoff Quality and Quantity from Permeable and Traditional Pavements in Semiarid South Texas. Journal of Environmental Engineering, 146(6), 05020001. https://doi.org/10.1061/(asce)ee.1943-7870.0001685
8. Ferrari, A., Kubilay, A., Derome, D., & Carmeliet, J. (2020). The use of permeable and reflective pavements as a potential strategy for urban heat island mitigation. Urban Climate, 31, 100534. https://doi.org/10.1016/j.uclim.2019.100534
9. Knabben, R. M., Trichês, G., Gerges, S. N. Y., & Vergara, E. F. (2016). Evaluation of sound absorption capacity of asphalt mixtures. Applied Acoustics, 114, 266–274. https://doi.org/10.1016/j.apacoust.2016.08.008
10. Antunes, L. N., Ghisi, E., & Severis, R. M. (2020). Environmental assessment of a permeable pavement system used to harvest stormwater for non-potable water uses in a building. Science of the Total Environment, 746, 141087. https://doi.org/10.1016/j.scitotenv.2020.141087
11. Zhao, L., Chen, R., Pang, L. X., Zhang, W., & Tan, X. (2019). Study on Photo-catalytic Efficiency and Durability of Nano-TiO2 in Permeable Concrete Pavement Structure. IOP Conference Series: Earth and Environmental Science, 371, 042011. https://doi.org/10.1088/1755-1315/371/4/042011
12. Ghimire, S. R., Johnston, J. M., Ingwersen, W. W., & Sojka, S. (2017). Life cycle assessment of a commercial rainwater harvesting system compared with a municipal water supply system. Journal of Cleaner Production, 151, 74–86. https://doi.org/10.1016/j.jclepro.2017.02.025
13. Zhu, Y., Xu, C., Yin, D., Xu, J., Wu, Y., & Jia, H. (2022). Environmental and economic cost-benefit comparison of sponge city construction in different urban functional regions. Journal of Environmental Management, 304, 114230. https://doi.org/10.1016/j.jenvman.2021.114230
14. Xu, C., Tang, T., Jia, H., Xu, M., Xu, T., Liu, Z., Long, Y., & Zhang, R. (2019). Benefits of coupled green and grey infrastructure systems: Evidence based on analytic hierarchy process and life cycle costing. Resources, Conservation and Recycling, 151(April), 1–10. https://doi.org/10.1016/j.resconrec.2019.104478
15. Nunes Carvalho, T. M., de Souza Filho, F. D. A., & Medeiros de Sabóia, M. A. (2020). Performance of rainwater tanks for runoff reduction under climate change scenarios: a case study in Brazil. Urban Water Journal, 17(10), 912–922. https://doi.org/10.1080/1573062X.2020.1846063
16. Martins Vaz, I. C., Ghisi, E., & Thives, L. P. (2020). Life cycle energy assessment and economic feasibility of stormwater harvested from pervious pavements. Water Research, 170, 115322. https://doi.org/10.1016/j.watres.2019.115322
17. Nnadi, E. O., Newman, A. P., Coupe, S. J., & Mbanaso, F. U. (2015). Stormwater harvesting for irrigation purposes: An investigation of chemical quality of water recycled in pervious pavement system. Journal of Environmental Management, 147, 246–256. https://doi.org/10.1016/j.jenvman.2014.08.020

18. ICPI. (2021). ICPI's Permeable Design Pro - Software for Permeable Interlocking Concrete Pavement Design. Permeable Design Pro Software V3. http://www.permeabledesignpro.com/
19. UFSC-Sustentável. (2021). Water consumption monitoring per consumer unit (Monitoramento do consumo de água por unidade consumidora, in portuguese). Federal University of Santa Catarina.
20. Ghisi, E., & Cordova, M. M. (2014). Netuno 4, programa computacional. LABEEE. Available at: http://www.labeee.ufsc.br/.
21. INMET. (2021). Banco de Dados Meteorológicos para Ensino e Pesquisa (BDMEP) do INMET. BDMEP. https://bdmep.inmet.gov.br/

Reinforcement Learning Based Strategies for Decision Support on Water Treatment Plants

Aída Álvarez Díez, Rocío Pena Rois, Santiago Muíños Landín, and Juan M. Fernández Montenegro

Abstract Treatments to be applied for water purification must be dynamically adaptable to any raw water conditions. Currently, treatments are applied based on standards that are generally correct but not optimized for the circumstances of each drinking water treatment plant (DWTP), neither for critical events. This work presents a methodology for the creation of an Artificial Intelligence (AI) water treatment model, based on reinforcement learning techniques, that provides suggestions about the most efficient treatments for various raw water conditions, increasing their resilience to climate and water related risks. The model has been developed, optimised and validated in a DWTP replica. The results and evaluation of the model are promising as a first approach of a decision support system for drinking water treatments suggestion to be applied to 4.0 DWTPs, although next versions may include more water quality parameters to characterize raw water.

Keywords Drinking water · Treatments management · Reinforcement learning · Decision support

1 Introduction

The water used for any domestic purposes must be wholesome. The requirements for treatment may vary according to rainfall, catchment activity or other reasons. Water from natural courses (rivers, lakes) is fully treated before being supplied to a distribution system from where it will go to consumers in drinking water supplies systems. Health risks from poor quality water will often be due to microbiological or chemical contamination [1]. Microorganisms can be associated with particles and turbidity in water [2], this physical contamination may represent a health risk and it makes disinfection more difficult. Furthermore, substances in water, which affect its

A. Á. Díez · R. P. Rois · S. M. Landín · J. M. F. Montenegro (✉)
AIMEN, O Porriño, 36418 Pontevedra, Spain
e-mail: juan.fernandez@aimen.es

J. R. da. C. S. Galvão et al. (eds.), *Proceedings of the 3rd International Conference on Water Energy Food and Sustainability (ICoWEFS 2023)*, Springer Proceedings in Earth and Environmental Sciences, https://doi.org/10.1007/978-3-031-48532-9_60

appearance, odor or taste, could make water objectionable to consumers and lead to rejection on aesthetic grounds.

Water treatments consists in sequential units to eliminate pollutants and pathogens. It usually includes pre-treatment, coagulation, flocculation and sedimentation, filtration and disinfection [3]. Most treatment systems are designed to remove microbiological contamination and those physical constituents, such as suspended solids (turbidity) that affect aesthetic acceptability or prevent effective disinfection.

Treatment models can help in understanding and predicting treatments effectiveness, especially in stream events when abstraction water changes, like during storms or droughts [4]. There are mathematical models that describe and simulate the most common treatment processes [5, 6]. These mathematical models are robust methods for the simulation of the drinking water treatment plants (DWTP) processes where the staff can check the outcomes of using different treatments. Although, this implies the staff to search for the best treatment using a try and error method (based on their expertise). Currently, DWTP staff usually utilize the treatment standards supplied by the treatments' providers and programs that recalculate the treatments based on the current water flow [7]. Nevertheless, none of their suggestions consider the peculiarities of the individual DWTP (small differences in water composition produced by local source water conditions and infrastructure materials). Here is when the data driven approaches are recommended [8]. Artificial Intelligence (AI), and in particular machine learning methods such as artificial neural networks (ANN), and reinforcement learning, are robust technologies that can handle the complex and dynamic nature of water treatment processes considering such individual particularities.

The goal of this research is the development of AI treatment models able to react to the events related to the quality of water (including water related risks) in a dynamical data driven approach, that allows to increase the performance and safety of the current systems. The AI models are meant to provide efficient and resilient treatment recommendations even when water quality changes occurred due to extreme events such as heavy rains and droughts. Events that are not frequent, and that are expected to happen more often due to climate change, so DWTP's staff are not prepared to react to them. Therefore, the outcome water treatment AI models are expected to help operators and managers in drinking water treatment plants to monitor and control the process. The next sections present the methodology (Sect. 2) and the results (Sect. 3). The conclusions are finally presented in Sect. 4.

2 Methodology

Machine learning techniques are key in Water 4.0. They are data driven methodologies therefore they are highly dependent in data quality and quantity. For this study it was considered different operation conditions in order to generate a resilient AI treatment suggestion model: (1) normal operation (low input turbidity and conductivity), (2) heavy rain events (high input turbidity) and (3) saline intrusion (high conductivity) events. Therefore, for each of those conditions it was necessary to

Table 1 Inlet water characterization for the three operation conditions considered in this study

Parameter	Normal condition	Heavy rain	Saline intrusion	Units
pH	7–7.5	7–8.5	7–7.5	
Turbidity	0–10	0–70	0–20	NTU
Conductivity	60–120	100–450	100–3500	μS/cm

obtain enough data points. The data for this study was obtained from two sources: a DWTP's processes simulator and a scaled replica. The inlet water was characterized by just pH, conductivity and turbidity to simplify the models in this study (see Table 1).

The outlet water was characterized by free chlorine and turbidity. It was considered that conductivity could be linked with chloride [9], and turbidity could be an indirect measurement of Total Dissolved Solids [10] and organic matter [11]. It was also considered that a content of 0.5 mg/L of free chlorine and turbidity below 5 NTU means no biological pollution is present in the outlet water, based on WHO recommendations [1]. Although, depending on the country these thresholds can be more relaxed.

The treatment processes considered in this study are four of the most common drinking water treatment processes: coagulation (with sedimentation), filtration and disinfection (chlorination).

The next sections describe the methodology followed during this study based on the aforementioned DWTP configuration. Section 2.1 provides more details about the two sources of data, which were also used for optimization and validation. Section 2.2 describes the AI modelling carried out to create the treatments suggestion model.

2.1 Testbed for Data Generation, Optimization and Validation

Drinking Water Treatment Plant Scaled Replica. The objective of the creation of a DWTP replica was the creation of data from a near to real environment for any event affecting raw water quality. Its second objective was the validation of the models' suggestions. Therefore, a DWTP replica was created with the most common processes for drinking water production (see Fig. 1). It consists of a reservoir tank with sensors for real time characterization of the inlet water (pH, turbidity and conductivity). Followed by a tank where water mixes with coagulants and flocculants, which then passes to a precipitation chamber. The next stage is a filtration unit filled with sand, finalizing with a chlorination section which applies the chorine and measures the free chlorine at the outcome water with a chlorine sensor. The outlet water is stored in a tank where a sensor is monitoring the turbidity. The DWTP replica was designed for a maximum flow rate of 50 L/d, although it would operate at 20 L/d.

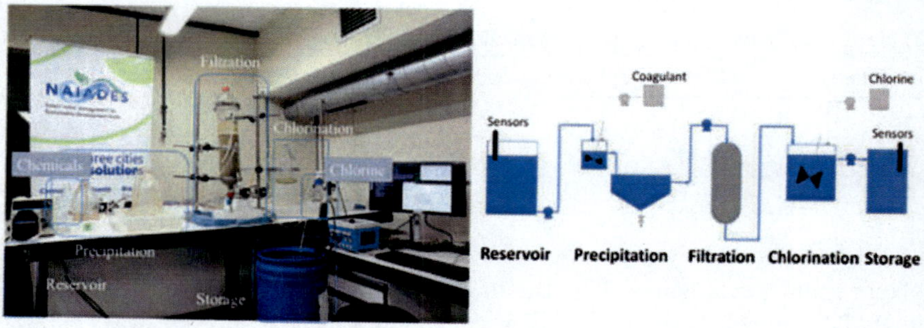

Fig. 1 DWTP replica on the left. Design of the replica on the right

For initial operation, a total resident time of 11 h was considered. A total of 5 sensors are utilized: 3 at the inlet (turbidity, pH and conductivity) and 2 at the outlet (Chlorine and turbidity). The turbidity is important to correlate solids and organic matter content of inlet water with the treatments' efficiency. A change of pH could be a signal of high values in other parameters, such as organic matter or nitrogen forms. Conductivity is important in salinity intrusion detection and other sources of pollution. Chlorine is necessary to ensure free Cl in the outlet water is in the desire range.

A coagulation/flocculation commercial product based in aluminum sulphate salts (25% in weight of $Al_2(SO_4)_3$) for drinking water was used in the precipitation process. The recommended dosage by the manufacturer (3 mg/L treat water) was used in normal operation conditions and it was adapted for extreme events conditions in order to obtain models and then correlate product dosage vs inlet water characteristics (mainly turbidity and conductivity). In order to facilitate the coagulant dosage, a solution with distilled water was made.

A chlorine solution was used in order to obtain a Cl concentration in the outlet to ensure drinking water microbiological quality. The objective was to obtain a Cl concentration less than 1 mg/L in the DWTP replica's outlet [1]. A commercial liquid solution (37% in weight of sodium hypochlorite) was selected because it is easier to manipulate.

As baseline, 1 mL of raw commercial product per liter was considered for each treatment, which was modified with operation conditions. The whole replica is controlled by a single board PC (Nvidia Jetson Nano). The proposed sequence described above (Fig. 1) was programmed into this PC.

Stimela. The gross of the data produced to train the AI models was simulated. The replica's residence time of 11 h limits the amount of data that can be generated; therefore, a simulator was necessary. Stimela is a mathematical model of drinking water treatment processes imbedded in Matlab/Simulink [12]. Stimela offers many models that can be interconnected to represent drinking water plants such as filtration modules, coagulation modules, etc. Stimela simulates the changes experimented in the water quality parameters when water is processed through the modules.

A Stimela schema was created considering the basic drinking water treatment processes defined previously. For parameters initialization to cover all experiment condition values while reducing computational load, ranges for each inlet raw water and treatments were specified.

From the raw water's parameters, the one that was determined to influence the most in the choosing of the treatments was the turbidity. If the turbidity of the raw water increases, the coagulant dosage must increase to maintain the quality of the treated water. If the coagulant dosage increases, the amount of chlorine must also be higher to maintain its effectivity. So, the ranges defined for the Turbidity were sampled n-times meanwhile the pH and conductivity were set to their median values for each range. For each sampled combination of (Turbidity, pH, Conductivity) all the possible combinations of treatment values were simulated (see Eq. (1)).

$$simulations = n_{samples} \cdot n_{Turb} \cdot n_{pH} \cdot n_{Conduct} \cdot n_{coag} \cdot n_{cl} \quad (1)$$

where $n_{samples}$ is the number of samples taken in each Turbidity's range, n_{Turb} is the total number of Turbidity's ranges, n_{pH} is the total number of pH's ranges, $n_{Conduct}$ is the total number of Conductivity's ranges, n_{coag} is the total number of coagulant dosage's options and n_{cl} is the total number of chlorine dosage's options.

2.2 *Reinforcement Learning Based Treatments' Suggestion*

Reinforcement Learning (RL) is part of machine learning approaches which consist of selecting the best Actions for an Agent in a defined Environment to reach States with the maximum Reward. In each event the Agent receives some representation of the environment's State, based on that, it selects an Action. As consequences of the action the Agent finds itself in a new state and receives a Reward depending on the goodness of this transition. It was decided to use RL due to its autolearning capabilities [13].

The versatility of RL makes it suitable for a lot of applications through different industries: algorithms that play games [14], self-driving cars [15], text generation [16], etc. In fact, RL algorithms have already been applied to water treatment plants processes with promising results [17, 18]. In one of those cases [17], the RL procedure was implemented for different objectives: such as the optimization of the environmental impact, based on a Life Cycle Analysis, and the cost reduction. Their method showed beneficial results for the WWTP, like the optimization of costs and energy by reducing the aeration and chemical dosages. On the other hand, Hernandez et al. [18] proposes a solution to make the RL systems to be trustable by the WWTP managers so the RL could be applied directly to control the plant minimizing the operational costs (costs derived from energy consumption and disposal of waste sludge).

A multiarmed bandit model was considered for this study. This specific type of RL has already been studied and suggested for some processes of WTPs in water purification in WWTP and finding the best treatment (they consider 3 different treatment technologies) [19, 20]. In the RL multiarmed bandits modelling there are a fixed limited set of resources. The Agent must pick between a set of choices which have a probability distribution that is partially known at the time of picking. There is a limited number of picks, the Agent must maximize the total reward. In our case, the Inlet Water Quality parameters define the Environment, the treatment combinations are the Actions, and the Outlet Water Quality parameters define the States. Different Environments were considered based on turbidity ranges.

Following this method, the transitions that happen for each combination of Environments and Actions were represented in a Transition Dictionary (TD). The TD is a dictionary that has a key for each Environment and the element associated with each key is the Transition Matrix (TM) calculated for the Environment. A TM is a matrix in which each element a_{ij}^k represents the probability of ending in the State i by taking the Action j (the superscript refers to the Environment $= k$) (see Eq. (2)).

$$\begin{aligned} a_{ij}^k &= \frac{P(State = i \cap Action = j \cap Environment = k)}{P(Action = j \cap Environment = k)} \\ &= \frac{rowswithState = i \wedge Action = j \wedge Environment = k}{rowswithAction = j \wedge Environment = k} \end{aligned} \tag{2}$$

Finally, the TMs of all the Environments were saved onto a dictionary, the TD.

The difficulty in the creation of the TD is the correct definition of the Environment, Actions and State. For example, for the TD related to the higher Turbidity values, the Turbidity's range was taken following an exponential distribution meanwhile Turbidity's range for the sets with lower input Turbidity follows a lineal distribution. This is because the relationship between the Turbidity and the coagulant dosage. For higher Input Turbidity, it is needed a higher dosage of coagulant to maintain the Output Turbidity constant. But the amount of coagulant increases in an exponential way in relation to the Input Turbidity, so the Turbidity's range must follow an exponential distribution. For the lower Turbidity cases, that exponential behaviour is not so significant, and a linear distribution creates the desired States. This is also the reason behind the fact that the range taken for the coagulant dosage for the high Turbidity's case also follows an exponential distribution meanwhile the coagulant dosage for the lower Turbidity's values follows a linear distribution. The chlorine dosage always follows a linear distribution because the free chlorine has a linear dependency with it. Finally, the States ranges are defined based on the legal values required. The reward function designed to for the multiarmed bandit is a combination of three different components (see Eq. (3)). The first one, α, accounts for the effectiveness of the action taken; this component is also a prohibitive one, if the action does not provide drinkable water, the agent will not be able to choose it. The second component, β, refers to the amount of product used. The third one, γ, is calculated based on the water's turbidity before entering the sand filter. Replacing the filter is an expensive procedure so minimizing the water's turbidity before it elongates the time in which it can be operational. All the different components are multiplied by a factor, c_i, , that normalizes the maximal input for each component so that every factor is equally accounted for. These factors can be tweaked if the user values one factor more important than other (i.e.: using less product before the longevity of the sand filter, then increases c_β and c_γ decreases)

$$REWARD = c_\alpha \alpha + c_\beta \beta + c_\gamma \gamma \tag{3}$$

Optimisation. The first treatments suggestion model was generated with data from the simulator. During the data generation process it was detected that the simulator was not capable of generating data with very high values in turbidity in the inlet water (values above 45–50 NTU). Therefore, a set of dedicated experiments was defined and executed to optimise the model. Furthermore, they were also used to check if the treatments suggested by the transition dictionary are correct and to optimise them when necessary.

Regarding the check of the transition matrix suggested treatments: there were defined a total of 36 In-States and 35 Actions per In-State. Ideally, all the 1260 transitions should be characterized but because of time limitations (taking into account the DWTP replica's operation time is 11 h); it was decided to characterize the Actions that would be chosen with higher. If the Action suggested for the In-State returned an undesired Out-State, the treatment would be changed until the desired output was achieved. This information would later be introduced in the Transition Dictionary to make it more accurate.

Once the probable Actions were characterized, experiments with a remarkably high turbidity were also performed, not only because Stimela has a problem to simulate those cases but also because our objective is to be able to predict the treatment for those extreme cases. These cases were also introduced in the Transition Dictionary to enhance accuracy.

3 Results

A set of experiments were formulated with the purpose of validating the dosages results in the tree operation conditions presented in Sect. 2. In Fig. 2 the three settings are depicted as the coloured regions. Once the experiments were defined, the trained model was used to calculate the coagulant and chlorine dosages. In total, seventeen valid experiments were done, four for normal conditions, seven for heavy rain conditions and seven for saline contamination conditions.

Initial Conditions (Turbidity, pH & Conductivity). In Fig. 2 the values of input turbidity and conductivity recorded in the laboratory are represented as the 'x' marks. The colour in each mark is related to the setting that the experiment was trying to reciprocate. The blue 'x' markers are not inside the blue region, this is because they have a higher conductivity than the planned experiments accounted for.

On the other hand, the pH values do not follow the guidelines. This may be caused by a change in the water source due to the drought that affected Galicia, where the

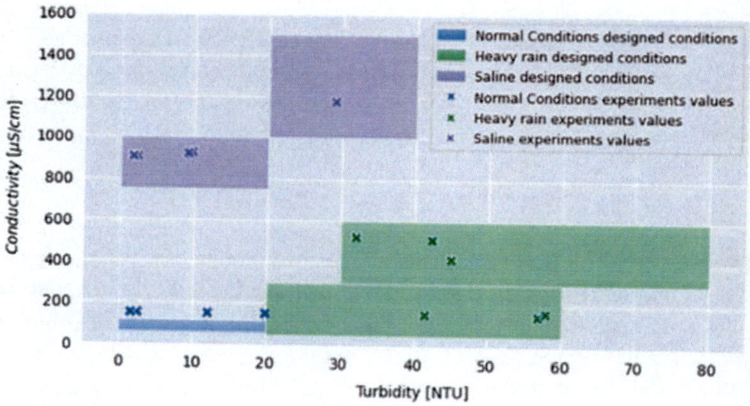

Fig. 2 Conductivity versus inputed turbidity

replica is located, during 17th of August & 4th of October of 2022. The flow of the river was dramatically reduced which could have been the main reason for the inlet pH increase. Nevertheless, the values are in the range of most drinking water catchments, so the validation was continued.

Recalculating Dosages. Because the pH levels differ so much from the initially proposed (between 7 & 7.5) the dosages calculated must be different from the ones the DSS would suggest if the actual parameters were introduced. In Fig. 3 the differences in the dosages are represented.

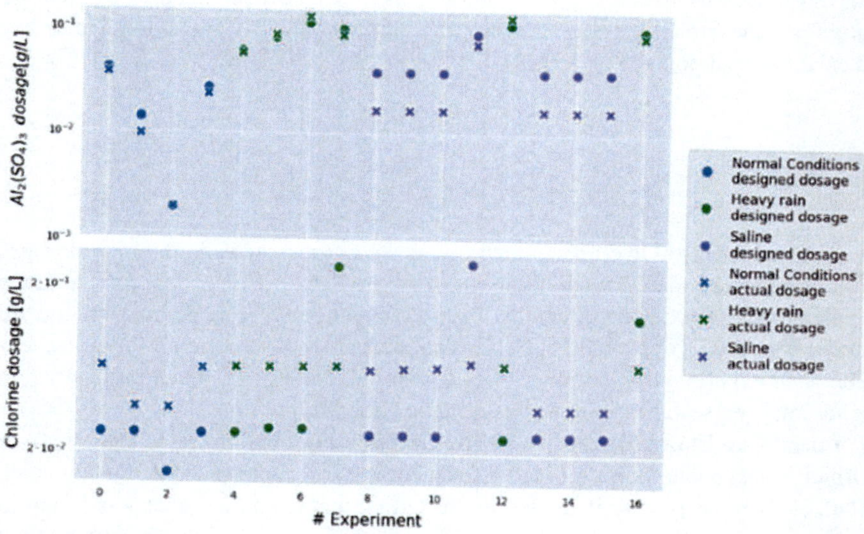

Fig. 3 Comparison between the dosage values used for the experiments ('o') & the values that the DSS would have recommended considering the water Turbidity, pH & Conductivity ('x')

As we can see in Fig. 3 the actual coagulant dosages (‘x’ marks) are quite like the ones that were used in the experiments (dots). On the other hand, the chlorine dosage is very different to the one used. It must be noted that the actual chlorine dosage is usually higher than the one used.

There could be multiple reasons for this behaviour, but the most probable ones are chloride presence or other pollutants in the raw water, a change in the characteristics in chloride product used, even an undetected failure in the replica.

Results. Two parameters were measured at the outlet: turbidity and free chlorine. The turbidity is related with the coagulant dosage. More coagulant dosage, less turbidity in the water leaving the DWTP & vice versa. The same can be said between the free chlorine and chlorine dosage.

Output Turbidity. The coagulant dosage used in the experiments is quite like the model’s suggestions, so the results of the experiments can be used to prove the validity of the coagulant dosage provided by the model suggestion. In Fig. 4 the turbidity of the water that leaves the replica is represented. The water is consistently under 1 NTUs. Reaching the higher values only with high levels of input turbidity, and never surpassing 2 NTUs. These results are satisfactory, even if the input water of the DWTP is of low quality the output water is very clear.

Residual chlorine. In this case, the recommended dosages tend to be higher than the ones used during the experiments. As we can see in Fig. 4, the residual chlorine tends to be higher than the desired maximum limit of 1 mg/L. Still, the vast part of the data is in the acceptable range for drinking water (up to 4 mg/L [21]).

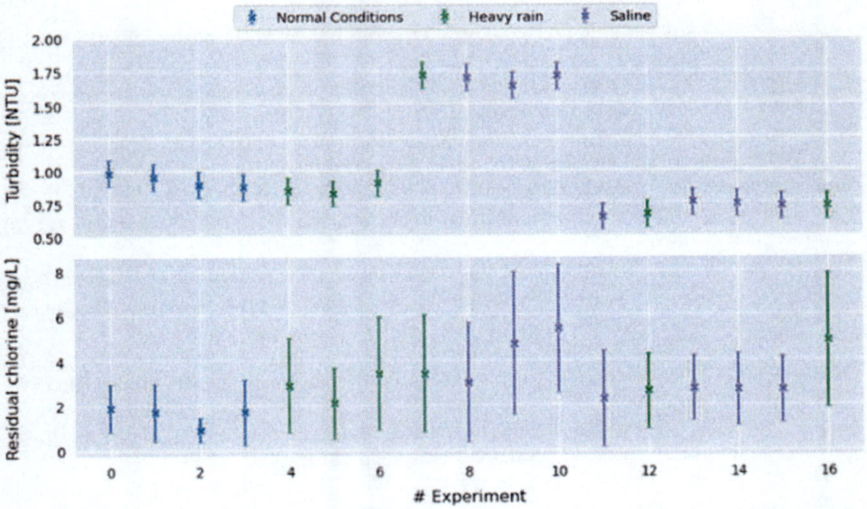

Fig. 4 Turbidity & Residual chlorine of the outlet for each experiment with the uncertainty. Water produced in experiments number 9, 10 and 16 is not acceptable for human consumption

4 Conclusion

In this study a methodology was presented to create an AI treatments suggestion model for basic 4.0 DWTP's management, this is, DWTPs with coagulation, sedimentation, filtration and disinfection processes. The inlet water was characterised by three water quality parameters (pH, turbidity and conductivity) to simplify the modelling and an AI approach similar the ones applied to wastewater treatment plant' processes were followed. An AI model based on RL multiarmed bandits was created using data generated from a treatments processes simulator and it was optimized and validated with data from a replica DWTP. As a result, it was demonstrated that the AI treatments suggestion model is capable of suggesting resilient and efficient treatment that guarantee drinking water at a replica scale for various scenarios (normal conditions and water related risk events). The future steps of this research should further optimize the models and include a validation with data from a real DWTP. Moreover, the introduction of biological factors in the characterization of the water should help to create more robust models.

Acknowledgements This project has received funding from the European Union's Horizon 2020 research and innovation programme under grant agreement No 820985.

References

1. World Health Organization (WHO): Guidelines for drinking-water quality. WHO chronicle, 38 (4), 104–108 (2011).
2. Schiperski, F.: Turbidity as an indicator of contamination in karst springs: a short review. Karst Groundwater Contamination and Public Health, 127–139 (2018).
3. Farhaoui, M., Derrazm, M: Review on optimization of drinking water treatment process. Journal of Water Resource and Protection, vol. 8, no. 8, pp. 777-786 (2016).
4. Aliashrafi, A., Zhang, Y., Groenewegen, H., Peleato, N. M.: A review of data-driven modelling in drinking water treatment. Reviews in Environmental Science and Bio/Technology, vol. 20, no. 4, pp. 985-1009, (2021).
5. Hidayah, E. N., Hendriyanto, O.: Hydrodynamic model of sedimentation and disinfection to predict water quality in water treatment plant. International Journal of Science, Technology and Society, vol. 2, no. 4, pp. 73-77 (2014).
6. Sun, Y., Vaidya, R., Khunjar, W. O., Rosenfeldt, E. J., Selbes, M., Wilson, C., Bott, C. B., Titcomb, M., Wang, Z. W.: Mathematical modeling of biologically active filtration (BAF) for potable water production applications. Water Research, vol. 167, p. 115128 (2019).
7. Rowe, J.: Chemical Feed Calculations in Water Treatment Processes. Workforce Libre Texts. https://workforce.libretexts.org/, last accessed 2023/03/27 (2020).
8. Yin, Y., Le Guen, V., Dona, J., de Bézenac, E., Ayed, I., Thome, N., Gallinari, P.: Augmenting physical models with deep networks for complex dynamics fore-casting. Journal of Statistical Mechanics: Theory and Experiment, vol. 2021, no. 12, p. 124012, (2021).
9. Cox, R. A., Culkin, F., Riley, J. P.: The electrical conductivity/chlorinity relationship in natural sea water. In Deep Sea Research and Oceanographic Abstracts. Elsevier, vol. 14, no. 2, pp. 203–220 (1967).

10. Serajuddin, M., Chowdhury, A. I., Haque, M., Haque, E.: Using turbidity to determine total suspended solids in an urban stream: a case study. In: Proceedings of the 2nd International Conference on Water and Environmental Engineering, STMCA, pp.148–154 (2019).
11. Bright, C., Mager, S., Particulate organic matter, suspended sediment and turbidity in Otago, New Zealand tussock grasslands. In: EGU General Assembly, Vienna, pp.102 (2018).
12. Van der Helm, A. W. C., Rietveld, L. C.: Modelling of drinking water treatment processes within the Stimela environment. Water Science and Technology: Water Supply, vol. 2, no. 1, pp. 87-93 (2002).
13. Sutton, R., Barto, A.: Reinforcement learning: An introduction, 2nd Edition, A Bradford Book, Cambridge, (2018).
14. Mnih, V., Kavukcuoglu, K., Silver, D., Graves, A., Antonoglou, I., Wierstra D., Riedmiller, M.: Playing Atari with deep reinforcement learning. arXiv preprint arXiv:312.5602 (2013).
15. Shalev-Shwartz, S., Shammah, S., Shashua, A.: Safe, multi-agent, reinforcement learning for autonomous driving. arXiv preprint arXiv:1610.03295 (2016).
16. Li, J., Monroe, W., Ritter, A., Galley, M., Gao, J., Jurafsky, D.: Deep reinforcement learning for dialogue generation. arXiv preprint arXiv:1606.01541 (2016).
17. Chen, K., Wang, H., Valverde-Pérez, B., Zhai, S., Vezzaro, L., Wang, A.: Optimal control towards sustainable wastewater treatment plants based on multi-agent reinforcement learning. Chemosphere, vol. 279, no. 130498 (2021).
18. Hernández-del-Olmo, F., Gaudioso, E., Dormido, R., Duro, N.: Tackling the start-up of a reinforcement learning agent for the control of wastewater treatment plants. Knowledge-Based Systems, no. 144, pp. 9-15 (2018).
19. Van Moffaert, K., Van Vaerenbergh, K., Vrancx P., Nowé, A.: Multi-objective X-armed bandits. In:International Joint Conference on Neural Networks, IEEE,pp.2331–2338(2014).
20. Martin, D. M., Johnson, F. A.: A Multiarmed Bandit Approach to Adaptive Water Quality Management. Integrated environmental assessment and management, vol. 16, no. 6, pp. 841-852 (2020).
21. Dery, J. L., Brassill, N., Rock, C. M., Minimizing Risks: Use of Surface Water in Pre-Harvest Agricultural Irrigation. Arizona University, Cooperative Extension (2019).

Policy Directed Information Technology Governing Network Welfare by Balancing the Water-Energy-Food Nexus and Human Perceptions (WEFH) Measures

S. Sorek, A. Peeters, and F. Yuval

Abstract The Water-Energy-Food (WEF) Nexus represents the most fundamental requirements for human existence. It is essential for social welfare and thus is a top strategic priority in the management of water related stakeholders, which enables security and sustainability of the WEF limited resources. In previous publications following Boltzmann entropy, we prove that the WEF resources are linked by information, across the regional administrative nodal network (municipalities, districts, and states) system. We suggest the Volume (normalized WEF expenditures product) as the nexus holistic measure from high to low Volume nodes. The node's Volume with its household income combined density allows for a decision-making policy tool on a node's Welfare Mass (WM—product of Volume and density). This enables the aiming at balancing regulation of the system welfare. The intervention to balance household WEF consumption inequality, in conjunction with the subjective attitude uses big data mining and analytics approach vis a vis water availability, yields a computer-based decision-support system (DSS) for defunding/subsidizing actions to over/under respective consumption leads to activating regulatory policy instrument. This WEFH inputs across a region welfare imbalanced nodal network, is quantified and visualized to derive recommendations for policy driven interventions to regional

S. Sorek
Zuckerberg Institute for Water Research, J. Blaustein Institutes for Desert Research, Ben-Gurion University of the Negev, 8499000 Midreshet Ben-Gurion, Israel
e-mail: sorek@bgu.ac.il

A. Peeters (✉)
TerraVision Lab, 8499000 Midreshet Ben-Gurion, Israel
e-mail: aviva@terravisionlab.com

Negev School of Architecture, Shamoon College of Engineering, 8410802 Beer Sheva, Israel

F. Yuval
Department of Public Policy & Administration, Faculty of Business & Management, Ben-Gurion University of the Negev, 8410501 Beer-Sheva, Israel
e-mail: fanyuval@som.bgu.ac.il

J. R. da. C. S. Galvão et al. (eds.), *Proceedings of the 3rd International Conference on Water Energy Food and Sustainability (ICoWEFS 2023)*, Springer Proceedings in Earth and Environmental Sciences, https://doi.org/10.1007/978-3-031-48532-9_61

ascribed nodes' WM source/sink terms. We show that Israel's administrative districts of temporal WEF expenditures follow the nexus features.

Keywords Information · Network · Policy · WEF nexus

1 Scientific Background

Water is crucial for sustainable socio-economic development, and its supply is linked to energy and food security. The Water-Energy-Food (WEF) Nexus represents the most fundamental requirements for human existence, essential for social welfare and thus is a top strategic priority in the management and oversight of water related stakeholders across administrative nodal network (municipalities and districts) system. The household WEF Volume (normalized expenditures product) the nexus holistic measure and argue, [1] and [2]; that it is linked by information from high to low Volume nodes. The node's Volume, its households' subjective weights towards the WEF components with its income combined density defines the node's scaled Welfare Mass (WM—product of Volume and density). The level of WM and subjective perceptions calibrated nodal network balanced model, yields a decision-making tool, for defunding/subsidizing actions to over/under respective consumption in some nodes can lead to activating regulatory policy instruments. The WEFH inputs (Fig. 1) involve cycles (Fig. 2) of decision-making strategy (DMS), that start with a decision-support system (DSS) the 1st of which is based on historical data mining, to map nodes' mean temporal Volumes. The 2nd DSS apply Ordinary Differential Equations (ODEs) to predict future corrective interventions.

2 Mathematical Model

The developed macro management tool involves cycles of DMS (Fig. 2), with big data disclosing unbalanced nodal network Volumes, for allocating investments based on spatiotemporal information (i.e., facts of nodes' WM values) leading to rectify knowledge for feedback control by ODEs simulations. First the node WEF expenditure components are normalized, so these different components' information can be mutually compared. Let $()_{\mathrm{i}}^{\mathrm{K}}$ denote a quantity at a time t^{K} ($\mathrm{K} = 0, 1, 2, \ldots, \kappa$) at a i node, $\mathrm{W}_{\mathrm{i}}^{\mathrm{K}}$, $\mathrm{F}_{\mathrm{i}}^{\mathrm{K}}$, $\mathrm{E}_{\mathrm{i}}^{\mathrm{K}}$ denote expenditures and $\hat{W}_i^K$, $\hat{F}_i^K$, $\hat{E}_i^K$ their normalized [i.e., $(\hat{})_i^K \equiv ()_{\mathrm{i}}^{\mathrm{K}}/()_{\mathrm{i}}^{0}$] values common to all network nodes.

The 'nexus' concept addresses mutual interactions between group of components such that feedback mechanisms cannot be reconciled by studying each component independently. Given the complex interconnected nature of the nexus' components: a change in one may have unintended repercussions in the others, as well as impact across the entire WEF related network. We thus refer to the product of the WEF

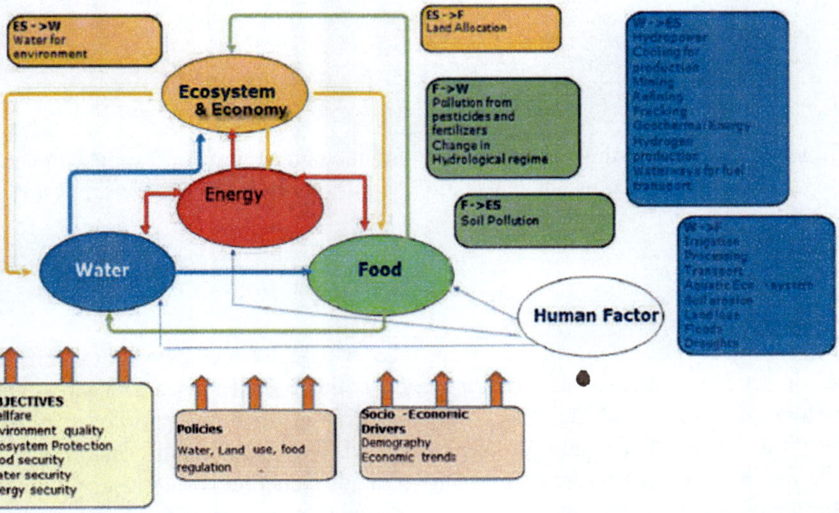

Fig. 1 Interplay between the WEFH measures mutually influencing a domain ecology reflected by economy stresses

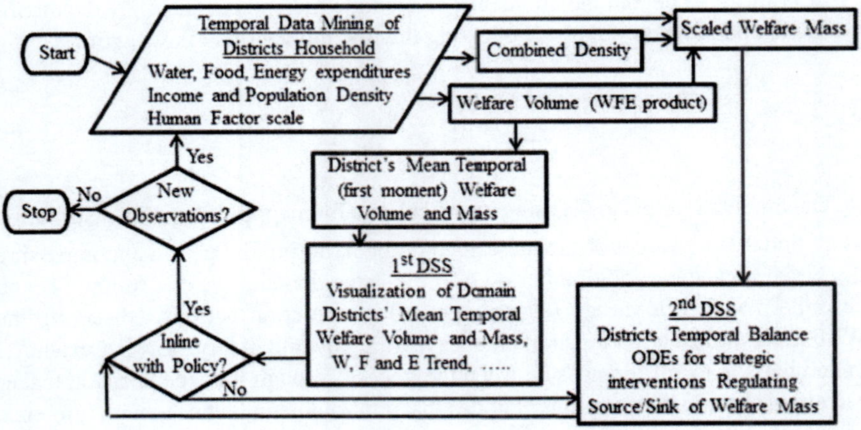

Fig. 2 Operations flowchart for WEF network nodes with two DSS processes

components as a one nexus holistic quantity, representative of its components. Let $\hat{U}_i^k = \left(\hat{W}\hat{E}\hat{F}\right)_i^k$ denote the normalized WEF Volume associated with WEF data at any time t^K and i node. The expenditure WEF components are (I) information-dependent, hence we obtain

$$\frac{1}{\hat{U}}\frac{d\hat{U}}{dt} = \beta\frac{dI}{dt}, \tag{1}$$

in which

$$\beta \equiv \frac{d\hat{W}/dI}{\hat{W}} + \frac{d\hat{F}/dI}{\hat{F}} + \frac{d\hat{E}/dI}{\hat{E}} = \frac{d}{dI}\left(\ln \hat{W} + \ln \hat{F} + \ln \hat{E}\right) = \frac{d\hat{U}/dI}{\hat{U}}. \quad (2)$$

in which β denotes summation over the Volume components of response to change in information. For constant β value, (2) conforms to Boltzmann entropy [3] that reads

$$\ln \hat{U} = \beta I, \quad (3)$$

suggesting directional information from nodes with high $\hat{U}$ values to nodes with lower $\hat{U}$ values. The transfer of information is a key notion because its transfer across nodes dictates the balance of their WEF componential information that allows for intervention, if needed. The novelty here is twofold. First, that this information follows (3) Boltzmann entropy, which entails two major facts that are utmost important to the nexus and the nodal network: (i) this information is valid trans all scales, and (ii) information that transfers between a small number of nodes, even just two, relies on the entire nodes in the network and will also affect them. Second, is our ability to influence this information by intervention, and control of the network sustainability.

Let $NK(\leq \kappa)$ denote the number of series time increments and $\tilde{\mathrm{U}}_{\mathrm{i}}^{NK}$ denote the i-node temporal mean Volume. Following the first moment notion we write

$$\tilde{\mathrm{U}}_{\mathrm{i}}^{NK} = \sum_{\mathrm{K}=1}^{\mathrm{K}=NK} \hat{\mathrm{U}}_{\mathrm{i}}^{\mathrm{K}}(\mathrm{t}^{\mathrm{K}} - \mathrm{t}^{0}) / \sum_{\mathrm{K}=1}^{\mathrm{K=NK}} (\mathrm{t}^{\mathrm{K}} - \mathrm{t}^{0}). \quad (4)$$

The first DSS, refers to (4) using historical data for mapping of spatial distribution of all nodal $\tilde{\mathrm{U}}_{\mathrm{i}}^{NK}$ values at the prescribed domain, during the period encompassing the NK time levels evolution period. For the same domain one can follow (4) and describe maps for a variety (e.g. sequential) of evolution periods. Observing the WEF nodal information transfer, enables learning quantitatively "to what extent … in comparison to" differences between the nodes. Through this, the decision maker can apply specific interventions at nodes that are Volume imbalanced with reference to others in the network. Let, θ_i^K, denote the subjective attitude for i node, $(\smile)_i^K$, per each of the WEF components and their normalized [i.e., $(\smile)_i^K/(\smile)_i^0$] values at an initial (k = 0) time level common to all domain nodes. Let N denote the number of households, the averaged θ_i^K value per N reads,

$$\theta_i^K = \frac{1}{N}\left(\alpha_W \sum_{n=1}^{N} \overset{\smile}{W}{}_n^K / \overset{\smile}{W}{}_N^0 + \alpha_F \sum_{n=1}^{N} \overset{\smile}{F}{}_n^K / \overset{\smile}{F}{}_N^0 + \alpha_E \sum_{n=1}^{N} \overset{\smile}{E}{}_n^K / \overset{\smile}{E}{}_N^0\right)_i;$$

$$(\alpha_W + \alpha_F + \alpha_E)_i = 1, \quad (5)$$

in which α_W, α_F and α_E denote assigned weights for the WEF components. Let $(IP)_i^K\left[=\left(\sum_{n=1}^{N}(IP)_n^K\right)_i\right]$ denotes the i node's income for its N households., Accounting for network relative nodal income becomes $\sum_i (IP)_i^K A_i/A_D$ in which A_i denotes the i node area and $A_D\left(=\sum_i A_i\right)$ denotes the domain area. Let the node income index be $I_i^K\left\{\equiv (IP)_i^K/\left[\sum_i (IP)_i^K A_i/A_D\right]\right\}$ then the i node equivalent households density reads

$$\rho_i^K \equiv (I\rho)_i^K.. \tag{6}$$

In view of (5), (6) and the normalized Volume $\hat{U}$, we define the i-node welfare-mass $(WM)_i^K = \left(\theta\rho\hat{U}\right)_i^K$ at the Kth time level. For the second DSS, considering the WEF Volume and the information flow across nodes suggested by (3), we write a system of temporal first ODEs for the network balance which reads,

$$\frac{d}{dt}(WM)_i = \sum_{j\neq i} \propto_{ij} \Delta_{ij} + Q_i, \quad \Delta_{ij} \equiv (WM)_j - (WM)_i, \tag{7}$$

in which Δ_{ij} denotes the transfers between node i and other j nodes, Q_i denotes the i-node source/sink term and α_{ij} denotes a transfer coefficient calibrated based on historic data. The Q_i, is the super imposed quantity, following a policy choice for the $k+1$ time level. By virtue of (7), we have the condition

$$\sum_i \frac{d}{dt}(WM)_i = \sum_i Q_i, \tag{8}$$

dictating the prevention of violating overall budget allocation.

In view of (7) solving for $(WM)_i$, let $\overline{U}$ define a variant of the Volume for node i outcome at the $k+1$ time level of (7) with $\Delta\hat{W}$, $\Delta\hat{F}$ and $\Delta\hat{E}$ denoting increments of the WEF normalized expenditure components at the k previous time level, namely

$$\overline{U} = \overline{WFE} = \left(\hat{W} + \Delta\hat{W}\right)\left(\hat{F} + \Delta\hat{F}\right)\left(\hat{E} + \Delta\hat{E}\right). \tag{9}$$

We rewrite (9) to read

$$f = \frac{\overline{U}}{\hat{U}} = \left(1 + \frac{\Delta\hat{W}}{\hat{W}}\right)\left(1 + \frac{\Delta\hat{F}}{\hat{F}}\right)\left(1 + \frac{\Delta\hat{E}}{\hat{E}}\right). \tag{10}$$

Expanding f as a Taylor series for the three variables $\frac{\Delta\hat{W}}{\hat{W}}$, $\frac{\Delta\hat{F}}{\hat{F}}$, $\frac{\Delta\hat{E}}{\hat{E}}$ in (10) around (1, 1, 1) reads

$$f \cong \frac{\overline{U}}{4\hat{U}} = \frac{\Delta\hat{W}}{\hat{W}} + \frac{\Delta\hat{F}}{\hat{F}} + \frac{\Delta\hat{E}}{\hat{E}} - 1. \tag{11}$$

Hence, in view of (11) the decision maker can regulate specific allocations per each of the WEF components across domain nodes at any time level.

With reference to W, we have

$$\lambda = \frac{\Delta \hat{W}}{\hat{W}}; \frac{\Delta \hat{F}}{\hat{F}} = \frac{\lambda}{2}; \frac{\Delta \hat{E}}{\hat{E}} = \frac{\lambda}{2}, \tag{12}$$

and in view of (11), resolve for λ from

$$\frac{\overline{U}}{4\hat{U}} = 2\lambda - 1. \tag{13}$$

3 Examples

For the sake of assessing possible outcome trends of balancing a nodal network WEF expenditures, let us consider the simplified version of (7) the nodes balance ODEs accounting for (5) the human factor subjective weight. We will consider a simplified simulation subject to $\theta_i^K = 1$ in (5), $\propto_{ij}= 1$ in (7) and zero sum of impulses of imposed sources and sinks for the *k + 1* time, namely $Q_i^{k+1} = Q_i^k + \Delta Q_i$; $\Delta Q_i \equiv Q_i^{k+1} - Q_i^k$, , which by virtue of (8) dictates the $\sum_i \Delta Q_i = 0$ condition. Let us consider annual data surveys by Israel central bureau of statistics (http://www.cbs.gov.il/reader).

Simplified simulation of the WM evolution trend over Israel's nodal network is described in Fig. 3. Simulation imply that subsidy imposed to nodes with initial relative inferior WM values ("Akko", "BeerSheva" and "ZefatKineretGolan in Fig. 4) will improve with no significant effect of a node with a much higher WM value ("RamatGan" in Fig. 4).

We note from Fig. 4 that the more urbanized districts [descending from nodes (7), (2), (9) and (0), corresponding, respectively, to labels 1, 2, 3 and 4 in Fig. 4, being the most rural] correspond to less expenses on energy in comparison to those of water or food. In all nodes depicted in Fig. 4, we note a "turning point" at the vicinity of the year 2007. This is aligned with international food price spike (Figs. 3 and 4 in the link mentioned in [4], reported in June 2014), occurred in 2007/2008. We maintain that the "turning point" delineated in Fig. 4 fit the nexus feature of inter dependencies between the components of the WFE expenditures and is a vivid manifestation of a network subject to causality driven directional information.

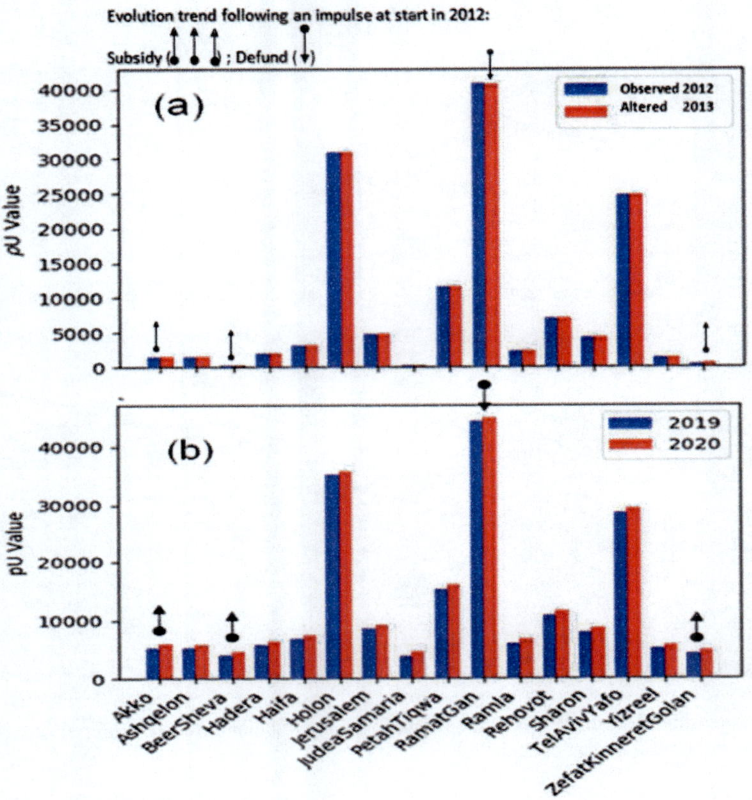

Fig. 3 Prediction of WM over Israel's nodal domain following one subsidy of 1000 units imposed to three different nodes, balanced by 3000 units defund imposed to one node: **a** Evolution at 2013 in comparison to initial observation at 2012, and **b** at 2020 in comparison to 2019. Data collected from Israel's central bureau of statistics

4 Conclusion

Our developed macro-policy-governed management strategy is associated with DSS that is directly related to the WEF expenditures at a node from the investigated domain, maintaining its network sustainability.

The domain nodal network historical data is addressed to map the nodes spatiotemporal WEF expenditures. Such mapping enables a managerial comparison pinpointing "to the extent of…in comparison with" of the domain nodes. Following this mapping, we implement simulations of WEF expenditure of the domain nodes, for future corrections predicting outcome under specific interventions. Such simulation is aimed at equalizing welfare in terms of fundamental needs of WEF components over an investigated domain.

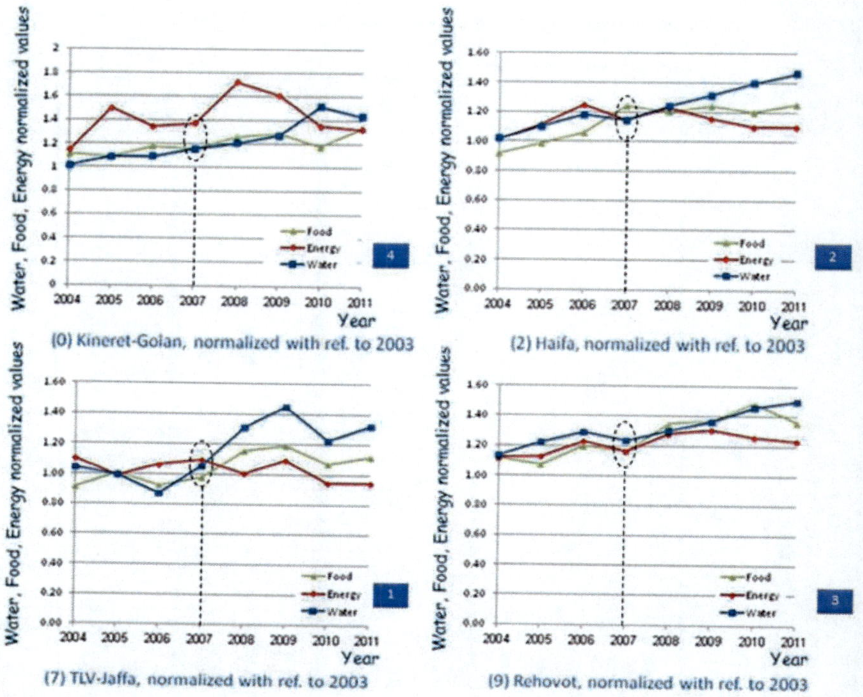

Fig. 4 Snapshots from the W-E-F normalized expenditures at four different districts. Data collected from Israel's central bureau of statistics

References

1. Y. Teitelbaum, A. Yakirevich, A. Gross and S. Sorek* (2020). Simulations of the Water Food Energy Nexus for policy driven Intervention, Heliyon, 6, 8, e04767. DOI: https://doi.org/10.1016/j.heliyon.2020.e04767, * Corresponding author.
2. S. Sorek, A. Peeters, F. Yuval, D. Savic (2022). Governance using the water-food-energy nexus and human-factor measures, PLoS ONE 17(1), 2022: e0261995. https://doi.org/10.1371/journal.pone.0261995.
3. C.G. Chakrabarti, D. Kajal (2000). Boltzmann-gibbs entropy: axiomatic characterization and application. Int. J. Math. Math. Sci.; 23: 243-251
4. G. Anania (2013). Agricultural export restrictions and the WTO: What options do policy-makers have for promoting food security? Issue paper No. 50, ICTSD, Geneva, Switzerland. Report

A Comparative Study for Water Reuse in an Urban WWTP with a Novel Non-thermal Plasma Technology

Luís Marinheiro, Inês Baptista, Ruben Jorge, Célia Manaia, and Stefan Löblich

Abstract Currently, the scarcity of potable water is a worldwide problem that makes it necessary to explore alternative sources for non-potable uses, such as the reuse of treated wastewater. The present study aimed to investigate and compare the application of several tertiary treatment technologies, and its combinations, in the production of reusable wastewater (RWW). A novel advanced oxidation technology in the water treatment field, Non-Thermal Plasma (NTP), was also studied and included in this comparative study. Pilot and industrial scale tests were carried out under real conditions in an urban WWTP. RWW were obtained with diverse qualities fitted to different urban uses. NTP applied in best operating optimized conditions allowed for a 3.2 log-reduction in *Escherichia coli* present in secondary effluent. A comparative cost analysis also showed that NTP is a competitive technology for disinfection in comparison with the other tested.

Keywords Water reuse · Disinfection · Non-thermal plasma

1 Introduction

Currently, water scarcity is a global concern. In southern Europe, where Portugal is located, around 20% of the population is currently subject to water shortages. Forecasts suggest that in 2030 this problem could affect around 50% of Europe's river basins, requiring measures to promote a more rational use of this resource [1,

L. Marinheiro (✉) · S. Löblich
AST-Ambiente, Rua do Bairro 400, 4485-010 Aveleda, Portugal
e-mail: lmarinheiro@ast-ambiente.com

I. Baptista · R. Jorge
Wedotech, Lda, Rua do Seixal 108, 4000-521 Porto, Portugal

C. Manaia
Universidade Católica Portuguesa, CBQF—Centro de Biotecnologia e Química Fina—Laboratório Associado, Escola Superior de Biotecnologia, Rua de Diogo Botelho 1327, 4169-005 Porto, Portugal

J. R. da. C. S. Galvão et al. (eds.), *Proceedings of the 3rd International Conference on Water Energy Food and Sustainability (ICoWEFS 2023)*, Springer Proceedings in Earth and Environmental Sciences, https://doi.org/10.1007/978-3-031-48532-9_62

2]. In this context, it is increasingly urgent to implement systems to produce reusable wastewater (RWW) for non-potable uses [3]. In Portugal, only 1.2% of wastewater was converted into RWW in 2020, while in neighboring Spain it was between 7 and 10% in 2016 [3, 4].

Conventional urban wastewater treatment plants are currently ill equipped to produce quality RWW compatible with the strictest reuse standards, and also to comply to the recent requirements included in the proposal for the revision of the wastewater treatment Directive, which includes micro-pollutant removal targets and antimicrobial resistance monitoring [5].

The wastewater treatment sector needs, therefore, complementary and cost-effective treatment solutions to upgrade the quality and flexibility of treatment at those facilities.

Recently, there has been a major focus on the research of advanced oxidation processes (AOP), as a tertiary treatment for conventional WWTP [6, 7]. Among these processes, the application of atmospheric non-thermal plasma (NTP) has gained visibility due to its potential to form a broad spectrum of oxidizing species physical and chemical phenomena: formation of oxidizing species, such as radicals (H•; O•; OH•) and other oxidizing species (H_2O_2, O_3, etc.), shock waves, ultraviolet radiation and electro-hydraulic cavitation [8, 9]. This diversity of effects provides a more diversified and effective action on the various contaminants to be treated [10]. Despite these advantages, the applications of this technology in a real environment context for wastewater disinfection are still very limited, although it has been shown to have great potential for *E. coli* removal, but still at lab scale with very small working volumes and artificial waters [11, 12].

The present study aimed to investigate and compare the application of different tertiary treatment technologies for RWW production, to obtain RWW with differentiated quality to serve different uses in an urban context, and to compare operating costs. This study was developed on site at a WWTP, with pilot and industrial scale treatment units, all tested in real conditions with secondary effluent from the WWTP. These included ozone, ultrafiltration, and reverse osmosis. Tertiary treatment already existing in the selected WWTP, with sand filtration combined with coagulation and UV disinfection, was also included in this comparative study. Additionally, in this work, the potential application of a NTP pilot scale unit was also investigated for RWW disinfection, and preliminary results are presented and compared with conventional tertiary treatment solutions.

There is a lack of practical pilot and industrial scale wastewater treatment studies performed under comparable conditions, with real effluents, and the findings of this study are a valuable contribution for the wastewater treatment sector, while highlighting the potential of a novel technology.

2 Materials and Methods

2.1 *Wastewater Treatment Plant*

This study was carried out in a municipal WWTP dimensioned for a population of 300,000 inhabitants, with an average effluent treatment flow of ca. 70,000 m^3/day. The WWTP has a treatment process with the following stages: preliminary treatment and primary settling with sand and grease removal, and secondary treatment with conventional activated sludge followed by secondary settling. There is also a tertiary treatment stage for RWW for internal use, consisting of sand filters and UV disinfection.

2.2 *Wastewater Treatment Technologies*

The multilayer sand filters (SF), existing in the WWTP, are made up of two units in series with a capacity of 100 $m^3 \cdot h^{-1}$. The filter media is composed of a 1st stage with anthracite and silica sand, and a 2nd stage with silica sand. This filtration is combined with the dosage of coagulant (100 $mg \cdot L^{-1}$ aluminum sulfate).

The ultraviolet (UV) disinfection unit is open channel, with low pressure mercury arc lamps in horizontal arrangement, with a capacity of 100 $m^3 \cdot h^{-1}$. This unit comprises a channel with 5 banks × 8 lamps, with a unitary UV emission of 26.7 W, and a dose of 190 $microWatts \cdot cm^{-2}$ at 1 m.

The containerized ultrafiltration (UF) unit, with a semi-industrial scale (AQUASMART—Water and Wastewater Treatment Solutions, Lda.), is equipped with 128 submerged UF modules, with a total active filtration surface of 768 m^2 and average processing flow of 10 $m^3 \cdot h^{-1}$. The UF membranes are high density polyethylene hollow fiber with an outside diameter of 0.4 mm and average pore size distribution of 20 nm, operating in outside-in mode.

The reverse osmosis (RO) pilot unit (AST—Soluções e Serviços de Ambiente, Lda.), has 3 spiral-wound type membrane modules, with a total active membrane area of 57 m^2, and an average processing flow of 2.2 $m^3 \cdot h^{-1}$.

The ozone pilot unit (De Nora) comprises an ozone generator, using ambient air, and a bubble column reactor where the ozone is mixed with the effluent to be treated. The average processing flow rate is 1.2 $m^3 \cdot h^{-1}$ and ozone dose applied was 88 $gO_3 \cdot m^{-3}$.

The NTP pilot unit operated with ambient air consisted of a plasma generator (model IXS-500-KISS, SFC) fed by an air compressor with a variable flow rate between 20 and 80 $L \cdot min^{-1}$. The power applied for gas ionization is variable between 7 and 40 W, which correlates with a pulse frequency range between 200 to 500 Hz. The plasma generated in the gaseous phase is distributed in a tank with a capacity of 50L for the effluent, through a diffuser hose installed at the bottom of the tank.

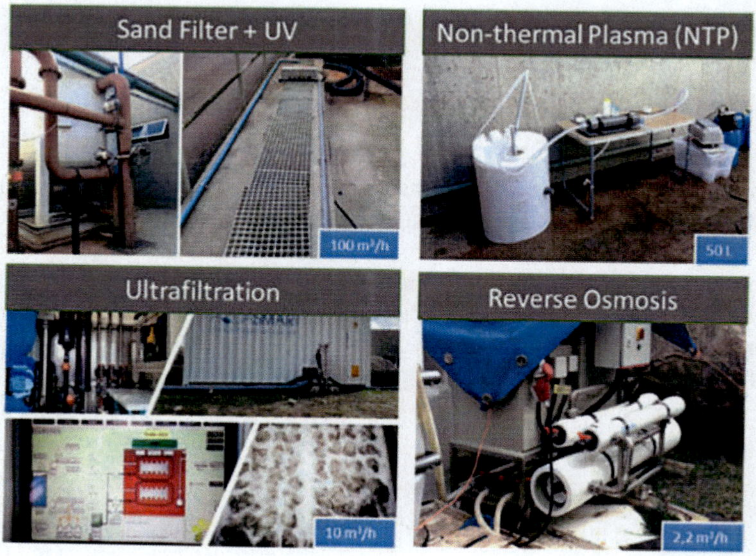

Fig. 1 Pilot units used in this study

All tests were performed in continuous mode, except for NTP which was operated in batch. Some of the pilot units used are presented in Fig. 1.

2.3 Sampling and Analysis

The WWTP effluent used in the tests was collected after secondary settling, in the effluent discharge point. Grab samples were collected from WWTP secondary effluent at the inlet of the tested technology and from its respective outlet. Physicochemical analysis of the inlet and treated effluent samples were performed by a certified laboratory using the Standard Methods' methodologies.

2.4 Results Analysis

The results presented for each configuration tested correspond to the mean ± standard deviation of at least 3 samples ($n \geq 3$) taken on different days, with the exception of NTP which study is being concluded. The RWW produced in this study were compared with the quality standards laid out in the Portuguese Decree-Law No. 119/2019, for wastewater reuse for irrigation, urban and industrial uses.

3 Results and Discussion

3.1 NTP Optimization

An initial optimization study was conducted with the NTP pilot unit. Experiments were carried out with secondary effluent in batch, and the *E. coli* disinfection was measured. The impact of operating conditions, such as gas flowrate (20, 50 and 80 $L{\cdot}min^{-1}$) and applied power for gas ionization (30 W and 40 W), were evaluated on the NTP disinfection efficiency. Also, the effluent pretreatment with sand-filtration was evaluated and compared to the raw effluent.

The preliminary results showed that at the same flowrate, better results were obtained with higher power. Also, at the same applied power, the lower flowrates provided the best disinfection results. Additionally, better *E. coli* removal was observed with effluent pre-treatment through sand-filtration, possibly due to suspended solid removal, such as sludge particles, which can improve radical action, as observed by others [7]. From this preliminary study, carried out in batch, the best *E. coli* removal of 3.2 log units (CFU/100 mL) was achieved with pretreated effluent after 90 min NTP treatment, applying a flowrate of 20 $L{\cdot}min^{-1}$ and 40W. This is an ongoing investigation, and further results will be shared in the presentation.

3.2 RWW Characterization

Comparative tests were carried out at the same WWTP with different technologies, at pilot or industrial scale, as tertiary treatment for RWW production. Table 1 presents results of the physical–chemical characterization for the different technologies or combinations tested, including the raw secondary effluent for quality comparison.

All RWW produced reported here had a biochemical oxygen demand (BOD) lower than the limit established for urban irrigation uses ($\leq 10\ mgO_2{\cdot}L^{-1}$). Regarding total suspended solids (TSS), RWW with quality superior to the most restricted limit ($\leq 10\ mg{\cdot}L^{-1}$) was obtained for all configurations tested. The same was observed for turbidity ($\leq$ 5 NTU). Regarding the total nitrogen (NT), all RWW obtained with the different combinations tested exceeded the recommended thresholds for application in irrigation ($NT \leq 10\ mg{\cdot}L^{-1}$), with the exception of reverse osmosis, although this parameter is facultative and does not hinder the RWW application for irrigation. This result for nitrogen is justified by the high concentration of nitrogen in the secondary effluent, and by the low removal capacity of this parameter for the tested technologies. Nitrogen reduction upstream of the tertiary treatment through nitrification improvement in the aeration tanks, would be the best strategy to reduce the nitrogen content in the RWW produced.

Figure 2 shows the *E. coli* removal achieved with the different technologies or combinations tested. The highest removal achieved was for the UF + RO, followed by SF + UV and UF. The RWW obtained from these technologies complied with

Table 1 Results of the physicochemical analysis of produced RWW

Parameter	Secondary Effluent	SF + UV	SF + OZ	UF	UF + OI	SF + NTP[a]
pH	7.6 ± 0.1	7.4 ± 0.1	7.4 ± 0.1	7.5 ± 0.2	6.6 ± 0.4	7.1 ± 0.3
BOD [$mgO_2 \cdot L^{-1}$]	13.6 ± 5.4	6.0 ± 4.0	5.8 ± 1.3	8.5 ± 1.1	< 3.0 ± 0.0	8.3 ± 4.0
TSS [$mg \cdot L^{-1}$]	20.4 ± 4.8	9.0 ± 1.0	3.3 ± 1.1	4.6 ± 2.0	2.3 ± 1.3	3.3 ± 1.2
Turbidity [NTU]	9.5 ± 2.5	6.7 ± 2.0	1.1 ± 0.1	0.9 ± 0.4	0.3 ± 0.1	1.3 ± 0.0
NH_4 + [$mg \cdot L^{-1}$]	50.7 ± 9.0	59.4 ± 0.7	52.1 ± 7.0	48.3 ± 2.5	2.9 ± 0.7	38.0 ± 0.9
NT [$mg \cdot L^{-1}$]	55.7 ± 8.2	61.5 ± 0.5	54.2 ± 5.9	51.3 ± 1.9	4.6 ± 2.8	38.0 ± 0.8
PT [$mg \cdot L^{-1}$]	2.7 ± 0.8	3.0 ± 0.5	0.6 ± 0.4	1.8 ± 0.3	< 1 ± 0.0	0.6 ± 0.0

[a] Batch tests results

the class A for irrigation (< 10 CFU/100 mL). The combination SF + NTP provided a class B RWW suitable for irrigation.

The SF + UV combination, that is installed at the WWTP, was not effective in reducing the microbiological load of *E. coli* and failed to provide RWW with microbial quality compatible for irrigation reuse or other urban applications, according to Portuguese regulations. The effluent low transmittance, and possibly the bad design of the UV channel, are the main reasons for this outcome. It should be noted that NTP was very effective in reducing *E. coli*, even in a complex matrix such as a secondary effluent, demonstrating its potential as a tertiary treatment for effluent disinfection and RWW production.

From the tests carried out so far, the application of membrane technologies, ozonation and NTP resulted in RWW production with superior microbiological quality, with compatible uses for irrigation and urban applications.

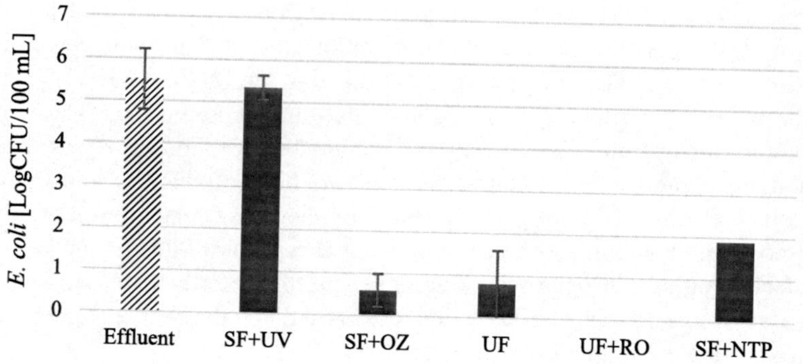

Fig. 2 Results for *E. coli* in produced RWW

A comparative operating costs analysis was performed, based on the pilot units used and conditions tested, englobing costs for energy, consumables, and reagents. This evaluation showed that NTP can be competitive, with an operating cost of 0.30 € per m^3 of produced RWW. This cost for NTP is lower than the one presented by Singh et al. [11] of 4.3$/$m^3$, for 3-log removal of *E. coli*, also the NTP energy consumption observed in our work was of 2.2 kWh/m^3, and they reported values between 20 and 86 kWh/m^3 [11]. In comparison, ozonation and reverse osmosis had higher running costs than NTP, although these delivered class A RWW for irrigation.

4 Conclusions

In this comparative study, several tertiary technologies were applied at pilot and industrial scale to produce RWW with differentiated quality. The highlight if this study was the optimization and application of NTP as a novel technology for effluent disinfection.

The microbial quality of the obtained RWW was the decisive factor for the RWW classification. Membrane technologies including UF, UF + RO and SF+OZ application produced RWW with the best quality for irrigation, followed by NTP. UV radiation failed to provide significant effluent disinfection.

The NTP application at the best optimized operating conditions, allowed for a 3.2 log-reduction of *E. coli* present in the secondary effluent, producing a class B irrigation RWW. Results obtained at pilot scale with the NTP showed that technology has potential for RWW production with quality for different uses. There are very limited studies performed with NTP at this scale, and with real effluents, and this study demonstrated its potential under normal operation conditions at WWTP.

Acknowledgements Project ApR-TEC (NORTE-01-0247-FEDER-047207) is funded by Sistema de Incentivos à Investigação e Desenvolvimento Tecnológico do Portugal 2020, Programa Operacional do Norte (NORTE 2020) and Fundo Europeu de Desenvolvimento Regional (FEDER). Authors also acknowledge SIMDOURO for providing access to the municipal WWTP were the pilot scale study was performed.

References

1. EEA (2016), Is Europe's freshwater use sustainable? European Environmental Agency. http://www.eea.europa.eu/highlights/world-water-day-is-europe, last accessed 2023/01/26.
2. EC (2012), Report on the Review of the European Water Scarcity and Droughts Policy. European Comission. https://ec.europa.eu/environment/water/quantity/pdf/COM-2012-672final-EN.pdf, last accessed 2023/01/26.
3. ERSAR. Relatório anual dos serviços de águas e resíduos em Portugal – Vol. 1 | 2020. (2020).
4. Jodar-Abellan, A., López-Ortiz, M. I., Melgarejo-Moreno, J. Wastewater treatment and water reuse in Spain. Current situation and perspectives. Water, 11(8), 1551 (2019).

5. EC (2022) Proposal for a DIRECTIVE OF THE EUROPEAN PARLIAMENT AND OF THE COUNCIL concerning urban wastewater treatment (recast). European Comission. https://eur-lex.europa.eu/resource.html?uri=cellar:fc078ec8-55f7-11ed-92ed-01aa75ed71a1.0001.02/DOC_1&format=PDF, last accessed 2023/01/26.
6. Poyatos, J. M., Muñio, M. M., Almecija, M. C., Torres, J. C., Hontoria, E., Osorio, F. Advanced oxidation processes for wastewater treatment: state of the art. Water, Air, and Soil Pollution, 205, 187-204 (2010).
7. Back, J. O., Obholzer, T., Winkler, K., Jabornig, S., Rupprich, M. Combining ultrafiltration and non-thermal plasma for low energy degradation of pharmaceuticals from conventionally treated wastewater. Journal of environmental chemical engineering, 6(6), 7377–7385 (2018).
8. Hijosa-Valsero, M., Molina, R., Montràs, A., Müller, M., Bayona, J. M. Decontamination of waterborne chemical pollutants by using atmospheric pressure nonthermal plasma: a review. Environmental Technology Reviews, 3(1), 71-91 (2014).
9. Jiang, B., Zheng, J., Qiu, S., Wu, M., Zhang, Q., Yan, Z., Xue, Q. Review on electrical discharge plasma technology for wastewater remediation. Chemical Engineering Journal, 236, 348-368 (2014).
10. Vanraes V., Nikiforov, A., Leys C. Electrical Discharge in Water Treatment Technology for Micropollutant Decomposition. *In* Plasma Science and Technology - Progress in Physical States and Chemical Reactions, chapter 16, Mieno, T. (Ed.). BoD–Books on Demand (2016).
11. Singh, R. K., Philip, L., Ramanujam, S. Continuous flow pulse corona discharge reactor for the tertiary treatment of drinking water: Insights on disinfection and emerging contaminants removal. Chemical Engineering Journal, 355, 269-278 (2019).
12. Murugesan, P., Moses, J. A., Anandharamakrishnan, C. Performance of an atmospheric plasma discharge reactor for inactivation of *Enterocococcus faecalis* and *Escherichia coli* in aqueous media. Journal of Environmental Chemical Engineering, 8(4), 103891 (2020).

Prior Indication of Agricultural Nature-Based Solutions (NbS) for Water and Food Security in the Context of Climate Change

Joyce Maria Guimarães Monteiro, Elaine Cristina Cardoso Fidalgo, and Ana Paula Dias Turetta

Abstract Impacts of climate change represent a serious challenge for the agricultural sector and expose the need for suitable solutions for its adaptation. Nature Based Solutions (NbS) are a key component to address climate change and challenges related to food and water security and biodiversity loss. The aim of this work is to present and test a methodological framework for prior indication of agricultural NbS for the water-food security nexus in the context of climate change. The study case is Nova Friburgo, a municipality located in the mountainous region of Rio de Janeiro State, highly vulnerable to the impacts of extreme weather events and an important center of food production. The method consists of an integrated environmental assessment that includes the synergies and trade-offs between ecosystem services and a water-food security nexus assessment, as well as the assessment of the implementation of NbS in public policies of the municipality of Nova Friburgo. The survey of NbS in the literature resulted in the selection of twenty-six NbS for agricultural management of water and food security (water-food security nexus). The most suitable NbS considering the integrated environmental assessment and the governance assessment were soil conservation practices, agroforestry/integrated crop-livestock-forestry system and organic agriculture. Agricultural zoning and agritourism received the highest score in terms of governance. The methodology was able to identify the most appropriated NbS for the study case. However, prior indication of NbS needs to be discussed with local decision makers, considering the local environmental, socioeconomic, and political factors.

Keywords Integrated environmental assessment · Governance · Sustainable agriculture

J. M. G. Monteiro (✉) · E. C. C. Fidalgo · A. P. D. Turetta
Embrapa Soil, Rio de Janeiro, Brazil
e-mail: joyce.monteiro@embrapa.br

J. R. da. C. S. Galvão et al. (eds.), *Proceedings of the 3rd International Conference on Water Energy Food and Sustainability (ICoWEFS 2023)*, Springer Proceedings in Earth and Environmental Sciences, https://doi.org/10.1007/978-3-031-48532-9_63

1 Introduction

Impacts of climate change include increased precipitation and extreme temperatures, floods and droughts that can affect society in unprecedented ways. Such impacts represent a serious challenge for the agricultural sector and expose the need for suitable solutions for its adaptation [23]. Nature-based solutions (NbS) address societal challenges such as climate change, food and water security, and disaster risk reduction through the protection, sustainable management, and restoration of natural and modified ecosystems, simultaneously providing human well-being and biodiversity benefits [10]. NbS are inspired and supported by nature, and simultaneously provide economic, environmental, and social benefits [13]. The 2030 agenda for Sustainable Development recognizes the need to address the negative global trends affecting society, the economy, and the environment. NbS can contribute to several Sustainable Development Goals (SDGs), generating various social, economic, and environmental co-benefits by promoting the delivery of bundles of ecosystem services [14, 17].

NbS include establishing practices such as forest landscape restoration, land conservation and management, integrated water resource management, ecosystem-based adaptation and mitigation, and ecosystem-based disaster risk reduction [9]. According Simelton et al. [33] in the agricultural sector, efforts related to NbS refer to the use of natural processes or elements to improve the ecosystem functions of environments and landscapes affected by agricultural practices and improve livelihoods and other social and cultural roles. This understanding finds support in ecosystem-based adaptation (EbA), as a set of nature-based methods for adaptation and food security that is closely associated with sustainable land management and water security [23]. Sustainable agriculture can be a solution integrated within the ecosystem service provision and water-food security nexus [34].

In fact, NbS for agriculture could maximize the ability of nature to provide ecosystem services and water and food security. These ecosystem services include the provision of services such as food and water; regulation services, such as the regulation of floods, droughts, soil degradation, and climate regulation; support services such as soil formation and nutrient cycling; and cultural services, such as leisure, spiritual, religious, and other intangible benefits [26].

Being prepared to face the impacts of climate change means adapting agricultural systems to become resilient and optimized, so that they contribute to water and food security and continue to provide benefits to all strata of society. The scientific literature widely addresses impacts of NbS interventions. There are several methods for assessing and quantifying NbS impacts on water and food security, biodiversity, and disaster reduction. It is applied for diverse environments, including urban, peri-urban, coastal or agricultural settings [3, 9, 22, 24, 27, 29, 33–35]. There are also several platforms available that present NbS and their impacts, such as Urban Nature Atlas (https://una.city/); Oppla (https://oppla.eu/); Panorama–Solutions for a Healthy Planet (https://panorama.solutions/en), Nature-based Solutions evidence platform (https://www.naturebasedsolutionsevidence.info/evidence-tool/).

This work proposes a methodology not to measure the impact of NbS interventions but to indicate NbS for agricultural management of water and food security (water-food security nexus) in order to face the impacts of climate change. The methodology was tested in Nova Friburgo, a municipality located in the mountainous region of the State of Rio de Janeiro, Brazil, highly vulnerable to the impacts of extreme weather events and an important center of food production, where these solutions are even more necessary.

2 Case Study

Nova Friburgo is a municipality in the mountainous region of the State of Rio de Janeiro, southeastern Brazil. The municipality is located in the mid north of the State of Rio de Janeiro, at 22° 16′ 55″ south latitude and 42° 31′ 52″ west longitude, approximately 136 km from the city of Rio de Janeiro (capital). The municipality has an average altitude of 985 m, with a total area of 933.4 km^2. In 2010, the population of Nova Friburgo was 182,082 inhabitants [19].

The climate is tropical at altitude (Cwa), with an average annual temperature ranging between 10 and 15 °C. The average annual precipitation is 1279.8 mm, with the wettest months from November to March, and the driest months from May to August [21].

The mountainous region of Rio de Janeiro is highly vulnerable to extreme weather events, while it has intense population density and food production areas [20]. Based on the last twenty-five years, in the region there were floods with landslides in 1987, 1988, 1999, 2000, 2003, 2007, 2008 and 2011, resulting in hundreds of deaths, missing persons, and homeless [16]. The natural geological vulnerability and the increase in population in areas with high risk of landslides in Nova Friburgo resulted in one of the biggest environmental catastrophes in terms of loss of human life that occurred in Brazil, with 918 deaths recorded and more than 8000 homeless, due to intense rains in 2011 [25].

Nova Friburgo has a diversified agricultural activity, with emphasis on the production of flowers and fruits, vegetables, and agrosilvopastoral activities. Nova Friburgo is an important vegetable production hub, responsible for supplying not only its consumption market in neighboring municipalities, but also the city of Rio de Janeiro, the second largest city in Brazil, contributing to the food security of a large contingent of the population [12].

Even though this municipality is a pioneer in organic agriculture in the State of Rio de Janeiro and even today has areas with organic agriculture, most of the agricultural system is conventional, based on the intensive use of synthetic inputs and mechanization. The lack of adequate soil management for slope environmental conditions, such as contour planting or no-tillage, is common. In general, plowing is done using a tractor on the steepest line, contributing to soil loss, silting up streams and transporting sediments. These facts have accentuated the local erosive processes. The continued use of lands for vegetable crops and pastures has impoverished and compacted the soils, and these areas remain productive due to the intensive use of fertilizers and agrochemicals [30].

There are around 1600 rural properties involved in agricultural production (mainly cattle raising and horticulture), occupying an area of 21,000 ha (average ≈ 13 ha/property). Of the more than 2400 producers, 90% are family farmers (properties up to 40 ha) [12]. Agritourism and ecotourism are important activities in the rural area, since the municipality attracts visitors and tourists to enjoy privileged nature (climbing, trails, and waterfall baths) [28].

3 Methodological Framework

We propose a method to help identify NbS interventions in agriculture to contribute to the decision of local actors before the NbS intervention.

The method consists of an integrated environmental assessment that includes the synergies and trade-offs between ecosystem services (ES) and the water-food security nexus (WFS) assessment, as well as the assessment of the implementation of NbS in public policies of the municipality of Nova Friburgo (PP).

The first step consisted of surveying and selecting the NbS whose interventions occur in the agricultural ecosystem to reduce the impacts of climate change, associated climate risks (climate variability, climate extremes, and extreme hydro meteorological events), and increase the benefits for water-food security nexus. The NbS were selected from several reference works, such as the Oxford University database (Nature-based solutions evidence platform -https://www.naturebasedsolutionsevidence.info/evidence-tool/), [3, 9, 18, 22, 29, 32–34].

The evaluation of the ES and WFS indicators relied on the participation of seven (7) top-level experts, including experts who work in Nova Friburgo. The experts participated in a meeting where the methodology was presented, and they received instructions for completing questionnaires organized in an Excel® spreadsheet.

3.1 *Synergies and Trade-Offs Between Ecosystem Services (ES)*

For each category of ecosystem services from the Millennium Ecosystem Assessment [26], we selected two ecosystem services directly or indirectly related to water and food security in the context of climate change from the work of [15].

The assessment of synergies and trade-offs of ecosystem services (ES) consists of experts' scores regarding the potential supply or demand for ecosystem services of each NbS intervention.

The suggested score value was − 1 (negative impact), zero (neutral impact) or + 1 (positive impact). The scale of values aims to reflect, respectively, the relationship of

Table 1 Representative model of the expert evaluation spreadsheet

	Categories	Provision		Regulation		Support		Cultural	
	Score	Impact (− 1; 0; + 1)		Impact (− 1; 0; + 1)		Impact (− 1; 0; + 1)		Impact (− 1; 0; + 1)	
	Ecosystem service	Water	Food	Erosion	Climate	Biodiversity	Soil quality	Leisure	Scenic beauty
NbS	1								
	2								
	3								

conflict, neutrality, and synergy between NbS and ecosystem services (ES). Table 1 represents the experts' evaluation worksheet, showing the categories of ecosystems, ecosystem services, and suggested scores for the evaluation of synergies and trade-offs of NbS interventions.

The ES score for each NbS was obtained considering the mode of the scores of the seven experts, which ranged from − 1 to + 1. To integrate these results into the analysis of water and food security, these values were transformed into positive, following the rule: replacement of assigned values equal to − 1 by zero; of values equal to 0 by 0.5; keeping values equal to + 1 unchanged. The average of the sum of eight transformed values resulted in normalized values in the range of 0–1, which represents the interval between negative (< 0.5) and positive (> 0.5) impact for the set of ecosystem services evaluated for each NbS.

3.2 *Water-Food Security Nexus (WFS)*

The assessment of the water-food security nexus (WFS) considers three water security sensitivity indicators and three food security sensitivity indicators accepted by the Brazilian Ministry of Science, Technology, and Innovation [1]. Water security indicators are the following: quantity and quality of water for human well-being; quantity and quality of water for productive activities; conservation of the environment to maintain the hydrological cycle, avoiding the risk of droughts and floods. Food security indicators are the following: food production and availability; promotion of sustainable food production systems; promotion of decentralized or diversified food production systems. Thus, the score for WFS is assigned to six indicators.

The specialists evaluated each NbS in relation to each indicator, scoring with value of zero (no impact) or one (with impact). The score for WFS for each NbS was obtained considering the mode of the scores of the seven experts. The final score for each NbS in relation to the water-food security nexus was the average of the sum of the six modal values, which resulted in normalized values in the range of 0 to 1.

3.3 *Governance Assessment (Public Policy—PP)*

Another stage of the methodological journey consisted of surveying public policies related to land use, food, water, and climate risk management in Nova Friburgo's municipal development plan. This search was carried out on the Nova Friburgo City Hall website (https://www.novafriburgo.rj.leg.br/leis) in 2021. Eight (8) policies were selected.

The assessment consisted of a textual search of each NbS selected in the content of public policies. Each NbS received a score of zero (0) when not included or one (1) when mentioned in the policy content. The value attributed to each NbS was summed, with the expected result ranging from zero (0) to eight (8).

In the sequence, each of the eight public policies was evaluated regarding the inclusion of governance criteria. The criteria considered were:

1. Being in force,
2. Inclusion of deployment forecast,
3. Inclusion of financial/compensation mechanisms (PES, fees, taxes, fines, credit lines),
4. Integration with the National Water Security Plan [2],
5. Integration with the National Food Acquisition Program (PAA, the acronym in Portuguese) [5],
6. Inclusion of a participatory process: citations of participatory committees, producer associations, among other instruments that include the participation of society,
7. Inclusion of training/dissemination process: communication instruments, training, environmental education, capacity building.

A value of zero or one was assigned, respectively, if the evaluated policy met or not a certain listed criterion. The sum of the scores for each policy against the governance criterion ranged from zero (0) to seven (7) according to the number of criteria met.

The score for the policies in relation to the governance criterion was obtained using the relative frequency of positive responses to the examined requirements, generating an index of compliance with the criterion listed for evaluating the policies. As an example, if the policy does not meet any criterion, the relative frequency is equal to 0/7; therefore the index is equal to zero. In the case of meeting one criterion, the relative frequency is equal to 1/7 and the index is 0.14, and so on, in the case of meeting all criteria, the index is equal to one.

The final score of the NbS in relation to the PP is given using the product of the score given to each NbS mentioned or not in the policy (0—case not mentioned and 1 case mentioned) and the index of compliance with the listed governance criteria of each public policy. The results (final score) were normalized into a [1, 1] range by applying the minimum–maximum formula (Eq. 1).

$$Z_{norm} = \frac{x - x_{min}}{x_{max} - x_{min}} \tag{1}$$

where min and max are the minimum and maximum values in x given its range.

3.4 *Prior Indication of NbS (Total ES + WFS) and the Comparison of PP*

The integrated environmental assessment for each NbS is the result of summing the values assigned to ES and the WFS. This result was normalized to be between zero (0) and one (1) by applying the minimum–maximum formula. The integrated environmental assessment was compared to the governance assessment (PP), also on a scale between zero (0) and one (1).

4 Results and Discussion

The survey and selection of NbS listed in the literature for agricultural management of water and food security (water-food security nexus) in order to face the impacts of climate change resulted in a list of 26 NbS. The list of NbS selected to test the proposed methodology was very broad and sometimes even repetitive, such as, for example, no-tillage (one kind of soil conservation practice) and soil conservation practices (in general).

If, on the one hand, this fact made the evaluation repetitive and increased the number of NbS to be evaluated, on the other hand, it facilitated the textual search of NbS in public policies, by increasing the possibility their citation in the content of the evaluated municipal public policies.

Figure 1 shows the result of the previous indication of selected NbS, highlighting the integrated environmental assessment (ES + WFS), its components (synergies and trade-offs between the provision of ecosystem services—ES and water-food security nexus—WFS), and the governance assessment (PP).

In terms of integrated environmental assessment, Fig. 1 shows that the most recommended set of practices are soil conservation practices, contour farming, strip cropping, agroforestry/integrated crop-livestock-forestry system, integrated livestock-crop system, cover crops and mulching, recovery of degraded areas, pasture management, contour buffer strip, organic agriculture, hydrological protection zone, and no-tillage.

From this set, the NbS soil conservation practices (in general) and agroforestry/integrated crop-livestock-forestry systems and organic agriculture had a good score in terms of governance. This indicates that the public policies analyzed include more general NbS than specific ones. The selection of a specific management (e.g., strip cropping, cover crops, and mulching) is often decided at the property level, according to the local situation and is not the object of municipal environmental planning.

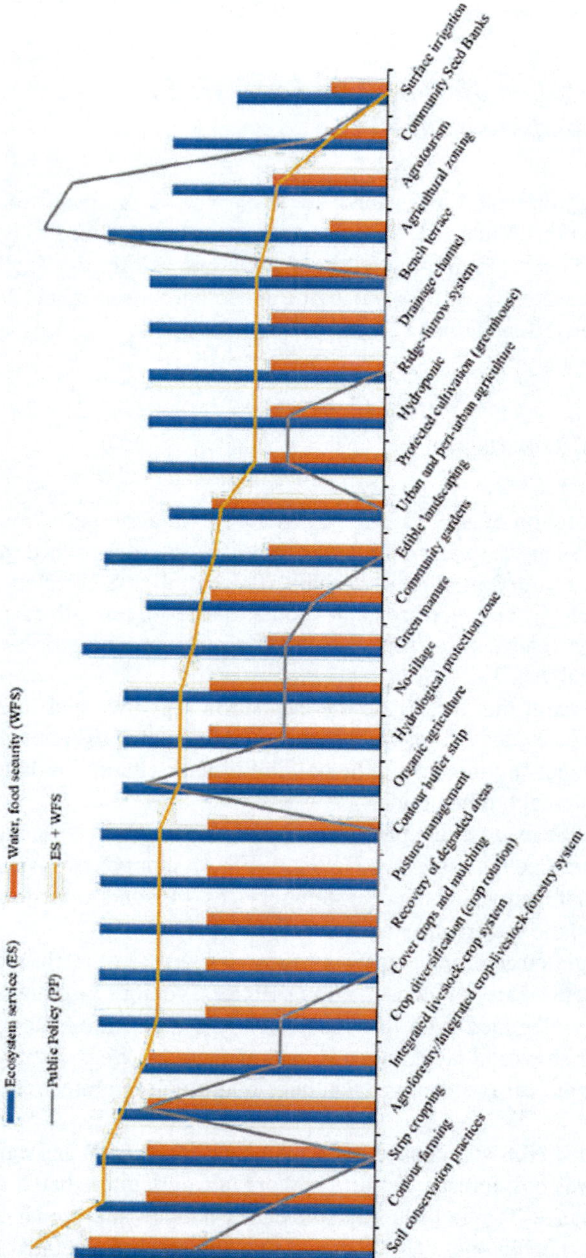

Fig. 1 Prior indication of agriculture NbS—case study Nova Friburgo, RJ—Brazil

It is important to emphasize that agroforestry/integrated crop-livestock-forestry are recommended and widespread practices in Brazil due to the essential role in the global challenge of meeting increasing global food demand while conserving and restoring natural ecosystems [6–8]. Nova Friburgo follows the same trend; perhaps this explains the good score in terms of governance for this NbS.

Agricultural zoning and agritourism, although considered by experts to be promising for providing ES, got low score in the water-food security nexus (WFS), resulting in a low integrated environmental assessment (Fig. 1). On the other hand, these NbS received the highest score in terms of governance and could be evaluated locally in terms of social and economic benefits that favor the strengthening and adaptability of the rural population to climate change.

When considering the most suitable NbS both for the integrated environmental assessment and for the governance assessment, the first three would be soil conservation practices, agroforestry/integrated crop-livestock-forestry system, and organic agriculture. These results seem to make sense due to the undeniable contribution of these agricultural NbS in the water-food security in the context of climate change [11, 33, 34, 36].

The prior indication is important considering the reality of agriculture in Nova Friburgo, represented by high vulnerability to extreme weather events and the impoverishment of soils used for vegetable crops and pastures, which contributes to soil compaction and erosion. On the other hand, agricultural activity contributes to the food security of a large contingent of the population, there is a vocation for organic agriculture, and agritourism is a promising activity in the rural area [4, 33].

However, it is fundamental to discuss these results with local decision makers, both to emphasize the integrated environmental benefits of some NbS without effective governance at the municipal level and to prioritize those that already have political arrangements for their implementation.

5 Conclusion

The focus of this work was to present and demonstrate the methodology, for the purpose of ex ante indication of municipal NbS and not to discuss the impact of NbS adoption, since most works in the literature present approaches to assess the impact of NbS adoption.

The proposed methodology seems promising for a prior assessment for the indication of NbS, since the approach includes the integrated environmental assessment, considering the provision of ecosystem services and the water-food security nexus, as well as the potential of implementation of NbS in Nova Friburgo (governance). The proposed methodology can be improved and/or adapted to different municipal realities.

The results of applying the methodology in the municipality of Nova Friburgo were satisfactory. However, a prior indication of NbS needs to be discussed with local decision makers, considering the local environmental, socioeconomic, and political factors that determine the adoption of NbS.

References

1. AdaptaBrasil Homepage, https://www.dropbox.com/s/weh5wvo9jj51z8o/Te%C3%B3rico-metodol%C3%B3gico_Recursos_H%C3%ADdricos_Seguran%C3%A7a_Alimentar_revisado_19_01_2022_vers%C3%A3o_1.1.pdf?dl=0, last accessed 2022/01/18.
2. ANA - National Water Agency. National Water Security Plan. Brasília, DF: ANA (2019). 112 p. Homepage: https://arquivos.ana.gov.br/pnsh/pnsh.pdf. last accessed: 2022/12/12.
3. Artmann, M.; Sartison, K. The role of urban agriculture as a nature-based solution: a review for developing a systemic assessment framework. Sustainability, v.10, n. 6, 1937 (2018).
4. Ballinas, M. B.-P., Arumugam, P., Verstand, D., Siegmund-Schultze, M. G. R., Keesstra, S. D., Garcia Chavez, L. Y., de Boer, R., van Eldik, Z. C. S., & Voskamp, I. M. Nature based solutions in Food Systems: Review of nature-based solutions towards more sustainable agriculture and food production. Wageningen University & Research. (2022) homepage https://edepot.wur.nl/578638, last accessed 2023/01/18.
5. Brazil. Law nº 10,696, of July 2, 2003. Provides for the renegotiation and extension of debts arising from rural credit operations, and other measures. Federal Government (2003).
6. Brazil, Law nº12.805 de 29 de April de 2013 - National Crop-Livestock-Forest Integration Policy (2013).
7. Brazil. Ministry of Agriculture, Livestock and Supply (MAPA). Sectoral plan for mitigation and adaptation to climate change for the consolidation of a low-carbon economy in agriculture: ABC plan (Low Carbon Emission Agriculture) / Ministry of Agriculture, Livestock and Supply, Ministry of Agrarian Development, coordination of Civil House of the Presidency of the Republic. – Brasília: MAPA, 173 p (2012).
8. Brazil. Ministry of Agriculture, Livestock and Supply (MAPA). Sector plan for adaptation to climate change and low carbon emissions in agriculture with a view to sustainable development (2020–2030): strategic vision for a new cycle / Secretariat for Innovation, Rural Development and Irrigation. – Brasilia: MAPA (2021).
9. Chausson, A.; Turner, B.; Seddon, D.; Chabaneix, N.; Girardin, C. A. J.; Kapos, V.; Key, I.; Roe, D.; Smith, A.; Woroniecki, S.; Seddon, N. Mapping the effectiveness of nature-based solutions for climate change adaptation. Global Change Biology, v. 26, n. 11, p. 6134-6155 (2020)
10. Cohen-Shacham, E.; Walters, G.; Janzen, C.; Maginnis, S. (Ed.). Nature-based Solutions to address global societal challenges. Gland: IUCN (2016).
11. Daryanto, S., Fu, B., Wang, L., Jacinthe, P.-A., and Zhao, W. Quantitative Synthesis on the Ecosystem Services of Cover Crops. Earth Sci. Rev. 185, 357–373 (2018)
12. EMATER-RIO. Relatório de atividades 2017. Rio de Janeiro, 70 p. (2018) Homepage http://www.emater.rj.gov.br/pdf/Relatorio_de_atividades2017.pdf, last accessed: 2022/12/12.
13. European commission. Towards an EU research and innovation policy agenda for nature-based solutions & re-naturing cities: final report of the Horizon 2020 expert group on 'Nature-based solutions and re-naturing cities'. Brussels: European Commission (2016).
14. Faivre, N.; Fritz, M.; Freitas, T.; Boissezon, B. DE; Vandewoestijne, S. Nature-Based Solutions in the EU: Innovating with nature to address social, economic and environmental challenges. Environmental Research, v. 159, p. 509-518, (2017).
15. Fidalgo, E. C. C., Prado, R. B., Turetta, A. P. D., Schuler, A. E. (ed.). Manual para pagamento por serviços ambientais hídricos: seleção de áreas e monitoramento. Brasília, DF: Embrapa, 78 p. (2017).
16. Freitas C. M., Carvalho, M. L, Ximenes, E. F., Arraes, E. F., Gomes, J. O. Vulnerabilidade socioambiental, redução de riscos de desastres e construção da resiliência - lições do terremoto

no Haiti e das chuvas fortes na Região Serrana, Brasil. Ciência e Saúde Coletiva, v. 17, n. 6, p. 1577- 1586 (2012).
17. Gómez Martín E., Giordano, R., Pagano, A. van der Keur, P., Costa, M. M. Using a system thinking approach to assess the contribution of nature based solutions to sustainable development goals, Science of The Total Environment, Volume 738 (2020)
18. Herzog, C. P., Rozado, C. A. Diálogo setorial UE-Brasil sobre soluções baseadas na natureza: contribuição para um roteiro brasileiro de soluções baseadas na natureza para cidades resilientes. Luxemburgo: Serviço das Publicações da União Europeia (2019). Homepage https://oppla.eu/sites/default/files/docs/Portuguese-EU-Brazil-NBS-dialogue-low.pdf last accessed 2022/12/12
19. IBGE. Censo demográfico 2010. Rio de Janeiro (2010). Homepage: http://censo2010.ibge.gov.br/resultados, last accessed 2022/11/11.
20. INEA - INSTITUTO ESTADUAL DO AMBIENTE (RJ). APA Estadual de Macaé de Cima: plano de manejo - análise da UC. Rio de Janeiro: módulo 3, p. 314. (2014).
21. INMET. Instituto Nacional de Meteorologia, 2013. 2013. Homepage http://www.inmet.gov.br/portal, last a: accessed: 2022/12/22
22. IUCN - International Union for Conservation of Nature and Natural Resources. Global Standard for Nature-based Solutions: a user-friendly framework for the verification, design and scaling up of NbS. Gland: IUCN (2020).
23. IPCC - Intergovernmental Panel on Climate Change. Climate Change and Land: special report on climate change desertification, land degradation, sustainable land management, food security, and greenhouse gas fluxes in terrestrial ecosystems P.R. Shukla, et al. (eds.) 43p.(2019).
24. Kabisch, N.; FrantzeskakI, N.; Pauleit, S.; Naumann, S.; Davis, M.; Artmann, M.; Haase, D.; Knapp, S.; Korn, H.; Stadler, J.; Zaunberger, K.; Bonn, A. Nature-based solutions to climate change mitigation and adaptation in urban areas: perspectives on indicators, knowledge gaps, barriers, and opportunities for action. Ecology and Society, v. 21, n. 2, 39 (2016).
25. Leite, C.; Marques, A. L.; Levy, W.; Wernke, A. Análise dos vazios urbanos infraestruturados: oportunidade de adensamento populacional e absorção de população em área de risco de deslizamento: cenários para Nova Friburgo (RJ) de 2015 a 2050. Arquitextos, ano 22, n. 253, 07, jun. (2021). Homepage https://vitruvius.com.br/revistas/read/arquitextos/22.253/8130. last accessed 2022/12/12.
26. MEA - MILLENNIUM ECOSYSTEM ASSESSMENT. Ecosystems and human well-being: current state and trends. Washington, DC: Island Press. v. 1. (2005)
27. Morri, E.; Santolini, R. Ecosystem Services Valuation for the Sustainable Land Use Management by Nature-Based Solution (NbS) in the Common Agricultural Policy Actions: A Case Study on the Foglia River Basin (Marche Region, Italy). Land 11, 57. (2022).
28. Pereira, M. da S.; Antonio, G. J. Y.; Machado, C. R. de L. (ed.). Plano Municipal de Desenvolvimento Rural Sustentável do Município de Nova Friburgo: 2020–2023. Nova Friburgo: Secretaria Municipal de Agricultura e Desenvolvimento Rural, 93 p. (2020). Homepage http://ainfo.cnptia.embrapa.br/digital/bitstream/item/222952, last accessed 2022/12/12.
29. Raymond, C. M., Frantzeskaki, N., Kabisch, N., Berry, P., Breil, M., Nita, M. R., GenelettI, D., Calfapietra, C. A. framework for assessing and implementing the co-benefits of nature-based solutions in urban areas. Environmental Science and Policy, v. 77, p. 15–24, Nov. (2017).
30. SEAPEC - Secretaria de Estado de Agricultura e Pecuária (RJ). Projeto Desenvolvimento Rural Sustentável em Microbacias Hidrográficas do Estado do Rio de Janeiro - Rio Rural / BIRD - Plano Executivo da Microbacia de Barracão dos Mendes. Rio de Janeiro (2011).
31. Seddon, N., Chausson, A., Berry, P., Girardin, C.A.J., Smith, A., Turner, B., Understanding the value and limits of nature-based solutions to climate change and other global challenges. Philosophical Transactions of the Royal Society B 375, 1794, 20190120 (2020).
32. Shiao, T.; Kammeyer, C.; Brill, G.; Feinstein, L.; Matosich, M.; VIgerstol, K.; Müller-Zantop, C. Business case for Nature-Based Solutions: landscape assessment. Oakland: United Nations Global Compact CEO Water Mandate and Pacific Institute (2020).
33. Simelton E, Carew-Reid J, Coulier M, Damen B, Howell J, Pottinger-Glass C, Tran HV and Van Der Meiren M. NbS Framework for Agricultural Landscapes. Front. Environ. Sci. 9:678367 (2021)

34. Sonneveld, B.G.J.S., Merbis, M.D., Alfarra, A., Ünver, O., Arnal, M.A. Nature-Based Solutions for agricultural water management and food security. FAO Land and Water Discussion Paper no. 12. Rome, FAO. 66 pp. (2018).
35. Xing, Y., Jones, P., Donnison I. Characterization of nature-based solutions for the built environment Sustainability, 9 (2017).
36. Van Noordwijk, M., Gitz, V., Minang, P.A., Dewi, S., Leimona, B., Duguma, L., Pingault, N., Meybeck, A. People-Centric Nature-Based Land Restoration through Agroforestry: A Typology. Land 9, 251. (2020).

Effectiveness of a UV Treatment Using Light Emitting Diodes (LED) for the Inactivation of Phytoplanktonic Organisms Present in Two Reservoirs of Cádiz (Spain)

Pablo Gento-Arazola, Leonardo Romero-Martínez, Enrique Nebot, Asunción Acevedo-Merino, Jóse Ángel Gálvez, and Javier Moreno-Andrés

Abstract The irradiation with ultraviolet (UV) light is a treatment which is often used for the inactivation of waterborne organisms. UV-C radiation impacts directly on the organisms DNA, hindering their reproduction. The objective in this study is determining the efficacy of a UV treatment, using LEDs as a source of UV-C light, for the inactivation of phytoplanktonic organisms present in two reservoirs of the Province of Cadiz (Spain). Samples of surface water were subjected to UV irradiation using a collimated beam reactor equipped with UV-LED lamps. The treatment efficacy was determined, calculating the growth inhibition by means of monitoring the fluorescence throughout an incubation period and, additionally, quantifying the inactivation of viable organisms by means of the most probable number (MPN) method. The results regarding the growth inhibition showed, in general, significant but weak efficacy for the treatment of green algae, requiring UV doses between 109 and 712 mJ cm^{-2} for inhibiting their growth a 90%. In the case of cyanobacterial growth, there was not, in general, significant effect of the UV treatment. On the other hand, by means of the MPN method, levels of inactivation greater than 2 log-reductions were determined with UV doses of 25 mJ cm^{-2}. The results obtained in this study indicate that the UV-LED treatment with high UV doses may be effective for the inactivation of green algae, but ineffective for the inactivation of cyanobacteria in the range of UV doses applied in this study.

P. Gento-Arazola (✉) · L. Romero-Martínez · E. Nebot · A. Acevedo-Merino · J. Moreno-Andrés
Department of Environmental Technologies, Faculty of Marine and Environmental Sciences, INMAR—Marine Research Institute, CEIMAR—International Campus of Excellence of the Sea, University of Cadiz, Cadiz, Spain
e-mail: pablo.gentoarazola@alum.uca.es

J. Á. Gálvez
Department of Biology, Faculty of Marine and Environmental Sciences, INMAR—Marine Research Institute, CEIMAR—International Campus of Excellence of the Sea, University of Cadiz, Cadiz, Spain

J. R. da. C. S. Galvão et al. (eds.), *Proceedings of the 3rd International Conference on Water Energy Food and Sustainability (ICoWEFS 2023)*, Springer Proceedings in Earth and Environmental Sciences, https://doi.org/10.1007/978-3-031-48532-9_64

Keywords Reservoirs · Harmful Algal Blooms · Cyanobacteria · LED Ultraviolet treatment · Growth inhibition · Most probable number

1 Introduction

The term "Harmful Algal Bloom" (HAB) was coined by the Intergovernmental Oceanographic Commission (IOC) of UNESCO [1] to designate the rapid growth of photosynthetic organisms, such as microalgae or cyanobacteria, that are perceived as harmful because of their adverse effects on the ecosystem or on the human health. These phenomena could be associated with the generation of biotoxins, presenting themselves as a potential health risk to humans and animals [2]. HABs are increasing in frequency and magnitude globally [3]. In a scenario of global change, where nutrient inputs to the natural environment are becoming more frequent and abundant, the likelihood of eutrophication episodes, together with unusual weather conditions, makes these episodes increasingly recurrent [2, 4].

HABs may occur in natural and artificial water bodies such as lakes and water reservoirs. In the case of the reservoirs, the impact of toxic cyanobacterial blooms on the drinking water supply system depends on many factors, such as the extent and nature of the bloom, the specific toxins and climatic conditions [5]. It should be noted that as the proliferation of toxic cyanobacteria in freshwater increases in frequency, protection of the water supply becomes more difficult. Additionally, reservoirs can also have a recreational-sporting use and be a landscape resource, forming an important element of tourism in the area where it is located. The recent European CYANOCOST [6] project estimated that 20% of bathing or recreational water areas in Europe were prone to cyanobacterial blooms. Recreational and tourist activities in waters with the potential for toxic blooms pose a significant health risk [7].

Ultraviolet (UV) light irradiation is a frequently used treatment for the inactivation of microorganisms in water. It is applied in purification processes, in wastewater treatment for reuse, in industrial facilities and in the treatment of ballast water from ships [8, 9]. UV-C irradiation in particular, with an emission wavelength between 240 and 280 nm, directly impacts the DNA of organisms, preventing their reproduction [10].

The traditional sources of UV light are mercury vapor lamps. In the medium term, these mercury vapor lamps will be replaced by new generation light-emitting diodes (LEDs), with emission in the UV range. LEDs have several advantages over traditional lamps, such as avoiding the use of mercury in their manufacture, absence of warm-up period, longer lifetime and flexibility in reactor configuration, given their individual character [11].

An essential parameter for the evaluation of the effects of UV treatment on organisms is the applied UV dose, understood as the total radiant energy passing through an infinitesimal unit of area [12]. This is calculated as the product of the average intensity (I_m) and the exposure time. The calculation of both parameters depends on the type of UV reactor used for the treatment. The collimated beam reactor (CBR) is a device in which both I_m and exposure time can be calculated with precision; however, its use is restricted to laboratory studies, prior to the use of continuous flow reactors [13].

The other essential parameter in the evaluation of the treatment efficacy is the response by the organisms. Despite some organisms may be killed directly by the UV exposure, the major effect by the UV-C treatment is damaging the DNA strands, preventing the organisms from reproducing; in this sense, the organisms remain alive but unable to reproduce. Due to this non-lethal effect on the organisms, it is not possible to directly determine the effect of the treatment, so growth-based techniques are required for this purpose. One such technique was proposed by the Organization for Economic Cooperation and Development (OECD) in 1984. This organization proposed a method to measure the inhibition of microalgae growth by comparing areas under the growth curves for a given period of time. Another alternative technique is the Most Probable Number method (MPN). The MPN procedure is widely used to estimate microbial densities in many matrices, including food and water. It derives from the original work of McCrady (1915) and consists of adding, to a culture medium, a volume of different serial dilutions of a sample to several tubes with at least five replicates. After incubation, the number of tubes showing growth at each inoculum level is recorded. This method has been recently adapted for determining viable phytoplankton on UV-treated samples [14].

The main objective in this study is to evaluate the feasibility of a LED UV-C treatment for the inactivation of microalgae and cyanobacteria in natural water samples collected from two reservoirs from the province of Cádiz (Spain).

In order to cover the main objective of this work, we intend to:

- Evaluate the growth inhibition of phytoplanktonic organisms caused by UV-LED treatment, distinguishing between the fraction associated with cyanobacteria and the rest of the green algae.
- Evaluate the inactivation of phytoplanktonic organisms by UV-LED treatment by means of the MPN technique.
- Determine the UV dose necessary to inactivate or inhibit the growth of 90% of the organisms in real samples.

2 Material and Methods

2.1 Study Areas and Sampling

The study areas selected for this work are the Los Hurones reservoir and the Arcos reservoir, both located in the province of Cádiz (Spain). Both reservoirs suffer from eutrophication processes, making them susceptible to HABs. The main use of the Hurones reservoir is to supply drink water to the province of Cádiz, and use of the Arcos reservoir is recreational.

Sampling in each reservoir was carried out biweekly from June to August 2022. The samples in Los Hurones were taken from the area close to the dam (36° 39′ 41″ N; 5° 33′ 43″ W), whereas the samples in Arcos were taken from a recreational pier (36° 45′ 53″ N; 5° 47′ 04″ W). In both cases, samples were collected from the water surface.

2.2 UV Reactor and Irradiation Procedure

Samples taken from both reservoirs were subjected to UV irradiation using a CBR (Photolab LED275-0.01/300-0.03/365-1cb, Apria Systems S.L., Spain). The CBR was equipped with a LED device with power of 10.5 mW and an emission wavelength between 265 and 285 nm, with a maximum at 275 nm. The UV dose applied to each sample was calculated as the product of the mean intensity and the exposure time [15].

The mean intensity of the CBR reactor was determined according to the spreadsheet associated with the protocol proposed by Bolton (2015) [16]. This protocol is based on reactor morphometry, direct measurement of irradiance at the center of the sample surface and the UV transmittance (T_W) of the target water. Irradiance was measured with a radiometer (HD 2102.1, Delta OHM, Padova, Italy) that was equipped with a probe capable of measuring UV-C and UV-B spectra (LP471UVBC, Delta OHM, Padova, Italy). T_w at 275 nm of the target water was measured with a spectrophotometer (Jenway 7315; Staffordshire, UK) immediately before sample irradiation. The I_m values in the different experiments were between 0.027 and 0.053 mW cm^{-2}, depending on the T_w in each case. The required exposure times were determined as the quotient between the desired dose and the I_m value in each case, for doses values of 25, 50, 75 and 100 mJ cm^{-2}.

The treatment was applied to samples in Petri dishes of 55 mm diameter and capacity for 60 mL of target water. 50 mL of the sample and a magnetic stirrer with a volume of 0.25 mL were added to the Petri dish to ensure continuous stirring during irradiation. After treatment, the samples were placed in 50 mL flasks.

2.3 Determination of the Treatment Efficacy

The effectiveness of the treatment was assessed using two separate techniques. According to the OECD's suggested method, growth inhibition was first computed by comparing regions on the growth curves, and then the MPN was used to measure the concentration of viable organisms.

For determining growth inhibition according to OECD protocol, nutrients (f/2 at a ratio of 1 mL of solution in 1000 mL of culture) were added to the samples and incubated at 25 ± 2 °C, illuminated with two lamps (Phillips 18 W LED tube, 1600 lm cold light) with an active radiation of 130 μEinstein m^{-2} s^{-1} (QSL-2100

Radiometer, Biospherical Instruments Inc., San Diego, CA, USA) and a 24-h continuous light cycle. Samples were monitored for culture growth by means of fluorescence measurements (AquaPen AP 110/C, Photon Systems Instrument, Czech Republic). These measurements were performed at two different wavelengths, at 455 nm and at 630 nm. The blue excitation light (455 nm) is focused on measuring chlorophyll concentration as an indicator of the green algae concentration. The red–orange excitation light (630 nm) is intended for excitation of phycobilins and is suitable as an indicator of the concentration of cyanobacteria. Fluorescence was measured periodically throughout the incubation up to 14 days. Growth curves for every sample were obtained by representing the fluorescence values against the incubation time. The area under the curve was calculated according to the Eq. (1), where A is the area and f is the fluorescence measured at each time (t). Once the different areas were calculated, they were normalized with respect to the control sample, representing the logarithm of this ratio (Log (A_t/A_c)). In this way, the growth inhibition profiles were obtained with respect to the UV dose applied.

$$A = \frac{(f_0 + f_1) * (t_1)}{2} + \frac{(f_1 + f_2) * (t_2 - t_1)}{2} + \frac{(f_{n-1} + f_n) * (t_n - t_{n-1})}{2} \quad (1)$$

On the other hand, phytoplankton inactivation through MPN was calculated following the IMO PPR 7 protocol. Successive dilutions of each sample were made and 1 mL of each dilution was added to 5 mL of autoclaved tap water added with f/2 in test tubes. Five tubes per dilution step were prepared. The tubes were capped and incubated under the same conditions as described for the OECD method. Fluorescence (AquaPen AP 110/C, Photon Systems Instrument, Czech Republic) emitted from each test tube was measured just after irradiation and remeasured after 14 days of incubation according to PPR protocol 7. Additionally, fluorescence was remeasured after 21 and 28 days of incubation to check for false negatives at 14 days due to too short incubation time. Subsequently, the tubes observed for determining existence of algal growth (positive tube) or absence of growth (negative tube). The criterion established in the PPR 7 protocol for a tube to be considered positive is that the fluorescence value obtained is at least four times higher than the standard deviation of the replicates of that dilution at the beginning of the incubation. The abundance present in each sample was calculated using a standardized spreadsheet [17] (https://standards.iso.org/iso/7218/) basing on the number of positive and negative tubes obtained for every dilution step.

3 Results and Discussion

3.1 Growth Inhibition

Inactivation curves represent the values of Log (A_t/A_c) against the UV dose applied (Fig. 1). Curves were represented for each experiment (1, 2 and 3) in each reservoir, determined with fluorescence either at 455 nm (green algae) or 630 nm (cyanobacteria).

In general, the growth inhibition of both kinds of organisms was not noticeable in the range of UV doses tested. The values of Log (A_t/A_c) followed in some cases a linear relationship with the UV dose applied, corresponding to first order inactivation kinetics (2), in which A_t is the area under the growth curve in the UV treated sample, A_c is the area under the curve of the control sample, k is the inactivation kinetics constant and D is the applied UV dose.

$$A_t = A_c \cdot e^{-k\,D} \tag{2}$$

The growth inhibition of green algae was significant ($p < 0.05$) in 4 out of 6 experiments, whereas the inhibition of cyanobacteria was significant in only 2 out of 6 experiments. In the cases with significant relationship, the UV dose required

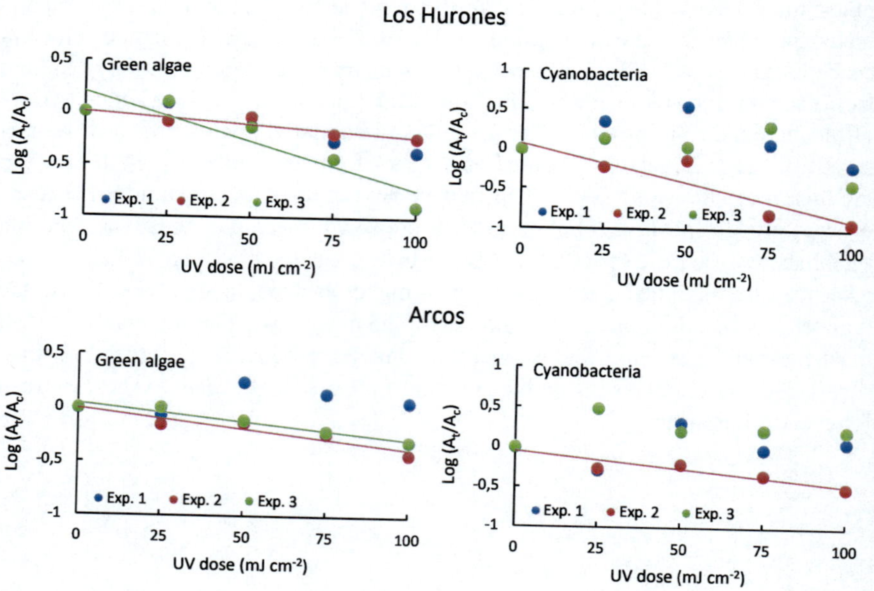

Fig. 1 Graphical representation of Log (A_t/A_c) for green algae and cyanobacteria after 14 days of incubation versus UV dose. Regression lines are included when significant ($p < 0.05$) relationship was determined

Table 1 Inactivation rates (k) and doses required to obtain 90% growth inhibition (D1) for the different types of organisms studied and both reservoirs. D1 was only calculated in the case of significant slopes at 95% confidence level (p-value < 0.05)

Reservoir	Experiment	Organism	k + SE (p-value) (cm^2 mJ^{-1})	R^2	D_1 (mJ cm^{-2})
Hurones	1	Green algae	0.0078 ± 0.0047 (0.193)	0.48	-
Hurones	2	Green algae	0.0208 ± 0.0025 (0.004)	0.96	109.1
Hurones	3	Green algae	0.0213 ± 0.0055 (0.030)	0.84	128.5
Hurones	1	Cyanobacteria	0.0061 ± 0.0092 (0.558)	0.13	–
Hurones	2	Cyanobacteria	0.0222 ± 0.0057 (0.030)	0.84	111.1
Hurones	3	Cyanobacteria	0.0064 ± 0.0084 (0.504)	0.16	–
Arcos	1	Green algae	-0.0032 ± 0.0036 (0.441)	0.21	–
Arcos	2	Green algae	0.0083 ± 0.0018 (0.018)	0.88	276.1
Arcos	3	Green algae	0.0074 ± 0.0011 (0.008)	0.93	325.3
Arcos	1	Cyanobacteria	-0.0039 ± 0.0068 (0.608)	0.10	–
Arcos	2	Cyanobacteria	0.0097 ± 0.0022 (0.023)	0.86	227.2
Arcos	3	Cyanobacteria	-0.0017 ± 0.0057 (0.783)	0.03	–

to achieve one log-reduction of the organism growth was calculated (D1), which represent a 90% of growth inhibition (Table 1). D1 values were always over the maximum UV dose that was applied in experiments of 100 mJ cm^{-2}; in this sense, greater UV doses are necessary of an effective treatment for growth inhibition of algae and cyanobacteria.

3.2 *Inactivation Determined with MPN*

MPN is a growth-based method focused to determine the concentration of viable (able to reproduce) organisms. The correct selection of the dilution steps to provide a suitable combination of positive and negative tubes is essential for obtain a numerical value of concentration as well as the 95% confidence interval. In the case of all positive tubes, more dilutions steps are required and only the lower limit of the 95% confidence interval is available. In the case of all negative tubes, only the upper limit of the 95% confidence interval is available. The analysis of natural samples, in which the initial cell concentration and the effect of UV irradiation are unknown, the selection of the proper set of dilutions is complex and requires certain expertise. The PPR7 protocol establishes an incubation period of 14 days; however, tubes were measured for positive or negative growth al at 21 and 28 days of incubation, with the purpose of checking for false negative with shorter incubations.

Figure 2 shows the concentration of viable organisms calculated through the MPN versus the UV dose applied to each sample corresponding to the Los Hurones reservoir, for incubation periods of 14, 21 and 28 days. The cases of all negative tubes are indicated with a downward arrow from the upper limit of the 95% confidence interval, indicating that the actual value is between zero and this value. The cases of all positive tubes are indicated with an upward arrow from the lower limit of the 95% confidence interval, indicating that the actual value is between this value and the infinity. In the experiment 1, all tubes in the dilution set for each sample were negative, except for the control and one sample measured the day 28. This implies that the dilution performed in this case were excessive for obtaining reliable results. In the rest of experiments, there are set of tubes that were all negative even with including a non-diluted set; in these cases, the actual concentration of viable organisms was below the limit of detection of the technique.

To quantify the inactivating by the treatment, the survival (S) was determined for every UV dose applied as the quotient between the concentrations of the viable organisms determined in the treated sampled and the concentration determined in the control ($S = MPN_{treated}/MPN_{control}$) (Table 2).

Incubations of 21 and 28 days provide, in general, greater values of viable organism concentration with respect to the incubations of 14 days (Fig. 2); this implies that determining the concentration of viable organisms requires longer incubations than the 14 days stated in the PPR 7 protocol. However, in relative terms of survival (Table 2), the level of inactivation increases with longer incubations in the most of cases. This is due to the increase of the viable organism concentration determined in the controls, that exceeds the increase observed in the treated samples.

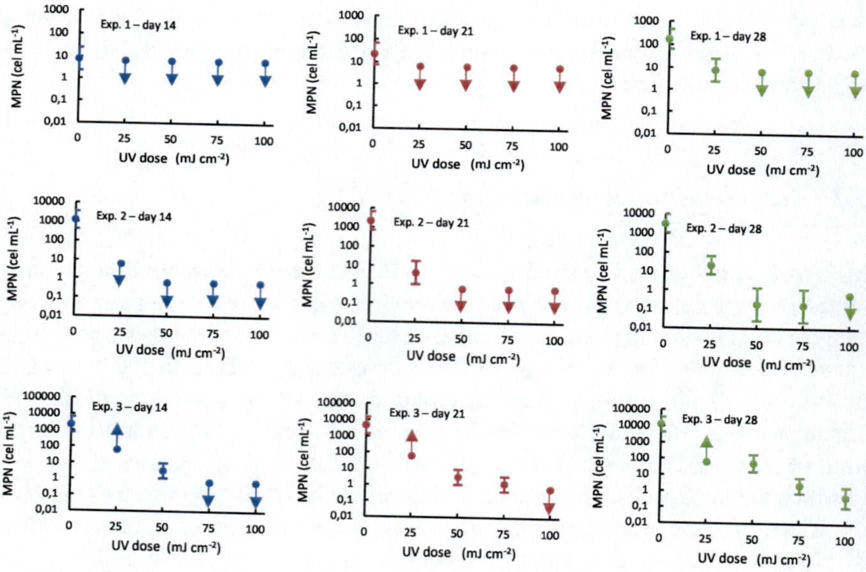

Fig. 2 Concentration of viable organisms calculated using MPN at Los Hurones. The up arrow indicates that the true value is between the point and infinity (all positive tubes), while the down arrow indicates that the true value is between the point and zero (all negative tubes)

Table 2 Logarithm of survival for the MPN at Los Hurones. In cases where a numerical value for the concentration of viable organisms could not be obtained (all tubes positive or negative), the lower or upper limit to the Log (S) value indicated is indicated. The values not indicated are due to the lack of numerical value of concentration in the control, preventing the calculation of survival in the treated samples

Experiment	UV Dose (mJ cm^{-2})	Log (S) day 14	Log (S) day 21	Log (S) day 28
1	25	≤ – 0.07	≤ – 0.54	– 1.34
1	50	≤ – 0.07	≤ – 0.54	≤ – 1.41
1	75	≤ – 0.07	≤ –0.54	≤ – 1.41
1	100	≤ – 0.07	≤ –0.54	≤ – 1.41
2	25	≤ – 2.29	– 2.71	– 2.16
2	50	≤ – 3.29	≤ – 3.54	– 4.22
2	75	≤ – 3.29	≤ – 3.54	– 4.26
2	100	≤ – 3.29	≤ – 3.54	≤ – 3.70
3	25	≥ – 1.55	≥ – 1.89	≥ – 2.30
3	50	– 2.84	– 3.17	– 2.38
3	75	≤ -3.54	– 3.58	– 3.75
3	100	≤ -3.54	≤ – 3.87	– 4.46

The effect by the UV treatment on the concentration of viable organisms is more evident in the case of using MPN as determination technique (Fig. 2) in comparison with using OECD growth inhibition technique (Fig. 1). More than 1 log-reduction (inactivation of the 90% of the viable organisms) was determined for the lowed UV dose applied of 25 mJ cm^{-2}, and more than 4 log-reductions (inactivation of the 99.99% of the viable organisms) were obtained by using UV doses of 100 mJ cm^{-2}. Similar results were obtained for the Arcos reservoir.

4 Conclusions

Two different methods were used to evaluate the inactivating efficacy of a LED UV-C treatment on natural populations of microalgae and cyanobacteria from two reservoirs.

The determination of the growth inhibition based on comparing the growth curves of treated and untreated samples reported absence of effect by the UV treatment in some cases, and slight inhibition in other cases, in the range of UV doses tested up to 100 mJ cm^{-2}. Absence of effect in cyanobacteria was more frequent with respect to green algae.

In the case of determining the treatment efficacy by means of the most probable number (MPN) method, the effect was noticeable even at low UV doses. UV doses of 25 mJ cm^{-2} rendered the inactivation of more than the 90% of the viable organisms, and inactivation greater than the 99.99% were reached by applying doses of 100 mJ cm^{-2}.

Results indicate that the treatment efficacy depends strongly on the method used for determining the response by the organisms to the UV irradiation.

Acknowledgements This work has been co-funded by the 2014–2020 ERDF Operational Programme and by the Department of Economy, Knowledge, Business and University of the Regional Government of Andalusia (Spain). Projects Ref.: FEDER-UCA18-108023 and FEDER-UCA18-105151. J. Moreno-Andrés acknowledges Grant IJC2020–042741-I funded by MCIN/AEI/ https://doi.org/10.13039/501100011033 and by the European Union NextGenerationEU/PRTR. Additionally the support received by the "Plan Propio - UCA 2022-2023" through the INICIA-INV grant: Initiation to the Research Master's Degree.

References

1. Sar, E.A., Ferrario, M.E., Reguera, B., 2002. Floraciones algales nocivas en el cono sur americano Bolton, J.R., 2000. Calculation of ultraviolet fluence rate distributions in an annular reactor: significance of refraction and reflection.
2. Sellner, K.G., Doucette, G.J., Kirkpatrick, G.J., 2003. Harmful algal blooms: Causes, impacts and detection. J Ind Microbiol Biotechnol. https://doi.org/https://doi.org/10.1007/s10295-003-0074-9.
3. Hallegraeff, G.M., 2003. Harmful algal blooms: a global overview, in: IOC-UNE-SCO (Ed.), Manual on Harmful Marine Microalgae. Monographs on Oceanographic Methodology. Paris, pp. 25–49.
4. Mantzouki, E., Campbell, J., …, Ibelings, B.W., 2018. A European Multi Lake Survey dataset of environmental variables, phytoplankton pigments and cyanotoxins. Sci Data 5. https://doi.org/10.1038/sdata.2018.226
5. Griffiths, D.J., Saker, M.L., 2003. The Palm Island mystery disease 20 years on: A review of research on the cyanotoxin cylindrospermopsin. Environ Toxicol. https://doi.org/https://doi.org/10.1002/tox.10103.
6. Bormans, M. (2012). European COST project CYANOCOST: Cyanobacterial blooms and toxins in water resources: Occurrence, impacts and management.
7. Jones, G.J., Olut, P.T., 1994. Release and degradation of microcystin following algicide treatment of a microcystis aeruginosa bloom in a recreational lake, as determined by hplc and protein phosphatase inhibition assay, Pergamon Wat. Res.
8. Hess-Erga, O.K., Moreno-Andrés, J., Enger, Ø., Vadstein, O., 2019. Microorganisms in ballast water: Disinfection, community dynamics, and implications for management. Science of the Total Environment. https://doi.org/https://doi.org/10.1016/j.scitotenv.2018.12.004.
9. Li, S., Tao, Y., Zhan, X.M., Dao, G.H., Hu, H.Y., 2020. UV-C irradiation for harmful algal blooms control: A literature review on effectiveness, mechanisms, influencing factors and facilities. Science of the Total Environment. https://doi.org/https://doi.org/10.1016/j.scitotenv.2020.137986.
10. Morató, J., Mir, J., Codony, F., 2003. Microbial response to disinfectant. Academic.
11. Chen, J., Loeb, S., Kim, J.H., 2017. LED revolution: Fundamentals and prospects for UV disinfection applications. Environ Sci (Camb). https://doi.org/https://doi.org/10.1039/c6ew00241b.
12. Bolton, J.R., 2000. Calculation of ultraviolet fluence rate distributions in an annular reactor: significance of refraction and reflection.

13. Sun, Z., Li, M., Li, W., Qiang, Z., 2022. A review of the fluence determination methods for UV reactors: Ensuring the reliability of UV disinfection. Chemosphere. https://doi.org/https://doi.org/10.1016/j.chemosphere.2021.131488.
14. IMO, 2019. PPR 7/INF.10. Revised guidance on methodologies that may be used for enumerating viable organisms. Description of the MPN serial dilution culture method + Motility method.
15. USEPA, D.M., 1986. Municipal Wastewater Disinfection. Cincinnati, Ohio.
16. Bolton, J.R., Beck, S.E., Linden, K.G., 2015a. Protocol for the determination of fluence (UV dose) using a low-pressure or low-pressure high-output UV lamp in bench-scale collimated beam ultraviolet experiments. IUVA News 17, 11–16.
17. Jarvis, B., Wilrich, C., Wilrich, P.T., 2010a. Reconsideration of the derivation of Most Probable Numbers, their standard deviations, confidence bounds and rarity values. J Appl Microbiol 109, 1660–1667. https://doi.org/https://doi.org/10.1111/j.1365-2672.2010.04792.x

The Challenge of Reducing the Sulfur Oxides Emission in the Maritime Transport: Risks Derived from the Discharge of the Scrubber Washwater into the Marine Environment

Leonardo Romero-Martínez, Lucía Ruiz-Salas, José Luis García-Morales, Asunción Acevedo-Merino, Javier Moreno-Andrés, and Enrique Nebot

Abstract The emission of sulfur oxides into the atmosphere is one of the major concerns derived from the maritime transport. To reduce their impact on the environment and the human health, the last update of the MARPOL Annex 6 established a limit of 0.5% of sulfur in in the fuel as of 2020. Nevertheless, ships can use fuel with higher sulfur content if they are equipped with an exhaust gas cleaning system, also known as scrubber. In the case of open-loop scrubbers, seawater is put in contact with the exhaust gas and then released into the ocean. This washwater absorbs the sulfur oxides efficiently, as well as other substances such as polycyclic aromatic hydrocarbon (PAHs), and increases int temperature by the contact with the hot exhaust gas. For this reason, the discharge of the washwater may concern to the aquatic environment and compromise the human activities. The objective in this study is to analyze the data recorded by the Continuous Emission Monitoring System of a ship during a series of commercial voyages, with the purpose of determining the potential threats derived from the discharge of scrubber washwater into the ocean. Data revealed that the washwater was discharged at pH below 4, a concentration of PAHs greater than 13 μg/L and temperature 7 °C greater than the recipient seawater. The discharge of washwater is especially concerning during the arrival and departure maneuvers, due to the releasing of acid, PAHs enriched and hot water into enclosed water bodies.

Keywords Sulfur oxides · Exhaust gas cleaning · Scrubber washwater

---

L. Romero-Martínez (✉) · L. Ruiz-Salas · J. L. García-Morales · A. Acevedo-Merino · J. Moreno-Andrés · E. Nebot
Department of Environmental Technologies, Faculty of Marine and Environmental Sciences, INMAR—Marine Research Institute, CEIMAR—International Campus of Excellence of the Sea, University of Cadiz, Cadiz, Spain
e-mail: leonardo.romero@uca.es

J. R. da. C. S. Galvão et al. (eds.), *Proceedings of the 3rd International Conference on Water Energy Food and Sustainability (ICoWEFS 2023)*, Springer Proceedings in Earth and Environmental Sciences, https://doi.org/10.1007/978-3-031-48532-9_65

1 Introduction

Maritime transport is nowadays the major means for moving goods around the world. It is an efficient method for moving freight, with values of CO_2 emission between 3.0 and 7.9 g t^{-1} km^{-1}, in comparison with trucks which emits 80 g t^{-1} km^{-1} and air freight, this one with 475 g t^{-1} km^{-1} [1].

Despite the efficiency related with the CO_2 emissions, maritime transport entails a series of environmental threats. For this reason, the International Maritime Organization (IMO) adopted the International Convention for the Prevention of Pollution from Ships, 1973 as modified by the Protocol of 1978 (MARPOL 73/78). MARPOL is divided in 6 Annexes, each one regulating a specific category of pollutants (Fig. 1). The most recent Annex 6 (2005) is focused on preventing the air pollution from ships.

Air pollution is one of the major concerns derived from the maritime transport. The different kinds of fuel used in maritime transport can be classified according to their content in sulfur: high sulfur fuel oil (HSFO) with maximum sulfur content of 3.5%, low sulfur fuel oil (LSFO) with maximum sulfur content of 1.0% and ultra-low sulfur fuel oil (ULSFO) with maximum sulfur content of 0.1%. In general, the price of the fuel increases as the content of sulfur decreases. The use of cheap fuel with high concentration of sulfur entails the emissions of high amount of sulfur oxides (SO_X) into the atmosphere. SO_X emission can be responsible of acid rain and human health threats on coastal areas [2]. In this context, the Annex 6 of the MARPOL establishes limits for the sulfur content in the fuel. The maximum sulfur concentration allowed has been decreased with the time (Fig. 2); the last update of the Annex 6 establishes a limit of 0.5% since 2020. The sulfur limits are even more restrictive in sulfur emission control areas (SECAs): the Baltic Sea, the North Sea, the most of US and Canadian coast and the US Caribbean currently. Compliance with

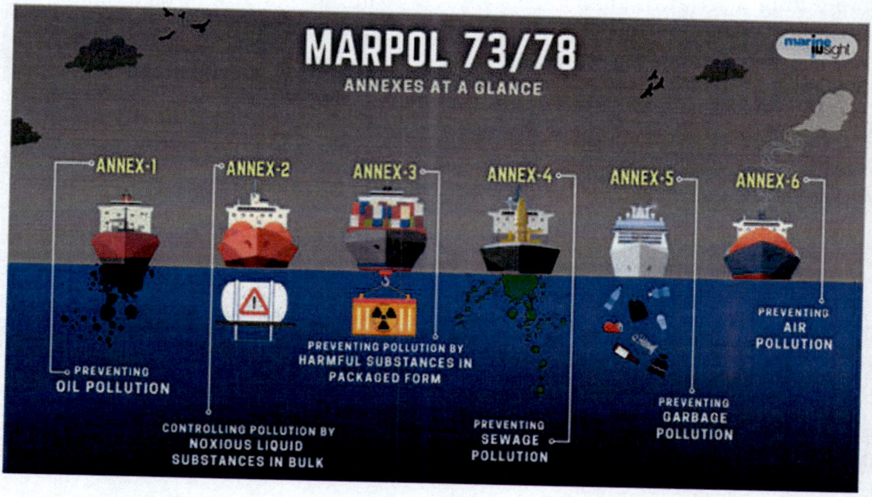

Fig. 1 Representation of the categories of pollutants regulated by the MARPOL 73/78

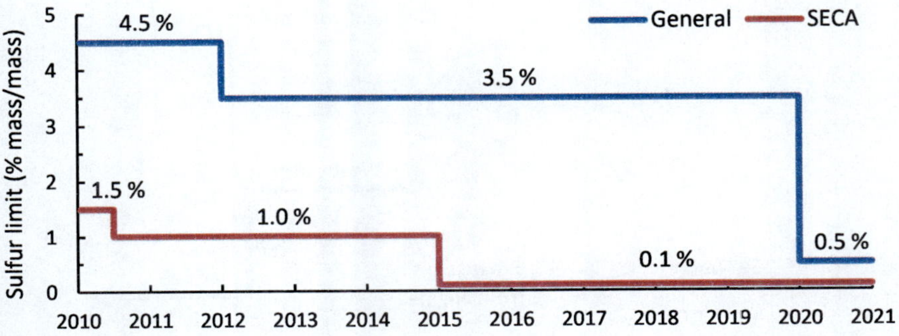

Fig. 2 Evolution of the sulfur limit in the fuel for maritime transport, in general and within the sulfur emission control areas (SECAs), according to the periodical updates of the MARPOL Annex 6

the Annex 6 regulations requires the adoption of strategies such as using ULSFO or liquefied natural gas (LNG); however, ships are allowed to using fuel with high content of sulfur if they are equipped with an exhaust gas cleaning system (EGCS), also known as scrubber.

Scrubbers are devices installed into the ships with the purpose of reducing the SO_X content in the exhaust gases. In general, their operation consists on contacting the exhaust gas with an alkaline substance which removes the sulfur from the gas. Scrubbers installed in ships uses commonly water as scrubbing material, named washwater. Seawater is commonly alkaline enough to efficiently remove SO_X from the exhaust gas; however, in the ship navigates on low alkaline water, such as rivers or the Baltic Sea, the water alkalinity requires to be enriched with chemicals, usually sodium hydroxide. According to the features of the washwater circuit, scrubbers can be classified into open-loop or closed-loop [3]. In open-loop scrubbers, the washwater is discharged back into the ocean after the contact with the exhaust gas (Fig. 3); optionally, the washwater can be subjected to a treatment for sludge removal. On the contrary, in closed-loop systems, the wash water is not discharged but recirculated and enriched in alkali such as sodium hydroxide to be used repeatedly.

The discharged washwater from open-loop scrubbers has low pH due to the SO_X absorbed, higher temperature due to the heat exchange with the exhaust gas, as well as pollutants from the exhaust gas such as polycyclic aromatic hydrocarbon (PAHs), heavy metals [4]. In this sense, the discharge of washwater may suppose an environmental risk [5]. In this context, the IMO establishes limits for some parameters of the discharged washwater in the Resolution MEPC 259(68) (Table 1) [6]. It is remarkable the absence of regulations on heavy metals, although they are present in the discharged washwater, especially vanadium, nickel and lead [7]. Temperature is neither regulated, although the values of inlet and discharge washwater temperature must be monitored and recorded.

The risks derived from the use of scrubbers for SO_X removal from the exhaust gas should be analyzed to determine the possible negative impact on the marine environment. The higher temperature, low pH, PAHs and other substances in the discharged

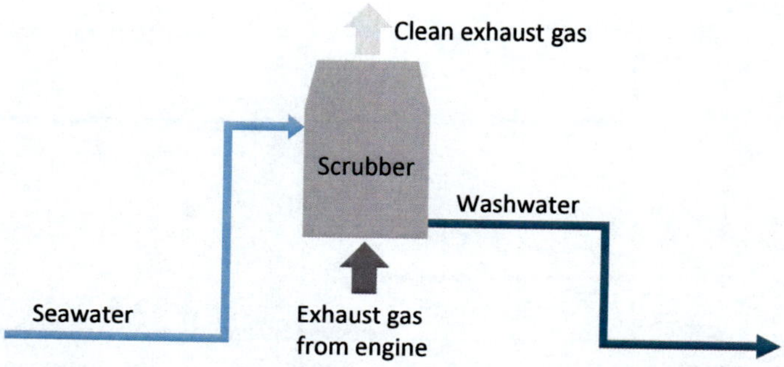

Fig. 3 Diagram of the washwater flow in an open-loop scrubber

Table 1 Compliance criteria for some parameter of the washwater discharge, according to the Resolution MEPC 259(68)

Parameters	Compliance criteria
pH	• No less than 6.5 measured at the ship's overboard discharge • Minimum pH 6.5 at 4 m from the overboard discharge point with the ship stationary
PAHs	• Not greater than 50 μg/L PAH_{phe} (phenanthrene equivalence) above the inlet water PAH concentration, normalized for a washwater flow rate of 45 t/MWh
Turbidity	• Not greater than 25 FNU (formazin nephlometric units) or 25 NTU (nephlometric turbidity units) or equivalent units, above the inlet water turbidity
Nitrates	• The washwater treatment system should prevent the discharge of nitrates beyond that associated with a 12% removal of NO_X from the exhaust, or beyond 60 mg/l normalized for washwater discharge rate of 45 tons/MWh whichever is greater
Heavy metals	• Not regulated

washwater may pose a hazard for the biological community in the recipient area, compromising the fishery, shellfishing and aquaculture activities, thus posing a risk for the food security.

The objective in this study is to analyze the composition of the scrubber washwater discharged by a merchant ship during its commercial voyage, and to assess the potential risks for the marine environment.

2 Material and Methods

This study focuses on a Ro-Ro ship that voyages between six Eastern Atlantic ports every week (Fig. 4). The ship is equipped with an open-loop scrubber system (Fig. 3). Data were obtained from the Continuous Emission Monitoring System (CEMS) as an Excel file. The file contained information about the geolocation of the ship,

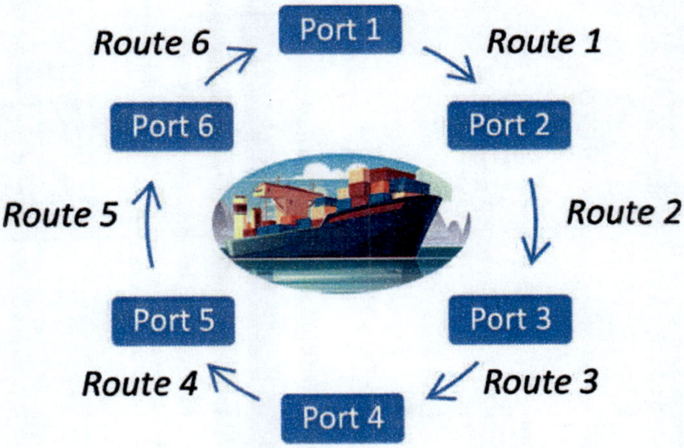

Fig. 4 Diagram of the commercial voyage followed by the studied ship

parameters on the operation of the engines and scrubbers, concentration of CO_2 (%) and SO_2 (ppm) and parameters relative to the scrubber washwater (pH, PAHs, inlet temperature, outlet temperature and turbidity). Data were collected every four minutes for two months, which corresponds to seven complete voyages.

Using the information on pH, PAHs, temperature and turbidity in the washwater, as well as geolocation and discharge flow rate, the potential risks for the marine environment derived from the washwater discharge were assessed.

3 Results and Discussion

Among the parameters registered by the CEMS, the most representative for assessing the environmental risk derived from the washwater discharged were pH, PAHs, temperature and turbidity.

The pH was stable at 3.7 ± 0.2 during the navigation between ports (Fig. 5a). In ports, where the engines operate at minimum power, the washwater was discharged at pH of 7.5 ± 0.4. During the arrival and departure, the pH was 5.3 ± 0.7 and 5.5 ± 0.4 respectively; both maneuvers represent the most concerning stages, since washwater with relatively low pH is released into harbor water, that is usually enclosed. Since pH in discharged water is below 6.5, it would be necessary determining the pH at 4 m from the overboard discharge point to assess compliance with the Resolution MEPC 259(68).

The PAHs concentration has a mean value of 13.7 ± 0.9 μg/L during the navigation. The concentration increases during the maneuvers of arrival and departure (Fig. 5b). During the stay in port, PAHs are discharged at a concentration of 8.4 ± 3.4 μg/L. According to the washwater flow rate values recorded by the CEMS, PAHs

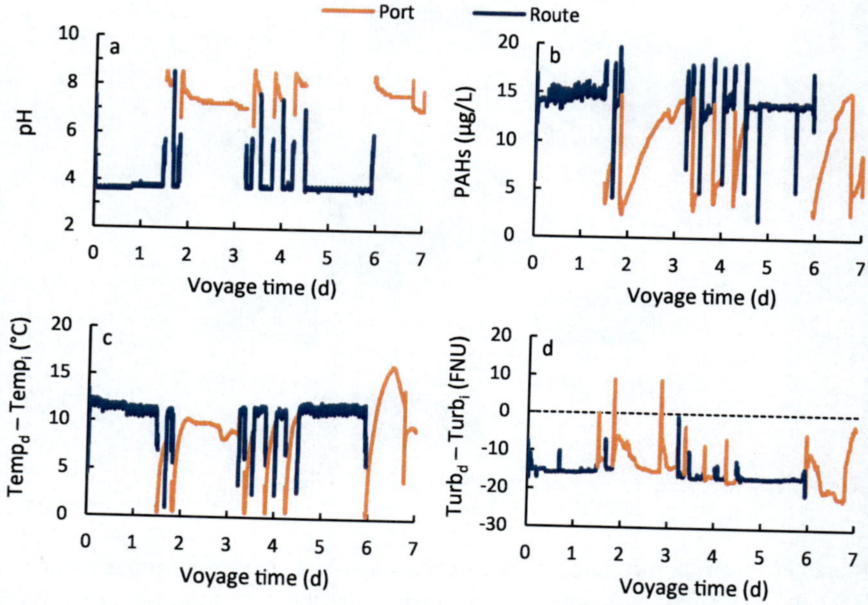

Fig. 5 Evolution of the pH (**a**), PAHs (**b**), difference between discharge ($Temp_d$) and inlet temperatures ($Temp_i$) (**c**) and difference between discharge ($Turb_d$) and inlet turbidities ($Turb_i$) (**d**) throughout one commercial voyage of the studied ship

in washwater are discharged at a rate of 8.52 mg/h during the navigation, 7.04 mg/h during the arrival and 5.79 mg/h during the departure; during the stay in port, PAHs are released at a rate of 0.07 mg/h due to the low flow rate of discharged washwater. The same as pH, the discharge of PAHs is concerning during the arrival and departure operations. Although the PAHs concentration is unknown in the inlet water, the concentration in the discharge does not exceed 50 μg/L, and thus the fulfills the Resolution MEPC 259(68) requirements.

The temperature in the released washwater is 11.6 ± 0.7 °C greater than the inlet water temperature during the navigation. The difference decreases during the arrival and departure (Fig. 5c) into values of 7.6 ± 1.6 and 6.0 ± 2.6 °C respectively. During the stay in ports, the differences are small at the beginning and increases as the ship is stopped; this fact is probably an artifact due to the low flow rate of discharge water in the analyzer when the ship is stopped. The environmental impact by the high temperature in the discharged water depends on the recipient water features, such as the renewal rate.

The turbidity of the discharged washwater is lower than the turbidity of the inlet water throughout all the navigation and almost all the time the ship is in ports (Fig. 5d). This paradoxical behavior implied that the Resolution MEPC 259(68) requirements are fulfilled, but there is likely an accumulation of materials within the washwater circuit.

The register of the pH, PAHs and temperature in the discharged washwater, together with the geolocation data recorded by the CEMS, allows determining the potential concerns derived from the discharged water features in relation with the characteristics of the recipient areas. The Fig. 6 shows the evolution of some features of the discharged washwater during the departure operation. The point 1 is the dock where the ship stops; the points 2 and 3 are within the harbor basin and thus the water renewal is limited. The rest of points are out of the harbor basin but close to the shore. Data indicate that water with pH below 5, PAHs concentration greater than 14 µg/L and temperature more than 7 °C greater is discharged into an enclosed water mass. The potential threat derived from the scrubber washwater discharge depends on the water renewal rate and the uses and activities performed in the areas neighboring the harbor.

Although the CEMS records an enormous amount of information, there are concerning substances that are not monitored, such as nitrates, sulfates and heavy metals. Determining the concentration of these substances in the discharged water, as well as a life cycle analysis (LCA), are recommended to determine whether the installation of a scrubber implies economic and environmental advantages with respect to using fuel with low sulfur content.

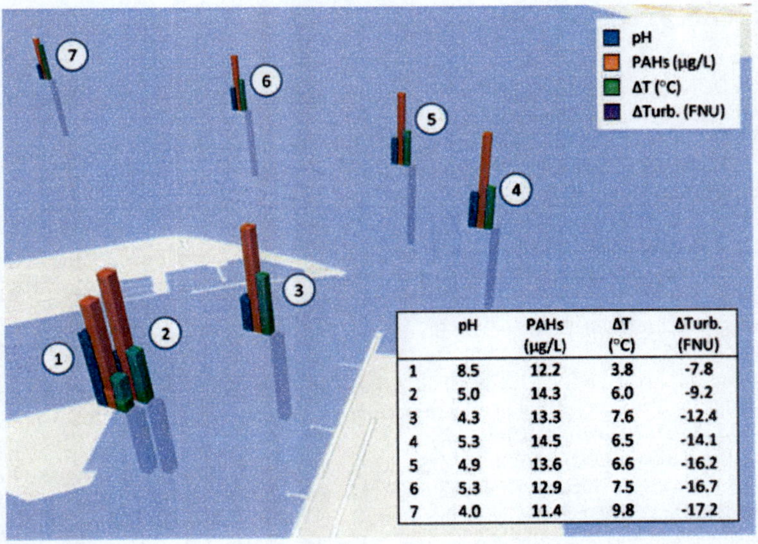

Fig. 6 Evolution of the discharged washwater features (pH, PAHs, difference between discharge and inlet temperature and difference between discharge and inlet turbidity) during the ship departure from a port

4 Conclusions

The scrubbers are devices installed into the ships to eliminate sulfur oxides from the exhaust gas, avoiding their releasing to the atmosphere, where might cause impacts on the environment and human health. In particular, open-loop scrubbers use seawater as scrubbing material; after the scrubbing process, this washwater is released into the ocean. The removal of sulfur oxides from the exhaust gases with a scrubber implies that they, as well as other harmful substances, are transferred into the aquatic environment instead to the atmosphere. The data analyzed in this study revealed that the discharged washwater has low pH, PAHs and a temperature greater than the recipient seawater. Washwater discharge may be particularly concerning during the arrival and departure maneuvers, in which the washwater is discharged into enclosed water bodies, thus may concern to the aquatic environment and the human activities.

Acknowledgements Project PID2021-123155OB-I00 funded by MCIN/AEI/https://doi.org/10.13039/501100011033 and by "ERDF A way of making Europe". Plan Propio—UCA 2022-2023 (MV2023-256).

References

1. ICS: Delivering CO_2 Emission Reductions: International Shipping is Part of the Solution (COP21). Int. Chamb. Shipp. London. (2015)
2. Sofiev, M., Winebrake, J.J., Johansson, L., Carr, E.W., Prank, M., Soares, J., Vira, J., Kouznetsov, R., Jalkanen, J.P., Corbett, J.J.: Cleaner fuels for ships provide public health benefits with climate tradeoffs. Nat. Commun. 9, 406 (2018). https://doi.org/https://doi.org/10.1038/s41467-017-02774-9
3. Tran, T.A.: Research of the Scrubber Systems to Clean Marine Diesel Engine Exhaust Gases on Ships. J. Mar. Sci. Res. Dev. 7, 243 (2017). https://doi.org/https://doi.org/10.4172/2155-9910.1000243
4. Hufnagl, M., Liebezeit, G., Behrends, B.: Effects of Sea Water Scrubbing. Final report, BP Mar. 145 pp. (2005)
5. Teuchies, J., Cox, T.J.S., Van Itterbeeck, K., Meysman, F.J.R., Blust, R.: The impact of scrubber discharge on the water quality in estuaries and ports. Environ. Sci. Eur. 32, (2020). https://doi.org/10.1186/s12302-020-00380-z
6. IMO: Resolution MEPC 259(68). 2015 Guidelines for exhaust gas cleaning systems. (2015)
7. Endres, S., Maes, F., Hopkins, F., Houghton, K., Mårtensson, E.M., Oeffner, J., Quack, B., Singh, P., Turner, D.: A new perspective at the ship-air-sea-interface: The environmental impacts of exhaust gas scrubber discharge. Front. Mar. Sci. 5, 139 (2018). https://doi.org/https://doi.org/10.3389/fmars.2018.00139

Drinking Water: Strategies for Sustainable Management and Water for Everyone

Ricardo Gomes, Cidália Pereira, João Galvão, and Vânia Ribeiro

Abstract Drinking water for human consumption is one of the most important natural resources that humanity must preserve. In different countries water management is carried out in different ways. While in countries with more scarce water resources the concern is water abstraction in others, where water is abundant, the concerns go further, from microbiological criteria to chemical parameters with increasingly reduced parametric values based on epidemiological studies. In low-income countries, the demand for drinking water, and its distribution, has revealed serious failures in water supply, with dire consequences in terms of public health for some communities. Now, these regions need a strategic plan for water management, from awareness raising among the poor on water saving to policy measures for improving aquifer water abstraction and investments in the water storage and distribution system. New technologies to obtain water suitable for human consumption are emerging, replacing conventional practices that can endanger the health of consumers, with solar thermal distillation standing out as a sustainable technique that can reach everyone in the most remote locations. Finally, building water networks should also be checked and maintained by consumers so that water from the public network, or other water storage systems, is fit for human consumption.

Keywords Water quality · Drinking water · Sustainability · Public health

---

R. Gomes · J. Galvão
R&D Unit, Institute for Systems Engineering and Computers—INESC, Coimbra, Portugal

C. Pereira · V. Ribeiro (✉)
CiTechCare, Centre for Innovative Care and Health Technology, School of Health Sciences, Polytechnic Institute of Leiria, 2411-901 Leiria, Portugal
e-mail: vania.ribeiro@ipleiria.pt

V. Ribeiro
GeoBioTec, GeoBioSciences, GeoTechnologies & GeoEngineering, Earth Sciences Department, Universidade NOVA de Lisboa, 2829-516 Caparica, Portugal

LSRE—Laboratory of Separation and Reaction Engineering and LCM—Laboratory of Catalysis and Materials, School of Management and Technology, Polytechnic Institute of Leiria, Morro do Lena-Alto do Vieiro, 2411-901 Leiria, Portugal

J. R. da. C. S. Galvão et al. (eds.), *Proceedings of the 3rd International Conference on Water Energy Food and Sustainability (ICoWEFS 2023)*, Springer Proceedings in Earth and Environmental Sciences, https://doi.org/10.1007/978-3-031-48532-9_66

1 Introduction

"Ensuring the availability and sustainable management of drinking water and sanitation for all" is presented as an objective of the SDGs of the United Nations (UN) [1], whose achievement depends on several factors, one of the most relevant being the sustainable management of water resources) [2]. The unrestrained use of water resources (mainly due to human pressure caused by urbanization, industry, and intensive agriculture) aggravated by climate change, pollution and biodiversity loss have contributed to water scarcity in some regions worldwide) [3]. Due to the significant population growth and economic expansion taking place on the African continent, which is expected to continue over the next few decades, serious problems are expected such as the shortage of potable water and the energy lack coming from the hydroelectric power stations) [2].

In addition to the issue of water access (rainwater, surface water, groundwater, or seawater), its quality from a chemical and microbiological point of view must also be considered to ensure the safety of the final consumer.

A recent study, covering more than 85 scientific publications, demonstrated that the impact of water pollution on human health is significant: more than 50 diseases caused by water pollution, the most common being diarrhoea, which is mainly transmitted by enteroviruses in the aquatic environment. Contaminated drinking water is responsible for 80% of illnesses and 50% of deaths in children worldwide [4].

In a time where the scarcity of drinking water affects several regions of the world, and it is not accessible for everyone, it becomes important to investigate strategies that allow sustainable water management and an efficient and effective use of this resource by the end user. Whether for human consumption (drinking, cleaning fruits and vegetables, cooking), cleaning, farming, industry, among the most varied needs of the populations of different countries, access to water will have to be done in different ways, considering the characteristics of each region.

2 Water Source and Water Quality

Water is a natural resource that exists in nature and can be found in liquid, solid or gaseous state. Under the effect of solar and wind action, water circulates naturally in nature and is in constant renewal/transformation [5]. Rainwater recharges surface and groundwater bodies depending on soil properties, climatic conditions, and subsurface interactions between rivers and aquifers. Its storage can be carried out in natural and artificial reservoirs. Access to drinking water in developing countries need an effective urban supply, which leads many households to invest in private supply or using a rainwater harvesting or boreholes [6]. In regions with a low drinking water, the use of reverse osmosis desalination is a solution for obtaining high quality water, attention must be paid to water chlorination to reduce the biofilms of mesophilic microorganisms that form may form due to temperature tolerance and poor nutrient in these waters [7]. In addition to surface and underground water sources, the following sources of alternative water stand out: desalination water (Cape Verde, Spain, Israel, United States, Chile, Australia), fog and dew water (Cape Verde, Australia, Ethiopia), reuse of domestic wastewater (Israel, Namibia, USA, Australia and Singapore) [8] and also the use of rainwater in buildings (Mexico) [9]. However, the important thing will always be to guarantee the potability of the water for the purpose for which it is intended.

3 Solutions for Water Abstraction, Storage, and Distribution

In regions where the supply water is not ensuring for 24 h and/or has failures, usually family's resort to storing water in cisterns or tanks on the roofs. This common practice in Guadalajara, Mexico, as well as in other Latin American countries, seems to be a solution whose cleaning and maintenance of tanks/cisterns should be improved as well as the water disinfection with chlorine to prevent a microbial load above acceptable limits [10]. In regions where access to drinking water is often lacking for populations, there are sometimes untapped resources for the production of renewable energies with potential for the production of drinking water from contaminated or salty resources, mainly in Africa where almost half of the total population lives who do not have access to drinking water [2].

In addition to the "physical scarcity of water", the lack of infrastructure in many African regions also limits access to drinking water, who populations suffer from "economic scarcity of water", which also ends up conditioning agriculture, production stable and sufficient food supply and therefore food security. To get around the problem of water scarcity, it is considered that the potential use of renewable energy sources (solar and wind) for water abstration and pumping for network distribution, as well as the use of sustainable desalination technologies (driven by renewable energies) in coastal zones with "physical water scarcity", could have enormous benefits [2]. Desalination require a lot of energy [11] that can be boosted by renewable energies, such as: (i) production of photovoltaic and wind energy to supply electricity to desalination systems by reverse osmosis or electrodialysis [12], and (ii) production thermal solar energy and the recovery of residual heat to feed stills driven by evaporation [13]. Recently new approaches that are more sustainable from the point of view of energy, materials, maintenance, and investment have been suggested, such as solar thermal distillation which is an excellent solution for a small-scale installation in places without access to drinking water [14, 15]. Drinking water distribution network can present complex environments that are difficult to predict due to water source, dynamic nutrient availability, pipe material, hydrodynamic conditions, water temperature and presence/absence of disinfection residues [16], which can compromise wholesomeness if the network conditions are not efficiently managed and maintained. Research into new coatings and their implementation may help to improve water quality. The use of cement with three-dimensional reduced graphene oxide (3D-rGO) has demonstrated a potential increase in the safety and durability of iron pipes in cement mortar coatings in drinking water distribution networks. This additive has advantages in inhibiting cement leaching, reducing disinfection by-products and reducing microbial load [17], which could be an asset to be considered in new hydraulic works for drinking water supply.

4 Drinking Water and Public Health

The water quality parameters to be evaluated to guarantee the drinking water quality, are constantly changing, and are established according to the technological development and studies on exposure to certain compounds present in water. With regard to European regulations, the Drinking Water Directive (Directive (EU), 2020/2184), which entered into force on January 13, 2023, which replaces the previous Directive (Directive (EU), 98/83) has as its main objective "to protect human health from the adverse effects of contamination of water intended for human consumption, ensuring its wholesomeness and cleanliness, and to improve access to water intended for human consumption" through the adoption of specific national standards in all Member States with the aim of providing high quality drinking water throughout the European Union [18]. This Directive highlights the introduction of new parameters and respective parametric values: bisphenol A (2.5 μg/l), chlorite (0.70 μg/l), chlorate (0.70 μg/l), haloacetic acids (HAAs) (0.60 μg/l), microcystine LR (1.0 μg/l), PFAS (0.50 μg/l) and uranium (0.30 μg/l) [19], whose studies demonstrated their toxicity to humans and the consequent need for regulation. Chromium and lead values have been updated, becoming more restrictive with a 50% reduction. Parametric values of chromium and lead were 50 μg/l and 10 μg/l according to WHO Guidelines and Directive (EU), 98/83 becoming 25 μg/l and 5 μg/l respectively in accordance with EU Directive 2020/2184 [18]. In addition to the chemical parameters, the adoption of a risk-based approach carried out through Water Safety Plans (article 7), the identification of possible emerging pollutants in the supply sources (catchment areas) (article 8), the evaluation of risks related to distribution, including in domestic installations between the water meter and the tap (Articles 9 and 10), and the request for effective and transparent communication to citizens about the quality of the water supplied (Article 14), which is a very important aspect to increase the final consumer's confidence in public network water [18, 19]. A Danish cohort study suggests an increased risk of an individual contracting colorectal cancer due to prolonged exposure to drinking water with nitrate concentrations above 3.87 mg/L [20], when the reference standard according to with the WHO is 50 mg/L [21]. Trihalomethanes (THM) are the class of disinfection by-products (DBP) that react with organic compounds present in water, most abundantly formed in water treated with chlorine (the most widely used water disinfectant) [22]. THM dichloroacetic acid and trichloroacetic acid have been classified by the IARC as probable human carcinogens (2B) [23]. Exposure to DBP (specifically trihalomethanes) has been associated with an increased risk of bladder cancer, as demonstrated by epidemiological studies. DBP levels in US public drinking water have declined significantly over the years, however a recent study linked 10% of bladder cancer cases with potential attribution to exposure to DBPs found in drinking water systems [24]. In developed countries, exposure to DBPs through the use of water from the municipal public network for domestic use (drinking, cooking, body hygiene, dishes and clothes or leisure swimming pools) which involves ingestion, inhalation, or dermal absorption is higher when compared to poor countries. The maximum level established for total THM varies according to the needs of countries, 250 μg/L in Australia, 100 μg/L in the European Union (EU) and 80 μg/L in the USA from underground water sources and advanced treatments without the application of chlorine are at the origin of the low levels of THM in some countries [22]. An Italian study demonstrated that floods could induce severe microbiological contamination with the occurrence of enteric viruses (hepatitis A virus, adenovirus, rotavirus, norovirus and enterovirus) of drinking water from wells at considerable distances (5–8 km) from the region of flooding and also in groundwater. The results highlight the need for additional water disinfection treatments before distribution to the public network. In addition, the study states that monthly controls on the occurrence of enteric viruses in well water should be imposed by a new water framework directive in semi-arid regions due to the vulnerability of karst carbonate aquifers to prolonged injections of water from flooding and contamination by enteric viruses [25]. If in more developed countries with regard to microbiological aspects, some parameters such as Pseudomonas are no longer covered by EU Directive 2020/2184, in other less developed countries it may be important to include them to verify the effectiveness of water treatment or sanitation of water reservoirs or dispensers, as well as potential environmental contamination [18]. Despite the disinfection levels of the public water network seem to be under control, it should be noted that the study and control of potential biofilms inside the pipes of the distribution network (which have been associated with a higher risk of opportunistic pathogens) should be considered for guaranteeing the potability of water [7].

5 Conclusions and Future Perspectives

New urban areas will have to be designed to become more efficient with regard to the management of drinking water. The study of water distribution networks, identification of alternative water sources and the use of new methodologies using renewable energies to obtain water suitable for human demand (such as solar thermal distillation) should be considered in order to make the systems more resilient and sustainable.

The existence of guidelines and legislation can help to ensure the safety of drinking water, particularly where binding legislation exists. The need for disinfection of drinking water to provide microbiologically safe water is unquestionable to prevent diarrheal disease and death, but there are measures to minimize the formation of BPD that must be checked in order to prevent potential cases of cancer.

Additionally, in each country there should be a detailed study of the impact of the various measures to be adopted that could have an impact on the health of consumers in the short and/or long term.

References

1. United Nations. 17 Goals to Transform Our World (2015) [Available from: https://www.un.org/sustainabledevelopment/]
2. De Angelis P, Tuninetti M, Bergamasco L, Calianno L, Asinari P, Laio F, et al. Data-driven appraisal of renewable energy potentials for sustainable freshwater production in Africa. Renewable and Sustainable Energy Reviews.**149**,111414 (2021)
3. Eslamain S, Eslamain FA. Handbook of Drought and Water Scarcity: Environmental Impacts and Analysis of Drought and Water Scarcity. 1st ed. United States: Boca Raton (2017)
4. Lin L, Yang H, Xu X. Effects of Water Pollution on Human Health and Disease Heterogeneity: A Review. Frontiers in Environmental Science. **10** (2022)
5. Oki T, Kanae S. Global Hydrological Cycles and World Water Resources. Science. **313**(5790):1068-72 (2006)
6. Raphael A, Ngaga Y, Lalika M. Watershed degradation and water provision in Morogoro Municipality, Tanzania. Ecohydrology & Hydrobiology (2022)
7. Farhat N, Kim L, Mineta K, Alarawi M, Gojobori T, Saikaly P, et al. Seawater desalination based drinking water: Microbial characterization during distribution with and without residual chlorine. Water Research. **210**,117975 (2022)
8. Munné A, Solà C, Ejarque E, Sanchís J, Serra P, Corbella I, et al. Indirect potable water reuse to face drought events in Barcelona city. Setting a monitoring procedure to protect aquatic ecosystems and to ensure a safe drinking water supply. Science of The Total Environment. **866**,161339 (2023)
9. Martínez-Castrejón M, Flores-Munguía EJ, Talavera-Mendoza O, Rodríguez-Herrera AL, Solorza-Feria O, Alcaraz-Morales O, et al. Water Efficiency Households Retrofit Proposal Based on Rainwater Quality in Acapulco, Mexico. Water. **14**(18), 2927 (2022)
10. Rubino F, Corona Y, Pérez JGJ, Smith C. Bacterial Contamination of Drinking Water in Guadalajara, Mexico. International journal of environmental research and public health. **16**(1) (2018)
11. Eke J, Yusuf A, Giwa A, Sodiq A. The global status of desalination: An assessment of current desalination technologies, plants and capacity. Desalination. **495**, 114633 (2020)
12. Pugsley A, Zacharopoulos A, Mondol JD, Smyth M. Global applicability of solar desalination. Renewable Energy. **88**, 200-19 (2016)
13. Reif JH, Alhalabi W. Solar-thermal powered desalination: Its significant challenges and potential. Renewable and Sustainable Energy Reviews. **48**, 152-65 (2015)
14. Chiavazzo E, Morciano M, Viglino F, Fasano M, Asinari P. Passive solar high-yield seawater desalination by modular and low-cost distillation. Nature Sustainability. **1**(12),763-72 (2018)
15. Dao V-D, Vu NH, Yun S. Recent advances and challenges for solar-driven water evaporation system toward applications. Nano Energy. **68**, 104324 (2020)
16. Prest EI, Hammes F, Van Loosdrecht MC, Vrouwenvelder JS. Biological stability of drinking water: controlling factors, methods, and challenges. Frontiers in microbiology.**7**, 45 (2016)
17. Liu Q, Zhuang Y, Shi B. Three-dimensional reduced graphene reinforced cement with enhanced safety and durability for drinking water distribution applications: Long-term experimental and theoretical study. Water Research. **230**, 119572 (2023)

18. Dettori M, Arghittu A, Deiana G, Castiglia P, Azara A. The revised European Directive 2020/2184 on the quality of water intended for human consumption. A step forward in risk assessment, consumer safety and informative communication. Environmental Research. **209**, 112773 (2022)
19. Directive (EU) 2020/2184 of the European Parliament and of the council, of 16 December 2020, on the quality of water intended for human consumption. Official Journal of the European Union (2020)
20. Schullehner J, Hansen B, Thygesen M, Pedersen CB, Sigsgaard T. Nitrate in drinking water and colorectal cancer risk: A nationwide population-based cohort study. International journal of cancer. **143**(1),73-9 (2018)
21. Guidelines for drinking-water quality, 4th edition, incorporating the 1st addendum, (2017) [Available from: https://www.who.int/publications/i/item/9789240045064]
22. Villanueva CM, Evlampidou I, Ibrahim F, Donat-Vargas C, Valentin A, Tugulea A-M, et al. Global assessment of chemical quality of drinking water: The case of trihalomethanes. Water Research. **230**, 119568 (2023)
23. IARC. Monographs on the Identification of Carcinogenic Hazards to Humans (2022) [Available from: https://monographs.iarc.who.int/list-of-classifications].
24. Weisman RJ, Heinrich A, Letkiewicz F, Messner M, Studer K, Wang L, et al. Estimating National Exposures and Potential Bladder Cancer Cases Associated with Chlorination DBPs in U.S. Drinking Water. Environmental health perspectives. **130**(8), 87002 (2022)
25. Masciopinto C, De Giglio O, Scrascia M, Fortunato F, La Rosa G, Suffredini E, et al. Human health risk assessment for the occurrence of enteric viruses in drinking water from wells: Role of flood runoff injections. Science of The Total Environment. **666**, 559-71 (2019)

Assessment of First Flush Roof Runoff Water Quality: A Case Study in Southern Brazil

Jéssica Kuntz Maykot and Enedir Ghisi

Abstract Rainwater harvesting systems may supply users with contaminated water if certain precautions are not taken. This study aims to quantify the levels of total coliforms and *Escherichia. coli* in rainwater at the beginning of runoff from a fibre cement roof located in Florianópolis, southern Brazil. Samples of roof runoff were collected, and physical–chemical parameters were quantified in addition to the microbiological ones. Finally, the existence of a possible correlation between these parameters was assessed. The results of laboratory tests were statistically described and submitted to correlation analysis. The analyses showed the existence of total coliforms in 100% of the samples and that *E. coli* can highly contaminate the roof runoff once some samples reach the test's detection limit (2419.6 NMP/100 mL). Furthermore, correlations between *E. coli* and the parameters pH, turbidity and apparent colour were observed. The study shows the importance of using first flush devices or techniques for rainwater treatment in rainwater harvesting systems to reduce users' health risks.

Keywords Rainwater quality · Rainwater harvesting systems · *Escherichia coli* · First flush

1 Introduction

One of the strategies that may be adopted to promote rational water usage is rainwater harvesting. The rainwater harvesting system, however, can be a source of water contamination if no proper care is taken, such as frequent cleaning of the catchment area and installation of screens or filters which prevent the passage of larger pollutants [1–3]. The use of devices to divert the initial portion of rainwater discharge (first

J. K. Maykot (✉) · E. Ghisi
Federal University of Santa Catarina, Florianópolis, Brazil
e-mail: jessica.maykot@posgrad.ufsc.br

E. Ghisi
e-mail: enedir.ghisi@ufsc.br

J. R. da. C. S. Galvão et al. (eds.), *Proceedings of the 3rd International Conference on Water Energy Food and Sustainability (ICoWEFS 2023)*, Springer Proceedings in Earth and Environmental Sciences, https://doi.org/10.1007/978-3-031-48532-9_67

flush), which corresponds to the volume of rain with the highest concentration of pollutants, also helps to improve water quality [4–6].

Even for non-potable purposes, harvested rainwater must have adequate quality for use. One of the water quality indicators is the concentration of thermotolerant coliforms, whose main representative species is *Escherichia coli.* The presence of colonies of these micro-organisms in water indicates the existence of human or animal faeces, which may contain pathogens, representing a risk to the user's health. Studies that investigated the coliforms concentration in the initial roof runoff show that there is high contamination in the first flush water, requiring some treatment prior to consumption [7–11].

In general, studies about the first flush quality show that there is an apparent roof-cleaning effect when comparing the microbial quality of the first flush sample with the microbial quality of the samples collected after the first flush diversion [12–14]. Investigating the quality of the first roof runoff discharge and the dependence on such quality during the roof runoff is important to enabling rainwater usage and user safety. For this purpose, this study aims to assess total coliforms and *E. coli* concentrations during the first flush of a fibre cement roof in Florianópolis, southern Brazil. In addition, correlation analyses between such indicators and the physical–chemical parameters (pH, conductivity, turbidity, total suspended solids and apparent colour) of water were performed.

2 Materials and Methods

In order to assess the first flush quality, runoff samples were collected from a fibre cement roof at the Federal University of Santa Catarina in Florianópolis, southern Brazil. The samples were submitted to laboratory tests, and the results were statistically described.

2.1 Study Location

Florianópolis is located at latitude 27.6 S and longitude 48.5 W [15]. According to the Köppen-Geiger classification, Florianópolis has a humid subtropical climate (Cfa), i.e. a climate with hot summers and cold winters, whose average temperature for the coldest month varies between 15 and 18 °C; without months of drought [16].

In order to perform the study, an unpainted fibre cement roof (a material widely used in roofs in Brazil) was sought, located close to trees and roads with heavy vehicle traffic. The collection area of roof runoff was limited and had 5 m^2, with a slope of 13.5°, facing northeast. The roof has asbestos in its composition, is 35 years old and has never been replaced.

2.2 Experimental Collector and Sample Collection

A PVC gutter with a grid attached (to prevent the entry of leaves, branches or debris) was installed to collect runoff from part of the roof. The first flush was conducted through pipes and connections to an experimental collector, consisting of five glass bottles arranged sequentially and coupled by connectors and a pipe inclined at 5° [17]. Such an inclination prevented the water transportation to posterior bottles before occupying the previous ones. In each bottle, an expanded polystyrene sphere blocked the water entrance when the bottle was full. In order to guarantee the bottles' sealing with the connections, fragments of butyl rubber were used around them. Each bottle collected approximately 1 mm of roof runoff. The accumulation of runoff between the surface of the bottle opening and the pipe connection (volume corresponding to 0.5% of the bottle's storage capacity) was not considered.

Before collecting the first flush samples, some procedures were complied with, such as cleaning, sterilizing, identifying and installing the bottles in the pipe connectors. The first flush collection bottles were rinsed with potable water and sterilized with boiling water. Subsequently, the spheres, before being inserted into the bottles, and the seal were rinsed (under potable water). The bottles were attached to the pipe connectors within 24 h before the beginning of the rainfall event.

Two types of containers were used to collect the first flush samples: polypropylene ones, previously cleaned with distilled water, used for sampling runoff submitted to the physical–chemical tests, and borosilicate glass ones, autoclaved for one hour at a pressure of 1.2 kgf/cm^2 and a temperature greater than 120 °C, used to collect samples for coliform tests.

A glass container, sterilized with boiling water, was also prepared to collect rainfall directly from the sky. The collection was performed manually until approximately one litre of rainfall sample was obtained. Samples were collected right after the end of the precipitation event. First, a small portion of the roof runoff collected from each bottle was discarded to homogenize the first flush, shaking them. After homogenization, a first flush portion was transferred to the autoclaved container, and another portion was transferred to the polypropylene one. This procedure of discarding the first flush portion and transferring it to the container samples was done successively until the containers were filled. Sample collection was performed for all five bottles (1, 2, 3, 4 and 5 mm). Container samples were closed and refrigerated until laboratory analyses could be conducted [18]. All analyses were performed within 24 h after collection. Figure 1 shows the experimental collector used in this study.

2.3 Laboratory Tests

Water quality analyses were performed at the Integrated Environment Laboratory (LIMA) at Federal University of Santa Catarina. The parameters pH, apparent colour, turbidity, total suspended solids, electrical conductivity, total coliforms and *E. coli*

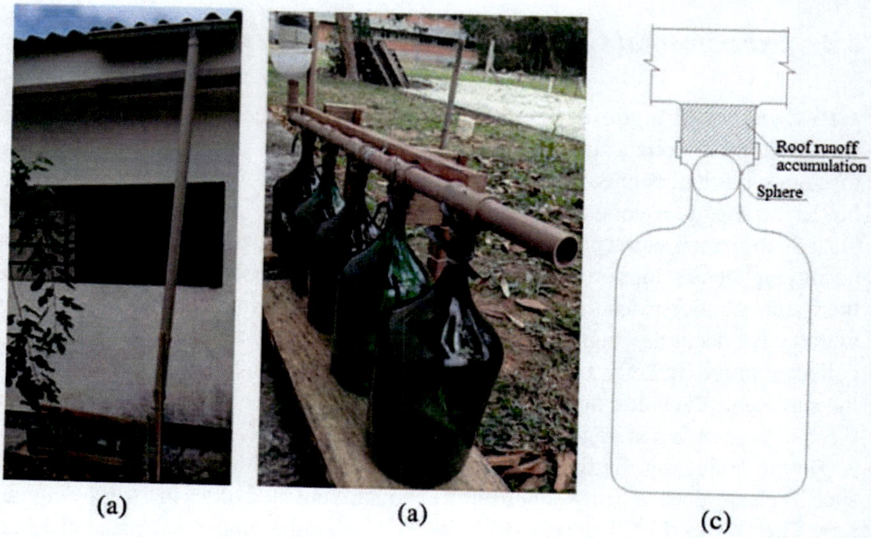

Fig. 1 Experimental collector. **a** View of runoff collector with gutter and pipes. **b** Pipe connections and bottles for runoff collection. **c** Representative scheme of a full bottle with sphere and runoff accumulation

were analysed. All the electronic devices used in the tests were calibrated, and more specific information about them is shown in Table 1. The pH was measured by inserting a pH meter into the sample, and the equipment was previously calibrated using pH 4, 7 and 10 buffer solutions. The apparent colour measurement was performed using a spectrophotometer (calibrated by a specialised technician) at 455 nm using the platinum-cobalt method.

The samples' turbidity was measured by the nephelometric method using a turbidimeter (manually calibrated by inserting standard formazin solutions into the

Table 1 Specifications of the equipment used to measure the quality parameters of rainwater

Equipment	Brand and model	Measurement ranges	Resolution	Accuracy
Portable pH meter	Kasvi K39-0014PA	0.0–14.0	0.1	± 0.1
Benchtop turbidity meter	Tecnopon TB1000	0–1 NTU	0.01 NTU	< 0.05
		0–10 NTU	0.01 NTU	
		0–100 NTU	0.1 NTU	
		0–1000 NTU	1 NTU	
Portable conductivity meter	Orion model 115	0.0 -199.9 μS 200–1999 μS 2.00–19.99 mS 20.0–199.9 mS	0.1 μS	± 0.5% of complete scale reading into each band
Spectrophotometer	Hach DR3900	320–1100 nm	1 nm	± 1.5

equipment at different concentrations and turbidity). The samples' conductivity was obtained using a conductivity meter, and its calibration was performed manually by inserting the sensor in a standard conductivity solution of 1413 μS/cm (± 0.5 %at 25 °C). The concentration of total suspended solids was obtained using a vacuum filtration system. The filtering membranes used were made of borosilicate glass fibre, free of binder resin and with a porosity of 0.6 μm.

Microbiological analyses were conducted using the Colilert test from the company Idexx. The solutions composed of substrate and sample, arranged in sealed aluminized trays, were incubated at 35 °C (± 0.5 °C) for 24 h. After incubation, quantification of microorganism colonies was performed using the most probable number method (MPN), with exposure of test trays to ultraviolet light to count *E. coli* colonies.

2.4 *Data Analysis*

The rainfall events from which samples were collected were characterized by rainfall indices, intensities, duration and dry weather periods between rainfall events. The results for turbidity, apparent colour, pH, total suspended solids, electrical conductivity, total coliforms and *E. coli* were statistically summarized.

Subsequently, the behaviour of the microbiological indicators during the first flush runoff was analysed using dispersion plots. Simple regression analyses were also conducted to verify whether it would be possible to approximate this behaviour to an exponential decrease function of the concentration of microorganisms during the first flush drainage, as shown by Doyle's [14] results. Regression analyses were performed with normalised data, using the concentration obtained for the first millimetre of runoff as a reference. The models were obtained using the least squares method.

Pearson's correlation analyses were performed between the microbiological indicators (total coliforms and *E. coli*) and the physical–chemical parameters measured. Subsequently, the pairs of variables whose Pearson correlation coefficients were higher than 0.70 (indication of strong linear correlation) were identified, and linear regression analyses were conducted. All statistical analyses were performed using the R language.

3 Results

First flush samples were collected from eleven rainfall events (58 samples). The samples' collections occurred between August 2020 and February 2021. The catchment of the first five millimetres in three rainfall events was not possible: four millimetres were collected in two of them, and only one millimetre could be collected in the other event. Table 2 summarizes the characteristics of the rainfall events and the quality of rainwater collected directly from the sky. The results indicate that, in general, rainwater has few pollutants.

The rainfall samples had a slightly acidic pH, possibly due to carbon dioxide and sulfate ions. In the atmosphere, when carbon dioxide is dissolved in an aqueous medium, it originates carbonic acid, and sulfate ions, in turn, originate sulfuric acid [19]. Another possibility is that the source of acidity in rainwater has a natural origin (organic acids) and exerts a constant influence on the value of this parameter [20].

Although in low concentration, the laboratory tests detected total coliforms in 100% of the rainfall samples. Only one sample showed a concentration of total coliforms greater than 50 MPN/100 mL. Regarding *E. coli,* the average concentration found in rainfall was much lower (0.9 MPN/100 mL) and in five precipitation events, there was no detection of colonies in rainwater.

NBR 15527 [21] is the Brazilian standard for rainwater harvesting, in which minimum requirements for rainwater quality for non-potable purposes are established. *E. coli* and turbidity concentrations were relatively low, and the quality parameters of most of the samples collected were within limits established in NBR 15527 [21] for non-potable uses. Regarding pH and *E. coli* concentration, all collected samples have complied with the limits established in the rainwater harvesting Brazilian standard. Only one event showed turbidity above 4.0 NTU, which may have occurred due to dust accumulation or transport of larger fragments of other pollutants in the container used for sample collection.

Table 3 summarizes the results of the laboratory tests to which the first flush samples from the fibre cement roof were subjected. When comparing the results between Tables 2 and 3, it is noted that the first flush runoff has a higher concentration of pollutants than the rainfall collected directly from the sky. Rainwater runoff from

Table 2 Statistical description of rainfall characteristics and quality parameters of rainwater collected directly from the sky

Parameters	Statistical measures			
	Mean	Standard deviation	Maximum	Minimum
Rainfall duration (h)	8.56	10.11	31.00	0.88
Rainfall index (mm)	16.90	15.50	57.25	4.25
Initial rainfall intensity (mm/h)	10.25	10.66	35.04	0.62
Antecedent dry weather period (days)	4.39	4.31	14.00	0.33
Total Suspended Solids (mg/L)	5.21	3.47	9.33	0.30
pH	6.6	0.3	7.0	6.1
Turbidity (NTU)	2.34	3.58	11.73	0.14
Electrical conductivity (μS)	36.8	36.4	113.5	6.6
Apparent colour (PtCo or uC)	17	15	50	0
Total Coliforms (NMP/100 mL)	31.4	45.3	137.6	2.0
Escherichia coli (NMP/100 mL)	0.9	1.4	3.1	< 1

Table 3 Statistical summary of first flush quality

Parameters	Statistical measures			
	Mean	Standard deviation	Maximum	Minimum
Total suspended solids (mg/L)	9.37	5.39	21.67	1.00
pH	8.2	0.3	8.8	7.3
Turbidity (NTU)	3.10	4.20	16.40	0.21
Electrical conductivity (μS)	86.3	50.8	278.0	17.1
Apparent colour (PtCo or uC)	70	58	291	24
Total Coliforms (NMP/100 mL)	1954.3	706.4	> 2419.6	461.1
Escherichia coli (NMP/100 mL)	375.8	692.8	> 2419.6	< 1

the fibre cement roof had a higher pH compared to rainwater captured directly, which was also observed in other studies [19, 22–24]. However, the pH values obtained in all samples meet the recommendations established in NBR 15527 [21]. As observed in this study, analyses of samples from fibre cement roofs in Poland also showed an alkaline pH [25]. Considering the long exposure time of the fibre cement roof, the occurrence of cementitious components' leaching is presumed, which neutralizes the acidity of the roof runoff, making it more alkaline.

Most of the first flush samples showed high concentrations of total coliforms, exceeding the upper limit of the test (2419.6 MPN/100 mL of sample). As for the *E. coli* concentration, most of the roof runoff water collected was between zero and 483.7 MPN/100 mL. Although the average concentration of such microorganisms in the first flush exceeded the limit established in NBR 15527 [21], approximately 66.7% of the samples showed concentrations below the standard limit. Mendez et al. [9] also found high values for the concentration of total coliforms in runoff samples from concrete tiles roof. The same authors, however, found concentrations of faecal coliforms lower than those in this study (mean lower than 25 CFU/100 mL and maximum concentration lower than 50 CFU/100 mL).

3.1 Behaviour of the Coliforms' Concentration During Roof Runoff

Figure 2 shows the behaviour of the concentration of total coliforms and *E. coli* during the first flush runoff. The fitted curve for the exponential model and the determination coefficients are also shown. Points with the same colour represent data from the same event sampled. Regression analysis shows no clear tendency in behaviour for the total coliforms and *E. coli* parameters as a function of the collected runoff. The models obtained for normalized data presented low coefficients of determination. This result indicates that the exponential decay model obtained for these variables does not

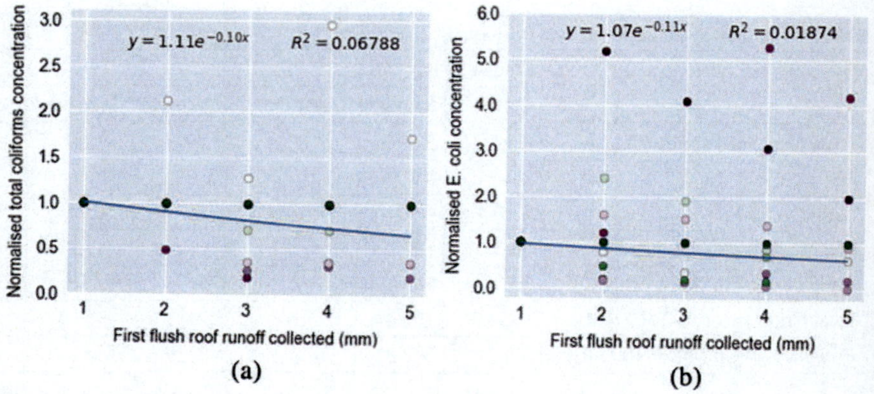

Fig. 2 Behaviour of normalized total coliforms concentration (**a**) and *E. coli* concentrations (**b**) as a function of the first flush roof runoff

reasonably fit the sampled data. Thus, one cannot predict total coliforms or *E. coli* concentrations considering only the deviation of the first flush runoff.

3.2 *Correlations Between Microbiological and Physical–Chemical Indicators*

Pearson's correlation analysis showed a strong negative correlation between the parameters *E. coli* and pH (r = − 0.76), in addition to strong positive correlations between *E. coli* and apparent colour (r = 0.84) and *E. coli* and turbidity (r = 0.79). No strong correlations were found between total coliform concentrations and the physical–chemical parameters analysed.

Shamsudin et al. [26] also found strong correlations between *E. coli* and physicochemical parameters of water quality, such as pH and turbidity (with correlation coefficients of 0.97 and 0.88, respectively). Different from these research findings, the correlation between pH and *E. coli* was positive. On the other hand, results of experiments conducted in a lagoon in Morocco [27] showed the existence of a strong negative correlation between pH and *E. coli* (r = − 0.98) and a strong positive correlation between turbidity and *E. coli* (r = 0.93). The increase in *E. coli* concentrations with the water turbidity may occur due to the reduction of the radiation level in the water. This environment becomes more favourable for the proliferation of these microorganisms [28].

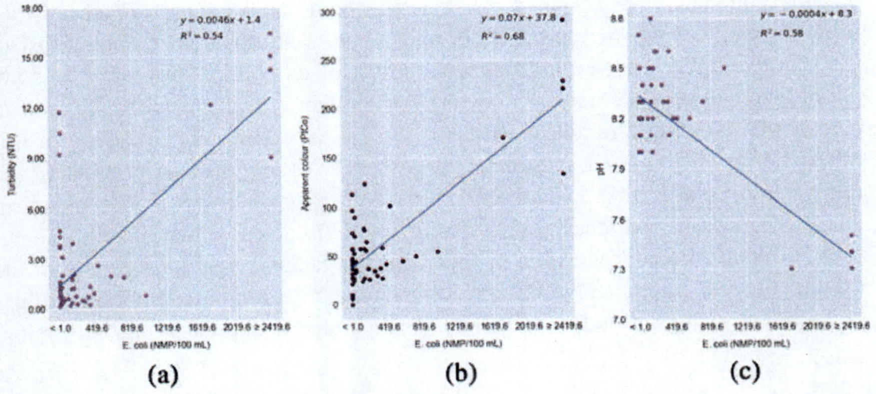

Fig. 3 Linear regressions between *E. coli* and turbidity (**a**), *E. coli* and apparent colour (**b**) and *E. coli* and pH (**c**)

3.3 Regression Analysis Between Correlated Parameters

Models that correlate microbiological parameters of water quality with physical–chemical parameters are helpful. Carrying out tests for microorganisms' quantification tends to be more expensive and generally demands special care to avoid contamination during the execution of the tests. Figure 3 shows the linear regression models obtained, their respective equations and coefficients of determination.

Through the linear model obtained with the correlation between the turbidity and *E. coli*, it is verified that turbidity greater than 2.32 NTU indicates concentrations of *E. coli* greater than 200 NMP/100 mL. However, with more accurate data observation, from 35 samples with turbidity levels below 2.32 NTU, 10 had a higher *E. coli* concentration than the recommended in NBR 15527 [21].

4 Conclusions

This research showed that a rainwater harvesting system's roof, gutters and pipes may be significant sources of contamination of the rainwater to be harvested. Significant increases were observed, concerning rainwater collected directly from the sky, of first flush roof runoff regarding the average concentration of total coliforms and *E. coli*.

When analysing the dispersion plots of the microbiological quality parameters as a function of the collected runoff, it was not possible to identify a clear trend of behaviour for the total coliform and *E. coli* indicators. This result was also observed by Yujie, Gouvello and Tassin [29], who did not observe a first flush effect for any microbiological parameter.

Linear models were obtained through regression analyses performed between turbidity and *E. coli*, pH and *E. coli*, and *E. coli* and apparent colour. Most of the runoff samples showed high concentrations of total coliforms and it was not possible to determine a first flush deviation value that guarantees the absence of *E. coli* or that the stored rainwater will not present excessive concentrations of this microorganism. Gikas and Tsihrintzis [10], when analysing the quality of harvested rainwater, concluded that using the first flush device improves the physicochemical quality of rainwater but does not prevent its microbiological contamination. Therefore, to reduce the concentration of these microorganisms, regular cleaning of the catchment area, gutters and pipes is recommended, in addition to using other types of treatment, such as filtration, for example, associated with the first flush diversion.

References

1. Abbasi, T.; Abbasi, S.A. Sources of Pollution in Rooftop Rainwater Harvesting Systems and Their Control. *Crit Rev Environ Sci Technol* **2011**, *41*, 2097–2167.
2. Domènech, L.; Heijnen, H.; Saurí, D. Rainwater Harvesting for Human Consumption and Livelihood Improvement in Rural Nepal: Benefits and Risks. *Water and Environment Journal* **2012**, *26*, 465–472.
3. Schets, F.M.; Italiaander, R.; van den Berg, H.H.J.L.; de Roda Husman, A.M. Rainwater Harvesting: Quality Assessment and Utilization in The Netherlands. *J Water Health* **2010**, *8*, 224–235.
4. Angrill, S.; Petit-Boix, A.; Morales-Pinzón, T.; Josa, A.; Rieradevall, J.; Gabarrell, X. Urban Rainwater Runoff Quantity and Quality – A Potential Endogenous Resource in Cities? *J Environ Manage* **2017**, *189*, 14–21.
5. Gould, J. Is Rainwater Safe to Drink? A Review of Recent Findings. In Proceedings of the 9th International Rainwater Catchment Systems Conference; Petrolina, 1999.
6. Sazakli, E.; Alexopoulos, A.; Leotsinidis, M. Rainwater Harvesting, Quality Assessment and Utilization in Kefalonia Island, Greece. *Water Res* **2007**, *41*, 2039–2047.
7. Coombes, P.J.; Kuczera, G.; Kalma, J.D.; Dunstan, H.R. Rainwater Quality from Roofs, Tanks and Hot Water Systems at Figtree Place. In Proceedings of the Hydro 2000: Interactive Hydrology; National Conference Publication: Barton, Australia, 2000; pp. 1042–1047.
8. Amin, T.; Han, M.Y. Microbial Quality Variation within a Rainwater Storage Tank and the Effects of First Flush in Rainwater Harvesting (RWH) System. *Aust J Basic Appl Sci* **2011**, *5*, 1804–1813.
9. Mendez, C.B.; Klenzendorf, J.B.; Afshar, B.R.; Simmons, M.T.; Barrett, M.E.; Kinney, K.A.; Kirisits, M.J. The Effect of Roofing Material on the Quality of Harvested Rainwater. *Water Res* **2011**, *45*, 2049–2059.
10. Gikas, G.D.; Tsihrintzis, V.A. Assessment of Water Quality of First-Flush Roof Runoff and Harvested Rainwater. *J Hydrol (Amst)* **2012**, *466–467*, 115–126.
11. Friedler, E.; Gilboa, Y.; Muklada, H. Quality of Roof-Harvested Rainwater as a Function of Environmental and Air Pollution Factors in a Coastal Mediterranean City (Haifa, Israel). *Water (Basel)* **2017**, *9*, 896.
12. Amin, M.T.; Kim, T.; Amin, M.N.; Han, M.Y. Effects of Catchment, First-Flush, Storage Conditions, and Time on Microbial Quality in Rainwater Harvesting Systems. *Water Environment Research* **2013**, *85*, 2317–2329.
13. Doyle, K.C.; Shanahan, P. Effect of First Flush on Storage-Reliability-Yield of Rainwater Harvesting. *Journal of Water, Sanitation and Hygiene for Development* **2012**, *2*, 1–9, https://doi.org/10.2166/washdev.2012.055.

14. Doyle, K.C. Sizing the First Flush and Its Effect on the Storage-Reliability- Yield Behavior of Rainwater Harvesting in Rwanda. Master's thesis, Massachusetts Institute of Technology: Cambridge (Massachusetts), 2008.
15. IBGE *Anuário Estatístico Do Brasil 1996*; Instituto Brasileiro de Geografia e Estatística: Rio de Janeiro, RJ, 1997.
16. Nimer, E. *Climatologia Do Brasil*; 2ª.; IBGE: Rio de Janeiro, 1989.
17. Yaziz, M.I.; Gunting, H.; Sapari, N.; Ghazali, A.W. Variations in Rainwater Quality from Roof Catchments. *Water Res* **1989**, *23*, 761–765.
18. APHA *Standard Methods for the Examination of Water and Wastewater*; 23ª.; American Public Health Association: Washington, DC, 2017.
19. Klein, L. de B. Controle Qualitativo e Quantitativo Do Escoamento Pluvial Em Diferentes Tipos de Coberturas. Dissertação de Mestrado, Universidade Federal de Santa Catarina: Florianópolis, SC, 2017.
20. Hoinaski, L.; Franco, D.; Haas, R.; Martins, R.F.; Lisboa, H. de M. Investigation of Rainwater Contamination Sources in the Southern Part of Brazil. *Environ Technol* **2014**, *35*, 868–881.
21. ABNT. *NBR 15527 - Água de Chuva - Aproveitamento de Coberturas Em Áreas Urbanas Para Fins Não Potáveis*; Rio de Janeiro, 2019.
22. Föster, J. Patterns of Roof Runoff Contamination and Their Potential Implications on Practice and Regulation of Treatment and Local Infiltration. *Water Science and Technology* **1996**, *33*.
23. Ibe, F.C.; Ibe, B.O. Roof Runoff Water as Source of Pollution: A Case Study of Some Selected Roofs in Orlu Metropolis, Imo State, Nigeria. *International Letters of Natural Sciences* **2016**, *50*, 53–61.
24. Sanjeeva, A.; Puttaswamaiah, S.G. Influence of Atmospheric Deposition and Roof Materials on Harvested Rainwater Quality. *Journal of Environmental Engineering* **2018**, *144*.
25. Polkowska, Ż. Examining the Effect of the Type of Roofing on Pollutant Content in Roof Runoff Waters from Buildings in Selected Districts of the City of Gdańsk. *Polish Journal of Environmental Studies*, **2004**, 13 (2), 191-201.
26. Shamsudin, S.N.; Rahman, M.H.F.; Taib, M.N.; Razak, W.R.W.A.; Ahmad, A.H.; Zain, M.M. Analysis between Escherichia Coli Growth and Physical Parameters in Water Using Pearson Correlation. In Proceedings of the 2016 7th IEEE Control and System Graduate Research Colloquium, ICSGRC 2016—Proceeding; Institute of Electrical and Electronics Engineers Inc., January 10 2017; pp. 131–136.
27. Mouna, H.; Ahmed, A.; Omar, A.; Doukkali, C.; Jadida, E. An Evaluation of Environmental Factors Affecting the Survival of Escherichia Coli in Coastal Area, Oualidia Lagoon. *Int.J.Curr.Microbiol.App.Sci*, **2014**, 3(10) 710-721.
28. Bell, C.; Albright, L. Attached and Free-Floating Bacteria in the Fraser River Estuary, British Columbia, Canada. *Mar Ecol Prog Ser* **1981**, *6*, 317–327.
29. Yujie, Q.; de Gouvello, B.; Tassin, B.; Gouvello Bernard, de; Bruno, T. *Qualitative Characterization of the First-Flush Phenomenon in Roof-Harvested Rainwater Systems*. LID (low impact development conference) 2016, June 2016, Beijing, China

Evaluation of Paper Pulp Industry Waste as a Conductivity and Water Colour Reducer

Paulo Mourão, José Gomes, and Sara Branco

Abstract This work consists in exploring the preparation, characterisation and application of adsorbents based on lignocellulosic materials obtained from waste from the paper pulp industry, namely "Knots and Rejects - KR" and "Primary Sludge - PS". These carbon materials were prepared by carbonisation under nitrogen flow, N_2, and by chemical activation with the activating agent phosphoric acid, H_3PO_4. The materials were characterised by FTIR and point of zero charge (pHpzc). Selected samples were applied to reduce the conductivity and colour of an effluent from the pulp industry (Biotek, S.A.—Vila Velha de Rodão). The application of these materials showed that some of the samples have the potential to reduce the conductivity of the effluent as well as to diminish its colour. It was possible to reduce the conductivity between 13.2% (KR600) and 20.3% (PS600), with the best results obtained with the samples prepared from PS waste. More interesting results were also observed in terms of colour attenuation of the effluent, with values ranging from 7.3% (KR600) to 69.1% (PS-H_3PO_4). Also in this case, the best results were obtained with the samples prepared from the PS waste. These results validate the strategy adopted and place it within a typical circular economy approach.

Keywords Paper industry · Lignocellulosic materials · Adsorbents · Waste valorisation · Conductivity · Colour

P. Mourão (✉)
Departamento de Química e Bioquímica, MED & CHANGE, IIFA, ECT, Universidade de Évora, Évora, Portugal
e-mail: pamm@uevora.pt

J. Gomes
Biotek, S.A., Vila Velha de Ródão, Portugal

S. Branco
Departamento de Química e Bioquímica, Universidade de Évora, Évora, Portugal

J. R. da. C. S. Galvão et al. (eds.), *Proceedings of the 3rd International Conference on Water Energy Food and Sustainability (ICoWEFS 2023)*, Springer Proceedings in Earth and Environmental Sciences, https://doi.org/10.1007/978-3-031-48532-9_68

1 Introduction

Throughout the paper production cycle, a large amount of inorganic (ash, dregs and grits) and organic (sludge) waste from the pulp and paper industry IPPI) is generated, as well as other by-products [1]. Residues from forest-based industries are defined as the leftovers from mechanical, physical or chemical processing that are not incorporated into the final product [2]. This fact drives this industrial sector, in collaboration with the scientific and technological community, to seek applications for these materials, either in their unaltered form (i.e. as produced) or after their subsequent modification by physical and/or chemical processes. The focus is therefore on the recovery of these wastes, while reducing the negative economic and environmental impact of having to dispose of them, sometimes in landfill, at the end of the production chain, transforming them in commercial products.

Among the various residues, those with the greatest potential are those of lignocellulosic origin, or those whose composition contains a higher percentage of these components, lignin and cellulose. In this case, knots, rejects and primary sludge are among the most interesting, not only because of their composition, but also because of the quantities produced during the industrial process. It should be noted that lignocellulosic materials form the essential part of the cell wall of plants and are an abundant and renewable source of carbon [3, 4]. They are composed of three macromolecular components, cellulose (between 40 and 60% of the total dry matter weight), hemicelluloses (20–40%) and lignin (10–25%), and low molecular weight components that include organic compounds, also called extractives, and inorganic compounds [3, 4]. The polysaccharides, cellulose and hemicelluloses, are linked to the lignin in the cell wall by covalent bonds and hydrogen bridges, giving the lignocellulosic biomass robustness, Fig. 1.

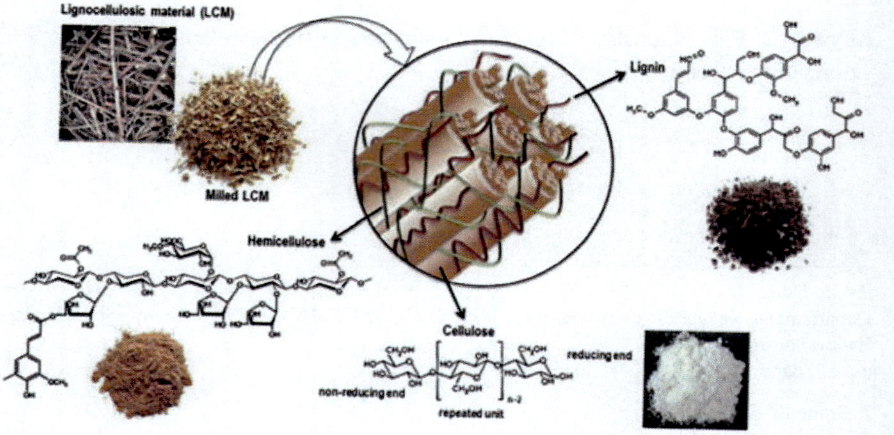

Fig. 1 Structure of the plant cell wall. Adapted from [3]

In this context, and from the perspective of a circular economy, the success of the approach will be greater the closer it is to the source, where this waste is generated, the application of the materials that are now being valorised [5]. Regarding the current problem related with water scarcity, well known to all mankind, solutions that can improve the quality of secondary water sources are very important, and the reduction and control of parameters such as conductivity and colour are key to an approach on the water cycle circularity within the pulp and paper industry [6].

2 Materials and Methods

2.1 Materials

The wastes used as work materials, "start materials", were generated throughout the paper industry process chain. The selected waste materials were "Knots and Rejects - KR" and "Primary Sludge - PS", Fig. 2. These two types of materials were thermal, and chemical modified and stabilized under a thermal chemical acid modification, with phosphoric acid. The studied range of temperatures were 400–600 °C. The carbonisation and activation steps were made in a horizontal tubular furnace TR-334/2018 (Termolab).

(a) (b)

Fig. 2 Representative samples of the paper pulp industry wastes materials: **a** "*Knots and Rejects - KR*" and **b** "*Primary Sludge - PS*"

All the prepared materials were evaluated from the chemical point of view by FTIR and point of zero charge (pzc). The FTIR analysis was made by attenuated total reflectance-ATR on a Perkin Elmer Spectrum I Two I FTIR spectrophotometer.

In second phase the samples modified were tested on the reduction of conductivity and colour parameters on simulated aqueous medium and on real effluents, provided by the Biotek unit of paper pulp production.

2.2 Methods

Activation and Modification of Wastes Materials. In this work the wastes materials were modified by three different ways: (i) by carbonization under nitrogen inert atmosphere at 400. 500 and 600 °C, (ii) acid modification with phosphoric acid, H_3PO_4, by the method of impregnation, and then submitted to activation under inert atmosphere of nitrogen at 450 °C, (iii) the previous carbonized samples were also acid modified with H_3PO_4, under similar conditions to previously one. The designation of samples follows the following sequence for: (i) *XT*, where *X* represent the waste material and *T* the temperature; (ii) and (iii) *XA*, where *X* represent the waste material or the carbonised sample, and *A* the activating agent.

Evaluation of Reduction of Colour and Conductivity Potential. All the selected samples chosen for potential colour and conductivity reduction tests were evaluated by colour measure (by UV-V is spectroscopy) and conductivity (by Contacting Method). The equipment used was the Nicolet Evolution300 UV-V is Spectrophotometer and an electrochemistry meter from Denver Instrument—model 20.

3 Results and Discussion

3.1 Waste and Modified Materials

The waste materials and the modified samples were characterized by FTIR-ATR and pHpzc in order to evaluate the initial characteristics of the wastes and the modification effects of the carbonization and chemical activation steps. The data presented on Table 1, shows that the waste had about one third of humidity when it was collected at the industrial plant, more precisely a value of 35.2 and 30.6% for KR and PS waste respectively. These values can be explained by the nature of the waste, in particular by the higher water absorption capacity of the lignocellulosic material, which is present in higher quantities in the case of KR waste. The yields of the different waste modification processes are quite different, but always higher than ~ 23%. More specifically, in the case of carbonisation, the values decrease slightly with increasing temperature, but always with yields higher than 25% for each of the precursors. Again, KR wastes are more sensitive to the effect of temperature, due to the nature of the starting material.

When analysing the chemical activation process of these wastes, the results are quite different. When the chemical activation was carried out on the original waste, the final yields were 29.4 and 22.9%, KR-H3PO4 and PS-H3PO4, respectively, for KR and PS waste. These results are in line with those already obtained in the carbonisation process. The higher amount of lignocellulosic matter and the lower amount of inorganic constituents lead to higher yields in the case of the KR-H_3PO_4 sample. On the other hand, when the material submitted to chemical activation was previously carbonised, the yields increase significantly to values around 80 and 99% for the KR600-H_3PO_4 and PS600-H_3PO_4 samples, respectively. The explanation for these very positive results lies essentially in the fact that these samples have been previously carbonised, so that we are dealing with physically and chemically stabilised materials, which make them less heterogeneous, after they have lost part of their volatile constituents.

In terms of chemical nature, the pHpzc values reflect a slightly basic character of both wastes, which is a result of the pulp production process, in particular of the chemical agents involved. When subjected to carbonisation, an increase in pHpzc is observed, which means an increase in the basic character of these samples, for pH values above 10. Once again, some inorganic contaminants, waste from the pulp production process, retained in the residues are responsible for this modification in the temperature range studied. On the other hand, the materials chemically activated with H_3PO_4 presented pH values in the acidic range, with values between 4.4 and 5.8. The lowest values were found in the samples prepared from KR waste. If we restrict the analysis to the samples of each precursor, slightly lower values always appear in the samples resulting from the activation of the direct wastes.

Table 1 Physical and chemical characteristics (Humidity and pHpzc) of the wastes and activated materials, and modification processes yield

Waste precursor	Modified samples	Humidity (%)	Yield (%)	pHpzc
KR	–	35.2		8.0
	KR400	–	30.6	–
	KR500	–	28.5	–
	KR600	–	27.9	11.0
	KR600-H_3PO_4	–	80.4	4.9
	KR-H_3PO_4	–	29.4	4.4
PS	–	30.6		7.4
	PS500	–	25.7	–
	PS600	–	25.2	10.1
	PS600-H_3PO_4	–	99.0	5.8
	PS-H_3PO_4	–	22.9	5.2

These values clearly reflect the effect of the acidic nature of the activating agent, H_3PO_4, and the lower resistance of the original waste to its action compared to the already carbonised samples.

These characteristics are also reflected in the FTIR spectra obtained for these samples. The FTIR profiles of these samples, Fig. 3, are similar to the typical spectra of carbon materials obtained by carbonisation and acid activation processes of ligno-cellulosic precursors [7]. In more detail, Fig. 3a–c, it can be seen that the carbonisation step induces some changes, particularly in the lower wavenumber range and an attenuation of the small bands in the higher range. On the other hand, when the chemically activated samples are compared with the starting materials, the differences are obvious along the whole spectrum, although with particular incidence in the lower wavenumber region.

3.2 *Conductivity and Colour Reduction Tests*

After preliminary experiments, selected samples were tested for the reduction of conductivity and colour parameters of the liquid medium. The first tests were carried out with simulated aqueous solutions to evaluate the ability of the samples to retain the species, namely ions, from the liquid phase. Based on these results, Fig. 4, the same samples were tested with a real effluent collected at the Biotek plant. From these latter tests, it can be concluded that the samples prepared from PS waste, both by carbonisation and acid activation, show even better performance.

The control and quantification of the colour variation was carried out by analysing the real effluent solution by UV-V is along the time of contact of this solution with the different samples. The results show that the colour removal process is relatively fast, reaching the maximum removal values after a contact time of 4 days for all samples. These results can be explained by the ability of these materials to retain, through processes such as adsorption, the coloured compounds responsible for the colour of the effluent. In terms of these parameters, the samples with the best performance are those prepared from the precursors by direct activation with phosphoric acid, KR-H_3PO_4 and PS-H_3PO_4, with values ranging from 59 to 69% respectively. These were followed by the carbonised and the subsequently chemically activated samples, PS600 and PS600-H_3PO_4. These results show that the samples prepared from PS waste have properties that enhance the removal of coloured compounds.

4 Conclusions

This work has demonstrated the feasibility of using waste from the pulping process, namely KR and PS, after mild acid modification with H_3PO_4 and thermal stabilisation in an inert atmosphere, in the temperature range between 450 and 600 °C, to produce new materials with potential application in the processes of conductivity and colour

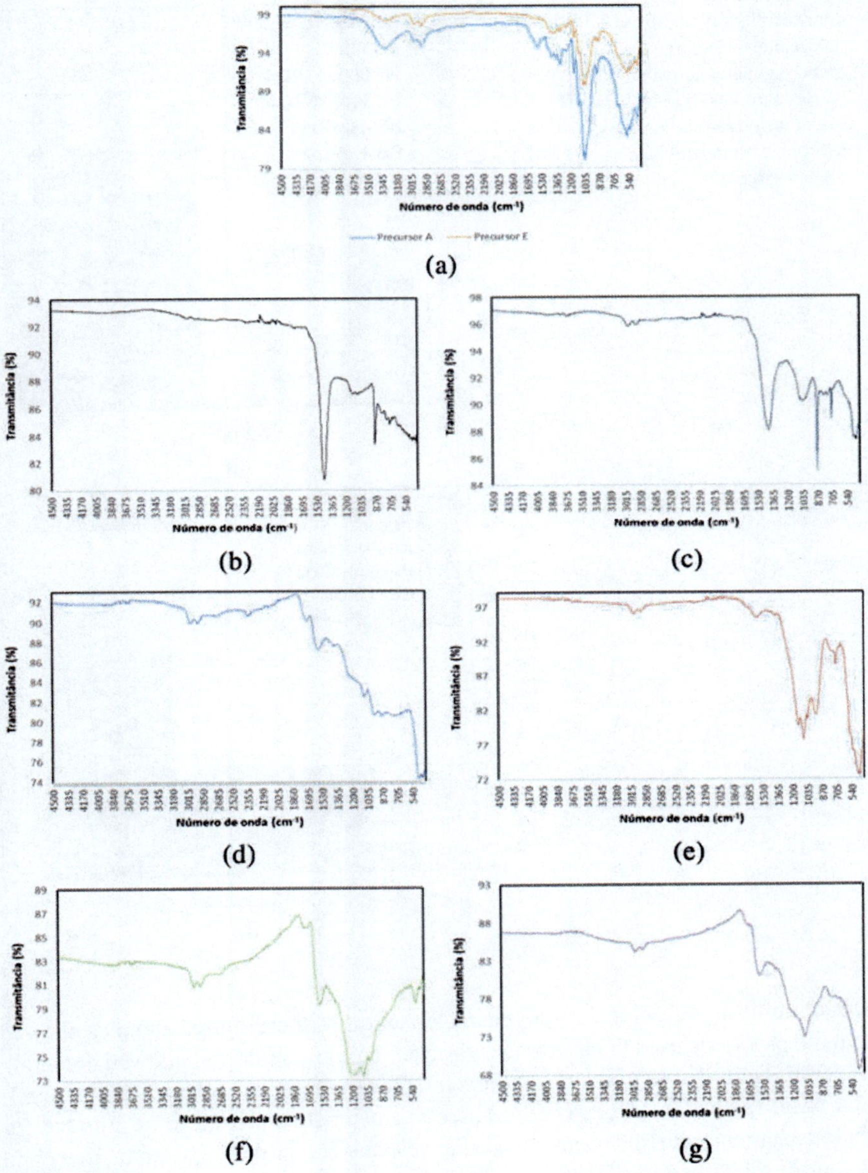

Fig. 3 Representative FTIR spectra of wastes and activated samples: **a** KR and PS, **b** KR600, **c** PS600, **d** KR600-H_3PO_4, **e** PS600-H_3PO_4, **f** KR-H_3PO_4 and **g** PS-H_3PO_4

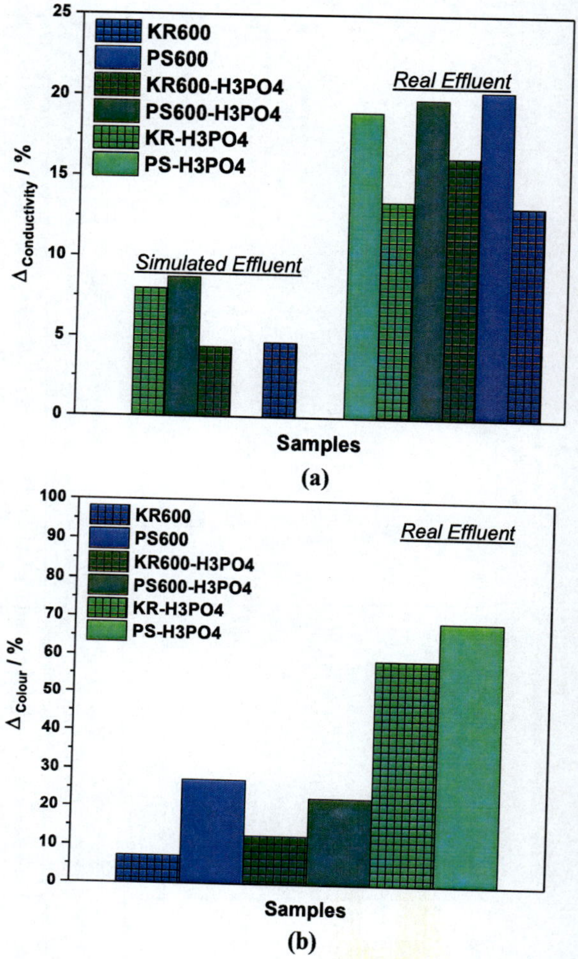

Fig. 4 **a** Reduction of conductivity on simulated and real effluents by carbonized and activated wastes samples, **b** reduction of colour on real effluent by carbonized and activated wastes samples

reduction in liquid effluents. The characterisation of the different samples showed that the processes used to modify the waste are quite versatile and allowed obtaining materials with distinct chemical characteristics, both acidic (pHpzc = [4.4; 5.8]) and basic (pHpzc = [10.1; 11.0]). These characteristics proved to be suitable for their use in removing the chemical species responsible for the high conductivity and colour of this type of effluent, even when tests were carried out with real effluents from the pulp mill itself. Conductivity was reduced by between 13.2% (KR600) and 20.3% (PS600), with the best results obtained with samples prepared from PS waste. The results are even more interesting in terms of colour reduction of the effluent, with colour reductions ranging from 7.3% (KR600) to 69.1% (PS-H_3PO_4). Also in this scenario, the best results were obtained with the samples prepared from PS waste. In conclusion, this approach of waste valorisation through the production and application route follows the best practices of an approach that can be classified as circular economy [8].

References

1. Bajpai, P.: Management of Pulp and Paper Mill Waste. Springer, Cham (2015).
2. Pinto, S., Almeida, M., Correia, A.M.S., Labrincha, J.A., Ferreira, V.M., K. Rosenbom, K.: Study on the environmental impact of lightweight aggregates production incorporating cellulose industrial residues. RILEM publications SARL, Barcelona (Spain), 2, 771–777, (2004).
3. Ballesteros, L.F., Michelin, M., Vicente, A.A., Teixeira, J.A., Cerqueira, M.Â.: Use of Lignocellulosic Materials in Bio-based Packaging. In: Lignocellulosic Materials and Their Use in Bio-based Packaging. Springer Briefs in Molecular Science. Springer, Cham. 65–85 (2018)
4. Smith, M.D.: An Abbreviated Historical and Structural Introduction to Lignocellulose. In: Understanding Lignocellulose: Synergistic Computational and Analytic Methods. ACS Publications, 1–15 (2019).
5. Gomes, J.A.F.L., Azaruja, B.A., Mourão, P.A.M., From MDF and PB wastes to adsorbents for the removal of pollutants. Appl. Surf. Sci. 380, 119-126 (2016).
6. Hubbe, M. A., Metts, J. R., Hermosilla, D., Blanco, M. A., Yerushalmi, L., Haghighat, F., Lindholm-Lehto, P., Khodaparast, Z., Kamali, M., Elliott, A.: Wastewater treatment and reclamation: A review of pulp and paper industry practices and opportunities. Bioresour. 11(3), 7953-8091 (2016).
7. Tchikuala, E., Mourão, P., Nabais, J.: Valorisation of Natural Fibres from African Bao-bab Wastes by the Production of Activated Carbons for Adsorption of Diuron. Procedia Eng. 200, 399-407 (2017).
8. Lin C.S.K., Kaur G., Li C., Yang X., Stevens C.V.: Waste Valorisation: Waste Streams in a Circular Economy. Wiley (2020).

Water Governance in Portugal and Brazil: A Comparative Analysis

Nádia Menegaz and **Margarida Queirós**

Abstract Faced with the globally installed water crisis, Portugal and Brazil have problems related to extreme hydrological events. The paper presents a first stage of research that comparatively analyzes the water resources governance systems in Portugal and Brazil, seeking to carry out a comparative diagnosis of the political systems and forms of territorial organization. In addition to differences related to territorial and demographic dimensions, Portugal and Brazil have different forms of state organization. The first is a unitary State and the second a federation, and this difference is the main starting point for this research. Conceptual, approach and arrangement differences between the two systems are pointed out. As a result, it is noticed that the differences found bring consequences to the water governance systems of each country. Between the differences appointed in both systems, the most significant is the level of the social participation specially during decision process of occurrence of extreme hydrological events.

Keywords Brazil · Portugal · Water governance

1 Introduction

The water crisis resulting from climate change, the risk society era [1–4] and the Anthropocene [5], has been identified as one of the main risks to humanity [6–10], and there has been a considerable increase in conflicts involving water, especially related to its scarcity [6, 11–13]. However, it is pointed out that the already installed crisis is associated with a governance crisis [8, 14–16]. Certainly, the risk of extreme hydrological events calls into question the governance systems of the institutions responsible for the political decision-making process, testing their strengths and weaknesses.

N. Menegaz (✉) · M. Queirós
Centre for Geographical Studies, Institute of Geography and Spatial Planning, Ulisboa, R. Branca Edmée Marques, 1600-276 Lisbon, Portugal
e-mail: nadiamenegaz@edu.ulisboa.pt

J. R. da. C. S. Galvão et al. (eds.), *Proceedings of the 3rd International Conference on Water Energy Food and Sustainability (ICoWEFS 2023)*, Springer Proceedings in Earth and Environmental Sciences, https://doi.org/10.1007/978-3-031-48532-9_69

This article comparatively analyzes the water resources governance systems in Portugal and Brazil, seeking to understand whether the respective models face and adequately respond to the challenges that involve the water resources sector in a context of climate emergency.

Since the 1970s, international conferences point out concerns about the future of natural resources and, annually, dozens of events and organizations continue to discuss the environmental directions of the planet. Governments have assumed a central role in these issues and, in principle, are responsible for solving them, especially those related to water scarcity, in the face of the strong economic pressures of the 70s and 80s in water resources sector, a period that can be called the "hydraulic mission". At that time, engineering was the driving force to solve problems related to the growing needs for the use of natural resources, based on Keynesian policies with strong State intervention and focused on economic development [17, 18].

Such a model, centralized and hierarchical, was questioned from the 1990s onwards, for being centered exclusively on command-and-control approaches (top-down) and, for this reason, was gradually replaced by public policies for decentralized and participatory water management [19, 20].

Currently, there are different approaches, depending on the context and the temporal, geographic or political scale [8, 21] and certainly the way governments organize themselves and their historical-cultural contexts are essential. And nowadays, water security is a political, technological, scientific, and socioeconomic challenge that involves several institutions and actors, which are not exclusively public, stimulating different forms of governance, and integrating multilevel and multiscale approaches [22, 23]. From a multilevel perspective, water management is both part of everyone's daily life, as well as that of society, and in all sectors, translating into a collective interest managed by different levels of competences. As for the multiscale feature, this natural resource cuts across cities, regions, nations, including water transfer at a global level—as addressed by the concepts of "virtual water" and "water footprint" [11, 24–27].

Despite the conceptual diversity regarding the topic of water governance, there is a consensus that studies on the subject, including comparative analyses, contribute to understanding political and structural phenomena in the field of water resource management, promoting changes in the scientific and governmental field, especially when they involve adaptive tools and policy learning processes [15, 28–31].

The issue of water in terms of the risks of scarcity and drought is vast and transdisciplinary and is studied both from the perspective of natural sciences (climate, hydrology, meteorology, ecosystem services, urban heat island, etc.) of social sciences (power conflicts, political economy, cultural values, management and governance, etc.), involving studies on risks to natural systems (drought, floods, soil permeability, biodiversity, resilience, etc.) and to the human systems (infrastructures, urbanization, resilience, etc.). The severe risks associated with water management solutions require transdisciplinary approaches and the emergence of a new water culture, based on ethical, democratic, and social equity principles [6, 31, 32].

In this context, the present study makes a comparative diagnosis of the systems and forms of territorial/administrative organization of political power regarding water management—which involve the risk of drought—in Portugal and Brazil. This raises clues to understanding the forms of organization of the States, the institutions of government responsible, the legal framework of the respective systems and the structures of civil society, and the implications for both systems of water governance—in aspects of greater or lesser suitability to the challenges of the climate crisis [33].

In addition to the differences related to territorial and demographic dimensions, both countries have different forms of State organization: the first is a "Unitary State" and the second a "Federated State". It is expected that federalism reflects more scales and levels of implementation and decision than 'unitarism', a fact that can translate the involvement of more stakeholders in decision-making processes. But a federated system also presupposes more complexity given that, in theory, it can lead to more conflicts regarding the allocation of responsibilities, in the balance of rights and duties between federated entities [14]. On the other hand, considering that water is a cross-cutting issue, that its management is associated with the natural territory of the river basin, a Unitary State may have a more holistic view of the issue [34].

2 Methodology

The methodology used to carry out this study was based on the review of public documents, including legislation, administrative structure and functioning of the decision-making process, especially related to the risk of extreme hydrological events.

In the case of Portugal, EU institutions were mapped, namely the European Commission, responsible for implementing legislation and common policies. The functioning structure of the Albufeira Convention was observed, in addition to considering the internal organization of the Portuguese government, namely the Ministry of the Environment and Climate Actions; APA-PT (Environmental Protection Agency); IPMA-PT (Portuguese Institute of Beach and Atmosphere); ANEPC-PT (National Emergency and Civil Protection Authority); DGADR-PT (General Directorate of Agriculture and Rural Development); DGEG-PT (Directorate-General for Energy and Geology). All other government bodies that make up the Commissions that are part of the governance process of the Portuguese political system were also mapped.

The main institutions observed in the case of the Brazilian political system were the Ministry of Regional Development; the ANA-BR (National Water and Sanitation Agency); the CNRH-BR (National Council Water Resources). and state water resources councils; state water resources management bodies; basin committees; the basin agencies (all described in National Water Resources Management System, SINGREH-BR). To seek some parity with the Portuguese system, for the purposes of the comparative analysis, the Ministry of the Environment, CONAMA-BR (National Environmental Council) and IBAMA-BR (Brazilian Institute for the Environment and Renewable Natural Resources), involved in environmental policy, and the Ministry of Mines and Energy and ONS-BR (National Electric System Operator), since a good part of the operation of water reservoirs in Brazil involves the electricity sector.

The brief detailing of the activities of each institution are presented in and Table 1.

Firstly, the focus was to understand the form of State (unitarism vs federalism) and political-administrative organization of each country, as it is considered that this stage is essential for a comparative analysis of the water governance systems. A second stage involved the analysis of the water legal framework of both nations, available on the websites of the main government institutions. Through this research, several points of divergence and similarities were identified, both under the conceptual approach, as under the multilevel and multiscale (arrangements), result of the form of political-administrative organization.

Qualitative literature research was also carried out in the main academic databases (Scopus, Google Scholar) and diversified repositories which allowed to deepen the contents and tracing a timeline of the systems and mapping their historical evolution. As a result, comparative tables of the political systems of water governance in both

Table 1 Portuguese and Brazilian policy water systems: comparative considerations

	Portugal	Brazil
Number of governance scales	3 (UE e PT, despite the existing of Reginal Administrations that works like a deconcentrate power of Portugal Central Government)	4 (Union, States, Municipalities, Watershed Committees)
Number of institutions involved (multiscale and multilevel process)	At least 30 (ministries, departments, agencies, councils, commissions, and others)	Considering the Federal scale, at least 23 (ministries, agencies, councils, committees, and others). But it depends on the size of the watershed and the number of states that cross it
Politic fragmentation between water and water resources	No	Yes
Social participation	Yes, only public consultation during the process of drawing up management instruments	Yes, public consultation during the process of drawing up management instruments and in the decision-making process through representation in deliberative bodies (Water Councils and Watershed Committees)
Social participation during the decision-making process in situations of risk	No, the decisions are made by the central Government	Yes, the Government discuss the situation with social representative institutions

Source [35–38]

countries were prepared, especially identifying the main government institutions involved and their multilevel and multiscale governance relationships.

3 Results

3.1 *Water Management: Political Systems in Portugal and Brazil*

In both countries, there is (from an early age) concern with the regulation of water, containing devices that point to the recognition of its access, either as a fundamental right, but also prioritizing the indispensable use for development. Also, in the case of Brazil and Portugal, there is an interventionist nature of the State in water matters, although with different models, presented in Table 1.

3.2 *The Organization of Portuguese Political and Territorial System of Water Management*

The Constitution of the Portuguese Republic states that Portugal is organized as a unitary State, in the form of a Semi-Presidential Constitutional Republic, arranged in an area of 92,152 km^2, distributed in 18 districts on the mainland (themselves grouped into 5 regions), 2 Autonomous Regions (archipelagos of Azores and Madeira), and 308 municipalities [39]. The Unitary State presents a single instance of political decision, a single legal-constitutional order [34], understood as the exclusive concentration of sovereign power in the central government; even though the Azores and Madeira archipelagos constitute autonomous regions endowed with political-administrative statutes and their own government bodies and local authorities (parishes and municipalities). It is inferred that a Unitary State tends to have a more centralized governmental organization, to ensure relations of hierarchization and subordination of the territorial units in relation to that government [40] which is presumed to have consequences for governance systems. According to Barraqué [19: 275–276], in 1976, those territorial levels did not play any role in the management of water resources, with the intermediate levels serving to collect information on the state of services and resources, for comparisons.

Since Portugal established as a nation it has had a significant percentage of watersheds shared with the neighbor nation (Spain), which implied, from the outset, the need to establish agreements between the two countries that, somehow, culminated in water management mechanisms. In 1297 Portugal and Spain signed a diplomatic agreement, the Treaty of Alcanises, with the definitive establishment of the borders between the kingdoms of Portugal and Castile, considering the rivers as geographical landmarks delimiting the territories of both countries [17, 41]. The Philippine

Ordinances (1603) portray the pioneering nature of water protection by providing the concept of pollution [42, 43]. But the current status of "public water domain" in Portugal began in 1864, with the publication of the Royal Decree that made sea water and respective beds and margins public [36].

In 1884, the first efforts were made to form a water policy with the creation of the "Plan and Organization of Hydrographic Services on Portugal mainland", characterized by the centralization of decisions at the state level, which, however, appeared as a first attempt to organize the territory, with the proposal to divide the country into hydrographic regions, using as the main criterion the grouping of the hydrographic basins of the respective rivers [42].

In 1919, already in the period after the implementation of the Portuguese Republic, the first consolidated normative is publicized, the Water Law (Decree 5787-IIII, of 10 May). With the fall of the dictatorial regime and the installation of the democratic regime in 1974, the current Constitution was approved in 1976 (having undergone some changes), bringing provisions on the organization of the theme of water [17, 44].

Today, as a member of the European Union, Portugal is subject to its guidelines, establishing more scales (and levels) in the governance process. And, as mentioned, Portugal has a good part of its hydrographic basins shared with Spain, implying a bilateral agreement for its management, the Albufeira Convention (Decree No. 182/99, of 17 August) and in force since 2000, regulating the regime of the flows (prevent risks that may affect the waters of common hydrographic basins). This convention deals with the establishment of a relationship of cooperation and mutual assistance in special situations [45].

The Water Framework Directive (Directive 2000/60/CE), currently the main instrument of water policy in the European Union (EU), was transposed into the national legal framework through the Water Law No. 58/2005, of December 29 (and amendments). According to Henriques [45], this Directive considers water management above all in terms of quality, with quantity seen in a subsidiary way. Other important provisions of Portuguese water policy are the Ownership Law (Law no. 54/2005, of 15 November), the Water Resources Use Regime (Decree-Law No. 226-A/2007, of 31 May) and Ordinance No. 1450/2007, of November 12th. Portuguese law also provides for Watershed Management Plans. The referred Water Law outlines that the national water authority is the Portuguese Agency (APA), representing the State. Representation of activity sectors and users of water resources is ensured through the following consultative bodies: a) the National Water Council (CNA), as a body of the government in matters of water resources (in which the public administration bodies and the most representative professional, scientific, sectoral and non-governmental organizations related to water are represented); b) the councils of the hydrographic region (CRH), as advisory bodies of the APA, in terms of water resources, for the respective hydrographic basins integrated therein (municipalities and the entities representing the main users related to the consumptive and non-consumptive use). The Coordination and Regional Development Commissions (CCDR) ensure the articulation of land use planning instruments with the rules and principles resulting from the law and the water plans and the integration of the water policy in the transversal environmental policies.

3.3 *The Organization of Brazil's Political and Territorial System of Water Management*

The connections between Portugal and Brazil are historical. The Brazilian territory was first claimed by Portugal due to the Treaty of Tordesilhas (1494) in the year 1500. In colonial times, the legal regime of Brazilian waters was basically characterized by the predominance of real rights, the use of rivers was common to the people, but the property belonged to the Royal Heritage [42].

In 1822, Brazil becomes independent. For Almeida [27], one can see the great influence of the Portuguese nation for the formation of history and Brazilian environmental legislation. But the territorial and demographic proportions between Brazil and Portugal are huge. In terms of scale, the comparative dimensions between the two countries are famous, i.e., we can mention the fact that only the city of São Paulo has a population greater than the whole of Portugal, that is, even what can be considered local/regional in Brazil, is equivalent to the entire territory of Portugal. Brazil has around 200 million inhabitants, while Portugal has around 10 million. Brazil is a Federated State, in the form of a presidential government, covering 8,510,345 km^2, organized into 26 Federated States, the Federal District and 5568 municipalities. The Brazilian Constitution of 1988 opted for 3 scales of power (Union, States, Municipalities).

A federation presupposes the sharing of power between federative entities [34, 46, 47]. The federative entities are endowed with government, autonomy, and legislative, administrative, and fiscal competences. In the case of Brazil, and more specifically regarding water, this brought up important points about the arrangement of the governance system. First, the domain of Brazilian waters is shared between the States and the Union, that is, the management of water in Brazil is sometimes up to the Union, or to the States, depending on the water body. In addition, there are many related issues that compete between the federative entities (environmental protection and natural resources; agriculture and food supply; housing and sanitation; exploitation of water and mineral), which it can imprint more complex arrangement of resource governance.

The drought was the driving factor for the creation of the first agency linked to water management, the "Inspetoria de Obras Contra a Seca" (Inspection of Works Against Drought), in 1909. Later, in 1930, the Code of Waters of Brazil appears, which deals with topics (among others) on irrigation and works against drought.

The current Brazilian Constitution, enacted in 1988, with the fall of the Brazilian Dictatorship, encompasses several devices that relate to water resources. In 1997, the National Policy of Water Resources (PNRH-BR) was instituted, through Law No. 9.433/1997, and is the main legal instrument that regulates the management of water resources in the country. In the case of the PNRH-BR, the law applies to bodies of water that belong to the Union, and each member state also has its own water resources policy, which must follow the PNRH-BR guidelines. The law provides for Water Resources Plans. The main institutions observed in the Brazilian political system are those registered in the SINGREH—BR, in addition to the bodies responsible for environmental and energy policy, as well as participatory decision-making forums for risk management of extreme hydrological events.

3.4 Conceptual and Approach Differences: Summary

Through the analysis of both government and legislative systems, we obtained findings that impact the current water governance system in each country, which can be divided into 2 blocks: (1) conceptual and approach; (2) arrangements/structure. The conceptual ones indicate definitions on some important points that diverge between the analyzed documentation and bring repercussions on the entire system. The approach concerns the scope of each policy, mainly on its fragmentation of cross-cutting themes that address water. And the arrangements illustrate the multilevel and multiscale distribution of the system itself.

Among the existing conceptual and approach divergences between the two analyzed systems, it is evident that water policy in Portugal, under the influence of the EU, focuses on concepts such as "water" and "watersheds" and that of Brazil on "water resources" and "water resources" plans. In Portugal, there is a concern with ecological quality and pollution of water bodies and other dimensions of water, such as groundwater, coastal and transitional water, in addition to addressing issues of reuse. In Brazil, all these themes are divided into other policies, such as environmental, soil conservation, maritime and sanitation, and the reuse themes are not mentioned, only the promotion of rational use is pointed out. Brazil clearly focuses on the theme of "uses". In the case of groundwater, these are the exclusive domain of the states, with a clear division between surface and groundwater in Brazilian politics. Although the PNRH-BR reinforces integrated and systematic management, highlighting the integration between water resources management and environmental management, articulation with land use management and integration with the management of estuarine systems and coastal zones, such guidelines still are little observed.

The cut that was made in Brazilian water policy can be associated with the organization of the State as a Federation, where the constituent power is understood to share among the federative entities themes that it considered of national, regional and local interest (Union, States and Municipalities). Because there is no true cooperation between federal entities and even at the vertical level, policy integration is not experienced in practice. Indeed, Souza [48] states that interstate relations in Brazil are highly competitive, both vertically and horizontally, and marked by conflicts, normally mediated by the Union. This shows that, despite Brazil having adopted the federal system more than a century ago, it is still characterized by a strong centralization in the federal sphere.

To exemplify the different ways of approaching the policies object of this study, we quote: Portuguese laws mentions the word 'quality' 34 times, 'pollution' 31 times, 'quantity' 7 times and 'use' 34 times. Brazilian laws, in contrast, refers to the word 'quality' 8 times, 'pollution' 1 time, 'quantity' 6 times, and 'use' 71 times. A conceptual point worth highlighting is the economic value of water in both countries. The Portuguese Water Law brings the concept of water as "a commercial product like any other, but a heritage that must be protected, defended and treated as such". Vital consequences result from this: the commercialization of water use titles is allowed (water market) meaning private waters. In Brazil, all water is for common use, that is, it cannot belong to a private entity and cannot be sold. Law No. 9.433/1997 states that "water is a limited natural resource, endowed with economic value". Therefore, it constitutes an economic and utilitarian dimension of water, but without a commercial value.

Such conceptual differences can be perceived in the economic-financial instruments of both policies. Both bring the user-pays and polluter-pays principles. However, in Portugal, larger uses (licenses and concessions for the use of water resources) are always subject to the payment of a water resource fee and a guarantee to ensure compliance with the obligations of the use itself. In Brazil, issuing grants for water resources is free of charge and, although charging for water resources is envisaged as one of the instruments of the PNRH-BR, it is not mandatory, and it is up to the Committee of the respective Hydrographic Basin to suggest it in the resource plan. In other words, one can clearly see how the commercial and economic character is more subtle on the Brazilian side.

3.5 *Differences in Organization*

The Brazilian Constitution divides competences (administrative, legislative, tax and jurisdictional) among the federative entities, a fact that has direct repercussions on the management and governance of water in Brazil. Portugal is a Unitary State meaning a deconcentration of power to the regional and local sphere, through the Regional Hydrographic Administrations (known as ARH-PT).

In Brazil, as the Water Resources Councils are deliberative bodies and present "chairs" to society (representatives of water users and civil organizations of water resources). In Portugal, the National Water Council and the Hydrographic Region Councils are advisory bodies for Government decisions, although there is only representation from scientific, economic, professional, and non-governmental organizations and Members invited for their recognized merit in the domain of water resources. In Brazil on the other hand, the PNRH-BR created decentralized decision-making entities, such as the River Basin Committees, which are SINGREH-BR entities, known for functioning as the "Water Parliament", where society and the public power discuss, deliberate, and share the management of water resources (there are currently 228 state committees and 10 interstate committees).

The Portuguese Water Law created the figure of the Association of Users that could be an analogous entity to the Brazilian Basin Committees, however, until today no association has been created. Another figure that the Portuguese system brings, by way of management decentralization, are the so-called Public Purpose Enterprises for which the management of multipurpose hydraulic infrastructure is granted by the State to one or more users of the main uses of water resources allocated to a project through a concession agreement. There are currently ten developments, with the assignment of the respective management still in the final stages [37]. Brazil encourages participation in the decision-making process, whether in the Water Councils or through the Basin Committees. It is also important to highlight that decisions involving situations of risk to extreme hydrological events are shared in participatory forums, from various instances and levels of government, water user sectors and civil society, through the so-called "Crisis Rooms" and the "Negotiated Allocation Meetings", coordinated by ANA-BR, the government entity responsible for implementing the PNRH-BR. Despite the strong participatory decision-making process involving water resources issues in Brazil, it is worth noting that the federal sphere continues to be a protagonist in the implementation and decisions on the subject [20, 46]. In Portugal, although the commissions, which are purely advisory bodies, hold seats for representatives of the society and are multisectoral, the most critical decisions are taken within the scope of the central government. Public consultation focuses on items such as watershed management plans, flood plans, among others. The decentralized and participatory power is perceived in the figures of the Multipurpose Enterprises and the Users' Association, both still in the incipient phase of implementation.

4 Conclusions

Apart of having a related history, Portugal and Brazil present considerable disparities when it comes to territorial, demographic, and hydrographic dimensions. They also exhibit different forms of government and state organization, bringing consequences for the water governance systems in each country.

The Brazilian legal framework presents overlapping jurisdiction between federative entities and levels of government (horizontal and vertical scales), resulting from the concurrent competences established by the Federal Constitution. This has strong consequences for water governance since issues such as pollution, coastal zones, and sanitation are addressed in different policies within the country's legal system. As an example, the pollution issues that are concurrent with the Union, states, and municipalities, is unrelated to the water resources policy, as this topic is not dealt with in the scope of the PNRH-BR, but in the National Environmental Policy (Law no. 6938/1981) or in state and municipal policies. The environmental impact assessment itself and the economic-financial analysis for the installation of water infrastructure projects that require the use of water are carried out by the Environmental Agency and not by the Water Resources Agency, highlighting the legal and administrative fragmentation.

For some authors, federalism is not the most important factor for the success or failure of public policies, when it can be the result of producing specific results or impacts of lesser importance [40, 48]. It is considered that it is not exactly federalism that imposes greater complexity on Brazilian politics, but its exacerbated vertical and horizontal legislative fragmentation, which brings overlaps and regulatory gaps. This can be corroborated by other cases of federated systems that present simpler and more objective arrangements, even with large territorial proportions like those of Brazil, as the United States and Canada [14, 49].

The legal framework surrounding the issue of water in Brazil is significantly more robust and complex than that of Portugal. It is believed that part of this, is due to the fact of the federative organization where many themes and competences were divided or shared between the federative entities. On the other hand, Portugal, being a Unitary State, also has a relatively complex network, especially in terms of the vertical aspect of its governance system. The horizontal aspects are observed since the country is bound by both the EU directives and the rules established in the Albufeira Convention.

Brazil has a more participatory and decentralized system than the Portuguese one, where even decisions involving risks of extreme hydrological events are debated with society. Consequently, the Brazilian system has more tools for a political learning process and adaptive governance [28, 50–52].

There are important conceptual differences between the two policies: the norm in Portugal is related to "water", and the Brazilian one to "water resources". The term 'water resource' links water to a specific destination, to an exchange and utility value; the concept 'water' values it, for its social and intrinsic value. Obviously, this has consequences for the culture that underlies each country's water governance system. The Portuguese system is strongly concerned with the water qualitative and ecological issues; quantitative management is little publicized, as well as the decision-making process in case of risk of extreme hydrological events, which is limited to the issuance of bulletins by the commissions involved. For Henriques [45], to better match the paradigms of qualitative and quantitative water management, Portugal should emphasize the regime of use, licensing and supervision of the water domain and the economic and financial regime of water uses, especially since the country is very dependent on flows originating in Spain. Although Portugal allows the water market, its legal framework brings a strong concern with ecological issues and the quality of water bodies and little dwells on the aspect of the use of the resource itself. Differently, in Brazil, where the commercialization of titles/grants of water resources is not allowed, the PNRH-BR clearly focuses on the regulation of use, but little is manifested by the ecology of water.

It is more evident that Portugal experiences a top-down governance model [53], multilevel and with fewer scales. Brazil, despite presenting a multi-scale and multi-level system and being organized in a form of federative government, still concentrates the main decisions at the federal level, especially when it comes to conflicts and overlapping competences. Finally, regarding the preparation to deal with risks to extreme hydrological events in the context of climate emergency, the Brazilian system has the advantage of having a more participatory system, with multiple stakeholders' discussion forums for decision making in crisis situations (i.e., risk of drought) such as the so-called 'Crisis Rooms' [35] and 'Water Allocation Meeting' [35]. In the Portuguese case, decisions taken for actions in risk events are basically centralized into government bodies and consultants representing society. The lack of public participation and the low levels of stakeholders' participation in decision-making process makes Portuguese system more vulnerable and less adaptable to a water crisis as extreme hydrological events.

References

1. Beck, U. (1992). Risk Society: Towards a New Modernity. Sage.
2. Beck, U. (1999). World Risk Society. Sage.
3. Beck, U. (2008). World at Risk. Polity Press.
4. Boyd, W., Beck, U., & Shrader-Frechette, K. S. (1993). Risk Society: Towards a New ModernityRisk and Rationality: Philosophical; Foundations for Populist Reforms. In Economic Geography (Vol. 69, Issue 4)
5. Crutzen, P. J. (2021). We Live in the Anthropocene, So Will Our Grandchildren. Herald of the Russian Academy of Sciences, 91(1), 13–16. https://doi.org/10.1134/S1019331621010020
6. Bogardi, J. J. (2021). Handbook of Water Resources Management: Discourses, Concepts and Examples. In A. Bogardi, J;Salamé, L.;Tingsanchali, T. Gupta, J.;ooijen, R.;Bhaduri, A;Nandalal, K;Kumar, N.;Kolechkina (Ed.), Handbook of Water Resources Management: Discourses, Concepts and Examples (1st ed.). Springer. https://doi.org/10.1007/978-3-030-60147-8
7. Melo, M. C. de, & Johnsson, R. M. F. (2017). O Conceito Emergente De Segurança Hídrica. Sustentare, 1(1), 72–92. https://doi.org/https://doi.org/10.5892/st.v1i1.4325
8. OECD. (2015). The OECD Principles on Water Governance. https://www.oecd.org/cfe/regional-policy/OECD-Principles-Water-portuguese.pdf
9. Steffen, W., Broadgate, W., Deutsch, L., Gaffney, O., & Ludwig, C. (2015). The trajectory of the anthropocene: The great acceleration. In *Anthropocene Review* (Vol. 2, Issue 1, pp. 81–98). https://doi.org/10.1177/2053019614564785
10. World Economic Forum. (2019). Global Risks Report 2019. Geneva Switzerland
11. Bleninger, T., & Kotsuka, L. K. (2015). Conceitos de água virtual e pegada hídrica: estudo de caso da soja e óleo de soja no Brasil. Revista Recursos Hídricos, 36(1), 15–24. https://doi.org/10.5894/rh36n1-2
12. Braga, B., & Kelman, J. (2020). Facing the challenge of extreme climate : the case of Metropolitan Sao Paulo. International Journal of Water Resources Development, 36(2–3), 278–291. https://doi.org/10.1080/07900627.2019.1698412
13. Candela, L., Domingo, F., Berbel, J., & Alarcón, J. J. (2008). An overview of the main water conflicts in Spain: Proposals for problem-solving. Options Méditerranéennes, A, 83(June 2014)
14. Garrick, D., Anderson, G., Connell, D., & Pittock, J. (2014). Federal rivers: A critical overview of water governance challenges in federal systems. Federal Rivers: Managing Water in Multi-Layered Political Systems, 3–19. https://doi.org/10.4337/9781781955055.00010
15. Pahl-Wostl, Claudia. (2017). An Evolutionary Perspective on Water Governance: From Understanding to Transformation. Water Resources Management, 31(10), 2917–2932. https://doi.org/10.1007/s11269-017-1727-1

16. Ribeiro, N. B., & Johnsson, R. M. F. (2018). Discussões Sobre Governança Da Água: Tendências E Caminhos Comuns. Ambiente & Sociedade, 21(2001), 1252. http://www.scielo.br/j/asoc/a/XRVK5sTy3ZxWmYqsGNmgj7b/abstract/?lang=pt
17. Ferreira, S. M., & Debeus, G. (2018). Avaliação dos Modelos de Gestão ao Longo da História de Portugal e Brasil: Um olhar acerca das tendências internacionais nas Políticas Hídricas. Revista Geografia Em Atos, 2(9), 22–43.
18. Setti, A. S. et. al. (2001). Introdução ao gerenciamento de recursos hídricos (3rd ed.). ANA; ANEEL Brasil
19. Barraqué, B. (1995). As Políticas da Água na Europa (I. Piaget (ed.)).
20. Miranda, G. M., & Reynard, E. (2020). Integrated water resources management in federations: The examples of Brazil and Switzerland. Water (Switzerland), 12(7), 1–15. https://doi.org/10.3390/w12071914
21. Jones, R. N., & Preston, B. L. (2011). Adaptation and risk management. Wires Climate Change, 2(2), 296–308. https://doi.org/10.1002/wcc.97
22. Kumar, P. (2015). Water Resources Research. Journal of the American Water Resources Association, 51(7), 5827–5838. https://doi.org/10.1111/j.1752-1688.1969.tb04897.x
23. Withanachchi, S. S., Ghambashidze, G., Kunchulia, I., Urushadze, T., & Ploeger, A. (2018). A paradigm shift in water quality governance in a transitional context: A critical study about the empowerment of local governance in Georgia. *Water* (Switzerland), 10(2), 1–27.https://doi.org/10.3390/w10020098
24. Graham, N. T., Hejazi, M. I., Kim, S. H., Davies, E. G. R., Edmonds, J. A., & Miralles-Wilhelm, F. (2020). Future changes in the trading of virtual water. Nature Communications, 11(1), 1–7. https://doi.org/10.1038/s41467-020-17400-4
25. Hoekstra, A. Y. et al. (2011). The Water Footprint Assessment Manual. In earthscan. https://doi.org/10.4324/9781849775526
26. Kundzewicz, Z. W., Krysanova, V., Benestad, R. E., Hov, Piniewski, M., & Otto, I. M. (2018). Uncertainty in climate change impacts on water resources. Environmental Science and Policy, 79(June 2017), 1–8. https://doi.org/10.1016/j.envsci.2017.10.008
27. Yawson, D. O., Mohan, S., Armah, F. A., Ball, T., Mulholland, B., Adu, M. O., & White, P. J. (2020). Virtual water flows under projected climate, land use and population change: the case of UK feed barley and meat, Heliyon, *6*(1), e03127. https://doi.org/10.1016/j.heliyon.2019.e03127
28. Huntjens, P. (2021). Towards a Natural Social Contract. In Towards a Natural Social Contract. https://doi.org/10.1007/978-3-030-67130-3_3
29. Huntjens, P., Pahl-Wostl, C., & Grin, J. (2010). Climate change adaptation in European river basins. Regional Environmental Change, 10(4), 263–284. https://doi.org/10.1007/s10113-009-0108-6
30. Özerol, G., Vinke-De Kruijf, J., Brisbois, M. C., Flores, C. C., Deekshit, P., Girard, C., Knieper, C., Mirnezami, S. J., Ortega-Reig, M., Ranjan, P., Schröder, N. J. S., & Schröter, B. (2018). Comparative studies of water governance: A systematic review. Ecology and Society, 23(4), 43. https://doi.org/10.5751/ES-10548-230443
31. Pahl-Wostl, C. (2016). Water security, systemic risks and adaptive water governance and management. In Handbook on Water Security.
32. Ostrom, E. (1990). Governing the commons: the evolution of institutions for collective action. In Governing the commons: the evolution of institutions for collective action. Indiana University
33. Pahl-Wostl, C., Gorris, P., Jager, N., Koch, L., Lebel, L., Stein, C., Venghaus, S., & Withanachchi, S. (2021). Scale-related governance challenges in the water–energy–food nexus: toward a diagnostic approach. Sustainability Science, 16(2), 615–629. https://doi.org/10.1007/s11625-020-00888-6
34. Loomis, J. J., de Oliveira, C. M. R., & Dziedzic, M. (2021). Environmental federalism in EIA policy: A comparative case study of Paraná, Brazil and California, US. Environmental Science and Policy, 122(February), 75–82. https://doi.org/10.1016/j.envsci.2021.03.015
35. ANA Homepage, https://www.gov.br/ana/pt-br, last accessed 2023/01/13.
36. APA Homepage, http://www.apambiente.pt, last accessed 2023/01/13.

37. Martins, J. P., & Godinho, F. (2021). A água e os rios no futuro: Contributos do CNA para decisões estratégicas no setor água. https://conselhonacionaldaagua.weebly.com/publicaccedilotildees-de-acesso-geral.html
38. Villar, Pilar and Graziera, M. L. Direito de águas à luz da governança (Vol. 1). 1sp edn. Agência Nacional de Águas e Saneamento Básico - ANA. Brasília (2020)
39. Portugal Homepage, https://www.parlamento.pt/Legislacao, last accessed 2023/01/21.
40. De Souza Machado, F. R. (2014). Federalismo e polÍticas públicas no brasil. In Cadernos de Saude Publica (Vol. 30, Issue 11). https://doi.org/10.1590/0102-311XRE011114
41. Martins, F. M. M. M. (2015). Portugal no contexto da Península Ibérica : sua dependência hídrica , demografia e segurança. Universidade de Évora
42. Costa, F.;Nossa, P.;Magalhães, S.;Magalhães, M. (2011). A Legislação dos recursos hídricos em Portugal e no Brasil - uma análise histórica comparativa. https://repositorium.sdum.uminho.pt/bitstream/1822/22593/1/AlegislaçãodosrecursoshidricosemPortugalenoBrasil_F.COSTAetal.pdf
43. Coutinho, F. S. D. N. (2014). LICENCIAMENTO AMBIENTAL E GESTÃO DOS RECURSOS NATURAIS NOS ASSENTAMENTOS RURAIS: uma compreensão hermenêutica a partir da realidade do semiárido paraibano
44. Pato, J. (2013). Políticas públicas da água em Portugal: Do paradigma hidráulico à modernidade tardia. Analise Social, 48(206), 56–79
45. Henriques, A. G. (2004). O Direito Internacional das Águas e a Convenção de Albufeira de 1998 sobre as Bacias Hidrográficas Luso-Espanholas. 7° Congresso da Água, 16
46. Empinotti, V. L., Gontijo, W. C., & de Oliveira, V. E. (2018). Federalism, water, and (de)centralization in Brazil: the case of the São Francisco River water diversion. Regional Environmental Change, 18(6), 1655–1666. https://doi.org/10.1007/s10113-018-1371-1
47. Linck, L. C., & Ianoni, M. (2022). Cooperative federalism in Brazil and the multilevel environmental management system. Desenvolvimento e Meio Ambiente, 60, 271–292. https://doi.org/10.5380/dma.v60i0.78983
48. Souza, C. (2005). Federalismo, desenho contitucional e instituições federativas no Brasil pós-1988. Revista de Sociologia e Política, 24, 105–121. https://doi.org/10.1590/s0104-44782005000100008
49. Fisch, J. (2018). The Case for Effective Environmental Politics: Federalist or Unitary State? Comparing the Cases of Canada, the United States of America, and the People's Republic of China. University of Michigan Journal of Law Reform, 51(51.4), 777. https://doi.org/10.36646/mjlr.51.4.case
50. Huntjens, P., Pahl-Wostl, C., Rihoux, B., Schlüter, M., Flachner, Z., Neto, S., Koskova, R., Dickens, C., & Kiti, I. N. (2011). Adaptive water management and policy learning in a changing climate: A formal comparative analysis of eight water management regimes in Europe, Africa and Asia. Environmental Policy and Governance, 21(3), 145–163. https://doi.org/10.1002/eet.571
51. Pahl-Wostl, Claudia, Arthington, A., Bogardi, J., Bunn, S. E., Hoff, H., Lebel, L., Nikitina, E., Palmer, M., Poff, L. R. N., Richards, K., Schlüter, M., Schulze, R., St-Hilaire, A., Tharme, R., Tockner, K., & Tsegai, D. (2013). Environmental flows and water governance: Managing sustainable water uses. Current Opinion in Environmental Sustainability, 5(3–4), 341–351. https://doi.org/10.1016/j.cosust.2013.06.009
52. Pahl-Wostl, Claudia, Becker, G., Knieper, C., & Sendzimir, J. (2013). How multilevel societal learning processes facilitate transformative change: A comparative case study analysis on flood management. Ecology and Society, 18(4). https://doi.org/10.5751/ES-05779-180458

53. Rowbottom, J., Graversgaard, M., Wright, I., Dudman, K., Klages, S., Heidecke, C., Surdyk, N., Gourcy, L., Leitão, I. A., Ferreira, A. D., Wuijts, S., Boekhold, S., Doody, D. G., Glavan, M., Cvejić, R., & Velthof, G. (2022). Water governance diversity across Europe: Does legacy generate sticking points in implementing multi-level governance? Journal of Environmental Management, 319(July). https://doi.org/10.1016/j.jenvman.2022.115598

Strategic Plan for Water Losses Management in Water Supply Systems

Sara Bona, Armando Silva-Afonso, Fernanda Rodrigues, and Ricardo Gomes

Abstract Water is at the core of sustainable development and is critical for human survival itself. However, due to failures in public supply networks, a significant amount of water is lost, highlighting the need for a reliable and robust leak detection and location system. Thus, it is important to find ways to improve efficiency in water use, being the reduction of water losses in water supply systems a decisive factor to achieve this goal, which implies the adoption of an active and responsible attitude, to ensure proper management and sustainability of this essential natural resource, increasingly scarce. This article has as its main goal to present an analysis of concepts and methods, through a literature review, and to propose a strategy for planning the reduction and control of water losses in water supply systems, which, admittedly, allows the achievement of economic gains and greater ecological efficiency, leading to an overall improvement of the service. In summary, the implementation of an integrated strategy for water losses management is a reliable option from several perspectives, which justifies its implementation by water companies, with the necessary contextual adaptations and specificities.

Keywords Water losses management · Water supply systems · Sustainability

---

S. Bona (✉) · A. Silva-Afonso · F. Rodrigues
Department of Civil Engineering, RISCO Research Center for Risks and Sustainability in Construction, University of Aveiro, 3810-193 Aveiro, Portugal
e-mail: sara.cbona@ua.pt

A. Silva-Afonso
Department of Civil Engineering, ANQIP—National Association for Quality in Buildings Services, University of Aveiro, 3810-193 Aveiro, Portugal

R. Gomes
Department of Civil Engineering, INESC Coimbra and SMAS de Leiria, Polytechnic of Leiria, Campus 2 Morro do Lena—Alto do Vieiro, 2411-901 Leiria, Portugal
e-mail: ricardo.gomes@ipleiria.pt

J. R. da. C. S. Galvão et al. (eds.), *Proceedings of the 3rd International Conference on Water Energy Food and Sustainability (ICoWEFS 2023)*, Springer Proceedings in Earth and Environmental Sciences, https://doi.org/10.1007/978-3-031-48532-9_70

1 Introduction

Hydraulic systems are of structural and essential importance for the progress of society and for ensuring public and environmental safety. Water scarcity caused by urban population growth and climate change is putting pressure on water companies to accelerate the development and diffusion of best practices for water losses management and, more generally, to define a strategic vision for the future of urban infrastructure to help the sector adapt to climate change in a world where natural resources are increasingly scarce. Particularly in urban environments, water losses are present and more significant in distribution networks [1, 2], even in systems with well-developed infrastructure and good operational practices [3]. Water losses are estimated to average 30% of the water supplied worldwide [4–6], but values in the order of 50% are recorded by most water companies [6, 7].

Leakage causes numerous undesirable consequences, besides the loss of the natural resource, financial losses [8], contamination risks [9] and unacceptable environmental risks, further aggravating the water scarcity situation [10]. For the water companies, this fact has serious implications both in the efficiency of these institutions and in the quality of service provided to their customers, directly impacting the increase of costs of service and the degradation of the quality of service provided to the community. To minimize such damages and optimize the water supply networks in all their amplitude, the development of alternative methods and the adoption of advanced technologies and methodologies for water losses management is inevitable [7, 10].

The inefficient management of water supply systems results not only in the depletion of water resources, but also corresponds to a direct increase in energy consumption and related CO_2 emissions, and, additionally, when treating larger volumes of water than necessary, there is an increase in the use of chemical reagents, representing consequent environmental and economic impacts [3, 11]. In the social scope, increasing efficiency in these systems allows net gains and added value for the consumer, namely through tariff moderation [9]. In addition to the above, it is worth mentioning that the success of a water losses management strategy corresponds, in practice, to the increase of water availability, allowing to avoid or postpone the need to resort to new water sources (for example construction of pipelines, reservoirs, dams or new boreholes), solutions that imply high investments and, therefore, much higher than those inherent to the implementation of a strategic plan for water losses management [12].

The implementation of a program for water losses management, which allows reducing the volume of water losses in a supply network to levels considered economically feasible and technically acceptable, should be considered one of the main strategic objectives of government agencies and water companies, as an alternative to mitigate ecological stress and as a means to achieve service sustainability. The environmental, economic/financial, social and public health importance of water losses in water supply systems is thus evidenced, as it is a global issue that has been assuming an increasingly relevant dimension and a central role in the concerns of the water companies of these systems. However, in recent years, there has been an increase, but the scarcity of research in this important area is still evidenced [8, 10, 13]. The present research outlined in this contribution emerged intending to understand a set of internationally recognized basic concepts and present an analysis based on practices, methodologies and tools appropriate and inherent to the theme of water losses management in supporting the optimization of performance and sustainability levels in water supply systems. With this objective, a study was conducted employing applied research with a qualitative approach around textual elements and considering the predefined methodology of content analysis [14].

2 Strategic Planning for Water Losses Management in Water Supply Systems

The creation of planning strategies for water losses management is a key step, which will establish the guidelines for the whole process. Strategic planning corresponds to the various sequential steps of actions outlined for the achievement of the established goals, aiming at the minimization of water losses in the system. In its elaboration, it is important to ensure that it is dynamic and sufficiently flexible so that it can be revised over time in response to changes arising from its implementation or new requirements. The suggested strategic model was developed based on the literature [1, 15–17], to assist in the rationale for decisions and actions to reduce real water losses in water supply systems (see Fig. 1).

2.1 *Problem Identification and Objective Definition*

In the first stage, the proposed methodology will focus on preliminary analysis for strategic decision-making, ensuring the identification of potential leakage problems and allowing the establishment of priorities in strategic leakage detection programs.

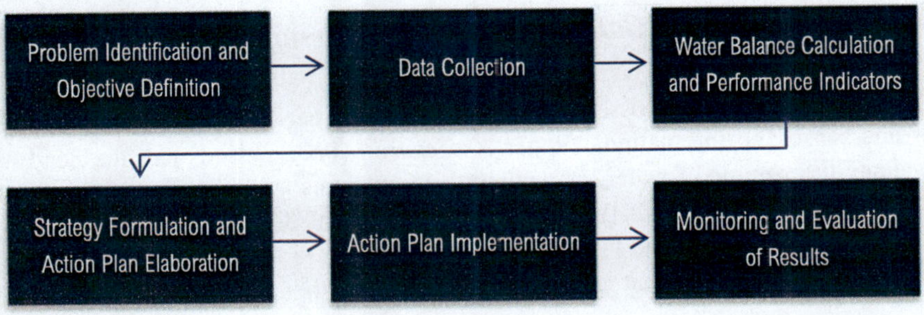

Fig. 1 Strategic planning methodology in the process of minimizing leaks and water losses

The definition of objectives and achievable goals is very important for the success of strategic planning. The characterization of the existing context and the assessment of the dimension of the problem is fundamental, and the leakage control strategy should be established based on this preliminary assessment.

2.2 Data Collection

To implement a strategy to reduce real water losses and consequently carry out the diagnosis, it is essential to analyze the current situation of the network systems. For the correct characterization of the network operation level, it is necessary to collect and compile a set of data regarding the water supply system of the respective locality under study, under the availability of the responsible water company [1], with emphasis on:

(a) *Network cadastre*: layout of the pipes, location of all network elements/accessories, diameter, age and material of each pipe;
(b) *Operational data and management reports*: history of repair of network burst, their frequency, duration and associated costs;
(c) *Flow measurement*: flow rates in strategic points of the network;
(d) *Consumption measurements*: demand by water users (for example fire departments, public irrigation, sidewalk washing and industries);
(e) *Pressure management*: pressures in the network throughout the day and at critical points in the water supply network;
(f) *Water cost (production and sale)*: water production/acquisition costs and the sale value to customers;
(g) *Staff and available technology*: experience and degree of knowledge of the technical staff, available technology and social, cultural and political influences.

In cases where there is insufficient data, the diagnosis should be complemented with information collected on-site to expand the necessary information and confirm the preliminary assessments, thus composing the diagnosis of existing systems [16].

2.3 Water Balance Calculation and Performance Indicators

The occurrence of water losses in a water distribution network is inevitable [3, 16]. In this sense, the International Water Association (IWA), faced with the need to evaluate the volume of water losses and its components and, in this way, allow an international comparison between the performances of different water companies (benchmarking), developed the tool—Water Balance (see Fig. 2)—that has become the basis of the analysis that is carried out around this theme [17, 18].

The main terms adopted correspond:

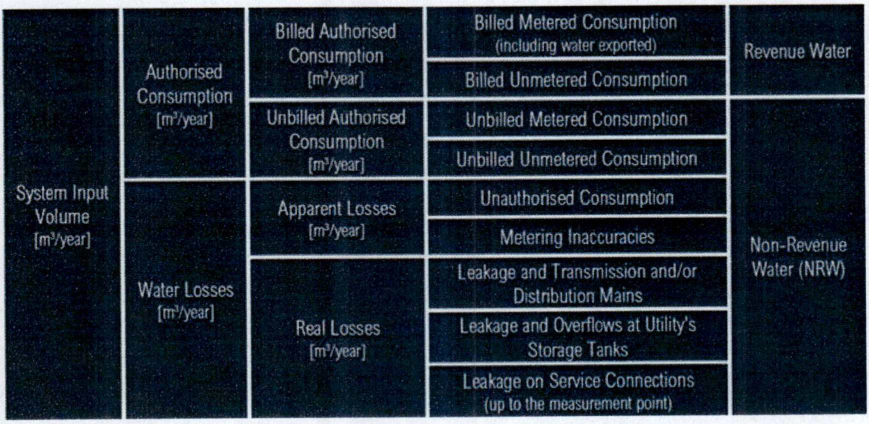

Fig. 2 Water balance. Adapted from [18]

(a) *System Input Volume*: this volume represents the sum of the volumes referring to the sources added to the volume of imported water, therefore, it represents the volume of water entering the distribution system.

(b) *Authorized Consumption*: volume of water, measured or unmeasured, but that was consumed by the registered customers/consumers, added with the volumes authorized by the company. It includes exported water, the volume of fire-fighting and social commitments made.

(c) *Apparent Losses*: correspond to the volume of water consumed but not accounted for by the water management entity, that is, they are caused by clandestine connections, theft or illegal use, hydrometer fraud, hydrometer measurement errors and commercial registry failures.

(d) *Real Losses*: are the physical leakages of water in the supply system, therefore, represent all the volume lost through leaks, bursts and overflows from pipes, reservoirs and branches, upstream of the customer's metering point.

(e) *Non-Revenue Water (NRW)*: corresponds to the difference between the volume of water introduced into the system and the authorized consumption that is billed. NRW represents real and apparent losses, as well as unbilled authorized consumption.

The calculation of the water balance, and the clear quantification of all the components, of a water supply system, is an indispensable step for the correct assessment of the size of the problem and its performance. The data required for an accurate water balance calculation are obtained by measuring the abstracted, produced, adduced and stored flows entering or leaving each distribution sector. Calibrated meters should be used whenever possible. In their absence, estimates based on other accessible data or other reliable engineering techniques should be used [12, 16].

Through the international method (IWA methodology), a set of performance indicators was identified, under the environmental, operational, and economic/financial aspects, used for the following components of the water balance: NRW, water losses,

Table 1 Main performance indicators related to water losses (IWA water audit methodology)

Water balance components	Performance areas	Base performance indicator	Detailed performance indicator
Non-revenue water	Financial	NRW (by volume): volume of NRW as a % of system input volume	NRW (by cost): value of NRW as a % of the annual cost of running the water system
Water losses	Operational	Volume/service connection/year	–
Apparent losses	Operational	–	Volume/service connection/year
Real losses	Water resources	Inefficiency of use of water resources: real losses as a % of system input volume	Volume/service connection/year

Adapted from [12, 17, 18]

real losses, and apparent losses [17] (see Table 1). These are widely used as a tool for the performance evaluation of water supply services [16].

2.4 Strategy Formulation and Action Plan Elaboration

This step comprises the identification of solutions to address existing problems in the public water supply network. The strategic action plan represents an essential tool that presents systematically, to the water companies, a vast set of information and actions that the responsible managers will use to support the decision of the measures to be implemented, to reduce water losses. This strategy should be defined after analyses to identify from which level of water losses it is cost-effective to implement the various technical solutions in the field [1, 19, 20].

At the European level, some water sector regulators have punitive powers, whereby water companies may face financial penalties if they do not achieve certain efficiency targets or, furthermore if they do not invest a certain percentage of their annual budget in water losses management and prevention measures [12]. Given the facts, there is no single, standardized solution that applies to all water companies. Each organization must evaluate the internal and external conditions to which it is subject, including the technical capacity of its professionals and its financial balance, and in this way, ensure, based on them, the balance between cost, performance and risk, in addition to defining the strategies to be adopted for water losses reduction.

In the sequence, Fig. 3 presents a model with a set of actions to help the feasibility and implementation of the strategic plan, being this tool developed according to the organization's goals. Besides being the document that will be a reference for the

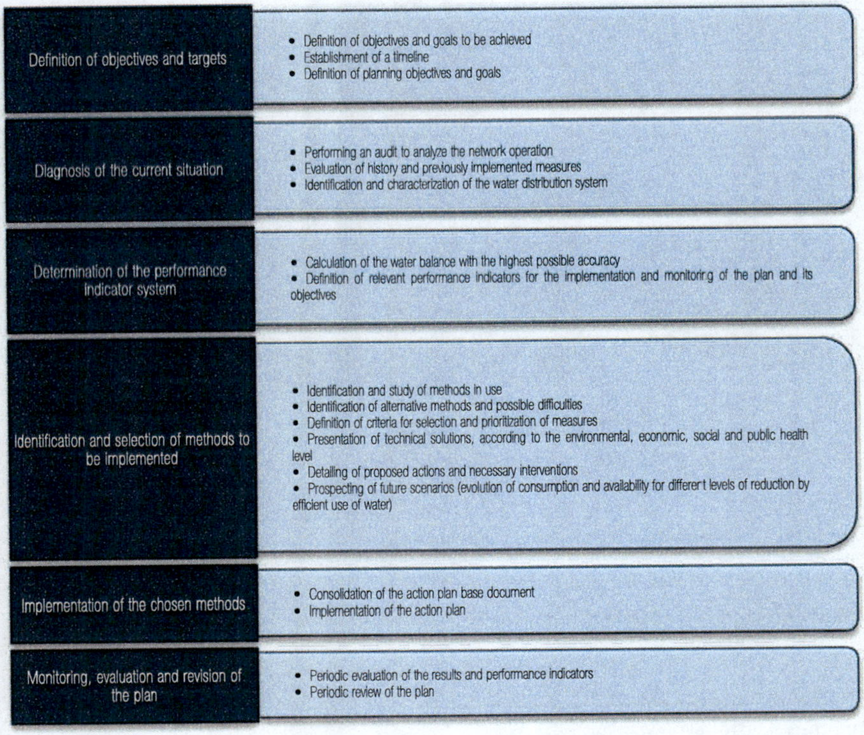

Fig. 3 Stages for the formulation of the strategy and action plan to reduce water losses. Adapted from [16]

implementation of actions to reduce water losses, this plan will allow for evaluation of the impact and help in the redefinition of new strategic action plans.

2.5 *Action Plan Implementation*

A significant part of the leaks in a water distribution network is not easily detectable and are usually caused by bursts in the pipes or associated components, and the lost water is absorbed by the soil surrounding the infrastructure or directed to aquifers, groundwater lines or even to other existing underground infrastructures [12]. Considering that the leaks in a water distribution network depend on several factors (such as age, material, soil characteristics and service pressure), its identification gains added relevance, because the results of the implementation of a specific technical solution will depend on it. The methodological approach commonly used to reduce the volume of real losses is presented in Fig. 4 [13, 17, 21].

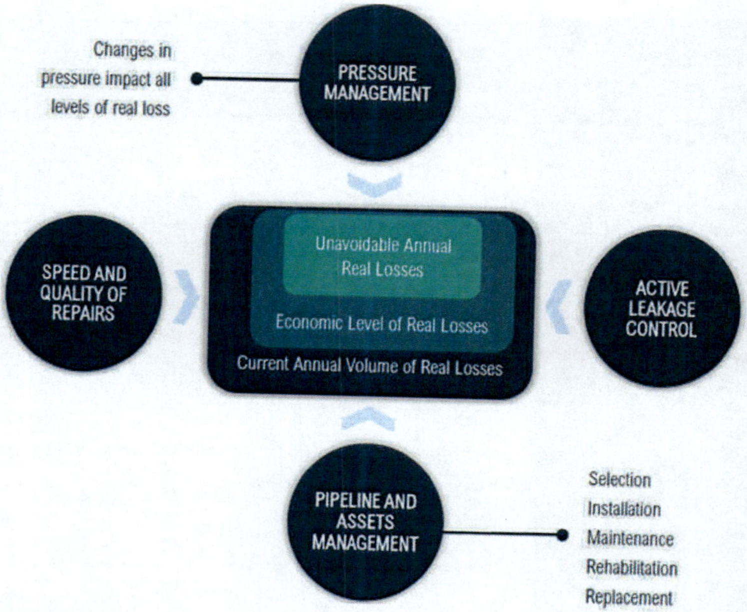

Fig. 4 Main strands of real losses reduction method. Adapted from [17, 21]

The service pressure in a water distribution network is a preponderant factor and unequivocally influences the occurrence of pipe bursts. The pressure management in water distribution networks presupposes the guarantee of a minimum level of service pressure, as low as possible, and ensures the stabilization of pressures in unfavourable situations of water demand at critical points of the network [19, 22–24]. The best way to control pressure in a supply system is to sectorize the system by establishing different pressure levels based on the network's local characteristics. This technique must be designed to ensure not only that service pressures are maintained in all parts of the network, but also that a stable pressure level is maintained. The benefits associated with correct pressure management are recognized and are highlighted by the potential reduction of water losses flow, reduction of consumption, stability of service pressure in the water distribution network, and reduction of the number of new bursts [2, 15].

Water losses are not of equal importance, since a small leak may represent a large volume of water losses if the identification and repair process is not immediate. Therefore, active leakage control presupposes procedures related to the speed and quality of repairs on the network: detection time, location time, and repair time. These procedures allow the volume of water losses to decrease and ensure that it remains at controlled levels. The identification of leaks in a water distribution network should use specific techniques and equipment, being advisable, when possible, to have monitoring systems that can automatically detect the existence of leakage, alerting the technical teams [3, 12, 25].

After the detection of a problem related to water losses in the network, the water company must start a process of investigation of the causes, considering two phases of approach in the intervention, the first inherent to the location by approximation, based on a process of narrowing down the areas to intervene, and the second the exact identification of the location of the leak, based on the application of specific techniques considered appropriate [12, 13]. The approaches should be complementary, with techniques and methodologies in identifying the problem, specifically [1, 12, 13]:

(a) Examples of techniques for detection and approximate location of leaks:

- *Sectorisation of the Water Distribution System (DMA—District Metered Area):* A tool that allows flow control by dividing the water distribution network into a set of zones with known and well-defined limits. With this division it is possible to measure the inflow and outflow of water in each zone, making it possible to obtain data on the water balance and consumption behaviour. Sector delimitation involves criteria that vary according to local characteristics, such as network topology, and population and branch density. The technique is applied when a consumption considered excessive in a given area is not compatible with the number and type of users in the isolated area. Flow rate measurement is usually carried out during the night.
- *Step Test:* This technique consists of the continuous flow measurement at the inlet point system (or the DMA) and the progressive closing of sectionalizing valves from downstream to upstream. As the isolated area increases, the flow measured at the inlet point system tends to zero. After closing the valves and stabilizing the flow at the inlet point system, if there is a significant drop in flow at the inlet of the non-compatible system in the isolated area, this may be related to the presence of leaks. The night period, when the consumption is lower and network pressure is higher, is the best time to perform these operations.

(b) Examples of techniques for detection and exact location of leaks:

- *Acoustic Probing:* Devices that detect the noise produced by leakage when it escapes through small holes in a pressurized pipe, as a result of phenomena that present dominant frequencies. These devices are widely used, but they are not the most appropriate when other noises overlap the noise emitted by leaks, pressures are very low, diameters are very high, or leaks occur in non-metallic conduits. Some of the equipment that incorporates acoustic sensors are acoustic loggers, listening rods, geophones and acoustic correlators.
- *Tracer Gas Injection:* This method is effective, although costly as it involves consumables (trace gas) and poses a risk of contamination. It is recommended when the pipe diameters are reduced and/or the noise emitted by the leak is insufficient for it to be detected by other equipment. In this case, the leak constitutes an exit point for the tracer that can be detected on the surface using a device that detects the presence of the injected gas.

- *Direct Observation and Visual Inspection:* The analysis of the system's behaviour allows the problem to be solved at an early stage. Due to its simplicity and low operational cost, the visual inspection of infrastructures is usually associated with the state of conservation of pipes, reservoirs and accessories. One of the techniques is surveillance of the pipes network to identify visible bursts and third-party works that may collide with the installed infrastructure. There are situations in which an analysis of the soil characteristics allows leak detection (irregular surface water outcrops, appearance of vegetation or mosses, or abnormal soil moisture).

2.6 Monitoring and Results Evaluation

Practical experience has shown that water losses management is a continuous task. Independent of the level of water losses that the system fits into, the permanence of actions is fundamental for the maintenance of good service performance. In this context, water losses management requires the presence of some primordial tools and solutions [12, 13], specifically:

(a) *Geographic Information System (GIS)*: Computational information tool for making available infrastructure cadastres, including georeferencing. This system should be constantly updated and include a representation of the terrain cartography. Interconnected to the GIS, the customer management information system, it is a system that aggregates all the relevant information of the managing entity's consumers.
(b) *Digital Elevation Model (DEM)*: This is the digital representation of the land surface elevation with respect to any reference datum. In other words, consists of a set of data in numerical support that, for a given area, allows associating to any water network point a value corresponding to its altitude.
(c) *Hydraulic Modelling*: A tool that allows the analysis, simulation and prediction of the hydraulic behaviour of the system, over time, from the characteristics of its components, its form of operation and the water demand.
(d) *Telemetering*: With this technological solution, the consumer flow meters are in communication with carefully located concentrator equipment, which, in turn, remotely sends the readings compiled to the central system, where the data is collected and processed.
(e) *Telemetry*: A system that provides real-time knowledge of the flow in the water distribution network and the service pressures at critical points in the network. It includes a computer infrastructure (Data Center) to support the Supervision System (SCADA System) and a Central Control Center. This method provides a set of information and tools that allow water companies, through centralized supervision, to take decisions that increase efficiency and reduce costs.

The final stage of the process refers to the evaluation of the results achieved. This is the stage that recommends the efficiency with which the resources were used, as well as the effectiveness of the measures put into practice, comparing the initial and final situations. It is essential to assess the potential gains between the costs that result from water losses and the benefits obtained with the implementation of the strategy, concerning water losses management, the improvement of the physical conditions of the infrastructure and the overall management of the supply system. In addition, the evolution of performance indicators should be analyzed, as well as an analysis of the causes of any deviations identified during the implementation of the plan [1, 16]. Therefore, the evaluation and control of the system's overall performance, in strategic planning, is an indispensable step, to adjusting new cycles accordingly, in a process of continuous improvement.

3 Conclusions

Water stress, water-related risks, and water quality pose growing challenges to water companies. However, inefficiency in water losses management by most water companies, coupled with the complexity of distribution systems, represent sources of very high rates of NRW. Therefore, traditional methods are no longer able to guarantee the high-performance indicators and sustainability of the systems. Currently, new investments and political priorities of governments have been directed to adequate water losses management in water supply systems, as a way to increase the levels of sustainability and ensure contributions to sustainable development, economic and financial, public health protection and social cohesion of communities.

Although it is increasingly important to raise awareness among the population to save and use water efficiently it is, also important to ensure that as little as possible of the resource is wasted during its abstraction, treatment, storage, and distribution. Software technology is an excellent tool to improve the efficiency of the service provided and strengthen values such as innovation, sustainability, efficiency, experience, optimization and savings for the water companies. It is worth pointing out that each water company must adopt the water losses management that best suits its reality (i.e. economic level of losses), establishing the best cost–benefit ratio, and aiming to optimize the water service levels to its customers and ensure greater added value for all stakeholders.

Acknowledgements This work was supported by the Portuguese Foundation for Science and Technology (FCT)—Aveiro Research Centre for Risks and Sustainability in Construction (RISCO), University of Aveiro, Portugal [FCT/UIDB/ECI/04450/2020]. This research was supported by the Doctoral Grant 2021 MPP2030-FCT (PRT/BD/152847/2021) financed by the FCT, under MIT Portugal Program, awarded to the first author (S.B.).

References

1. SIGA, Plano Estratégico de Redução de Perdas/Deteção e Controlo de Fugas, http://pintobraz.com/Ap_SIGA_web_dupla_face.pdf, last accessed 2022/12/09.
2. J. Sousa, J. Muranho, A. S. Marques, and R. Gomes: Optimal Management of Water Distribution Networks with Simulated Annealing: The C-Town Problem, J Water Resour Plan Manag, vol. 142, no. 5, p. C4015010, Oct. 2015. https://doi.org/10.1061/(ASCE)WR.1943-5452.0000604.
3. R. T. Fritz, J. de C. Gimenes, and A. C. de Pina Filho: A study of automation for loss reduction in water distribution system. Brazilian Journal of Development, vol. 6, no. 8, pp. 56408–56416, 2020. https://doi.org/10.34117/bjdv6n8-161.
4. S. G. Vrachimis, S. Timotheou, D. G. Eliades, and M. M. Polycarpou: Leakage detection and localization in water distribution systems: A model invalidation approach, Control Eng Pract, vol. 110, p. 104755, May 2021. https://doi.org/10.1016/J.CONENGPRAC.2021.104755.
5. Q. S. Fang, J. X. Zhang, C. L. Xie, and Y. L. Yang: Detection of multiple leakage points in water distribution networks based on convolutional neural networks, Water Supply, vol. 19, no. 8, pp. 2231–2239, Dec. 2019. https://doi.org/10.2166/WS.2019.105.
6. Sistema de Informação de Indicadores de Perdas de Água - Associação Portuguesa de Distribuição e Drenagem de Águas, https://app.powerbi.com/view?r=eyJrIjoiNGYwN2E4YWYtNWU0NS00Y2 RmLWJmMTctYmJlOD-JiOGRkNDQ5IiwidCI6Ijc5YTBlYjk3LWM0YWEtNGI2Zi05NDAyLTJhZmFjNGQ3ZDU5MCIsImMiOjh9, last accessed 2023/01/07.
7. Y. X. Feng, H. Zhang, S. Rad, and X. Z. Yu: Visual analytic hierarchical process for in situ identification of leakage risk in urban water distribution network, Sustain Cities Soc, vol. 75, p. 103297, Dec. 2021. https://doi.org/10.1016/J.SCS.2021.103297.
8. T. K. Chan, C. S. Chin, and X. Zhong: Review of Current Technologies and Proposed Intelligent Methodologies for Water Distributed Network Leakage Detection, IEEE Access, vol. 6, pp. 78846–78867, 2019. https://doi.org/10.1109/ACCESS.2018.2885444.
9. T. AL-Washali, S. Sharma, F. AL-Nozaily, M. Haidera, and M. Kennedy: Modelling the Leakage Rate and Reduction Using Minimum Night Flow Analysis in an Intermittent Supply System, Water 2019, vol. 11, no. 1, p. 48, Dec. 2018. https://doi.org/10.3390/W11010048.
10. S. Tariq, B. Bakhtawar, and T. Zayed: Data-driven application of MEMS-based accelerometers for leak detection in water distribution networks, Science of The Total Environment, vol. 809, p. 151110, Feb. 2022. https://doi.org/10.1016/J.SCITOTENV.2021.151110.
11. M. D'Ercole, M. Righetti, R. M. Ugarelli, L. Berardi, and P. Bertola: An Integrated Modeling Approach to Optimize the Management of a Water Distribution System: Improving the Sustainability While Dealing with Water Loss, Energy Consumption and Environmental Impacts, Procedia Eng, vol. 162, pp. 433–440, Jan. 2016. https://doi.org/10.1016/J.PROENG.2016.11.085.
12. F. S. Serranito and A. Donnelly: Controlo Ativo de Perdas de Água, EPAL, Empresa Portuguesa das Águas Livres S.A., Lisboa, 2015.
13. R. de J. Gomes: Modelação matemática como ferramenta de gestão e exploração de sistemas de distribuição de água - Dissertação apresentada para a obtenção do Grau de Doutor em Engenharia Civil na Especialidade de Hidráulica, Recursos Hídricos e Ambiente, Universidade de Coimbra, 2011.
14. L. Bardin: Análise de Conteúdo. rev. e atual, 70th ed., vol. 3. Lisboa, 2009.
15. C. Bozkurt, M. Firat, A. Ateş, S. Yilmaz, and Ö. Özdemir: Strategic water loss management: Current status and new model for future perspectives, Sigma Journal of Engineering and Natural Sciences, vol. 40, no. 2, pp. 310–322, Jun. 2022. https://doi.org/10.14744/sigma.2022.00035.
16. S. de T. M. Bezerra and P. B. Cheung: Perdas de água: tecnologias de controle, vol. 1. João Pessoa: UFPB, 2013.
17. AWWA Water Loss Control Committee, Committee report: Applying worldwide BMPs in water loss control, J Am Water Works Assoc, vol. 95, no. 8, 2003. https://doi.org/10.1002/j.1551-8833.2003.tb10430.x.

18. IWA International Water Association: Losses from Water Supply Systems: Standard Terminology and Recommended Performance Measures, Aug. 2000.
19. R. Gomes, A. S. A. Marques, and J. Sousa: District Metered Areas Design Under Different Decision Makers Options: Cost Analysis, Water Resources Management, vol. 27, no. 13, pp. 4527–4543, Oct. 2013. https://doi.org/10.1007/S11269-013-0424-Y/FIGURES/5.
20. R. Gomes, A. S. Marques, and J. Sousa: Decision support system to divide a large network into suitable District Metered Areas, Water Sci Technol, vol. 65, no. 9, pp. 1667–1675, 2012. https://doi.org/10.2166/WST.2012.061.
21. R. S. McKenzie and A. Lambert: ECONOLEAK Economic Model for Leakage Management - Development of an Economic Model for Evaluating Leakage in South Africa. Pretoria, South Africa: Water Research Commission (WRC Project No. 1145), 2002.
22. R. Gomes, A. S. Marques, and J. Sousa: Estimation of the benefits yielded by pressure management in water distribution systems, vol. 8, no. 2, pp. 65–77, Apr. 2011. https://doi.org/10.1080/1573062X.2010.542820.
23. R. Gomes, A. Sá Marques, and J. Sousa: Identification of the optimal entry points at District Metered Areas and implementation of pressure management, vol. 9, no. 6, pp. 365–384, Dec. 2012. https://doi.org/10.1080/1573062X.2012.682589.
24. R. Gomes, J. Sousa, and A. S. Marques: The influence of pressure/leakage relationships from existing leaks in the benefits yielded by pressure management, Water Utility Journal, vol. 5, pp. 25–32, 2013.
25. E. Vivas, P. Leite, and S. Pinto: Utilização de ferramentas de inteligência artificial no apoio à deteção de fugas em sistemas de abastecimento de água, 2021.